PROGRAMMING LANGUAGES:

History and Fundamentals

Prentice-Hall Series in Automatic Computation

George Forsythe, editor

ARBIB, *Theories of Abstract Automata*
BATES AND DOUGLAS, *Programming Language/One*
BAUMANN, FELICIANO, BAUER, AND SAMELSON, *Introduction to ALGOL*
BLUMENTHAL, *Management Information Systems*
BOBROW AND SCHWARTZ, editors, *Computers and the Policy-Making Community: Applications to International Relations*
BOWLES, editor, *Computers in Humanistic Research*
CESCHINO AND KUNTZMANN, *Numerical Solution of Initial Value Problems*
CRESS, DIRKSEN, AND GRAHAM, *Fortran IV with Watfor*
DESMONDE, *A Conversational Graphic Data Processing System: The IBM 1130/2250*
DESMONDE, *Computers and Their Uses*
DESMONDE, *Real-Time Data Processing Systems: Introductory Concepts*
EVANS, WALLACE, AND SUTHERLAND, *Simulation Using Digital Computers*
FIKE, *Computer Evaluation of Mathematical Functions*
FORSYTHE AND MOLER, *Computer Solution of Linear Algebraic Systems*
GOLDEN, *Fortran IV: Programming and Computing*
GOLDEN AND LEICHUS, *IBM 360: Programming and Computing*
GORDON, *System Simulation*
GREENSPAN, *Lectures on the Numerical Solution of Linear, Singular and Nonlinear Differential Equations*
GRISWOLD, POAGE, AND POLONSKY, *The SNOBOL4 Programming Language*
GRUENBERGER, editor, *Computers and Communications—Toward a Computer Utility*
GRUENBERGER, editor, *Critical Factors in Data Management*
HARTMANIS AND STEARNS, *Algebraic Structure Theory of Sequential Machines*
HULL, *Introduction to Computing*
LOUDEN, *Programming the IBM 1130 and 1800*
MARTIN, *Design of Real-Time Computer Systems*
MARTIN, *Programming Real-Time Computer Systems*
MARTIN, *Telecommunications and the Computer*
MARTIN, *Teleprocessing Network Organization*
MINSKY, *Computation: Finite and Infinite Machines*
MOORE, *Interval Analysis*
SAMMET, *Programming Languages: History and Fundamentals*
SCHULTZ, *Digital Processing: A System Orientation*
SNYDER, *Chebyshev Methods in Numerical Approximation*
STERLING AND POLLACK, *Introduction to Statistical Data Processing*
STROUD AND SECREST, *Gaussian Quadrature Formulas*
TRAUB, *Iterative Methods for the Solution of Equations*
VARGA, *Matrix Iterative Analysis*
VAZSONYI, *Problem Solving by Digital Computers with PL/1 Programming*
WILKINSON, *Rounding Errors in Algebraic Processes*
ZIEGLER, *Time-Sharing Data Processing Systems*

PRENTICE-HALL INTERNATIONAL, INC., *London*
PRENTICE-HALL OF AUSTRALIA, PTY. LTD., *Sydney*
PRENTICE-HALL OF CANADA, LTD., *Toronto*
PRENTICE-HALL OF INDIA PRIVATE LTD., *New Delhi*
PRENTICE-HALL OF JAPAN, INC., *Tokyo*

PROGRAMMING LANGUAGES:
History and Fundamentals

JEAN E. SAMMET

Programming Language Technology Manager
Federal Systems Division
IBM Corporation

PRENTICE-HALL, INC.
ENGLEWOOD CLIFFS, N. J.

Concept of the tower of BABEL to represent a large set of programming languages is due to the *Communications of the ACM*, a publication of the Association for Computing Machinery, Inc. The illustration appears on the front endpaper.

© 1969 by Prentice-Hall, Inc.
Englewood Cliffs, New Jersey

All rights reserved. No part of this book
may be reproduced in any form
or by any means without permission
in writing from the publisher.

Current printing (last digit):
10 9 8 7 6 5 4

Library of Congress Catalog Card No. 68-28110

Printed in the United States of America

6. And the LORD said, Behold the people is one, and they have all one language; and this they begin to do; and now nothing will be restrained from them, which they have imagined to do.
7. Go to, let us go down, and there confound their language, that they may not understand one another's speech.
8. So the LORD scattered them abroad from thence upon the face of all the earth; and they left off to build the city.
9. Therefore is the name of it called Babel; because the LORD did there confound the language of all the earth; and from thence did the LORD scatter them abroad upon the face of all the earth.

Gen. xi

PREFACE

The primary purpose of this book is to serve as a reference for an overall view of higher level languages. The book brings together in one place, and in a consistent fashion, fundamental information on programming languages, including history, general characteristics, similarities, and differences.

A second purpose of the book is to provide specific basic information on all the significant, and most of the minor, higher level languages developed in the United States.

The third purpose of the book is to provide history and perspective for this particular aspect of the programming field. Comments on both are the responsibility of the author and are not necessarily accepted by all the people concerned. Because of the rapidly changing nature of this type of work, new languages appear daily (literally) and so this book represents a snapshot of—and an (indirect) explanation of how we arrived at—the situation at a given point in time, namely the fall of 1967. In a few instances, major happenings of 1968 which could be inserted into galley or page proofs were included, but in general the text and bibliography cover the period through 1967.

The most well known language (FORTRAN) is merely *one* of approximately *120* languages described in this book. (Of this total, approximately 20 are completely dead or on obsolete computers, about 35 are receiving very little usage, about 50 are for specialized application areas, and about 15 are widely used and/or implemented.) No major attempt has been made to include languages which are known or used only within a single organization. Most of those discussed here have been described in published literature. However, a few languages discussed only in reports issued by the developing organization have been included.

Other purposes are to provide an extensive bibliography of relevant material, to show various philosophies of language design, to describe a number of the key factors involved in choosing a language, and to provide the reader with enough information so that he can decide which languages he wishes to examine in detail.

It is *not* the purpose of this book to teach how to program in any of the languages described, nor is the purpose to provide specific or detailed comparisons of related languages, nor is it meant to provide a cookbook for selecting a language for a particular application. A discussion of implementation techniques is also outside the scope of this book. Except for a few special cases, only languages which have been developed in the United States, *and* which have been implemented, are described. This restriction applies also to comments of the type "there has not been anything of this kind done"; such remarks apply to U.S. work only and might be invalid when considering other countries. Furthermore, some of these remarks are very time dependent and because of the rapidly changing nature of the field become invalid when considering work done after 1967.

Since even the very definition of a programming language is debatable, it is clear that inclusion or exclusion in the book is based on my view of the meaning of the phrase *programming language*. This is discussed in Chapter I. The amount of space given to each language in the book is usually dependent on both the complexity of the language and the author's judgment of its importance (either past, present, or future). Every effort has been made to ensure that the descriptions are accurate and not misleading. (See the Acknowledgments.)

The reader is assumed to have had experience, or at least one course, in programming. In many places, more than this minimum is needed for full understanding although the basic points should be comprehensible to readers with little experience.

Although not written as a textbook, this book could be used as the basic source for course I2 (Programming Languages) in *Curriculum 68* as described by the ACM Curriculum Committee on Computer Science. It might be used for background reading in courses B1 (Introduction to Computing), I1 (Data Structures), I5 (Compiler Construction), A1 (Formal Languages and Syntactic Analysis), A8 (Large-scale Information Processing Systems), and A9 (Artificial Intelligence and Heuristic Programming). Since this book was not available at the time that the committee made its report, obviously the book could not appear in any of the bibliographies suggested for the courses.

Chapter I provides a general introduction to the subject of programming languages, advantages and disadvantages, various classifications, and factors involved in the choice of a language. Chapters II and III discuss respectively the functional (i.e., non-technical) and technical characteristics of programming languages. Most of the language descriptions are based on the outline and concepts established in those two chapters, and so a careful reading of Chapters II and III is required in order to understand the rest of the book. This admittedly has the disadvantage of not necessarily being the best way to describe a specific language, but has the advantage of providing some consistency throughout the discussions. However, some of the flavor and style of the individual language descriptions have been allowed to creep in, and serve as a preview of what would be encountered in a more detailed study. In a very few instances, minor inexact statements have been made because the accurate language description would require details beyond the scope of this book.

The remaining chapters and sections are relatively independent, and in most cases a specific language description can be read without knowledge of any other languages. Chapter IV deals with languages used primarily for numerical scientific problems. Chapter V discusses those used for business data processing. Chapter VI discusses the list and string processing languages, and Chapter VII describes languages used for doing formal algebraic manipulation. Chapter VIII describes languages which can be *effectively* used in more than one of the areas covered in the preceding four chapters. Sample programs for most of the languages in Chapters IV through VIII have been included; they are meant solely to illustrate the syntactic style of the languages, and they are not guaranteed to be either correct or efficient. A few problems have been coded in several different languages to provide an easy comparison.

Chapter IX describes about 50 languages which are used in more specialized application areas. Because greater specific knowledge of those areas is needed to appreciate the languages, the discussions are very brief and superficial. The criteria for judging these languages is less stringent than those used for languages in preceding chapters, so that some of the languages included are not conceptually very different from older ones which were deliberately omitted, and they do not necessarily satisfy all the defining characteristics of Chapter I. The criteria have been reduced to make clear the great need for specialized languages.

Chapter X discusses a few significant unimplemented concepts. Finally, Chapter XI contains the author's personal views on future long range developments in programming languages.

The philosophy and arrangement of references is described in Appendix A, which also contains a list of authors of included citations. *The beginning of this Appendix should be read before looking at any of the reference lists.* Appendix B contains a list of each language described in this book, the meaning of its acronym, a very brief description, and the one or two best references for it.

The book outline was constructed with great care and is shown completely in the Table of Contents. A significant amount of fundamental information is contained there and it should serve as a basic outline to the subject as viewed by the author.

The process of preparing this manuscript for printing actually required the solution of some technical problems in language description, not all of which I successfully solved. In order to make the reading of the text easier, it was necessary to settle on a distinctive type style to represent actual words in a language. Thus, specific statements which would be legal in a program, such as normal arithmetic and control statements, are written as A = B + C and GO TO HECK. However, in those cases where the format of a statement was being described, it was necessary to distinguish between the fixed words in the language, and the variable names which are to be supplied by the programmer. This was settled by writing GO TO *statementlabel* or MOVE A TO B where *A* and *B* represent variables. However, there is clearly a difference between the two occurrences of the phrase "GO TO statement" in the following sentence, and the distinction is handled as follows: The GO TO statement in FORTRAN is of the form GO TO *statement*. More specifically the style used in the book is the following: Fixed words appearing in a definition of the format of a command, and all words in a specific example, are set THIS WAY or this way. Words or letters representing characters to be supplied by the programmer (metalinguistic variables as described in Chapter II) are set *THIS WAY* or *this way*. (Thus the letters A and B would be set as A, B if used in an example, and as *A, B* if used in the definition of a format.) In a discussion *about* a statement or a *list* of statement names, the name of the statement is set *THIS WAY* or *this way*. The use of upper and lower case letters have no significance and are merely those most commonly used in descriptions of the particular language involved. In ALGOL based languages, the tradition of boldface was used with the above concepts, resulting in **this style** or ***this style***. The character ' has been used to represent the prime or apostrophe available on printer chains or typewriters.

Another problem which exists is the variable size of the characters and spaces used in setting this book. In input/output media for computers, all characters require the same amount of space, and there is either a blank between them, or there is not. To simulate this a specific size space was used to represent the computer *blank* character; however, this space could not always be maintained in programs where vertical alignment of columns was critical. Thus, in material that represents specific examples, the spacing is critical and is shown as well as possible; on the other hand, it is essential to realize that in many languages the presence of one, many, or no spaces is immaterial and the reader should not conclude that because a space was present or absent that this is a requirement of the language. The language description specifies whether or not the blank character is significant.

While these ideas may appear confusing here, it is hoped that the type styles will be clear in the text. However, the really careful reader will find that my scheme breaks down in many minor places, and it is left as an exercise for the reader to find the actual places and to propose a solution.

The subject of programming languages is quite controversial and even includes debate on what should be included or excluded. Therefore, this book reflects the author's personal opinions to a much larger extent than would a book on a more stable or well defined area. It should be clearly understood that the specific views expressed in the text, and the implied views represented by the selection and arrangement of the languages, are solely those of the author.

Jean E. Sammet
Cambridge, Massachusetts

ACKNOWLEDGMENTS

As stated in the preface, every effort has been made to ensure that the language descriptions are accurate and not misleading. In order to accomplish this, virtually every individual language writeup was sent to one or more *experts* in that particular language. Because a large majority of these people spent significant time in reviewing more than one draft and making constructive comments, it seems appropriate to list the names of the languages or sections with the individuals who commented on them, and to take this opportunity to express my deep thanks. It must be emphasized that any errors, omissions, or misleading statements in the text are completely and solely my responsibility, and not those of the people listed here. They are also not necessarily in agreement with the views I have expressed about various aspects of the language or its contributions to the technology.

Languages marked with an asterisk were also included in a review of a larger section. Languages that are marked with an asterisk but for which no reviewer is listed were reviewed only as part of a larger section.

Language	Reviewer(s)	Language	Reviewer(s)
A-2 & A-3*	Hopper, G. M.	APL	Iverson, K.
ADAM*	Lafferty, E.	APL/360	Falkoff, A. D.
AED*	Ross, D. T.	APT*	Ross, D. T.
AESOP*	Hazle, M.	BACAIC*	
AIMACO	Hopper, G. M.; Jones, J. L.	BASEBALL*	
		BASIC	Kurtz, T. E.
ALGOL	Ingerman, P. Z.	C-10*	Lafferty, E. L.
ALTRAN	Brown, W. S.	CLIP*	
AMBIT*	Christensen, C.	CLP*	
AMTRAN	Seitz, R. N.	COBOL	Betscha, R. F.

xii ACKNOWLEDGMENTS

Language	Reviewer(s)	Language	Reviewer(s)
COGENT*		LISP 1.5*	Abrahams, P. W.;
COGO*	Logcher, R.		Edwards, D.
COLASL	Carter, G. L.	LISP 2	Abrahams, P. W.;
COLINGO*	Lafferty, E. L.		Barnett, J. A.
COMIT*	Bennett, J. L.;	MAD	Galler, B. A.
	Yngve, V.	MADCAP	Wells, M. B.
Commercial		Magic Paper	Clapp, L. C.
Translator	Goldfinger, R.	MAP	Brackett, J.
CORAL*		MATHLAB	Engelman, C.
CPS	Rochester, N.;	MATH-MATIC*	Hopper, G. M.
	Schroeder, D. A.	META 5*	
DEACON*		MILITRAN*	
DYNAMO*		MIRFAC	Gawlik, H. J.
473L Query*		NELIAC	Halstead, M. H.
FACT	Clippinger, R. F.	OPS-3*	Jones, M.
FLAP	Morris, A. H., Jr.	PL/I	Beech, D.; Cundall, P.;
FLOW-MATIC	Hopper, G. M.		Hankam, E. V.;
FORMAC	Tobey, R. G.		Mitchell, R.
Formula ALGOL	Perlis, A. J.	PRINT*	
FORTRAN	Heising, W.;	Protosynthex*	
	Ridgeway, R.	QUIKTRAN	Morrissey, J.
FORTRANSIT*		Short Code*	
FSL*		SIMSCRIPT*	
GAT*	Galler, B.	SIMULA*	
GPSS*	Krasnow, H.	SNOBOL*	Bennett, J. L.;
GRAF	Johnson, C.		Griswold, R. E.
ICES*	Logcher, R.	SOL*	
IPL-V*	Newell, A.	Speedcoding*	
IT	Galler, B. A.;	STRESS*	Logcher, R.
	Perlis, A. J.	Symbolic Math.	
JOSS	Shaw, J. C.	Lab.	Martin, W. A.
JOVIAL	Perstein, M.;	TMG*	
	Shaw, C. J.	TRAC*	Mooers, C.
Klerer-May	Klerer, M.	TRANDIR*	
L⁶*	Knowlton, K.	TREET*	Haines, E. C.
Laning and Zierler*		UNICODE*	Hopper, G. M.
Lincoln Reckoner	Wiesen, R.		

Section	Reviewer(s)	Section	Reviewer(s)
IV.1 and IV.2	Hopper, G. M.	IX.3.1	Jones, M.; Krasnow, H.
Chapter VI	Bobrow, D. G.; Raphael, B.	3.2	Walker, D.
IX.2.1	Ross, D. T.	3.4	Ross, D. T.
2.2	Logcher, R.	3.5	Walker, D.
2.5	Cheatham, T. E., Jr.	3.6	Walker, D.

In addition to the above, Patricia Cundall and Peter Ingerman carefully reviewed most of the manuscript and made valuable comments. I am particularly grateful to Robert Tabory not only for his review of several manuscript drafts but also for many constructive discussions on detailed and general points. Other people too numerous to cite individually, including students in various courses who saw successive drafts, made constructive suggestions on various sections.

Each sample program not taken from a printed source was coded by a person with much experience in the language, although no guarantees of either correctness or efficiency are made. I am grateful to the following people for the coding of one or more of the sample programs: J. Andrada, F. Bequaert, J. Bennett, C. Engelman, H. J. Gawlik, E. C. Haines, W. Harrison, M. H. Halstead, K. Knowlton, T. Kurtz, C. Mooers, M. Perstein, B. Raphael.

Most of the material used in figures and sample programs came from publications of the Association for Computing Machinery, Inc., the International Business Machines Corporation, and various Proceedings of the AFIPS Joint Computer Conferences. I particularly appreciate the cooperation of these organizations, as well as all others who permitted the use of copyrighted material.

My appreciation to the IBM Corporation—specifically the Systems Development Division and the Federal Systems Division—is unbounded, both for providing me with the majority of the time I used for preparing this book, and for supplying all the typing support, as well as the computer time and telephone lines for the use of ATS (Administrative Terminal System) which was used for typing and correcting most of the drafts. I am specifically very grateful for the continued support and encouragement of my managers, Nathaniel Rochester and Joel D. Aron.

Last but not least, a devoted crew of typists struggled with my manuscript and having to learn an unfamiliar system of typing and correcting (namely, ATS). They include Josephine Auterio, Margaret Mahoney, Carrie Jo Clausen, and Dorothy Pearlman. The last two did a Herculean task in helping prepare the initial reference lists. Finally, my secretaries Beatrice Roffman and Carolyn Willet cheerfully coped not only with the book typing but with a myriad of other chores as well.

CONTENTS

GENERAL INTRODUCTION   1

1. Machine Language Programming   2
2. Symbolic Assembly Language Programming   2
3. Early Development of Better Tools   3
   3.1. Specific needs to be met   3
   3.2. Brief history of early efforts   5
4. Definition of Programming Languages   8
   4.1. Definition problem   8
   4.2. Defining characteristics   9
      1. Machine code knowledge is unnecessary   9
      2. Potential for conversion to other computers   9
      3. Instruction explosion   10
      4. Problem-oriented notation   10
   4.3. Basic terminology   11
      1. Source program   11
      2. Object program   11
      3. Compiler   12
      4. Interpreter   12
      5. Automatic coding   13
      6. Automatic programming   13
   4.4. Difference between programming language and application package   13
5. Advantages and Disadvantages of Higher Level Languages   14
   5.1. Advantages   14
      1. Ease of learning   14
      2. Ease of coding and understanding   15
      3. Ease of debugging   15
      4. Ease of maintaining and documenting   16
      5. Ease of conversion   16
      6. Reduce elapsed time for problem-solving   17

5.2. Disadvantages		17
1. Advantages do not always exist		17
2. Time required for compiling		17
3. Inefficient object code		18
4. Difficulties in debugging without learning machine language		18
5. Inability of the language to express all needed operations		18
5.3. Overall evaluation		19
6. Classifications of Programming Languages and Proposed Definitions		19
6.1. Procedure-oriented language		19
6.2. Nonprocedural language		20
6.3. Problem-oriented language		21
6.4. Application-oriented language		21
6.5. Special purpose language		21
6.6. Problem-defining language		21
6.7. Problem-describing language		22
6.8. Problem-solving language		22
6.9. Reference language		22
6.10. Publication language		22
6.11. Hardware language		23
7. Factors in Choice of a Language		23
7.1. Suitability of language for problem area and projected users		23
7.2. Availability on desired computer		24
7.3. History and evaluation of previous use		25
7.4. Efficiency of language implementation		25
7.5. Compatibility and growth potential		25
7.6. Functional (= nontechnical) characteristics		26
7.7. Technical characteristics		26
References		26

II FUNCTIONAL CHARACTERISTICS OF PROGRAMMING LANGUAGES   30

1. Description of the Concept of Functional Characteristics	30
2. Properties of Languages	31
3. Purpose of Language	32
3.1. Application area	33
3.2. Type of language	34
3.3. Type of user	34
3.4. Physical environment	36
4. Conversion and Compatibility	36
4.1. Types of compatibility	37
1. Machine independence	37
2. Compiler independence	38
3. Dialects and language L-like	39
4. Subsetting and extensions	40
5. Relation to language definition	41

4.2. Ease of conversion	41
1. Based on compatibility	41
2. Ease of SIFTing to another language	42
3. Ease of translating to another language	42
5. Standardization	43
5.1. Purposes	43
5.2. Problems	43
1. Conceptual problems	44
2. Technical problems	44
3. Procedural problems	46
5.3. Method of establishing standards	46
5.4. Overall status	48
6. Types and Methods of Language Definition	48
6.1. Administrative	49
1. Who designed the language?	49
2. What were the objectives of the language?	49
3. Who implemented the language?	50
4. Who maintains the language?	50
6.2. Technical	51
1. Syntax, semantics, and pragmatics	51
2. Formalized notation	52
6.3. Types of documentation	57
7. Evaluation Based on Use	58
7.1. Availability on differing computers	58
7.2. Evaluation of language versus evaluation of compiler	59
7.3. Usage relative to objectives	59
7.4. Advantages	60
7.5. Disadvantages	60
7.6. Mistakes to be avoided in the future	61
References	61

III TECHNICAL CHARACTERISTICS OF PROGRAMMING LANGUAGES   65

1. Description of Concept of Technical Features	65
1.1. Introduction	65
1.2. Major parts of language	66
1. Data and its description	66
2. Operators	66
3. Commands	67
4. Declarations	67
5. Compiler directives	68
6. Delimiters	68
7. Program structure	68
2. Form of Language	68
2.1. Character set	69
2.2. Types of basic elements (= tokens)	70
1. System-defined	70
2. User-defined and restrictions	71

2.3. Identifier definition 72
    1. Types of identifiers 72
    2. Formation rules 72
    3. Use of reserved words 73
    4. Data names for aggregates (subscripts, qualification) 74
2.4. Definition and usage of other basic elements 77
    1. Operators 77
    2. Delimiters 77
    3. Use and meaning of punctuation 78
    4. Significance of blanks 79
    5. Use of noise words 79
    6. Literals 80
2.5. Type of input form used 80
    1. Physical input form 81
    2. Conceptual form 82

3. Structure of Program 82
  3.1. Types of subunits 83
    1. Nonexecutable: declarations and compiler directives 83
    2. Smallest executable unit 83
    3. Sets of smallest executable units 84
    4. Loops 85
    5. Functions, subroutines and procedures 85
    6. Comments 86
    7. Interaction with the operating system and environment 87
    8. Inclusion of other languages 87
    9. Complete program (including sequencing rules) 88
  3.2. Characteristics of subunits 88
    1. Methods of delimiting 89
    2. Recursive 89
    3. Parameter passage and differing types 90
    4. Embedding 92

4. Data Types and Units and Computations with Them 92
  4.1. Types of data variables and constants 93
    1. Arithmetic 93
    2. Logical (= Boolean) 93
    3. Character 93
    4. Complex 93
    5. Formal (= algebraic) 94
    6. String 94
    7. List of pointer 94
    8. Hierarchical 94
    9. Others 95
    10. Combinations of variable and constant types 95
  4.2. Accessible data units 95
    1. Hardware data units 95
    2. Language data units 97
  4.3. Types of arithmetic 97
    1. Integer, fixed point, mixed number 98
    2. Floating point 98
    3. Rational 98

	4. Complex numbers	99
	5. Double or multiple precision	99
	6. Logical	99
	7. Other	100
	8. Higher level data units	100
4.4.	Rules on creation and evaluation of arithmetic and logical expressions	100
	1. Intermingling rules	100
	2. Conversion rules	101
	3. Precision and computation rules for various modes	102
	4. Precedence and sequencing rules	102
4.5.	Scope of data	103

5. Executable Statement Types   104
   5.1. Assignment   104
      1. Methods of specifying computation   105
      2. Conversion rules for results   105
   5.2. Alphanumeric data handling   106
      1. Editing statements   106
      2. Conversion statements   106
      3. Sorting statements   106
   5.3. Sequence control and decision making   107
      1. Control transfer statements   107
      2. Conditional statements   108
      3. Loop control statements   109
      4. Error condition statements   110
   5.4. Symbolic data handling   111
      1. Algebraic expression manipulation statements   111
      2. List-handling statements   111
      3. String-handling statements   112
      4. Pattern-handling statements   112
   5.5. Interaction with operating system and/or equipment   112
      1. Input/output statements   113
      2. Library reference statements   113
      3. Debugging statements   114
      4. Storage and segmentation allocation statements   114
      5. Operating system and machine feature statements   115
   5.6. Others   115

6. Declarations and Nonexecutable Statements   115
   6.1. Data description   116
   6.2. File description   117
   6.3. Format description   118
   6.4. Storage allocation   118
   6.5. Environment or operating system descriptions   119
   6.6. Procedure, subroutine, function declarations   119
   6.7. Compiler directives   119
   6.8. Others   120

7. Structure of Language and Compiler Interaction   120
   7.1. Self-modification of programs   120
   7.2. Self-extension of the language   121
   7.3. Ability to write the compiler for a language in that language   121

7.4. Effect of language design on implementation efficiency	122
1. Compile time versus object time efficiency	122
2. Generality versus restrictions	123
3. Specific features with significant effect	124
4. Storage allocation requirements	124
5. Possibility for providing choice of tradeoffs	124
7.5. Debugging aids and error checking	125
8. Other Features Not Included	125
References	126

IV LANGUAGES FOR NUMERICAL SCIENTIFIC PROBLEMS — 128

1. Scope of Chapter	128
2. Languages of Historical Interest Only	128
2.1. Very early systems	129
1. SHORT CODE	129
2. Speedcoding	130
3. Laning and Zierler system	131
4. A-2 and A-3	132
5. BACAIC	133
6. PRINT	134
2.2. More widely used systems	134
1. MATH-MATIC (AT-3)	135
2. UNICODE	137
3. IT, FORTRANSIT, and GAT	139
4. ALGOL 58 (cross-reference only)	143
3. FORTRAN	143
3.1. History of FORTRAN	143
3.2. Functional characteristics of ASA (USASI) FORTRAN and Basic FORTRAN	150
3.3. Technical characteristics of ASA (USASI) Basic FORTRAN	157
3.4. Technical characteristics of ASA (USASI) FORTRAN	165
3.5. Significant contribution to technology	169
3.6. Significant extensions of FORTRAN	170
1. Proposal Writing language	170
2. FORMAC (cross-reference only)	171
3. QUIKTRAN (cross-reference only)	172
4. GRAF (cross-reference only)	172
5. DSL/90 (cross-reference only)	172
4. ALGOL	172
4.1. History of ALGOL	172
1. ALGOL 58	172
2. ALGOL 60	175
3. Revised ALGOL 60	177
4. ALGOL 6X	178
4.2. Functional characteristics of Revised ALGOL 60—Proposed ISO Standard	178
4.3. Technical characteristics of Revised ALGOL 60—Proposed ISO Standard	182

4.4. Significant contribution to technology	192
4.5. Extensions of ALGOL	194
1. Formula ALGOL (cross-reference only)	194
2. LISP 2 (cross-reference only)	195
3. AED (cross-reference only)	195
4. SFD-ALGOL (cross-reference only)	195
5. SIMULA (cross-reference only)	195
6. DIAMAG	195
7. GPL	195
8. Extended ALGOL	196
5. Languages Motivated by ALGOL 58	196
5.1. NELIAC	197
5.2. MAD	205
5.3. JOVIAL (cross-reference only)	215
6. On-Line Systems	215
6.1. Introductory remarks	215
6.2. JOSS	217
6.3. QUIKTRAN	226
6.4. BASIC	229
6.5. CPS	232
6.6. MAP	240
6.7. Lincoln Reckoner	245
6.8. APL/360 and PAT	247
6.9. Culler-Fried System	253
6.10. DIALOG	255
6.11. AMTRAN	258
7. Languages with Fairly Natural Mathematical Notation	264
7.1. Introductory remarks	264
7.2. COLASL	265
7.3. MADCAP	271
7.4. MIRFAC	281
7.5. Klerer-May system	284
8. Miscellaneous	294
8.1. CORC	294
8.2. OMNITAB	296
8.3. More nonprocedural languages	299
References	300

V LANGUAGES FOR BUSINESS DATA PROCESSING PROBLEMS   314

1. Scope of Chapter	314
2. Languages of Primarily Historical Interest	316
2.1. FLOW-MATIC (and B-0)	316
2.2. AIMACO	324
2.3. Commercial Translator	324
2.4. FACT	327
2.5. GECOM	328

3. COBOL	330
3.1. History of COBOL	330
3.2. Functional characteristics of COBOL	334
3.3. Technical characteristics of COBOL	345
3.4. Significant contribution to technology	375
4. File Handling	376
4.1. Extensions of COBOL	376
1. IDS	376
4.2. General (cross-reference only)	376
References	377

VI STRING AND LIST PROCESSING LANGUAGES — 382

1. Scope of Chapter	382
2. Languages of Historical Interest Only	388
3. IPL-V	388
3.1. History of IPL-V	388
3.2. Functional characteristics of IPL-V	389
3.3. Technical characteristics of IPL-V	393
3.4. Significant contribution to technology	400
4. L^6	400
5. LISP 1.5	405
5.1. History of LISP 1.5	405
5.2. Functional characteristics of LISP 1.5	407
5.3. Technical characteristics of LISP 1.5	410
5.4. Significant contribution to technology	416
6. COMIT	416
6.1. History of COMIT	416
6.2. Functional characteristics of COMIT	417
6.3. Technical characteristics of COMIT	421
6.4. Significant contribution to technology	435
7. SNOBOL	436
7.1. History of SNOBOL	436
7.2. Functional characteristics of SNOBOL	436
7.3. Technical characteristics of SNOBOL	438
7.4. Significant contribution to technology	448
8. TRAC	448
9. Languages Not Widely Used	454
9.1. AMBIT	454
9.2. TREET	457
9.3. Others	461
1. CLP	461
2. CORAL	462
3. SPRINT	462
4. LOLITA	464
References	464

VII FORMAL ALGEBRAIC MANIPULATION LANGUAGES — 471

1. Scope of Chapter — 471
2. Languages of Historical Interest Only — 473
   2.1. ALGY — 473
3. FORMAC — 474
   3.1. History of FORMAC — 474
   3.2. Functional characteristics of FORMAC — 475
   3.3. Technical characteristics of FORMAC — 476
   3.4. Significant contribution to technology — 490
4. MATHLAB — 491
   4.1. History of MATHLAB — 491
   4.2. Functional characteristics of MATHLAB — 491
   4.3. Technical characteristics of MATHLAB — 493
   4.4. Significant contribution to technology — 501
5. ALTRAN — 502
6. FLAP — 506
7. Systems Requiring Special Equipment — 510
   7.1. Magic Paper — 510
   7.2. Symbolic Mathematical Laboratory — 514

References — 520

VIII MULTIPURPOSE LANGUAGES — 523

1. Scope of Chapter — 523
2. Languages of Historical Interest Only — 524
3. JOVIAL — 524
   3.1. History of JOVIAL — 524
   3.2. Functional characteristics of JOVIAL — 526
   3.3. Technical characteristics of JOVIAL — 530
   3.4. Significant contribution to technology — 539
4. PL/I — 540
   4.1. History of PL/I — 540
   4.2. Functional characteristics of PL/I — 542
   4.3. Technical characteristics of PL/I — 548
   4.4. Significant contribution to technology — 582
5. Formula ALGOL — 583
6. LISP 2 — 589

References — 598

IX SPECIALIZED LANGUAGES 603

1. Scope of Chapter 603
2. Languages for Special Application Areas 605
   2.1. Machine tool control 605
      1. APT 605
      2. Others 606
   2.2. Civil engineering 610
      1. COGO 611
      2. STRESS 612
      3. ICES 615
   2.3. Logical design 620
      1. APL (Iverson) 620
      2. LOTIS 621
      3. LDT 621
      4. Langage for simulating digital systems 622
      5. Computer Compiler 623
      6. Computer Design Language 623
      7. SFD-ALGOL 623
   2.4. Digital simulation of block diagrams 627
      1. Introduction 625
      2. DYANA 628
      3. DYSAC 629
      4. DAS 631
      5. DSL/90 632
   2.5. Compiler writing 633
      1. Introduction 633
      2. CLIP 635
      3. TMG 636
      4. COGENT 638
      5. META 5 638
      6. TRANDIR 640
      7. FSL 641
      8. AED (cross-reference only) 641
   2.6. Miscellaneous 642
      1. Matrix computations: Matrix Compiler 642
      2. Cryptanalysis: OCAL 642
      3. Movie creation: Animated Movie Language and BUGSYS 644
      4. Social science research: DATA-TEXT 646
      5. Equipment checkout: STROBES, DIMATE 647
3. Specialized Languages across Application Areas 650
   3.1. Discrete simulation 650
      1. Introduction 650
      2. DYNAMO 651
      3. GPSS 653
      4. SIMSCRIPT 655
      5. SOL 656

	6. MILITRAN	657
	7. SIMULA	657
	8. OPS	660
3.2.	Query	662
	1. Introduction	662
	2. COLINGO and C-10	664
	3. 473L Query	665
	4. ADAM	667
	5. BASEBALL	668
	6. DEACON	668
	7. Protosynthex	669
	8. AESOP	670
3.3.	Graphic and on-line display languages	674
	1. GRAF	674
	2. PENCIL	677
	3. Graphic	677
	4. DOCUS	678
	5. AESOP (cross-reference only)	678
3.4.	Computer-aided design	678
	1. General	678
	2. AED	680
3.5.	Text editing and processing	684
3.6.	Control languages for on-line and operating systems	687
References		693

X SIGNIFICANT UNIMPLEMENTED CONCEPTS — 707

1. Scope of Chapter — 707
2. UNCOL — 708
3. Information Algebra — 709
4. APL (Iverson) — 712
5. English — 715
6. Hardware Implementation of Programming Languages — 717
References — 719

XI FUTURE LONG-RANGE DEVELOPMENTS — 722

1. Introduction — 722
2. Theory-Oriented Category — 723
   2.1. Language definition, translation, and creation — 723
   2.2. Next major conceptual step — 725
   2.3. Nonprocedural languages — 726
   2.4. Problem-describing languages — 726
   2.5. Use of mathematical concepts — 727

3. User-Oriented category	727
3.1. User-defined languages	727
3.2. Use of natural language	729
3.3. Communication with hardware and software	732
3.4. Languages for new application areas	733
3.5. Languages for writing software	734
4. Interrelationships among Some of These Concepts	734
5. Conclusions and Summary	736
References	736

A BIBLIOGRAPHY ARRANGEMENTS AND AUTHOR LIST   **738**

B LANGUAGE SUMMARY   **753**

NAME AND SYSTEMS INDEX   **765**

SUBJECT INDEX   **776**

LIST OF FIGURES AND SAMPLE PROGRAMS

CHAPTER I

I-1 List of automatic programming systems (1959)   6–7

CHAPTER II

There are no figures in Chapter II.

CHAPTER III

III-1 Example of tree and data layout   76

CHAPTER IV

IV-1 List of BACAIC facilities	133
IV-2 MATH-MATIC commands	136
IV-3 List of UNICODE instructions	138
IV-4 Table of FORTRAN I statements for the IBM 704	145
IV-5 Summary of FORTRAN II statements	146
IV-6 List of FORTRAN statements implemented on IBM computers, circa 1961	149
IV-7 List of intrinsic functions in Basic FORTRAN	162
IV-8 List of basic external functions in Basic FORTRAN	162
IV-9 Rules for assignment of e to v in FORTRAN	167

IV-10	List of statements in Proposal Writing Language	171
IV-11	Program in Proposal Writing Language	172
IV-12	Format for changing operators and modes in MAD	215
IV-13	Short summary of JOSS I	220
IV-14	JOSS II commands and functions	224–225
IV-15	Summary of BASIC statements	231
IV-16	Summary of CPS facilities (one-page summary with notes).	234–235
IV-17	Example of the use of Least Square command in MAP	243
IV-18	Standard scalar operators in APL	249
IV-19	Summary of DIALOG facilities	257
IV-20	List of AMTRAN operations	262–263
IV-21	The COLASL alphabet	266
IV-22	Example of COLASL program including heavy commentary	268
IV-23	Expressions in COLASL	270
IV-24	Two forms of the same MIRFAC program	283
IV-25a	Page 1 of the Klerer and May Reference Manual	286
IV-25b	Page 2 of the Klerer and May Reference Manual	287
IV-25c	"Addendum" to the Klerer and May Reference Manual	288
IV-26	Example of expressions with ambiguities in Klerer-May system	292
IV-27	Interpretation of ambiguities in expressions shown in Figure IV-26	293

CHAPTER V

V-1	Eight different ways to add three numbers	315
V-2	FLOW-MATIC verb formats	317–322
V-3	FLOW-MATIC program	323
V-4	List of Commercial Translator commands	326
V-5	List of FACT verbs	328
V-6	Schematic diagram showing the structure of pUSASI COBOL	342
V-7	Formats of COBOL verbs	353–359
V-8	Flowchart represented by a single COBOL sentence	361
V-9	Data description (=Record description) skeleton in COBOL	366–367
V-10	Report description (*RD*) skeleton in COBOL	367
V-11	Report group skeleton in COBOL	368–369
V-12	File description (*FD*) skeleton in COBOL	370
V-13	Sort file description (*SD*) skeleton in COBOL	370
V-14	*OBJECT—COMPUTER* format in COBOL	371
V-15	*SPECIAL—NAMES* format in COBOL	372
V-16	*FILE—CONTROL* format in COBOL	373
V-17	*I—O—CONTROL* format in COBOL	374

CHAPTER VI

VI-1	Illustration of list	383
VI-2	Inserting element in a list	384
VI-3	List structure, i.e., list containing list as element	385

xxviii LIST OF FIGURES AND SAMPLE PROGRAMS

VI-4	Fairly complete list of basic processes in first Information Processing Language	390–391
VI-5	List of IPL-V basic processes	398
VI-6	Summary of all L^6 instructions	403–404
VI-7	Syntactic definition of LISP S- and M-expressions	409
VI-8	Example of CORAL statements	463

CHAPTER VII

VII-1	7090/94 FORMAC verb formats with examples	481–484
VII-2	Example of use of MATHLAB on the AN/FSQ-32	492
VII-3	Typical executive control characters in first Magic Paper	511
VII-4	Format control characters in first Magic Paper	512

CHAPTER VIII

VIII-1	JOVIAL usage and compilers	526–527
VIII-2	Executable PL/I statement formats	558–561
VIII-3	List of *On-conditions* in PL/I	565
VIII-4	List of *Built-in* functions in PL/I	568

CHAPTER IX

IX-1	Example of APT program for specific part to be cut	607
IX-2	APT vocabulary list	608–610
IX-3	Small COGO program for figure shown	611
IX-4	List of 7090 COGO command names	612
IX-5	Specifications for a particular COGO command	613
IX-6	STRESS program for analysis of space truss shown in the diagram	614–615
IX-7	Partial listing of STRESS commands	616
IX-8	Some of the CDL commands in ICES	619
IX-9	Portion of LOTIS program defining a hypothetical computer	621
IX-10	Relationship between information flow on a block diagram and LDT input	622
IX-11	System layout and data paths in a sample computer, and part of computer compiler program	624
IX-12	Computer Design Language program to define multiplication	625
IX-13	SFD-ALGOL program to describe action of pushdown stack shown in diagram	626–627
IX-14	Mechanical system and corresponding DYANA program	628
IX-15	DYSAC components	629
IX-16	DYSAC statements for input data sections	629

IX-17	Problem, diagram, and corresponding DYSAC program	630
IX-18	Mathematical problem, block diagram, and corresponding DAS program	631
IX-19	Examples of some DSL/90 functional blocks, switching functions, and function generators	632
IX-20	Differential equation, diagram, and corresponding DSL/90 program	633
IX-21	Example of part of TMG program	637
IX-22	Example of part of COGENT program	639
IX-23	META program to convert JOVIAL constants to PL/I constants	639
IX-24	Example of part of TRANDIR program	640
IX-25	Example of part of FSL program	642
IX-26	OCAL program for solving cryptanalysis problem	643
IX-27	Part of movie animation program	644
IX-28	Example of part of BUGSYS program	645
IX-29	Example of DATA-TEXT input data	646
IX-30	Example of defining new variables based on conditions	647
IX-31	DATA-TEXT program	647
IX-32	Example of STROBES program	648
IX-33	Program for testing equipment	649
IX-34	DYNAMO example of model of retail store	652
IX-35	Some of the GPSS block types and corresponding operations	653
IX-36	Example of harbor arrival problem and GPSS solution	654
IX-37	Names of SIMSCRIPT commands and phrases	655
IX-38	SIMSCRIPT program for order in machine shop	656
IX-39	Complete SOL program for the multiple console on-line communication system shown in the diagram	657–658
IX-40	List of MILITRAN statements	659
IX-41	MILITRAN program for finite length queue	660
IX-42	Skeleton SIMULA description of job shop system	661
IX-43	OPS-3 program for multi-server queuing model	662
IX-44	List of basic commands in COLINGO	665
IX-45	Portion of a C-10 program	666
IX-46	Punctuation characters and key words in 473L Query language	667
IX-47	Examples of control statements, questions, and answers from Protosynthex II	670
IX-48	Communication tree in AESOP	671
IX-49	AESOP commands used on typewriter	672–673
IX-50	Small GRAF program	674
IX-51	List of PENCIL primitives	675
IX-52	PENCIL program for diagram shown	676
IX-53	Example in Graphic language	677
IX-54	GOL, DOL, and PIL operations in DOCUS	679
IX-55	General structure of AED-1 compiler	683
IX-56	Examples of editing statements in ES-1	684
IX-57	List of DATATEXT commands	685
IX-58	Sample of DATATEXT usage	686
IX-59	Names of commands in CTSS context editor	686
IX-60	Sample of actual CTSS session	690
IX-61	Uses of Job Control Language: (a) catalogued procedure, (b) changes to catalogued procedure, and (c) result of changing catalogued procedure	692

CHAPTER X

X-1 Use of UNCOL to reduce the number of compilers in going from M languages to N machines 708
X-2 System information for payroll problem in Information Algebra 710
X-3 Payroll program written in Information Algebra 711
X-4 Partial list of APL notation 713
X-5 m-way merge sort written in APL 714

CHAPTER XI

There are no figures in Chapter XI

SAMPLE PROGRAMS

ALGOL 60	178
ALTRAN	502
AMBIT	455
AMTRAN	259
APL/360	248
BASIC	230
COBOL	336–337
COLASL (Fig. IV-22)	268
COMIT II	418
CPS	233
FLAP	507
FORMAC (PL/I-FORMAC)	477
Formula ALGOL	583
FORTRAN	151
IPL-V	392
JOSS II	218
JOVIAL	528
Klerer-May System	285
L^6	401
LISP 1.5	407
LISP 2	591
MAD	207
MADCAP	272
MAP (Fig. IV-17)	243
MATHLAB 68	499
MIRFAC	282
NELIAC	199
OMNITAB	297
PL/I	544–545
PL/I-FORMAC	477
Proposal Writing Language (Fig. IV-11)	172
QUIKTRAN	227
SNOBOL3	437
Symbolic Mathematical Laboratory	515
TRAC	449
TREET	458

1 GENERAL INTRODUCTION

Programming languages have become the major means of communication between the person with a problem and the digital computer used to help solve it. In fact, it would be impractical to solve most problems if the computer had to be instructed in machine language. This has come about because most machines tend to operate in binary, and this is clearly an unsatisfactory method of communication for humans; hence the primary interface between the computer user and the computer itself has become the programming language used. In this context, *language* has the broadest possible meaning and includes not only the description of the problem to be solved but also the needed instructions to the operator or operating system. It should be noted that throughout this book the terms *programming language* and *higher level language* will be used synonymously, with the former being my preferred term. In particular, as noted later in this chapter, I do not consider an assembly language (even a very sophisticated one) to be a programming language. This view differs from that held by some people who maintain that anything in which programs are written is a programming language.

The main function of this chapter is to motivate the need for programming languages, and to define and characterize them. One section discusses the advantages and disadvantages of programming languages; however, in spite of the disadvantages, the net evaluation is that programming languages are here to stay. Finally, there is a lengthy list of types of programming languages, together with some proposed definitions. This provides one way of classifying programming languages. Many of these types are overlapping, i.e., a language can fall into several categories simultaneously.

2 GENERAL INTRODUCTION

I.1. MACHINE LANGUAGE PROGRAMMING

Every computer has a specific set of instructions which it can execute once the instruction is placed into the appropriate part of the machine. The actual set of symbols which the hardware can interpret for execution is the direct machine language. Since most computers are designed so that their storage locations and registers contain binary characters (i.e., bits), the most common machine language is actually binary. Thus the sequence

$$011011\ 000000\ 000000\ 000000\ 000001\ 000000$$

might mean *place the contents of storage location* 64 *in the accumulator*. To write one instruction, let alone many of them, in this form is clearly impractical, and this was recognized very rapidly in the early days of computers. A partial step to alleviate this problem involved the use of mnemonic codes to represent the instruction, while the rest of the information was left in binary. Thus, the sequence

$$\text{CLA}\ 000000\ 000000\ 000000\ 000001\ 000000$$

might have the same meaning as the binary string given earlier. While this was a partial improvement, it was still far from easy to write even one instruction correctly. The next step forward came when the numbers (representing the storage locations or registers in the computer) were allowed to be written in decimal form. Thus the sequence CLA 0 0 0 0 64 might have the same meaning as the earlier strings.

The border line between machine language and symbolic assembly language is not well defined. Some people would choose to refer to the format given just above as *assembly language*. At the present time, it is not worth debating the merits of either view.

I.2. SYMBOLIC ASSEMBLY LANGUAGE PROGRAMMING

The biggest disadvantage to machine language as described in the previous section, even in the form CLA 0 0 0 0 64, was that the insertion or elimination of a single instruction (or piece of data) caused many—if not all—of the addresses in other instructions to be incorrect. This situation could be improved somewhat by a scheme of *relative addressing* or *regional addressing*, in which the program was divided into sections, each of which started in a fixed location. Addresses within each section were given relative to the starting location. Thus CLA 0 0 0 0 R64 might refer to the 64th location within the R section of code.

While this was an obvious improvement, it was the development of completely symbolic notation and addressing for both instructions and data that freed the programmer from worrying about changing all occurrences of R64 to R63. A preliminary step in this direction was the work done at MIT on Whirlwind in 1952, and by Rochester [RT53] which used numeric *symbolic addresses*. These numbers had no mnemonic or numerical significance but were merely used as symbols for addresses. The culmination of this early work was the use of mnemonic symbols for both instructions and data, thus permitting the user to write CLA TEMP where TEMP stands for the location in memory of the value of a variable, e.g., temperature. A whole generation of programmers thus learned to use the IBM 704 by writing programs in SAP (Symbolic Assembly Program) [XY56].

I.3. EARLY DEVELOPMENT OF BETTER TOOLS

I.3.1. Specific Needs to Be Met

The histories of automatic coding, programming languages, and the development of better tools to assist programmers are almost—but not completely—synonymous. As noted in the ensuing discussion, virtually all the significant problems arising in the *early* days were, or have been, solved by the increasing development of higher level languages (which are defined and discussed in Section I.4). It was not until the existence of second generation computers (circa 1959) that the speed, cost, and difficulty of manually changing jobs began to force the development of operating systems; this latter is probably the one major area that is entirely distinct from programming languages which has had, and is having, a major impact upon the overall computing community. (Time-sharing is considered to be in the broad category of operating systems.) One can go even further and say that the development of operating systems was really to help the installation managers rather than the individual programmers; the latter seldom see a direct benefit from the operating system unless it greatly reduces their turnaround time (which is not always true!). Time-sharing of course attempts to bring to the programmer of third generation computers the advantages of the on-line debugging which the user of the first computers usually had. However, even this inherently nonlanguage development requires consideration of the language with which the user will address the system.

The desirability of having a symbolic code forced the development of symbolic assembly languages. However, that was not sufficient to meet the growing demands of programmers. For one thing, programmers wanted the ability to use other people's code wherever it was appropriate. This could not always be done because of differences in notation and lack of an effective way to link the pieces together. One of the main motivations for using other people's code was that certain programs were being written over and over again. For example, square root and trigonometric routines were being written by the dozens. In some cases, this proliferation was justifiable because one person was interested in saving space and therefore wrote as short a subroutine as he could, while somebody else wanted to save as much time as possible and therefore removed loops even at the expense of using more storage locations. In another case, people had differing requirements for precision, and this caused another whole set of routines to be developed with varying degrees of precision. However, eventually the individual effectiveness of a particular program became less important and was subjugated to the overall effectiveness of a group of programmers. Thus there developed the need for effective library facilities and, in particular, library routines—many of them parameterized—that could be invoked very easily by a programmer.

Another area where a need rapidly became apparent was in routines which differed not in concept but only in specific cases or which had many input parameters, not all of which were numbers. The best example of this was the early sort routines, which used the same techniques but differed in the coding because the key might be in the first word of the record or the fifth or the ninth, and the key might be three characters long or eight, etc. The early work of F. Holberton [HF54] in developing sort generators for UNIVAC had a significant impact on this type of problem because she provided a set of routines which would partially write themselves once given the necessary input parameters.

Programmers not only wanted the ability to use other people's code, but they wanted the capability of easily bringing together small sections of a program. One of the earliest and most significant efforts along these lines was the development of the subroutine library for the EDSAC as represented and described by Wilkes, Wheeler, and Gill [WI51].

Finally, there was an increasing demand for being able to write shorthands of various kinds. Once people had written sequences of code, they were interested in finding a shorthand way to write the same or similar information, and calling the material from the library was not always appropriate. In addition, people wanted better and better notation, where they implicitly defined "better" as "more natural".

All these needs were attacked by different people in different ways. In my opinion, Dr. Grace Hopper probably did as much as any other single person to sell many of these concepts from an administrative and management, as well as a technical, point of view. See, for example, Hopper [HP55].

One of the first meetings held to discuss the subject that was then called *automatic programming* was sponsored by the Office of Naval Research in May, 1954 and reported in [DN54]. At that time, a number of interesting systems were described, some of which are covered in later sections. Probably the most significant ideas that were mentioned at that early meeting and that are not covered in this book are the concept of code generation discussed by Holberton, the editing generator of Waite and Elmore, the analytic differentiator of Kahrimanian, and the grandiloquent objective (not yet achieved) of the universal code described by Gorn; all these are described in [DN54].

The next significant meeting was the symposium on "Advanced Programming Methods for Digital Computers" held under the joint sponsorship of the Navy Mathematical Computing Advisory Panel and the Office of Naval Research in June, 1956 [DN56]. About the only paper presented at that meeting which had any significance as far as programming languages are concerned was the paper by Thompson [TM56] which discussed OMNICODE. This is sufficiently similar in principle to the PRINT system discussed in Chapter IV so that it will not be mentioned here (although the reader should realize that the details are significantly different).

The next meeting of major importance was the symposium on "Automatic Coding" held in January, 1957 at the Franklin Institute in Philadelphia [FK57]. The major items covered there were PRINT I, B-$\emptyset$, IT, and the Matrix Compiler; they are described elsewhere in the book.

I.3.2. Brief History of Early Efforts

A number of systems were developed in the early years (defined to be prior to 1957) which made significant contributions to the development of higher level languages. Chief among these, in approximate chronological order, are Short Code (UNIVAC), Speedcoding (IBM 701), Laning and Zierler system (Whirlwind), A-2 and A-3 (UNIVAC), BACAIC (701), and PRINT (705). These are all described in Chapter IV, and the references are given there. The early work of Rutishauser in Switzerland is also mentioned there, even though this book is only attempting to deal with American developments.

These systems generally provided some type of mathematically oriented operation (e.g., addition, computation of sines) and control functions, together with either fixed or variable operands. In each case, the information written by the programmer in one line or statement was either interpreted or was directly equivalent to several lines of actual machine code. However, most of these systems had a fixed format and, in particular, did not permit the writing of mathematical expressions in anything resembling natural notation. Only the Laning and Zierler system and BACAIC had this latter facility.

Figure I-1 is a list of the *automatic programming systems* of 1959. Note that many of them would not be considered programming languages by the criteria established later in this chapter.

No attempt has been made in this book to include those systems which contributed strongly to the development of either better symbolic assembly

Computer	In ACM Library	Do Not Have
704	AFAC CAGE CORBIE FORTRAN KOMPILER 3 MYSTIC NYAP PACT IA REG-SYMBOLIC SAP	ADES FORC
701	BACAIC DUAL-607 FLOP JCS-13 KOMPILER 2 PACT I QUICK SEESAW SHACO SPEEDCODING 3	BAP DOUGLAS GEPURS LT-2 QUEASY SO 2 SPEEDEX
705	ACOM AUTOCODER ELI PRINT I SOHIO SYMBOLIC ASSEMBLY	FAIR
702	AUTOCODER ASSEMBLY SCRIPT	
650	ADES II APT BACAIC BELL BELL L2, L3 CASE SOAP III DRUCO I EASE II ELI FAST FOR TRANSIT FORTRUNCIBLE IT IT 3 MYSTIC RELATIVE RUNCIBLE SIR SOAP I SOAP II	BALITAC ESCAPE FLAIR KISS MITILAC OMNICODE SPEEDCODING SPUR
650 RAMAC	GAT-2	
NORC	NORC COMPILER	
7070	BASIC AUTOCODER	

Figure I-1. List of automatic programming systems (1959). Source: *Comm. ACM*, Vol. 2, No. 5 (May 1959), p. 16. By permission of Association for Computing Machinery, Inc.

Computer	In ACM Library	Do Not Have
1103, 1103A	APT BOEING CHIP FAP FLIP-SPUR MISHAP MYSTIC RAWOOP-SNAP UNICODE USE	TRANS-USE
UNIVAC I, II	A0, A1, A2 ARITHMATIC (A-3) BIOR FLOWMATIC (B-0) GP GPX (II ONLY) MATHMATIC (AT-3) MATRIX MATH NYU, OMNIFAX SHORTCODE UNISAP X-1	MJS RELCODE
DATATRON 201 204 205	APX III DUMBO PURDUE COMPILER SAC SIMPLE UGLIAC	ANCP BELL DATACODE I DOW COMPILER SHELL SPAR STAR 0
G-15	DAISY 201 FLIP INTERCOM 101 INTERCOM 1000	POGO
WHIRLWIND	COMPREHENSIVE SUMMER SESSION	ALGEBRAIC
FERUT	TRANSCODE	
JOHNNIAC	EASY FOX	
ILLIAC	ILLIAC	
LGP-30	ERFPI JAZ SPEED	
MIDAC	EASIAC MAGIC	
LARC		K5 SAIL
FERRANTI MERCURY	AUTOCODING MAC (NORWAY)	
FERRANTI PEGASUS	AUTOCODE	

programs or file-handling techniques. In particular, such systems as PACT (see papers by Baker [BK56] and others in the same journal, and Steel [ST57]), BIOR [RR55a], and SURGE [NM00] are deliberately excluded.

I.4. DEFINITION OF PROGRAMMING LANGUAGES

I.4.1. Definition Problem

The ASA (now USASI) standard *Vocabulary for Information Processing* [AA66b] defines a *programming language* on page 23 as "A language used to prepare computer programs". The IFIP-ICC Glossary [IF66] defines on page 79 a language as "A general term for a defined set of *symbols* and rules or conventions governing the manner and sequence in which the symbols may be combined into a meaningful communication," with a note that "An unambiguous language, intended for expressing *programs*, is called a PROGRAMMING LANGUAGE." This glossary also states that the term *pseudocode* has been used in England to denote a programming language which is not a computer language, but this usage is deprecated by the IFIP-ICC Glossary.

While these definitions may be true from an overall abstract point of view, they do not—in my opinion—reflect actual current usage. Furthermore, neither glossary includes the term *higher level language*. It is intuitively clear that there is a significant difference between symbolic assembly languages and the languages which are discussed in this book. However, not only is there a lack of a specific term for the items in this book but, furthermore, a symbolic assembly program with a very powerful macro facility can certainly be made to look very much like what is frequently called a *programming* or *higher level* language. (See for example the XPOP system of Halpern [HL64].) For the purposes of this book, and admittedly contrary to the opinion of many, these two terms will be used interchangeably. One of the prime differences between assembly and higher level languages is that to date the latter do not have the capability of modifying themselves at execution time. In one instance—namely, LISP 1.5 (see Section VI.5)—an equivalent result can be achieved because the program is represented internally in the same form and can be acted on as data. However, no language in this book has the facility for changing, e.g., a GO TO to an IF. The lack of this capability has not proved much of a handicap and is cited merely because it is one of the clear-cut distinctions between an assembly language and a programming language.

Because there is no satisfactory definition, it seems more effective to try to define a programming language through its characteristics rather than by a specific definition. Thus, in my opinion, a programming language is *a set of characters with rules for combining them which have the following characteristics.*

I.4.2. DEFINING CHARACTERISTICS

1. *Machine Code Knowledge Is Unnecessary*

A programming language requires no knowledge of the machine code by the user. In other words, the user need only learn the particular programming language and can use this quite independently of his (perhaps nonexistent) knowledge of any particular machine code. Thus he need not learn about what registers are available on the computer, nor the specific hardware instructions that are required to activate the computational and logical processes. In many cases he can also remain ignorant of the internal representation of numbers; thus he can avoid worrying about whether numbers are represented internally as binary, hexadecimal, decimal, etc. However, this should not be interpreted to mean that the user can completely ignore the actual computer if he wants to obtain maximum (or even reasonable) effectiveness from it. For example, he may wish to take advantage of certain machine facilities (e.g., mass storage devices) which are known to him and which can provide more efficient programs; even more specifically he obviously cannot use input/output equipment which does not exist on a particular computer configuration. He might conceivably wish to concern himself with whether numbers were represented in binary or decimal fashion because this could affect certain points of computational precision that might be of concern to him.

In summary, the first characteristic of a programming language is that the user can write a program without knowing much—if anything—about the physical characteristics of the machine on which the program is to be run. This same comment does not apply if he wishes to obtain maximum efficiency.

A further constraint is that the user should be unable to affect directly the machine registers and memory. This rules out such ideas as Wirth's PL360 [WT68] from being considered a higher level language (nor does he claim it is).

2. *Potential for Conversion to Other Computers*

Since the first characteristic states that the user need not know the details of the particular computer on which his program is to be run, it follows that a programming language must have some significant amount of machine independence. The whole question of compatibility and conversion is discussed at some length in Section II.4. It is sufficient to say here that a major characteristic of a programming language is that there must be a reasonable potential of having a source program written in that language run on two computers with different machine codes without rewriting the source program. Again, we are dealing with both absolute and relative quantities. In the absolute sense, the program may be able to be moved from one machine to another with no rewriting. In most programming languages, some—but often very little—rewriting of the source program is necessary.

3. *Instruction Explosion*

When a source program [see the definition in Section I.4.3(1)] written in a programming language is translated to the actual machine code, there is normally more than one machine instruction created for each statement in the programming language. For example, a statement in a programming language might be something of the form A = B + C * D or MOVE A TO B. Normally each of these phrases requires more than one machine instruction to execute it, and this is the major difference between a symbolic assembly language and a higher level language. In fact, many compilers actually translate the source program to a symbolic assembly language.

To be considered a programming language, there should be no need for the user to write any sequence of machine code. This provision causes the exclusion of macro assemblers from the category of programming languages (by assuming that *some* user must write the machine instructions for the macro).

4. *Problem-Oriented Notation*

A programming language must have a notation which is somewhat closer to the specific problem being solved than is normal machine code. It usually permits a relatively free format. Thus, for example, the first illustration given in Section I.4.2(3) might be translated into a sequence of instructions such as

$$\begin{array}{l} \text{CLA C} \\ \text{MPY D} \\ \text{ADD B} \\ \text{STO A} \end{array}$$

which is clearly less understandable than the programming language form A = B + C * D. Again, this notational question is a relative one because what is considered problem-oriented and relatively free in one case might be considered quite rigid and unnatural in another. However, as in each of the preceding discussions, the comparison that is being made is with a symbolic assembly language and not between two types of higher level language.

To fall within the spirit of this concept of problem-oriented notation, a programming language must *not* require that each statement type or executable unit be specifically identified or flagged in a standard terminology and location. Furthermore, a fixed format, or the existence of a form to be filled out, is not considered problem-oriented notation. The first of these requirements rules out the Autocoder series, e.g., [IB61a], from being considered programming languages. The second rule excludes the class of report generators, e.g., [IB65d], and decision tables, e.g., [EP66], [KV60], [UC61]. The exclusion of report generators, and to a lesser extent the exclusion of the Autocoder languages, runs contrary to much of the popular nomenclature. Very specifically, Report Program Generators (RPG) are commonly referred to as *programming languages*. However, I believe my classification is justified by the rigidity and lack of flexibility of the normal RPG "programs" which consist primarily of filling in a preprinted form. *This in no way implies any failing or lack of importance of systems of this type; it merely excludes them from the class of languages considered in this book.*

I.4.3. Basic Terminology

1. *Source Program*

The actual program written in a higher level language is called the *source program*. This is the material that is put into the computer by the user for the purpose of obtaining results. The source program contrasts with the object program (which does not always exist), defined in the next paragraph.

2. *Object Program*

A source program can usually be translated to an *object program*. [Note the differences between compiler and interpreter in Sections I.4.3(3) and I.4.3(4) below.] The object program can actually exist in many forms, depending on the particular system involved. It can exist in pure binary form, or it could actually exist in a fairly complex symbolic assembly language form. The phrase *object program*, strictly speaking, relates only to the final binary form that can be executed by the computer, but in common conversation it is often used to denote the result of translating the source program at least down to an assembly level.

3. Compiler

A compiler is a program, not a piece of hardware. A *compiler* is simply a program which translates a source program written in a particular programming language to an object program which is capable of being run on a particular computer. A compiler is therefore both language and machine dependent. The most important characteristic of a compiler is that its output is a program in some form or another and not an answer of any kind. This contrasts with the interpreter defined in Section I.4.3(4). The first completed compiler seems to be the A-0 system developed by Dr. Grace Hopper and her staff at Remington Rand in 1952; see [HP53] and [HP53a].

A compiler must perform at least the following functions: Analysis of the source code, retrieval of appropriate subroutines from a library, storage allocation, and creation of actual machine code. In current systems, some or all of these functions (except the first) may actually be performed by another part of the general operating system (e.g., a loader), but these functions are conceptually part of the compiling process. Thus the compiler acts very much as an executive routine to obtain and combine the necessary pieces of information to produce a machine-executable program.

The word *translator* has been in and out of vogue for years as a synonym for compiler. In my opinion, translator is too general a term to use for the specific process of turning a source program written in a higher level language into machine code.

4. Interpreter

An interpreter is a program which executes a source program, usually on a step-by-step, line-by-line, or unit-by-unit basis. In other words, an interpreter will usually execute the smallest possible meaningful unit in the programming language. The output of an interpreter is an actual answer, i.e., the result of performing the actions designated in the program.

The greatest disadvantage of an interpreter is that certain phases of work and analysis must be done repeatedly. In particular, the scan of a statement which is to be executed for varying values of a particular parameter must take place each time that a new value is to be used. This contrasts with the compiler, which performs this translation function only once. On the other hand, the disadvantage to the compiler is that it does not produce answers; as soon as a change in the program is made, a recompilation must be made.

The originally clear-cut distinctions between compilers and interpreters have become quite blurred. Some systems (e.g., QUIKTRAN, see Keller, Strum, and Yang [KR64]) compile partway, i.e., translate the source program to some other form and then interpret that information. This is an attempt to obtain the advantages of both concepts, while minimizing the disadvantages of both.

5. *Automatic Coding*

In the very early stages of work in this area, the phrase *automatic programming* was used to mean the process of writing the program in some higher level language. As time went on, it became clear that this encoding was only part of the entire process of programming since there were phases of analysis, documentation, debugging, testing, etc. Hence, the term *automatic coding* began to apply to the portion of the overall programming effort that related specifically and only to the process of actually writing the source program and having it translated to a form where it could be run on a computer.

6. *Automatic Programming*

The term *automatic programming*, which as stated above was originally used to cover anything to do with higher level languages, is defined on page 10 of the USASI Glossary [AA66b] as "The process of using a computer to perform some stages of the work involved in preparing a program". Thus, automatic coding is a particular subset of automatic programming, which is as it should be since coding is one of the many facets of programming.

I.4.4. DIFFERENCE BETWEEN PROGRAMMING LANGUAGE AND APPLICATION PACKAGE

In the past few years there has been an increasing number of special application packages developed. One of the earliest and most significant of these was the work done on linear programming. More recent areas involve type composition [IB00f], demand deposit accounting [IB00e], traffic control [IB66c], inventory management [IB00] and [IB00d]. However, it is important to realize that an application package and a programming language are not the same. An application package tends to be a set of routines which are heavily parameterized, so that an individual user can supply the specific information which is needed for his particular direct usage. The information is often supplied through tables or filling in a form. File formats are usually specified by the application package. In some cases the execution sequence of the routines is predetermined, e.g., student scheduling [IB66j]. In others, the user decides which routines he needs and what the sequence should be, e.g., bill of material processing [IB66k]. In the latter situation, the user sometimes has to write a control program in an assembly or higher level language to set up and call the necessary routines. A programming language, on the other hand, provides flexibility in the way in which information is conveyed and, more importantly, provides the tools with which the subroutines or packages can be built-up. An application package is limited to use in a narrow area. A programming language usually involves a wider potential range of applications, although the languages discussed in Chapter IX are designed for very specific—and sometimes quite narrow—applications.

I.5. ADVANTAGES AND DISADVANTAGES OF HIGHER LEVEL LANGUAGES

As with any item, it is impossible to obtain something for nothing; therefore, there are both advantages and disadvantages to programming languages, where the alternative is some type of assembly language. It is essential to realize that the comparison is being made between a symbolic assembly language (which might but does not necessarily have macros) and some type of higher level language which has the defining characteristics given in Section I.4.2. Furthermore, the comparison is being made between an assembly and higher level language of roughly equivalent orders of complexity within their given classes. Thus, in examining the advantages and disadvantages of a powerful (very simple) programming language, it is tacitly being compared to a powerful (very simple) assembly language. This point will be critical in several of the advantages given below. Furthermore, the programming language must be appropriate to the task; thus a language with notation well suited to scientific problems is not likely to be much help in business data processing (although this has actually been done with FORTRAN; see Robbins [RM62]).

I.5.1. Advantages

1. *Ease of Learning*

A very significant advantage to a higher level language is that it is easier to learn than a machine-oriented language. This is probably the main place in which the relative aspect referred to above is significant. An extremely powerful programming language might be harder to learn than an assembly language with only a dozen instructions. However, given programming and assembly languages of approximately the same complexity in their relative classes, the programming language will be easier to learn. This ease of learning actually has two facets to it. The programming language may itself be complex, but its ease of learning often comes because the notation is somewhat more related to the problem area than is the machine code—this is essentially the fourth defining characteristic given on page 10. The second facet is that more attention can be paid to the language and the logic of the program rather than to the idiosyncracies of the physical hardware which are significant when one deals in machine code.

A further element of comparison in the ease of learning is that learning a subset of a complex programming language may be, and probably will be, very much easier than learning a subset of a complex assembly language. Furthermore, the subset of the programming language will probably be more useful and powerful than an equivalent subset of the assembly language. Thus, although some programming languages have extremely thick manuals, this is because they provide all the very detailed definitions that are needed for writing compilers and sophisticated programs; the user who does not wish to learn (or have) all the power available to him need not be bothered with the full language.

2. *Ease of Coding and Understanding*

Because the notation is considerably more problem-oriented, the actual coded program is generally easier to write. This is exemplified not only by the case of algebraic expressions given on page 10 but by such things as IF C IS GREATER THAN A+B, GO TO ALPHA OTHERWISE GO TO BETA. which is easier to write than an equivalent symbolic form which might look like the following:

CLA A	
ADD B	(Calculate A+B)
SUB C	(Calculate A+B−C)
TRN ALPHA	(Transfer control to ALPHA if A+B−C is less than 0, i.e., if C is greater than A+B)
JMP BETA	(Transfer control to BETA)

The other half of the advantage is the ease of understanding the program once it is written. These two aspects reflect the differences between actually writing the program and trying to understand an already existing program (either one's own or, more likely, someone else's). The higher level language is clearly easier to read and understand, as seen from the example above.

In addition, the complexities of today's large computers make it very difficult to learn to program them at all, let alone effectively.

3. *Ease of Debugging*

A problem written in a programming language is generally easier to debug than one written in a symbolic assembly language, for two major reasons. First, there tends to be less material written because of the explosion factor given as the third defining characteristic of a programming language. Thus, in comparison with a program written in assembly language, the source program will generally be physically shorter. Since the number of errors is roughly proportional to the length of the program, obviously there will be fewer errors. In some cases it might turn out that the program written in the higher level language was actually longer if measured by the number of characters actually written. (This happened in the example given on p. 15.) However, the program is easier to debug because the notation is so much more natural; more attention can be paid to the logic of the program with less worry about the details of the machine code. For example, although there might be more characters involved in writing **READ NEXT RECORD FROM TAPE ALPHA** than in **REDABC, ALPHA**, the former is easier to understand, particularly when there may be a whole sequence of six-letter instructions which differ by at most one letter.

4. *Ease of Maintaining and Documenting*

One of the greatest advantages to a programming language is the fact that it provides certain documentation automatically because of the notational advantages; it is also considerably easier to maintain. There are very few programs which last very long without requiring some changes, and a combination of reasonably natural notation plus shortness of program make the higher level language quite advantageous. In addition, one of the great difficulties in changing a program written in assembly language is to make sure that a change in one instruction does not have major (and unpleasant) ramifications elsewhere. This factor applies not only to the logic of the change (which must also be considered when dealing with the higher level language) but, more significantly, to various tricky coding techniques which might be forgotten by the time the change was made and result in incorrect code.

5. *Ease of Conversion*

Since the second defining characteristic of a programming language is the potential for conversion to other computers, it is not surprising that this is considered an advantage. Since by now it is clear that programming costs equal or exceed hardware costs, it is not surprising that the problem of conversion is a very major one. In many cases, companies have been unable to acquire new computers because of the enormous cost of converting their existing programs to the new machines. This has forced the manufacturers to pay much more attention to compatibility among the computers they offer to their customers and to provide technically graceful ways of converting programs from one machine to another. However, since programming languages are relatively machine independent, the ease of conversion becomes an extremely important advantage. The various types of conversion, and their significance, are discussed in Section II.4.

I.5. ADVANTAGES AND DISADVANTAGES OF HIGHER LEVEL LANGUAGES 17

6. *Reduce Elapsed Time for Problem-Solving*

Probably the greatest single overall advantage to a programming language is that it usually reduces the total amount of elapsed time from inception of the problem to its solution. This is particularly true for one-shot problems—problems in which a single or only a small number of cases need to be run. Higher level languages have cut this elapsed time from months to weeks in some cases and from days to hours in other cases. Although sometimes one particular facet of the overall process might be worse in a higher level language (specifically, the compilation time, discussed in Section I.5.2.2), the overall problem solution time is greatly reduced. This is somewhat less of an advantage for long-term production runs, such as payroll. In that case, the advantages of ease of maintenance and documenting probably overshadow the elapsed time advantage, although the latter is still available.

I.5.2. Disadvantages

1. *Advantages Do Not Always Exist*

There is a subtle point that the advantages stated above do not always exist in specific cases, and a person might be worse off; however, this would only tend to arise in a comparison of a complex and powerful programming language versus a simple assembly language. Thus the programming language might be extremely difficult and hard to learn; and unless proper attention is paid to the compiler and other facets of the overall system, the other advantages may not themselves accrue. Fortunately, this seldom occurs.

2. *Time Required for Compiling*

A very obvious disadvantage to the use of a higher level language is that the additional process of compilation requires more machine time than the straight assembly process; the compilation time might, in fact, require more than the machine time saved from easier debugging. This additional machine time is most easily observed by recognizing the fact that a very common compiling technique is to translate the source program to an assembly language which already exists for the given computer and letting the standard assembly program create the final object code. (Naturally techniques have been developed to avoid this particular difficulty, but they are not always applicable.) Compilation time is a particular disadvantage on one-shot problems in which the compilation time sometimes exceeds the time actually required to produce the answers. Another disadvantage associated with compilation time is the necessity of *re*compiling every time a change in the source program is made. However, sometimes modern assemblers are so complex that they take "longer" than the translation of an equivalent program in a higher level language.

3. *Inefficient Object Code*

A disadvantage which significantly affects production runs occurs when the compiler produces inefficient object code. When a program is to be run repeatedly, it is important that the final program be efficiently coded because of the constant repetitive use. The counterargument to this, of course, is that compilers nowadays generally produce code that is at least as good as the average programmer, and there are only a limited number of really expert programmers who can write the most efficient machine code. A further counterargument is that it is usually possible to take very critical routines, which are generally quite short and encapsulated, and code them as efficiently as possible in machine code.

A disadvantage in this area which is sometimes unjustly blamed on the compiler occurs when the programmer writes inefficient source programs in the higher level language and obtains inefficient object programs as a result. Although it is easier to code in a higher level language than in machine code, there is still a difference between good and poor coding. A program that has been written inefficiently (e.g., unnecessary control transfers and extra computations) with respect to the programming language will produce inefficient object code regardless of how good the compiler is.

4. *Difficulties in Debugging Without Learning Machine Language*

If a person does not know machine code, and the compiler does not provide the proper type of diagnostics and debugging tools, the program may actually be harder to debug than an assembly language program which the user understands. A person who must look at an octal memory dump will have a lot more trouble debugging his high level source program than he would if he had written it in assembly language. Thus a compiler which does not provide proper attention to this aspect may greatly reduce the advantages of a higher level language or cause them to disappear entirely.

5. *Inability of the Language to Express All Needed Operations*

In some problems there are operations to be performed which cannot be expressed in the programming language, or if they are available, they will be so awkward as to be almost useless. Thus, to handle individual bits in a language designed only to manipulate numeric quantities is virtually impossible, and certainly inefficient. The user may find himself trapped by being unable to do certain manipulations without resorting to machine code. This usually occurs when he has chosen the language unwisely for his particular application. However, a more common problem is the poor match between the older (and more popular) languages and third generation hardware. For example, a language with no facilities for dealing with random access memories requires the user to either ignore his equipment or resort to machine language to deal with it.

I.5.3. Overall Evaluation

In spite of the fact that higher level languages have been with us for over 10 years, there has been relatively little quantitative or qualitative analysis of their advantages and disadvantages. One very small study is given by Shaw [SH66] and some information is given by Nelson *et al.* [NE65].

In spite of this paucity of definitive information, the current milieu calls for the use of higher level languages. People who use assembly code are—if not in an actual minority—considered somewhat archaic or old-fashioned. The fact that there is a tremendous proliferation of languages (as witnessed by all those described, plus others not even mentioned, in this book) indicates that we have not yet solved the problem of knowing what is really needed by the user. Some comments about possible future directions are given in Chapter XI. However, the net overall evaluation appears to be that higher level languages have proved their worth and are definitely here to stay.

I.6. CLASSIFICATIONS OF PROGRAMMING LANGUAGES AND PROPOSED DEFINITIONS

As indicated earlier, it is very difficult to define a programming language. However, it is a little easier to propose definitions for classes of programming languages. The terms to be defined are the following: Procedure-oriented and nonprocedural; problem-oriented, application-oriented, and special purpose; problem-defining, problem-describing, and problem-solving; hardware, publication, and reference. Note that some of these are overlapping and that a particular language may fall into more than one of these categories.

I.6.1. Procedure-Oriented Language

A *procedure-oriented* language is one in which the user specifies a set of executable operations which are to be performed in sequence; the key factor here is that these are definitely executable operations, and the sequencing is already specified by the user. FORTRAN, COBOL, and PL/I are examples.

I.6.2. NONPROCEDURAL LANGUAGE

The term *nonprocedural language* has been bandied about for years without any attempt to define it. It is my firm contention that a definition is not really possible because *nonprocedural* is actually a relative term meaning that decreasing numbers of specific sequential steps need be provided by the user as the state of the art improves. The closer the user can come to stating his problem without specifying the steps for solving it, the more nonprocedural is the language. Furthermore, there can be an ordered sequence of steps, each of which is "somewhat nonprocedural," or a set of executable operations whose sequence is not specified by the user. Both cases contribute to "more nonproceduralness". Thus, before the existence of such languages as FORTRAN, the statement

$$Y = A + B * C - F / G$$

could be considered nonprocedural because it could not be written as one executable unit and translated by any system. Right now, the sentences *CALCULATE THE SQUARE ROOT OF THE PRIME NUMBERS FROM 7 TO 91 AND PRINT IN THREE COLUMNS* and *PRINT ALL THE SALARY CHECKS* are nonprocedural because there is no compiler available that can accept these statements and translate them; the user must supply the specific steps required. Another type of nonprocedural statement is a higher level primitive operation, e.g., integration. Note that there is a fundamental language difference between writing INTEGRATE F(X) FROM A TO B USING SIMPSON'S RULE and CALL SIMP (F(X), A, B) although the same subroutine could be used for both. In cases where subroutines do not exist (as in the earlier two examples), then obviously the detailed steps must be specified.

As compilers are developed to cope with increasingly complex sentences, the nature of the term changes. Thus, what is considered nonprocedural today may well be procedural tomorrow. The best examples of currently available nonprocedural systems (not really languages) are report generators and sort generators in which the individual supplies the input and the output without any specific indication as to the procedures needed.

Specific attempts to raise the level of nonproceduralness in different ways are discussed by Wilkes [WI64], Rice and Rosen [RI66], Klerer and May [KL67], and Schlesinger and Sashkin [QL67]. General discussions of some of the issues are given by Young [YJ65] and Whiteman [WF66].

I.6.3. PROBLEM-ORIENTED LANGUAGE

The term *problem-oriented* has been used in many ways by different people, but it seems that the most effective use of this term is to encompass any language which is easier for writing solutions to a particular problem than assembly language would be. Any current programming language illustrates this; thus, the term *problem-oriented* is a general catchall phrase.

I.6.4. APPLICATION-ORIENTED LANGUAGE

The term *application-oriented* seems to apply best to a language which has facilities and/or notations which are useful primarily for a single application area. The best illustrations of this are such things as APT for machine tool control and COGO for civil engineering applications, both of which are discussed in Chapter IX. Notice that both of these are of course problem-oriented languages. On the other hand, FORTRAN and COBOL are problem-oriented but much less application-oriented than APT or COGO. Here again, the term is somewhat relative because FORTRAN is suitable for applications involving numerical mathematics, whereas COBOL is obviously suited for business data processing and the overlap between these is relatively small. The wider the application area, the more general the language must be.

I.6.5. SPECIAL PURPOSE LANGUAGE

A *special purpose* language is one which is designed to satisfy a single objective. The objective might involve the application area, the ease of use for a particular application, or pertain to efficiency of the compiler or the object code.

I.6.6. PROBLEM-DEFINING LANGUAGE

A *problem-defining* language is one which literally defines the problem and may specifically define the desired input and output, but it does *not* define the method of transformation. There is a significant difference among a problem (and its definition), the method (or procedure) used to solve it, and the language in which this method is stated. The best current illustrations are report and sort generators, although none of these involves languages in the sense of Section I.4.

I.6.7. Problem-Describing Language

A much more general type of language classification is that referred to as *problem-describing*, in which the objective is described only in very general terms, e.g., *CALCULATE PAYROLL*. All this does is cite, in the most general way, the problem which is to be solved but gives no indication of its detailed characteristics, let alone how to solve it. We are an extremely long way from this!

I.6.8. Problem-Solving Language

Finally, a *problem-solving* language is one which can be used to specify a complete solution to a problem. Like the term *nonprocedural*, this is a relative term which changes as the state of the art changes. All procedure-oriented languages are problem-solving languages.

I.6.9. Reference Language

A *reference* language is the definitive character set and form of a language. It usually has a unique character for each concept or character in the language, is one-dimensional, and need not be suitable as computer input. In some cases, English is the reference language; in other cases, a fixed set of symbols is provided. The concept of having a reference language, as distinguished from a publication or hardware representation language (discussed below), was introduced by the ALGOL committee in their first report [PR58]. In fact, ALGOL is the only language in this book with these three forms. The reference language need not be particularly easy to read.

I.6.10. Publication Language

A *publication* language is some well-defined variation of the reference language which is suitable for publication. It is designed to be suitable for printing and/or writing; therefore, it would have reasonable rules and characters for such things as subscripts, exponents, spaces, and Greek letters. The publication language would normally be the means of communication between people (using printed media). There can be many publication languages and they can contain different characters, but there must be a well-defined mapping between the publication and reference languages. An illustration of this is the use of an up arrow ↑ to denote exponentiation in the ALGOL reference language, but the use of a raised symbol in the publication language, e.g., A ↑ 2, becomes A^2.

I.6.11. Hardware Language

A *hardware* language, sometimes called a *hardware representation*, is a mapping of the reference language into a form which is suitable for direct input to a computer. The number and type of characters used must be that accepted by the computer involved. A hardware language must have a well-defined mapping between itself and the reference language, e.g., ✷✷ might be a hardware representation of the ↑ in the reference language.

I.7. FACTORS IN CHOICE OF A LANGUAGE

Assuming that a decision has been made *not* to use assembly language (see, e.g., Shaw [SH66]), there is currently no scientific, or even logical, way to choose the best programming language for a particular situation. Part of the difficulty stems from the fact that the situation itself is usually defined poorly, and potential for change in the application area is a factor which must be taken into consideration. It is definitely *not* the purpose of this book to provide *all* the information needed by a potential user to choose the programming language most suited for his purposes. However, it *is* one of the purposes to supply *some* of this information and to indicate the factors which should be considered. The reader is cautioned to be very careful in applying the items discussed in this section to a particular case. Not all factors are relevant in all situations, nor are they all equally important. In virtually all cases, no single language will be ideal for a particular application, let alone for a particular installation, and probably not for an entire company.

An increasing amount of work is being done to develop some fairly specific *methods* for evaluating languages *and* their compilers. Scientific evaluations have seldom been made, and documented even less often, and the few attempts to date seem to be without any quantitative measurements. Questionnaires and comparisons have been developed by Shaw [SH62] and Budd [QH66]; although the latter pertains only to FORTRAN and COBOL, it is quite detailed for those languages. General discussions are given by Haverty [HV64], Chapin [CZ65], and Schwartz [SC65]. A number of unpublished papers on evaluations for specific military applications also exist.

Some of the terms and/or concepts used below are defined and discussed in some detail in Chapters II and III, particularly the former.

I.7.1. Suitability of Language for Problem Area and Projected Users

The most important factor in the choice of a language is whether it contains the elements needed to solve the particular class of problems for which it is being considered. In the simplest case, a language which provides good facilities for handling equations may not provide the character handling and input/output facilities needed to process a payroll. Conversely, a language which is *too* large, i.e., has many more facilities than are needed, is not necessarily desirable since the user will be paying a heavy price because of less efficiency in his specialized area. While these points are fairly obvious at a gross level, there are other elements in the language suitability issue. For example, if there is to be much array handling, then the type and amount of subscripting which is permitted may be significant. Another case might involve the types of data names which are permitted; for example, if the application involves inventory and all the stock items are identified by numbers, then it might be more convenient if these were allowed as names in the program.

In addition to the capabilities of the language, the type of actual users must be considered. There is an obvious difference among experienced programmers, professionals in other fields, novice programmers, open shop versus closed shop, etc. The amount of formalism or naturalness in the language relative to the projected users is of vital importance.

In summary, the potential user must first examine the language at a gross level to see whether it supplies the general capabilities he needs. Then he must determine whether individual features which might be very important in a particular situation are available. (See also Section I.7.7.) Finally, he must consider the *style* of the language relative to the intended users.

I.7.2. Availability on Desired Computer

The most obvious question which must be asked (and which is also raised in Section II.7.1) is whether there is an implementation of the language on the desired computer (configuration). It is obviously useless to decide on a superb language for a particular application and then find there is no way to obtain running programs. Of course in some cases the language may be deemed so worthwhile that a particular installation would choose to finance a compiler if there was not one existing already.

If there is a compiler available, then a particular point to watch out for is the exact computer configuration which it requires. It does not help to find an excellent language and an efficient compiler if the latter requires twice as much memory capacity as the installation possesses. Again, in this case, if other factors warrant it, then there might be justification for obtaining the extra memory.

I.7.3. History and Evaluation of Previous Use

Once the user has found what he considers a suitable language and there is a compiler available on his computer, then he should consider the history of usage of this language. He should investigate such items as the reactions of previous users, users' views on its applicability in actual practice, the efficiency of the implementation (see Section I.7.4.), its potential for expansion into other (and probably unforeseen) application areas, ease or difficulty of training and effectiveness of documentation, and problems of conversion and compatibility. In short, he should consider the language based on the practical experience of others with regard to the factors in Chapter II.

I.7.4. Efficiency of Language Implementation

In choosing a language, it is essential to understand the difference between a language and a specific implementation of it. However good the former may be, a very bad compiler may render the language almost useless. The prospective user must investigate this situation very thoroughly. There may be elements in the language (some are discussed in Chapter III) which would prevent a good compiler from ever being developed. On the other hand, the first compilers for a new language almost always tend to be inefficient and remain that way until better implementation techniques are found and finances and time permit them to be used. Similarly, a language may be very difficult to implement on a particular computer (configuration), although it might have an excellent compiler on another. While this latter point is obvious in considering small versus large computers, there are other more subtle points which are relevant (e.g., type of input/output and type of indexing permitted).

The user who finds a language which is well suited to his purpose may choose to suffer the (presumably temporary) inconvenience of an inefficient compiler for the sake of long-range benefits.

I.7.5. Compatibility and Growth Potential

The meaning of compatibility and its applicability to problems in programming languages is discussed in Section II.4. The prospective user must understand what types of compatibility and conversion are available, and how important they are to him. In addition, the potential use of the language in new and unforeseen areas must be considered. While this is

26 GENERAL INTRODUCTION

obviously impossible in detail (since it would be self-contradictory to consider unforeseen areas), some thought can be given to the matter. For example, a scientific installation might consider whether it might ever be involved with data processing. A large command and control project might consider whether the application would grow into other areas. Finally, the language should be viewed from the point of possible extensions to meet other needs.

In addition to looking ahead, the user may need to look behind if there are existing applications. The consideration of a new language may involve problems of compatibility with the old one.

I.7.6. Functional (= Nontechnical) Characteristics

There has tended to be much confusion in the past due to lack of consideration of the difference between the nontechnical and the technical characteristics of a programming language. It is my hope that the delineation of these issues and a detailed discussion of them in two separate chapters will alleviate this difficulty. It suffices to point out here that the prospective user must consider the nontechnical characteristics (as discussed in Chapter II) as carefully as he considers its technical elements in order to arrive at a proper judgment.

I.7.7. Technical Characteristics

While the nontechnical characteristics of a programming language may tend to prevent it from being used in a particular application, an affirmative choice can only be made if the language contains the necessary technical features. Some relevant factors were mentioned in Section I.7.1. A careful study of Chapter III should provide a complete checklist to be used against a specific language. The importance of particular elements in a given situation is a value judgment to be made by the prospective user.

REFERENCES

[AA66b]   *American Standard Vocabulary for Information Processing*, X3.12, American Standards Association [now United States of America Standards Institute], New York, 1966.

[AD54]   Adams, C. W. and Laning, J. H., Jr., "The M.I.T. Systems of Automatic Coding: Comprehensive, Summer Session, and Algebraic", *Symposium on Automatic Programming for Digital Computers*, Office of Naval Research, Dept. of the Navy, Washington, D.C. (1954), pp. 40–68.

[BK56]   Baker, C. L., "The PACT I Coding System for the IBM Type 701", *J. ACM*, Vol. 3, No. 4 (Oct., 1956), pp. 272–78.

[CZ65]	Chapin, N., "What Choice of Programming Languages?", *Computers and Automation*, Vol. 14, No. 2 (Feb., 1965), pp. 12–14.
[DN54]	*Symposium on Automatic Programming for Digital Computers*, Office of Naval Research, Dept. of the Navy, Washington, D.C. (1954).
[DN56]	*Symposium on Advanced Programming Methods for Digital Computers* (Washington, D.C., June 28, 29, 1956). ONR Symposium Report ACR-15, Office of Naval Research, Dept. of the Navy, Washington, D.C. (Oct., 1956).
[EP66]	"How to Use Decision Tables", *EDP Analyzer*, Vol. 4, No. 5 (May, 1966), pp. 1–14.
[FK57]	*Automatic Coding* (Proceedings of the Symposium on Automatic Coding held January 24–25, 1957 at the Franklin Institute in Philadelphia), *Jour. of the Franklin Inst.*, Monograph No. 3, Philadelphia, Pa. (Apr., 1957).
[FR63]	Ferguson, H. E. and Berner, E., "Debugging Systems at the Source Language Level", *Comm. ACM*, Vol. 6, No. 8 (Aug., 1963), pp. 430–32.
[HF54]	Holberton, F. E., "Application of Automatic Coding to Logical Processes", *Symposium on Automatic Programming for Digital Computers*, Office of Naval Research, Dept. of the Navy, Washington, D.C. (1954), pp. 34–39.
[HL64]	Halpern, M. I., "XPOP: A Meta-Language Without Metaphysics", *Proc. FJCC*, Vol. 26, pt. 1 (1964), pp. 57–68.
[HM66]	Homer, E. D., "An Algorithm for Selecting and Sequencing Statements as a Basis for a Problem-Oriented Programming System", *Proc. ACM 21st Nat'l Conf.*, 1966, pp. 305–12.
[HP53]	Hopper, G. M., "Compiling Routines", *Computers and Automation*, Vol. 2, No. 4 (May, 1953), pp. 1–5.
[HP53a]	Hopper, G. M. and Mauchly, J. W., "Influence of Programming Techniques on the Design of Computers", *Proc. IRE*, Vol. 41, No. 10 (Oct., 1953), pp. 1250–54.
[HP55]	Hopper, G. M., *Automatic Coding for Digital Computers* (Talk presented at The High Speed Computer Conference, Louisiana State University, Feb. 16, 1955), Remington Rand Corp., ECD-1 (1955).
[HV64]	Haverty, J. P., *Programming Language Selection for Command and Control Applications*, RAND Corp., P-2967, Santa Monica, Calif. (Sept., 1964).
[IB00]	*Management Operating System: Inventory Management and Materials Planning—Detail*, IBM Corp., E20-0050-0, Data Processing Division, White Plains, N.Y.
[IB00d]	*Retail IMPACT—Inventory Management Program and Control Techniques*, IBM Corp., E20-0188, Data Processing Division, White Plains, N.Y.
[IB00e]	*Demand Deposit*, IBM Corp., 1140-FB-03X, Data Processing Division, White Plains, N.Y.
[IB00f]	*Type Composition*, IBM Corp., 1130-DP-04X, Data Processing Division, White Plains, N.Y.
[IB00j]	*The Bank Central Information System—Locate File*, IBM Corp., E20-0138, Data Processing Division, White Plains, N.Y.

28 GENERAL INTRODUCTION

[IB61a] *IBM 7070 Series Programming Systems: Autocoder* (Reference Manual), IBM Corp., C28-6121-0, Data Processing Division, White Plains, N.Y. (1961).

[IB65d] *IBM System/360 Operating System Report Program Generator Specifications*, IBM Corp., C24-3337, Data Processing Division, White Plains, N.Y. (1965).

[IB66c] *1800 Traffic Control System: Application Description*, IBM Corp., H20-0212-0, Data Processing Division, White Plains, N.Y. (1966).

[IB66j] *Student Scheduling System/360, Application Description*, IBM Corp., H20-0202-0, Data Processing Division, White Plains, N.Y. (1966).

[IB66k] *System/360 Bill of Material Processor (360-ME-06X), Programmer's Manual*, IBM Corp., H20-0246-0, Data Processing Division, White Plains, N.Y. (1966).

[IF66] *IFIP-ICC Vocabulary of Information Processing* (First English language edition). North-Holland Publishing Co., Amsterdam, Netherlands, 1966.

[KL67] Klerer, M. and May, J., "Automatic Dimensioning", *Comm. ACM*, Vol. 10, No. 3 (Mar., 1967), pp. 165–66.

[KR64] Keller, J. M., Strum, E. C., and Yang, G. H., "Remote Computing: An Experimental System Part 2: Internal Design", *Proc. SJCC*, Vol. 25 (1964), pp. 425–43.

[KV60] Kavanagh, T. F., *"TABSOL—A Fundamental Concept for Systems-Oriented Languages"*, *Proc. FJCC*, Vol. 18 (1960), pp. 117–36.

[LA54] Laning, J. H. and Zierler, W., *A Program for Translation of Mathematical Equations for Whirlwind I*, M.I.T., Engineering Memorandum E-364, Instrumentation Lab., Cambridge, Mass. (Jan., 1954).

[NE65] Nelson, E. A. et al., *Research into the Management of Computer Programming: Some Characteristics of Programming Cost Data from Government and Industry*, System Development Corp., TM-2704/000/00, Santa Monica, Calif. (Nov., 1965).

[NM00] *SURGE: A Data Processing Compiler for the IBM 704*, North American Aviation, Inc., Columbus, Ohio.

[PR58] Perlis, A. J. and Samelson, K. (for the committee), "Preliminary Report—International Algebraic Language", *Comm. ACM*, Vol. 1, No. 12 (Dec., 1958), pp. 8–22.

[QH66] Budd, A. E., *A Method for the Evaluation of Software: Procedural Language Compilers—Particularly COBOL and FORTRAN*, MITRE Corp., (DDC) AD 651142, Commerce Dept. Clearinghouse, Springfield, Va. (Apr., 1966).

[QL67] Schlesinger, S. and Sashkin, L., "POSE: A Language for Posing Problems to a Computer", *Comm. ACM*, Vol. 10, No. 5 (May, 1967), pp. 279–85.

[RI66] Rice, J. R. and Rosen, S., "NAPSS—A numerical analysis problem solving system", *Proc. ACM 21st Nat'l Conf.* 1966, pp. 51–56.

[RM62] Robbins, D. K., "FORTRAN for Business Data Processing", *Comm. ACM*, Vol. 5, No. 7 (July, 1962), pp. 412–14.

[RR55a] *BIOR (Business Input-Output Rerun) Compiling System*, Remington Rand Corp., ECD-2 (1955).

REFERENCES

[RT53]   Rochester, N., "Symbolic Programming", *Trans. IRE Professional Group on Electronic Computers*, Vol. EC-2, No. 1 (Mar., 1953), pp. 10–15.

[SC65]   Schwartz, J. I., "Comparing Programming Languages", *Computers and Automation*, Vol. 14, No. 2 (Feb., 1965), pp. 15–16, 26.

[SH62]   Shaw, C. J., *An Outline/Questionnaire for Describing and Evaluating Procedure-Oriented Programming Languages and Their Compilers*, System Development Corp., FN-6821/000/00, Santa Monica, Calif. (Aug., 1962).

[SH64a] Shaw, C. J., "More Instructions ... Less Work", *Datamation*, Vol. 10, No. 6 (June, 1964), pp. 34–35.

[SH66]   Shaw, C. J., "Assemble or Compile?", *Datamation*, Vol. 12, No. 9 (Sept., 1966), pp. 59–62.

[ST57]   Steel, T. B., Jr., "PACT IA", *J. ACM*, Vol. 4, No. 1 (Jan., 1957), pp. 8–11.

[TM56]   Thompson, C. E., "Development of Common Language Automatic Programming Systems", *Symposium on Advanced Programming Methods for Digital Computers* (Washington, D.C., June 28, 29, 1956). ONR Symposium Report ACR-15, Office of Naval Research, Dept. of the Navy, Washington, D.C. (Oct., 1956), pp. 7–14.

[UC61]   Grad, B., "Tabular Form in Decision Logic", *Datamation*, Vol. 7, No. 7 (July, 1961), pp. 22–25.

[WF66]   Whiteman, I. R., "New Computer Languages", *Internat'l Science and Technology* (Apr., 1966), pp. 62–68.

[WI51]   Wilkes, M. V., Wheeler, D. J., and Gill, S., *The Preparation of Programs for an Electronic Digital Computer*. Addison-Wesley Press, Reading, Mass., 1951.

[WI64]   Wilkes, M. V., "Constraint-Type Statements in Programming Languages", *Comm. ACM*, Vol. 7, No. 10 (Oct., 1964), pp. 587–88.

[WT68]   Wirth, N., "PL360, A Programming Language for the 360 Computers", *J. ACM*, Vol. 15, No. 1 (Jan., 1968), pp. 37–74.

[XY56]   "UA SAP 1 and 2", *SHARE Distribution No. 36* (Feb., 1956).

[YJ65]   Young, J. W., Jr., "*Non-Procedural Languages*", Presented at *ACM S. Calif. Chap. Seventh Annual Technical Symposium*, Mar. 23, 1965.

II FUNCTIONAL CHARACTERISTICS OF PROGRAMMING LANGUAGES

II.1. DESCRIPTION OF THE CONCEPT OF FUNCTIONAL CHARACTERISTICS

This chapter concerns itself with the functional characteristics of programming languages. The term *functional characteristics* is used to refer to those aspects of programming languages which are primarily nontechnical and/or which are not part of the language specifications themselves. The functional aspects normally relate to economic and political factors and also to those aspects of compilers which affect the use of the language in a significant way. The actual elements of the language are considered *technical characteristics* and are described in Chapter III.

Although the primary characteristics of a programming language are its technical facilities and the way in which they are provided, these are far from being the only features in determining the use and usability of a language. Just as in the case of computer hardware selection there are factors that transcend the physical characteristics, so there are multitudinous and interlocking issues which apply to programming languages. For example, originally the selection of a computer depended primarily on the speed of individual instructions such as addition, multiplication, etc. After a while, it became clear that the amount of time required for memory access was very significant; still later it became apparent that speed of input/output, sizes of secondary storage units, and the interrelationship of all these hardware features were very important. Finally it became clear that the selection of hardware depended not only on the hardware itself but on its relationship to the software; so the concept of *thruput* became of paramount importance. Thus, just as the total amount of productive work which could be done using a particular piece of hardware and its associated software became the prime criterion in computer selection, so in the case of programming languages there are factors beyond the immediate language definition which significantly affect its selection and use. It is the function of this chapter to try to describe these functional characteristics and to indicate their importance. The items to be discussed include general properties of a language, the purpose of a language, issues of conversion and compatibility, standardization, types and methods of language definition, and evaluation.

II.2. PROPERTIES OF LANGUAGES

There are a number of language properties which are difficult to define and which often appear to be subtle aesthetic qualities rather than tangible characteristics. Although it is not possible to provide rigorous definitions for these qualities, it is nevertheless worth trying to provide some brief intuitive feel for each of them.

Two properties which may occur either in parallel or at opposite ends of a scale are generality and/or simplicity. Generality really means a wide scope, i.e., ability of the language to apply directly and effectively to a wide class of problems (see Section II.3.1). Simplicity usually refers to ease of learning, use, and implementation. These properties are at opposite ends of the spectrum because putting a large number of capabilities into one language, thus making it general, causes loss of simplicity by requiring many different facilities to be learned. On the other hand, a very simple language cannot provide too many facilities because in so doing it will lose that characteristic. It will tend to provide a few very powerful primitives. The only way in which generality and simplicity can exist together is when the ability to handle a large number of differing application types is achieved by providing a simple framework and allowing (and requiring) the user to build up the larger capabilities that he needs. This is sometimes referred to as the *core-language* concept. It is often difficult to separate the concept of generality from the availability of many special purpose features.

Two properties which are often at opposite ends of a spectrum are succinctness and naturalness. An example of naturalness might be FIND THE SQUARE ROOT OF 17 USING NEWTON–RAPHSON ITERATION, whereas its succinct equivalent might be SQR (17, NR). To a person well trained in formal notation, the succinct notation may even be more natural. Such a case clearly arises in comparing the sentence ADD A TO B AND MULTIPLY THAT RESULT BY C TO PRODUCE D with the equation $D = C * (A + B)$. Both of these are clearly within current technology. The choice is usually based *both* on the type of intended users and the personal choice of the language designers.

The notational properties of languages play at least as great a role as any other characteristic. A language notation can be succinct and/or natural and/or formal. There is a significant difference between the facilities a language provides and the notation by which they are invoked.

Consistency is a property which programming languages should have, but often cannot. In this instance, consistency means the constant application of the same rules in the same way throughout the entire language. While it may seem both easy and obvious that this can and should be done, sometimes achieving this objective is not worth the sacrifice. (The change of optional key words in different divisions of COBOL illustrates such a case. See Section V.3.)

The property of efficiency is seldom applied to a language, but it is an appropriate one nevertheless. Unfortunately, the criteria for efficiency are as widespread as the people who use or implement the language. For example, efficiency could mean the number of pencil strokes required to write a program or the ease of use by novice programmers or a language design which permitted rapid compilation or the provision of a number of compiler aids to provide optimal object code. There is no single measure of efficiency, and the language examiner should be careful of what facet is being measured in attempting to ascertain the efficiency of the language. It is also essential to realize that efficiency of a language and a compiler are not the same thing; the latter usually cannot be achieved without some appropriate language design, but the best language in the world can have a very inefficient compiler (see Section II.7.2).

Another very general property of a language is whether it is easy to write and/or whether it is easy to read. These are not necessarily coexistent in a single language, and one may in fact tend to militate against the existence of the other. Thus a language which is extremely easy to read (e.g., some of the languages discussed in Section IV.7) might be difficult to prepare for computer input. A related property is whether the average user will be very error prone. If the language has many specific and strict rules about spacing and punctuation, there is more of a tendency for error in writing the program. Finally, while one of the avowed advantages of programming languages is that they are easier to learn than assembly languages, some higher level languages may be designed to be very easy to learn while others do not have that as a characteristic or objective. Being easy to learn is definitely not necessarily the same as being easy to read, write, or avoid errors.

II.3. PURPOSE OF LANGUAGE

In looking at a language, the first and most important characteristic is its purpose. It is futile and foolhardy to look at languages and complain about them for not accomplishing some particular task, when their avowed purpose was quite different. Defining the design objectives of a programming language requires specifying the type of applications, the type of language, and the type of potential user.

II.3.1. APPLICATION AREA

Generality often breeds inefficiency, just as familiarity breeds contempt. Thus, a language which is designed to be all things to all people will probably be less successful than a language with somewhat narrower objectives, unless the design is very carefully done. We must consider the application area for which a language is designed; it may be aimed at a very narrow range of endeavor such as machine tool control, or it may be designed for a wider class of problems—for example, numerical scientific computations—or it may be designed to cover the whole gamut of all problems to be run on a computer. (There are not likely to ever be languages which satisfy the last objective.) To date, most languages have dealt with application areas such as numerical scientific computations (e.g., FORTRAN) and then more recently nonnumerical scientific computations (e.g., FORMAC), business data processing (e.g., COBOL), simulation (e.g., SIMSCRIPT), or machine tool control (e.g., APT). Some languages (see Chapter VIII) were designed to be very wide in scope and encompass several of the items in the preceding list. However, even in some of these cases—notably JOVIAL and PL/I—the types of applications envisioned were fairly standard scientific and data processing.

It is essential to distinguish between the basic application area for which the language was designed and the actual usage to which it may be put. There are numerous examples of languages which were aimed at coping with a given class of problems but which eventually were used for many other things. The best example of this is FORTRAN, which was originally designed for use in numerical scientific work but has been used for subjects as widely separated as logical design and payroll writing. The important factor in viewing the issue of application area is not so much what the language has been or can be used for but what it is really designed to be good at. To the extent that one extends beyond the hard core of the basic objective, one finds the language may be more general and may be useful to a larger class of people than originally intended. Any inefficiencies which result from such extended usage should not be blamed on the language.

The objectives of a language are usually stated in the terminology of the intended users. Thus COBOL is described as *business oriented*, although it is restricted to administrative and financial areas of business. An operations researcher might expect that COBOL would be useful in solving scientific problems associated with business planning, whereas COBOL was never intended for use in that class of problems. The person who is concerned with choosing the best language for his problem or his installation is strongly advised to consider not only the stated objectives but the frame of reference for the terminology. This can usually be achieved by considering the technical features in the language.

II.3.2. Type of Language

In addition to being concerned with the application area for which a language is designed, it is necessary to consider the type of language within the classifications described in Section I.6. Obviously, a language can fall into more than one of these categories. In fact, a person interested in categorizing a language can, in the manner of a Chinese menu, choose one from each of columns A, B, and C and choose any number from column D, although there do not exist languages for every possible combination.

A	B	C	D
procedure-oriented	problem-defining	hardware	problem-oriented
nonprocedural	problem-describing	publication	application-oriented
	problem-solving	reference	special purpose

II.3.3. Type of User

In designing a language, considerable attention must be given to the kind of user for whom the language is designed. We can separate two very broad classes—namely, professional programmers and, in contrast, people who have a problem to be solved and must program it but consider their profession to be the field in which the problem arose. If the objective is to help the latter category (who will be called nonprofessional programmers), then considerable effort must be made to make the languages easy to learn and to use. Various tricks and quirks relative to the machine or even relative to the language itself should be minimized, because the nonprofessional programmer is more concerned with being able to state his problem easily than he is with obtaining the maximum efficiency from a particular machine. For a nonprofessional programmer, the distinction between writing the program and reading it or using it after it is written is significant. It is well-known that two of the major problems in administering any activity involving programming are the need for program maintenance and the troubles arising from personnel turnover. Thus one objective might be to make it very easy for nonprofessional programmers (or for that matter, even professional programmers) to pick up and understand somebody else's program. For example, one can envision a situation in which very succinct information is fed into a compiler and much more elaborate and detailed information is put out, so that the program becomes easily understandable to a wide variety of people. (This has actually been done by developing the Rapidwrite system for COBOL—see Humby [HY62], [HY63].)

In the case of a language designed for use by a professional programmer, a major characteristic is to provide maximum capability. In other words, the programming language can and should aim at relieving the professional programmer of many annoying details but still provide him with great flexibility. Thus, for example, he should have some very nice way of stating the beginning and ending points of loops and the increment to be used, but he may want a large number of ways of specifying or controlling the loops (e.g., incrementing or decrementing, varying several parameters in one statement). In the case of the nonprofessional programmer, he may be satisfied with one relatively simple way of handling this particular facility. Similarly, the professional programmer will almost always want to be able to get at the machine code. No programming language to date has been designed so well that the professional programmer has been completely satisfied with it; there are always things that he wants to do that seem to require resorting to machine code. This facility does not generally interest the nonprofessional programmer.

One other feature in considering this distinction between the professional and the nonprofessional programmer is in the type of debugging aids that are made available. These are discussed in somewhat more detail in Sections III.5.5.3 and III.7.5, but it should be pointed out here that a programming language which requires a nonprofessional programmer to understand machine language in order to debug his higher level language program is not much help. Only if the debugging can take place at the source language level is he really aided. On the other hand, in very tricky cases the professional programmer may want the ability to get memory dumps and to examine contents of index registers. This is particularly true if the language does not provide really good debugging aids.

In attempting to aim a language at a nonprofessional programmer, one can give strong arguments for making the language as natural as possible. In other words, if the user is concerned only with solving the problem, he will presumably prefer to communicate with the computer in the language which is most natural to him. He is not necessarily concerned with all the fine points that the professional programmer wishes to be able to control. The issue of what is meant by *natural* and how much is desirable and feasible is a hotly debated one. (See Sammet [SM66b], Halpern [HL66], and Dijkstra [DJ63] for further discussions of this point.)

An issue of vital concern to the nonprofessional is the amount of "nonlanguage" material he must learn. Since the compilers are usually part of an operating or time-sharing system, the user can seldom just "write his program". He is often required to worry about such things as control cards, form of his object deck, etc. A discussion of these problems is beyond the scope of this book.

II.3.4. Physical Environment

A major defining purpose of a language is the physical environment in which it is to be used. The most significant distinctions existing today are between batch and on-line systems. However, on-line systems can be further subdivided into those which are *truly* conversational and thus permit significant man-machine interaction and those which are merely time-sliced and provide only some additional facilities to an inherently batch-oriented language. (There are actually many levels of gradation of these concepts.) There is also a possibility that the language may be intended for use in real-time situations. The actual size of the computer may be relevant since some languages are clearly aimed at maximum effectiveness on large computers, while others may be intended for small machines. The possibility of multiprocessing configurations could also affect the language.

II.4. CONVERSION AND COMPATIBILITY

Of all the characteristics of programming languages about which there has been great confusion, the subject of compatibility and its associated factor of conversion rank very high. These characteristics are of prime importance from a management point of view, although they may be of very little concern to the programmer himself. In some cases, the characteristics provide the deciding factor in determining what languages should be used. The reason for the importance of compatibility and conversion is easy to understand as soon as one realizes that the investment in programs for a particular machine may run into millions of dollars. In particular, by now the costs of programming tend to equal or, in some cases, even exceed the actual cost of the hardware. Thus, it is no light matter to ignore the question of what happens to the programs if one wants to change machines. Hardware technology does not stand still and is continually improving. This means that users can generally improve their economics by obtaining new equipment which permits them to do the same jobs faster or cheaper, or both. However, the decision to obtain new machines is usually influenced very strongly by the prior investment in programming. Thus, if a large amount of money has been invested in programs which cannot be run on a new machine, it becomes necessary to think very long and hard before obtaining new equipment, even though the new machines could certainly do the job faster and cheaper. The timing cycles of hardware and software development are such that by the time an installation has its programs running satisfactorily on one machine, the manufacturers have usually come up with new and better hardware. There has been a great deal of misinformation and misunderstanding on what is meant by compatibility and what types of conversions are possible and meaningful. It is the purpose of this section to try to clarify these issues.

II.4.1. Types of Compatibility

1. *Machine Independence*

The first type of compatibility that people are concerned with is compatibility across machines, i.e., how dependent on a particular machine or class of machines is a given programming language. Clearly, if the programming language makes reference to hardware that is unique to a given machine (e.g., sense lights, backward-reading tapes, and discs), then there is no hope that a program written in this language can be directly handled on a machine without these features, unless they are simulated; this is usually prohibitive in cost. Similarly, if the language—as a language—makes particular use of the fact that the machine is fixed word length versus variable word length, binary versus decimal, or has a particular number of bits or characters per word, then again there is no chance of having the program directly transferable to another machine. A partial solution to this problem is to allow the user to state in his program the precision he requires. (This is done in PL/I.) However, this is a double-edged sword because the user may pay a heavy penalty for the inefficiency caused by a precision which is grossly disparate from the word size, e.g., specifying 11-digit precision on a computer with 10 decimal digits per word. If the user is aware of these factors, he can make a more intelligent choice.

Clearly if a language makes use of the hardware characteristics of a specific computer, programs cannot possibly be *directly* compatible, i.e., directly usable on another machine. There might be exceptions to this but they would depend on very clever programming on the part of the compilers, and this has not yet been done. The true definition of machine compatibility is the ability to take a deck of cards, or whatever other input media is used, insert it into a different type of computer (i.e., not one "in the same family"), and have the program run and produce the same answers. Anything less than that capability is a partial or pseudo type of compatibility. We have not yet achieved this facility for the languages, let alone for the extra information required by the operating system.

Two of the machine features which tend to "ruin" compatibility most are word size and collating sequence; actually both of these could be corrected by the compiler—but at prohibitive cost. The word size affects the precision and sometimes even the actual results of numeric calculations because numbers are usually stored in one or two machine words. Thus, unless the actual number of characters (or bits) to be used is specified (and implemented by the compiler), the arithmetic results will differ from machine to machine whenever the word lengths are different. In many cases this does not have a practical bad effect, but the potentiality is certainly there. In the case of the collating sequence, the situation is actually worse because incorrect results are easily obtained as a result of branches which operate differently. If the collating sequence on one computer places letters before numbers, but this is reversed on another machine, then any test of data based on this sequencing information will produce opposite results in going from one machine to another. Again, this could be corrected by having the language specify the collating sequence and require the compiler to turn out the correct code, but nobody has yet been willing to do this because of the tremendous cost at object time.

Other facets of the data base problem, such as wordmarks, fixed versus variable length words, and general record layouts, cause incompatibility. This difficulty exists independently of the language characteristics.

2. *Compiler Independence*

It is clear that when one talks about machine independence, there is an implied reliance on the ability of compilers to do the same things on different machines. In other words, a statement in the programming language that causes an addition to be performed must be translated into the proper instructions on all machines. That is quite obvious; what is not so obvious is the amount of incompatibility which can actually be engendered by the compilers themselves even on the same computer. One of the best examples of this is the situation in which a compiler accepts and correctly handles a statement which is not really legal in the language, but which is certainly meaningful to anybody using the language; e.g., one of the early FORTRAN compilers correctly translated a certain type of implied multiplication. What happens in cases like these is that people tend to write programs knowing the characteristics of their particular compiler, and they are in for a rude shock when the same problem is translated by another compiler.

A second kind of incompatibility caused by compilers is much more subtle and, therefore, much more difficult to track down. Because of the lack of precision in defining programming languages, there are often ambiguous rules relative to the meaning of certain statements in the language, and every compiler writer must make a decision on how to interpret such statements. This is bad enough, but what makes it even worse is that in many cases the ambiguity is not even recognized as such. Thus two people looking at a statement or sequence of statements in the language definition may, in all good faith and in all clear conscience, come up with two entirely different views of what is meant. On top of that, neither person may even recognize that an alternative view is possible, until it is pointed out to him. This causes the compilers to be incompatible in the sense that two different compilers may accept the same source statement and not only produce different object code but, more importantly, cause the source programs to produce different results. Unfortunately there is no way around this incompatibility until better means of defining languages are developed. It is because of this problem that many people have taken the view that the only complete definition of a language is a compiler for the language. My personal view is that this is so impractical that I prefer the unpleasant alternative of admitting that we do not yet know how to define programming languages rigorously.

3. *Dialects and Language L-Like*

One of the most difficult problems in the question of compatibility has to do with the existence of dialects. A *dialect* means a minor variation on a particular language. These variations may exist for any number of reasons. One group may feel that they can obtain a more efficient compiler if they simply make a minor change in the rules. An illustration of this involves naming conventions, whereby the language definition may not require data names and/or statement labels to start with a letter, but some particular compiler writer may decide that his efficiency can be improved by an order of magnitude if he imposes such a restriction. In other cases, minor deviations may occur because one group does not like the actual notation used by the language designers and substitutes a different one. The most common reason for dialects is that for any language there is almost always somebody who feels he can improve the language by making certain additions and/or changes. A more laudable motive is the creation of modifications to meet the needs of a particular application. It is important to notice that the difficulty usually arises more from changes than from additions or restrictions, although the latter two also present problems and are discussed in the next section.

The phrase *language L-like* is frequently heard; it usually refers to a language which is similar in spirit and notation to *language L*, but differs from it markedly enough not to be considered a dialect. The deviations usually involve (1) some changes in notation, (2) some omissions of features or some restrictions, and (3) some additions. As an illustration, LISP 2 (see Section VIII.6) is described as being an extended *ALGOL-like* language.

The prime distinction between being a dialect and being *language L-like* is one of degree. If there are only minor variations, then the word *dialect* is appropriate. Unfortunately, there is seldom universal agreement on how minor the variations really are.

4. Subsetting and Extensions

The issue of subsetting differs significantly from that of the dialects, in the sense that dialects involve changes, whereas subsets imply incompleteness but presumably no changes. Strictly speaking, a subset is a type of deviation but usually one that is less severe in its implications for compatibility.

A language S is considered a proper subset of a language L if (1) there are some programs which can be legally written in L which cannot be legally written in S, (2) all legal S programs are legal L programs, and (3) the results from a program written in S when executed with an S compiler are the same as the results obtained from an L compiler on the same machine, except for those aspects which are implementation dependent. Subsetting obviously permits upward but not downward compatibility. That is, by definition, programs written in S must run on L compilers, but the converse is not true.[1]

Subsetting may take a number of forms. One is simply the nonability to handle a certain class of features in the language. For example, if a language allows both double and single precision, a subset of the language might not allow double precision. Another form is to omit certain special cases in a general feature; e.g., the subset might omit double-precision integers but not double-precision floating points. Another type of subsetting involves placing additional restrictions that the language itself does not have, such as requiring data names to begin with a letter, although the language may not require this.

The primary motivations for subsetting are cost and time. Subsets permit smaller compilers, which can be developed more cheaply and/or more rapidly. Furthermore, subsets tend to compile faster.

Problems with regard to compatibility arise when nonnested subsets exist. In other words, if there are several subsets of a given language, and there is no hierarchy among them, then there is chaos for the user who tries to move from one subset compiler to another. Clearly, if the overall language contains features A, B, C, and D, and Compiler 1 eliminates feature A and Compiler 2 eliminates features A and B, then a hierarchy exists which permits upward compatibility. On the other hand, if Compiler 1 eliminates feature A, whereas Compiler 2 eliminates only feature B, then there is no relationship between those two compilers. They can only be related back to the main compiler which is implementing the entire language. Thus, nonnested subsets will always lead to lack of compatibility among implementations of each other.

One interesting facet of subsetting occurs when the language is implemented by *bootstrapping*, which means that a translator for a subset of the

---

[1] This definition was essentially suggested to me by E.F. Codd.

language is coded in machine language and the compiler is written in this subset of the language. This can be done only for certain languages. Sometimes more than one level of subset is required to create the full compiler.

A language E is an extension of a language L if L is a subset of E. Types of extensions might be the provision of additional facilities, such as new variable types and commands to handle them or removal of restrictions (e.g., on the ways in which data names can be defined). If L is the prime language under consideration, then the existence of its extension, namely E, is of no concern to the users of L. If E is a proper extension of L, then the compiler for E should accept legal programs written in L and produce the correct results. Unfortunately this is seldom true in practice and, after extending L to produce E, restrictions are usually placed on L programs, regardless of whether or not they use the additional facilities of the E language (compiler). This happens because extending a language is seldom easy and almost always requires some change—albeit minor—in the original language.

A common occurrence is to start with a language L, create a subset of it (called S), allow some minor deviations (say S'), and then put in some extensions which are not in L (say $S'+$). The result is an *L-like* language. If $S'+$ is significantly smaller than L, then it is really an *L-like extended subset* and this term will be used throughout this book.

5. *Relation to Language Definition*

Many of the problems of compatibility are caused by the current inability to define languages in a complete and accurate fashion. A good start has been made on defining the syntax of the language, but only a little effective work has been done in defining semantics and virtually no work in defining pragmatics. These terms will be discussed in more detail later. The crucial point is that the lack of compatibility across compilers and very often across machines is related to the fact that the language definition may not have been completely rigorous or understandable.

II.4.2. Ease of Conversion

1. *Based on Compatibility*

As indicated earlier, there is great motivation to ease the conversion of programs from one computer to another. The best way to do this is to maintain complete compatibility between a language acceptable to one machine and the same language handled on another machine. *Acceptability* can be achieved by hardware or software or a combination of both (i.e., emulation). In that case, the conversion problem is negligible. Achieving compatibility is one of the strongest motives for writing programs in a higher level language. Unfortunately, languages tend to change somewhat when implemented on new machines for reasons indicated above, thus reintroducing the dialect problem. The relevant factor becomes the amount of difficulty that is involved. If only small changes are necessary to make the program run on a different machine, then there has been a large amount of compatibility preserved and the conversion is very easy. On the other hand, if major changes and difficulties are encountered, then the conversion is difficult. (This is not meant to imply that conversion of *programs* from one machine to another is the only factor in changing machines. However, it is the only aspect under discussion here.[2])

2. *Ease of SIFTing to Another Language*

The term *SIFT* stands for *S*hare *I*nternal *F*ORTRAN *T*ranslator and was first used in connection with the program to go from FORTRAN II to FORTRAN IV on the 709/90/94 (see Allen, Moore, and Rogoway [AX63]). In the development of FORTRAN IV, great attempts were made to make FORTRAN II a proper subset of FORTRAN IV. However, there were cases in which this was not possible, either because it would place too great a restriction on the new version of the language or, in other cases, because people had taken strong advantage of what the compilers of FORTRAN II would do and these nonlanguage facilities were not applicable to the larger and newer compilers. The term *sift* became used fairly generally to refer to the partial translation of one higher level language to another one which is fairly similar. This normally means the automatic conversion of equivalent language elements and flagging the others for manual conversion. This is a type of conversion, which again is dependent for its ease on the amount of sifting which can be done. A particular illustration—namely, of ALTAC to FORTRAN II—is described by Olsen [OL65].

3. *Ease of Translating to Another Language*

In the worst case, one may be faced with the problem of trying to have one language, which has been implemented for a particular machine, translated into the form of another language for another machine. It is of course assumed that such a translation will preserve the *high level* characteristics of the original program and will not cause severe degradation of the eventually resulting object code. An almost useless translation (from an efficiency viewpoint) occurs when a less powerful language is translated on a statement-by-statement basis to a more powerful one. This has actually occurred in translating from powerful assembly programs (Autocoder)

---

[2] See *Datamation*, Vol. 12, No. 6 (June, 1966) for several papers on this subject.

to COBOL. It is an interesting—and as yet unsolved—problem as to what general characteristics of languages are needed to permit one to be translated into the other automatically without severely losing the efficiency of the original source program. The ease of conversion is dependent upon how easily—if at all—this translation can be made.

One of the reasons for wanting an effective translation is that the newer machine may not have available on it a compiler for the earlier language. A second reason occurs when the installation managers wish to have everything coded in the newer language and, therefore, want to have the old programs translated automatically.

II.5. STANDARDIZATION

One of the key factors in the definition and use of a programming language is the role played by standardization. The purpose of this section is to describe the purposes and problems in standardizing programming languages and the procedures that are involved and to give a brief status report. More details about the latter are shown in the individual language descriptions.

II.5.1. PURPOSES

The basic purpose of standardizing programming languages is to achieve compatibility, which in turn reduces costs. Compatibility in programming languages permits savings in training personnel because they do not need to learn a new language. It also permits savings in documentation because the number of new manuals that must be written is sharply reduced. Standardization also minimizes—although it does not eliminate completely—the problem of converting to new computers. (This assumes that a standard language is implemented for a new set of machines.)

Even assuming a language standard exists, there is a management problem in enforcing the standard. This is not significantly different from the problem of enforcing *any* standard or set of conventions in a programming organization.

II.5.2. PROBLEMS

There are three main problem areas in standardization: Conceptual, technical, and procedural. It should be recognized that the conceptual and procedural problems are not unique to programming languages; they apply to most technology.

1. Conceptual Problems

The first conceptual problem is one of timing; i.e., when should standardization of a language take place. Unless this is given careful consideration, it is likely to come too soon or too late. If it is too soon, then the standardization is premature; it is not clear what is needed, and there is a risk of standardizing on a number of things that really are not very good. On the other hand, if standardization is delayed too long, then there are dialects—admittedly some of them very minor changes—and this in turn creates a number of vested interests which are reluctant to accept a standard which deviates from their particular version.

A second conceptual problem is the risk of stifling progress. Somehow the standardization process must avoid eliminating or preventing technical progress. This is extremely difficult because there is no easy way of coping with new and bright ideas if they come in after the standard is established, or even while it is in the process of being established. An excellent example of this arose in one subcommittee meeting which suggested a somewhat better method for handling the proposed revised ASCII code. Unfortunately, too much work had already been done by too many people to permit the change, even though several groups agreed it was an improvement.

2. Technical Problems

The first technical problem in standardization is one of definition. We do not yet know how to define a programming language rigorously. No completely formal method exists, even for the purely syntactic definitions, although tremendous strides have been made along these lines. There are only beginning attempts at defining semantics rigorously, and no effort has been made toward coping with the problem of pragmatics. (These terms are defined in Section II.6.2.) A verbal description is inadequate (although used) because the English language is ambiguous and it is impossible to spell out every possible contingency or interpretation. Some people would cope with this problem by accepting the processor (i.e., the compiler) as the basic definition of a language. This might work satisfactorily if there were only one processor per language, but that clearly is not the case. It is certainly not feasible to say that the first compiler written will be the formal definition of a language. Even if that were done, or some other compiler were chosen, there would still arise the problem of requiring everybody to investigate the details of the compiler coding to find what a particular issue meant. In some cases this would still not provide a complete definition for the entire language.

A second technical problem is to try to determine when a compiler (or a program) actually meets the standards. Since we do not have a completely rigorous definition of the language, we clearly do not have a rigorous way of testing whether or not a given compiler meets that language specification. Even if we accept the unfeasible alternative that a particular compiler will define a language, this still does not tell how to determine whether another compiler actually meets the language specification. The use of test problems is definitely not the answer because a particular compiler could easily be designed to meet the test problems but still be very far from the standard.

A third technical problem is to determine how to do maintenance in an orderly way and still not invalidate the compilers. This is tied in with the problem of the language definition because most of the maintenance involves clarifying unclear points. The difficulty that arises here is the one pointed out in Section II.4.1.2—namely, two different groups may have implemented a particular point differently without even realizing that there was another possible view on what they were doing. Once there is a large amount of money invested in the implementation, it is very difficult to persuade any one group to change its view on what should be done. Since in many cases maintenance also involves extensions, these have to be looked at very carefully in the light of present implementations. Certain extensions could invalidate all compilers written for a particular language, even though the extension was extremely desirable.

A fourth technical problem is the one of subsetting, which was discussed earlier in Section II.4.1.4. Since a standard must achieve wide acceptance in order to fulfill its purpose, a highly complex language may reduce the number of groups which can implement the standard. On the other hand, reducing the level of the standard to the smallest computer will lower the value of the standard considerably. The best solutions for this problem seem to be controlled subsetting and/or modularity of features.

The last technical problem is the multiplicity of standards for programming languages. It is preposterous at this point and in the near future to consider standardizing one language for all programming. The best we can hope for is one language for each major application area and some languages (e.g., PL/I) which cover more than one application area. However, it is important to notice that FORTRAN and ALGOL were both standardized; yet they covered very similar application areas, namely, the solution of scientific numerical problems. The reason for the two standards was quite simple; there were large investments in both languages, and neither group was willing to retreat and disclaim all interest in having its language become standardized. Thus, there has been a necessary regression from the mythical ideal of one programming language standard to one for each major application area, and a further regression to merely standardizing *any* "suitable" language to prevent dialects of it from being developed and used.

3. Procedural Problems

The procedural problems in establishing standards are enormous, but this is necessary to prevent the promulgation of undesirable standards. In this context, *undesirable* merely means *not acceptable* by virtually all the groups to whom the standard will apply. The complexity of the procedures—which have been established to protect the rights of all those involved—of necessity delays the establishment of a standard. This often causes difficulty to those groups who are at a stage in their technological or manufacturing development where they are ready to implement the standard, but it does not exist officially and may yet be changed.

II.5.3. METHOD OF ESTABLISHING STANDARDS

Most standards are adoptions or rework of existing practices. Some come into being through a specific committee which does developmental work and announces at the outset that their result is to be a standard of some kind. (This was done with COBOL.) Other standards become what are called *de facto standards*—i.e., they are so commonly used that by general agreement and general practice they are a standard, even though no formal mechanism whatsoever has been used to establish them as such. In most of these cases, however, although there may be widespread agreement on the basic item, there are almost always deviations which must be eliminated from an actual standard. (FORTRAN is an illustration of this situation.) There is a very formal and specific procedure for establishing official standards, and this section will discuss this procedure in some detail.

The authority for *industrial* standardization in the United States is vested in the United States of America Standards Institute (USASI), which replaced the American Standards Association (ASA) in August 1966. Obviously any group, e.g., government, professional societies, and user groups, can (and does) standardize anything, but USASI is recognized as *the* central and official source of activity for any type of industrial standardization in the United States. Unlike European countries, standardization is a voluntary process in the United States. Thus, nobody is obligated to obey a standard just because it exists; whereas, in many European countries, once a standard exists, it is a government regulation and must be followed. There are a number of factors which are relevant to the standardization process under USASI and which are independent of programming language standardization per se. It is worth noting these, so that the problems and procedures for programming language standardization can be seen in perspective. (A more detailed description of the procedures is given by Goodstat [GS67] and Steel [ST67].)

The USASI provides an elaborate structure with built-in checks to prevent "railroading" of anything as a standard; a broad basis of participation is required both to do the work in establishing the standard as well as to approve it. A consensus among all interested parties is required before something is approved as a standard, and a consensus is much more than a mere majority. If a significant-sized minority objects to a standard, then it is normally sent back for rework. It is characteristic of USASI in particular (and most organizations in general) that if they provide an elaborate structure of the kind just indicated, then of necessity the committee procedures and regulations will be long and complicated. In addition, there is also an international standards organization which has different rules from USASI, and groups in the United States usually wish to satisfy both standards organizations.

The USASI normally asks some group to sponsor work in a particular area. This is usually a trade association or similar group. In the case of the computing industry they asked BEMA (Business Equipment Manufacturers Association) to provide sponsorship. Thus, BEMA established the sectional committee X3, which in turn established seven technical working committees as follows: (1) Optical character recognition, (2) coded character sets and data formats, (3) data transmission, (4) common programming languages, (5) glossary, (6) problem description and analysis, and (7) magnetic character recognition.

The charter of X3.4 (which was formed in 1960) is "Standardization of common programming languages of broad utility through standard methods of specification, with provision for revision, expansion and improvement, and for definition and approval of test problems". At the time of this writing, X.3.4 has established eight subcommittees. A list of these follows, with a brief indication of the function and purpose of each subcommittee.

X3.4.1. Language theory. This committee has been dormant for a long time but was responsible initially for investigating some of the technical problems associated with standardization.

X3.4.2. Language specifications. This committee is concerned with miscellaneous activities, which includes deciding what languages are appropriate candidates for standardization and the criteria involved. These tasks are not as easy as they sound due to the need for being concerned with a large number of vested interests. This committee also has the responsibility for reviewing an actual proposed language standard for X3.4.

X3.4.3. FORTRAN. This committee defined the standard FORTRANs and is responsible for their maintenance.

X3.4.4. COBOL. This committee is responsible for the definition of the standard COBOL and for its maintenance.

X3.4.5. ISO/TC97/SC5 secretariat and USA participation. This committee handles interaction with the international standards organization: SC5 is roughly the equivalent of X3.4 at the international level.

X3.4.6. Glossary. This committee is responsible for determining and/or reviewing glossary items which are particularly relevant to the subject of programming languages.

X3.4.7. Machine tool control. This group was actually the latest formed; it did not come into existence until the latter part of 1964. It is concerned with the development of standards for machine tool control.

X3.4.8. ALGOL. This was actually a sub-subcommittee under X3.4.2 and was eventually formed into a subcommittee in its own right.

The main work of X3.4 has been in deciding what languages to try to standardize and then actually attempting to do it. Because the maintenance and definition are different for each language, the procedures need to be different.

Once X3.4 has created a proposed draft standard, it is submitted to the parent body, X3, which arranges for its publication and wide distribution. A period of approximately 6 months is then allowed for commentary by any person or organization whatsoever. Following (and sometimes during) this period, a ballot is taken according to USASI rules and procedures and, based on that ballot, either the proposed standard is sent back to the committee for rework or it is submitted to the Information Processing Systems Standards Board (IPSSB) for its determination that the proper procedures were used and a consensus really exists. In almost all cases, IPSSB provides final approval of the standard. (There is a still higher group, but it is seldom needed.) Once the standard becomes promulgated, it is then recognized as an American standard. Again it must be emphasized that adherence to this standard is completely voluntary on the part of any organization. Experience to date has shown that such standards do play a very significant role in the activities of computer manufacturers.

II.5.4. OVERALL STATUS

The descriptions of each language indicate the status of the standardization for that language and the process that was involved.

II.6. TYPES AND METHODS OF LANGUAGE DEFINITION

Fortunately or unfortunately, language definition is an administrative as well as a technical issue. Many factors discussed below play an important role in the creation, development, and usage of the language. These aspects tend to be ignored or misunderstood but they play a vital role in the overall consideration of the language.

II.6.1. ADMINISTRATIVE

1. *Who Designed the Language?*

The first administrative question to be asked about any language is: Who designed the language? Also, how was the group constituted? Who was the sponsor or directing authority? What kind of pressures were they under? Several languages have been designed by committees, where the committee consisted of participants from a number of organizations. This is not necessarily bad since even when a language is designed within one organization, it is normally designed by more than one person and this group could also be called a committee. It is not at all clear whether a committee composed of people from different organizations fares significantly worse than one formed solely within an organization. The main reason for this is that current and past language design has been based very much on personal opinion, rather than just on fact or objectivity. Many of the properties described in Section II.2 mean different things to different people, and certainly the method of applying them is nebulous. Language design is an art, not a science. Furthermore, as in any endeavor, language designers also tend to use their past experience even though it is not always applicable to the current situation. The one factor that pervades *inter*company language design which generally does not affect *intra*company work is a number of political considerations. In particular, an intercompany committee may have on it people who are under directives from their organization to try to place into the language those features which are helpful to their equipment (and possibly harmful to others) and, of course, to prevent the converse from happening. These are all unfortunate facts of life which must be taken into account in considering any language.

2. *What Were the Objectives of the Language?*

In examining any language, it is necessary to know the objectives. Just as it would be silly to complain that an automobile is not a good device for crossing an ocean, it is equally foolish to say that a language is a poor one because it does not satisfy the person examining it. A language designed for use by nonprogrammers may seem very loquacious or inefficient when viewed by a professional programmer. Conversely, terminology or techniques that are useful to a person with considerable programming experience may be confusing or meaningless to a person who just wants to find answers quickly.

There are two legitimate questions which can be asked about a programming language and its objectives. The most important is: Does the language satisfy the objectives? The second is: Were the objectives worthwhile? The first question is a very good one if it is applied honestly, and the prime criterion of a good language is whether it achieves the goals specified for it. The question about worthiness of objectives is a dangerous one. Using the earlier analogy, a device good for crossing an ocean may be a silly idea to someone who has no interest in moving off dry land. Too many criticisms of programming languages tend to be made by people who have no knowledge of, or interest in, the problem area; they insist that the objective is bad when in reality they do not understand or care about it.

3. *Who Implemented the Language?*

The question of who implemented the language is another administrative facet which cannot be ignored. If the language is implemented by the same people who designed it, then there is the greatest chance of success because the language can be modified as the needs of the implementation demand. Of course, a poorly conceived implementation design should not be allowed to ruin the language by forcing unnecessary restrictions. There are more difficulties when the implementation group differs significantly from the language group and the latter must be consulted on every change in the language. Making sure that the right kinds of interactions occur in both cases is clearly an administrative problem. As mentioned earlier, very often the definition of a language is not completed until the compiler is completed.

Implementation is normally done either by a group within one company (usually a computer manufacturer, but sometimes a user with its own language) or an outside software group (which is charged with the responsibility for preparing a compiler for a particular machine or class of machines). Even here, there are difficulties that depend on whether the implementation for a given class of machines is under direct control at a low enough organizational level to be effective. Thus, if a company has a class of machines which are either similar or purportedly compatible in some sense, then the question of how compatible the compilers are becomes another administrative and management problem.

4. *Who Maintains the Language?*

The maintenance of a language is not the same as the maintenance of a compiler or a program. The language maintenance is by far the stickiest of the administrative problems. In some cases, the group who originally designed the language retains the responsibility for its maintenance. This maintenance has many facets, starting from answering the questions of the implementers who do not understand a particular language specification to responding to requests for changes on features that are difficult to implement and, ultimately, making improvements and/or extensions to the language. As was true with the question of who designed the language, the maintenance is sometimes done by an interorganization group and sometimes within a single organization. However, when the maintenance of the language is divorced from the implementation, a certain amount of chaos is likely to arise. This occurs because the implementors usually need an immediate decision on what a particular point means; those who are maintaining the language may not be ready to meet that week to answer the question; yet coding must continue. Similarly, people who are pressing for improvements and/or extensions to the language are apt to find a very responsive chord in the maintenance group, but an unresponsive chord in the implementation group. The latter will certainly resist improvements to the language if it invalidates their compilers. Thus, if the maintainers of the language are significantly separated from the implementers, they may make changes and/or decisions and/or improvements which seriously affect the implementation. Even if the two groups coincide, the thorniest of all the administrative problems is to decide when to allow the language to be significantly improved, at the cost of much compiler rewriting.

II.6.2. TECHNICAL

The technical issues in language definition are, of course, the very heart of determining what the language actually is, i.e., what its specifications are. These issues are often mixed up with the notation (metalanguage) of the definition, i.e., the actual way in which the language definition is written down on one hand, and the questions of the rigorousness of the definition of the syntax, semantics, and pragmatics on the other hand. In my opinion, too much of the discussion of the actual features and qualities of a language centers around the way in which the language is defined. While obviously a poor and unrigorous definition makes it difficult if not impossible to determine what the language specifications really are, it should be kept in mind that the language and the means of defining the language are not the same thing. It is for this reason that the discussion of the technical methods of language definition are included in this chapter, even though they are definitely *technical* and this chapter is purportedly concerned with nontechnical characteristics of programming languages. I would go even further and say that many of the nontechnical problems exist because the computing community has not yet solved satisfactorily the technical problems of defining programming languages rigorously.

1. *Syntax, Semantics, and Pragmatics*

The three characteristics of a language definition are syntax, semantics, and pragmatics. (These are discussed specifically by Zemanek [ZE66] and were largely the subject matter of the 1964 IFIP working conference, whose proceedings appear in Steel [ST66a].

By *syntax* we mean a rigorous statement of what sequences of characters are considered correct in the language and, ultimately, what character sequences constitute a (syntactically) legal program. Thus, the syntax could specify that the sequence A + B is legal; whereas the sequences +AB or A+B are not allowed. On the other hand, a different language might say that the second or third (or both) of these was legal; whereas the first was not. In any case, the syntax simply specifies the legitimate strings in the language. The meaning of the string is determined by the semantics. Thus, for example, the string A + B might mean addition if A and B were numbers; whereas it might mean union if A and B were sets or logical conjunction if A and B were truth values. Clearly, a single legal string can have a great many meanings; the collection of all these meanings for each legal string is called the *semantics* of the language. The *pragmatics* is the relationship of these strings and their meanings to the user. Thus, the user himself must understand and appreciate what is meant by arithmetic, set union, and logical conjunction. Furthermore, there must be agreement between his intended use of a string of symbols and its actual semantic interpretation by a compiler.

The following statements appear to be true: (1) There is sometimes a hazy line between what is syntax and what is semantics; e.g., the rule that the number of subscripts on a variable in FORTRAN must agree with the information in the DIMENSION statement can be considered both syntactic and semantic, although it is primarily syntactic. (2) There is no notation yet developed which will express completely unambiguously all the syntax of a programming language, even if there were agreement on what was purely syntax. (3) Little work has been done on formalizing semantics, although the work of the IBM groups in Hursley, England and Vienna, Austria has made a good start on PL/I (see the reference lists at the end of this chapter and Chapter VIII for numerous reports). (4) Nothing has been done about formalizing pragmatics. Thus, the problem of rigorously defining a language—assuming there is an intuitive idea of what the language should be—is one in which a large amount of technical work needs to be done. However, significant work in providing formal notation for syntax has been done and has helped the language definition problem enormously. See Floyd's survey [FL64] and the other items in the list of references at the end of the chapter.

2. *Formalized Notation*

Since the English language permits numerous ambiguities, it is desirable to provide a formal or rigorous method for defining programming languages. Considerable work has been done to provide such formalism for the syntax,

II.6. TYPES AND METHODS OF LANGUAGE DEFINITION

but very little work has been done for the semantics; hence, the latter will therefore not be discussed at all.

A complete discussion of the formalized notations used for describing programming languages is beyond the scope of this book. However, the basic principles can be stated rather simply. This whole area is the major interface point between artificial languages and the work of linguists concerned with natural languages. Further details can be found in the references in Floyd's paper [FL64].

To define a language, some language must be used for writing the definitions. This latter is called a *metalanguage*. It is a general term which can include any formal notation or even English itself. Metalanguage is a relative term since it is itself a language which must be defined, and that requires a *metametalanguage*. For the languages discussed in this book, we need only be concerned about the single level of metalanguage.

The first, and still the most significant, contribution made in this area was by John Backus in his paper [BS60] describing IAL (later called ALGOL). After an informal description of the proposed language, Backus states (page 129) "There must exist a precise description of those sequences of symbols which constitute legal IAL programs . . . . For every legal program there must be a precise description of its 'meaning', the process or transformation which it describes, if any . . . ." The second part of this objective has not yet been carried out completely and successfully, although significant work is well underway. The prime elements of the metalanguage are the concepts of a *metalinguistic formula* or *expression* composed of *metalinguistic variables* (whose values are strings of symbols), a *metalinguistic equivalence symbol*, and *metalinguistic connectives*. The metalinguistic variable (which is also called a *syntactic unit*) normally has mnemonic meaning, although this is not required; thus *integer* is a metalinguistic variable whose values are the digits 0, 1, 2, 3, 4, 5, 6, 7, 8, or 9. (Angular brackets < > are a commonly used notation for syntactic units.) The most important connectives are *or*, *concatenation* (i.e., adjoining two strings to make one string), *choice*, and *optional*. Not all these connectives are used in each metalanguage; it is largely a matter of (1) personal choice and (2) structure of the language being defined, as to which combinations are used. The concepts of recursion within definitions and repetition of syntactic units are also widely used; these are illustrated later.

The most common (although by no means the only) combinations of symbols are those which have been used for the ALGOL 60 report (Naur [NA60] or [NA63]) and for the COBOL report [US65].[3] In the former, commonly referred to as BNF for Backus Normal Form or Backus-Naur Form, the metalinguistic symbols and their meanings are

---

[3] Citations are given in the reference lists at the ends of Chapters IV and V, respectively.

Symbol	Meaning
:=	equivalence
< >	surround metalinguistic variable
juxtaposition	concatenation
|	or

In the COBOL report,

Symbol	Meaning
small letters	metalinguistic variable
juxtaposition	concatenation
{ }	choice
[ ]	optional
upper-case letters	optional fixed words in language
upper-case letters underlined	required fixed words in language
...	repeat previous syntactic unit

As a simple example using BNF (i.e. the "ALGOL metalanguage"), consider the definition of an integer. We start by defining a digit by writing

$$<digit> := 0\,|\,1\,|\,2\,|\,3\,|\,4\,|\,5\,|\,6\,|\,7\,|\,8\,|\,9$$

$$<integer> := <digit>\,|\,<integer>\,<digit>$$

The first line specifies that a metalinguistic variable called *digit* is one of the characters 0, 1, 2, 3, 4, 5, 6, 7, 8, or 9. The second line illustrates *recursion* as part of the definition because it says that an *<integer>* is either a *<digit>*, or an *<integer>* followed by a *<digit>*. A negative integer would be defined by saying

$$<negint> := -<integer>$$

The following are integers (by the definition above):

    3   32   0045   000000   2598600002100900

Note that there is no limit stated on the number of digits allowed. From the definition of negative integer, examples are

    −3   −32   −000000   −05290600

but not

    −32−   −3−2

As a more abstract illustration, suppose

$$<ab> := (\,|\,*\,|\,<ab>\,)\,|\,<ab>\,<d>$$

$$<d> := A\,|\,B\,|\,C\,|\,D\,|\,E$$

II.6. TYPES AND METHODS OF LANGUAGE DEFINITION

Then the following are legitimate values for <ab>:

(	(A
*)))	*C
()E)	()E))
()	*)))ABCDE)ABCDE)

The following are *not* legitimate values for <ab>:

A	A)(
A(	ABC)
((*	*(
)*	)E

Using the "COBOL metalanguage," consider the following abstract example:

$$\text{integer K } \begin{Bmatrix} \text{bibble} \\ \text{A}\underline{\text{B}} \end{Bmatrix} \text{ [bull] } \ldots$$

where *integer* has the expected meaning, *bibble* represents a letter, *bull* represents a digit, and the three dots ... indicate repetition of the immediately preceding syntactic unit, namely, *bull*, i.e., digits. Note that it is the syntactic unit which is repeated, not necessarily the individual value of the unit. Then the following are legitimate values for the metalinguistic expression above (which is not actually given a name):

3K5	5KC3333
5B3259	4KAB
2KL	2B

Since the K is not underlined, it is optional. Note that in the first case it is impossible to tell whether the K has come from the specific K, or from *bibble*. In the last case, the B can be from either the *bibble* or from the A$\underline{B}$. A language with the characteristic that its strings can be broken apart in only one way is called *uniquely deconcatenable*, and the example above defines a language which is not uniquely deconcatenable.

To show the difference between these two examples of metalanguages more fully, each formula will be written both ways. The first one can be written as

$$\left\{ \begin{matrix} (\\ * \end{matrix} \right\} \left[\left\{ \begin{matrix} d \\ ) \end{matrix} \right\} \cdots \right]$$

and the second as

    <partial> := <integer> K <bibble> <bull> | <integer> K <bibble> |
    <integer> K A B <bull> | <integer> K A B |
    <integer> K B <bull> | <integer> <bibble> <bull> |
    <integer> <bibble> | <integer> A B <bull> |
    <integer> A B | <integer> B <bull> | <integer> B |
    <integer> K B

    <full> := <partial> | <full> <bull>

The primary advantage to metalanguages similar to those used in the ALGOL report is their ability to name a metalinguistic variable and use it in a formula. The metalanguages similar to those used in the COBOL report do not have that facility. This often makes it very difficult to define certain metalinguistic variables. On the other hand, in most cases where any complicated choice is involved, the COBOL approach is simpler. However, the COBOL approach involves two dimensions, while the ALGOL metalanguage requires only one. A more detailed discussion of the differences between the two general approaches is given in Sammet [SM61a]. A discussion of the problems of two-dimensional syntax is given in Rochester [RT66].

While some readers may feel that such notation introduces undesirable formalism, it certainly serves to eliminate a number of ambiguities. For example, the following definition appears on page 5 of the *COMIT Reference Manual* [MT61]:

> A name consists of a string of twelve or less characters chosen from the letters of the alphabet, the numbers, and period and hyphen in medial position:
>
> Characters for use in names:
>
>   A B C ... Z
>
>   0 1 2 ... 9
>
>   . —   except as first or last character

The question left unanswered by this definition is whether more than one period and/or hyphen can appear in a name. Thus, it is not clear whether or not A.B.C.D and A.B — C are legal names.

II.6. TYPES AND METHODS OF LANGUAGE DEFINITION   57

The illustrations of metalanguages above should not be thought to include all the major concepts. For other ways to define artificial languages, see Gorn [GO61a] and Floyd [FL64]. However, the two above, and minor variations of them, have proved to be most useful. They have also given rise to the whole compilation technique known as *syntax-directed compiling*. Very briefly, this is a method whereby languages are defined by providing tables of their syntax and tables of the operations (e.g., convert to machine code) which are to be performed on different syntactic units. Among the earliest works along these lines were the independent efforts of Irons [IR61], Glennie [GC60], and Brooker and Morris [BX62]. An overall description of the technique of syntax-directed compilation is given by Cheatham and Sattley [CH64].

In my opinion, if there is ever to be any hope of allowing users to define their own artificial languages, it will most likely occur through the use of formal methods of language description and processors which can accept these definitions and either translate them to running code or interpret them to produce answers directly.

II.6.3. TYPES OF DOCUMENTATION

It is a truism that a language or a program is only as good as its documentation. Without written specifications for an artificial language, there is no language. The real problems exist in determining what type of documentation should exist.

There are essentially four types of documentation for a programming language. The first is the reference manual containing the exact specifications, using whatever metalanguage (including English) has been agreed upon by the language designers. It is in this document that the real technical troubles usually fall since, as discussed in Section II.6.2, there are as yet no satisfactory techniques for defining programming languages rigorously.

The second type of document is a user's manual, which can be tutorial or introductory. Such manuals are usually replete with examples and often omit many of the trickier points of the language. This usually causes the individual who wishes to know all about the language to refer to the specification manual, which may be very difficult to read. In such a case, the tutorial description has served its purpose, namely to allow individuals to learn to use the language in a reasonable way but not necessarily with all the fine points. Ideally, the tutorial manual would exist in stages, providing first the most basic information and then progressing toward the most complex, so that all points are covered.

The third type of document is written for a specific implementation and often combines elements of the other two. Although ideally there should be no need for a new language description manual for each compiler, in practice this has turned out to be necessary. Minor differences in implementation techniques or machines cause differences in such points as numbers of variables allowed, precision of the arithmetic, special cases not handled, etc. Such manuals often contain information on how to write programs most efficiently for the particular version involved. Sometimes the individual implementation manuals are based on a more general manual which is assumed to be the basic information. (See, for example, the IBM FORTRAN manuals listed in the references.)

The fourth type of document is some form of summary or very short (ideally 1–2 pages) document to be used as a ready reference by those familiar with the language and needing only to refresh their memories. (See, for example, the CPS summary in Section IV.6.5.)

For a general discussion of the problem of documenting programming languages and the ways in which seven languages (ALGOL, COBOL, COMIT, FORTRAN, IPL-V, JOVIAL, NELIAC) were documented, see the series of articles edited by Yngve and Sammet [YN63a] and their specific comments [YN63]. In addition, there is a series of individual language bulletins which have appeared independently and/or under the auspices of the ACM SICPLAN Notices. The latter is an informal "news and notices" bulletin edited by C.J. Shaw and has appeared monthly.

II.7. EVALUATION BASED ON USE

It is characteristic of the computer business that systems are often evaluated on theory and personal preferences rather than on the basis of practical usage. This is advantageous if the system is so obviously bad that nobody ought to even try using it. Unfortunately, nobody has yet devised a foolproof way of making such judgments. It is always very easy—and much more fun—to examine a language in an abstract condition that is independent of its usage. This tends to relieve people of the problem of obtaining facts to back up their contentions, and it allows them to operate continuously in the realm of opinion. However, this is not the most effective way to proceed. It is essential that work be done to determine valid criteria for evaluation based on *usage*, rather than on whim. We need to understand the advantages and disadvantages of specific systems—evaluated against specific objectives—so that mistakes can be avoided in the future.

II.7.1. AVAILABILITY ON DIFFERING COMPUTERS

The most obvious question for a prospective user is whether the language has been implemented for his computer. The answers can range from *yes* to *in process now* to *never will be*. Part of the evaluation of a language is its availability and usage on one or more machines. If it has been widely implemented, then there is more accumulated experience for both users and implementers. There is also strong indication that the language has been used successfully. If it has been available only on large machines and now is to be used on a small computer for the first time, then certain new problems will arise.

As discussed below, the usefulness of the language must be judged independently of the compilers which implement it.

II.7.2. EVALUATION OF LANGUAGE VERSUS EVALUATION OF COMPILER

There are two ways of looking at a language—one is on paper and the other is as implemented on a machine. In the first instance, an individual can examine the language and decide whether or not it is easy for him to solve his problem using that language. In making such an evaluation, he uses such criteria as ease of learning, ease of writing, and applicability to his class of problems. When he attempts to evaluate the implementation, however, he has other characteristics he must be concerned with, such as rapidity of compilation and effectiveness and efficiency of the object code which is produced. Unfortunately, there are too many instances in which the evaluation of the language is based primarily on the evaluation of the compilers. All too often people say *language X is no good*, when what they really mean is *the compiler they are using for that language is very poor*. Once the compiler is improved, then their view of the language changes. It is extremely important to separate these two aspects. (There are cases in which new languages received semipermanent black marks because the first compiler(s) for the language was so bad.)

The two greatest criticisms of compilers are slow compilation and poor object code. The latter can be considered bad because of slow running time or large storage requirements or both. Secondary objections can be raised about the diagnostics at compile or object time or both, inadequate listings from the compiler, unavailability of load-and-go (i.e., compile and immediately execute), and poor debugging facilities. The *language* should not be deemed poor unless it can be shown that its features would permanently cause one or more of these faults. (This point is discussed in Section III.7.)

II.7.3. USAGE RELATIVE TO OBJECTIVES

The most important factor in evaluating a language is to compare its achievements against its objectives. It is therefore necessary that the objectives of the language be well understood before the language design begins.

It is equally essential that prospective users understand the avowed objectives of the language so that they do not try to use a language for the wrong purpose. Unfortunately, the purposes are seldom clearly stated. Either they are not realized when the language development is started or the designers try to claim too much for the language or else they try to claim a more *sellable* set of objectives than are actually intended or implemented. It is certainly fair to consider whether the objectives are worthwhile, but it is not fair to complain about a language for not meeting some objective that was never intended.

There are cases in which languages have been known to exceed their objectives. One way in which this can occur is when a language becomes useful outside its primary application area. The widespread use of FORTRAN for a variety of problems that are not numerical scientific makes it the outstanding example of this additional factor.

II.7.4. Advantages

Only after a language has been in use for a while can its advantages be ascertained. The first thing to determine is whether or not it met its objectives. If so, then the language can be considered to be successful. (The question of whether the objectives were worthwhile is a separate issue and should not be combined with the evaluation of the language.) However, there are two other possible advantages which might exist. The first is that the language may exceed its objectives by being useful for areas which were not originally intended or by being particularly easy to implement or to learn (if these were not part of the original objectives). A second advantage may occur if the language has certain special features which turn out to be very valuable and can be used in other areas. The concept of list processing is a good illustration of this; the basic list processing languages (IPL-V and LISP—see Chapter VI) showed the value of list processing so successfully that it became important for inclusion in newer languages (e.g., PL/I).

The most important thing to realize is that the full advantages (see Section I.5.1) of a language cannot be determined without actually using it on a computer. This is not true about the disadvantages, which can often be found before going near a computer. Those advantages which can be ascertained without actually using a machine are the ease in learning, ease in coding in, documentation it provides, and ease in transferring a program from one person to the next.

II.7.5. Disadvantages

Obviously, the most important disadvantage to a particular language is that it does not meet its objectives. This can sometimes be determined before actually running on a computer, but there must be some honest attempt to try using the language. For example, if one objective of the language is to make it easy for nonprogrammers to use it, then a failure of this aspect can be determined after appropriate training and attempts at program writing. Similarly, if efficient compilation is an objective, the implementers may discover the disadvantages very early in their work.

One important thing to keep in mind is that one cannot measure the disadvantages of a language in a vacuum; one must consider them in the light of the objectives. If the purpose of the language is to solve *numerical* scientific problems, then one cannot say that the language has disadvantages because it cannot do formal differentiation or integration.

The main disadvantages that can be discovered without actually using a computer are that it fails to have the advantages cited in Section II.7.4 and it is not possible to express all the needed operations in the language.

II.7.6. Mistakes to be Avoided in the Future

Only after the language has been in use for a considerable period of time can one determine what mistakes have been made. These mistakes might be in the actual design objectives, in the sense that they were either too narrow or too broad and, therefore, incapable of achievement; or the mistakes might be involved with the relationship between the language and the implementation; or it may be that the language was not suitably designed to meet its objectives. Again, all these factors can be determined only after actual usage.

REFERENCES

II.1.—II.3.

[DJ63] Dijkstra, E. W., "On the Design of Machine Independent Programming Languages", *Ann. Rev. Automatic Programming*, Vol. 3 (R. Goodman, ed.), Pergamon Press, New York, 1963, pp. 27–42.

[DT62] "The RAND Symposium: 1962, pt. 1", *Datamation*, Vol. 8, No. 10 (Oct., 1962), pp. 25–32.

[DT62a] "The RAND Symposium: 1962, pt. 2", *Datamation*, Vol. 8, No. 11 (Nov., 1962), pp. 23–30.

[HL66] Halpern, M. I., "Foundations of the Case for Natural-Language Programming", *Proc. FJCC*, Vol. 29 (1966), pp. 639–49.

[HY62] Humby, E., "Rapidwrite—COBOL Without Tears", *Symbolic Languages in Data Processing*, Gordon and Breach, New York, 1962, pp. 573–83.

[HY63] Humby, E., "Rapidwrite", *Ann. Rev. Automatic Programming*, Vol. 3 (R. Goodman, ed.), Pergamon Press, New York, 1963, pp. 299–310.

[SM66b] Sammet, J. E., "The Use of English as a Programming Language", *Comm. ACM*, Vol. 9, No. 3 (Mar., 1966), pp. 228–30.

II.4. CONVERSION AND COMPATIBILITY

[AX63]   Allen, J. J., Moore, D. P., and Rogoway, H. P., "SHARE Internal FORTRAN Translator", *Datamation*, Vol. 9, No. 3 (Mar., 1963), pp. 43–46.

[HL65]   Halpern, M. I., "Machine Independence: Its Technology and Economics", *Comm. ACM*, Vol. 8, No. 12 (Dec., 1965), pp. 782–85.

[OL65]   Olsen, T. M., "Philco/IBM Translation at Problem-Oriented, Symbolic and Binary Levels", *Comm. ACM*, Vol. 8, No. 12 (Dec., 1965), pp. 762–68.

II.5. STANDARDIZATION

[AT64]   Alt, F. L., "The Standardization of Programming Languages", *Proc. ACM 19th Nat'l Conf.* 1964, pp. B.2-1–B.2-6.

[GS67]   Goodstat, P. B., "Standards in Data Processing", *Data Processing Magazine*, Vol. 9, No. 3 (Mar., 1967), pp. 22–25.

[ST67]   Steel, T. B., Jr., "Standards for Computers and Information Processing", *Advances in Computers*, Vol. 8 (F.L. Alt and M. Rubinoff, eds.), Academic Press, New York, 1967, pp. 103–52.

II.6. TYPES AND METHODS OF LANGUAGE DEFINITION

[AL67]   Alber, K., *Syntactical Description of PL/I Text and Its Translation into Abstract Normal Form*, IBM Corp., TR 25.074, Vienna Lab., Vienna, Austria (Apr., 1967).

[AN66]   Allen, C. D. et al., *An Abstract Interpreter of PL/I*, IBM Corp., TN 3004, Hursley, England (Nov., 1966).

[BA67]   Bandat, K., *On the Formal Definition of PL/I*, IBM Corp., TR 25.073, Vienna Lab. (Mar., 1967).

[BC66]   Beech, D. et al., *Concrete Syntax of PL/I*, IBM Corp., TN 3001, Hursley, England (Nov., 1966).

[BC66a]   Beech, D., Nicholls, J. E., and Rowe, R., *A PL/I Translator*, IBM Corp., TN 3003, Hursley, England (Oct., 1966).

[BC67]   Beech, D. et al., *Abstract Syntax of PL/I*, IBM Corp., TN 3002 (Version 2), Hursley, England (May, 1967).

[BS60]   Backus, J. W., "The Syntax and Semantics of the Proposed International Algebraic Language of the Zurich ACM-GAMM Conference", *Proc. 1st Internat'l Conf. Information Processing, UNESCO, Paris, 1959*, R. Oldenbourg, Munich and Butterworth, London, 1960, pp. 125–32.

[BU65]   Burkhardt, W. H., "Metalanguage and Syntax Specification", *Comm. ACM*, Vol. 8, No. 5 (May, 1965), pp. 304–305.

[BX62]   Brooker, R. A. and Morris, D., "A General Translation Program for Phrase Structure Languages", *J. ACM*, Vol. 9, No. 1 (Jan., 1962), pp. 1–10.

REFERENCES

[CH64] Cheatham, T. E., Jr. and Sattley, K., "Syntax Directed Compiling", *Proc. SJCC*, Vol. 25 (1964), pp. 31–57. (Also in [RO67].)

[FL64] Floyd, R. W., "The Syntax of Programming Languages—A Survey", *IEEE Trans. Elec. Comp.*, Vol. EC-13 (Aug., 1964), pp. 346–53. (Also in [RO67].)

[GC60] Glennie, A. E., *On the Syntax Machine and the Construction of a Universal Compiler*, Tech. Report No. 2, Carnegie Inst. Tech. Computation Center (AD-240512) (July, 1960).

[GO61] Gorn, S., "Some Basic Terminology Connected With Mechanical Languages and Their Processors", *Comm. ACM*, Vol. 4, No. 8 (Aug., 1961), pp. 336–39.

[GO61a] Gorn, S., "Specification Languages for Mechanical Languages and Their Processors, A Baker's Dozen", *Comm. ACM*, Vol. 4, No. 12 (Dec., 1961), pp. 532–42.

[IB62] *IBM 1620 FORTRAN* (Reference Manual), IBM Corp., C26-5619-0, Data Processing Division, White Plains, N.Y. (1962).

[IB64a] *IBM Operating System/360: FORTRAN IV*, IBM Corp., C28-6515-2, Data Processing Division, White Plains, N.Y. (1964).

[IB66] *FORMAL DEFINITION OF PL/I*, IBM Corp., TR 25.071, Vienna Lab., Vienna, Austria (Dec., 1966).

[IB66h] *IBM 7090/7094 IBSYS Operating System-Version 13: FORTRAN IV Language*, IBM Corp., C28-6390-3, Data Processing Division, White Plains, N.Y. (Apr., 1966).

[IR61] Irons, E. T., "A Syntax Directed Compiler for ALGOL 60", *Comm. ACM*, Vol. 4, No. 1 (Jan., 1961), pp. 51–55. (Also in [RO67].)

[IV64] Iverson, K. E., "A Method of Syntax Specification", *Comm. ACM*, Vol. 7, No. 10 (Oct., 1964), pp. 588–89.

[LW64] Landweber, P. S., "Decision Problems of Phrase-Structure Grammars", *IEEE Trans. Elec. Comp.*, Vol. EC-13, No. 4 (Aug., 1964), pp. 354–62.

[MT61] *COMIT Programmers' Reference Manual*, M.I.T. Research Lab. of Electronics and the Computation Center, Cambridge, Mass. (Nov., 1961).

[PU67] Pursey, G., *Concrete Syntax of Subset PL/I*, IBM Corp., TN 3005, Hursley, England (Feb., 1967).

[RT66] Rochester, N., "A Formalization of Two Dimensional Syntax Description", *Formal Language Description Languages for Computer Programming* (Proc. of the IFIP Working Conference on Formal Language Description Languages). (T. B. Steel, Jr., ed.), North-Holland Publishing Co., Amsterdam, 1966, pp. 124–38.

[SM61a] Sammet, J. E., *A Definition of the COBOL 61 Procedure Division Using ALGOL 60 Metalinguistics*. Summary in *Preprints of 16th Nat'l Meeting of the ACM* (Sept., 1961), pp. 5B-1 (1)–(4).

[ST66a] Steel, T. B., Jr. (ed.), *Formal Language Description Languages for Computer Programming* (Proceedings of the IFIP Working Conference on Formal Language Description Languages). North-Holland Publishing Co., Amsterdam, 1966.

[UE67] deBakker, J. W., *Formal Definition of Programming Languages With an Application to the Definition of ALGOL 60*. Mathematical Centre Tract 16 (Mathematisch Centrum), Amsterdam (1967).

[YN63] Yngve, V. H. and Sammet, J. E., "Toward Better Documentation of Programming Languages: Introduction", *Comm. ACM*, Vol. 6, No. 3 (Mar., 1963), p. 76.

[YN63a] Yngve, V. H. and Sammet, J. E. (eds.), "Toward Better Documentation of Programming Languages", *Comm. ACM*, Vol. 6, No. 3 (Mar., 1963), pp. 76–92.

[ZE66] Zemanek, H., "Semiotics and Programming Languages", *Comm. ACM*, Vol. 9, No. 3 (Mar., 1966), pp. 139–43.

TECHNICAL CHARACTERISTICS OF PROGRAMMING LANGUAGES

III.1. DESCRIPTION OF CONCEPT OF TECHNICAL FEATURES

III.1.1. Introduction

In Chapter II there was a discussion of those characteristics of programming languages which were distinct from the detailed specifications of the language itself. Many of those factors were avowedly nontechnical, including economic and political aspects. This chapter is devoted to a discussion of the fundamental technical characteristics in programming languages. The main functions of this chapter are (1) to describe briefly most—if not all—of the salient features that are likely to be present in the common types of programming languages and (2) to provide a consistent framework for discussion of individual languages. It must be emphasized that not all languages have all the features mentioned here, nor is this list absolutely complete; however, it should definitely serve as a checklist for comparing and describing the languages. There does not appear to be any really major attempt at such a classification anywhere in the literature. Some superficial attempts at a breakdown into a few broad categories are given by Perlis [PR65] and Raphael [RA66]. A questionnaire, which contains many of the points listed in this chapter, and was, in fact, a starting point for the development here and in Chapter II, was developed by C. J. Shaw [SH62]. The (unpublished) questionnaire developed by the ACM SICSAM Subcommittee on Language Comparison in developing its report by Raphael *et al.* [RA67] also provides a gross way of dividing the major language elements.

III.1.2. Major Parts of Language

In considering a programming language, there are seven major component parts. These are not mutually exclusive, nor is this the only possible way of dividing a language into its elements. For purposes of this book, however, this particular set of categories seems to be the most useful. The categories are (1) the data and its description, (2) operators, (3) commands, (4) declarations, (5) compiler directives, (6) delimiters, and (7) program structure. Each of these will now be described from an overall point of view to show how they interrelate. Details will be given in later sections.

1. *Data and Its Description*

The purpose of a program is to accomplish some type of computation, where *computation* is *not* limited to numerical calculations. The elements on which the computation is to be performed are called the *data*. This might consist of numerical quantities, lists of names and addresses, mathematical formulas, or just an arbitrary string of characters. The data might even be generated completely internally from the program. In most cases there arises the need for the concept of data variables whose values are to be determined during the execution of the program. Because of the multiplicity of data types which can be used, there is a need for descriptions of them. The methods of describing the data vary considerably, ranging from implicit assumptions to specific declarations [see Section III.1.2 (4)].

2. *Operators*

The use of operators is one of the ways of combining or acting on data elements. Operators generally fall into the *computational, relational,* or *logical* category, although there are other miscellaneous possibilities, e.g., *find first element on a list* and *find third bit*. The distinction between operators and commands is not clear-cut; the most common difference is that operators generally appear in expressions and do not *themselves necessarily* cause permanent results (e.g., writing IF A = B + C does not create a result B + C), while commands precede a set of parameters and cause direct execution.

The common *computational* operators are addition, subtraction, multiplication, division, and exponentiation. These can be represented by any symbols chosen by the language designers, including specific words. Thus, one language might permit the user to write A + B, while another requires A PLUS B. The relational operators, e.g., GREATER THAN, EQUAL TO, LESS THAN, and varying combinations of these, are commonly used to compare arithmetic quantities but the result is (at least implicitly) a *logical value*. Common operators for logical data (data which can have only the values TRUE and FALSE) are AND, OR, NOT. Operators need not necessarily be written *between* variables; when they are, then the notation is called *infix*; when they appear before or after the variables, the notation is called *prefix* or *postfix* (=*suffix*), respectively.

3. Commands

The heart of a language is the set of executable actions that can be performed on the data elements. Each command performs a specific task as specified by the language designers, e.g., *assign a new value to a variable* or *transfer control to another command*. The major types of commands are described in some detail in Section III.5. In some cases the commands in a particular language are defined through a specific set of formation rules but usually the individual commands are listed along with syntactic rules on how to specify the data they are to operate on.

4. Declarations

Under Section III.1.2.1 it was pointed out that data elements had to be described so that the system would know on what it was working. One technique of providing this information is through the use of explicit declarations. These *declarations* do not cause action to be taken directly at object time, but rather they supply information to the compiler. One simple but common illustration is the controlling of arithmetic precision by including somewhere in the program an indication that double-precision arithmetic is to be performed on certain variables. Declarations can take the form of separate statements; they can be associated with the commands themselves, or they can simply be associated with a description of the variable. For example, one could write

DOUBLE PRECISION X, Y, Z

or

Z = X + Y (DOUBLE)

or

X(DOUBLE), Y(DOUBLE), Z(DOUBLE)

The concept is the vital issue here and not the exact form in which the information is conveyed. Declarations can also be used to convey information about storage requirements or even about equipment (e.g., equating a sense switch with a variable).

In some cases, declarations provide information about what is to be done and leave it to the compiler to figure out how to do it, e.g., some of the pattern-matching statements in COMIT and SNOBOL. Further discussion on declarations is given in Section III.3.1.1.

Declarations are a special case of a more general concept called *compiler directives*, but they are sufficiently important to warrant this separate discussion.

5. Compiler Directives

The parts of the language which are directly associated with executable object code are the commands and the data. There are numerous cases, however, in which it is not possible for the compiler to translate such material without having more information. This latter is normally supplied through *compiler directives*, of which the declarations described above are a special case. Other types of information which might be supplied to a compiler relate to the environment in which the system is being used, to specific input/output facilities, or to efficiency criteria, etc.

6. Delimiters

The delimiters are a part of the language which serves only the syntactic purpose of helping define the various other parts of the language. The delimiters might include such things as punctuation marks, blanks, or even key words. They can be token separators (e.g., + in A+B) or terminators for larger units (e.g., . in GO TO ALPHA.). This is discussed further in Section III.2.4.2.

7. Program Structure

Assuming that the language contains the six elements discussed above, there must be a meaningful way of combining these to produce some desired action. The way in which this is done is the *program structure*. This concept involves the rules needed for combining sets of commands and the data on which they operate. It also provides rules for building larger programs from smaller ones. This is discussed in more detail in Section III.3.

III.2. FORM OF LANGUAGE

There is a difference between the form of the *language* and the form of the *program* written in the language; the latter is discussed in Section III.3. The form of the language can be considered to consist of the following major constituents: (1) The character set, (2) the basic elements (=tokens), (3) identifier definition, and (4) definition and usage of other basic elements. The identifier definition is logically a part of the general usage of the basic elements, but it has been shown as a separate topic because of its importance.

III.2.1. CHARACTER SET

The fundamental constituent (although fundamental only in a trivial way) of a programming language is the character set which it uses. There may actually be three character sets, corresponding to the publication, hardware, and reference languages described in Section I.6. Depending on the particular language involved, one or more of these may be involved. The readability of the language, as well as many other features, is heavily dependent on the character set used. For example, if there is a <, i.e., a *less than* sign in the character set, then this eliminates the *necessity* (although not necessarily the desirability) of having a string of letters to represent this operator. Conversely, and more likely to occur, the absence of specific characters for relational operators forces the use of some representation for them. This is usually done by using some appropriate letter string, e.g., **LESS THAN** or **.LT.**.

The character set for the *language* is not necessarily the same as that allowed for the data. The latter can be much larger (or smaller, although this is less likely). The program can therefore operate on more characters than are available for actually writing the program.

Character sets for computer input are obviously constrained by the hardware available and, as a result, the most common classes are those which use the 47 (or 48) characters of the key punch and those which use the characters on a typewriter. There is no single standard set for either class, however, since the hardware can provide certain choices. For example, two common sets on the IBM 026 key punch are the "FORTRAN character set", which includes the following in addition to the letters and digits:

$$+ \quad - \quad * \quad / \quad) \quad (\quad . \quad , \quad \$ \quad = \quad ' \quad \text{blank}$$

and the "commercial set", which uses the following nonalphanumeric characters:

$$\& \quad . \quad - \quad \$ \quad * \quad , \quad \% \quad / \quad \# \quad @ \quad \text{blank}$$

One way of extending a limited character set is by means of an *escape character*. In this case, one specific character is used for this purpose and no other. When the escape character precedes other characters, they take on a second meaning. Thus, for example, if the dollar sign were an escape character, then A $+ B might mean *A is greater than B*, whereas A + B has the normal arithmetic meaning. In some cases, concatenation of operators is used even without an escape character to denote a single operator (which really should be a single character), e.g., >= means ≥, i.e., *greater than or equal to*.

As of this writing, there is not enough widespread use of the typewriter for any definite character set to emerge as the most common, although the PL/I set is quite likely to do so. The probable increasing use of ASCII will also begin to have a significant effect on the choice of character set for a programming language.

Although language design can proceed without a fixed determination of the character set, I consider it undesirable. There is a significant difference between designing a language from a hardware language rather than from a reference language. (See Sections I.6.9 and I.6.11.) The former usually imposes many more constraints than the latter. In general, if one starts with the reference language and then specifies the hardware language later, the result will be quite different than if one starts with the hardware language at the beginning. In my opinion, it is much better to work directly from the hardware language because in that way the maximum effectiveness for the given physical character set is achieved. The effect of the character set is most heavily shown in the rules for naming, the choice of operators with or without word equivalents, and the punctuation (rules) used. It is obviously desirable—although equally obviously not technically essential—that characters retain their normal meaning when there is one. Thus, it would be ineffective to have a plus sign + mean *equality* and have an equal sign = mean *greater than*.

III.2.2. Types of Basic Elements (=Tokens)

The word *tokens* is used to refer to the basic elements in the language. In this context, the elements are *atomic*, i.e., they have no possible further subdivisions. The definition of token depends on the language; in one case it might be a single character, while in another it could be a sequence of characters surrounded by spaces. While the *types* of tokens and many specific ones are system-defined, some individual instances of tokens (e.g., names) can be user-defined. In the latter case, there are restrictions imposed by the system.

1. *System-Defined*

The system tokens are the graphic operators, the key words, and the graphic punctuation symbols. The graphic operators are those characters which are in the character set for the language and which have a defined semantic meaning as an operator. The most common occurrences of these are the +, −, *, and / signs. For those character sets containing them, the >, <, =, and combinations of them are normally used to designate the relational operators. The key words are those which have fixed meaning in the language. They may be used as commands, operators, compiler directives, delimiters, or punctuation. Finally, the punctuation characters are defined by the system from among the available graphics. The punctuation characters (whether individual graphics or key words) serve as delimiters.

2. *User-Defined and Restrictions*

There are categories of tokens which the user defines (or, more precisely, creates in his program) within the restrictions imposed upon him by the language designer. The most important of these are the identifiers, but the existence of constants, literals, and comments also must be discussed.

For any program, the concepts of data and variables exist in some form. As mentioned earlier, the data may consist of numerical quantities, alphabetic quantities, strings of characters, or anything else permitted by the language. This data, however, must be able to be referred to in some general way. This is done by giving it a name, and the name is more rigorously called an *identifier*. Similarly, the concept of a variable—i.e., a quantity whose value changes during the program—exists, and it must be named or *identified*. There is a significant difference between an identifier and the item it is naming. The identifier may refer primarily to a storage location or to a whole hierarchy of data elements or to a formal variable which never receives any value. It may also refer to elements of the program structure. Possible ways of defining such identifiers are discussed in Section III.2.3.

Most programs require the use of some fixed quantities during the course of the computation. The quantities are most usually numbers, although they can also be logical, or character string, constants. A *constant* is one of the user-defined basic elements in a programming language. In this case, the term *user-defined* means that the programmer decides which values to use. However, he is bound by the restrictions of the language, which may allow some kinds of constants but not others. For example, he might be able to use fixed point numbers but not floating point numbers. He might be allowed numeric constants but not logical constants. The most common restrictions are on the size of the constants; these rules tend to reflect the computer(s) on which the programming language is expected to be used. The presence or absence of a decimal point is significant in some languages, i.e., 2. is not necessarily treated the same as 2 in 2. + 2 * A.

A special type of constant is known as a *literal*. A number of cases arise in which one wishes to use the string ABC to mean a data name (i.e., an identifier for an element of data). In this case it simply represents a location somewhere which contains information which is desired. On the other hand there are many times when one wishes to use the string ABC to mean exactly itself. This latter usage is the meaning of the word *literal*. In other words, a literal is a string of characters which represents itself and not something else. Thus there is a difference between the *number* 23 and the *literal* 23; the latter has no numeric significance. The problem in the language design arises in specifying the means of identifying literals. This is discussed in Section III.2.4.6.

Since one of the advantages of a programming language is to provide better documentation of the task being performed, it is essential that there be a means of providing *comments* in the program. Comments are one of the possible types of user-defined tokens. Most programming languages provide a method by which the user can intersperse comments into his program. These must have appropriate *flagging* so that the compiler will not attempt to translate them.

III.2.3. IDENTIFIER DEFINITION

1. *Types of Identifiers*

There are two major types of identifiers: *Data names* and *program unit labels*. The former can be individual data elements or records or files or aggregates of data. The latter are more commonly called *statement names* or *statement labels*, but these terms are misleading because the language may not have *statements* or it may be able to name several different parts of the program. A program unit label may itself be treated as data in certain types of commands and may also be used to identify nonexecutable parts of the program (e.g., declarations).

2. *Formation Rules*

There are a number of different ways in which data and/or statement names can be created. For example, it is possible to specify that data names can consist only of alphabetic characters, numeric characters, a single alphabetic character, or alphabetic and numeric characters in any sequence. Other common alternatives include allowing letters and numerals to be intermingled providing the first character is a letter, and/or placing a limit on the number of characters. Finally, it is possible to allow punctuation marks or other characters as part of data names and statement labels, with or without specific restrictions to go with them. (However, the use of a hyphen or its equivalent is intuitively reasonable, whereas the use of semicolons in the middle of a word is not.)

One of the key features that must be decided in the formation of rules for identifiers is whether there is any difference between the rules for a variable name (i.e., the name of a piece of data) and the program unit label (called statement label for short). Some languages use the same rules for both, whereas others provide for some distinguishing characteristic between them. In any case, clearly one name cannot be used to represent two different items *at the same time* unless it is always clear from context which is meant. An interesting problem in establishing rules for naming variables is connected to the method of representing multiplication. In ordinary algebra, we write xy and mean the product of two variables. However, in a programming language, if data names have more than a single letter, it becomes very difficult, or logically impossible, to distinguish between the product of two variables and a single data name with two letters. Thus, most programming languages which allow more than a single letter for the variable name are forced into providing a specific operator (usually the asterisk) to indicate multiplication. Conversely, if the language is to permit multiplication by merely indicating juxtaposition, then it usually restricts names to a single letter. Some of these problems could be handled by appropriate use of blanks, but it is usually not worth the trouble.

3. *Use of Reserved Words*

A language can contain *key words* which are merely character strings having a specific meaning in the language. Some, all, or none of these may be defined as *reserved words*, which are forbidden for use as either data names and/or statement labels, or their beginnings. For example, FORTRAN has no reserved words, although it does have key words (e.g., DO, DIMENSION). On the other hand, all the key words in COBOL (e.g., PERFORM, RECORD, READ) are reserved words and cannot be used as either data names or statement labels. Most key words in PL/I are not reserved. The advantage of refusing to allow the programmer to use key words for naming variables or statements is that the scanning of the source program becomes considerably easier. The disadvantage of disallowing reserved words for the use of the programmer is that he must always have in front of him a list of these reserved words and make sure that he does not use them. Furthermore, if he is choosing a language for use in an existing installation, he must make sure that the words he has already used for his files do not conflict with reserved words in the language. In some cases, there is even a more severe restriction which says that a data name or statement label cannot start with any letters which coincide with one of the reserved words. It can become even more confusing when, as in the case of COBOL, reserved words vary from division to division (see Section V.3) and the user must keep this in mind. In PL/1 there are built-in function names which have specialized rules. This is one of the characteristics that is very significant as far as implementation efficiency is concerned, but it is at the expense of the user's convenience.

Some reserved words may be used as noise words (see Section III.2.4.5).

4. *Data Names for Aggregates* (*Subscripts, Qualification*)

In most practical problems, data is grouped together into some meaningful form of aggregate. The most common types of aggregates are sets of items of the same type, normally called *arrays*, and sets of items of distinct types grouped together into some type of hierarchy, normally called *hierarchical* or *structured data*. It is logical to have arrays of hierarchies or hierarchies containing arrays as elements.

There are many cases in which one wishes to give a single name to a list of elements and then refer to an individual element in this by a *subscript*. In other words, one might have a list called A with 12 elements in it; then it is normal to want to refer to these as A_1, A_2, ... , A_{12}. To do this in a programming language, it is necessary to introduce the concept of subscripts as part of the data name. Because almost all the input is in one dimension, the subscripts can seldom be written below the line as done in normal mathematical notation; some other notation must be used and this fact usually becomes a significant problem in language design. (This is one of the key places in which the publication language will differ from the hardware language, as discussed in Section I.6.) Once we have established the principle of desiring to refer to an element in a list by its position designator, then there arises the question of a two-dimensional array; this is normally coped with by allowing two subscripts. Similarly, the position of an element in an N-dimensional array is denoted by N subscripts. The most common notation for this is the use of parentheses adjacent to the variable name, e.g., A(3,2) would refer to the item in the third column and second row or in the third row and second column of a two-dimensional array. One of the key points in a language design is the number of subscripts which will be allowed. From a language point of view, there is usually no reason to impose any limit, but restrictions are placed for implementation reasons.

An additional characteristic of subscripts is the amount of flexibility used in defining them. For example, a language could permit only constants but this would be rather pointless. An almost equally severe restriction is to permit only a fixed point variable to specify the value of the subscript. The next most flexible rule is to allow arithmetic expressions involving addition and multiplication of fixed point variables. From there, generality can be increased to allowing any combination of fixed point variables (with rules required for division) and, finally, to allowing any combination of variables, including floating point, Boolean, or anything else for which some rule can be specified that will end with an integer. This latter allows statements such as IF A = 5 THEN 3 ELSE 7 to be used as subscripts in some languages, e.g., ALGOL. A fairly common practice is to allow any arithmetic expressions, including floating point numbers, and then to truncate the result to produce the integer which is needed to obtain the position in the array. Other design questions involve the allowed range (i.e., negative or zero, as well as positive) and whether or not subscripts can themselves be subscripted and, if so, in what form and to what depth.

Hierarchical data occurs when a particular data item has subitems to which names should be given. Alternatively, there may be data items which can be grouped into a larger unit which can then be named by an identifier. Consider a complete name and address as an illustration. Suppose this is of the form JOHN DOE, 7777 OCTAL ROAD, CITY, STATE, ZIPCODE. Depending upon the purpose, we might wish to reference just the name, any single one of the other items, the city and state, or any combination of these fields. Suppose that we could assign a name to each piece and to each meaningful group of pieces. Then we might have something of the following form:

A) NAME—AND—ADDRESS

B) NAME

C) ADDRESS

D) STREET

E) AREA

F) CITY

G) STATE

H) ZIPCODE

This is really a representation of the tree and data layout shown in Figure III-1. In another file, we might also have a data item called **AREA**, and the problem becomes one of specifying which occurrence of the name (and corresponding data) is meant. The technique which is used for this is normally called *qualification*. By this is meant the usage of enough names in the hierarchy to uniquely identify the desired data name. In a fairly common case, suppose that the **NAME—AND—ADDRESS** data item appears in both an

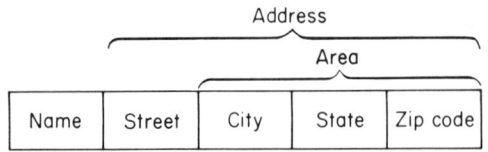

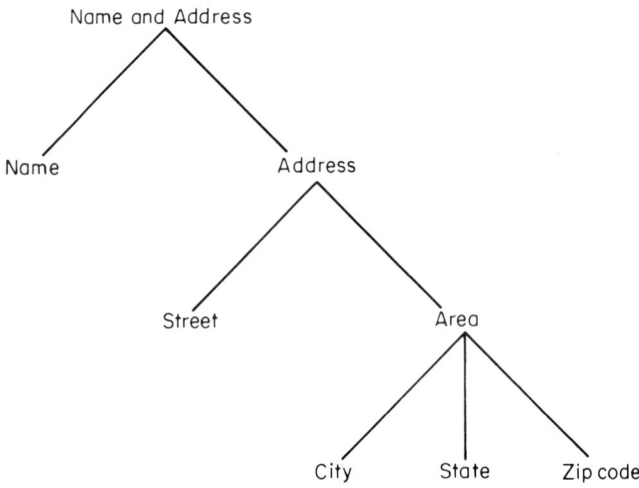

Figure III-1. Example of tree and data layout.

INPUT file and an OUTPUT file. Then in order to uniquely identify which was meant, it would be necessary to say NAME—AND—ADDRESS IN INPUT. Similarly, if the AREA was used as a data name in several items, it would be necessary to use some higher level name to identify it. For example, if AREA also appeared in a data element called SALES—RECORD, then the user would write AREA IN SALES—RECORD or AREA IN NAME—AND—ADDRESS to identify the one he wanted. The actual notation used to specify this qualification differs in each language. These additional names can be attached at the beginning or the end of the relevant data name and are then called *prefixes* or *suffixes*, respectively. Various notational devices which are used to indicate these are periods, hyphens, specific key words (e.g., IN as used above), etc. A variety of rules can be used for the identification, ranging from the most severe which says *all names up to and including the topmost one which gives uniqueness must be shown*, to the most liberal which says *you need only show enough to make it clear to the compiler which piece of data you are referencing*.

Various combinations of subscripting and qualification are permitted in those languages which allow both individually.

III.2.4. DEFINITION AND USAGE OF OTHER BASIC ELEMENTS

Assuming the basic elements (=tokens) of the language as described above, these can be combined or used in numerous ways. This section discusses a number of the factors in both the syntax and the semantics of these combinations.

1. *Operators*

The operators in a language are one of the two categories which can cause things to be done (the other being the commands). Operators provide means of combining or relating data types; hence they can be defined as connectives for variables. Thus, for arithmetic quantities, the ordinary arithmetic operations are essential and the relational operators, such as **GREATER THAN** and **EQUAL TO**, are fairly standard in any language which permits comparison of arithmetic quantities. (Note that the use of relational operators with arithmetic variables does *not* yield an arithmetic result.) Operators are also defined for nonarithmetic data types. For example, Boolean (i.e., logical) variables are often combined by logical operators such as **AND, NOT,** and **OR**; while **DIFFERENTIATION** is an operation performed on formal expressions. It is important to note that operators may be represented by symbols, such as + and =, or by any legal combination of other characters in the language, e.g., **PLUS** and **EQUALS**. As mentioned before, it would certainly be possible to attach nonnormal meanings to operators but this would be of value only when the normal meanings were not needed at all. In some cases, normal *operator functions* are actually represented as commands; e.g., instead of writing **A + B**, one writes **ADD A TO B**. In the first case, the + is considered an operator, while in the second the **ADD** is considered a command. However, the distinction between operators and commands is not clear-cut. The former tend to indicate actions which *need* to be performed, while the latter specify actions which *must* be performed. Thus, **A + B** *implies* that an addition will take place while **ADD A TO B** *requires* it to take place. There are counterexamples to both these conceptual definitions.

2. *Delimiters*

The concept of delimiter was mentioned in Section III.1.2.6. In this context it is merely necessary to note that a delimiter can be any combination of tokens that the language designers feel desirable. For example, a delimiter can be a key word, a particular punctuation symbol, a blank, or varying combinations of these. The prime purpose of the delimiter is to define the beginning and/or end of elements in the language. Thus, a delimiter might be needed to define the end of an identifier, the beginning of some unit in the program, or the beginning and end of a literal. In some cases the delimiter exists as a concept rather than a specific entity since the termination of one element might be determined only by the beginning of another, e.g., the end of a paragraph in COBOL is identified by the occurrence of the beginning of another paragraph.

3. *Use and Meaning of Punctuation*

The word *punctuation* in English usually refers to the use of characters such as commas, semicolons, periods, hyphens, and parentheses. For our purposes here, we shall limit attention to the first three, although many of the points to be made are valid or relevant independently of how many characters are specifically classified as being punctuation characters. In many cases, punctuation characters are used as delimiters.

Because many of the languages attempt to make things natural (or at least not unnatural) for the user, a period or semicolon is often used to end a sentence or statement (see Sections III.3.1 and III.3.2). It is essential to notice that an *end of statement* marker is not required in the case of fixed format input (discussed in Section III.2.5) because the beginning of a new card (which does not have a continuation mark on it) is an automatic termination of the previous statement. Thus the problem of statement termination really becomes an issue only in the case of continuous string input or where one wants to permit a second statement to begin on the same line (or card) as the ending of the first statement.

Another area in which punctuation is needed is in listing a string of parameters; normal English usage requires commas between them and this concept is followed in almost all programming languages. For example, if one wants to print the three characters A, B, and C, a very natural notation would be PRINT A, B, C. The use of commas in a situation like this is intricately tied in with the use of significant blanks. (See Section III.2.4.4.) Thus, if blanks are considered significant, then a space between the A and the B is sufficient to denote that these are two different names. On the other hand, if the blanks are not significant, then some marker must be used to indicate that AB is not the name. Thus, the nonuse of significant blanks forces more punctuation than would be necessary if blanks are critical. It should be noted that even where blanks are critical, punctuation is still required in certain cases.

The most common uses of punctuation in a programming language are the use of the period or some similar character (often a dollar sign is used) to indicate the end of some level of executable unit and the use of a comma or some other mark to separate items in a list. The next most significant usage occurs with a semicolon which is often used to delimit other executable subunits. Sometimes the semicolon and the period play different roles, in the sense that the semicolon is used for an end of statement mark, while the period is not used for punctuation. The reason for the latter is that the actual period mark is required in numbers, e.g., 3.14159. In some cases where a significant difference exists between 2. and 2 there is a problem if 2. appears at the end of the unit that the period terminates. Since obviously a period should be used with numbers, great care must be given to the syntactic rules for embedding periods in the language.

Other punctuation marks are sometimes used for more specialized purposes; the most prevalent is the use of parentheses in mathematical formulas where the normal rules of mathematical notation apply. Problems arise, however, when parentheses are also used to designate subscripts (which is done in almost all languages) and also to represent functions. It is not obvious whether F(I) represents the function F with the parameter I, the variable F with the subscript I, or even the variable F multiplied by the variable I.

4. *Significance of Blanks*

The *blank* or *space* character usually plays a special role in programming languages, even if in a negative sense. One characteristic of the word formation rules is whether or not blanks are significant. *Being significant* means that A♭B and AB are not the same thing (where ♭ means the blank character). The advantage to having blanks be *non*significant is that the user does not have to worry at all about where he puts them. This presumed advantage tends to be more than counterbalanced by the double virtues of providing an extra character to use for technical purposes and permitting a person's natural tendency to use blanks as separators to be indulged. Thus, since English is written with a blank space *following* (but not within) each word, it is quite natural to require a blank at the end of a word in an artificial language. However, there tend to be special rules for blanks near operators. Thus it might be legal or illegal to write A♭+♭B, or it might not matter.

5. *Use of Noise Words*

The term *noise words* refers to character strings which can be inserted or omitted in a program at the user's option without changing the meaning of the program. There are several different rules which can be established within this general principle. For example, the most flexible rule (it is not used in any language in this book except COLASL—see Section IV.7.2) is that between any two fixed words there can be any number of arbitrary words. The prime difficulty with such a general rule is the difficulty (or potential impossibility) of distinguishing noise words from legitimate names. A much more restricted (but far more reasonable) rule which does exist is that certain fixed words may be present or absent in certain fixed places; this was done in COBOL, for the express purpose of improving readability. For example, in the sentence READ PAYROLL RECORD INTO INPUT—AREA, the word RECORD is a noise word and can be omitted. Between these extremes there are a number of possibilities.

Noise words are often key words. They can also be reserved words so that the programmer is restricted from using them as identifiers.

6. *Literals*

Obviously, there must be some way of indicating whether ABC is a name or the three-letter string (i.e., literal). It might appear at first glance that this could be determined from context. In some cases this is possible, but in many other cases it is not. For example, if we write PRINT ABC, then the compiler is at a loss to know whether we want the three-letter string ABC printed out or the quantity whose name is ABC. Thus it is necessary to put some kind of a marker around the literals. This can be done in a variety of ways: One way is to take some single character from the character set and designate this as a beginning and ending delimiter for a literal. If, for example, one chooses the dollar sign, $, then writing ABC would designate the literal string of characters. Another alternative (which is practical only if there is an extremely large character set) is to choose two characters, one of which is used to designate the beginning and the other, the end of the literal. The prime advantage to this second method is that it makes it easier to solve the problem of how to represent the literal delimiter as a literal. If one uses $ for the beginning and end of a literal, then one might require that $$$ be written to designate the literal $ itself. Thus $AB$$CD$ is really two literals, AB and CD, while $AB$$$CD$ is the string AB$CD. Another alternative to representing the literal delimiter itself is to choose a particular fixed word, such as QUOTE, to designate the literal value. Actually, any fixed word, or pair of them, could be used as literal delimiters instead of characters, but this is less convenient for the user and still requires rules for terminating the literal or embedding the terminal character in the string. Still another possible way of delimiting literals is to have one character to mark the beginning and then a count of the number of literal characters following. This is a poor choice because the user frequently miscounts the length of his literal if it is more than a few characters. There are various other techniques which can be used.

III.2.5. TYPE OF INPUT FORM USED

There are several ways in which to consider the amount of form required in a particular programming language. The word *form* in this case means both the way in which legitimate sequences of characters are placed on the input media and the philosophy associated with the types of sequences which are used.

1. *Physical Input Form*

Physical input can be in one or two dimensions. The latter means that syntactic significance is given to more than one physical line simultaneously, e.g., the use of subscripts and/or superscripts as in A_i^3. Because the most common input media for program preparation are punched cards and paper tape, all programming languages using that type of standard hardware have a one-dimensional string form as input.[1] (Note that since paper tape is prepared through the use of a typewriter, the direct use of a regular keyboard does not change the basic principles involved.) There are languages, however, which permit two-dimensional input through the use of special hardware; they are described in Section IV.7.

Once we assume a single dimension, then the primary distinction is between fixed format and a continuous string. The term *fixed format* is a relative one since programming languages vary in their requirements. They run the gamut from a continuous string (e.g., PL/I) to some requirements about card columns (e.g., FORTRAN).

Since many of the earlier programming languages assumed the use of punched cards for input, it is quite natural that they took some advantage of the fixed columns to represent specific items. The advantage to using fixed columns is that the compiler need not scan every single character to find the one that is wanted. Thus, if there is a requirement that a statement label should start in a particular column, it is not necessary to continue scanning until you find something that looks like the beginning of a statement label. In the case of higher level languages, information often extends over more than one card. For that reason, it is convenient to permit early columns in the card to contain some kind of symbol to indicate that the current card is a continuation of a previous one. Space can then be left for specific commands or, more likely, for a label, followed by specific commands and the operands. One of the significant differences between the format on a punch card for a higher level language and for an assembly program is that the latter usually has a fairly restricted amount of space and format allowed for the operands, whereas the higher level language usually permits them to be written continuously following the command.

Assuming continuous string input (which of course can be punched on cards), it is necessary to scan every character to reach any future one, and the meaning of symbols can only be determined from context. Thus, the string

---

[1] The MAC-360 system which appeared too late for a thorough discussion is an exception to this. See Section IV.7.6

ABC can be determined to be either a statement label or the name of a variable (or something else) only by seeing what precedes and what follows it. Delimiters (including punctuation) play a much more important role in a continuous type of format because new statements can often be recognized only as coming immediately after the end of a statement and not on a new card or fixed location.

Techniques more suitable to punched cards can be used for paper tape (by using tabs) and vice versa. The important point is the continuity versus fixed format aspect of the input string rather than the physical media being used.

Graphic display devices with input by light pen or keyboard have not yet become a significant input media, although they undoubtedly will become so. They may require new developments in specifying the physical form of a program.

2. Conceptual Form

The *conceptual form* of the language is completely independent of the input media. At the extremes are the concepts of an "English-like" language (e.g., COBOL) versus a highly symbolic one (e.g., LISP). A number of arguments pro and con can be given for both of these views. At one extreme, a language can be designed to be as close to natural English (for the commands allowed) as the designers can make it and still be able to implement it. COBOL was an early attempt to follow this philosophy. At the other extreme is the philosophy that a programming language should be as succinct and formal (not necessarily natural) as possible for the class of problems it handles. No really good illustration of this exists among the fairly common languages, although LISP 1.5 has some of this flavor. This latter viewpoint is often expressed by proponents of ALGOL (although ALGOL programs are not difficult to read after a little training). A formal notation is not necessarily harder to read or write than a more natural notation; the decision on ease of use is almost entirely a matter of personal taste.

Obviously, the physical input form of the language must be based on the conceptual form, and certain combinations are inherently meaningless. For example, it would be impractical to design a language which is "English-like" and then use a rigid input format.

The definition of the conceptual form of a language is really rather intuitive and is being left to the reader to formulate after consideration of the languages in this book.

III.3. STRUCTURE OF PROGRAM

The previous section discussed a number of items relative to the form of the language, in particular, those items which are very significant in determining what strings are legal, such as rules for naming and punctuation. This section involves a similar type of discussion, but from the point of view of units and subunits of a program. In other words, merely a long list of syntactically correct character sequences is not going to necessarily produce a meaningful program, and all languages have types of subunits which must be combined properly. For that reason, this section discusses in some detail the types of subunits that are permitted and their characteristics. The characteristics are mentioned briefly where appropriate with the subunit description, with a full discussion given in Section III.3.2.

III.3.1. Types of Subunits

There are a number of different types of subunits which go into making a complete program, and there are two ways in which a discussion like this can be handled: One is to define a program and then successively work down until the smallest unit is defined; the alternative is to define smaller units and work upward until a program is defined. The latter seems to be a little easier to cope with in terms of normal understanding and will be done here. With the exception of the declarations and comments, all the subunits are executable, i.e., they will directly cause some action to take place at object time. It is important to realize that the subunits discussed represent concepts, not detailed ways in which they are handled. Those are discussed with each of the languages. Furthermore, not all these subunits appear in every language.

1. *Nonexecutable: Declarations and Compiler Directives*

There are many instances in which information must be given to the compiler to permit it to compile correct code or, in some cases, to compile more efficient code. This information is nonexecutable and is usually—but not always—given explicitly as declarations. In some cases, the information is given explicitly or implicitly as part of an executable unit. Declarations are used primarily to supply information about variables or data; common examples are the dimensions for a particular variable, fixed or floating point, the arithmetic precision desired in a calculation, etc. More generally, declarations usually indicate which characteristics a variable has from among those listed in Section III.4, or they supply information about storage allocation or interaction with the environment. The category *compiler directive* includes declarations; in addition, some languages make a subtle distinction by including some things in the former category but not the latter.

2. *Smallest Executable Unit*

The smallest executable unit (SEU) is a general name for what is usually a single command and its operands. There must always be a clear-cut way of determining its end, either through the use of a special symbol or through the syntax of the command. A SEU might be something of the form

$$Y = 3$$

or

$$Z = A * X + B$$

or of the form

PERFORM ALPHA THROUGH BETA FOR Y VARYING FROM 1 TO 17 IN STEPS OF 2

The most common word used for SEU's is *statement*. A sequence of these statements can be used to make up a program. Note that the SEU cannot necessarily be given a name. Furthermore, SEU does *not* refer to the smallest unit for which the compiler turns out code since in the executable unit Y = A * X + B it is necessary for the computer to do two arithmetic operations and a store operation.

There are a number of different types of commands which can be used as SEU's, and these are discussed in Section III.5.

3. *Sets of Smallest Executable Units*

One SEU by itself will not accomplish very much, unless it is being used in an on-line *desk calculator* system (e.g., QUIKTRAN). Hence it is necessary to be able to group these together in a way which permits assigning a name for identification and cross-reference purposes and which permits them to be treated as a single unit in other ways. There are a number of ways of doing this; for example, in ALGOL there are blocks, in COBOL there are paragraphs and sections, and in PL/I there are several other levels of combining. The use of the same term, e.g., *block* or *compound statement*, in different languages does not mean that they are defined or used the same way since there is no standardization at all for these ideas. A block, however, is the name usually given to a sequence of executable units which can be treated as a single nameable executable unit. That is, it can be referenced from a control transfer statement; it can be used as a single statement in the range of a loop; or it can be used as a single statement wherever a single statement is valid in the language. A block normally has some kind of designator to indicate its beginning and end; the words **BEGIN** and **END** are, in fact, often used. A block often contains declarations about the variables used in the block. The main function of a block is to permit the handling as a unit of a number of individual statements. The word *block* is sometimes used for other items and, conversely, is not always used for the concept above.

It is also possible to combine SEU's to create a subunit which has fewer properties than those normally associated with a block, e.g., a sentence in COBOL. In such a case, the ability to name the subunit is usually the first characteristic to be eliminated.

4. *Loops*

A common and essential part of a programming language is the capability of repeating a certain sequence of executable units for more than one value of a particular parameter (or set of parameters). This concept is called a *loop* and has four constituents: Range, the value(s) of the parameter(s), the terminating condition for the loop, and the place to which control is transferred when the execution of the loop is finished. The first three elements are normally in one statement (which is of course an executable unit), but they need not be. Any language with a conditional statement of the form IF ... THEN can be used to write a loop, although many statements will be required in this case. The range is simply the set of statements which are to be repeated for the differing values of the parameter(s). Depending on the particular language, these may be sets of contiguous statements immediately following the loop statement, they may be widely scattered statements connected by control or conditional transfers, or they may be contiguous sets of statements elsewhere in the program. In some cases, the range of a loop may consist of one or more sets of subroutines. Regardless of how the range is determined, there must always be a way to designate the beginning, ending, and intervening values for the loop parameter(s). This point is discussed more fully in Section III.5.3.

5. *Functions, Subroutines, and Procedures*

It was discovered many years ago that it was inefficient and very difficult to check out a very large number of statements as a single unit. Furthermore, it was very wasteful to have many people writing the same routines over and over again, and this occurred very frequently with commonly needed computations, e.g., finding $\sin x$ or finding the roots of a polynomial. Thus was born the idea of a *subroutine*, which is simply a self-contained set of statements to perform a particular task. Although some subroutines are independent of the values of any input parameters, most are written to perform their task for the particularly supplied value(s). Thus, a subroutine may have zero, one, or many input values and one or many results. (It could conceivably have none, but this usually would be indicated by an error condition, which is a form of result.) Most large programs are built up from a series of subroutines.

The special case of a subroutine which has a single result is usually called a *function*. Routines for computing the elementary mathematical functions (e.g., sine, cosine, exponential, logarithm) are usually included in languages designed for solving mathematical problems. Because there is only a single item as a result, functions can usually occur in the same places as numbers or variables. Thus, Y = A + SIN(B) * C would normally cause the actual value of the sine of B to be multiplied by C. An alternative—but obviously more inconvenient—way of doing this is to write

$$Z = SIN(B)$$

$$Y = A + Z * C$$

Although there is usually a real choice between using open or closed subroutines in assembly languages, most subroutines used in programming languages are closed. The lack of macros in most programming languages discourages the use of open subroutines.

I consider procedures to be the same as subroutines, but some people feel there is a subtle difference, in which the procedure is a more meaningful unit than the subroutine. This is primarily a historical difference, with the problem of parameter passage (see Section III.3.2.3) and declarations more closely associated with the use of the word *procedure* than with *subroutine*. In any case, the terms *subroutine*, *procedure*, and even *function* are relative because what may be a complete program in one case is merely part of a program (i.e., a subroutine) in another. For example, the code to invert a matrix might be the entire program in one instance but merely part of a more complicated program in another case. In modern languages, a program is usually written as a procedure.

There is a significant difference between the *use of* (commonly called *invoking*) the function, subroutine, or procedure and the information which defines the function, subroutine, or procedure. The definition is usually written as a self-contained unit, called the *body*, outside the main control flow of the program and invoked from within the program. The process of invoking requires supplying parameters (unless the function, subroutine, or procedure does not allow them).

6. Comments

Although it may seem strange to include a category called *comments* with all the other subunits in this section, it is an item which must be considered. Since one of the key problems in programming is to provide adequate documentation of programs, any means to assist in this is worthwhile. A device which exists in virtually all programming languages is a specific way of indicating comments which are meaningful to people but not to the compiler. This is most often done by specifying some delimiter(s) for the beginning (and end) of the comments. When the compiler encounters these flags, it knows that it should ignore the following material until the beginning of the next subunit which requires compiler action. This permits the user to write anything which will be helpful to him or others.

The liberal use of comments, together with the *problem-oriented* notation should supply most of the documentation for a particular program.

7. Interaction with the Operating System and Environment

In modern computers, an operating system is quite common, and the compiler must operate under it. It is therefore often—but not always—necessary for the user to be able to communicate with the operating system in order to obtain or put out data, to know when parts of his problem are finished, to take advantage of the specific hardware that might be involved, etc. Thus, there are potentially in the language a number of statements or subunits which have to do with the operating system. Included among these are procedures for recovery from tape errors or other computer malfunctions, where the programmer feels he can recover or he wishes to take specific action based on a particular error message.

The types of things which may (or must) be written to provide interaction with the operating system and the environment may actually take the form of a number of the subunits previously discussed. For example, they could be declarations, which really provide information to the operating system rather than the compiler but which must be translated by the compiler to provide the proper interface. Alternatively, they can be executable statements which are called into action by either the compiled object code or by the operating system. In fact, for very large and complex operating systems, it is necessary to create a special language to provide the necessary interface between the user and his program on one hand and the operating system and hardware on the other. This point is carried to its extreme limits in the case of the control statements and operations in a time-sharing system. In such a situation, the language by which the user communicates with the system tends to be more important or more complicated than the language he uses to solve his problem. (See Section IX.3.6.)

8. Inclusion of Other Languages

It is an unfortunate fact that very few programming languages are so self-sufficient that the user does not wish he had some capability from another language available to him. In some instances, the language does provide the ability for the user to insert—directly or indirectly—a different form of language. This must be preceded by some kind of declaration which informs the compiler that it must switch out of its present scanning mode and bring in some other mechanisms. If this were not done, then the compiler would have to scan and translate statements of which it had no knowledge.

One of the key types of languages which users like to be able to intermingle with their higher level language programs is machine language. At the very minimum they wish to insert machine-language subroutines which already exist; even more importantly, they would like to be able to include machine instructions in the midst of the higher level language code. (This is particularly important for small sections which should be coded as efficiently as possible.) However, there is a significant difference between being able to invoke a machine language subroutine which already exists, and the inclusion of machine language in line. The former requires the user to make sure that the interface of information about variables is handled properly, whereas the latter requires the compiler to do it. Relatively few languages include this facility (which is really a requirement on the compiler). The earliest (and in fact the only) publicly available *language* to require this facility in great generality was MATH-MATIC [RR60]. (A version of FORTRAN with this facility, namely FORTRAN III, was distributed for use on the 704.)

The facility to invoke a subroutine which has been coded in another language is definitely *not* considered an instance of being able to *include another language*.

9. *Complete Program* (*Including Sequencing Rules*)

A complete program is usually created by combining any or all the preceding subunits according to specified rules. Whether this produces the answers desired by the user is a pragmatic question.

The legal ways of sequencing and concatenating executable units of various kinds, and combining these with the declarations, must be defined for each language. There are a few common practices, however, and almost all these are imposed to make the implementation easier; very few have any inherent meaning for the language or the program itself. For example, declarations are sometimes required to precede the items which they are describing. In other cases, all the declarations must be together and must be labeled as such. In still other cases, all the subroutines or equivalent subunits must follow the main part of the program. Thus, the issue of intermingling executable and nonexecutable subunits is a significant characteristic in the overall form of the program.

III.3.2. CHARACTERISTICS OF SUBUNITS

Some types of subunits have characteristics which have a significant effect on programs written in the language. Not all these characteristics are meaningful for each type of subunit; with the description of each characteristic, the units it applies to are indicated.

1. *Methods of Delimiting*

All subunits—without exception—must have some way of indicating their beginning and end, although in some cases the only way of determining the end of one unit is by recognizing the beginning of another unit. In the cases of single executable units or declarations, this is usually controlled either by fixed format (see Section III.2.5) or by the syntax of the individual unit. It is in a situation like this that the concept of reserved words introduced in Section III.2.3.3 becomes particularly significant. Thus, if there is a declaration with an unlimited number of parameters following it (e.g., FIXED A, B, C, D ...), then there must be either a specific way of terminating the list (e.g., using a period) or else a way of determining the beginning of the next statement. This *beginning* can be defined in a number of ways, but it is most commonly done by using a reserved word (e.g., COMPLEX G, H, ...).

In larger subunits, special symbols or words are introduced. A very common way of ending smallest executable units (defined in Section III.3.1.2.) is by either semicolons or periods. Sequences of executable units are often delimited by key words at the beginning and end.

In addition to delimiting the actual subunit itself, there must be a method of specifying the name of the subunit; the name sometimes performs a delimiting function, but it also presents other syntactic problems. Methods of handling these things will be seen in the individual languages.

Another facet of the delimiting problem is the scope issue. This involves a determination of what characteristics of variables or other program information is relevant in different parts of the program. (See Section III.4.5.)

2. *Recursive*

A subunit is said to be *recursive* if it can be used in, or referenced from, itself. This means that if subroutine XYZ is invoked, say by writing

CALL XYZ, (*parameter list*)

then subroutine XYZ will be said to be recursive if CALL XYZ, (*parameter list*) can appear within the XYZ subroutine itself or in another routine which is invoked from XYZ, although normally with a different set of parameters. This particular characteristic is extremely important because some problems require this kind of capability and others are stated most easily by using recursion. To clarify what is meant by recursion, the following example of the factorial is the most easily undersood:

SUBROUTINE FACTORIAL (N)

FACTORIAL 1 = 1

IF N > 1, FACTORIAL N = N * FACTORIAL (N − 1)

END SUBROUTINE

In this case, the FACTORIAL subroutine is called repeatedly from within itself to find the value of N!. Note that writing SQRT(SQRT(X)) is *not* an example of recursion, because the normal ways of writing square root routines do not require the routine to call itself. The term *recursive* is applied only to those subunits for which there are parameters involved. Thus, functions, subroutines, and procedures can be considered as being recursive.

3. *Parameter Passage and Differing Types*

The basic nature and concept of functions, subroutines, and procedures require that they have parameters. In other words, these subunits are sets of commands whose objective is to produce specified results for each set of inputs. The inputs can be simple (e.g., an identifier) or complicated (e.g., an expression or an invocation of a function). The invocation of a subroutine (or function or procedure) requires that the values of the variables for which the subunit is to be executed be supplied. (This process is called *passing the parameters*.) The variables which appear in the subunit itself are usually called *formal parameters*. The word *value* in this case is somewhat misleading because there are three primary types of parameter passage which exist. The first is referred to as *call by value*, the second is the *call by name*, and as the third has no standard terminology, I use *call by location*. (This is the same as the *call by simple name* proposed by Strachey and Wilkes [SQ61] and is the one used in FORTRAN.) While the concept of *location* does not appear in programming languages as such, the rules given for handling the parameter passage use this implementation factor directly or indirectly as part of the language definition. In the *call by value* case, the actual value of each parameter called by value is assigned to the corresponding formal parameter at the time of invocation. In the *call by name* case, the name of the particular parameter involved is inserted into the code of the subunit before executing it. In those languages for which the input parameter can be an expression, the distinction is more significant; the *call by value* causes the expression to be evaluated before entering the subunit, whereas in the *call by name* case the occurrence of the parameter in the subroutine is replaced by the code to evaluate the expression (or by a call to a subroutine to perform the evaluation). The *call by location* applies only to a single variable and not to an expression; in this instance the occurrence of the parameter in the subroutine is replaced by the code to access the location, thus permitting the subroutine to destroy constants by storing over them.

As an illustration of these concepts, consider the following example (assuming there is no concept of global or local variables in the language): Suppose we have a subroutine called INC with an input parameter S and an output parameter R. Suppose that the body of the subroutine consists only of the two statements:

$$A = 4$$

$$R = S + 1$$

Assume that the subroutine is invoked by writing CALL INC (S, R), where S and R represent the input and output parameters, respectively. Then

$$A = 3$$

$$\text{CALL INC } (A + 1, B)$$

will give B the value 5 in the *call by value* case, and the value 6 in the *call by name* case. The reason for this is that in the call by value, the expression A + 1 is computed (yielding 4) *before* entering the subroutine, and so B = 4 + 1 = 5. However, in the call by name case, the subroutine actually becomes

$$A = 4$$

$$R = A + 1 + 1$$

Hence the result is 6. In both these cases, the original value of A prior to invoking the subroutine (namely 3) is preserved.

Any reader who studies this matter further and then wants to test his understanding of the concepts should see Weil [WL65].

To show the call by location case, suppose the subroutine is called ADDONE and has a single formal parameter R used for both input and output, and it has the body

$$R = R + 1$$

Then if we write

$$A = 3$$

$$\text{CALL ADDONE } (A)$$

the result is that the value of A has been reset to 4. The same thing happens in the call by name case. However, in call by value, the assignment of the result is only to the formal parameter R and there is no change in A.

4. *Embedding*

Embedding (sometimes called *phrase substitution*) means that one or more subunits can be inserted in, or used as part of, another subunit. Looking at the same idea from another angle, this means that within a given subunit other subunits can be identified which may be of the same or different types. One of the most common examples of this is the inclusion of statements within IF ... THEN ... ELSE ... sentences. Thus, we can write IF A = B THEN X =: Y ELSE X =: Y + Z and the relatively simple executable units X =: Y and X =: Y + Z and the relation A = B are all embedded in the larger unit. In this case, not everything shown is executed because a branch is definitely indicated here; in other cases, everything shown is executed. Embedding can also occur when declarations are included in larger units. A common case of embedding is the inclusion of functions in place of variables in statements like Y = A + COS B. Subroutines often contain other subroutines, sometimes going down many layers. A more interesting case that is seldom allowed (although it is permitted in ALGOL 60) is the inclusion of conditional statements for arithmetic variables. For example, the sentence A = 4 + (IF X = Y THEN 5 ELSE 8) assigns the value 9 to A if X = Y and the value 12 to A if X ≠ Y.

It is clear that only certain types of subunits can be embedded in others. Thus, the examples given above are perfectly reasonable, as in the situation in which A = B + LOG (IF A = B THEN C ELSE D). A case which is intuitively unreasonable, however, is A = B + PRINT XYZ. The rules for embedding always exist in the language definition, although they may be hard to identify as such. This issue is related to the problem of the sequence in which executable units are carried out. The main difference is that embedding implies (and requires) an immediate replacement of a variable by a more complex unit, whereas sequencing involves the order in which parts of the program can be written.

III.4. DATA TYPES AND UNITS AND COMPUTATIONS WITH THEM

The key features of a programming language are the types of data it can handle and the ways in which it can operate on that data. The major aspects are the specific types and characteristics of data variables, the types of data units either in the machine or in the language which are available relative to the commands in the language, the types of arithmetic which are permitted, how expressions are created and evaluated, and how data elements are defined relative to the whole program. This section discusses all these points.

III.4.1. Types of Data Variables and Constants

There are a number of different types of data variables, and each requires (or permits) different types of computation and operations to be performed upon them. More specifically, certain types of executable statements can meaningfully operate only on certain types of data variables.

1. *Arithmetic*

The first and most obvious type of data variable is the real (i.e., not complex) arithmetic, sometimes called *numeric*. The data name represents a number; therefore, to say that A + B = C where A and B are numbers means that C will be a number. The types of arithmetic that can be performed, however, are discussed in Section III.4.3.

Every programming language has rules about the size of constants and the ways in which they can be written in the program. These will *not* be described for each language.

2. *Logical (=Boolean)*

The second most common type of data variable is the logical (=Boolean) variable; this is simply a variable which can take on only two values, normally designated as *true* or *false*. These are usually represented in the machine by 0 and 1, 0 and non-0, or any other two distinct numbers. Boolean variables are usually the direct or indirect operand of an *IF clause*; see Section III.5.3.2.

3. *Character*

A character data type is really one which is nonnumeric. The two most important special cases of character data types are *alphanumeric* (where the data consists of letters and/or numbers) and *bit* (where the only elements are 0 and 1). The former is usually introduced into languages which are concerned with handling data processing applications to provide a data type that *excludes* characters other than letters and numbers.

4. *Complex*

Some languages permit complex numbers and variables. In such cases, the complex number is made up of a real and an imaginary component. Presumably, although it is not logically necessary, a language which allowed the representation of complex numbers would automatically perform arithmetic on them. (See Section III.4.3.4.) This is not required, however, and a language might permit the user to declare certain variables as complex so as to allow the proper amount of storage for them and then require the user to do the arithmetic himself.

5. *Formal* (=*Algebraic*)

A type of data variable is the *formal* or *algebraic* which stands only for itself or for an expression, such as Z or $A^2 + B^2$, where these are not numeric. A formal data variable has no value in either the numeric or logical sense. The operations performed on it are usually those related either to string handling, or more usually to formal mathematics or algebraic manipulation. Languages emphasizing this data type are discussed in Chapter VII.

6. *String*

Strings and lists are types of data variables which represent similar but somewhat different concepts. A string is a type of data variable that consists of one or more characters concatenated, and it will be operated on as such, e.g., ABC or JOHN Y. DOE. Depending on the intended purpose, either the individual characters have meaning and the whole unit does not or vice versa. Among the operations to be performed on strings are concatenation, deconcatenation, and replacement. The special case in which the string consists only of bits is significant because this can usually be interpreted as some type of logical variable and operated on accordingly.

7. *List or Pointer*

A list or pointer data variable conveys information either directly or indirectly—about the location of another data item. A *list variable* is definitely not the same as a *string*. A string can be represented internally in a computer as a list but this does not change the conceptual view of the string. Operations on lists include adding and deleting elements and pushing down and popping up. The pointer variable itself is a specific data type.

8. *Hierarchical*

A data type that is actually the superstructure or concatenation of a number of other data variables is called hierarchical (see Section III.2.3.4). Looking from the top down, a hierarchical variable is one which has identifiable subparts, each of which might be of a different data type. The simplest example is the zip code, which consists of three codes concatenated. In that illustration the subordinate variables are all the same data type, but this is not essential to the concept. For example, the complete identification of a person in a payroll file might consist of his name, address, and social security number. In this case, some of the subordinate data variables are alphabetic, while others are purely numeric.

The operations performed on hierarchical data variables are usually limited to movements within storage and to operation on the subunits that is appropriate to their type.

9. *Others*

While individual languages may introduce other variable types, they are not sufficiently general to justify their inclusion here.

10. *Combinations of Variable and Constant Types*

A number of the types above can be combined to form new variable types. The most common are expressions, vectors, arrays, records, and files.

A string of variables, all of the same data type, is called a *vector*. A sequence of vectors is called a *matrix*. Note that the elements of the vector or array can be of any type; theoretically they need not all be of the same type but, practically speaking, they usually are. The operations which can be performed on vectors and arrays are normally the same ones which are performed on their individual elements, except that they are applied to the larger set.

Combinations of arithmetic, logical, or formal variables, together with appropriate operators, are called *arithmetic*, *logical*, or *formal expressions*. Some of the issues involved in defining the rules for creating these expressions are given in Section III.4.4.

III.4.2. ACCESSIBLE DATA UNITS

The commands in a language may operate on, and the declarations may describe, different data units. There are two main aspects of this data unit consideration: One is the essential hardware of the machine itself, and the other is the set of variable types allowed in the language.

1. *Hardware Data Units*

The main issue is which of the subunits of data which exist in a particular computer are accessible to some or all the commands in a programming language. The criteria for inclusion or exclusion usually involve implementation, efficiency, and compatibility. If it were not for these problems, it would not matter whether a variable occupied a single bit, a character, or a full word.

In a binary machine, obviously there are some machine instructions which are able to access and operate on individual bits. This is particularly useful for logical operations in which there are only two values of the variable. The problem arises when the programming language tries to include commands which operate on individual bits without also sinking to what would essentially be the level of machine instructions. One way of handling this is through the use of declarations and implementation techniques. In such a situation, a variable is defined as occupying (or requiring) a certain number of bits. The command specifies only the variable, without regard to its internal representation; the compiler must bridge the gap by turning out the necessary machine instructions to manipulate the bit(s).

On binary and character, as well as fixed and variable, word length machines, there is a basic unit size which contains a single character. On some machines there may be several such units to represent different types of characters. For example, to represent 63 characters, obviously 6 bits are needed; alternatively, one can save space (at the cost of time and other complexity) by allowing different amounts of space for the representation of letters and numbers. The programming language might cope with these various possibilities. As in the case of the bits, it is often more effectively handled by allowing the commands to operate solely on a variable and letting the compiler take care of obtaining the right information from storage.

On a fixed word length machine, most machine instructions deal either with the complete word or with specifically defined subsets of the word. These latter often contain addresses of storage locations or various registers. Obviously the commands in a programming language are not basically concerned with such subdivisions. Some languages, however, have been specifically designed around certain machine word structures and their associated hardware (most noticeably LISP), but they have been successfully implemented on very different types of computers.

In summary, a language which permits the user the ability to access any bit in the machine is clearly more flexible than one which permits him to access only words. On the other hand, he is paying a very heavy penalty for this capability because he must supply much more information about his data.

One of the biggest disadvantages to allowing programming language commands direct access to bits or even larger units is the difficulty of maintaining machine independence. This problem arises in trying to switch programs and their associated data structures from one machine to another. If the commands have been designed to be independent of the specific data unit, then there will be relatively little difficulty in the executable portion of the program, but significant changes are probably needed in either the declarations or the file layout or, most likely, both.

2. Language Data Units

Section III.4.1 discussed a number of different variable types that can exist in a programming language. There is a strong interaction between the existence (and relative importance) of these data types and the actual executable commands which exist. For example, it would be inconsistent to permit logical variables if there were no command (or operator) capable of recognizing *truth* or *falsity* and acting on this recognition. Similarly, having strings as a legitimate data type would be relatively useless unless there were commands to operate on them. (However, strings can actually serve a useful purpose without special commands, by being used for printouts.) Thus, for each variable type there should be one or more commands which permit that type as an operand; conversely, each command must clearly specify which of the allowable variable types in the language are legitimate operands of the command.

Referencing of arrays and hierarchies is normally handled by subscripting and qualification rather than by specific commands.

III.4.3. TYPES OF ARITHMETIC

There are a number of types of arithmetic which are available in programming languages. The significant point is what arithmetic is provided by the language and not what happens to be available in the hardware. The two most common types of arithmetic (both in the language and the hardware) are of course fixed point and floating point. There is no requirement that the arithmetic available in the programming language must also be available in the hardware; conversely, the programming language may not make use of all the hardware available, e.g., COBOL does not provide floating point capability in the language, even though many machines have that facility (and some implementations have provided it). The only requirement for permitting varying kinds of arithmetic in the language is that the user have some method of specifying in his source program what type of arithmetic he wants performed. This is usually done through the use of declarations or sometimes through a combination of declarations and commands. It is better to specify the type of arithmetic desired solely by the use of declarations since this permits one to have an **ADD** or **COMPUTE** command which is independent of the data type. Thus, the most common way of handling this matter is to have a specific declaration in which a variable is defined to be of the desired arithmetic type. However, by doing it this way there becomes a need to establish rules for what forms of intermingling of arithmetic data types are permitted in one command. This point is discussed in Sections III.4.4 and III.5.1.

1. Integer, Fixed Point, Mixed Number

In considering the various types of arithmetic, the simplest and most obvious is the *integer* arithmetic, which means exactly what it says: Namely, the addition of the integers 2 and 3 yields exactly 5. Integer arithmetic is needed to give complete accuracy, particularly in counting. Unfortunately, on a binary computer if one adds 0.1 to itself 10 times, the result is not necessarily the number 1; it might be 0.99999 or 1.0001, depending on the machine and the conversion programs that have been written. It is clear that there are a number of cases in which the exact arithmetic is needed. On the other hand, integers clearly do not suffice for most scientific problems and so floating point numbers are used.

Although the terms *fixed point* and *integer* are often used interchangeably, this is not really accurate. The term *fixed point* is actually a more general one and can be used for numbers which are not actually integers because they require a decimal point, but they are not floating point numbers because they are not in exponent-mantissa form. Thus, the addition of 3.2 and 4.6 to produce 7.8 can be considered *fixed point*. The distinction is between the ways in which the numbers are represented internally; a more accurate description of the concept just cited is the phrase *mixed number* arithmetic. Few computers permit this type of arithmetic because the radix point is usually at the left or right end of the word. Hence, it is usually a capability provided by the programming language itself.

2. Floating Point

Floating point numbers are used to eliminate the need for scaling and/or to provide for the use of a wide range of numbers. Through the use of the so-called scientific notation, both large and small numbers can be contained in a single word (although with some loss of precision in the fractional part). This is done through the use of an exponent and mantissa; thus, the number 3.14159 is represented as 0.314159×10, or 31.4159×10^{-1}. Depending on the machine, there are differing ways of representing floating point numbers, but they all reduce to separating the number into two parts, one representing an exponent of a fixed quantity (usually 2 or 10, depending on whether the machine is binary or decimal) and the other representing the decimal part of the number. The programming language itself does not usually distinguish between the actual representation; this is handled by the compiler in translating to machine code.

3. Rational

Rational arithmetic is a relatively newer concept in practical programming usage than either integer or fixed or floating point. *Rational* arithmetic means the ability to take two numbers of the form A/B and C/D and add them to produce the result (AD + BC)/BD. The need for this arises particularly in scientific problems, where one wishes to do very precise arithmetic of this kind. For example, in most implementations of programming languages, if one computes $\frac{1}{3} + \frac{1}{3}$ in fixed point or integer mode, the result may well be 0 or 1, depending on whether the language specifies or the compiler chooses to truncate or round up the result. In floating point form, the result would clearly be 0.66667, whereas the real answer that is desired is $\frac{2}{3}$. Since no machine known to me has this capability built into the hardware, it must be supplied by the compiler.

4. *Complex Numbers*

While it would be possible for the programmer to handle complex arithmetic by separating the real and imaginary parts, he would certainly prefer not to have to do this. Through the use of declarations, or some other similar technique, the compiler becomes aware of the fact that a particular variable is a complex number; the compiler then becomes responsible for doing the arithmetic on the real and imaginary parts separately. Obviously, there is a potential storage allocation problem here because it is no longer possible for the variable to be contained in one machine word.

5. *Double or Multiple Precision*

Regardless of the amount of arithmetic precision inherent in a given computer, there are always problems requiring more. As a result, many languages (and compilers) include the capability for double, higher, or variable precision. (The fact that a particular compiler might provide variable precision arithmetic does not make it a language feature.) The actual precision is obviously a relative concept because double precision on one computer may be a single or triple precision on another computer. This is a problem that has not been satisfactorily solved with regard to compatibility. It is very easy—and has been done frequently—to include declarations, or even to modify commands as necessary, to notify the compiler that additional precision is desired. However, if this source program is run on another computer, the arithmetic results may be quite different. (See Section II.4.)

6. *Logical*

The arithmetic performed on logical variables obeys the rules of logic (with the most common precedence rules described in Section III.4.4.4), and produces a logical quantity as a result.

7. *Other*

Although I do not share the view, some people consider operations on strings to be a form of arithmetic. Common operations include concatenation and deconcatenation, counting the number of characters, searching for a pattern, and transforming the string.

While no other types of arithmetic seem to be present in programming languages, there is always the possibility for such.

8. *Higher Level Data Units*

Languages that include vectors and arrays sometimes include special commands, or permit modification of regular commands, to allow arithmetic on these higher level data units. In most cases this simply means performing the indicated arithmetic on each element in the vector or the array. When the language permits data hierarchies, then there are sometimes commands which can be applied to the hierarchy, with the actual action being performed on all the subunits in the hierarchy.

III.4.4. RULES ON CREATION AND EVALUATION OF ARITHMETIC AND LOGICAL EXPRESSIONS

Because of the variety of types of arithmetic and types of data variables, there are a number of rules about the creation and evaluation of arithmetic and logical expressions which must be made. The word *mode* is used to apply to both the type of data variable and also to the types of arithmetic. The first set of rules is needed to specify how much intermingling of different types of arithmetic is to be permitted; the second issue involves rules for converting numbers and data types from one form to another; the third problem involves rules for precision. Finally, precedence rules for operators and sequencing rules for evaluation are needed.

1. *Intermingling Rules*

In early versions of some languages it was not legal to add an integer and a floating point number; this is called the *mixed mode* case. This restriction still holds in a number of cases (e.g., FORTRAN), but it has been removed in more recent languages (e.g., PL/I). Similarly, there are often rules preventing the use in a single arithmetic expression of variables requiring different precision. There is nothing inherently logical in these restrictions, and they are usually imposed just to make the implementation easier and/or to avoid specifying the rules described in Sections III.4.4.2 and III.4.4.3. A problem does exist, however, if one tries to add arithmetic, Boolean, and formal variables together; it is not at all obvious what form the result should take. Thus, in any programming language a decision must be made as to what types of intermingling are permitted and what conversion rules will be applied to such combinations. Obviously, it is impossible to establish intermingling rules without simultaneously specifying the meaning of the combinations.

2. *Conversion Rules*

Once it has been determined what types of data variables can be intermingled, it is necessary to determine what kinds of rules are involved in such intermingling. Conversely, in order to say it is legal to add an integer and a floating point number or an arithmetic and a Boolean variable, it is necessary to specify the rules used to determine the result. Thus, if one adds 3 and 2.5, the answer could be either 5, 5.5, or 6, depending on whether one wanted an integer value rounded up or down, a floating point value, or a fixed point noninteger. Normally when floating point and integer numbers are added, the result is floating point because the precision from the integer calculation has already been lost and the floating point number itself may be too large to fit into an integer format. (It is often, but not always, true that if variables of a single type, such as floating point or logical, are combined, then the result would be of the same type.)

If one tries to add arithmetic and logical variables, then there is another problem because it is hard to define what variable type the results should be. As indicated earlier, logical variables normally take on one of two specific machine values; therefore, it is certainly possible to produce a number as the result of adding an arithmetic and Boolean data variable; however, this is a fairly meaningless number. Interestingly enough, use of this concept with arithmetic multiplication can be both meaningful and useful if the values 0 and 1 are used to represent the logical variables internally. If A is an arithmetic variable and L is a logical variable, then A * L will be either 0 or A, depending on the truth value of L.

Other conversion rules are not so obvious. For example, if one tries to add a logical and a formal variable, what is the result? Similarly, combining strings and other types of variables produce very many questions. There are numerous rules of thumb and arbitrary guidelines but no very definitive way of determining what the result should be, aside from the wishes of the language designers. In cases where one mode *includes* another, there is usually automatic conversion to the more general case, e.g., integer $\subset$ rational $\subset$ floating point, or arithmetic $\subset$ formal.

3. Precision and Computation Rules for Various Modes

Once it has been established what variable types can be combined and what type of variable will be produced as a result, there are still a number of rules which must be specified. Unless the rules are carefully defined, each compiler will produce a different answer for variables of differing kinds. Thus, if we have A = B + C * D + E / F, where B is single-precision fixed, C is double-precision floating point, D is a logical variable with a numerical representation, and E and F are rational numbers, there must be precision rules as to how to do the arithmetic. The language with the most flexibility on this point is PL/I; it therefore has the most complicated rules.

4. Precedence and Sequencing Rules

There is a normal precedence rule for the five arithmetic operators, namely exponentiation, then multiplication and division, and then addition and subtraction. (Cases involving minus signs and exponents sometimes require special definitions.) This means that the expression A+B**C*D−E is interpreted as A+(B**C)*D−E. Naturally these precedence rules can be overridden by the use of parentheses for grouping purposes. Thus although 3 + 4 × 5 is evaluated as 23, the expression (3 + 4) × 5 yields the result 35. However, implicit multiplication (shown by juxtaposition, e.g., AB) is seldom allowed unless data names are restricted to single letters. It is either impossible or difficult (requiring other complicated rules) to distinguish between the data name AB and the product of variables A and B. Furthermore, the precedence rules are insufficient to deal with an expression of the form A/B*C. The normal sequencing rule is left to right; thus, A/B*C is evaluated as (A/B)*C, subject to whatever conversion and precision rules have been defined. Without some sequencing rule, the user would probably obtain different numeric results from different compilers even when using the same computer. (This happens even when all these rules are spelled out because of differences in conversion routines.)

The most common precedence rule for logical variables is to apply first the unary NOT operator, then the AND, and last the OR. Thus X AND NOT Y OR Z is evaluated as (X AND (NOT Y)) OR Z.

In expressions containing both arithmetic and logical operators, the latter are normally evaluated first.

It should be assumed that all languages obey the precedence rules given above unless stated otherwise.

Since expressions can usually contain functions, and these might have parameters called by location, there is a possibility that the value of a variable in an expression might change while the expression was being calculated; e.g., if A + DBL(A) + A is to be calculated and DBL(A) doubles the value of A, it is unclear what value of A is being used where. This (and some related confusing situations) is often referred to as the *side effects* problem.

III.4.5. SCOPE OF DATA

The problem of determining the scope (i.e., the meaning in different portions of a program) of data is a very complex one. In its simplest occurrence, the same data name might be used in a subroutine and also in the main program. This is usually handled in a simple way by specifying that any variable used in a subroutine body is protected from the rest of the program; i.e., it has no meaning outside its use as a formal parameter in the subroutine. However, there are times when it is desirable to use the same variable in a subroutine and in the main program or in two or more subroutines. The technique of COMMON used in FORTRAN helped solve this problem. However, with the advent of the block structure in ALGOL, the problem became much more complex. When the main program itself is broken down into smaller subunits which can be nested, then the scope of the type and value of a variable must be clearly defined. Thus if we have executable subunits A, B, C, D, E arranged in the overall program as follows,

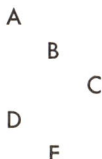

and a variable V is defined prior to starting the execution of A, then what happens to it in B and C, and how does this affect its value when entering D? Suppose there is an occurrence of V in one of these subunits; what rules or restrictions on this should exist? Inherently there are two concepts which are needed: Local and global. In the former case, a variable is considered defined only within the specific subunit in which it is declared; the variable has the same value and characteristics in any smaller subunits unless it is redeclared. Furthermore, if a local variable is referenced outside the unit in which it is defined, this will be an error since it is undefined (unless it is redeclared). The intuitive definition of global is that the variable has the same characteristics and means the same thing each time it is used in the entire program. Readers who are interested in pursuing the details and ramifications of these points should see the Revised ALGOL 60 report [NA63] and

104   TECHNICAL CHARACTERISTICS OF PROGRAMMING LANGUAGES

the PL/I manual [IB66b] since these are the main languages in which this is an issue.

III.5. EXECUTABLE STATEMENT TYPES

The two main components of a programming language are the data types and the executable statements permitted. Depending on the view of the person involved, one of these can be considered the main facet of the language and the other the secondary, or they can be considered equally important. The view that the data types allowed are the major factor is expressed by Perlis [PR65]. I hold the alternative view—namely that the most important factor of a programming language is the list of commands that it can perform. Probably the most reasonable view is that this is a "chicken and egg" situation and both components are equally necessary.

This section provides a categorization of the multiplicity of executable statement types which are relatively commonly used. Obviously, all languages omit certain specific ones. The exact details are discussed under the individual languages.

Executable statements are sometimes referred to as *imperatives* as contrasted with the *declaratives* which supply information about data or aspects of the program.

III.5.1. ASSIGNMENT

Assignment statements have the function of assigning a name to the result of some operation—usually but not always an arithmetic computation. The operation that is involved may be as fundamental as renaming or doing basic arithmetic or as uncommon as differentiation or concatenating strings. In some types of assignment statements, the results are capable of being assigned to many variables; i.e., a single result can be given several different names.

There are two classes of rules that are needed in connection with the assignment statement, and they affect the way in which the operation is carried out. One involves the actual method of specifying the action to be performed and the other involves the rules which are needed in converting a result into a form which the *name of the result* requires.

It is unfortunate that the most common notation used for the assignment statement is the equal sign, =, although some languages (e.g., ALGOL) do use a separate symbol. In virtually every case, the = is used to mean that the value of the expression on the right-hand side is to be assigned the name on the left-hand side. This results in such mathematically unpleasant statements as I = I + 1, where the meaning is that some variable I is to have the quantity 1 added to it and the result is to be given the name I. This naturally causes the original value of I to disappear.

1. *Methods of Specifying Computation*

In some languages the indicated computation on the right-hand side is controlled by the data types involved (see Section III.4.1), the arithmetic performed on them, and the operators and/or commands involved, as well as the rules on conversions and modes (see Section III.4.4). Thus, the assignment statement A = B + C is meaningless until we know all the data types involved, not just those on the right-hand side. If A, B, and C are numbers, then the action and result are obviously numeric, although there are problems of precision, mode, etc. However, if B and C are strings, then the operation indicated by + might be concatenation. The actual result is based on conversion rules (discussed in the next section).

In contrast to the principle of using just data types to indicate computation, it is possible to specify other desired actions. In this case, it is an actual command which is being used on the right-hand side, and this command falls into another category. For example, in FORMAC one writes LET A = SUBST X**2 + Y**2, (X, Y+1). Here the actual assignment is caused by the LET ... = and the SUBST or equivalent command is the operation being performed.

2. *Conversion Rules for Results*

Most of the previous discussions have pertained to variables which are in one arithmetic statement. However, it is perfectly possible—and often necessary from the problem solution viewpoint—to provide entirely different types of variables on the two sides of the *assignment sign*. It is necessary to consider conversion rules *across* the *assignment sign*. In other words, if we write Z = X + Y * A, then once the right-hand side is computed it is necessary to specify the value of the left-hand side. It might seem obvious that the value of Z should be exactly the same as the computed value of X + Y * A, but this is not always desired. The differences can vary from as little as the arithmetic mode (e.g., fixed point on one side, floating point on the other) to the very significant difference of having a string on one side and a number on the other. In the example of A = B + C given earlier, it might be assumed that B and C were strings to be concatenated, and it is still perfectly possible for A to be numeric. The conversion rule involved might require that the string on the right-hand side be converted to its equivalent numeric value; obviously, in order to do this, the right-hand side would have to be capable of being converted to a numeric quantity. For example, the string 4527 is, but the string A3+,45 is not unless some special interpretation is given, which of course can be done.

III.5.2. ALPHANUMERIC DATA HANDLING

Most data processing applications involve the use of *alphanumeric* (often abbreviated as *alphameric*) data. The primary types of manipulations on this data are editing (which has many facets), converting, and sorting. The form of the data may be the same here as in Section III.5.4, but the operations to be performed (i.e., the commands) are significantly different.

1. *Editing Statements*

The term *editing* covers a number of areas and applies primarily to the preparation of data for use outside the computer. In the simplest case, it may involve actions as basic as zero suppression or insertion of dollar signs where needed. This type of editing involves primarily a single variable or piece of data. At the other extreme, we can consider report generation as a form of editing. In this case, many pieces of data are being examined, possibly changed in format, and moved around to prepare them in a particular form relative to a printed page. Even arithmetic computation may be involved for the purpose of obtaining totals on a report.

Depending on the language, the various facets of editing may be called into play by a direct command; more often editing is accomplished through the implicit invocation of the necessary routines through another command and/or through data descriptions.

2. *Conversion Statements*

The use of conversion—at least in this context—is primarily an internal matter. There is often a need for data to be transformed from binary to EBCDIC or from purely numeric form to alphanumeric form. As with the editing, this can be accomplished through direct commands or through data declarations combined with other commands. Thus the statement MOVE A TO B in early specifications of COBOL called for A to be converted to the form specified for B and stored there.

3. *Sorting Statements*

Sorting is seldom thought of as a command in a programming language, but it does exist in at least one—namely COBOL—and was considered for inclusion in PL/I. Most sort routines are run as independent programs by using a proper set of control cards. However, the disadvantage to that process is that the user may wish to perform actions on his files either before the sorting takes place or after, or both. If there is a sorting command as an integral part of the language, then the other mechanisms of the language can be used prior to and after the execution of the sort command. This concept is really the reverse of the situation where some sort generators permit hand-coded routines to be included in the sort package.

III.5.3. SEQUENCE CONTROL AND DECISION MAKING

All programming languages must have statements to change the sequence of control of execution and also to make certain decisions with alternative choices based on the result. If this were not true, it would be impossible to stop most of the operations which were started and it would be impossible to have any branches in a program. Sequence control statements are usually conditional and unconditional transfers, where the latter can be considered to include procedure calls. Decision making facilities can take many forms; the most significant are the conditional statements, the loop control statements, and the error condition statements.

1. *Control Transfer Statements*

For each language there is a *normal flow of control* which specifies what the next executable unit is. This is usually, but definitely not always (e.g., COMIT), the next statement in sequence. Naturally there must be a way of changing this normal flow of control.

The simplest type of control transfer is the unconditional jump, in which the next statement to be executed is specifically named rather than automatically being the next one in sequence. The conditional jump is based on some choice or test being made. The conditional jump statement is really just the simplest type of conditional statement, normally consisting only of a test and a new location to transfer control to, e.g., IF A > B GO TO ALPHA; if the condition is not satisfied, then the next statement in sequence is executed. The invocation of a subroutine is a special type of control transfer which provides an unconditional jump and also some type of automatic return jump to the next executable unit in sequence after the subroutine call.

There are also switch-control statements which the programmer can use to cause control to transfer to one of a number of possibilities depending on existing or prior conditions. The choices can be made in elementary or complex ways, depending on the language, but all have the objective of specifying (at object time) the place to which control should be transferred. The most common is the *computed GOTO*, which usually is of the form

GOTO ($S_1, S_2, \ldots, S_n$), J where the value of J designates which S is used (e.g., $J = 2$ means control is transferred to S_2).

2. *Conditional Statements*

Although the purpose of a conditional statement is to permit alternative actions to occur, there are many forms of conditional statements and they do not always cause a control transfer. In fact, a significant aspect in judging the power of a programming language is the strength and flexibility of the conditional statements which it contains. Conditional statements usually start with some key words, such as IF or WHEN, followed by some condition which is to be tested and an action to be taken if the condition is satisfied. Thus, for example, a very common type of statement is to say "IF A IS LESS THAN B do something and/or transfer control somewhere". The power of the conditional statement depends on its two different parts: One is the amount of flexibility in the condition that can be stated (and tested) and the second is the amount of nesting of conditionals and/or other statement types which is permitted. An example of a complicated condition which can be tested is A > B + C AND B = SIN(X) OR X = Y * Z AND A = B = C. Obviously, precedence rules are needed to control the evaluation of the truth value of such an expression. (See Section III.4.4.) In considering the amount of nesting permitted, note that the simplest *sentence form* is

IF condition THEN statement ELSE statement

(The case of IF condition GOTO is too trivial to discuss.) In other words, a test is supplied and, depending on its result, either of two statements is selected. If either (or both) of the statements can themselves be conditional statements, or contain more than one command, then increased power and flexibility is obtained. For example, the single sentence (written in separate lines for clarity)

IF A > B + C OR C = SIN(X)
THEN IF X < Y THEN PUT Z = A + C
ELSE PERFORM ALPHA THRU BETA
ELSE IF A > B − C
THEN Z = A − C
ELSE Z = A * C

permits the user to state a variety of conditions and actions in one sentence. Obviously, rules of matching clauses, precedence, and existence of key words must all be specified very carefully. A discussion of this problem and existing and proposed solutions is given by Abrahams [AH66].

The situation in which a conditional statement can appear as part of an expression was discussed in Section III.3.2.4 in connection with embedding.

3. *Loop Control Statements*

One of the most powerful things in a programming language is its ability to state easily that a loop is to be executed. There are four parts to a loop control statement. One is the range of the loop; this indicates just which statements are to be executed under its control. A second part is the set of parameters which are to vary to actually create the loop; for each parameter, there are a set of values which that parameter will assume. These may be either individually stated values or, more commonly, a sequence of numbers with a constant increment, i.e., an initial value with a fixed increment to reach a final value. It is possible to have one or many parameters in a loop control statement, and they can vary simultaneously or sequentially. A third major feature is the termination criterion which specifies a rule for determining when the loop is ended. This is usually based on reaching the last set of values of all the parameters, but a loop can also be terminated when a particular condition (which may be stated independently of the parameters) is reached. The final part of the loop control statement is the specification of where control should be transferred after the loop is finished. In some cases it will be to the statement immediately under the loop control statement itself; in others it will be following the range of the loop; in still others it will be to some place designated by the loop control statement itself. One of the important facets of the range of the loop is to determine whether transfers within or outside that range are permitted and under what circumstances. In particular, if a control transfer out of the range of the loop is permitted before the loop is finished, then there must be a whole set of rules on what has happened to the parameters and what will happen in the future to the component parts of the loop control statement.

Finally, a significant characteristic is what type of nesting of loop control statements is permitted; i.e., can a loop control statement be within the range of another loop control statement? Usually, such nesting is permitted provided it is total; i.e., the range of the second loop is completely contained within the range of the first.

Language specifications sometimes tend to be somewhat vague in specifying *all* the necessary information needed in connection with a loop control statement. This is less likely to happen nowadays, but it is interesting to note a few of the problem areas because they exist in some earlier manuals and languages. For example, an often unspecified rule is whether or not the range must be executed at least once if the termination criterion is already satisfied at the time the loop-control statement is given. This could easily happen if a variable J was to vary from 1 to 10 until K was greater than 50, and K was greater than 50 when the loop was supposed to start. The results will differ markedly, depending on the implementation, unless the rule is stated very clearly. Another common case is lack of precision on the termination criteria. This is particularly prevalent on a binary computer, where a variable is supposed to vary from 0.1 to 1 in steps of 0.1. Depending on the conversion technique used (let alone any other implementation-dependent factors), the range might be executed 9, 10, or 11 times unless the criterion is stated very explicitly. A third disastrous example occurs when it is unclear whether the parameter is to equal or exceed the final value to terminate the execution of the range.

A loop control statement can almost always be written in terms of other statements in the language (i.e., using IF ... THEN and *assignment statements*). Hence one technique sometimes used in the language design is to define the meaning of the loop control statement in terms of other more basic statements. By doing this, virtually all ambiguities (about the loop control statement) can be eliminated.

4. *Error Condition Statements*

There are two major types of error condition statements which can be included in programming languages. One refers to the errors which may be caused by the data or committed by the programmer (and which he can test for), and the other involves errors which may be committed by the hardware (although in theory this never happens). Some of the types of errors which the programmer might make or be responsible for correcting are such things as overflow when it is not expected, trying to store a number in a field which is too small to contain it, incorrect data, or getting into a loop and not getting out. In some languages, direct testing statements can be written at the place in the program the user thinks an error might occur. In other cases, he may provide somewhere in his program a command that says "if such an error occurs, then do the following". In this latter situation, the generated code will cause an automatic interruption of the normal sequence and transfer control to the "corrective action" that the programmer has specified. In still other cases, the programmer may simply request that a flag be set when some condition occurs, and he retains the option of testing for it whenever he feels it is necessary.

There are certain machine errors which could occur (most notably in the input/output area) for which the programmer would like to test or provide corrective action. In some languages he can do this directly, although usually this is a function of the compiler and/or the operating system.

III.5.4. Symbolic Data Handling

A whole class of commands can be subsumed under the general category of symbolic data handling. These commands deal with algebraic expressions, lists, strings, or general patterns. Some commands may deal with more than one of these items. These commands can in turn be used or embedded in other commands, particularly the assignment statements.

1. *Algebraic Expression Manipulation Statements*

It was shown in Section III.4.1.5 that one of the possible types of data variables was a formal (= algebraic) one, which in turn leads to the possibility of algebraic expressions. Therefore, it is necessary to have certain kinds of statements for the manipulation of formulas and/or expressions. Among the most common of these manipulations are the abilities to substitute for a particular variable either another expression or a number, to differentiate, and to apply the distributive law to remove parentheses. Other types of statements which might be desired are the abilities to match expressions, integrate, to find the greatest common divisor of two expressions, etc. From a purely syntactic view, these facilities can be provided in a number of ways, e.g., as statements, as functions, or through interpretation of the data description. Thus, writing Y = (A−B)*(A+B) will have a very different meaning if A, B, and Y are formal variables rather than numbers and no new command is needed. On the other hand, *differentiation* requires a specific executable concept to be named; it might be defined as a function, an operator, or a command. The methods of handling these facilities depend highly on the personal taste of the language designers. (See Chapter VII.)

2. *List-Handling Statements*

Normally list-handling statements appear primarily in list processing languages, although there are a few exceptions to this. In any case, the types of statements that are normally involved are those which add to the beginning or the end of a list, insert information in the middle of a list, delete information, and create and delete common sublists. Sometimes these statements are interrelated with storage allocation statements. As in Section

III.5.4.1, these facilities can be provided in different ways, with executable statements as one of the most likely.

3. *String-Handling Statements*

In languages which have strings, sets of statements (or some other syntactic form) are needed to cope with them. Among the more common types of operations involved are concatenating and deconcatenating strings, inserting strings between others, and removal and deletion of portions of a string.

4. *Pattern-Handling Statements*

Some languages provide the ability to scan a string of text for patterns. This is extremely valuable in a wide variety of applications such as language translation, formula manipulation, and even compilation itself. The most common type of statement is one which specifies a pattern to be found and then demands some action to be taken after finding the pattern. Thus, a statement might require that all occurrences of the pattern **AB?CD** (where ? represents any arbitrary string) be replaced by the string **THEN**. Pattern-handling statements tend to be intermingled with string-handling statements because the normal reason for searching for a particular pattern is to do something to it or with it after it has been found.

III.5.5. INTERACTION WITH OPERATING SYSTEM AND/OR EQUIPMENT

No programming language can be used in a vacuum, and the user, compiler, and language must all interact with each other and with the physical and operating environment which exists. Thus, any language which is actually to be run on a computer and produce results of some kind must have some type of input/output statements. In order to make use of programs already in existence, there is a need for library referencing facilities. Debugging statements which aid the programmer can be officially part of the language, but they serve only a secondary purpose. The language must provide—or the compiler must handle automatically—storage and segmentation allocation facilities. Finally, some languages must— or do— provide facilities for interacting with the operating system and/or with special machine features which exist. (Obviously, these latter preclude the possibility of the language being machine independent.)

Each language must have a means of terminating the execution of the program either by stopping the computer or by giving control to the operating system. Note that this is an executable command based on the program logic and it differs from whatever symbol is used to indicate the physical end of the program as input to the compiler.

1. *Input/Output Statements*

The input/output statements are those commands which relate to getting data in and out of the computer. They fall into a variety of classes. One of the important distinctions is the difference between a physical unit of information on some storage media such as tape and a logical unit of information which may bear little or no relation to the physical unit.

The input/output statements usually refer to the logical unit, where the compiler (and/or operating system) has the responsibility for making the necessary information available from the physical units. Thus, one of the characteristics of good input/output statements is that they are either relatively or completely independent of the physical media that is being used. In other words, it should not really matter whether information is coming in from punch cards, paper tape, magnetic tape, disk, or drum. Of necessity, this goal cannot always be achieved, particularly if random access is required; but it is a desirable one to strive for because when the physical medium is changed, the program need not be.

The input/output statements usually, but not always, include separate facilities for initiating actions (e.g., checking or writing labels) on some external media and then asking for (or writing) some fixed amount of data. The amount of data being called for may be controlled by the programmer or by the data itself. In the first case, the user may specify how much information is to be brought in, whereas in the latter the data itself contains some kind of a delimiter and the input command is told to bring in enough information until the delimiter is reached.

2. *Library Reference Statements*

In the earlier discussion on subroutines it was pointed out that one of the reasons for the heavy development of subroutine libraries was to permit people to make use of work done by others. This benefit can and should accrue in the use of higher level languages, but this can only be accomplished by including appropriate statements in the language. Essentially, two facilities are needed—one to store a program in the library and the other to bring it from the library when needed. (The former capability is not usually included in the language itself.) In the case of programs or subroutines with no input parameters, the task is fairly simple; but when there are variables which must be assigned in order to use the routine, then there arise language (and implementation) problems of parameter transmittal. One characteristic of the language that affects the implementation significantly is whether or not the library routines are stored in their original (i.e., higher level language coded) form or in a translated form. (In the latter case, sometimes programs originally written in assembly language can be included in the library.) Depending upon these and other factors, the subroutines may be brought in at compile or object time.

There is also a need for library facilities for languages with complicated data descriptions. These facilities may be provided in the same way as for the subroutines.

3. *Debugging Statements*

Although one of the avowed purposes of higher level languages is to minimize the debugging problem, it is paradoxical that sometimes programming languages actually increase this problem. The reason is that the user of a higher level language tends not to know machine language and the higher level language often fails to provide him with information which is vital to his debugging in a form which he can understand. Thus, if a FORTRAN program stops and the only thing the user receives is a core dump, he may be completely unable to find his error. Fortunately, better language (and implementation) techniques are being developed to aid this problem. See for example the survey by Evans and Darley [EV66].

There are specific debugging statements which can be included as an integral part of the language. There is a difference between debugging statements *in the language* and debugging facilities provided by the compiler. Foremost among the former are various traces and snapshots. The user can include in his program statements which allow him to obtain printouts (in a form he understands) of various variables at specified times in his program. In other cases he is allowed to declare that he wants a trace with varying degrees of detail. In still other cases, he may ask for a dump of his variables, but with the proper names associated. One other technique sometimes used is to tie various debugging statements in with tests for error conditions (as mentioned in Section III.5.3.4).

It is essential to realize that even though a language may not contain any debugging statements, a particular compiler might provide such facilities.

4. *Storage and Segmentation Allocation Statements*

Because so very much of the problem in getting particular applications run on computers is involved with storage allocation, it is often necessary to provide information about storage requirements directly in the source program. Most of these actually exist as declarations rather than specific executable commands. However, there are other statements which control the amount of storage that is used, eliminate information from storage, indicate what things must be saved and what can be disposed of, etc. While these statements appear in some mathematical or data processing programming languages, they are absolutely essential to any list processing language. In some of the latter there are commands which provide control of the *garbage collection* (i.e., handling of the free list).

Information about overlays and program segments which can be stored individually is usually provided through declarations rather than as executable commands.

5. *Operating System and Machine Feature Statements*

Because of the increasing importance and complexity of operating systems, the compilers for programming languages are being interwoven more and more with the operating system. Whereas previously the same subroutine might be called into storage separately for each program, now this can be handled by the operating system to minimize storage requirements. While most of these aspects continue to be primarily implementation problems rather than direct language problems, there is nevertheless a carry-over into the language itself. Thus, the user often must know whether or not the data he wants can be made available by the operating system, whether a program which is being handled on another processor is ready, or how much longer he has on the machine, etc. He must also be able to time certain actions so that he can control the results if the process apparently takes too long. Another illustration of possible interaction with the operating system or machine itself is the language specifications which involve overflow and underflow, or attempts to divide by zero. Finally, the STOP (or its equivalent) statement, which indicates termination of a program, is included in this category.

III.5.6. OTHERS

There are statements in some languages which do not fall into any of these categories, but they are not general enough to warrant discussion here.

III.6. DECLARATIONS AND NONEXECUTABLE STATEMENTS

As indicated in several earlier sections, there are a great many situations in which information either can or should be supplied to the compiler. In some cases, this is necessary for the logic of the situation, whereas in others it helps significantly with the implementation, resulting in better object code. The most important statements of these types are those which describe or declare information about data items, files, data formats, storage allocation, and the hardware or operating system environment.

III.6.1. Data Description

Section III.4 discussed a wide variety of data types and sizes which are used in programming languages and a number of different ways of operating upon the data. In order for each executable command to perform its task, it must have all the information about the data which is logically required. Thus, for example, if numbers are stored in binary and are to have arithmetic performed upon them, the machine instructions needed will be rather different than if the numbers are stored in a character form. The primary method for doing this is to supply data declarations or descriptions which provide all the needed characteristics of the data. Thus, numeric data for mathematical problems will normally be characterized as being fixed or floating point; real or complex; single, double, or multiple precision; etc. With the exception of the precision, the amount of space occupied by each type of data is usually considered standard for a particular language or at least for a specific compiler. On the other hand, business applications require data which has more widely varying characteristics. In the first place, it is both alphabetic and numeric, and different parts require different sizes and internal formats. For example, a person's name in a payroll file may take a large number of characters, whereas a mark to indicate whether the person is male or female is clearly going to take the smallest possible unit which can be represented in the machine. Because of the need for this wide variety of type and size of data, it is essential that the compiler find out just how the data is being stored so that the appropriate machine language commands can be generated to handle it. All this information must be provided in the data description declarations either directly or indirectly. In a few cases the compiler will make a determination. In addition, it is characteristic of business as well as other types of applications that the data is usually in some type of hierarchical form. Again a payroll file is an excellent illustration because one can consider that a person's address consists of a street location, followed by a town, a state, and a zip code. Clearly, the entire address consists of all four of those items, whereas one might be concerned with only the zip code, the state and zip code, the town, or almost any other combination of these fields. For that reason, it is essential that the compiler knows what type of hierarchy has been used for the data so that it knows what machine instructions to generate.

Languages containing strings, lists, and/or arrays whose size is determined at object time must contain enough descriptions to permit compilation.

One of the key problems with regard to any complex data description is whether the information is being described in its internal representation (for a particular computer) or whether it is being described as it appears in a logical fashion on some external media. Thus, if the identification number for a person in a large organization is six characters, this is quite independent of the internal representation; the question of whether it is stored internally in character or binary form is essential information for the compiler but irrelevant for the user. Various compromises for this issue have been devised and will be described in the relevant languages.

One interesting philosophical point in creating the data declarations is where and how they are to be grouped. The earliest viewpoint—exemplified by FORTRAN—is that all data with a given characteristic should be shown together, e.g., DOUBLE PRECISION X, Y, Z and REAL X, Y, Z. A more recent view—exemplified by PL/I—is that all the declarations for a particular variable could be shown together, e.g., X, DOUBLE PRECISION, REAL, Y, DOUBLE PRECISION, REAL. (This is not the actual notation for PL/I or any other language.) This matter appears to be based considerably on personal taste and implementation techniques.

The requirement for data declarations is double-edged, in the sense that, applied as stated above, every variable would need to have a complete set of descriptions associated with it. Thus, each variable in a mathematical problem would have to be defined as being SINGLE or DOUBLE PRECISION, REAL or FLOATING, etc. This is undesirable because it requires far too much writing on the part of the user. Hence, the concept of *default* declarations has existed for some time (although there is a counterargument which says that errors are prevented by requiring the complete information to be written by the user). The concept simply means that associated with certain types of variables are certain characteristics, and these will be automatically assumed by the compiler unless some other information is specifically supplied. An early illustration of this is the fact that in FORTRAN any variable beginning with one of the letters I, J, K, L, M, N was automatically assumed to be an integer unless it was specifically declared as floating point.

III.6.2. File Description

A file description usually applies to external and logical characteristics of large amounts of data and how this is connected with the physical hardware units which are being used. For example, the same logical file may be stored on tape and disks in entirely different ways. This type of information must normally be made available to the compiler to avoid object time inefficiencies.

Among the characteristics usually included in a file description are the names of the different logical records it contains, information about the header and trailer blocks, and relation between logical and physical records. Descriptions of the format play a key role in query languages (see Section IX.3.2).

III.6.3. Format Description

In order to put out data, the user may definitely wish to have control over the format in which it appears. For example, if he wishes numbers to appear in three columns, he must be able to have a way of specifying this. Similarly, if he is using the computer to write a payroll check, it is essential that the amount is placed onto the right part on the check or else the check will be invalid. Thus, there needs to be a number of statements to the compiler which actually describe the format of the data which is to be put out. Similarly, information which is on some external media may have a format which the compiler must know about in order to be able to interpret the information coming in. When this formatting of output data is carried to a very high degree of complexity, it actually becomes a report generator.

Format descriptions often supply page and line controls, as well as the information necessary to permit the editing and converting discussed in Section III.5.2.

The details of the format descriptions available in each language will *not* be given as they require far too much space.

III.6.4. Storage Allocation

It was noted in Section III.5.5.4 that much of the control of storage allocation was done through the use of declarations rather than through executable commands. One prime example of this is the declarations which supply information about the dimensions of a data array; usually they give the number of dimensions and the maximum number of elements in each dimension. The compiler needs this information to allocate storage at either compile time (if the dimensions are fixed) or at object time (if the dimensions are allowed to vary).

Control of segmentation is usually done through declarations, either by indicating the appropriate places to segment a program that was too long to fit in storage directly or by indicating parts of the program which can be considered segments to overlay others.

III.6.5. Environment or Operating System Descriptions

It is perfectly possible to have a language describe the environment (hardware or software) in which it is to be run. This would allow the user to indicate for a particular program just what machine (configuration) he wanted it to run on. Ideally he could be allowed to compile *on* one machine *for* another. There is a significant difference, however, between having the language facility to describe a different object machine and having a compiler which will create code for it! Because of the increased importance of operating systems, and the sometimes hazy line between them and the hardware, descriptions involving the operating system are possible and/or needed. There are also situations in which information about data is supplied to permit (or require) the operating system to take action.

The only programming languages which have included this information in any significant way are COBOL, which provides a description of the environment and some information about the operating system, and PL/I, which deals heavily with the operating system.

III.6.6. Procedure, Subroutine, Function Declarations

In Section III.5.3.1 it was noted that the invoking or usage of procedures, subroutines, and functions was a particular type of control transfer. The executable part of the program contains only CALL (or whatever equivalent word is used) together with the necessary parameters. However, the body of the procedure, subroutine, or function must be given somewhere in the program. This is really a form of declaration, since the procedure, etc., is not executed where it is written. In other words, it is necessary to have some type of heading which indicates that the following piece of program is a procedure, etc.; this heading must also convey the required information about the parameters. These parameters are usually referred to as *dummy arguments* or *formal parameters* since they will be replaced when the procedure is actually invoked. This is then normally followed by the body of the procedure, subroutine, or function itself. Although the body is composed of executable code (together with any necessary declarations), it is never executed unless it is invoked from the program itself.

III.6.7. Compiler Directives

As mentioned earlier, compiler directives exist for several purposes: One is to improve the efficiency of either the compilation process itself or the object code that is turned out; another is sheer logical necessity, in the sense that the compiler cannot obtain the information any other way; a third is to cause action *not* to take place, as with comment statements included in a source program. These directives generally pertain to action that the compiler is to take when a certain situation is encountered at compilation time or to produce code to take care of these situations if they exist at execution time. One example of a compiler directive is some kind of a flag indicating that a macro is about to be used and must be handled separately. Another is an indication that the next part of the source program is written in another language.

III.6.8. OTHERS

As with the executable statements, there are other types of declarations in specific languages, but none are significant enough to justify discussion here.

III.7. STRUCTURE OF LANGUAGE AND COMPILER INTERACTION

This section attempts to describe certain characteristics of the language which relate to its overall structure and its interaction with the compiler. Some of these points are not considered to be 100 percent technical and were noted in Chapter II; however, they require either restatement or further amplification in this particular context. Other factors affecting implementation have already been mentioned briefly when they were particularly significant.

III.7.1. SELF-MODIFICATION OF PROGRAMS

Probably the greatest single difference between a program coded in either assembly language or machine language, and one coded in a higher level language, is the fact that the former usually has the ability to modify itself, whereas the latter does not. Thus, in a machine code or assembly language, one can normally modify either the operator or the operand and thus include in the program the code to change an instruction which used to be **ADD 509** into one which says **SUBTRACT 604**.

There is no higher level language known to me which has this characteristic. However, a similar effect can sometimes be achieved by having the programs look like data to themselves. Only a very few languages permit this, LISP being the main one. In my opinion, this is desirable from the user's point of view, although it does sometimes seriously limit the efficiency of the compiler that can be written for that particular language.

III.7.2. SELF-EXTENSION OF THE LANGUAGE

Since most languages are never quite as good as their designers intend them to be, there is always a desire on the part of the user to extend the language in a number of directions. These directions include the definition of new terminology in a program at the user's option, the need to extend capabilities—both data types and commands—and the wish to be able to abbreviate frequently used program strings. Note that there is a significant difference between the mere existence of a subroutine and an actual (although temporary because it lasts only for the program involved) extension of the language. The latter allows considerable user control over the format of what is written, whereas the subroutine does not. Subroutines which provide additional facilities (e.g., matrix-handling operations) are *not* considered language extensions if they are invoked through the normal subroutine calling mechanism.

The degree to which a language can be extended within the framework of the language itself, without having the compiler modified, is a measure of the self-extendability of the language. The most common way of handling this is through macros, which can be similar to those used in assembly programs. The only languages which attempted to provide any capabilities along these lines are COBOL and PL/I, and only the (macro) facility in PL/I has been implemented. (There was a *DEFINE* verb in COBOL for a long time but it was never implemented and was finally removed from the language.) A very simple type of self-extension occurs in languages which permit the user to name and define functions in a source program and use them in expressions as if they were system-defined. Most languages used for scientific problems have this capability.

Several suggestions for ways to provide macro facilities have been proposed but not necessarily implemented; see, e.g., the references at the end of the chapter. One of the earliest discussions of this subject was given by McIlroy [ML60].[2]

III.7.3. ABILITY TO WRITE THE COMPILER FOR A LANGUAGE IN THAT LANGUAGE

The concept of judging a language (even partially) by its ability to be used for writing its own compiler is one which I feel has been greatly overrated. The writing of a compiler is a particular type of computing application. (See Section IX.2.5.) Some languages are well-designed for that

---

[2] The interest in macros and/or self-extension of higher level languages has grown significantly during the time period in which this book was written. The concept is more important and further developed than the brief discussion here would seem to indicate.

particular application, whereas others are not. Of course, it is desirable to permit a compiler to be written in the language it is compiling because this is an excellent way of testing it. By writing the compiler for a language in that language and then passing it through itself, there might be a better chance of checking out the compiler than trying any large number of artificially constructed programs. However, it seems completely unreasonable to expect that a language which is very suitable for writing scientific applications or for doing simulation should also be capable of writing a compiler for the language. Even if it can be done logically, it may lead to poor object code (although this is often the fault of the compiler design).

If a language can be used to write its own compiler, this is an additional bonus, but no language should be criticized for not having that capability (unless, of course, that was its avowed purpose).

III.7.4. Effect of Language Design on Implementation Efficiency

Complaints about a language are really very often complaints about the compiler. (This was discussed in more detail in Section II.7.2.) This is a two-way street, however, in the sense that the type of design criteria that go into the language can have a serious impact on the efficiency of the implementation. Thus, it is possible to design a language for which it is virtually impossible to create an efficient implementation. Conversely, it is possible to design a language in such a way as to increase the efficiency of the compilation; the characteristics affecting the compilation may be major or minor facets of the language itself.

1. *Compile Time Versus Object Time Efficiency*

The decision of whether a compiler is to be most efficient at compilation time or to produce optimal object code is one which can usually be made technically by the implementer, although it must be made administratively by those people responsible for determining the ultimate method of usage. In particular, there is a major difference between a university (or possibly a scientific) installation which expects many small jobs, each to be run only once or twice, and an organization concerned with business data processing (or even scientific) production runs. Certain features in the language may tend to encourage compile time versus object time efficiency or vice versa. Generally speaking, the more special cases, flexibility, and power the language has, the less efficient will be the compilation process; however, sometimes these are introduced for the express purpose of permitting the creation of good object code.

There is invariably a tradeoff between compilation and object time efficiency, simply because it takes time to create efficient object code. In other words, if the compiler scans the object code produced to eliminate redundant instructions, then computer (compilation) time will be needed to accomplish this task.

There do not seem to be any language facilities which can be used to make the compilation process more efficient and *simultaneously* cause production of efficient object code. The converse is not true. Specific features which provide information to make it possible for a compiler to create efficient object code often require additional compilation time. An early example of an attempt at this was the FREQUENCY statement in the first FORTRAN (see [IB56] in Section IV.3); it was later dropped because it was found not to be worth the trouble it cost. In other cases there is no particular language feature provided but the compiler attempts to produce very good object code from specific parts of the language. The best examples of this are storage allocation and loop control. These obviously require compilation time.

2. *Generality Versus Restrictions*

As mentioned before, the greater the size, flexibility, and power of the language, the harder it will be to compile. From this, one might draw the conclusion that it would help the implementation to impose more restrictions. Unfortunately, this does not always work, particularly if the restrictions are imposed on top of some features providing great generality. For example, a language that permits names of any length causes certain problems in table design; this difficulty is compounded if the language says that certain types of names, e.g., statement labels, must be no more than six characters long. The generality causes a problem with the storage allocation at compile time but the restriction tends to slow down the scanning and processing because it involves a special case. A second problem arises from this type of situation if the compiler actually *checks* to make sure that the restriction is obeyed. (See Section III.7.5 for a discussion of error checking at compilation time.)

In many situations, restrictions permit more rapid compilation. For example, difficulties in scanning individual sequences of characters can be greatly reduced by putting restrictions on naming conventions. If a language is designed so that a data name can contain only alphabetic characters, then the compilation may be much more efficient than if the name can contain any characters. Even if one wants to permit numbers as part of the data name, then the restriction of saying that the first character must be a letter helps the compiler because when it encounters a digit, it knows that it should look for a numerical quantity rather than considering this as the potential beginning of a data name.

A restriction on the use of reserved words (see Section III.2.3.3) is another facet that improves compiler efficiency in the scanning. The term *reserved words* usually means the concept that a certain number of words in the language are fixed and cannot be used for data names or statement labels. In some languages there are *fixed words* but they can also be used for data names; this requires a great deal more investigation by the compiler of the context in which the word is used. All kinds of variations or further restrictions can be placed on this concept, even going so far as to say that no data name can begin with letters which might look like a reserved word.

3. *Specific Features with Significant Effect*

As an example of a minor characteristic of a language which has a great impact on compiler efficiency, one can consider the placement of declarations. If there is a rule that declarations about a variable must precede the first use of the variable, then the compiler might be able to generate good code on the first pass.

The issue of recursion is another one which has a severe effect on the efficiency of the implementation, but primarily it affects the object code. If procedures are allowed to be recursive but are not required to be defined as such, then the compiler is required to turn out code which will provide for all procedures to be recursive. This produces inefficient object code for those procedures which are not recursive. Thus, the inclusion of a declaration stating that a procedure or a subroutine is recursive permits the compiler to provide the mechanism only for those particular subroutines rather than for all of them.

4. *Storage Allocation Requirements*

For languages with large amounts of data to handle or for any language on a small machine, the storage allocation problem is a critical one. For this reason, any information which can be given to the compiler is helpful. Types of useful information include possible segmentation points, portions of the program which can be overlaid, and maximum expected size of variable-sized data and arrays.

5. *Possibility for Providing Choice of Tradeoffs*

Some features in a language provide the user directly or indirectly with some choice of tradeoff between compile and object time efficiencies. In most cases this occurs by default; i.e., if a user does *not* include some feature which improves his object time efficiency, he may (but does not always) save compile time. However, a more interesting situation would arise if the language contained definite provisions for the user to specify what type of efficiency he most desired.

III.7.5. DEBUGGING AIDS AND ERROR CHECKING

The line between debugging aids as part of the language and as part of the compiler is sometimes unclear. We must also distinguish between debugging at compile time and at object time. Some languages include statements which assist debugging at object time, such as *on error condition* statements which cause automatic transfers at object time under certain circumstances. In other cases the languages provide specific statements requesting traces of previously executed statements in order to obtain information. These kinds of debugging aids which are specifically inserted into the language are rather different from certain debugging aids and error checking which may be provided by the compiler itself. For example, a good compiler will normally do a great deal of error checking when scanning the source program. The simplest and most common type of error, at least in a large class of problems, is mismatched parentheses. A good compiler—and even in many cases a mediocre compiler—will specify that parentheses are mismatched. A good compiler will then attempt to indicate where the difficulty is most likely to be. Other ways in which a compiler can aid debugging are to list statement labels which are never referenced or give cross-references of statements which refer to data names.

There is a difference between debugging aids which exist in the language and are primarily for the purpose of handling object time errors and those checks which the compiler itself performs primarily to find syntactic (or even semantic) errors in the source code. A type of error which may be considered either syntactic or semantic is to have a number of subscripts associated with a variable which differs from the dimension declaration (or its equivalent). Most compilers will detect such errors at compile time and provide an error message to this effect. Far fewer compilers will insert checks into the object code to see that the maximum value of the subscript does not exceed the specified limit. One of the tradeoffs to be decided is how many of these object time error checks will be inserted, considering the amount of computer time they require. Ideally the user should decide; PL/I actually permits him to do so in many cases.

III.8. OTHER FEATURES NOT INCLUDED

Although this chapter attempts to list all the significant technical characteristics of programming languages, it cannot possibly list everything. Any particular features a certain language has that have not fallen in the categories above will be discussed under the particular language description.

REFERENCES

III.1.—III.6.

[AH66] Abrahams, P. W., "A Final Solution to the Dangling *else* of ALGOL 60 and Related Languages", *Comm. ACM*, Vol. 9, No. 9 (Sept., 1966), pp. 679–82.

[BG64] Bergin, G. P., "Method of Control for Re-entrant Programs", *Proc. FJCC*, Vol. 26 (1964), pp. 45–55.

[EV66] Evans, T. G. and Darley, D. L., "On-Line Debugging Techniques: A Survey", *Proc. FJCC*, Vol. 29 (1966), pp. 37–50.

[IB56] *The Fortran Automatic Coding System for the IBM 704 EDPM* (Programmer's Reference Manual), IBM Corp., 32-7026, New York (Oct., 1956).

[IB66b] *IBM System/360 Operating System: PL/I Language Specifications*, IBM Corp., C28-6571-4, Data Processing Division, White Plains, N. Y. (Dec., 1966).

[NA63] Naur, P. (ed.), "Revised Report on the Algorithmic Language ALGOL 60", *Comm. ACM*, Vol. 6, No. 1 (Jan., 1963), pp. 1–17. (Also in [RO67].)

[PR65] Perlis, A. J., "Procedural Languages", *Information System Sciences: Proceedings of the Second Congress*. Spartan Books, Washington, D.C., 1965, pp. 189–210.

[QM67] Standish, T. A., *A Data Definition Facility for Programming Languages*. Ph. D. Thesis, Dept. of Computer Science, Carnegie Inst. Tech. (May, 1967).

[RA66] Raphael, B., "The Structure of Programming Languages", *Comm. ACM*, Vol. 9, No. 2 (Feb., 1966), pp. 67–71.

[RA67] Raphael, B. *et al.*, "A Brief Survey of Computer Languages for Symbolic and Algebraic Manipulation", *Symbol Manipulation Languages and Techniques, Proceedings of the IFIP Working Conference on Symbol Manipulation Languages* (D.G. Bobrow, ed.). North-Holland Publishing Co., Amsterdam, 1968, pp. 1–54.

[RR60] *MATH-MATIC* (Remington Rand Automatic Programming System), Remington Rand Univac, U-1568 Rev. 1 (1960).

[SH62] Shaw, C. J., *An Outline/Questionnaire for Describing and Evaluating Procedure-Oriented Programming Languages and their Compilers*, System Development Corp., FN-6821/000/00, Santa Monica, Calif. (Aug., 1962).

[SQ61] Strachey, C. and Wilkes, M. V., "Some Proposals for Improving the Efficiency of ALGOL 60", *Comm. ACM*, Vol. 4, No. 11 (Nov., 1961), pp. 488–92.

[WL65] Weil, R. L., Jr., "Testing the Understanding of the Difference Between Call by Name and Call by Value in ALGOL 60", *Comm. ACM*, Vol. 8, No. 6 (June, 1965), p. 378.

[YM67] Mealy, G. H., "Another Look at Data", *Proc. FJCC*, Vol. 31 (1967), pp. 525–34.

III.7. STRUCTURE OF LANGUAGE AND COMPILER INTERACTION

[CH66] Cheatham, T. E., Jr., "The Introduction of Definitional Facilities into Higher Level Programming Languages", *Proc. FJCC*, Vol. 29 (1966), pp. 623–37.

[GA67] Galler, B. A. and Perlis, A. J., "A Proposal for Definitions in ALGOL", *Comm. ACM*, Vol. 10, No. 4 (Apr., 1967), pp. 204–19.

[LR67] Leroy, H., "A System of Macro-Generation for ALGOL", *Proc. SJCC*, Vol. 30 (1967), pp. 663–69.

[LV66] Leavenworth, B. M., "Syntax Macros and Extended Translation", *Comm. ACM*, Vol. 9, No. 11 (Nov., 1966), pp. 790–93.

[ML60] McIlroy, M. D., "Macro Instruction Extensions of Compiler Languages", *Comm. ACM*, Vol. 3, No. 4 (Apr., 1960), pp. 214–20.

[UH67] Brown, P. J., "The ML/I Macro Processor", *Comm. ACM*, Vol. 10, No. 10 (Oct., 1967), pp. 618–23.

[WA67] Waite, W. M., "A Language-Independent Macro Processor", *Comm. ACM*, Vol. 10, No. 7 (July, 1967), pp. 433–40.

[ZX67] Cohen, L. J., "COMMEN: A New Approach to Programming Languages", *Proc. SJCC*, Vol. 30 (1967), pp. 671–76.

IV LANGUAGES FOR NUMERICAL SCIENTIFIC PROBLEMS

IV.1. SCOPE OF CHAPTER

The scope of this chapter is almost self-evident. All the languages noted here had their objectives either directly as, or deeply rooted in, the solution of scientific problems by numerical techniques, using a digital computer. These problems tended to be characterized by small requirements for input/output and vast amounts of computation. Matrix inversion or the evaluation of mathematical formulas for regularly changing sets of values of the variables are prime examples of this. Such problems contrast with the data processing problems which tend to have much input/output but relatively little calculation. This distinction has become considerably less clear over the past years, but it still has some validity and certainly did at the time these languages were developed.

The availability of computers, and reasonable languages to use on them, helped make a major field out of numerical analysis. Prior to the advent of computers, relatively few people were familiar with the subject; with the existence of proper equipment, and the apparent inability of such equipment to handle problems analytically, numerical analysis techniques developed and flourished, virtually causing analytic solutions to disappear from practical working situations. Attempts to reverse this trend are described in Chapter VII.

IV.2. LANGUAGES OF HISTORICAL INTEREST ONLY

It is quite natural that most of the early work in language development, both in quantity and significant quality, was concerned with handling numerical scientific problems. The relative standardization and simplicity of mathematical notation—particularly in the writing of expressions—made it very natural for people to try to input this directly into a computer. The first two very early systems—namely SHORT CODE (UNIVAC) and Speedcoding (IBM 701)—did not make any attempts at this. It was Laning and Zierler at M.I.T. (on Whirlwind) who seem to have developed the first system in the United States which would allow a fairly natural mathematical expression as direct input to a computer. The work of Rutishauser in Switzerland had a similar motivation and is discussed below. The A-2 and A-3 systems provided scientific facilities on a data processing machine (UNIVAC), while PRINT did the same on the 705. The BACAIC system on the 701 was another—although less effective—early attempt to permit some type of mathematical notation as computer input.

These very early and primitive systems were followed by a number of others which were more widely used but did not survive, either for technical inadequacies or because of the exigencies of the marketplace. However, each contributed something to either the need at the time or to the technology, so they are worth discussing.

Since this book is defined to cover basically only developments in the United States, there is no listing of the early work which was done in Europe. Anything purporting to be a history of programming languages, however, would be incomplete if it did not at least mention the early work of Heinz Rutishauser of the Swiss Federal Institute of Technology, Zurich, Switzerland. As early as 1952, he described methods for allowing the input of mathematical expressions in a fairly natural form to a computer. Furthermore, Rutishauser described a system for translating them, i.e., a compiler, as contrasted with the interpretive system of Laning and Zierler. The early compiling work done in the United States by Dr. Grace Hopper initially involved very artificial pseudocodes rather than mathematical notation. In addition to permitting mathematical expressions as input, Rutishauser also allowed loop-control statements which look very much like the ones which eventually became commonly used, e.g., for $k = 1$ (1) 10.

IV.2.1. VERY EARLY SYSTEMS

1. *SHORT CODE*

The earliest document that I have seen which purports to describe a higher-level language (relative to that point in time) was the October, 1952 description of the SHORT CODE for UNIVAC [RR52] suggested by Dr. John Mauchly in 1949 and programmed by R. Logan, W. Schmitt, and A. Tonik. It was originally coded for the BINAC by W. Schmitt. In

[RR52] it is stated (in the preface) that Dr. Mauchly's "suggestion was, in effect, to have a program which would accept algebraic equations as originally written . . . ". While SHORT CODE did not achieve that goal (nor has any other system except those discussed in Section IV. 7), it was a remarkable objective for that point in time.

The basic principle involved was to use a 2-character code to designate either an operation or a variable and to use six of these codes at a time (because the UNIVAC had a 12-character word). Thus the problem of evaluating the equation A = B + C was written as

$$00 \; S0 \; 03 \; S1 \; 07 \; S2$$

where S0, S1, S2 represent the quantities A, B, C, respectively, and 03 and 07 stand for the operations of equality and addition, respectively; the 00 was the line number. Admittedly this looks primitive by today's standards, but then so does the hardware of 1952!

There were about 30 operations provided, including such things as floating point arithmetic operations, bracket indicators for evaluation of expressions, finding integral roots, tests of size, mathematical functions, and input/output operations. The system was interpretive.

The value of this system is in its objective, not in its execution, although the programmed floating point was an enormous help to the user.

2. *Speedcoding*

Work on the Speedcoding System for the IBM 701 was started in January, 1953 under the supervision of John Backus and the general direction of John Sheldon. Those who worked on the project were H. Herrick, D. Quarles, S. Skillman, J. Pulos, and L. Siegel. The first official manual [IB53] was dated September, 1953.

The basic principle of Speedcoding was to create two sets of operations, designated *OP1* and *OP2*; the first category contained three addresses, while the second contained only one. These operations were not part of the hardware, and were selected for their utility to the mathematician. The card format permitted one of each operation type, as well as a location field, in a single card. Thus

$$523 \; \text{SUBAB} \; 100 \; 200 \; 300 \; \text{TRPL} \; 500$$

shows an instruction in location 523 which subtracts the absolute value of the contents of 200 from the value in 100 and puts the result in 300; then it tests to find the sign of the result in 300 and transfers to 500 if it is positive.

There were about 45 *OP1* operations, including 10 arithmetic operations,

5 mathematical functions, and about 35 input/output instructions. (The large number of the latter is necessary because each tape number required a separate instruction.) The arithmetic was done in programmed floating point.

The *OP2* operations included testing, address modification, and control transfer instructions and, in particular, programmed index registers.

The system was interpretive.

3. *Laning and Zierler System*

An interpretive algebraic coding system was developed by J. H. Laning and W. Zierler in 1952 and 1953 at M.I.T. on the Whirlwind Computer [LA54]. It appears to be the first system in the United States to permit the user to write his mathematical expressions in a notation resembling normal format. Variables were represented by single letters, and thus multiplication could be indicated by juxtaposition rather than a specific operator. Paper tape on a Flexowriter was used, and upper-case numbers appeared as exponents, thus making it possible to write $a = b^2$. The user wrote his assignment statements in a very natural form, e.g.,

$$c = 0.0053(a-y)/2ay,$$
$$y = 5y,$$

Such statements could be numbered.

Normal precedence rules were used in evaluating the expression, so that $a + bc$ was handled correctly as $a + (bc)$. Both numerical and symbolic subscripts were permitted; the former were denoted by a vertical bar (which existed on the Flexowriter), followed by a superscript number; e.g., $u|^3$ represented u_3. Symbolic subscripts used the vertical bar and the letter; e.g., $v|j$ represented v_j. Floating point arithmetic was programmed.

Both unconditional and conditional control transfers were permitted. The former is designated by SP *n* where *n* is the equation number. Writing CP *n* caused a transfer to *n* if the previously computed quantity was negative. A switch control could be used by writing SP *x* where *x* represents a variable; control was transferred to the equation whose number was the value of *x*. Closed subroutines could be executed by writing SR *n*.

Although there were no loop control statements originally, they apparently were added later.

Loops could be controlled either by sequences of values or by fixed increments; e.g.,

$$g|N = 1, 1.2, 1.4, 1.6, 1.8, 2, 3, 4, 5,$$

$$g|N = 1(.2)2(1)5,$$

are both equivalent. The flexibility of the second form has not yet appeared in any significant programming language since then.

Over 20 common mathematical functions were available in a library, and they were invoked by writing F with a superscript identification number.

A *PRINT* statement and a *STOP* performed the obvious functions.

The rules of arithmetic, conversion, etc., were controlled by the CS system in use on Whirlwind at the time (see Adams and Laning [AD54]).

There was also a routine to solve differential equations.

This is truly an impressive system and, in my opinion, probably the most significant of all the early work.

4. *A-2 and A-3*

Although Remington Rand had the earliest automatic coding systems which were really compilers (namely A-0 and A-1), the first one to really receive extensive usage (starting in 1955) was A-2 [RR55]. It was developed by F. M. Delaney, M. H. Harper, M. Koss, J. E. McGarvey, and R. K. Ridgeway under the direction of Dr. Grace Hopper.

A-2 was a three-address code, specifically tailored to the 12-character word of UNIVAC. Thus it provided such instructions as

ADD A B C

AAL x1   del x   lim x
1CN ≠ opn #
2CN = opn #

where the first instruction meant *add the values of the variables* A *and* B *and call the result* C; the second meant *increase* x1 *by* del x *and if the result was less than* lim x, *then go to the address* (i.e., opn #) *designated in the* 1CN *line; if it equaled* lim x, *then go to the* 2CN *address*. The arithmetic was done in programmed floating point.

A-3 (also called ARITH-MATIC) was an improvement of, but not completely compatible with, A-2. It also provided a number of additional facilities which were not in A-2. However, A-3 never received much usage as such because it was made available at the same time as the AT-3 (later called MATH-MATIC) system, which was the Remington Rand conceptual equivalent of FORTRAN. (See Section IV.2.2.1.) The MATH-MATIC source program was translated to A-3 as an intermediate language, and this was then in turn translated to machine code. However, for those operations which could not conveniently be performed using MATH-MATIC, the user actually could write A-3 code in his MATH-MATIC program and have the linkage and translation performed correctly and automatically.

5. BACAIC

An interesting system called the *B*oeing *A*irplane *C*ompany *A*lgebraic *I*nterpretive *C*oding System (BACAIC) for the IBM 701 is described in a report by M. Grems and R. Porter [GR55] dated July, 1955. It permitted the writing of mathematical expressions in a fairly natural notation, except that no constants could be included; they had to be replaced by something of the form Ki, for i an integer. Multiplication was indicated by a centered dot, and an assignment statement by an asterisk at the right. Thus,

$$B \cdot B - K4 \cdot A \cdot C \cdot \ast D$$

caused the assignment to the variable D of the value of the expression $B^2 - K4 \cdot A \cdot C$. A complete list of the symbols for writing expressions is given in Figure IV-1. Note that a modified form of square brackets rather than parentheses was available on their key punch machine. It was not only permitted but required to number expressions and refer to them by

Symbols	How Used	Explanation
A thru Z	A + B	Refer to all parameters by the letters A thru Z, (except K).
K1 thru K99	K1 + B	Refer to all constants by a K-number.
1 thru 50	1 + B	Refer to the value (computed or estimated) of an expression by its expression number. An arbitrary limit of the number of expressions is 50.
[or $	[A + B	Front bracket for a term.
] or ,	A + B]	Back bracket for a term.
+	X + Y	Addition.
−	X − Y	Subtraction.
·	X · Y	Multiplication.
/	X / Y	Division.
SIN	SIN A	Sine of angle A. A is in radians.
COS	COS A	Cosine of angle A. A is in radians.
ASN	ASN A	Arcsine A, where the angle will be in radians.
ACN	ACN A	Arccosine A, where the angle will be in radians.
EXP	EXP X	$(e)^x$, exponential to the X.
LOG	LOG X	The *natural* logarithm of X.
PWR	X PWR N	$(X)^N$, the quantity X raised to the power N.
SRT	SRT X	$\sqrt{X}$, the square root of the quantity X.
SQR	SQR [X + Y]	The quantity following this symbol is squared.
∗	A − B ∗ Y	A substitution symbol. Compute the quantity on the left side of the symbol ∗, and substitute it for the constant or expression number on the right side.

Figure IV-1. List of BACAIC facilities.
Source: Grems and Porter [GR55], p. 6.

that number in other statements. Thus **WHN A GRT B USE 6** meant that if the value of **A** was greater than or equal to the value of **B**, then expression number 6 was computed next. Otherwise, the expression that followed was computed.

Logically unnecessary parentheses could be excluded; thus the expression **A + B · C** was handled correctly. They even allowed multiple assignment statements, such as

$$A + B * S * T * U$$

which meant assign the value of **A + B** to **S, T,** and **U**. This particular facility did not appear again in any major language until ALGOL 60.

6. *PRINT*

PRINT, which stands for *PR*e-edited *INT*erpretive system, was designed to meet the scientific computing needs of those people with an IBM 705. It was an interpretive system, which simulated floating point instructions. Coding was started at IBM in February, 1956, and the first customer tried the system in July, 1956; thus it was actually completed before FORTRAN.

PRINT provided a series of operation codes with variable fields, such as **RPT** n +−i +−j +−k which performed the next instruction n times, indexing its first, second, and third address by i, j, and k word lengths, respectively. The general form of the command was

OpCode Variable Field

where the variable field contained one to four variables, depending on the *Op Code*. The operations provided included the arithmetic ones, a few mathematical functions, testing, and input/output commands; these could all be indexed. Operations which could not be indexed included tests, index commands, and some input/output.

The main reason for including any mention of PRINT is the fact that it was the other significant attempt (besides A-2 and A-3) to provide facilities to handle scientific problems on a machine designed for use in data processing applications.

IV.2.2. MORE WIDELY USED SYSTEMS

For differing reasons, the systems discussed in this section received much wider use than those previously mentioned. Although none of them survived, they all helped the development of programming languages.

1. *MATH-MATIC (AT-3)*

Because of the logical sequence of material in the chapter, the information on the MATH-MATIC system will actually precede that of FORTRAN. However, a number of the comments to be made about MATH-MATIC will assume some knowledge of FORTRAN. Therefore the reader who is completely unacquainted with FORTRAN is advised to read Section IV.3 before trying to understand this particular system.

The work on MATH-MATIC, which was originally known as AT-3, was started around 1955 by a group at Remington Rand UNIVAC in the department under Dr. Grace Hopper. A preliminary manual for the MATH-MATIC system (Ash *et al.* [AS57]) was available in April, 1957. It was prepared by a group under the technical direction of C. Katz, with key participants being R. Ash, E. Broadwin, V. DellaValle, M. Greene, A. Jenny and L. Yu. The objective of this system was exactly the same as that of FORTRAN, namely to reduce the time and effort required to solve numerical scientific problems. This system was designed for use on the UNIVAC I, whose two major characteristics that affect this system are that it had no floating point arithmetic and only a 1000-word memory. It will be shown that these played a major role in the development of the system.

A list of the acceptable statements is given in Fig. IV-2. There were a number of the elementary mathematical functions available. A modified Unityper (which was the device used to prepare magnetic tape for input) had a keyboard which provided numerical superscripts which could be used as exponents. However, variable exponents had to be designated using the POW operator.

Data names could be subscripted, and the subscript could involve the four basic arithmetic operations; the limitation on the number of subscripts was dependent on the size of the UNIVAC word: The subscripted variable name, including the parentheses and the subscripts themselves, could not exceed 12 characters. Since UNIVAC was a fixed point machine, the floating point arithmetic had to be done by subroutines, and each number required two words of memory. (This obviously tended to have a catastrophic effect on an already small memory.)

Like any other system, this survived an evolutionary process so that some of the items shown in Fig. IV-2 were not in the original version. Even the early package was surprisingly strong however, and contained a number of features which did not find their way into other languages until considerably later. Among the more interesting or unusual commands in MATH-MATIC were the following: The *EXECUTE* statement; the flexibility of the *IF* statement, since any number of *IF clauses* could be included; the facility for having the range of the loop specified by the loop control statement itself, and the ability to have several variables varying within

CONTROL SENTENCES

    (n) CONTAIN $X(m,n)$.
    (n) CONTAIN $X(m,n,p)$.
    (n) EXECUTE SENTENCE F .
    (n) EXECUTE SENTENCES F THRU L .
    (n) IF $X > Y$ JUMP TO SENTENCE F .
    (n) IF $X < Y$ JUMP TO SENTENCE F .
    (n) IF $X = Y$ JUMP TO SENTENCE F .
    (n) IF $X > = Y$ JUMP TO SENTENCE F IF $V < W$ JUMP TO SENTENCE G .
    (n) IF $X = < Y$ JUMP TO SENTENCE F IF $V = W$ JUMP TO SENTENCE G IF $P > Q$ JUMP TO SENTENCE H .
    (n) IGNORE .
    (n) JUMP TO SENTENCE F .
    (n) PRINT-OUT A B C . . . N .
    (n) SET TO *number* A B C .
    (n) STOP .
    (n) TYPE-IN A B C . . . N .
    (n) VARY X X_0 (X_i) X_L SENTENCES F THRU L .
    (n) VARY X X_0 (X_i) X_L Y Y_0 (Y_i) Y_L SENTENCES F THRU L .
    (n) VARY X X_0 (X_i) X_L Y Y_0 (Y_i) Y_L Z Z_0 (Z_i) Z_L SENTENCES F THRU L .
    (n) VARY X X_0 X_1 X_2 . . . X_n SENTENCES F THRU L .
    (n) VARY X Y X X_0 Y_0 Z_0 X_1 Y_1 Z_1 X_2 Y_2 Z_2 . . . X_n Y_n Z_n SENTENCES F THRU L .

INPUT/OUTPUT SENTENCES

    (n) WRITE-LABEL X . . . X FOR SENTENCE F .
    (n) TITLE FOR SENTENCE F X . . . X .
    (n) TITLE FOR SENTENCE F X . . . X HEADINGS A . . . A B . . . B C . . . C .
    (n) HEADINGS FOR SENTENCE F A . . . A B . . . B C . . . C .
    (n) CHECK-LABEL X . . . X FOR SENTENCE F .
    (n) CHECK-COUNT SENTENCE F IF EXCEED X . . . X JUMP TO SENTENCE L .
    (n) READ A B C .
    (n) READ-ITEM $X(m,p)$ LABEL X . . . X .
    (n) READ A B C IF SENTINEL RESET AND JUMP TO SENTENCE F LABEL X . . . X .
    (n) READ A B C IF SENTINEL REWIND AND JUMP TO SENTENCE F LABEL X . . . X .
    (n) PRE-READ A B C .
    (n) READ-ARRAY $X(I,J)$.
    (n) WRITE A B C .
    (n) WRITE-ITEM $X(m,n,p)$.
    (n) WRITE EDIT X Y Z .
    (n) WRITE-ITEM EDIT $X(m,p)$.
    (n) WRITE CONVERT TO n DECIMAL X Y Z .
    (n) WRITE-ITEM CONVERT TO n DECIMALS $A(m,p)$.
    (n) WRITE CONVERT X Y Z .
    (n) WRITE-ARRAY CONVERT TO n DECIMALS $X(m,n,p)$.
    (n) CLOSE-INPUT SENTENCE F .
    (n) CLOSE-INPUT AND REWIND SENTENCE F .
    (n) CLOSE-OUTPUT SENTENCE F .

Figure IV-2. MATH-MATIC commands. The assignment statement and mathematical operators are not included in the figure. Note that the Input/Output sentences include only commands involving tape. Other I/O commands are included under control sentences. The *(n)* represents the statement number.

Source: [RR60], extracts from pp. 16–34.

the loop control statement. All these facilities were later picked up in one way or another in either COBOL or ALGOL.

There are two unusual features of MATH-MATIC which are very important but which have not really been supported in any major language since then. The first is the ability to handle lower level languages *in line*. MATH-MATIC was able to accept and handle statements written in both UNIVAC machine code (commonly called C-10) and also in A-3, which was a three-address intermediate language with its own compiler. (See Section IV.2.1.4). The user could use the same variables in all three languages (subject to some reasonable conventions) simply by listing these variable names in a dictionary.

In my opinion, the most interesting feature of MATH-MATIC was its implementation of automatic segmentation; this is a facility that appears not to have been implemented since then, although a great many people have talked about it and made claims that they were trying to do it but say they have been prevented because of the difficulty. The MATH-MATIC system provided completely automatic segmentation, in the sense that any object program which was too large to fit in one memory load would automatically have inserted into the object code (by the compiler) the necessary control transfers and input/output statements to reload memory as many times as necessary. Thus the compiler created an object program which brought into memory that part of the program which was to be executed next. This in itself is not very difficult to do, but it is naturally quite inefficient. The MATH-MATIC system went still further, by examining the code for loops and attempting to put them into a single segment. Thus, if the normal segmentation caused part of a loop to be in one memory segment and part in another, the compiler would create a shorter segment preceding the loop and put the loop all in one segment. Obviously if the number of statements within a loop could not fit into one memory load, the compiler could not do much about it; but in that case it looked to see if there was a subloop within the larger one and, if so, it would put that into a single memory segment.

It is interesting to speculate whether if this system had been implemented on a machine with fewer limitations than UNIVAC or with wider market acceptance such as the IBM 704 or 705, it would not have become the major language that FORTRAN became.

2. *UNICODE*

The UNICODE system was developed at Remington Rand UNIVAC around 1957–1958 for the 1103A and 1105. In many ways it is a hybrid language, falling somewhere between MATH-MATIC and FORTRAN. It tends to look very much like MATH-MATIC because it was obviously to Remington Rand's advantage to have their two scientific languages be the same; on the other hand, the 1103A and the 1105 were obviously machines much better suited to the solution of scientific problems than the UNIVAC was. In addition, from a timing point of view, UNICODE was able to pick up some of the facilities of FORTRAN which are of value. A list of the allowable statements is shown in Fig. IV-3.

COMPUTE X .

X=A .

VARY X P(Q)R SENTENCES N THRU M THEN JUMP TO K .
VARY X P(Q)R SENTENCE N .
VARY X P(Q)R WITH Y S(T)U SENTENCES N THRU M .

JUMP TO SENTENCE K .

IF X=Y JUMP TO SENTENCE K .
IF X NOT=Y JUMP TO SENTENCE K .
IF X<Y JUMP TO SENTENCE K .
IF X>Y JUMP TO SENTENCE K, IF X<=Y JUMP TO SENTENCE M .
IF X<Y JUMP TO SENTENCE K, IF X=Y JUMP TO SENTENCE M,
  IF X>Y JUMP TO SENTENCE N .
IF X>=Y JUMP TO SENTENCE K .

RESUME K .

LIST X(I,J), TAPE L, ((Title)) .
LIST F(X,Y,Z,T), X, Y, Z, T, TAPE L, ((Title)), (Column Heading),
  (Col. Hdg), (Col. Hdg), (Col. Hdg), (Col. Hdg) .

TYPE X(I,J), Y, - - -, F(X), Z .

READ A .
READ A, IF END OF DATA, JUMP TO SENTENCE K .

PRINT - - - - - - - - .

START .

STOP .

END OF TAPE .

DIMENSION X(- -,- -), Y(- -,- -,- -,- -), - - -, Z(- -,- -,- -) .

Figure IV-3. List of UNICODE instructions. The assignment statement and mathematical operators are not included in the figure.
Source: [RR59], p. 60 (slightly modified).

One of the concepts from FORTRAN that UNICODE adopted was that variables beginning with the letters I, J, K, L, or M were fixed point and all others were floating point. They could have up to four subscripts. In both UNICODE and MATH-MATIC it was possible to have a numerical superscript because the modified Unityper allowed them. However, variable exponents had to be designated by the operator POW. Similarly, the modified Unityper allowed the relational symbols >, <, and =.

Because of its somewhat hybrid nature, UNICODE cannot be said to have contributed anything significant to the improvement of scientific languages since it introduced no new concepts of its own.

3. *IT, FORTRANSIT, and GAT*

The IT (*I*nternal *T*ranslator) system was developed for the IBM 650 by A. J. Perlis, J. W. Smith, and H. V. Zoeren, based on a version developed for the Datatron by these people and M. Koschman, J. Chipps, and S. Orgel at Purdue University. The system is described by Perlis *et al.* [PR57] and [PR57a]. IT was designed primarily to handle numerical scientific problems. It had two enormous shortcomings: One was the hardware language, which was forced on the designers by the 650, and the other was the scanning technique used, which forced unnatural (and essentially incorrect) evaluation of mathematical expressions. In spite of these difficulties, however, IT was a significant step in compiler development. Most importantly, IT showed that an algebraic language could be implemented on a small machine (2000 words) with a small effort; it required only about two man-years to develop. This language was the forerunner of several others (e.g., RUNCIBLE, GATE, CORREGATE, and GAT which is discussed later) which had more reasonable hardware and used better compiling techniques. Thus a major contribution of IT was to inspire some aspects of compiler research at Case Institute of Technology and the University of Michigan. In terms of other work being done at that time, it is worth contrasting the notation for IT with that used for MATH-MATIC (see Section IV.2.2.1), which was far more natural and also ran on a machine with limited storage (originally UNIVAC I, with 1000 words). However, MATH-MATIC had the advantage of a reasonable character set and magnetic tapes for external storage, thus placing it in the FORTRAN category.

In current terminology, the reference language consisted of the digits and letters, punctuation characters and operators, and some other symbols; the *hardware representation* is the *single letter* shown in the right-hand column:

Symbol	Name	Representation
(	Left parenthesis	L
)	Right parenthesis	R
.	Decimal point	J
←	Substitution	Z
=	Relational equality	U
>	Greater than	V
≥	Greater than or equal	W
+	Addition	S
−	Subtraction	M
×	Multiplication	X
/	Division	D
exp	General exponentiation	P
,	Comma	K
''	Quotes	Q
	Type	T
	Finish	F

Floating point variables were represented by $Yi\ldots In$, which means the subscripted variable $Yiii\ldots n$, e.g., Y3, Y147, Yll26. The letter C could be used in place of Y for floating point variables, thus giving the programmer two mnemonic classes of variables. Fixed point variables were represented by $III\ldots In$, e.g., I3, Ill26. These variables were used primarily as indices. Limited subscripting and mixed mode arithmetic were permitted.

Subroutines were considered operands and represented as "$n\ E,\ v1,\ v2,\ \ldots,\ vj$" which means the subroutine number n which is a function of the variables $v1, v2, \ldots, vj$, e.g., "21E, "42E, Y1 + Y2"" would represent $\log_{10}(\sin(Y1+Y2))$ if subroutines 21 and 42 represented the logarithm and sine routines, respectively.

Each statement must be numbered using an integer less than 626 (the reason for this particular value is unknown to me), but the execution sequence is determined by the physical ordering and not by these numbers.

Statement types allowed included assignment; unconditional control transfers, where the address could be assigned or computed; conditional transfers involving relations between operands but not expressions (i.e., parentheses around all pairs were required), e.g., the statement numbered k

$$k:\ G\ I3\ IF\ (Y1\ +\ Y2)\ =\ 9$$

would cause the statement with the identifier value of I3 to be executed if Y1 + Y2 = 9. But,

$$k:\ G\ I3\ IF\ Y1\ +\ Y2\ =\ 9$$

is illegal. Additional statements included a halt, an input statement (READ), and an output statement of the form

$$k:\ T\ v1\ T\ v2\ T\ v3\ T\ v4$$

which has the effect of punching out a single card containing the names and current values of the (up to four) variables. A conditional output statement was provided.

Loop control was written as

$$k:\ j,\ v1,\ v2,\ v3,\ v4$$

where the range was down through statement j, with the parameter $v1$ varying from $v2$ to $v4$ in increments of $v3$. A subroutine call was accomplished by writing

$$k:\ "nE,\ \ldots\ "$$

The compilation technique used involved a right to left scan with no hierarchy of operators, except for parentheses. Thus the expression Y1 × Y2 + Y3 was translated as Y1 × (Y2+Y3).

The greatest difficulty with IT was that each nonalphanumeric character was represented as a single letter, as shown above. This caused such completely unreadable programs as

```
1    READ                    F
2    Y2 Z OJ                 F
3    4K I1K 11K M1K 1K       F
4    Y2 Z CI1 Y1 X Y2        F
5    H                       FF
```

which is to evaluate the polynomial

$$y = \sum_{i=0}^{10} a_i x^i$$

whose actual *reference language* version is

```
1:    READ

2:    Y2 ← 0

3:    4, I1, 11, −1, 1,

4:    Y2 ← CI1 + Y1 × Y2

5:    H
```

It has always been my contention that, from the user's viewpoint, the difficulties of the hardware representation completely outweighed any advantages gained by the fundamental concepts in the language, which were quite reasonable considering the machine involved and the year in which the language was developed. On the other hand, IT was an early example of the implementation of a programming language for scientific problems on a small machine, and a number of students and scientists at the cited universities made effective use of it. (Although the developers naturally claim otherwise, it is not obvious to me that these same users would not have had equivalent success with the 650 SOAP (*S*ymbolic *O*ptimum *A*ssembly *P*rogram).)

The FORTRANSIT system was developed (apparently) to have the best of both worlds, namely FORTRAN and IT. FORTRANSIT was merely a very simple subset of FORTRAN which the user could write; this was translated into IT, which in turn was translated to the SOAP assembler on the 650, and then to machine code. The FORTRAN statements acceptable to FORTRANSIT were

$a = b$ (arithmetic statement)

GO TO n

GO TO $(m_1, m_2, \ldots, m_3), i$

IF $(a)\ n_1, n_2, n_3$

PAUSE

STOP

DO $n\ i = m_1, m_2$

DO $n\ i = m_1, m_2, m_3$

CONTINUE

READ n, list

PUNCH n, list

DIMENSION $V, V, V, \ldots$

Certain restrictions and extensions of FORTRAN were imposed and permitted (respectively). For example, names could be only five characters instead of six and there were no built-in functions. On the other hand, mixed mode expressions were permitted, with a note of caution that they would be evaluated in floating point and would be incompatible for FORTRAN on the 704.

GAT was a system developed at the University of Michigan by R. Graham and B. Arden, who described it in [GM00]. GAT was based strongly on IT, but it managed to overcome the strongest disadvantage of the latter by having a 650 which permitted additional characters rather than just letters and digits. It used a different implementation technique than IT did, so arithmetic expressions were evaluated normally. Many of these techniques were used in the development of MAD (see Section IV.5.2).

Some of the features which were put into GAT and which were not fundamentally part of IT were the ability to handle alphanumeric strings of five or less letters by enclosing them within $ delimiters, additional letters used for fixed and floating point variables, correct mathematical handling of arithmetic expressions which were not parenthesized, *DIMENSION* statement, and use of function calls as operands within an expression.

GAT was also implemented for the 1105 and used at the University of North Carolina.

Extensions of GAT, called GATE and CORREGATE, were developed.

4. *ALGOL 58*

Although by now ALGOL 58 is a "language of historical interest only", its historical and technical development played such a major role in the creation of ALGOL 60 that it has been discussed within the general ALGOL description. See Section IV.4.1.1.

IV.3 FORTRAN

IV.3.1. HISTORY OF FORTRAN DEVELOPMENT

The history of the development of FORTRAN is almost equivalent to, or certainly parallels, the overall development of programming. It is not my intention to give a complete description of either; hence this section will describe only the main highlights of FORTRAN development.

The earliest significant document that seems to exist is one marked "PRELIMINARY REPORT, Specifications for the IBM Mathematical FORmula TRANslating System, FORTRAN", dated November 10, 1954 and issued by the Programming Research Group, Applied Science Division, of IBM. The first sentence of this report states "The IBM Mathematical Formula Translating System or briefly, FORTRAN, will comprise a large set of programs to enable the IBM 704 to accept a concise formulation of a problem in terms of a mathematical notation and to produce automatically a high-speed 704 program for the solution of the problem." It is interesting to note that the authors (who are not identified in the document) felt a need to justify such a development. They devoted several pages to a discussion of the advantages of such a system. They cited primarily the virtual elimination of coding and debugging, reduction in elapsed time, doubling of machine output, and the feasibility of investigating mathematical models.

The first manual for FORTRAN was the reference manual [IB56]. A primer [IB57] was issued later. The first page of the reference manual listed the working committee as the following people, all of whom worked for IBM except those designated otherwise: J. W. Backus, R. J. Beeber, S. Best, R. Goldberg, H. L. Herrick, R. A. Hughes (University of California, Radiation Laboratory), L. B. Mitchell, R. A. Nelson, R. Nutt (United Aircraft Corporation), D. Sayre, P. B. Sheridan, H. Stern, and I. Ziller.

The leader of this effort was John Backus, who thus deserves a major share of the credit for the initial development of what has become the most widely used higher level language in the world.

The 704 FORTRAN system was issued early in 1957 by the Programming Research Department of IBM. As far as professional presentation is concerned, the first paper appears to be the one given at the 1957 Western Joint Computer Conference by Backus et al. [BS57]. It is interesting to compare the preliminary specifications of November, 1954 and the finally issued manual of October, 1956. There is surprisingly little difference, although some interesting changes occurred. The preliminary specifications only allowed for names with one or two letters but defined function names as an alphabetic character followed by two or more characters. Then there were three significant deletions from the preliminary specifications: (1) Mixed number expressions were allowed, and these did not reappear until ALGOL, nor within the FORTRAN family itself except for FORTRANSIT (see Section IV.2.2.3), and then much later in FORMAC (see Section VII. 3). (2) The DO statement allowed the range to be explicitly stated as a pair of statement numbers. (3) The IF statement allowed a comparison between two variables, rather than merely a test against zero. The list of FORTRAN statements in this first system is given in Fig. IV-4. Note that a considerable number of these statements are heavily machine dependent; in particular, those relating to the sense switches, the overflows and divide check, and of course the references to tape and drum.

Although FORTRAN is considered quite commonplace now, it was not readily or easily accepted at that time. Customers raised many objections, foremost among them was that the compiler probably could not turn out object code as good as their best programmers. A significant selling campaign to push the advantages of such systems was underway at that time, with the spearhead being carried for the numerical scientific languages (i.e., FORTRAN) by IBM and for the "English-language-like" business data processing languages by Remington Rand (and Dr. Grace Hopper in particular).

In June, 1958 a new version of FORTRAN with significant language additions was released as FORTRAN II for the 704 [IB58]. A summary list of the FORTRAN II statements is given in Fig. IV-5. From a technical point of view, the following are the most significant additions of FORTRAN II to FORTRAN I: The subroutine concept exemplified by the *SUBROUTINE, CALL,* and *RETURN* statements and the *FUNCTION* statement; the *COMMON* statement was added to provide communication between subroutines; the *END* statement was added to avoid putting an end of file mark to indicate the end of the program; also, the use of subprograms permitted the linkage to assembly-coded programs (i.e., SAP).

Statement	Normal Sequencing
$a = b$	Next executable statement
GO TO n	Statement n
GO TO n, $(n_1, n_2, \ldots, n_m)$	Statement last assigned
ASSIGN i TO n	Next executable statement
GO TO $(n_1, n_2, \ldots, n_m)$, i	Statement n_i
IF (a) n_1, n_2, n_3	Statement n_1, n_2, n_3 as a less than, $=$, or greater than 0
SENSE LIGHT i	Next executable statement
IF (SENSE LIGHT i) n_1, n_2	Statement n_1, n_2 as Sense Light i ON or OFF
IF (SENSE SWITCH i) n_1, n_2	" " " as Sense Switch i DOWN or UP
IF ACCUMULATOR OVERFLOW n_1, n_2	Statement n_1, n_2 as Accumulator Overflow trigger ON or OFF
IF QUOTIENT OVERFLOW n_1, n_2	Statement n_1, n_2 as MQ Overflow trigger ON or OFF
IF DIVIDE CHECK n_1, n_2	Statement n_1, n_2 as Divide Check trigger ON or OFF
PAUSE or PAUSE n	Next executable statement
STOP or STOP n	Terminates program
DO n $i = m_1, m_2$ or DO n $i = m_1, m_2, m_3$	Next executable statement
CONTINUE	" " "
FORMAT (Specification)	Not executed
READ n, list	Next executable statement
READ INPUT TAPE i, n, list	" " "
PUNCH n, list	" " "
PRINT n, list	" " "
WRITE OUTPUT TAPE i, n, list	" " "
READ TAPE i, list	" " "
READ DRUM i, j, list	" " "
WRITE TAPE i, list	" " "
WRITE DRUM i, j, list	" " "
END FILE i	" " "
REWIND i	" " "
BACKSPACE i	" " "
DIMENSION $v, v, v, \ldots$	Not executed
EQUIVALENCE $(a,b,c,\ldots), (d,e,f,\ldots), \ldots$	" "
FREQUENCY n $(i,j,\ldots), m(k,l,\ldots), \ldots$	" "

Figure IV-4. Table of FORTRAN I statements for the IBM 704. The spacing is not significant.
Source: [IB56], p. 50. Reprinted by permission from *The FORTRAN Automatic Coding System for the IBM 704 EDPM*. © 1956 by International Business Machines Corporation.

FORTRAN systems for the 709 and 650 were officially released late in 1958. In 1960, FORTRANs for the 1620 and 7070 were released, and in 1962 FORTRAN IV was released on the 7030 (STRETCH).[1] The first apparent implementation of FORTRAN (using that name) by a manufacturer other than IBM was the version of FORTRAN I for the UNIVAC Solid State 80 which apparently was run as early as January, 1961.[2] Later that year, an augmented version of FORTRAN II was developed for the Remington Rand LARC by Computer Sciences Corporation.[3] Although not using the name FORTRAN, the ALTAC system developed for the Philco 2000 was an extended FORTRAN II running even earlier, in April, 1960.[4] By 1963 virtually all manufacturers had either delivered or committed

---

Arithmetic statements (*arithmetic formulas and function definitions*)

$a = b$

Control statements

GO TO n
GO TO n, (n$_1$, n$_2$, . . . , n$_m$)
ASSIGN i TO n
GO TO (n$_1$, n$_2$, . . . , n$_m$), i
IF (a) n$_1$, n$_2$, n$_3$
SENSE LIGHT i
IF (SENSE LIGHT i) n$_1$, n$_2$
IF (SENSE SWITCH i) n$_1$, n$_2$
IF ACCUMULATOR OVERFLOW n$_1$, n$_2$
IF QUOTIENT OVERFLOW n$_1$, n$_2$
IF DIVIDE CHECK n$_1$, n$_2$
PAUSE or PAUSE n
STOP or STOP n
DO n i=m$_1$, m$_2$ or DO n i=m$_1$, m$_2$, m$_3$
CONTINUE
CALL name (argument list)
RETURN
END (i$_1$, i$_2$, i$_3$, i$_4$, i$_5$)

Input/Output statements

FORMAT (specification)
READ n, list
READ INPUT TAPE i, n, list
PUNCH n, list
PRINT n, list
WRITE OUTPUT TAPE i, n, list
READ TAPE i, list
READ DRUM i, j, list
WRITE TAPE i, list
WRITE DRUM i, j, list
END FILE i
REWIND i
BACKSPACE i

Specification statements

DIMENSION v, v, v, . . .
EQUIVALENCE (a, b, c, . . .), (d, e, f, . . .), . . .
FREQUENCY n(i, j, . . .), m(k, l, . . .), . . .
SUBROUTINE name (argument list)
FUNCTION name (argument list)
COMMON a, b, c, . . .

Figure IV-5. Summary of FORTRAN II statements. The spacing is not significant.
Source: [IB58], pp. 59–60. Reprinted by permission from *FORTRAN II for the IBM 704 Data Processing System.* © 1958 by International Business Machines Corporation.

---

[1] Heising [HE63], p. 85.
[2] [CC63], p. 96.
[3] [CC63], p. 96.
[4] [CC63], p. 96.

themselves to producing some version of FORTRAN. Oswald [OS64] cites the existence of 43 FORTRAN compilers and compares 16 of them. Obviously many more have been written since then.

Because of the widespread use of FORTRAN, several things happened (understandably due to naivety and lack of foresight with respect to compatibility and growth problems). The methods of implementation differed not only between manufacturers but within the same manufacturer (i.e., the same features were handled differently even on IBM machines). It is important to note that the differences being referenced here are those which have—or at least potentially could have—some effect on the user; the significant concern is not the fact that different techniques of implementation create different types of efficiencies or inefficiencies but merely with the kind of differences which can happen in the end result. For example, different compilers handled the DO loop quite differently on a drop through the first time and sometimes even produced different values of the variable at the end of the execution of the DO loop. This is not very surprising since this type of specification was never written, or for that matter never intended to be written, into the language itself. A second effect of the widespread usage was that many people and groups found themselves wishing for improvements and changes. "The SHARE FORTRAN Committee... went on record in March 1961 as favoring a new FORTRAN language which did *not* contain all of FORTRAN II as a subset."[5] A FORTRAN III was developed by I. Ziller of IBM and used internally; its main characteristics were the addition of a Boolean algebraic statement, various devices to handle alphabetic information, an *external* capability to pass subprograms as arguments, and the inclusion of machine language instructions *in line*. It should be emphasized that this latter differs significantly from the ability to call a subroutine which happens to have been written in an assembly code. Because of the timing and other considerations, some of these features found their way as additions to FORTRAN II, others appeared in FORTRAN IV, and some others never were considered further. (The in-line machine coding facility was released in one version but later dropped because it was felt that this would completely ruin attempts at compatibility, and furthermore the differences between the 704 and the 709 played havoc with that particular feature.)

A gradual series of improvements or extensions were made to the 709/90 FORTRAN, including the provision of such facilities as double-precision and complex arithmetic.

In an attempt to stem some of the confusion arising from the multitude of implementations, IBM issued a General Information Manual on FORTRAN in 1961 [IB61]. It included a list of the available FORTRAN statements,

---

[5] Allen, Moore, and Rogoway [AX63], p. 46.

together with an indication of which machines the statements were being implemented on. This is shown as Fig. IV-6. Several things are worth noting from that manual: First, an attempt was made to indicate the differences in facilities provided by the various implementations, of which this table gives only partial information; secondly, the 1401 version which was released later is not included (probably the most significant aspect of the 1401 version is the fact that a reasonable version of FORTRAN was actually put onto such a small machine); a third very interesting fact is that neither [IB61] nor, for that matter, quite a few other manuals used the phrase FORTRAN II in their title, even though that is what they actually were. In other words, FORTRAN II was issued so relatively soon after FORTRAN I that the distinction rapidly became blurred and to some extent was even dropped, although it was clear in *Program Library* references.

During this entire period of time, FORTRAN was becoming more and more widely used. In some sense, its introduction caused a partial revolution in the way in which computer installations were run because it became not only possible but quite practical to have engineers, scientists, and other people actually programming their own problems without the intermediary of a professional programmer. Thus the conflict of the open versus closed shop became a very heated one, often centering around the use of FORTRAN as the key illustration for both sides. This should not be interpreted as saying that all people with scientific numerical problems to solve immediately sat down to learn FORTRAN; this is clearly not true but such a significant number of them did that it has had a major impact on the entire computer industry. One of the subsidiary side effects of FORTRAN was the introduction of the FORTRAN Monitor System [IB60]. This made the computer installation much more efficient by requiring less operator intervention for the running of the vast number of FORTRAN (as well as machine language) programs.

As stated earlier, SHARE went on record as favoring an improved version of FORTRAN; in 1962 a preliminary bulletin was issued to describe what eventually became known as FORTRAN IV, which would run under IBSYS-IBJOB on the IBM 7090/94. A number of significant features were added to FORTRAN II, including the following: Type statements (*LOGICAL*, *DOUBLE PRECISION*, *COMPLEX*, *REAL*, *INTEGER*, and *EXTERNAL*), logical expression as argument of an *IF*, function and subroutine names passed as arguments in references to other functions and subprograms, and *DATA* and *BLOCK DATA*. Some of these facilities were available in specific implementations of FORTRAN II but they only became official parts of the language in FORTRAN IV. Dropped from FORTRAN II were the machine dependent statements involving sense lights and switches, overflows, and the use of the words *TAPE* and *DRUM* in connection with the *READ* and *WRITE* statements. The *FREQUENCY* statement which had been included

Statement	COMMON	650	650 FORTRANSIT	1620	705	Basic 7070/7074	7070/7074	704	709/7090
ACCEPT n, list				X					
ACCEPT TAPE n, list				X	X		X	X	X
ASSIGN i TO n					X		X	X	X
BACKSPACE i					X	X	X	X	X
CALL NAME $(a_1, a_2, \ldots, a_n)$							X	X	X
COMMON $(a_1, a_2, \ldots, a_n)$							X	X	X
CONTINUE	X	X	X	X	X	X	X	X	X
DIMENSION $v_1, v_2, \ldots, v_n$	X	X	X	X	X	X	X	X	X
DO n $i = m_1, m_2, m_3$	X	X	X	X	X	X	X	X	X
END $(l_1, l_2, l_3, l_4, l_5)$		1	1	2		2	2	2	2
END FILE i					X	X	X	X	X
EQUIVALENCE $(a, b, c, \ldots), (d, e, f, \ldots), \ldots$				X		X	X	X	X
FORMAT $(s_1, s_2, \ldots, s_n)$					X	X	X	X	X
FREQUENCY $n(i, j, \ldots), m(k, l, \ldots), \ldots$							3	X	X
FUNCTION name $(a_1, a_2, \ldots, a_n)$								X	X
GO TO n	X	X	X	X	X	X	X	X	X
GO TO n, $(n_1, n_2, \ldots, n_m)$					X		X	X	X
GO TO $(n_1, n_2, \ldots, n_m), i$	X	X	X	X	X	X	X	X	X
IF ACCUMULATOR OVERFLOW n_1, n_2					X	X	X	X	X
IF DIVIDE CHECK n_1, n_2						X	X	X	X
IF QUOTIENT OVERFLOW n_1, n_2						X	X	X	X
IF (a) n_1, n_2, n_3	X	X	X	X	X	X	X	X	X
IF (SENSE LIGHT i) n_1, n_2					X		X	X	X
IF (SENSE SWITCH i) n_1, n_2				X	X		X	X	X
PAUSE n	4	5	5	4	5	5	5	5	5
PRINT n, list				X	X	X	X	X	X
PUNCH n, list	6	6	6		X	X	X	X	X
PUNCH TAPE n, list				X					
READ n, list	6	6	6	X	X	X	X	X	X
READ DRUM i, j, list					3			X	X
READ INPUT TAPE i, n, list					X	X	X	X	X
READ TAPE i, list					X	X	X	X	X
RETURN							X	X	X
REWIND i					X	X	X	X	X
SENSE LIGHT i					X		X	X	X
STOP n	4	4	5	4	X	X	X	X	X
SUBROUTINE name $(a_1, a_2, \ldots, a_n)$								X	X
TYPE n, list				X		X	X		
WRITE DRUM i, j, list					3			X	X
WRITE OUTPUT TAPE i, n, list					X	X	X	X	X
WRITE TAPE i list					X	X	X	X	X

1. l_i are not permitted.
2. l_i are optional and may be ignored.
3. May be included but will be ignored.
4. The n is not permitted.
5. The n is optional and may be ignored.
6. The n is optional and is ignored.

Figure IV-6. List of FORTRAN statements implemented on IBM computers, circa 1961.
Source: [IB61], p. 65. Reprinted by permission from *FORTRAN* (General Information Manual). © 1961 by International Business Machines Corporation.

to provide information useful for object time optimization was also dropped. FORTRAN IV was definitely not a compatible extension of FORTRAN II. One of the interesting results of this FORTRAN IV creation was the development of the SIFT program (discussed in more detail in Section II.4.2.2).

In May, 1962, the ASA X3.4.3 (=FORTRAN) Committee to develop an American Standard FORTRAN was formed and eventually produced two standards, known officially as FORTRAN and Basic FORTRAN, which correspond *roughly* to FORTRAN IV and FORTRAN II, respectively. (However, Basic FORTRAN *is* a proper subset of FORTRAN.) This will be discussed in more detail in Section IV.3.2 but it is essential to realize that FORTRAN—and in fact two of them—have the distinction of being the first programming languages that were actually standardized through the normal procedures of the USASI (then called the American Standards Association).

IV.3.2. FUNCTIONAL CHARACTERISTICS OF ASA (USASI) FORTRAN AND BASIC FORTRAN

It is actually rather difficult to characterize FORTRAN[6] according to the language properties that were discussed in Section II.2. FORTRAN is not very general, but that is not an entirely accurate statement considering some of the projects which have been accomplished using FORTRAN. Similarly, FORTRAN has fairly natural notation for algebraic expressions but tends toward succinctness in most other aspects. It is fairly consistent internally but has nothing particular in the language to cause or prevent great efficiency. It is easy to read and write and easy to learn, but its use is somewhat error prone.

The original objective in the first FORTRAN manual is worth quoting, both for historical interest and because it is still essentially valid today:

> The FORTRAN language is intended to be capable of expressing any problem of numerical computation. In particular, it deals easily with problems containing large sets of formulae and many variables and it permits any variable to have up to three independent subscripts.
>
> However, for problems in which machine words have a logical rather than a numerical meaning it is less satisfactory, and it may fail entirely to express some such problems. Nevertheless many logical operations not directly expressible in the FORTRAN language can be obtained by making use of provisions for incorporating library routines.[7]

---

[6] It should be clear, in the ensuing text, when FORTRAN is being used in a very general sense and when it refers specifically to a single language standard.

[7] [IB56], pp. 2–3. Reprinted by permission from *The FORTRAN Automatic Coding System for the IBM 704 EDPM (Programmer's Reference Manual)*. © 1956 by International Business Machines Corporation.

SAMPLE PROGRAM—FORTRAN

Problem: Construct a subroutine with parameters A and B such that A and B are integers and $2 < A < B$. For every odd integer K with $A \leq K \leq B$, compute $f(K) = (3K + \sin(K))^{1/2}$ if K is a prime, and $f(K) = (4K + \cos(K))^{1/2}$ if K is not a prime. For each K, print K, the value of $f(K)$, and the word PRIME or NONPRIME as the case may be.

Assume there exists a subroutine or function PRIME(K) which determines whether or not K is a prime, and assume that library routines for square root, sine, and cosine are available.

Program:

```
      SUBROUTINE PROBLEM (A, B)
      INTEGER A, B
      J = 2*(A/2) + 1
      DO 10 K = J, B, 2
      T = K
      IF (PRIME(K) .EQ. 1) GO TO 2
      E = SQRT (4.*T + COS(T))
      WRITE (1, 5) K, E
      GO TO 10
    2 E = SQRT (3.*T + SIN(T))
      WRITE (1, 6) K, E
   10 CONTINUE
    5 FORMAT (I6, F8.2, 4X, 8H NONPRIME)
    6 FORMAT (I6, F8.2, 4X, 5H PRIME)
      RETURN
      END
```

From the quotation above, it is clear that FORTRAN staked out a claim to handling problems in numerical computation. It has actually turned out to be used for a wide variety of other things, as discussed later, but its primary objective started and remains as the effective solution of numerical scientific problems. Of the classifications given in Section I.6, FORTRAN is definitely procedure-oriented, problem-oriented, and problem-solving; it is simultaneously a hardware, publication, and reference language; i.e., it is defined in a manner which makes it immediately acceptable as a hardware input language, and there are no other versions. Although not stated anywhere explicitly, it was clearly aimed at helping the nonprogrammer, i.e., the engineer or scientist. It was designed as a batch system—which is hardly surprising considering the time at which it was done!

When the first work on FORTRAN was accomplished in 1954ff, there was no real thought of making the language machine independent. This is clearly exemplified by the statements shown in Figure IV-4, which include references to sense lights, sense switches, accumulator, etc. By the time that FORTRAN IV was developed, however, these machine dependent characteristics were eliminated specifically to help in achieving such an objective.

With reference to the two standards, the languages are relatively machine independent. The major exceptions are the actual precision of the arithmetic which is being done (and which of course depends completely—in the practical world—on the size of the machine word) and some of the input/output statements.

The situation is also fairly good relative to compiler independence. The standard contains rigid rules about some of the normally tricky areas, e.g., ordering of array elements and special cases with DO loops.

In discussing the question of dialects, we must distinguish between dialects of the standards and dialects which have existed as a historical tradition. The former, by definition, should be either eliminated entirely or certainly minimized since that is definitely the purpose of the standard. For several reasons the existence of dialects has actually been somewhat less of a problem than with other languages, most notably ALGOL. First, within IBM it was possible to control the language specifications and so no dialects in the sense of *language deviations* really appeared (although there were differences based on implementation); secondly, the major reason that other manufacturers implemented FORTRAN was to permit their customers to transfer FORTRAN programs originally written for IBM equipment, so of course the manufacturers would go to great lengths to avoid dialects; thirdly, dialects would defeat the purpose of the standard; a fourth reason is that since FORTRAN was designed for direct input on a computer, there have not been transliteration problems nor any particular reasons to deviate from that set of specifications. These comments about dialects, however, do not apply at all when considering extensions. Historically, the greatest deviations among FORTRAN systems were extensions rather than dialects. As stated above, other manufacturers were highly motivated to retain the exact IBM FORTRAN notation for their customers; on the other hand, they were competitively motivated to provide additional features beyond those available in IBM compilers. Thus whole classes of particular implementations added new features of one kind or another and in a few cases FORTRAN II was extended to include some of the features of FORTRAN IV but in a compatible fashion. In some cases, extensions to the standards are being implemented. Probably the greatest difficulty in obtaining compatibility stems from implementation of slightly different sets of features; this is a chronic problem with all languages, although the COBOL standard has been created in a way which will recognize and define this problem. (See Section V.5.3.2.) Discussions of some of these issues are given by Heising [HE64a], McCracken [MR65], Oswald [OS64], and Wright [WR66].

As part of the subsetting characteristics, it should be noted that I have no first-hand (or even reliable second-hand) knowledge of FORTRAN being used to bootstrap its own compiler. What has been done in a few cases was to add some character and string-handling subroutines (coded in machine language) to FORTRAN and use these together with FORTRAN to do some bootstrapping. By defining extensions to FORTRAN, it can compile itself; but FORTRAN itself does not contain the necessary character-handling facilities to obtain even moderate efficiency.

Interestingly enough, in spite of the much more rigorous definition of ALGOL, there seems to have actually been less incompatibility in FORTRAN based on misunderstanding of the language definition. Most of the incompatibilities in FORTRAN have stemmed from idiosyncracies of individual compilers. Both FORTRAN standards have been written in straight narrative prose, without any attempt at formalization.

In a preprint, Rosen [RO61] describes a few of the problems in converting from FORTRAN on the 704 to ALTAC on the Philco 2000. He says there "To the best of my knowledge this is the first time that a compiler has assumed the major burden of transition from a large scale computer of one manufacturer to an even larger scale computer of another manufacturer."[8] I also believe that this is certainly one of the first, if not the first, serious and practical attempt at doing this. (In December, 1960, two COBOL programs were run on UNIVAC II and the RCA 501, with only minor changes required. This was a demonstration, however, rather than a practical attempt to do this on a large scale. See Section V.3.2 for further discussion of this.) Rosen describes some of the incompatibilities that actually arose as follows: Some which were based on idiosyncracies of the 704 FORTRAN compiler, e.g., handling negative integers; differences among the subroutines, e.g., attempting to find the square root of a negative number; permission (or lack thereof) of entering *FORMAT* statements at object time; differences caused by the primary use of on-line card readers and printers in the 704 versus off-line equipment with ALTAC; and larger range of exponents in the 2000 which affected the *E* conversion and the *FORMAT* statements. More recent work in this area is described by Olsen [OL65]. It seems to me that a number of these problems still remain, even with the existence of the standard. Nevertheless, in spite of this, there have undoubtedly been more FORTRAN programs converted to run from one machine to another than all other languages put together.

The word SIFT, standing for *S*hare *I*nternal *F*ORTRAN *T*ranslator, has become a somewhat generic term, as indicated in Section II.4.2. However, its original usage was the program which would do some translation between FORTRAN II and FORTRAN IV described in Allen, Moore, and Rogoway [AX63]. As those authors point out, most of the incompatibilities between FORTRAN II and FORTRAN IV could be resolved by simple transliteration. There were three areas, however, which required more analysis: The *EQUIVALENCE - COMMON* interaction, double-precision and complex arith-

---

[8] Rosen [RO61], p. 2B–2(1).

metic declarations, and Boolean statements. SIFT was a program primarily written in FORTRAN which would provide the necessary changes in a FORTRAN II program to make it work correctly as a FORTRAN IV program. It apparently worked successfully as a practical tool.

Various attempts have been made to translate FORTRAN to other languages but the tendency has been more toward the other way; in other words, because FORTRAN had been so popular and implemented on so many other machines, there is much more of a need to translate other languages to FORTRAN than vice versa. In particular, a number of ALGOL programs have been hand-translated to FORTRAN in order to test them when no ALGOL compiler was available. The most outstanding example of translating FORTRAN was to MAD (see Section IV.5.2), using a program called MADTRAN. This was done to take advantage of the fast MAD compiler.

FORTRAN has the distinction of being the first programming language to be standardized through the normal ASA (now USASI) procedures. For that reason, it is worth describing in some detail just how this was accomplished. The following material is taken intact from the writeup by W. P. Heising [HE64] which appeared preceding the publication of the proposed FORTRAN and Basic FORTRAN specifications [CC64].

> The American Standards Association (ASA) Sectional Committee X3 for Computers and Information Processing was established in 1960 under the sponsorship of the Business Equipment Manufacturers Association. ASA X3 in turn established an X3.4 Sectional Subcommittee to work in the area of common programming language standards. On May 17, 1962, X3.4 established by resolution a working group, X3.4.3-FORTRAN to develop American Standard FORTRAN proposals.
> RESOLVED:
> That X3.4 form a FORTRAN Working Group, to be known as X3.4.3-FORTRAN, with the
>
> *Scope.* To develop proposed standards of FORTRAN language.
>
> *Organization.* Shall contain a Policy Committee and a Technical Committee. The Policy Committee will be responsible to X3.4 for the Working Group's mission being accomplished. It will determine general policy, such as language content, and direct the Technical Committee.
>
> *Policy Committee Membership.* Will be determined by the X3.4 Steering Committee subject to written guidelines which may be amended later and including the following:
>
> > a. For each FORTRAN implementation in active development or use, one sponsor voting representative and one user voting representative are authorized.
> > b. A representative who is inactive may be dropped.

c. Associate members, not entitled to vote but entitled to participate in discussion, are authorized.

Technical Committee. Will develop proposed standards of FORTRAN language under the Policy Committee direction. The Technical Committee will conduct investigations and make reports to the Policy Committee.

On June 25, 1962 invitations to an organizational meeting of X3.4.3 were sent to manufacturers and user groups who might be interested in participating in the development of FORTRAN standards. The first meeting was held August 13–14, 1962 in New York City. X3.4.3 decided to proceed because (1) FORTRAN standardization was needed, and (2) a sufficiently wide representation of interested persons was participating.

A resolution on objectives was adopted unanimously on August 14, 1962.

"The objective of the X3.4.3 Working Group of ASA is to produce a document or documents which will define the ASA Standard or Standards for the FORTRAN language. The resulting standard language will be clearly and recognizably related to that language, with its variations, which has been called FORTRAN in the past. The criteria used to consider and evaluate various language elements will include (not in order of importance):

a. Ease of use by humans,
b. Compatibility with past FORTRAN use,
c. Scope of application,
d. Potential for extension,
e. Facility of implementation, i.e., compilation and execution efficiency.

"The FORTRAN standard will facilitate machine-to-machine transfer of programs written in ASA Standard FORTRAN. The Standard will serve as a reference document both for users who wish to achieve this objective and for manufacturers whose programming products will make it possible. The content and method of presentation of the standard will recognize this purpose."

It was the consensus of the group that (1) there was a definite interest in developing a standard corresponding to what is popularly known as FORTRAN IV, and (2) there was interest in developing for small and intermediate computers a FORTRAN standard near the power of FORTRAN II, however suitably modified to be compatible with the associated FORTRAN IV. Accordingly, two Technical Committees, designated X3.4.3-IV and X3.4.3-II, respectively, were established to create drafts. Most of the detailed work in developing drafts has been done by technical committees.

The X3.4.3-II Technical Committee completed and approved a draft in May, 1963. A Technical Fact Finding Committee was appointed and reported in August, 1964 on a comparison of the X3.4.3-II approved draft and an approved working draft of the X3.4.3-IV Technical Committee. This brought to light stylistic, terminological, and content differences and conflicts. In April, 1964 the X3.4.3-IV Technical Committee completed a draft of FORTRAN. In June, 1964 X3.4.3 received and compared the two drafts and (1) resolved conflicts in content, and (2) resolved the conflicting style and terminology. This was accomplished by recasting the X3.4.3-II document to reflect the style of the X3.4.3-IV document while retaining the original content. To reduce confusion, X3.4.3 decided to call the languages Basic FORTRAN and FORTRAN.[9]

The standards were approved March 7, 1966 and became ASA Standards X3.9-1966 [AA66] (=FORTRAN) and X3.10-1966 [AA66a] (=Basic FORTRAN). Within a year after approval, significant questions of interpretation had arisen, and so the X3.4.3 Committee had to be reactivated to deal with them.

An international standard was accepted for most practical purposes in October, 1965 but its final official approval has been delayed by administrative problems and errors. It added another subset (based on the ECMA work) to the two existing levels.

As indicated under the history, FORTRAN was initially designed by a group of individuals (who were listed earlier), most of whom worked for IBM; the objectives of the language were also stated earlier. The initial implementation of the language was done by the people who designed it, and they also did much of the implementation of FORTRAN II. Since then, an enormous number of people have become involved, both in and outside IBM. FORTRANs have been implemented by a large number of different people in IBM, by most computer manufacturers, and by virtually every independent software house. The maintenance of the language (prior to the standard) was done for IBM by various groups and for those outside of IBM in no central place. As indicated earlier, however, most manufacturers attempted to follow the IBM specifications.

FORTRAN was initially defined simply through the use of English prose and examples. When it came time to define the standards, those who were most heavily involved and had the strongest influence had no interest in using any type of formalized notation. Thus both standards have been written in narrative English with formats shown where appropriate. Some attempts (e.g., Rabinowitz [RN62] and Burkhardt [BU65]) have been made to provide a formal definition of FORTRAN but these are of necessity somewhat incomplete. Since here we are defining the characteristics of the Standard FORTRANs, there is only one form of documentation, namely [AA66] and [AA66a]. Readers who are interested in obtaining a somewhat better understanding about the documentation of the earlier IBM versions are referred to Heising's article [HE63].

Some version of FORTRAN has been made available on virtually every

---

[9] Heising [HE64], p. 590. By permission of Association for Computing Machinery, Inc.

computer ever made—see [CC63] for a list in 1963. Fortunately, by now FORTRAN has been clearly evaluated on the basis of its language facilities rather than on the basis of its compilers. The obvious advantages to FORTRAN are its practical effectiveness for solving numerical scientific problems and its subsequent widespread use with reasonable compatibility and conversion facilities. Its largest disadvantages stem from attempts to use it outside of the realm for which it was intended, namely for any type of alphanumeric data handling. (Even there it has occasionally succeeded—see, e.g., Fimple [FP64] and Robbins [RM62].) Thus, FORTRAN's largest disadvantages actually stem from its popularity; since so many people are used to it and like to program in it, they would like to use it for everything and when it does not supply their needs, it is (unjustly) criticized. The biggest disadvantage to FORTRAN is that it could not truly be extended in any type of clean way to provide the additional facilities that the state of the art now permits. Thus, although the original hope, when the development of (the now-called) PL/I was started, was to extend FORTRAN IV to provide the facilities that were needed, this rapidly turned out to be impossible. In other words, both Basic FORTRAN and FORTRAN suffer from the fact that they are based on a language developed in 1954 which did not have all the facilities that were desired 10 years later. What is perhaps most amazing is that, although it was developed in 1954, FORTRAN (in various forms) is still quite acceptable and widely used.

History and more recent developments have taught us some of the mistakes to avoid in the future.

IV.3.3. Technical Features of ASA (USASI) Basic FORTRAN

The character set used in Basic FORTRAN (as defined in [AA66a]) consists of the 26 capital letters, the 10 digits, and the following 10 symbols:

$$+ \quad - \quad * \quad / \quad) \quad (\quad = \quad . \quad , \quad \text{blank}$$

Octal digits are used in the *STOP* and *PAUSE* statements.

A data name consists of a letter followed by zero to four alphanumeric characters. Statement labels consist of from one to four digits. There are no reserved words so that any string of characters (subject to the definition above) can be used as a data name. One or two subscripts are allowed on data names; they are shown in parentheses, separated by commas, e.g., X(I, J), and they can be of the form (*constant* * *variable*) ± *constant*. There are no nonnumeric literals permitted except in the *FORMAT* statement.

The only operators are the five arithmetic ones, where a double asterisk denotes exponentiation. There are no delimiters. The only punctuation is a comma, which is simply used to separate lists of items and, of course, the parentheses, which are used in specified places. Blanks have no significance except in a few special cases. There are no noise words permitted.

The input form is highly card-oriented. The actual standard refers to the significant unit of a *line*, which is a string of 72 characters containing character positions called *columns*. A statement label can only be placed in columns 1 to 5. The statement appears in columns 7 to 72. The physical format permits continuation lines so that a statement can consist of an initial line optionally followed by up to five ordered continuation lines. A continuation line is any line that has a character other than zero or blank in column 6 and does not contain the character C in column 1.

The language is not highly formalized nor is it strongly English-like.

Basic FORTRAN has a number of declarations which are described later. The smallest executable unit is a single statement. In the sense used in Section III.3.1, there are no groups of smallest executable units. Loops are handled by the DO statement or by the tests in an IF statement.

There are actually four categories of procedures defined in Basic FORTRAN: *Statement, intrinsic*, and *external functions*; and *external subroutines*. The first three are normally called *functions* or *function procedures*. The statement function is defined internal to the program unit in which it is referenced and consists of a single statement of the form $F(a_1, a_2, \ldots, a_n) = E$ where F is the function name, E is an expression, and the a_i are the dummy arguments. This statement function must precede the first executable statement of the program unit and must follow the declarations. Aside from the dummy arguments, the expression E can only contain constants, variables, references to intrinsic functions and previously defined statement functions, and external functions. Intrinsic functions are the specific functions of *absolute value, float, fix*, and *transfer of sign*. An external function is defined externally to the program that references it.

Comments are designated by having the letter C in column 1. There is no interaction with the operating system or the environment specified in the Basic FORTRAN standard. There is no provision for direct inclusion of machine language but external procedures can be written in other languages.

All language units (i.e., declarations and executable units) are assumed to start with an "initial line" which requires a zero or blank in column 6. A statement or declaration can have up to five ordered continuation lines. There is no definitive end of a unit—rather it is determined by recognizing the beginning of the next unit. There is no type of recursion permitted. Parameters are called by location. Functions can be embedded into assignment statements but that is the only type of embedding permitted.

Declarations must precede statement function definitions, which must precede the executable statements. The former must be in the order

DIMENSION, COMMON, EQUIVALENCE. A complete program consists of an initial line, any number of statements, and an *END* line.

There is no provision for recursion in the language. A programming technique to achieve some of this effect is described by Ayers [AY63].

The only data type permitted is arithmetic, and this can be either integer or real (i.e., floating point). The type of a variable is determined by its name; any name beginning with one of the letters I, J, K, L, M, or N, denotes an *INTEGER* type; while all others imply type *REAL*. The executable commands are able to access only the variables, and it is generally assumed that these will take up a single machine word.

The only arithmetic done is integer or floating point. Use of real *and* integer variables (or constants) in the same expression is not permitted; thus there is no need for any conversion rules. The standard specifically disclaims all intent to specify precision or range of numerical quantities; however, there are some rules indicating sequence of evaluations.

The scope of a data name is the entire program, except that dummy arguments with names duplicating others can be used in function and subroutine definitions. There is often a need to use the same variable, however, in several subroutines or in the main program and a subroutine. In order to accomplish this, the COMMON declaration is used. This has the form COMMON $a_1, a_2, \ldots, a_n$ where each a_i is a data name. For each given COMMON statement, the items on this list are declared to be in COMMON storage and thus accessible by any part of the program, including all subroutines.

There is exactly one assignment statement, of the form v=e where v is a variable name and e is an arithmetic expression. Conversion is automatically performed if the v and e are of different types; if e is real, the result is truncated to create the necessary integer; while if e is integer, then it is floated. For example, the statements

$$I = 5$$
$$A = I + 2$$

would cause A to be assigned the value 7 in floating point form.

There are no character data-handling statements.

The normal flow of control is to the next physical statement in sequence. Thus the statement numbers provide labels only and have no inherent meaning.

The primary unconditional control transfer statement is the GO TO statement, which can be either unconditional of the form GO TO k, where k is a statement label, or can be a switch control. The latter is called a *computed* GO TO statement and is of the form GO TO $(k_1, k_2, \ldots, k_n)$, i where

k_i is a statement name and i is an integer. In this case, it is assumed that i has been computed independently; and if it has the value j, then control is unconditionally transferred to k_j. Thus, writing J=3 followed by GO TO (17, 20, 2, 5), J would cause control to be transferred to statement label 2.

Subroutines are invoked by writing CALL $s(a_1, a_2, \ldots, a_n)$ or CALL s where s is the name of a subroutine and the a's are actual arguments. These arguments can be variables, names of arrays, a single element in an array, or an expression. They must agree in order, number, and type with the corresponding dummy arguments. However, the subroutine can actually define or redefine one or more of these arguments, so as to return results, which can have the net effect of destroying the original value that existed before calling the subroutine. This is essentially the *call by location* discussed in Section III.3.2.3. A subroutine has its logical end marked by a RETURN statement, which causes control to return to the next executable statement after the CALL.

The CONTINUE statement only causes continuation of the normal execution sequence. It is used as the last statement in the range of a DO loop to permit alternate paths within the range to terminate at the end of the range.

The only conditional statement is the *arithmetic* IF, which is of the form IF (e) k_1, k_2, k_3 where e is an arithmetic expression and the k's are statement names. Control is transferred to k_1, k_2, or k_3 as the value of e is less than zero, zero, or greater than zero, respectively. For example,

$$A = 5$$
$$B = A - 7$$
$$\text{IF (B) 9, 15, 20}$$

causes control to go to statement 9.

The loop control in Basic FORTRAN is handled by the DO statement which is of the form DO n i = n_1, n_2, n_3 where the , n_3 can be omitted if it is equal to the integer 1. The range of the DO consists of the sequence of statements physically following the DO, through and including the statement with the label n. There is a single parameter, namely the one designated as i, which is varied by assigning it first value n_1 and incrementing by n_3 after control reaches statement n. The loop is considered finished only when the parameter exceeds the value designated by n_2. The terminal statement cannot be a GO TO of any form nor an IF, RETURN, STOP, PAUSE, or DO statement. It is permitted to leave the range of the DO statement by executing a control transfer of some kind; in this case the parameter is defined and is equal to the most recent value obtained. It is not legal, however, to cause control to be passed into the range of the DO from outside its range. It is legal to have a DO statement nested within another DO statement; in fact, they can have the same range, which then causes the net effect of looping on several parameters in succession. Overlapping of ranges is not allowed. For example, the program

```
         DO 7 M = 2, 9, 2
            A (M) = M ** 2
            DO 4 J = 1, 100
    4       B(J, M) = J + M
    7       C(M + 1) = 3 * M
```

would cause B(1,2), B(2,2), B(3,2),..., B(100,2), B(1,4), B(2,4),..., B(100,4),..., B(1,8), B(2,8), ..., B(100,8) to be assigned the values 3, 4, 5,..., 102, 5, 6, ..., 104, ..., 9, 10, ..., 108 respectively. Furthermore, A(2)=4, A(4)=16, A(6)=36, A(8)=64 and C(3)=6, C(5)=12, C(7)=18, and C(9)=24 would be assigned. It would be illegal to have the statement numbers 4 and 7 interchanged because then the ranges would be overlapping.

There are two types of input/output statements: The READ and WRITE statements, and the auxiliary input/output statements, namely BACKSPACE, REWIND, and END FILE. The READ and WRITE statements themselves can involve either *formatted* or *unformatted* records. The formatted READ statement is one of the forms READ (u, f)k or READ (u, f) where k is a list (defined below). The formatted WRITE has an identical format, except for using the word WRITE, of course. In both cases, the information is converted as specified by the FORMAT declaration (identified by the label f) described under declarations. The unformatted READ is one of the forms READ (u)k, where the k can be omitted. In this case, the next record is read from the input unit and if there is a list of names, the values read are assigned to the sequence of elements specified by the list. The unformatted WRITE must have a list associated with it; otherwise there would be no way of knowing what information was to be put out. The list specifies names of variables and array elements and can be a simple list, e.g., *a, b, c*, a simple list enclosed in parentheses, e.g., (*a, b, c*), a DO *implied list*, or two lists separated by a comma. A DO *implied list* is a list followed by a comma and then followed by something of the form $i = m_1, m_2, m_3$ where the m_i are defined the same as in the DO statement; if m_3 is omitted, it is assumed to be 1. The range of this DO specification is the set of names preceding it and the elements are specified for each cycle of the implied DO. Thus (A, B, I = 1, 3) means $A_1, B_1, A_2, B_2, A_3, B_3$.

The REWIND statement causes the specified unit to be positioned at its initial point; the **BACKSPACE** causes backspacing to the preceding record of the specified unit (if it is already positioned at its initial point, the statement has no effect); the **END FILE** statement causes an end of file record to be put on the specified unit.

A FORTRAN processor is required to provide two different types of library functions, called *intrinsic* and *basic external functions*. They are shown in Figs. IV-7 and IV-8, respectively. Each can be used in an arithmetic expression, where the types must agree. Other than these fixed functions, there is no provision for library references except through the mechanism of subroutines.

There are no specific debugging or storage-allocating statements in FORTRAN, although there are some declarations relating to storage. There

		Number of Arguments	Symbolic Name	Type of:	
Intrinsic Function	*Definition*			Argument	Function
Absolute value	$\|a\|$	1	ABS	Real	Real
			IABS	Integer	Integer
Float	Conversion from integer to real	1	FLOAT	Integer	Real
Fix	Conversion from real to integer	1	IFIX	Real	Integer
Transfer of sign	Sign of a_2 times $\|a_1\|$	2	SIGN	Real	Real
			ISIGN	Integer	Integer

Figure IV-7. List of intrinsic functions in Basic FORTRAN. Source: [AA66a], p. 19.

		Number of Arguments	Symbolic Name	Type of:	
Basic External Function	*Definition*			Argument	Function
Exponential	e^a	1	EXP	Real	Real
Natural logarithm	$\log_e(a)$	1	ALOG	Real	Real
Trigonometric sine	$\sin(a)$	1	SIN	Real	Real
Trigonometric cosine	$\cos(a)$	1	COS	Real	Real
Hyperbolic tangent	$\tanh(a)$	1	TANH	Real	Real
Square root	$(a)^{1/2}$	1	SQRT	Real	Real
Arctangent	$\arctan(a)$	1	ATAN	Real	Real

Figure IV-8. List of basic external functions in Basic FORTRAN. Source: [AA66a], p. 20.

are two statements which involve interaction with the operating system. The first is the *STOP* statement, which may be followed by an octal digit string containing from one to four digits; execution of this statement causes termination of the execution of the program. The second executable statement in this category is the *PAUSE*, which also can be followed by an octal digit string; the execution of the *PAUSE* statement causes the program to stop temporarily but execution must be resumable; the decision to do this is not under the control of the program itself but rather under the control of the operating system or the operator. If execution is resumed without otherwise changing the state of the processor, then the normal execution sequence is continued.

There are no data or file descriptions. There is a complicated *FORMAT* description which is used in conjunction with the *READ* and *WRITE* commands. This provides information on the form of the data, e.g., use of exponents, number of decimal places, and number and width of columns.

Storage allocation is partially controlled by the *DIMENSION*, *COMMON*, and *EQUIVALENCE* statements. The *DIMENSION* statement is of the form *DIMENSION* v(i) or *DIMENSION* v(i, j) where v is a data name and i and j are the maximum values that the subscript of that variable can assume, and any number of variables can be listed, e.g., *DIMENSION* A(10), C(5,9), M(2,8). Interestingly enough, the (Basic) FORTRAN standard actually specifies the way arrays are to be stored by defining their subscript values as follows: Assuming dimensions A and E, then subscript (x, y) has the value $x + A(y - 1)$, considering the array as a linear string.

The *EQUIVALENCE* statement has the form *EQUIVALENCE* (k_1), (k_2), ..., (k_n) where each k_i is a list of the form $a_1, a_2, ..., a_n$ where each a_i is a data name. The *EQUIVALENCE* statement is used to permit sharing of storage by two or more variables (or arrays). Thus each element in a given list is assigned the same part of storage by the compiler. When two variables share storage because of the *EQUIVALENCE* statement, the names cannot both appear in *COMMON* statements in the same program unit. It is important to note that the primary difference between *EQUIVALENCE* and *COMMON* is that the former permits several variables to share the same storage location, whereas the latter simply makes the designated variables accessible to all parts of the program.

There are different ways of declaring the four types of procedures which are permitted. The simplest is the statement function, which is defined internally to the program unit in which it is referenced. It is defined by a single statement of the form $f(a_1, a_2, ..., a_n) = e$, where f is the function name, e is an expression, and the a_i are dummy arguments. Note that since statement function definitions must precede the first executable statement in the program, it is logically possible to distinguish this from an ordinary assignment statement.

An external function is written as **FUNCTION** $f(a_1, a_2, \ldots, a_n)$ where f is the name of the function and the a_i are dummy arguments. This is then followed by the code for the procedure, which must contain the function name as a variable name, and at least one **RETURN** statement. The value of the former at the time of execution of any **RETURN** statement is the value of the function.

An external subroutine is defined similarly by writing **SUBROUTINE** $s(a_1, a_2, \ldots, a_n)$ or **SUBROUTINE** s. The name of the subroutine cannot be used in the body of the subroutine.

Basic FORTRAN does not permit any self-modification of the program nor any self-extension of the language. There are no specific compiler directives.

With regard to the ability to write a compiler for FORTRAN (any level) in FORTRAN (any level), this is utterly impractical to do because FORTRAN has no facilities for handling parts of words or for doing character manipulation. However, some people have written subroutine packages to do these things, and then used FORTRAN for the rest of the work and for the logic. For example, the SIFT program described earlier was written primarily in FORTRAN.

FORTRAN is a language which has tended to sacrifice compile time efficiency in order to obtain object time efficiency. An early illustration of this (which has long since disappeared) was the **FREQUENCY** statement that appeared in FORTRAN I and II; this was dropped in FORTRAN IV and in the creation of Basic FORTRAN. This statement required a tremendous amount of analysis at compile time and it did not yield enough benefits at object time. FORTRAN has no restrictions on naming of variables and simultaneously does not have any significant blanks; these facts make compilation quite inefficient in some cases (the former is the more significant problem). For example, if the compiler encounters something of the form DO 5 J = 7 it cannot tell whether that is an assignment statement or really a DO loop; in fact, this is only determined by the presence or absence of a comma after the number following the equal sign. In spite of this, several compilers which concentrated on rapid compilation have been written e.g., Rosen, Spurgeon, and Donnelly [RO65] and Schantz, *et al.* [SY67].

There are no debugging aids or error checking required in the language.

Actually, there has been no opportunity to use Basic FORTRAN outside its primary application area because it is relatively new. Earlier versions of FORTRAN (often with special machine language coded subroutine packages) have been used, however, for just about anything that one would care to name, e.g., list processing, polynomial handling, data processing, and phases of compiler writing. (See, respectively, the work described by Weizenbaum [WZ63] and Sakoda [SA65], Fowler and MacMasters [FH64], Robbins [RM62], Allen, Moore, and Rogoway [AX63].) Thus there is no doubt about FORTRAN's versatility, whether forced or inherent, in the language.

Comments about the distinction between Basic FORTRAN and FORTRAN are given in the next section.

IV.3.4. Technical Features of ASA (USASI) FORTRAN

It should be explicitly understood that all the characteristics and properties of Basic FORTRAN apply also to FORTRAN unless specifically stated otherwise. A discussion and summary of the restrictions on Basic FORTRAN are at the end of this section.

The $ is added to the character set as a currency symbol.

Data names can have six characters and three subscripts instead of two. Statement labels can have five digits. The first significant addition to FORTRAN is the set of relational and logical operators as follows:

.LT.	Less than
.LE.	Less than or equal to
.EQ.	Equal to
.NE.	Not equal to
.GT.	Greater than
.GE.	Greater than or equal to
.OR.	Logical disjunction
.AND.	Logical conjunction
.NOT.	Logical negation

Since there are now relational operators, they are used between two arithmetic expressions to produce the values *true* or *false* for Boolean (logical) variables. In this case, either arithmetic expression may be of type real or double-precision, or both arithmetic expressions may be of type integer. (See the description of new data types below.)

The restriction on the sequence of the declarations which exists in Basic FORTRAN is removed here.

In addition to the four types of arguments for procedures permitted in Basic FORTRAN, one can also use the name of an external procedure and a Hollerith constant; the latter is an exception to the rule requiring agreement of type. If an actual argument is an external function or subroutine name, the corresponding dummy argument must be used as an external function or subroutine name, respectively.

The most significant additions to Basic FORTRAN are in the area of data variables and the arithmetic performed upon them. Boolean (logical), complex, and Hollerith (i.e., alphanumeric) variables are permitted in FORTRAN. As a result, complex and Boolean arithmetic are performed.

In addition, "double-precision" floating point variables and constants are permitted; hence double-precision arithmetic is done. However, this is not required to be *twice* the precision of floating point but merely greater. There is no relaxation on the rules of not permitting integers and real variables in the same expression. However, real numbers can be combined with double-precision or complex numbers, with either of the latter resulting.

Boolean arithmetic is done in the normal manner by using the operators OR, AND, and NOT, resulting in a logical variable.

More flexibility on scope of variables is permitted since it is possible in a COMMON statement to have a block name assigned to a set of variables which are to be in a COMMON block. This block name is defined locally only for the variables it is associated with and has no meaning other than that. Hence the same block name can appear more than once in a COMMON statement.

In FORTRAN there are now three types of assignment statements: *arithmetic, logical,* and *go to*. The first is the same as in Basic FORTRAN, except that more complicated conversion rules for the results are required; these are shown in Fig. IV-9. The logical assignment statement is of the form v=e, where v is a logical variable and e is a logical expression; the right-hand side is evaluated and its value assigned to the logical variable v. The *go to* assignment statement is of the form ASSIGN k TO i, where k is a statement name and i is an integer variable name. The net effect of this statement occurs only for subsequent execution of any *assigned go to* statement (which is a new and additional type of control transfer added in FORTRAN). An assigned go to statement is of the form GO TO i, (k_1, k_2, ..., k_n) where i is a variable of type integer and the k_i are statement labels. The effect of this is to cause the transfer of control to that variable k_i which has the exact value that has been assigned to i. Thus if we wrote ASSIGN 17 TO M and then somewhat later executed GO TO M, (15, 902, 17, 21), there would be an unconditional transfer to statement 17.

A logical IF statement has been added and it is of the form IF(e)s where e is a logical expression and s is any executable statement except a DO statement or another logical IF statement. If the value of e is *true*, then s is executed; if e is false, then s is executed as if it were a CONTINUE statement.

Further flexibility on transfers back into a DO nest are permitted in FORTRAN. For details, see Section 7.1.2.8 of the FORTRAN standard [AA66].

There are more functions provided in the library.

One of the primary additions to FORTRAN is the *type* statements, which are of the form $tv_1, v_2, \ldots, v_n$ where t is one of the following: INTEGER, REAL, DOUBLE PRECISION, COMPLEX, or LOGICAL and each v is the name of a variable, an array, a function, or an array declaration. Most significant of all, type statements can be used to override the implicit typing (i.e., variables beginning with the letters I, J, K, L, M, or N are

If v Type Is	And e Type Is	The Assignment Rule Is
Integer	Integer	Assign
Integer	Real	Fix & assign
Integer	Double precision	Fix & assign
Integer	Complex	P
Real	Integer	Float & assign
Real	Real	Assign
Real	Double precision	DP evaluate & real assign
Real	Complex	P
Double precision	Integer	DP float & assign
Double precision	Real	DP evaluate & assign
Double precision	Double precision	Assign
Double precision	Complex	P
Complex	Integer	P
Complex	Real	P
Complex	Double precision	P
Complex	Complex	Assign

1. P means prohibited combination.
2. Assign means transmit the resulting value, without change, to the entity.
3. Real Assign means transmit to the entity as much precision of the most significant part of the resulting value as a real datum can contain.
4. DP Evaluate means evaluate the expression according to the rules of 6.1 (or any more precise rules) then DP Float.
5. Fix means truncate any fractional part of the result and transform that value to the form of an integer datum.
6. Float means transform the value to the form of a real datum.
7. DP Float means transform the value to the form of a double precision datum, retaining in the process as much of the precision of the value as a double precision datum can contain.

Figure IV-9. Rules for assignment of e to v in FORTRAN.
Source: [AA66], p. 13.

integers, while the others are real). The array declaration is of the form v(i), where (i) itself represents the subscript and can be composed of one, two, or three expressions. Thus the type statement can also include the DIMENSION information, e.g., LOGICAL I (3, 5). The DIMENSION statement (or the equivalent information which can appear in a type statement or in a COMMON statement) is more flexible. An array with an integer variable name can appear in a subroutine; the variable names are called *adjustable dimensions*. The dummy argument list of the subroutine must contain these two items. Values of the actual dimensions must be defined prior to calling the subroutine and cannot be redefined during the execution of the subroutine.

The COMMON statement permits the naming of different blocks and is of the form COMMON/$x_1/a_1/.../x_n/a_n$ where each x_i is a symbolic name or empty and each a_i is a nonempty list of variable names, array names, or array declarations.

An EXTERNAL statement is of the form EXTERNAL $v_1, v_2, \ldots, v_n$ where each v_i is an EXTERNAL procedure name. The basic purpose of an EXTERNAL procedure is to write in languages other than that of the standard. If an EXTERNAL procedure name is used as an argument to another EXTERNAL procedure, it must appear in an EXTERNAL statement program unit in which it is so used.

A FUNCTION declaration is of the form t FUNCTION f $(a_1, a_2, \ldots, a_n)$ where t is one of the type declarations.

A data initialization statement has been provided and is of the form DATA $k_1/d_1/, k_2/d_2/, \ldots, k_n/d_n$ where each k_i is a list containing data names and each d_i is a list of constants, any of which may be preceded by $j*$, where j is an integer constant and means that the constant is to be specified j times. The purpose of the data initialization statement is to define initial values of variables or array elements. There must be a one-to-one correspondence between the list of specified items and the constants.

A statement of the form BLOCK DATA can appear as the first statement of subroutines that are called *block data subroutines* and that are used to enter initial values into elements of labeled common blocks. This special subroutine contains only type statements and EQUIVALENCE, DATA, DIMENSION, and COMMON statements. If any entity of a given common block is being given an initial value in such a subroutine, a complete set of specification statements for the entire block must be included, even though some of the elements of the block do not appear in DATA statements. Initial values may be entered into more than one block in a single subroutine.

The following list is taken directly from Appendix C of USA Standard FORTRAN [AA66];[10] it summarizes the principal differences between the two Standards:

USA Standard Basic FORTRAN (as compared to USA Standard FORTRAN) has:

1. A maximum of five continuation cards (instead of 19 continuation cards).
2. A maximum of five characters in a symbolic name (rather than six).
3. Neither logical type, logical nor relational expressions, logical IF statement, nor "L" format descriptor.
4. No "$" in its character set.
5. Neither complex type, double precision type, type-statement,

---

[10] USA Standard FORTRAN [AA66], p. 35.

double precision and complex constants and expressions, nor "D" and "G" format descriptors.
6. No EXTERNAL statement.
7. No 3-dimensional arrays, subscripts.
8. A prohibition on FUNCTION subprograms, in that they may not define nor redefine any of their arguments nor any entity in common.
9. No array declarator permitted in a COMMON statement.
10. No labeled common blocks.
11. No ASSIGN nor assigned GO TO statements.
12. No DATA statement nor BLOCK DATA programs.
13. A maximum of four (rather than five) octal digits in the PAUSE statement.
14. No print carriage control for formatted output records.
15. No Hollerith datum nor the "A" format descriptor and therefore no FORMAT can be read in during execution.
16. No provisions in a FORMAT statement for (a) scale factor, (b) data exponent on input for "F" descriptor, (c) second level parentheses.
17. A restriction on external functions that they may not alter variables in common or variables associated with common via an EQUIVALENCE statement.
18. A requirement that all DIMENSION statements must precede all COMMON statements, which must in turn precede all EQUIVALENCE statements.
19. A statement label may contain only 4 digits rather than 5.

IV.3.5. SIGNIFICANT CONTRIBUTION TO TECHNOLOGY

FORTRAN has probably had more significant impact on computing than any other single development. However, the most significant contribution made by FORTRAN is its usage rather than its technology. Because it was designed so early, better ways have been found to do almost everything that is currently in FORTRAN. Lest this be considered too cynical an attitude, the most important technological contributions seem to be (1) the development of a language which could be used on available hardware, (2) the use of *EQUIVALENCE* statements to give the programmer some control over storage allocation, (3) the nondependence on blanks (which might also be considered a hindrance), and finally (4) the relative ease of learning the language and its palatability to a large group of people. In addition to these, considerable thought was given to the possibilities of compiler optimization of object code, and the language design showed it. For example, the inclusion of the *FREQUENCY* statement to facilitate flow analysis, the fixed limitation on the number of subscripts, and the various concerns with storage allocation all contributed to potential compiler efficiency. The fact that the first two of these examples became unimportant is *not* a reflection on the value of the initial technological contribution.

IV.3.6. SIGNIFICANT EXTENSIONS OF FORTRAN

Because FORTRAN is the most widely used language, it is not surprising that people have extended it for use in areas other than that for which it was originally intended. Some of these extensions are minor in concept and character, while others are far-reaching in both practical usage and implication. It should be emphasized that what is being considered here are specific extensions of the *language* and not of facilities which can be handled through the use of subroutines (which included things as different as character handling, list processing, recursion, and packages for doing data processing. Specific citations for these were given in Section IV.3.3).

The only items being included here are actual language extensions to FORTRAN. There are surprisingly few of these, primarily because most of the *additions* were made as subroutine packages. In some of these cases the extensions were implemented by writing a preprocessor to translate the additional statements to FORTRAN (e.g., proposal writing language and FORMAC); while in other cases new translators have been written (e.g., QUIKTRAN).

1. *Proposal Writing Language*

A rather unusual extension to FORTRAN is the proposal writing language developed by Carleton, Lego, and Suarez [CT64]. They have added the 12 statements shown in Fig. IV-10 to FORTRAN II as it existed for the IBM 704 in 1959. All FORTRAN statements are used, except for the *STOP* which is modified somewhat to control the termination of the proposal writing process. The system is implemented by a preprocessor which converts all new statements to *CALL*s to appropriate subroutines.

Many of the statements are self-explanatory, in particular *LEFT MARGIN, RIGHT MARGIN, TABULATE, SINGLE SPACE, DOUBLE SPACE, RESTORE PAPER,* and *END PARAGRAPH*. The meanings of the others are as follows:

ALPHABETIC INPUT causes the computer to read from card format, searching for as many alphabetic variables as are named in the list. The *m characters of Hollerith information* is interpreted by having the characters within the set of parentheses inserted into the text immediately following the material previously prepared for printing. A similar result occurs when using *ALPHABETIC INSERT*, except that in this case the statement provides the names of variables whose values are inserted into the text. The *NUMERIC INSERT* causes the insertion of the value of the

```
ALPHABETIC INPUT list of alphabetic variable names
RIGHT MARGIN m
LEFT MARGIN m
TABULATE m
SINGLE SPACE
DOUBLE SPACE
RESTORE PAPER
ALPHABETIC INSERT alphabetic variable
   (m characters of Hollerith information)
NUMERIC INSERT FORTRAN variable, constant or expression,
           Conversion type, n FIGURES, COLUMN m
PARAGRAPH n, list of variables
END PARAGRAPH
PREPARE PARAGRAPH n, list of arguments
STOP
```

Figure IV-10. List of statements in Proposal Writing Language. Source: Extracts from Carleton, Lego, and Suarez [CT64].

FORTRAN variable, constant, or expression into the current line of the proposal. The numbers can be converted to integers or to the FORTRAN E type floating point format. The optional clause *n* **FIGURES** specifies the number of digits desired to the right of the decimal point and the *COLUMN* can be used to line up a decimal point under previous numbers.

The *PARAGRAPH* statement allows the user to define a subprogram; the latter is generally used to produce a single paragraph of text describing individual items such as motors. The *n* is simply the identification number for the subprogram, and the *list of variables* provides the parameters. However, information cannot be passed back to the main program from a paragraph subprogram.

The *PREPARE PARAGRAPH* serves to invoke the subprogram and add its text to the body of the proposal, with the appropriate substitutions made for the parameters.

The *STOP* statement initiates a completion phase of the proposal writing system.

A simple illustration of a program written in this language is given in Fig. IV-11.

2. *FORMAC* (cross-reference only)

A significant extension of FORTRAN to do formal algebraic manipulation on the computer is FORMAC. It is described in detail in Section VII.3.

```
ALPHABETIC INPUT DESCPT, INPUT3, BETA
RIGHT MARGIN 72
LEFT MARGIN 15
TABULATE 40
SINGLE SPACE
DOUBLE SPACE
RESTORE PAPER
(DC, SERIES WOUND)
ALPHABETIC INSERT DESCPT
NUMERIC INSERT HP, DECIMAL, 2 FIGURES, COLUMN 14
PARAGRAPH 7, ID, COMNT, XC
END PARAGRAPH
PREPARE PARAGRAPH 7, ITEMNO+1, DESCPT, X+Y
STOP
```

Figure IV-11. Program in Proposal Writing Language. Source: Carleton, Lego, and Suarez [CT64], p. 460.

3. *QUIKTRAN* (*cross-reference only*)

An on-line version of FORTRAN was developed, initially for the IBM 7040. The primary extensions were in the area of the control functions for the time-sharing system and debugging facilities for the user. The overall features of the system and the debugging facilities are described in Section IV.6.3.

4. *GRAF* (*cross-reference only*)

An extension of FORTRAN to handle graphics is defined in GRAF. A new data type called a *display variable* is added, and various commands and declarations are defined which apply to this. GRAF is described in Chapter IX.3.3.1.

5. *DSL/90* (*cross-reference only*)

An extension of FORTRAN to simulate block diagrams is defined in DSL/90. See Section IX.2.4.5.

IV.4. ALGOL

IV.4.1. HISTORY

1. *ALGOL 58*

There are two major nontechnical *firsts* contributed to the computing community by ALGOL: (1) It was the first major language to be designed by a committee of people from different organizations, and (2) this committee was in fact international. (The PACT system was designed by an intercompany committee, but it is not considered either major or up to the level of the programming languages in this book. References are given in Chapter I.)

The following is a description of the creation of what was originally called IAL (*I*nternational *A*lgebraic *L*anguage) and subsequently became known as ALGOL 58 (*ALGO*rithmic *L*anguage). It is quoted directly from the report by Perlis and Samelson [PR58].[11] Although this report was defined as preliminary and the committee anticipated preparing a more complete description of the language for publication, they never did (until the ALGOL 60 report). However, the ALGOL 58 (nee IAL) report spawned a number of significant languages and implementation technique developments.

> In 1955, as a result of the Darmstadt meeting on electronic computers, the GAMM (association for applied mathematics and mechanics), Germany, set up a committee on programming (Programmierungsausschuss). Later a subcommittee began to work on formula translation and on the construction of a translator, and a considerable amount of work was done in this direction.
>
> A conference attended by representatives of the USE, SHARE, and DUO organizations and the Association for Computing Machinery (ACM) was held in Los Angeles on 9 and 10 May 1957 for the purpose of examining ways and means for facilitating exchange of all types of computing information. Among other things, these conferees felt that a single universal computer language would be very desirable. Indeed, the successful exchange of programs within various organizations such as USE and SHARE had proved to be very valuable to computer installations. They accordingly recommended that the ACM appoint a committee to study and recommend action toward a universal programming language.
>
> By Oct 1957 the GAMM group, aware of the existence of many programming languages, concluded that rather than present still another formula language, an effort should be made toward unification. Consequently, on 19 Oct 1957, a letter was written to Prof. John W. Carr III, president of the ACM. The letter suggested that a joint conference of representatives of the GAMM and ACM be held in order to fix upon a common formula language in the form of a recommendation.
>
> An ACM Ad-Hoc committee was then established by Dr. Carr, which represented computer users, computer manufacturers, and universities. This committee held three meetings starting on 24 Jan 1958 and discussed many technical details of programming language. The language that evolved from these meetings was oriented more towards problem language than toward computer language and was based on several existing programming systems. On 18 April 1958 the committee appointed a subcommittee to prepare a report giving the technical specifications of a proposed language.
>
> A comparison of the ACM committee proposal with a similar proposal prepared by the GAMM group (presented at the above-mentioned

---

[11] Perlis and Samelson [PR58], pp. 8–9. By permission of Association for Computing Machinery, Inc.

ACM Ad-Hoc committee meeting of 18 April 1958) indicated many common features. Indeed, the GAMM group had planned on its own initiative to use English words wherever needed. The GAMM proposal represented a great deal of work in its planning and the proposed language was expected to find wide acceptance. On the other hand the ACM proposal was based on experience with several successful, working problem oriented languages.

Both the GAMM and ACM committees felt that because of the similarities of their proposals there was an excellent opportunity for arriving at a unified language. They felt that a joint working session would be very profitable and accordingly arranged for a conference in Switzerland to be attended by four members from the GAMM group and four members from the ACM committee. The meeting was held in Zurich, Switzerland, from 27 May to 2 June 1958 and attended by F. L. Bauer, H. Bottenbruch, H. Rutishauser, and K. Samelson from the GAMM committee and by J. W. Backus, C. Katz, A. J. Perlis, and J. H. Wegstein for the ACM Committee.

It was agreed that the contents of the two proposals should form the agenda of the meeting, and the following objectives were agreed upon:

I. The new language should be as close as possible to standard mathematical notation and be readable with little further explanation.
II. It should be possible to use it for the description of computing processing in publications.
III. The new language should be mechanically translatable into machine programs.

Although not stated explicitly in the report, it was intended by the designers that the language should be a standard. The publication of the IAL report created a significant stir in the computing community. A number of groups in Europe started to implement it and found a number of omissions, ambiguities, etc. A few implementations were started in the United States but only one really became successful and widely used—namely the Burroughs version for the 220, known as BALGOL [QG61]. (See comments on usage in Forest [FS61].) A large series of dialects and derivatives began to spring up, e.g., CLIP, JOVIAL, MAD, and NELIAC; these are discussed in other sections. Springer-Verlag (a publishing firm) announced a plan to publish a series of books on numerical computation in which all the algorithms would be written in ALGOL. (However, they did not release anything until the fall of 1967, and that volume dealt only with the definition and translation of ALGOL 60.) SHARE announced support of IAL (ALGOL) and formed a working committee.

Articles and correspondence by individuals and committees appeared, indicating areas of difficulty. See, e.g., Green [GT59], Kanner [KF59], Irons and Acton [IR59], and the SHARE committee report [CC59]. In March, 1959 the first issue of the *ALGOL Bulletin* was issued at Regnecentralen,

Copenhagen, with Peter Naur as the editor. The impetus for this bulletin came from a meeting held in Copenhagen in November, 1958, where about forty interested people from several European countries held an informal meeting to discuss implementation. A group was formed to implement ALGOL for several machines, with agreement to be reached on everything down to and including the paper tape code used; this later became known as the ALCOR (*AL*gol *CO*nverte*R*) group. The *ALGOL Bulletin* was initially used for communication primarily by Europeans, while Americans sent their comments to the *ACM Communications* for publication. This bulletin continues, although it stopped for 2 years after Naur's resignation; it was revived in 1964 by IFIP with Fraser Duncan as editor. The *ALGOL Bulletin* serves as a very effective means of communication among people strongly interested in ALGOL.

Among the more intriguing technical features of ALGOL 58 were its essential simplicity; the introduction of the concept of three levels of language, namely a reference language, a publication language, and hardware representations; the **begin** ... **end** delimiters for creating a single (compound) statement from simpler ones; the flexibility of the procedure declaration and the **do** statement for copying procedures with data name replacement allowed; and the provision for empty parameter positions in procedure declarations. While ALGOL 58 is not an exact subset of ALGOL 60, the only items of significance which are in the former but not the latter are the **do** which was removed as a concept (although the word was used for something else) and the empty parameter positions. Because of this major carry-over, specific technical description of ALGOL 58 is not necessary.

The avowed purpose of ALGOL 58 was "to describe computational processes". For this reason, there were no input/output facilities provided, and this situation was not remedied in the ALGOL 60 report.

2. *ALGOL 60*

At the UNESCO sponsored International Conference on Information Processing held in Paris in June, 1959, several noteworthy events occurred. There was an open discussion of the weaknesses of ALGOL, and the now famous paper by John Backus [BS60] appeared. This paper presented a formal method of defining syntax—later referred to as Backus Normal Form (BNF)— and gave the proposed definition for ALGOL, using this technique. Although the paper created some excitement, its full significance for language definition and interface point with computational linguistics became obvious only later. That paper marked the beginning of a more rigorous approach to programming languages and is one of the major landmarks in the field.

It was agreed that there should be an international meeting in January, 1960 for improving the language and preparing a final report. At a European

ALGOL conference in Paris in November, 1959, the following people were selected to attend the January, 1960 conference: F. L. Bauer, P. Naur, H. Rutishauser, K. Samelson, B. Vauquois, A. van Wijngaarden, and M. Woodger. They represented the following organizations: Association Francaise de Calcul, British Computer Society, Gesellschaft für Angewandte Mathematik und Mechanik, and Nederlands Rekenmachine Genootschap. The seven individuals held a final preparatory meeting at Mainz in December, 1959.

In the United States, the ACM Committee on Programming Languages met in November, 1959 to consider all the comments on ALGOL which had been sent to the *ACM Communications*. See for example the items cited earlier and those given in [CC61a]. The following representatives were selected to attend the January, 1960 conference and they held a final preparatory meeting in Boston in December, 1959: J. W. Backus, J. Green, C. Katz, J. McCarthy, A. J. Perlis, J. H. Wegstein, and W. Turanski (killed just prior to the January, 1960 conference).

The 13 representatives from Denmark, England, France, Germany, Holland, Switzerland, and the United States met in Paris from January 11 to 16, 1960. Prior to that meeting, P. Naur had prepared a completely new draft report from (1) the preliminary report, (2) recommendations, and (3) Backus' notation. The committee strove for agreement on each item of the draft report, and the "Report on the Algorithmic Language ALGOL 60" [NA60] "represents the union of the Committee's concepts and the intersection of its agreements."[12]

Thus, IAL became ALGOL, which became ALGOL 58, and eventually disappeared into ALGOL 60, which was a substantial improvement over ALGOL 58, although still retaining much of the flavor of the original version.

In February, 1960 an "Algorithms" section appeared in the *ACM Communications*. For a few months the ALGOL 58 language was used, but from the time of the issuance of the ALGOL 60 report that language was used. Although FORTRAN has been much more widely used in the United States than ALGOL, it was not until September, 1966 that algorithms written in FORTRAN were considered acceptable for publication in this section. (By the end of 1967 no FORTRAN algorithms had been published.) The section itself has proved to be a useful catalog of a number of procedures and techniques, many of which have been tried on computers and certified as being correct (or had the errors indicated). In 1966 the ACM collected them and issued a notebook [AC66] which is updated periodically. It is interesting to note that in many cases the only practical method of computer checking available was to write a FORTRAN program to correspond to the ALGOL algorithm and check that; this situation occurred (and still does)

---

[12] Naur [NA60], p. 299.

because relatively few installations in the United States have ALGOL compilers.

3. *Revised ALGOL 60*

During the early part of 1960, an informal working group on ALGOL existed in the United States primarily for the purpose of discussing implementation techniques. This group was greatly enlarged and organized itself as an ACM ALGOL Maintenance Group. This committee contained about 60 members from 28 organizations and concerned itself with interpretation and the philosophical (and technical) issues of changes to the ALGOL 60 report. Most of its work was done by mail, and the results were also communicated through the *ALGOL Bulletin*.

At the same time, considerable work on implementation was being done in Europe. This served to highlight ambiguities and develop new implementation techniques. Few compilers attempted to implement the entire language; in fact, the battle as to who had implemented more, or at least more efficiently, became a favorite game of ALGOLers.

In January, 1962 a rather detailed questionnaire was included in *ALGOL Bulletin* No. 14. Its purpose was to solicit opinions on a number of technical ambiguities and also on the philosophy of specific proposed extensions and subsets.

On April 2 to 3, 1962, the following authors of the ALGOL 60 report were present at a meeting in Rome: F. L. Bauer, J. Green, C. Katz, R. Kogon (representing J. W. Backus), P. Naur, K. Samelson, J. H. Wegstein, A. van Wijngaarden, and M. Woodger. Also present were W. L. van der Poel (as an observer) and the following people as advisors: R. Franciotti, P. Z. Ingerman, P. Landin, M. Paul, G. Seegmüller, and R. E. Utman. The purpose of the meeting was to correct known errors, attempt to eliminate apparent ambiguities, and to provide other needed clarification of the ALGOL 60 report. There was no consideration of extensions. The results of the questionnaire in *ALGOL Bulletin* No. 14 were used as a guide.

There were two main results from this meeting: First, the issuance of "Revised Report on the Algorithmic Language ALGOL 60" (Naur [NA63]) and, secondly, the transferal by the authors of any collective responsibility they might have had with respect to the development, specification, and refinement of the ALGOL language to the IFIP Working Group on ALGOL (WG 2.1). The revised (i.e., Rome) report was reviewed by IFIP TC 2 (Technical Committee on Programming Languages) in August, 1962 and was approved by the IFIP Council. The (Rome) report itself completely incorporated the ALGOL 60 Report, with only an editor's footnote to indicate the places in which the new report differed from the 1960 one. In addition, the Rome report contained a brief description of the April, 1962 conference (edited by M. Woodger), indicating that there were still five areas which required further study.

The *ALGOL Bulletin* continues to be used for discussions of ALGOL 60 (revised), although the attention of WG 2.1 has been devoted more toward various extensions of ALGOL.

4. ALGOL 6X

For years the ALGOL community has heard about the possibility of a new version of ALGOL, presumably to be issued during the 1960's and hence known as ALGOL 6X. In the spring of 1968, a draft report describing ALGOL 68 (van Wijngaarden [VW68]) was issued. Its short and long range fate are unknown.

IV.4.2. Functional Characteristics of Revised ALGOL 60— Proposed ISO Standard

ALGOL is a moderately general language with as much succinctness and similarity with mathematics as is reasonable. It is internally consistent and has a general "cleanliness." It is easy to read and write (except for some of the very subtle points). It is easy to learn and is not particularly error prone.

SAMPLE PROGRAM—ALGOL 60

Problem: Construct a subroutine with parameters A and B such that A and B are integers and $2 < A < B$. For every odd integer K with $A \leq K \leq B$, compute $f(K) = (3K + \sin(K))^{1/2}$ if K is a prime, and $f(K) = (4K + \cos(K))^{1/2}$ if K is not a prime. For each K, print K, the value of $f(K)$, and the word PRIME or NONPRIME as the case may be.

Assume there exists a subroutine or function PRIME (K) which determines whether or not K is a prime, and assume that library routines for square root, sine, and cosine are available.

Program:
```
        procedure problem (a, b);
            value a, b; integer a, b;
            begin integer k; real e;
                for k := 2 × (a ÷ 2) + 1 step 2 until b do
                begin
                    e := if prime (k) then sqrt(3 × k + sin(k))
                                      else sqrt(4 × k + cos(k));
                    if prime (k) then putlist(k, e, ' prime')
                                 else putlist(k, e, 'nonprime')
                end
        end
```

The stated purpose of the language is to describe computational processes. It has also turned out to be very useful for teaching an introduction to the concepts of computational processes. It is definitely a problem-solving language and uses the concept of all three language types: Publication, reference, and hardware. In fact ALGOL was the first language to introduce this trichotomy and the only one in this book to use these distinctions. Since the primary objective of the language is to state algorithms, somewhat less attention was paid to the definition of the proposed user; any person with a computational process to describe who wishes a computer-oriented higher level language can use ALGOL with varying degrees of effectiveness. It is definitely designed for use in a batch environment.

ALGOL provides a significant amount of compatibility through the use of its reference language. Conversions to a particular machine tend to be completely incompatible because of differences in the hardware representation. There is also a problem about arithmetic precision because no specifications are given relative to the amount to be carried in any computation, and there is no provision for double-precision arithmetic. Like any other higher level language, ALGOL is compiler dependent for those areas in which the language specifications are somewhat unclear and also for those areas in which the result is left undefined in the language specifications. However, the user should only employ these at his own risk.

ALGOL has probably spawned more major outgrowths and fewer major dialects than any other language. This is partly because some of the "dialect deviations" really show up in the hardware representation and partly because of the historical development described earlier. There were several significant outgrowths from ALGOL 58 (primarily MAD, NELIAC, and JOVIAL which are described elsewhere). It is my firm contention that these are *not* dialects of ALGOL in any reasonable meaning of the word. They were motivated by ALGOL 58 but differ so markedly from it (and of course from ALGOL 60) that they should not be called dialects any more than PL/I is a dialect of ALGOL 60. There has been amazingly little of the minor kind of dialect, and the dialects have been caused almost entirely because of the problem of hardware representation.

It is in the area of subsetting prior to the existence of the proposed standard that the major differences exist. Virtually no two implementations handled the same subset of the language. The criterion for excluding features usually was based on machine limitations or compiler design, which makes certain features hard to handle, or the desire for great efficiency of either object code or compilation time. The varying subsets in general were not nested. The major features that have been omitted from a significant number of compilers are recursive procedures, integer labels, and *own* variables. These subsets also have extensions for input/output, most differing from each other.

The four major subsets which have been defined over a period of time are SMALGOL, the ALCOR, IFIP, and ECMA subsets. These are defined, respectively, in [CC61b], [CC63a, CC64d], [CC64c], and [CC63b]. There are three subsets defined in the ISO standard: Levels 1 and 2 are the ECMA subsets with and without recursion, respectively, while level 3 is the IFIP subset. Each subset is wholly contained within the one of next higher number.

Some early ALGOL compilers have been partially bootstrapped, but it would be logically impossible to do this completely because of the lack of input/output facilities in the language. In some cases, procedures for character handling and input/output were coded in machine language and these were used with ALGOL to accomplish bootstrapping.

Because of the essential simplicity of ALGOL, and more importantly because of the formalism of the syntactic definitions, there has been less of a problem of compatibility based on incomplete language definition than in some other languages. This does not mean that the language interpretation problem is nonexistent, as a wide variety of correspondence in the *ALGOL Bulletin* and elsewhere proves otherwise.

The problem of converting an ALGOL program from one machine to another is basically a problem in different hardware representations, as well as normal difficulties accruing from machine and compiler differences. There have been no significant attempts at translating ALGOL programs to another language, except by hand to FORTRAN primarily for the purpose of checking out algorithms.

Because ALGOL was created by an international committee and, in fact, has received wider usage and attention in Europe than in the United States, it is only natural that the standardization effort in the United States would subordinate itself to that of the international organization, ISO. The basic policy adopted by X3.4 was to wait until an international standard was developed and then to see whether or not this was appropriate as an American standard. However, the X3.4.2 (and then X3.4.8) subcommittees of X3.4 contributed to the ISO standard in several ways. They proposed solutions of the issues left unresolved by the Rome meeting and took strong positions on the necessity for a subset and for input/output. The ACM Programming Languages Committee sponsored the committee to produce a specific proposal for standard input/output (see [CC64a]). This was adopted, along with the independent IFIP proposal [CC64b]. The international standard was accepted for most practical purposes in October, 1965, but its final official approval has been delayed by administrative problems and errors.

The designers of the language were indicated in the historical section. The essential sponsorship for this effort came from professional computer societies in the United States and Europe. This contrasted with COBOL, where the development was by an American committee, heavily dominated by computer manufacturers initially, under essentially Department of Defense sponsorship. As stated earlier, the basic objective of ALGOL was to allow the specification of computational processes. For this reason, very little concern was given to the problem of inputting the language directly to a computer. The greatest concession to this problem was in the recognition of the concept of a hardware language. However, a direct translation of the reference language character set was far beyond any existing or even proposed equipment available at the time that the language was defined. Hence various techniques had to be devised to permit the definition of a hardware language on punched card and paper tape equipment.

There have been a large number of implementations of significant and small subsets of the language. No up-to-date list of these is available. (See [CC63] for an early list.) By the end of 1967 there was no implementation of the standard known to me. The maintenance of the language has been traced in the historical development and currently resides with IFIP Working Group 2.1.

It is in the area of the technical definition of the language that ALGOL shines. This is the first language in which the syntax was defined with a formal notation, and this has given rise to a number of very significant developments, in particular the syntax-directed compilers. While many people interested in the problems of language definitions thoroughly appreciated the formal notation used in the ALGOL report, a significantly large class of people who were only concerned with using the language found the notation difficult to read. Thus, the very value of the formal definition contributed to a lesser usage of the language simply because it discouraged people. This unfortunate situation is being remedied over a period of time. It must be noted that even with the formalism of the syntax, there were a few inconsistencies, errors, and ambiguities, and of course the semantics were no better defined for ALGOL than for any other language. Interestingly enough, no *formal* definition of a *program* appeared in the 1960 version, although an *informal* definition was given. This omission was corrected in the revised report.

There are five major types of documentation that have existed. The first is the set of official reports (Naur [NA60] and [NA63]). The second is the *ALGOL Bulletin*, which is completely informal and extremely valuable. The third are various attempts at pointing out ambiguities and/or clarifying them; see some of the references at the end of the chapter. Fourth, there have been a number of proposals for extensions and/or changes, e.g., Strachey and Wilkes [SQ61], Haynam [HN65] and several by Wirth. The last two categories appear primarily in the *ALGOL Bulletin* and/or *ACM Communications*. Finally, a number of descriptive or tutorial articles have appeared, e.g., Schwarz [QN62] and Bottenbruch [BH62], as well as several books, e.g., Baumann *et al.* [BN64]. A general discussion of ALGOL documentation appears in Naur [NA63a]. A number of references not specifically cited in the text—although by no means a complete list—is given at end of the chapter.

In my opinion, the evaluation of ALGOL has taken somewhat the opposite course from many other languages. The problem that exists in many other situations is that the faults of the compiler are blamed on the language. In the case of ALGOL, very often faults of the language are actually blamed on the compiler. Praise for the language has been much higher than praise for the compiler, although the latter has generally tended to be fairly good, particularly as better implementation techniques were gradually developed.

ALGOL was defined as being primarily useful in the area of numerical mathematics and certain logical processes and it has certainly proved its use in these areas, but not outside these areas. Since the phrase *logical processes* covers a number of widely different areas, e.g., sorting and compiling, ALGOL has actually been used to specify solutions to many differing types of problems. However, because the input/output facilities were not defined for many years after the basic language, ALGOL in its pure definition was not a computer usable language.

The primary advantage to ALGOL seems to be its universality and the effectiveness for stating a very wide class of algorithms for numerical mathematics and for some logical processes (e.g., sorting). It has also proved valuable as a method of teaching basic computing processes. The primary disadvantages are its lack of ability to handle alphanumeric data or complicated data structures and the (at least initial) lack of input/output specifications. It was originally argued by devotees of ALGOL that the programmer is free to write his own input/output procedures; while this is true, he is also free to write in assembly language all the other features that might happen to be in ALGOL and so this seems to be a very specious argument. The primary mistakes to be avoided in the future are in fact the two disadvantages just cited, namely the lack of adequate data-handling facilities and the lack of specified input/output. The second of these has already been corrected in the ISO proposed standard and the former will probably be corrected in a later version of ALGOL (presumably ALGOL 68). A further facility will need to be added to carry out list and string-handling processes.

IV.4.3. Technical Features of Revised ALGOL 60— Proposed ISO Standard

The language which is being defined here is the proposed ISO Standard [IO65], of which official copies were not generally available at the time of this writing. Since [IO65] is itself primarily based on, or a concatenation of, many documents, primarily Naur [NA63], [CC63b], [CC64a], [CC64b], and [CC64c], the basic information is actually available. Since ALGOL is conceptually composed of the three languages—reference, publication, and hardware, that which is being described here is the reference language.

The character set is composed of the 52 upper-and lower-case letters, the 10 digits, the logical values true (**true**) and false (**false**), and the following characters, which are called delimiters in ALGOL:

$$
\begin{array}{llllll}
+ & - & \times & / & \div & \uparrow \\
= & \neq & < & \leq & > & \geq \\
\equiv & \vee & \wedge & \neg & \supset & \\
(&) & [&] & ' & ' & . & , & : & ; & := & {}_{10} & \sqcup
\end{array}
$$

go to	**for**	**own**	**switch**
if	**step**	**Boolean**	**string**
then	**until**	**integer**	**label**
else	**while**	**real**	**value**
begin	**comment**	**array**	**procedure**
end	**do**		

Note that although many of the items look like words, e.g., **step**, **begin**, and **label**, they are really considered single characters. This means that from the point of view of the compiler there is no difference between the single letter **a**, a period . and the (apparent) word **if**; each is considered a single character. (Key words in ALGOL are customarily printed in boldface to emphasize this point.)

Because most items which would intuitively be considered key words are defined as single characters, e.g., **if**, there are no fixed *words* in the language. This causes no trouble in the reference language but it is a source of difficulty in creating a hardware representation and implementing it. The graphic operators for the categories arithmetic, relational, and logical are shown in the first three lines in the list of characters.

Both data names and program unit labels can consist of an unlimited string of letters and/or digits, but the first character must be a letter. As far as reserved words are concerned, there is (only) a strong recommendation that the following identifiers be reserved for the standard mathematical functions; these identifiers would be expressed as procedures and could be used without explicit declaration: *abs*, *sign*, *sqrt*, *sin*, *cos*, *arctan*, *ln*, *exp* and *entier*. A variable can have subscripts; they are written within square brackets and are separated by commas. There are no restrictions on the form of subscripts; in particular, subscripts can be arithmetic expressions and conditional expressions; they can have subscripts; and there is no limit on the number or depth of subscripts allowed, eg., ALPHA [X+2, 3∗YZ[2+I[3+J[2]]]]. Since there is no data structure in ALGOL, there is no qualification needed or permitted.

In normal terminology, there are three kinds of operators (*arithmetic, relational,* and *logical*), although ALGOL defines a fourth called *sequential*. There are actually six arithmetic operators—the normal five plus the division sign, ÷, which has a separate meaning. The relational operators are $<, \leq, =, >, \geq, \neq$. The logical operators are $\supset, \vee, \wedge, \neg, \equiv$. Although ALGOL defines a great many *delimiters* (including what are called operators here), the only ones within the meaning of Chapter III are the **begin**, **end**, and *punctuation symbols*.

Literals are defined as strings which can be any sequence of basic symbols except the delimiters ' and ' which must be properly paired.

Blanks have no significance outside of strings. There are no noise words in ALGOL.

There are several punctuation symbols used (and referred to in ALGOL as *separators*), namely the comma, colon, semicolon, and colon equals, :=, which is considered a single character. The semicolon is used only to separate different statements or declarations from each other (e.g., **integer** x; **real** y; x := y + 2). The comma is used in some cases to separate items in a list. The colon is used to identify statements by preceding them with an identifier and a colon, e.g., **case** 2 : x := 5 and also to separate identifiers from their semantic descriptions in the parameter list of a procedure declaration, e.g., **procedure** name (a, b) result: (y); it also separates the upper and lower bounds of the subscript values in an array declaration, e.g., **array** a, b[5 : 8, 2 : n]. The := is used primarily to designate assignment statements, e.g., a := b + c + 5, and then to apply an assignment concept in a **for** list and a **switch** list.

The physical input form is generally considered to be a continuous string; thus there are no concepts of card columns or images. The primary difficulty is the one which has been mentioned before; namely the hardware representation can be very awkward relative to the actual reference language. This occurs primarily, although not exclusively, from the use of what are really English words as basic symbols. One way of handling the hardware transliteration of these is to provide a specific escape character that is used before every occurrence of one of these basic symbols; another way is to restrict the use of labels and identifiers so as to exclude these words and to use the concept of reserved words in the hardware representation. Some of the specific hardware representations that have been used or advocated are given in [QG66], [CC63a and CC64d], and [IB66i]. The conceptual form of ALGOL is essentially one of simplicity, one of reasonable correspondence to mathematics, and one with as much flexibility as the designers felt was needed for the intended classes of applications.

There are a number of nonexecutable statements in ALGOL, ranging from those which essentially relate to the type of arithmetic to be done with the variable (e.g., **real** and **integer**) to those involving procedures. The complete list of declarations (used in the sense of this book, not necessarily agreeing with ALGOL terminology) permitted is as follows: **own**, **Boolean**, **integer**, **real**, **array**, **switch**, **procedure**, **string**, **label**, and **value**. These declarations apply only to the block in which they appear (and also to blocks included within that block).

The smallest executable unit is a single statement of the form x := y. From this simple statement a complex structure can be built up. Individual statements can be combined into compound statements by enclosing the sequence in the delimiters **begin**, **end**, e.g., begin x := y + z; p := r + 1 **end**. The **begin-end** pair causes the groups of individual statements to be treated as a single statement. A larger type of subunit is a *block* which consists of a series of declarations and statements, again completely enclosed within the **begin-end** pair, e.g., begin real x, y; integer z; x := y **end**. A block is itself a statement, and thus one or more of the statements which constitute a block may themselves be blocks.

Loops in ALGOL can be handled by the conditional statements and by the **for** statement.

ALGOL provides for both functions and procedures, and the procedures are really the backbone of the practical use of the language. Most algorithms are written as procedures so that they can be invoked from other programs. As is standard, the functions are a special kind of procedure, namely one in which there is a single numerical or logical result.

The symbols ;, **begin**, and **end** can be replaced by the following, respectively, to permit writing comments:

 ; **comment** < any sequence not containing ; > ;
 begin comment < any sequence not containing ; > ;
 end < any sequence not containing **end** or ; or **else** >

Comments can therefore also be put into the procedure declaration.

ALGOL does not include any interaction with the operating system or the environment. There is no provision for references to other languages.

A program is a block or a compound statement which is not contained within another statement and which makes no use of other statements not contained within it. Declarations appear in a block immediately after the **begin** symbol and can then be followed by any number of statements until the **end** symbol designates the termination of the block.

A statement in ALGOL is normally ended by the use of a semicolon. A compound statement and a block are both delimited by the symbol **begin** at the beginning and the symbol **end** at the end. Declarations (considered to include the identifiers to which they apply) are delimited at the end by a semicolon. A procedure declaration is preceded by the symbol **procedure** and normally ended by the symbol **end** which refers to the block within the procedure which is accomplishing the desired task.

ALGOL is almost the only (and was the first) language to allow recursive procedures. There is no requirement that the procedure be defined specifically as being recursive; this leads to inefficiency at object time in many implementations because the compiler must prepare to cope with a procedure which might be recursive but in fact is not.

ALGOL introduced the concept of two different types of parameter passage for procedures—*call by name* and *call by value*. It also permits considerable embedding, most notably, **if** statements in arithmetic expressions.

The language allows arithmetic and logical variables and arrays of them. String variables are defined, but there are no operations defined on them. In addition, there are two other elements which ALGOL defines as data types, namely **label** and **switch**.

There are no hardware data units accessible as such in an ALGOL program; this is quite consistent with the concept of machine independence and a reference language. All the variable types can be either declared and/or operated on, except for strings.

Two types of numerical arithmetic are provided in ALGOL, namely integer and real (i.e., floating point.) A special division operation designated by $\div$ is defined only if the variables involved are of type integer. Boolean arithmetic is also provided.

Real and integer numbers can be intermingled, and the result of the arithmetic operations is an integer (only) if the operands are integers.

Boolean expressions can be embedded in arithmetic ones. The normal precedence rules for arithmetic and logical operations apply in the evaluation of arithmetic and Boolean expressions, respectively.

General arithmetic expressions can include **if** clauses in which one out of several simple arithmetic expressions is selected on the basis of the actual values of the Boolean expressions. In this case, the Boolean expressions are evaluated one by one in sequence from the left until the one having the value *true* is found. The value of the arithmetic expression is then the value of the first arithmetic expression following this Boolean expression. Thus **if** Ab < C **then** 17 **else** q [**if** w < 0 **then** 2 **else** n] + r is a meaningful arithmetic expression.

There are two special functions, *sign* and *entier*, which yield integer results. The *sign* (e) equals 1, 0 or −1 if e is greater than 0, equal to 0, or less than 0, respectively. The value *entier* (e) is defined as the largest integer not greater than e. There are no rules given for precision. There are a number of rules relative to the computation in various modes, and in particular there is a real problem relative to side effects if procedures involving call by name are used in expressions.

ALGOL was the first language to introduce and define significant scope rules for data variables. The main unit considered for this purpose is the block. The basic principle is that the data named by an identifier occurring within a block is usually specified to be local to the block and is always local to the block if declared within the block. Thus the (data's) identifier has no existence outside the block, and conversely this identifier can be used elsewhere but is not accessible inside the block. At the time of exit from a block, all local identifiers lose their local significance, and in particular their values are not available at the next reentry to the block. All identifiers except labels and formal parameters of procedure declarations must be declared. If the user desires to retain the meaning and value of an identifier throughout significant portions of a program, this must be declared as an **own** variable, which means that when the block is reentered, the previous values are available. Nonlocal variables are those which are used in a block but declared in a larger (i.e., containing) block. Global variables are those defined only in the outermost block.

The assignment statement in ALGOL can be both single and multiple. Thus A := B := C := D + E; means that the variables A, B, and C, are all assigned the value D + E. It is assumed that in the case of a multiple assignment statement the type associated with all variables and procedure identifiers on the left must be the same. If the type is **Boolean**, the expression must also be **Boolean**; if the type is **real** or **integer**, the expression must be arithmetic. If the types of the arithmetic expressions on the left and right do not match, then the appropriate transfer functions are understood to be automatically invoked to go from the right to the left.

There is no character data handling.

Normal sequence of control is from one statement to the next. There is an unconditional control transfer designated by the single symbol **go to**. It is also possible to have this symbol **go to** followed by a subscripted identifier, e.g., **go to** K[I], where K has been previously identified as a switch by means of a switch declaration. The value of the subscript designates which of the possible labels is chosen. Designational expressions are rules for obtaining statement labels as values and can be written as subscripts. Designational expressions can be **if** statements, e.g.,

goto if A = 0 **then** ALPHA **else** BETA

causes control to transfer to ALPHA if A equals 0 and to BETA otherwise.

Functions are invoked by writing them, with their actual parameters, wherever the value is needed, e.g., in expressions. Procedures are invoked by writing the procedure name, with the actual parameters, where the procedure body is to be executed. The procedure given on p. 191 is invoked by writing

Spur (ALPHA) ORDER: (5) Result to: (ISH)

and the function might be used as follows:

$$y := z + \text{Step(line)} \times \text{Value}$$

A conditional statement in ALGOL can have one of three forms: The simplest is just **if** B **then** U, where B is a Boolean expression and U is an unconditional statement; the second form is **if** B **then** U **else** V, where V is any kind of a statement and in particular can itself be a conditional statement; the third form is **if** B **then** F, where F is a **for** statement which is described below. Any of these conditional statements can have labels. In the latter case it is possible to have something as complicated as the following:

if $s < 0 \lor p > Q$ **then** ALPHA: **begin if** $a > v$ **then** $a := v/s$
                                                                    else $y := a + b$ **end**
    else if $v < s$ **then** $a := v + q$
        else if $v = s - 1$ **then go to** S;
$y := a - b$

In somewhat more general terms, a conditional statement can have the form:

if B_1 **then** S_1 **else if** B_2 **then** S_2 **else if** B_3 **then** S_3 **else if** B_n **then** S_n; S_m

where the B_i are Boolean expressions, the S_i are unconditional statements, and there is no limit on i. The statement is executed by evaluating the Boolean expressions in sequence from left to right until one yielding the value **true** is found. Then the unconditional statement following this Boolean is executed and, unless this statement defines its successor explicitly, the next statement to be executed is (in the example above) S_m, i.e., the statement following the complete conditional statement.

The basic loop control statement in ALGOL is the **for** statement; this consists of a **for** list, followed by a **do**, followed by a statement. The **for** statement can be labeled. The range is the statement following the **do**. The **for** list consists of a single parameter which is to be varied, followed by the list over which it is being varied, possibly followed by more parameters and their values with termination criteria, e.g., a complete loop control statement might be

for $i := 2*j$ **step** 2 **until** 71, $3 + k$ **while** $k = p$ **do** $X[i] := Y[i] + 5$;

The list over which a variable may assume values can be expressed in one of three forms: It can be just a regular arithmetic expression; it can be something of the form A **step** B **until** C where A, B, and C are arithmetic expressions, or finally the form E **while** F, where E is an arithmetic and F is a Boolean expression. The last two of these can be described, respectively, most concisely in terms of additional ALGOL statements as follows:

>   V := A;
> L1: **if** (V − C) × sign(B) > 0 **then go to** *Element exhausted*;
>   *statement* S;
>   V := V + B;
>   **go to** L1;

and

>   L3: V := E;
>   **if** ⌐ F **then go to** *Element exhausted*;
>   *statement* S;
>   **go to** L3;

where in both cases V is the controlled variable, and *Element exhausted* points either to the evaluation using the next element in the **for** list or to the next statement in the program if it is the last element of the list.

There are no provisions in ALGOL for error condition statements. There are also no facilities for handling symbolic data of any kind.

The proposed ISO standard provides for two types of input/output procedures—one is a very small subset of the other. The difference in philosophy is that the user of the small subset is assumed to program most of the things that he wants himself, whereas the larger system should provide virtually everything that is needed.

The small subset assumes that the following primitive procedures are available to the user: *insymbol, outsymbol, length, inreal, outreal, inarray*, and *outarray*. Communication between the external media and the variables of the program is provided by the procedures *insymbol* and *outsymbol*. The appropriate correspondence is established between the basic symbols given in a string parameter of those procedures with the variables given in an integer parameter. The *length* procedure defines the value of the length of the string as equal to the number of basic symbols of the string enclosed between the outermost string quotes. The procedures *inreal* and *outreal* cause the appropriate correspondence to take place between the next value on the external medium and the destination parameter. The procedures *inarray* and *outarray* cause the transferal of the numbers forming the value of the array in the procedure declaration. The elements of the array are transferred in an order which corresponds to the lexicographic order of the values of the subscript.

The major and quite complete input/output facilities given in the proposed ISO standard can be subdivided into two major categories: Formats and input/output procedures. The former bear a strong resemblance in concept, and even in some places in detail, to the picture description in COBOL.

Whenever input or output is done, certain standard operations are assumed to take place unless otherwise specified. These nonstandard operations are specified through the use of a *layout* procedure which provides for horizontal and vertical control; although it is assumed that the output is for a page prepared by a high-speed printer, the concepts are applicable to other devices. The descriptive procedures *format, h end, v end, h lim, tabulation,* and *no data* set one of seven "hidden variables" to a particular value, and the total description is provided by this set.

There are some list procedures which describe the sequence of items which are to be transmitted for input or output. The actual transmission between the external medium and the program variables are handled through the input/output procedures and calls which were mentioned in the abbreviated version. However, these input/output procedures and calls are extended to include the ability to specify layouts and lists as provided in this more elaborate input/output system.

ALGOL has no provision for library references other than through the mechanism of procedures and the "standard" functions. There are no debugging statements in the language, nor are there any direct statements for allocating storage or segmentation. However, because the size of arrays can vary at object time, because of scope problems, and because of the provisions for recursive procedures, the compilers themselves must generate fairly elaborate object time storage allocation facilities. This has generally tended to be the area in which the most experimentation in ALGOL implementation has been done, particularly on small computers where this is critical.

The language contains no provisions for interaction with the operating system or for making effective use of specific machine features; to do this would be definitely contrary to the fundamental spirit of ALGOL, which is to be machine independent and serve as a good communication mechanism for algorithms.

There is no separate data description as such. The data types allowed in the language were described earlier. The actual specification for each type is given as part of the block head in which the variable is to be used. The **array** declaration provides for dimension information on data. The upper and lower bound for each dimension are separated by colons, and the differing dimensions are separated by commas; all this is contained within square brackets, e.g.:

array A[7 : n, 2 : m], B[−2, m + 5]

real array m[**if** c < 0 **then** 2 **else** 3 : 20]

There is no separate file description since files as such are not an allowable entity in ALGOL. There are some specific format descriptions included in the large input/output specifications. There are no declarations in the language specifically about storage allocation, although information of this type is conveyed to the compiler through **array** declarations and is implied in the block structure. There are no separate compiler-directing statements.

In order to give the reader some flavor of formalism used to define ALGOL, the following definition of a procedure declaration is copied from Section 5.4.1 in [IO65] or [NA63], using *their* notation:

⟨formal parameter⟩ ::= ⟨identifier⟩
⟨formal parameter list⟩ ::= ⟨formal parameter⟩
  | ⟨formal parameter list⟩ ⟨parameter delimiter⟩
    ⟨formal parameter⟩
⟨formal parameter part⟩ ::= ⟨empty⟩ | (⟨formal parameter list⟩)
⟨identifier list⟩ ::= ⟨identifier⟩ | ⟨identifier list⟩, ⟨identifier⟩
⟨value part⟩ ::= **value** ⟨identifier list⟩; | ⟨empty⟩
⟨specifier⟩ ::= **string** | ⟨type⟩ | **array** | ⟨type⟩ **array** | **label**
  | **switch** | **procedure** | ⟨type⟩ **procedure**
⟨specification part⟩ ::= ⟨empty⟩ | ⟨specifier⟩ ⟨identifier list⟩;
  | ⟨specification part⟩ ⟨specifier⟩ ⟨identifier list⟩;
⟨procedure heading⟩ ::= ⟨procedure identifier⟩
    ⟨formal parameter part⟩; ⟨value part⟩ ⟨specification part⟩
⟨procedure body⟩ ::= ⟨statement⟩ | ⟨code⟩
⟨procedure declaration⟩ ::= **procedure** ⟨procedure heading⟩
    ⟨procedure body⟩ | ⟨type⟩ **procedure** ⟨procedure heading⟩
    ⟨procedure body⟩

Examples of such declarations (taken from [IO65] or [NA63]) are as follows:

procedure Spur (a) Order: (n) Result: (s); **value** n; **array** a;
integer n; **real** s;
begin   body of procedure   **end**

integer procedure Step (u); **real** u;
  Step := **if** $0 \leq u \wedge u \leq 1$ **then** 1 **else** 0

The procedure body always acts like a block even if it does not have the form of one. A function declaration has the same form as a procedure declaration except that in the body of the former there must be an assignment statement with the procedure identifier on the left.

The **switch** declaration is specifically defined. It consists of a sequence of values (which can be defined by general arithmetic expressions) which are called the *switch list*; e.g.,

switch A := 3, S1, Q[4+m], **if** v=3 **then** S2 **else** 4

A switch is used as the argument in a **go to** statement. Thus, writing

go to A[3] would cause control to be transferred to the switch referenced by Q[4 + m].

ALGOL does not have the facility for modifying itself. There are no provisions for self-extension of the language, although proposals for doing this have been put forth. (See, for example, Galler and Perlis [GA67] and Garwick, Bell, and Krider [GW00].) With regard to writing ALGOL in itself, this was impossible in the strict sense prior to the time of the proposed ISO standard because there was no input/output; in addition, there were no character-handling facilities. However, a number of compilers, or specific algorithms needed in compiling, have been written in ALGOL with the necessary machine language procedures added, e.g., Huskey and Wattenburg [HU61], Stone [TS67].

Relatively little of the language design contributes directly to implementation efficiency. This is truer of the object code than of compilation. Such facilities in the language as integer labels, **own** variables, recursive procedures, and arrays whose size is not known until object time all tend to lower the efficiency of the object program and also to hamper maximum compilation speed because of the need to cope with these things. Thus although ALGOL has great generality in many of its specifications, some of these had to actually be restricted in order to provide reasonable implementations. For example, while the language specifications do not put an upper limit on the size of an identifier, most implementations do; similarly, most implementations must put restrictions on the numbers of variables, and some implementers have chosen to omit some of the features which cause the most difficulty. Fortunately, new techniques which minimize these difficulties are continually being developed.

Like the compilers for any other language, some implementers chose to put in good error checking at compile time, while others ignored or minimized this problem.

ALGOL was defined as being primarily useful in the area of numerical mathematics and certain logical processes, and it has certainly proved its use in these areas, but not outside these areas. Since the phrase *logical processes* covers a number of widely different areas, e.g., sorting and compiling, ALGOL has actually been used to specify solutions to many differing types of problems. However, because the input/output facilities were not defined for many years after the basic language, ALGOL in its pure definition was not a computer usable language.

IV.4.4. SIGNIFICANT CONTRIBUTION TO TECHNOLOGY

ALGOL has made a large number of significant contributions to the technology; among the more important are (1) block structure and defining the scope of variables; (2) formal language definition; (3) recursive procedures; (4) significant embedding capability for differing subunits; (5) a general simplicity combined with power for stating computational processes; (6) concepts of separate reference, publication, and hardware languages; (7) a requirement for the development of better implementation techniques; and (8) spawning a significant number of languages as outgrowths. Although in a somewhat different category, the collection of algorithms written in ALGOL [AC66] is certainly a significant contribution.

The block structure was the first occurrence of the concept of unlimited levels of nesting of executable units. For example, FORTRAN had none, COBOL allowed three, but ALGOL permitted any number. This, of course, led to a series of problems about the scopes of variables which had to be solved.

Probably the greatest contribution accruing from ALGOL is something which is actually not directly related to the language, namely the use of a formal syntactic definition for the language and the publication of a report in that form. This in turn has led to increased studies in methods of language definition and, in a parallel fashion, to the development of syntax-directed compilers.

Recursive procedures were introduced by ALGOL. They certainly should be considered a significant contribution to the technology, but it is not clear how great a one. The advocates of this facility claim that many important problems cannot be solved without it; on the other hand, people continue to solve numerous important problems without it and even in a few cases manage to handle (sometimes in an awkward way) some of the problems which the recursion proponents claim cannot be done.

The fourth point is a general one, namely that ALGOL is a clean language of great power for expressing algorithms to solve a wide class of problems. This is quite different from the practicality of ALGOL as a language for use on a computer. Because of the significant difference between the appearance and usage of the reference and hardware languages and because for many years there were no input/output facilities, the contribution seems to be much more in the area of expressing algorithms than in direct use on a computer.

The introduction of the concept of a reference language with transliterations to publication and hardware languages is considered by me to be quite significant, although it has not actually been used to any real extent by any language development since then. However, the development of a program to go from a hardware representation to the reference language, with the latter on a paper tape for controlling a photocasting device, has been done by von Sydow [VS67] and might prove to be of major significance.

Finally, many of the features in ALGOL forced the development of new or better implementation techniques. Problems such as handling recursive procedures, **own** variables, dynamic storage allocation, and the distinctions between *call by name* and *call by value* have all been solved to varying degrees of efficiency. Furthermore, the implementation of ALGOL on a number of small computers has created a body of useful compilation techniques for this class of machine.

At least as significant as ALGOL itself is the fact that ALGOL 58 engendered enough interest to cause the creation of several other languages. While this is a disadvantage in adding to the proliferation, each of the *direct outgrowths* (JOVIAL, MAD, NELIAC) has itself contributed to the technology. Furthermore, the specific features mentioned in the first paragraph of this section have been at least considered by almost every language designer since 1960. (Of course this does not mean that all the features can or should be put into new languages.)

ALGOL has had an impact on programming language development that far exceeds its direct practical usage in the United States. It has been used far more in Europe than in the United States, but even in Europe it appears that FORTRAN is more widely used for practical problem-solving.

IV.4.5. EXTENSIONS OF ALGOL

Just as some practical language extensions, and even more sets of subroutine packages, have been added to FORTRAN, so there has been continuing discussion of extensions to ALGOL. These discussions have tended to range from "it would be helpful if ALGOL added improvements such as" string handling (e.g., Wegstein and Youden [WG62]), list processing (e.g., Peck [PK67]), and minor improvements (e.g., Strachey and Wilkes [SQ61]), to major full-fledged additions or rewrites, e.g., EULER (Wirth and Weber [WT66a and WT66b]), subcommittee reports, etc., and of course ALGOL X and ALGOL Y. These terms have been used for years by the IFIP ALGOL committees to denote future versions. It is possible that by the end of 1968 a new version of ALGOL, which will include some character-handling facilities, will have been defined and the information officially disseminated.

The major extensions of ALGOL (where the term extension is interpreted very loosely) are cited and cross-referenced to where they are covered in detail. A few minor extensions of ALGOL are described very briefly and/or cross-referenced. Only implemented systems are included.

1. *Formula ALGOL* (*cross-reference only*)

Formula ALGOL is discussed in detail in Section VIII.5. Basically, it adds to ALGOL not only string and list processing facilities, but also the concept of a *formal* variable on which algebraic manipulation can be performed. Languages with only the concept of formal variable type (e.g., FORMAC) are discussed in Chapter VII.

2. *LISP 2 (cross-reference only)*

LISP 2, which is described in detail in Section VIII.6, is a curious anachronism in that its name does not give any indication of its relationship to ALGOL. It is really an ALGOL-like extension, with as much LISP facility at the outer language level as possible. It is the inner level, as well as the philosophy and structure which are heavily LISP-oriented. At an open meeting, I asked one of the active workers on LISP 2 why the system was called LISP 2 instead of having ALGOL in the title. The answer received was that if the name involved ALGOL, the reader or user would expect certain things; whereas if it were LISP, the same person would expect something quite different.

3. *AED (cross-reference only)*

The addition of string and list processing facilities to ALGOL, plus the concepts of plex structures and other facets of the computer-aided design work, is represented in the AED System. It is discussed in Section IX. 3.4.2.

4. *SFD-ALGOL (cross-reference only)*

The SFD-ALGOL (*S*ystem *F*unction *D*escription-ALGOL) is a language which permits the user to describe synchronous systems. It is discussed in Section IX.2.3.7.

5. *SIMULA (cross-reference only)*

SIMULA is an extension to ALGOL to provide facilities for doing discrete simulation. It is discussed in Section IX.3.1.7.

6. *DIAMAG*

The DIAMAG system is an on-line version of ALGOL which adds a number of language elements to permit communication with the time-sharing system. See Auroux, Bellino, and Bolliet [AU67].

7. *GPL*

The GPL (*G*eneral *P*urpose *L*anguage) manual by Garwick, Bell, and Krider [GW00] appeared too late to permit a thorough discussion. The language has some additions to, and deviations from, ALGOL, e.g.,

BYTE and pointer (*PTR*) data types, more general block structure, and user ability to define a *MACRO* (for in-line object code). However, the most interesting aspect of this language is its facility to allow the user to define new operators and new data types.

8. Extended ALGOL

The Extended ALGOL for the Burroughs B5500, as described in [QG66], is quite powerful and has added many useful features to ALGOL. Among the major ones are the following:

1. The ability to access a partial word by giving the beginning and ending bits.
2. A *TIME* function which can yield either the current date, or various elapsed times.
3. A *concatenate* expression which provides an efficient method of forming a primary (or Boolean primary) from selected bits of two or more primaries (or Boolean primaries).
4. A *FILL* statement which permits assignment of specified values to one row of an array.
5. A *DOUBLE* statement which causes assignment of the double-length result of operations on double-length variables.
6. *STREAM* procedures which manipulate words, characters, and bits.
7. Edit and move statements.
8. Specific hardware and efficiency-oriented input/output statements.
9. An *ALPHA*betic data type.
10. A *DEFINE* declaration to permit use of a single identifier for a longer legal program string.
11. *SORT* and *MERGE* statements.

This language appears to have a great many facilities which are useful in business data processing and in character-handling applications, e.g., compiler writing. It is definitely a multipurpose language and has apparently been used to write its own compiler.

IV.5. LANGUAGES MOTIVATED BY ALGOL 58

Although ALGOL 58 itself had only a very short life-span, its importance transcends its actual usage. Aside from its obvious role in the development of ALGOL 60, it motivated the creation of several other languages which have become quite significant. Two of these, namely NELIAC and MAD, are described in this section since their basic purpose, flavor, and general facilities fall primarily in the class of languages for numerical scientific problems. However, JOVIAL is an outgrowth of the CLIP effort, which itself was based on ALGOL 58, and both are far more general. JOVIAL is discussed in Section VIII.3 and CLIP in Section IX.2.5.2.

The role played by ALGOL 58 in the development of each of the languages is described within the discussion of the history. It will presumably be clear, however, that although each of these languages may have started out being based on ALGOL 58, the end results bear little resemblance; if it were not for the historical statements, I doubt whether the reader would recognize any connection with ALGOL 58. This comment is in no way meant to be a criticism of any of these efforts; it merely points out that objectives and achievements are often far from motivations.

IV.5.1. NELIAC

The language known as NELIAC (*N*avy *E*lectronics *L*aboratory *I*nternational *A*LGOL *C*ompiler) was developed concurrently with the creation of ALGOL 58. The work started at the Navy Electronics Laboratory in San Diego, California in the summer of 1958 because NEL was expecting delivery of some large computers for which there were no compilers. Professor Harry Huskey served as a consultant to a NEL group headed by Dr. Maurice Halstead. An attempt was made to follow ALGOL 58 as it was developing, but since the people concerned with NELIAC were anxious to get a system running and could not wait for the official specifications, they put in facilities or syntactic features of their own.

The key references for this language are Halstead's book [HS62], Halstead's article on documentation [HS63], and the article by Huskey, Love, and Wirth [HU63]. Without indicating specific quotation marks except where direct quotes are being made, the reader should assume that the information described here is obtained from one of these three sources.

In July, 1958 the work on the first compiler was started and it was finished within 6 months. Table IV.1 shows Halstead's list of the implementation status in 1962. Since then, at least the following have been developed: A version called the BC NELIAC was implemented on the 7094 and versions on the UNIVAC 1107/1108, CDC 3100, 3600, 3800, Burroughs D825, and IBM System/360. Associated with one or more of these implementations or other aspects of the NELIAC developments were R. Johnson, W. Landen, K. S. Masterson, Jr., R. McArthur, C. B. Porter, S. W. Porter, R. Rempel, R. T. Stelloh, J. B. Watt, and W. Wattenburg. In spite of the large number of versions that exist, there is really no standard definition of NELIAC. There was a NELIAC users conference held in January, 1963 which appointed a committee to develop such a standard, but it never came into being. As a result, and due to the relative ease of writing and documenting NELIAC compilers (about which more will be said later), there are as many versions as there are physical compilers. The language described will be that discussed in Halstead's book [HS62]. Certain improvements to this language were made, however, by Huskey et al. [HU63].

During much of 1962 and part of 1963, there was a great deal of con-

Table IV.1. MACHINES HAVING NELIAC COMPILERS*

Machine	Date Operational	Implementor
RRU M-460	February 1959	NEL
CDC 1604 (Basic)	February 1960	NEL-USNPGS
Burroughs 220	May 1960	NEL
IBM 704	September 1960	NEL-UC
IBM 709	November 1960	NEL-AEPG, Ft. Huachuca
IBM 7090	March 1961	UC-Lockheed
†Philco CXPQ	April 1961	NEL
†PB 250	September 1961	NEL
CDC 1604 (Expanded)	October 1961	NEL
†RW AN/UYK	April 1962	NEL
RRU M-490	May 1962	NEL-NOTS
†Philco BasicPac	May 1962	AEPG, Ft. Huachuca
†CDC 160A	June 1962	NEL
Sylvania Mobidic	In process	AEPG, Ft. Huachuca
NRL NAREC	August 1962	NEL-NRL
†RRU SS80	In process	Frankfort Arsenal
IBM 7070	In process	NOL-NEL
Burroughs 825	In process	NEL

*Halstead [HS63], p. 92. By permission of Association for Computing Machinery, Inc.
†Note—Compilers marked with a dagger are of the intermediate type, running on a larger machine.

troversy centering around NELIAC and its potential competition with JOVIAL. Some comparisons which I feel were not too meaningful were made, and a significant political battle ensued when a decision was made (and carried out) by the Navy to use JOVIAL for some of its command and control problems.

The language is general, with notation that is both formal and succinct. It is easy to learn, read, and write, but its usage appears to me to be error prone because of the large number of specialized punctuation rules.

The language was designed to be used for mathematics and engineering problems, business problems, command and control problems, and systems programming. It is procedure-oriented, problem-oriented, and problem-solving and simultaneously a hardware, reference, and publication language if special typewriters are available. It was meant to be used by all types of programmers, both professional and nonprofessional. It was definitely a batch system but requires typewriters with the defined character set or 029

SAMPLE PROGRAM—NELIAC

Problem: Construct a subroutine with parameters A and B such that A and B are integers and $2 < A < B$. For every odd integer K with $A \leq K \leq B$, compute $f(K) = (3K + \sin(K))^{1/2}$ if K is a prime, and $f(K) = (4K + \cos(K))^{1/2}$ if K is not a prime. For each K, print K, the value of $f(K)$, and the word PRIME or NONPRIME as the case may be.

Assume there exists a subroutine or function PRIME(K) which determines whether or not K is a prime, and assume that library routines for square root, sine, and cosine are available.

Program:
```
SAMPLE NELIAC (A,B,K=0000;F=00000.000):
{A(0→0)=D:A+1→K;A→K;
K=K(2)B{PRIME(K)=1:SQRT(3×K+SINI(K))→F,{PRINTER<K,F<PRIME>>};
                 SQRT(4×K+COSI(K))→F,{PRINTER<K,F<NON−PRIME>>};}}
```

key punches which then require a hardware representation for some of the characters.

The compatibility of NELIAC is something about which one could write an entire book (and to some extent Halstead did). It was meant to be machine independent, particularly since the NELIAC compilers were written in NELIAC except for the initial bootstrapping. However, the ability to allocate variables to particular bits or characters in words makes this impossible. Furthermore, differing input/output facilities made NELIAC obviously dependent upon particular machines. Since the compilers were readily changed, it is not clear if the results obtained on one compiler could necessarily be obtained on the other; there does not appear to be any significant substantive data on this. There were definitely subsets created to permit bootstrapping, but much less was done in the way of providing direct clearly identifiable extensions; most of the latter appeared in the form of removing certain restrictions, which of course can be considered a form of extension. There is no doubt but that a NELIAC program can be very easily converted, if not directly at least with a very small amount of effort, to run on another machine with a NELIAC compiler. NELIAC can be transliterated fairly easily into ALGOL publication language, but in my view its external appearance does not resemble ALGOL 58 very much.

Any consideration of NELIAC compatibility must take into account the basic philosophy that the language is meant to be easy to modify for any particular implementation or application. In Halstead's words, "In short, it is a dynamic language, with compatibility preserved by the ease with which desirable features can be added to any implementation as it becomes useful to do so."[13] While recognizing the validity of such an approach for solving problems, I do not feel that this philosophy addresses itself to compatibility

---

[13] Private communication, 1967.

200 LANGUAGES FOR NUMERICAL SCIENTIFIC PROBLEMS

difficulties in any realistic way, where compatibility is meant to ensure ease of running a program on more than one computer.

As indicated before, there was no official group standardization effort, and the closest thing to such a standard is Halstead's book. Unfortunately that does not contain a complete description of the language, except through the actual listings of one of the compilers. The language was designed by a group at the Naval Electronics Laboratory under the guidance of Huskey and Halstead, with the objectives indicated above. It has been implemented by a number of different people, but the language is not being maintained in an official way. According to Halstead's description, the official documentation for any given compiler is the listing of the NELIAC statements comprising that compiler. The primary source document for the language is Halstead's book [HS62]. Other references are shown in the bibliography at the end of the chapter. One of the interesting facets of the documentation is that Halstead's book serves not only to define the language but also to cover the methods in writing NELIAC compilers as well.

NELIAC has been implemented in varying degrees of completeness on many computers (which were listed earlier). The language has apparently been most effective for relatively small programs but less so for extremely large ones; however, programs containing 100,000 words of object code have been successfully written. NELIAC has proved easy to use and the compilers are fast because of the simplicity of the language.

The character set consists of the 52 upper- and lower-case letters, the 10 digits, and the following 26 characters:

$$
\begin{array}{cccccc}
, & ; & : & . & & \\
(&) & [&] & \{ & \} \\
+ & - & \times & / & & \\
\rightarrow & \uparrow & | & & & \\
= & \neq & < & > & \leq & \geq \\
\cap & \cup & & & & \\
_8 & & & \text{(subscript)} & &
\end{array}
$$

Various hardware representations exist.

There are key words but in most cases they are optional. The graphic operators and punctuation are shown above.

Identifiers must begin with a letter of the alphabet; they can contain only letters, spaces, numbers; they must be uniquely determined within the first 15 characters. Upper- and lower-case letters are interchangeable, i.e., ABC and abc represent the same name. The words IF, IF NOT, DO, GO, TO, and FOR are both reserved and noise words. They can be omitted in certain cases, but when present, they must be preceded and followed by at least one space. They cannot be used as identifiers. The six single letters I, J, K, L, M, N, when standing alone, are used as counters and cannot be used for identifiers. An identifier can have one subscript which consists of one of those letters or an integer, or a letter with an integer added to it; some compilers permit the use of a variable as a subscript. Generally only one subscript is permitted.

The operators consist of the five arithmetic operators, the absolute sign, and the six relational operators. There are also two Boolean operators. Punctuation is extremely critical and used as the delimiters in most cases, as seen from the following rules.

(1) The comma is used to delimit executable statements. (2) A semicolon can be used for this purpose also except in conditional statements, where the true and false parts must end with either semicolons or periods. (3) A semicolon also delimits the dimension statement. (4) A comma following a single identifier denotes a subroutine call. (5) A period following a single identifier denotes an unconditional control transfer. (6) A double period denotes the end of the program. (7) A colon is used to delimit a statement label and also to delimit the comparison in a conditional statement. (8) Braces are used to enclose a subroutine, to delimit the range of a loop, and to delimit either alternative of a comparison statement. Braces are also used to specify parts of a word in both the dimension statement and the executable statements.

As indicated in the rules for naming, blanks are significant. Although in many cases both fixed words and punctuation *may* be written, it is the punctuation which is critical and the words can be omitted. In general, noise words are not permitted, except for the cases of this type. Literals are enclosed in quote marks " which appear in some hardware representations, but they are often limited to one machine word.

Since there are differing devices—although primarily paper tape in the earlier systems—used for input, the physical input form depends on the device somewhat. The conceptual form is certainly free form to the largest extent possible.

There is really no smallest executable unit considered as a separate entity because the language is completely based on the concept of a triplet consisting of two operators and an operand between them. Thus there is no grouping. Looping is controlled by either conditional statements or by the FOR statement. Both subroutines and functions can be defined with input and output parameters; they are called by location. Delimiting is normally accomplished by the specification of the meaning of each triplet. However, function and subroutine definitions must be enclosed in braces. Functions can be embedded in expressions. There is no language provision for recursion. The noun list which contains the dimensioning information, and can contain initial values, must appear at the beginning. Other languages can be entered through the use of the *crutch* operator, which serves as a flag to the NELIAC compiler that some other language is to be processed.

Only arithmetic variables, and vectors of them, are permitted. Perhaps the most interesting aspect about NELIAC is the facility to assign names to parts of computer words. Thus the programmer can pack several variables within a single word, and this is shown in the noun list, e.g., Unit Cost: {cost per apple (0→5), cost per orange (6→11), cost per lemon (12→17),}. These character limits can overlap. In addition to that, even if the programmer has not specified a partial word in his noun list, he can nevertheless refer to a part of any named noun by the use of parentheses and an arrow; e.g.,

Code (0→10) → Form (15→25),
List A[I] (3→7) → B,
C(15→20) + D (25→30) → C(0→6),

Note that this reduces or eliminates compatibility between binary and decimal machines, or even between machines of different word lengths.

Only fixed and floating point numbers are allowed. All variables are considered integers by default, unless they appear in the noun list with some floating point number to set the mode. Boolean arithmetic is done with the relational operators and the use of the *cap* and *cup* characters.

In some (but not all) of the compilers, integer and floating point numbers are permitted in the same expression, and the computation is done in floating point. In some compilers, arithmetic expressions cannot contain grouping parentheses, but normal precedence is used.

Scope is handled in a rather interesting way. If a variable is to be used in several programs, the most recent value will be taken unless an absolute sign (i.e., vertical bar) is placed after the first letter of the noun or verb; in this case it has only local significance. For nouns the absolute sign is used only in the *Noun List* or *Dimensioning statement*. For verbs or verb phrases, it should be used only at the point where the verb is being defined.

There is a single type of assignment statement which assigns the value of an arithmetic expression to a variable or to parts of a variable. Thus writing A→B (1→5) would store the contents of A in positions 1–5 of word named B and leave the rest untouched. However, A(1→5) → B would set all of word B to zero before carrying out the assignment. It is possible to have assignment statements concatenated; hence A[I] + B[J] → C → D would store A[I] + B[J] in both C and D. In most systems, converting variable types across the arrow is not automatic.

There is no provision for handling alphanumeric information.

An unconditional transfer is actually shown by means of punctuation; the rule is that whenever a name is preceded by a nonarithmetic operator and followed by a period, control will pass to the indicated name. However, if a comma follows the name then this serves as a subroutine jump. Thus for example, writing ,ABC. or A + B → C; ABC. will cause an unconditional control transfer to the statement named ABC. By writing ,ABC, FIND NEXT WORD: X1 − Y → Z, the instructions in subroutine ABC will be executed and then the statement X1 − Y → Z will be executed. Subroutines and functions are invoked by writing the parameter list in parentheses after the name, separating the arguments by commas, and following this with a comma, e.g., SUM(I, GAMMA, 4),. The comma is not used when the function is included in an expression.

The conditional statement is written in the form

IF *relation: statement-for-true-path;* IF NOT , *statement-for-false-path.*

Several points need to be made about this general form. First, the relation can have only a single variable on the right of the relation sign. Thus, the following are legitimate:

IF A(2→7) < B < C:
IF A + (B/C) → U = V:

whereas

IF A + B = B + 1:

is incorrect. In some compilers it is possible to include Boolean *and*s and *or*s, thus permitting

IF A < B ∩ C = D ∪ A = D:

Second, the words IF NOT are not required; the semicolon or period following the statement for the true path is the defining character for the negative path. Third, the executable statements can themselves be conditional, provided they are stated completely within either the true or false path. Finally, the actual sequencing rules are controlled by punctuation. Since either a period or a semicolon can terminate the true or false path, these paths can contain subroutine calls or unconditional transfers. For example, one can write

IF Y < 2: Y + 1 → Y, ROUTINE 1, ROUTINE 2, ETC. IF NOT , Y − 1 → Y, CASE 1, CASE 2;

Loop control is handled by the FOR statement, which has the following format:

FOR *variable = initial value (increment) final value (range)*

As an example, the user can write

FOR I = 0(1)20 (NR[I] + COST[I] → TOTAL COST[I], COMPUTATION: A + B[I] → C[I],)

In some compilers there was a restriction that the final value must be reached precisely by adding increments to the initial value.

Since ALGOL did not have any input/output statements, it is not surprising that NELIAC did not originally. Many of the early compilers just created subroutines or borrowed existing input/output packages. However a semi-machine-independent technique has been implemented on a few of the NELIAC compilers. This technique makes use of the < and > symbols to serve as quotation marks. Three forms exist, of which the first specifies the output of headings, the second the output of numerical data, and the third calls for the reading of data into the computer, as shown below:

    Form 1.   , A (B << C >>)
    Form 2.   , A (B < C >)
    Form 3.   , A (B > C <)

The letter A can be replaced by any appropriate comment, B refers to a particular piece of peripheral equipment, and C represents names of nouns or noun lists. If B is either omitted or calls for equipment not available, some fixed equipment designated by each installation will be used. Thus for example, the user can write

    , Print Headings (Flex << List >>)
    , Print Result (HSP < Answer >)
    , Read Data (> Temperatures <)

There are no library facilities or built-in functions, debugging statements, storage-allocating statements, nor interaction with the operating system, although certain systems provide some of these facilities.

The only declaration is the noun list which appears at the beginning of the program and in which every variable to be used in the program must be listed. Initial values can be given here, and dimensioning is handled by simply writing the dimension enclosed in parentheses after the name of the variable, e.g., COST (3) = 10, 15, 8. Also within those declarations are specified the part of the word that the user is applying to the variable, e.g., LENGTH (0→5). The dimension and word portion can be given at the same time.

Subroutines and functions are defined by following the name with a colon, then a parameter list in parentheses, and then the body of the routine enclosed within braces and terminated by a comma, e.g., FUN: (X, Y) {body}. Types and numbers of parameters depend on a particular implementation.

The strongest feature of NELIAC is the ease with which it can be used to write its own compiler. There has been a very definite interaction between the language design and the implementation. In fact the syntax of the language is built entirely on the implementation technique, which is what they call the Co-No (current operator—next operator) Tables. There are actually 676 combinations in the Co-No Table, but of course some of these are errors. Furthermore, a number of the combinations require the same code generation.

The self-compilation feature produces an interesting effect on object code efficiency. As the compiler is improved, it may improve both the object time and compile time efficiency simultaneously because the compiler is run through itself. Hence, any differences in the relative effect on compile time and object code efficiency would tend to be caused by the logic of the compiler rather than by the code generation phase.

One of the applications for which NELIAC has been used is the rather interesting Blood Bank Program described by Singman *et al.* [SI65]. An older application was its use to write a very simple compiler for the 1401 (see Watt and Wattenburg [WJ62]). An ALGOL system at the University of California at Berkeley has been written in NELIAC. A successful command and control application was the Command Ship Data System for the Navy, whose object program contained over 100,000 instructions. NELIAC has also been used to help program a search procedure (see Halstead, Uber, and Gielow [HS67]). Unfortunately, public documentation of actual applications of NELIAC is very scarce and the descriptions seem to be internal documents of the using installation. As somebody noted facetiously, the main use of NELIAC has been to write other NELIAC compilers.

NELIAC has made several contributions to the technology. It was the first language used consistently to create all its own compilers, and it was used several times to compile programs on one machine, producing object code for another. For example, the NELIAC on the CDC 1604 produced code for the CDC 160A. The use of the partial word designations in the executable statements is still unique to NELIAC, except for the JOVIAL *BIT* and *BYTE*, which are not as powerful or flexible. The use of the clearly defined tables of operators and operands formalized the compiling technique but simultaneously caused a more stilted language.

IV.5.2. MAD

The *M*ichigan *A*lgorithm *D*ecoder (MAD) is a system which was developed at the University of Michigan originally for the IBM 704 and then for the 709/90/94. The work was started in the spring of 1959 by R. Graham, and subsequently he, B. Arden, and B. Galler produced the first version of MAD; this required about two man-years of work. Their original intent was to implement ALGOL 58. However, they found there were things they did not like in ALGOL 58, so some changes were made. In the summer of 1959 they felt a fast compiler was needed for their IBM 704. A system which was very much along the spirit of GAT (see Section IV.2.2.3) was finished around February, 1960 and given to SHARE. It was designed to be a fast translator for the 704 and expected to be used primarily by students. The main objectives were speed of translation, generality, ease of use, few rules to learn, and ease of adding to the language.

As with many other such systems, it is still being improved by the University of Michigan, and there have been numerous versions (each containing the previous one as a subset). It has also been heavily used at M.I.T., but that version was frozen in a manual dated November, 1964. The source document used for this description is the 1966 MAD manual [UM66].

It is interesting to see that the MAD compiler has served a rather inverted purpose relative to compatibility. Because a number of people were interested in using the FORTRAN language and yet wanted to obtain the speed of the MAD compiler, a system called MADTRAN (written in MAD) was developed. MADTRAN was simply a translator from FORTRAN to MAD, which then of course compiled machine code. It was distributed through SHARE and used at several places. MADTRAN assumes correct FORTRAN programs, so it provides no diagnostics; thus if the programmer has errors, he is in trouble. It is believed, however, that any correct FORTRAN program will be correctly translated by the MADTRAN system.

The basic purpose of MAD was stated by Arden, Galler, and Graham [AR61a] as being "developed for the specific purpose of training large numbers of university students and handling the large volume of university research problems. The primary motivation for writing this rapid translator may be traced directly to the special environment of a university computing center." They then go on to say that "ALGOL 58 provided the basic pattern for the language and to the extent that ALGOL 58 is like ALGOL 60, MAD is an ALGOL translator."[14] In my opinion, MAD is no more of an ALGOL translator than a FORTRAN translator. MAD was motivated by ALGOL 58, but it does not resemble ALGOL 58 in any significant way.

The original language was primarily designed and implemented by the people noted at the start of this section. Maintenance and improvements have been made by these and a few other people at the University of Michigan since then. The primary documentation has been a sequence of official manuals, each including some extensions which were originally documented as addenda and minor revisions to the many printings of the first edition [UM60]. The source used for the description in this book is the August, 1966 version [UM66]. Because of the way these manuals were printed, no

---

[14] Arden *et al.* [AR61a], p. 27.

SAMPLE PROGRAM—MAD

Problem: Construct a subroutine with parameters A and B such that A and B are integers and $2 < A < B$. For every odd integer K with $A \leq K \leq B$, compute $f(K) = (3K + \sin(K))^{1/2}$ if K is a prime, and $f(K) = (4K + \cos(K))^{1/2}$ if K is not a prime. For each K, print K, the value of $f(K)$, and the word PRIME or NONPRIME as the case may be.

Assume there exists a subroutine or function PRIME(K) which determines whether or not K is a prime, and assume that library routines for square root, sine, and cosine are available.

Program:

```
              EXTERNAL FUNCTION (A,B)
              ENTRY TO PRINTER.
              INTEGER A, B, K, HOL
              BOOLEAN PRIME
              THROUGH LOOP, FOR K = 2*(A/2)+1, 2, K.G.B
              WHENEVER PRIME.(K)
            1     F = SQRT.(3*K+SIN(K))
            2     HOL = $ $
              OTHERWISE
            1     F = SQRT.(4*K+COS(K))
            2     HOL = $NO$
              END OF CONDITIONAL
   LOOP       PRINT FORMAT ALPHA, K, F, HOL
              FUNCTION RETURN
   ALPHA    1 VECTOR VALUES = $3H K=, I5, 3H,F=, F10.5, 4H        ,
            2 C2, 5HPRIME$
              END OF FUNCTION
```

individual(s) is shown as an actual author, although indications of those involved are given on the title page.

Variations of the language have been implemented for at least the IBM 7040, Philco 210-211, and the UNIVAC 1107. These are not necessarily compatible with the version described in the manual or in this section.

The character set consists of the 26 upper-case letters, the 10 digits, and the following special characters:

$$+ \quad - \quad * \quad / \quad = \quad) \quad (\quad . \quad , \quad \$ \quad ' \quad \text{blank}$$

There are a number of key words.

Constants can be either integer, floating point, logical (1B = *true* and 0B = *false*), octal (written as up to 12 octal integers followed by the letter K), or alphabetic. The alphabetic constants are normally called literals and consist of up to 6 characters of any kind preceded and followed by the dollar sign, $, e.g., ABC, $D + 3 = $, etc. Any of the constants above can be declared to be a mode (e.g., integer, logical, etc.) other than the normally declared one (floating point) by following it with the letter M and a code number.

Both data names and statement labels have the same form, namely a letter followed by zero to five letters or digits. The same name cannot be used for a variable name and a statement label. Both variable and statement labels can be subscripted, and the latter are enclosed without parentheses. There is no limit on the number of subscripts. For one- or two-dimensional arrays, data name subscripts can consist of any arithmetic expression and can themselves be subscripted. However, the expressions for subscripts in an array with more than two dimensions must be integers. There are a few reserved words, primarily some function and system names. The bulk of the key words in the language are *not* reserved words since they are longer than six letters and the other symbols are surrounded by periods. There is, however, a standard set of abbreviations which can be used to replace the longer words. This consists of the first and last letters with an apostrophe between them, such as W'R for WHENEVER and D'N for DIMENSION.

All operators except those represented by single nonletter characters are surrounded by periods. The operators include the five arithmetic ones, + − * / and .P. (for exponentiation); the absolute value .ABS.; negation −, and the relational and logical operators. The relational operators include *equal, not equal, greater, greater than or equal, less than*, and *less than or equal*; these are designated, respectively, by .E., .NE., .G., .GE., .L., and .LE.. The Boolean operators are .NOT., .OR., .AND., .THEN., .EQV., and .EXOR., where the last three represent, respectively, *implies, equivalence*, and *exclusive or*. There are also operations for full words, consisting of *bitwise negation* .N., *logical and* .A., *logical or* .V., and *exclusive or* .EV.. Punctuation as such has no meaning, except for commas to separate lists of items. Blanks are not significant except in literals. There are no noise words.

The input form is designed for cards, with the statement labels anywhere in columns 1 to 10, the statements beginning anywhere in columns 12 to 72, and a continuation symbol permitted in column 11. A statement or declaration must start on a new card. In my opinion, the conceptual form is a mixture of ALGOL and FORTRAN.

There are several declarations and they are described later. The smallest executable unit is any of the specific statements that exist. There is no way of grouping them. There is loop control by means of the THROUGH statement and by the use of conditional statements.

There are two main types of functions, the *internal* and *external*. A MAD *procedure* is merely a function with multiple inputs and/or outputs. Statement labels and function names may be used as arguments. The external functions are normally called *subroutines* or *procedures*; i.e., they are headed by the declaration *EXTERNAL* and the statements that follow are to be translated independently of the main program in which they are used. Internal functions are translated as they occur relative to the other parts of the program. A one-sentence definition of internal functions is permitted; e.g., the user can write

INTERNAL FUNCTION COSUM. (A,B,C) = A*B+B*C+A*C

Comments are indicated by the letter R in column 11. Declarations and comments can appear anywhere in the program. The program must end with the statement *END OF PROGRAM*. There is a provision for an *ERROR RETURN* statement which will then transfer control to the operating system.

Declarations and statements are delimited only by their card position, i.e., new statements and declarations must start on a new card. Recursive procedures and functions are permitted, although the user must do some of the saving and restoring work himself. Parameters are passed by value and can be expressions.

There are arithmetic, Boolean, statement label, function name, and alphanumeric variables. There is also a facility for performing certain operations on the bitwise representation of integers. The arithmetic done is integer, floating point, and Boolean. Fixed and floating point variables and constants can be combined in a single expression. When they are, the expression is considered to be in floating point mode. However, parts of the computation may be performed in integer arithmetic; as a consequence, the final results may differ from those which would be obtained by having the entire computation done in floating point.

There is no problem with the scope of data because there are only single statements in the language, i.e., no groups of executable units.

The basic assignment statement consists of a variable on the left followed by an equal sign and an expression on the right, e.g., ALPHA=X+3+F.(X,Y). These must be of the same type, except that an integer or floating point expression on the right will be converted to a floating point or integer expression on the left if necessary.

The basic control transfer is of the form *TRANSFER TO S* where S is any statement label or any expression which defines the label of an executable statement. In particular, it is legitimate to write *TRANSFER TO BETA* (K + 2) since statement names can be subscripted. A subroutine is invoked either by direct use of its name or by using the word *EXECUTE*, with or without arguments, e.g., EXECUTE SORT. (A, B, C) or SORT. (A, B, C). A function can be used in an expression, e.g., Z = A + COS.(X)/SIN.(X−.5).

The *CONTINUE* statement serves as a function point in the program but causes no computation to take place.

There are two types of conditional statements. The first is the simple conditional of the form WHENEVER B, Q where B is a Boolean expression and Q is any executable statement except the END OF PROGRAM, another conditional, an iteration, or a function entry; e.g., WHENEVER X .LE. 1, Y = A + B is legal but WHENEVER X .LE. 1, WHENEVER Y .LE. 2, A = B + C is not. This is an illustration of a place where a comma must be written, as is true in other statements. A compound conditional is of the following form:

S1	WHENEVER B1
	executable statements
S2	OR WHENEVER B2
	executable statements
...	...
	OR WHENEVER Bn
	executable statements
Sn	END OF CONDITIONAL

Often the last condition is one for which the condition is always true, and in this case the programmer may write OTHERWISE. The S_i are optional statement labels. The executable statements can be simple or compound conditional statements. Testing starts with the first B_i, and as soon as something with the value *true* is encountered, then the set of statements following it up to the next S(j + 1) are executed. Computation then continues from the first statement after the END OF CONDITIONAL statement. Note that this means only one of the alternative computations is performed.

The loop control is handled by the THROUGH statement which has one of the two following forms:

THROUGH S, FOR VALUES OF V = $E_1, E_2, \ldots, E_n$
THROUGH S, FOR V = E_1, E_2, B

where S is a statement label. In the second case, B is a Boolean expression. Thus we can have something of the form THROUGH POLY, FOR J = N, −1,J.L.0 which means that N is to be decreased by 1 until it is less than 0. These loops can all be properly nested, and there are no restrictions on jumping into or out of the statements in the range of the iteration. A form of the iteration statement may be assigned a value and embedded into expressions.

The only error condition statement is the ERROR RETURN, which can be put into function, subroutine, and procedure definition programs.

The following is a list of input/output statements, where F is the name of a format specification vector or a literal giving the format itself, N is an expression whose value is a tape number, L is an input/output list, and the square brackets designate an optional clause:

PRINT FORMAT F, L
PRINT ON LINE FORMAT F, L
PUNCH FORMAT F, L
READ FORMAT F, L
READ BCD TAPE N, F, L
WRITE BCD TAPE N, F, L
READ BINARY TAPE N, L
WRITE BINARY TAPE N, L
LOOK AT FORMAT F, L
REWIND TAPE N
END OF FILE TAPE N
BACKSPACE RECORD OF TAPE N [, IF LOAD POINT TRANSFER TO S]
BACKSPACE FILE OF TAPE N [, IF LOAD POINT TRANSFER TO S]
SET LOW DENSITY TAPE N
SET HIGH DENSITY TAPE N
UNLOAD TAPE N

There are also some simplified input/output statements, namely

READ DATA
where the variable names and their values are punched on cards in the form $V_1 = n_1, V_2 = n_2, \ldots, V_n = n_n$ or even as values in an array written $V(j) = n_1, n_2, \ldots, n_n$
READ AND PRINT DATA
PRINT COMMENT $string starting with carriage control$
PRINT RESULTS L
PRINT BCD RESULTS L
PRINT OCTAL RESULTS L

If any type of error occurs during an input/output statement, the subroutine ERROR is automatically entered; the subroutine sets a flag and returns control to the operating system. There is also a way for the programmer to obtain control. Flexible format specifications are provided; they bear a strong resemblance to those of FORTRAN. The list L can be a single variable or array name, a list defined by an iteration statement including some logical expressions, or various other items.

There are no facilities for library references, except the normal use of subroutines, nor any debugging statements. There are a few built-in functions of rather special kinds, e.g., SETDIM, which updates dimension information; SYMM and TRANSP, which involve matrices. There apparently is no set of basic mathematical functions officially defined in the language. There are a number of system subroutines available to the user which provide some connection either with the operating system or the input/output system.

There are a number of interesting statements which are included to facilitate the writing of recursive functions and procedures. These statements cause the designation and actual use of a vector for the temporary storage of data and function returns. The statements are as follows:

SET LIST TO $U[,E]$ U is an array element name which designates the initial temporary storage location. The optional E is an expression defining the upper limit of the length of the list.

SAVE DATA L L is a list of variables or arrays and the list is stored in order in the current temporary storage vector.

SAVE RETURN   This causes the reentry point to the calling program to be stored as the next available element of the current temporary storage vector.

RESTORE DATA L The most recent n elements become the values of the n variables in the list L and are available for use by a *SAVE* statement.

RESTORE RETURN   This restores the reentry point to the program calling the function. The last element of temporary storage is then made available for use by successive *SAVE* statements.

The data description declarations are handled by assuming all variables and function values have *normal* mode unless declared otherwise, and this normal mode is considered floating point unless stated otherwise. Any of the other modes may be specified as the *normal* one by writing the declaration NORMAL MODE IS M where M is one of the following: INTEGER, BOOLEAN, STATEMENT LABEL, FUNCTION NAME, FLOATING POINT or n, where n is a code number. Only one such declaration can appear in a program and it applies to the entire program. Anything which is to override this normal mode must be put into a specific declaration. Format descriptions are similar to those of FORTRAN.

Storage allocation declarations are *DIMENSION*, *EQUIVALENCE*, *ERASABLE*, and *PROGRAM COMMON*. The *DIMENSION* statement is quite flexible. It permits negative subscripts. Among other things, it also allows changing the dimensions at object time; it permits the programmer to vary the location of the initial element in an array within the overall block which has been set aside for the array; and it allows an arbitrary storage mapping to be specified.

The *EQUIVALENCE* declaration has essentially the same meaning as the one in FORTRAN; i.e., the named variables are to occupy the same storage locations. The *ERASABLE* declaration assigns the indicated variables to a special storage area which eliminates the effective previous assignments to this separate storage area. There is also a *PROGRAM COMMON* declaration which allows a main program and several *EXTERNAL* function programs to refer to variables and arrays by the same name. There is no interaction among *EQUIVALENCE* and these other declarations.

The general form of a function declaration is

$\begin{Bmatrix}\text{INTERNAL}\\ \text{EXTERNAL}\end{Bmatrix}$ **FUNCTION** *name. (parameter list)*
    body of the routine
    FUNCTION RETURN *expression*
    END OF FUNCTION

An alternate way of writing the first two lines is by omitting the function name on the first line and writing **ENTRY TO NAME.** on the second. A procedure which provides several output values as parameters omits the expression. Functions can also have multiple entry points. The same function definition can be used to define any number of functions and/or procedures. They must use exactly the same set of arguments, however. Thus, the functions **CPADD. (X,Y,A,B)** and **CPMPY. (X,Y,A,B)** can be defined by one definition, but the functions **SIN. (A)** and **COS. (B)** would require separate definitions.

An interesting compiler directive is the *PARAMETER* declaration, which permits the assignment of values to symbols at compile time. The user writes **PARAMETER** $A_1(B_2), A_2(B_2), \ldots, A_n(B_n)$ where A_i is an identifier and B_i is an identifier or a constant. The effect is to replace A_i by B_i in any later occurrence in the program. This is of course a special case of the more general macro facility concept.

A rather unusual declaration (for its time) in MAD is the presetting of vectors; this is done by writing **VECTOR VALUES** $A(i) = C_0, C_1, \ldots, C_n$, where A is an array with each element's position reduced to a single subscript value. The values of the C_i are assigned. The C_i can be constants, statement labels, or character strings.

Probably the most interesting facility in MAD is its ability to allow some language extension by permitting the user to redefine existing operators or define new operators or modes for a particular program. This can be done because of the basic philosophy that is used in the implementation of MAD, namely the concept of an operator code number followed by modes of the variables and an address where the instruction sequence for that particular operator number begins. The exact format is as follows:

    operator no.   mode a   mode b   mode r2   mode r1   address

where $r1$ and $r2$ represent results (although there is normally only one).

Operators have a precedence associated with them, and this must be defined for a new operator. This provides the necessary information to the translator so it can understand how to handle something with a new operator in it. Three new modes may be added to the existing five: *Floating Point, Integer, Boolean, Function Name,* and *Statement Label*. It is also possible to change the meaning of some of the existing operators. To help clarify this, note the above display which shows a single entry in the operator-mode table. The relevant concept here is that an additional *mode* is introduced to facilitate the operation of conversion wherever this is necessary. For example, if A + B is to be computed, where A is an integer and B is a floating point number, then a preliminary conversion must be undertaken to make the operands compatible for addition. The table indicates *r1* and *r2* as new values of A and B, respectively, where one or both may be in modes different from their original form. The actual sequence of instructions that correspond to this table can be changed in certain cases. Thus any of the arithmetic operators, as well as some of the other operators, can be redefined to have a different meaning. As an example (taken from [UM66],[15] suppose that .EV. was not in the language. In order to define a binary operation .EV. with integer operands, where the result is a bitwise *exclusive or* of the two operands, the programmer would write

```
DEFINE BINARY OPERATOR .EV., PRECEDENCE SAME AS .V.
MODE STRUCTURE 1 = 1 .EV. 1
```

This would be followed by the sequence of instructions necessary to carry this out. Another possibility would be to define a binary operator which had double-precision operands and which produced a double-precision product. By means of an *INCLUDE* statement, definition *packages* may be called from a disc for matrix, complex, and double-precision arithmetic.

The general formats for adding or changing operators and modes are shown in Figure IV-12.

There are two main reasons for the speed of the MAD compiler. The first is its rather unique requirement that the key words all contain more than six letters, thus making it fairly obvious when a data name is encountered. The use of abbreviations in the form indicated on p. 208 makes it easier for the programmer to write and still permits the compiler to retain its speed. Abbreviations are expanded in the compiler-produced listing. Other reasons depend specifically on the internal techniques that are used.

MAD has proved to be quite successful in the universities because it provides some interesting features that are not available in other languages and the speed of compilation makes it a very useful tool for running a large number of student problems. It has not received much industrial usage.

One contribution to the technology made by MAD is that it provided a language which made a very fast translator easy to develop. However, the more significant contributions are some of the features which have been

---

[15] [UM66], p. 118.

(1) DEFINE $\begin{Bmatrix}\text{UNARY}\\\text{BINARY}\end{Bmatrix}$ OPERATOR $\begin{bmatrix}\text{defined}\\\text{op}\end{bmatrix}$, PRECEDENCE $\begin{Bmatrix}\text{SAME AS}\\\text{LOWER THAN}\\\text{HIGHER THAN}\end{Bmatrix}$ $\begin{bmatrix}\text{existing}\\\text{op}\end{bmatrix}$

MODE STRUCTURE $\begin{bmatrix}\text{mode}\\\text{no.}\end{bmatrix}$ = $\begin{bmatrix}\text{mode}\\\text{no.}\end{bmatrix}$ $\begin{bmatrix}\text{defined}\\\text{op}\end{bmatrix}$ $\begin{bmatrix}\text{mode}\\\text{no.}\end{bmatrix}$

defining sequence

(2) DEFINE $\begin{Bmatrix}\text{UNARY}\\\text{BINARY}\end{Bmatrix}$ OPERATOR $\begin{bmatrix}\text{defined}\\\text{op}\end{bmatrix}$, PRECEDENCE $\begin{Bmatrix}\text{SAME AS}\\\text{LOWER THAN}\\\text{HIGHER THAN}\end{Bmatrix}$ $\begin{bmatrix}\text{existing}\\\text{op}\end{bmatrix}$

MODE STRUCTURE $\begin{bmatrix}\text{mode}\\\text{no.}\end{bmatrix}$ = $\begin{bmatrix}\text{mode}\\\text{no.}\end{bmatrix}$ $\begin{bmatrix}\text{defined}\\\text{op}\end{bmatrix}$ $\begin{bmatrix}\text{mode}\\\text{no.}\end{bmatrix}$, SAME SEQUENCE

AS $\begin{bmatrix}\text{mode}\\\text{no.}\end{bmatrix}$ $\begin{bmatrix}\text{existing}\\\text{op}\end{bmatrix}$ $\begin{bmatrix}\text{mode}\\\text{no.}\end{bmatrix}$

(3) MODE STRUCTURE $\begin{bmatrix}\text{mode}\\\text{no.}\end{bmatrix}$ = $\begin{bmatrix}\text{mode}\\\text{no.}\end{bmatrix}$ $\begin{bmatrix}\text{existing}\\\text{op}\end{bmatrix}$ $\begin{bmatrix}\text{mode}\\\text{no.}\end{bmatrix}$

defining sequence

(4) MODE STRUCTURE $\begin{bmatrix}\text{mode}\\\text{no.}\end{bmatrix}$ = $\begin{bmatrix}\text{mode}\\\text{no.}\end{bmatrix}$ $\begin{bmatrix}\text{existing}\\\text{op}\end{bmatrix}$ $\begin{bmatrix}\text{mode}\\\text{no.}\end{bmatrix}$, SAME SEQUENCE

AS $\begin{bmatrix}\text{mode}\\\text{no.}\end{bmatrix}$ $\begin{bmatrix}\text{existing}\\\text{op}\end{bmatrix}$ $\begin{bmatrix}\text{mode}\\\text{no.}\end{bmatrix}$

Figure IV-12. Format for changing operators and modes in MAD. Braces indicate that a choice is to be made from the enclosed material. Square brackets have no logical significance.
Source: [UM66], p. 114.

put into other languages (although of course in a different form), notably *VECTOR VALUES* and the *PARAMETER* declarations; one feature (definitions of new operators and data types) is just beginning to be appreciated (see Section III.7.2).

IV.5.3. JOVIAL (*cross-reference only*)

JOVIAL is based on CLIP, which in turn was motivated by ALGOL 58. Both these languages (only JOVIAL has actually survived) are far more general than the others in this chapter and therefore are not described here. JOVIAL is discussed in Section VIII. 3 and CLIP in Section IX.2.5.2.

IV.6. ON-LINE SYSTEMS

IV.6.1. INTRODUCTORY REMARKS

This section considers languages that are designed to be used in on-line systems to do primarily numerical scientific work. In most cases the language is embedded in a stand-alone system rather than as part of a general purpose time-sharing system. In each case, the language has been designed to provide the user at a remote terminal with some facilities for doing numerical computations. The scope of these ranges from something slightly above a desk calculator (the original JOSS[16]) to small programming languages (BASIC, CPS) to a virtually full-fledged version of FORTRAN (QUIKTRAN) to systems involving special keyboards and scopes (Culler-Fried, AMTRAN) to systems providing high level language primitives (MAP, Lincoln Reckoner) to a system with many unusual operations (APL/360) and one using a stylus, push button, and scope but no input typewriter (DIALOG).

The only parts of these languages which are being discussed here are those which relate to the actual computations and/or the debugging facilities that are available to the user. Any aspects associated with the language which are primarily (or constitute a major set of) control functions to the time-sharing system itself are dealt with in Section IX.3.6. If the language can coexist with other languages or systems, this factor is considered irrelevant in this section. Languages originally designed for use in a *batch* environment which might be available under general time-sharing systems (e.g., MAD is available under CTSS – see Crisman [ZR65]) are not considered *on-line languages*.

A description of one organization's experience with a variety of systems is given by O'Sullivan [OU67].

The differences in the philosophy of the first four systems discussed in this section must be emphasized. In the case of JOSS and BASIC, the idea was to produce as simple a system as possible to do useful work. However JOSS was slightly more oriented to an industrial environment with working engineers and scientists, while BASIC was created for use on a college campus with a multitude of students doing small jobs that might only be experiments. There is a provision for execution of lines immediately when entered in JOSS, but there must be a full program in BASIC. There was no concern in either case for connection with other languages. There are a number of versions of JOSS implemented in varying places, but they will not be included here because they are not significantly different. A partial list of these dialects appears in Section IV.6.2.

QUIKTRAN represents the opposite end of the spectrum, in the sense that it took an *existing* and widely used language, namely FORTRAN, and adopted it for use on a remote terminal. This was done by omitting a few of the language features which made it hard to implement in the environment used and adding a number of features to facilitate debugging in a remote console type operation. A further factor in QUIKTRAN was the need to maintain compatibility with FORTRAN, so that programs checked out at the terminal could be compiled with a regular compiler for production runs.

---

[16] JOSS is the trademark and service mark of The RAND Corporation for its computer program and services using that program.

CPS (and RUSH) adopted a philosophy which was somewhere between these two extremes. CPS is a PL/I-like small extended subset of PL/I which attempted to retain the ease of learning and usage of JOSS but the syntax of PL/I. Much more so than in the QUIKTRAN case, it was found impossible to maintain strict compatibility with PL/I and still provide a language to meet the other objectives. However, it provides the user with a fundamental grounding in PL/I syntax, and from this he should be able to go on easily to learn more of PL/I.

A good comparison of the four systems, AMTRAN, Culler–Fried, Lincoln Reckoner, and MAP, is given by Ruyle, Brackett, and Kaplow [RU67]. MAP provides some high level mathematical operations to the user and is the only system in this section (except for the Lincoln Reckoner) which runs under a general purpose time-sharing system. The Lincoln Reckoner provides high level operations to the user and concentrates on a concept of coherent programming, which makes it practical for many people to create and use each others' programs. The Culler-Fried work pioneered in the use of display scopes and specially designed push-button keyboards to build up operations. AMTRAN has a number of features from different stylistic approaches, e.g., push buttons to build up high level operations, some FORTRAN-like programming facilities, and the use of the scope.

DIALOG involves the use of a display scope and input stylus very heavily, but it has a fairly conventional algebraically oriented language with some display features.

APL/360 (and PAT) are implemented subsets of the more general language defined by Iverson (see Section X.4).

IV.6.2. JOSS

JOSS is a system which was first developed by J. C. Shaw, T. O. Ellis, I. Nehama, A. Newell, and K. W. Uncapher to run on the JOHNNIAC (now in a museum) at the RAND Corporation. JOSS has been in daily use since January, 1964. (An earlier but simpler version was running during most of 1963.) A description appears in Shaw [JC64]. A later version was implemented on the PDP6 in 1965 and is described in Baker [BK66]. While officially the name JOSS applies to both versions, in practice JOSS I and JOSS II are used. This discussion will use the numeral only where it is necessary to distinguish them. Because of the significance of JOSS I as the first on-line system of such simplicity, it will be described; the additional facilities of JOSS II will then be given.

There are a number of versions of JOSS implemented in varying places, with differing amounts of deviation. A partial list of the names of these dialects is as follows (references have been included where possible): CAL, CITRAN, ESI, ISIS, PIL/I, and TELCOMP.

SAMPLE PROGRAM—JOSS II†

Program	Explanation
*Program to find the mean (M) and standard * deviation (S) of N numbers. Let M=sum(i=1(1)N:p(i))/N. Let V=sum(i=1(1)N:p(i)*2)/N — M*2. Let S=sqrt(V). N=10	$M = [\sum_{i=1}^{N} p(i)]/N$ $V = [\sum_{i=1}^{N} p(i)^2]/N - M^2$ $S = \sqrt{V}$
3.1 Demand p(i).	

Do part 3 for i=1(1)N.
     p(1) = 7.5
     p(2) = 3.5
     p(3) = 8.1
     p(4) = 7/3
     p(5) = 10
     p(6) = 11
     p(7) = .8*
     p(7) = 8.2
     p(8) = 5.5
     p(9) = 7
     p(10) = 8.2
	The easy way to type in many values.
	
	 Cancelling an error.

Type M,S.
     M =   7.13333333
     S = 2.55812431	

*We can also get the geometric mean, if we * want it: Type prod[i=1(1)N:p(i)]*(1/N). **prod[i=1(1)N:p(i)]*(1/N) = 6.53141414**	$[\prod_{i=1}^{N} p(i)]^{1/N}$

Comment: Boldface characters represent information typed out by the computer.

†Informal memorandum by D. C. McGarvey, March 24, 1966.

The goal of the JOSS experiment (JC64] was to demonstrate "the value of on line access to a computer via an appropriate language".[17] "It was designed to give the individual scientist or engineer an easy, direct way of solving his small numerical problems without a large investment in learning to use an operating system, a compiler, and debugging tools, or in explaining his problems to a professional computer programmer and in checking the latter's results."[18] It has unquestionably served this purpose.

The main published documentation of JOSS I is Shaw [JC64]. Numerous talks have been given on the subject, and much of the training of new people

---

[17] Shaw [JC64], p. 456.
[18] Shaw [JC64], p. 455.

in its use is done by either a movie (whose scenario is in Baker [BK64]), examples or direct on-line use. There was no single document which gave a complete description of the language, although the one-page summary shown in Figure IV-13 served as a good checklist.

The input/output device is a specially designed typewriter. The character set consists of the 26 upper-case letters, the 10 digits, and the following 27 characters:

$$
\begin{array}{cccccccccc}
+ & - & \cdot & / & * & & & & \text{blank} & \\
' & '' & \# & \$ & \leq & \geq & < & > & = & \neq \\
(&) & [&] & . & , & ; & : & ? & | & -
\end{array}
$$

The centered dot is used for multiplication and the pointed star for exponentiation. Parentheses and square brackets are interchangeable wherever grouping is needed.

Data names consist of a single upper- or lower-case letter which can have two subscripts separated by commas. The values of subscripts must be between 0 and 99.

Statement labels consist of numbers which can have decimal points in them to allow for insertion of lines since every line must be numbered. When referring to them, the word *step* is used for a single line, e.g., 3.2, while *part* refers to all lines with the designated number at the left of the decimal point, e.g., 3.1, 3.3, or 3.4. A period is used to denote the end of a statement. Blanks are significant in some places.

The input is free form. However, a step cannot exceed a single line, nor can there be more than one step on a line. The first word on a line must be capitalized.

Only arithmetic variables are permitted; they can be either integers or mixed numbers. However JOSS represents all numbers internally as floating point numbers to the base 10.

Statements are executed immediately after they are typed, or they may be stored.

The basic assignment statement in JOSS is designated as

$$\text{Set } var = expression$$

with an optional period after the expression. Thus the user writes

$$\text{Set } Y = 123 \cdot 456.$$

or

$$\text{Set } Y = C \cdot \exp(-x * 2/2).$$

Arbitrarily complex expressions can be used on the right-hand side. The

DIRECT or INDIRECT   DIRECT (only):

   Set x=a.   Cancel.

   Do step 1.1.   Delete step 1.1.
   Do step 1.1 for x = a, b, c(d)e.   Delete part 1.
   Do part 1.   Delete form 2.
   Do part 1 for x = a(b)c(d)e, f, g.   Delete all steps.
                                         Delete all parts.
   Type a,b,c,_.   Delete all forms.
   Type a,b in form 2.   Delete all.
   Type "ABCDE".
   Type step 1.1.   Go.
   Type part 1.
   Type form 2.   Form 2:
   Type all steps.   dist. = ... accel. = __.___
   Type all parts.
   Type all forms.   x=a
   Type all values.
   Type all.   RELATIONS:
   Type size.
   Type time.   = ≠ ≤ ≥ < >
   Type users.
                                      OPERATIONS:
   Delete x,y.
   Delete all values.   + - · / * () [] ||

   Line.   CONDITIONS:

   Page.   if a<b<c and d=e or f≠g

INDIRECT (only):   FUNCTIONS:

   1.1 To step 3.1.   sqrt(a)   (square root)
   1.1 To part 3.   log(a)   (natural logarithm)
                                         exp(a)
   1.1 Done.   sin(a)
                                         cos(a)
   1.1 Stop.   arg(a,b)   (argument of point [a,b])
                                         ip(a)   (integer part)
   1.1 Demand x.   fp(a)   (fraction part)
                                         dp(a)   (digit part)
                                         xp(a)   (exponent part)
                                         sgn(a)   (sign)
                                         max(a,b)
                                         min(a,b,c)

PUNCTUATION and SPECIAL CHARACTERS:

   . , ; : ' "   ≠ $?
   __.___ indicates a field for a number in a form.
   .... indicates scientific notation in a form.
   ≠ is the strike-out symbol.
   $ carries the value of the current line number.
   * at the beginning or end kills an instruction line.
   Brackets may be used above in place of parentheses.
   Indexed letters (e.g. v(a), w[a,b]) may be used above in place of x, y.
   Arbitrary expressions (e.g. 3·[sin(2·p+3)-q]+r) may be used above
     in place of a, b, c, ...

            Figure IV-13. Short summary of JOSS I.
            Source: Shaw [JC65], p. 10.

word Set can be omitted in an abbreviated form of direct input, but it is required in the stored program, e.g.,

$$\text{Set } x = \text{sqrt}(z) + y.$$

or

$$x = \text{sqrt}(z) + y.$$

or

$$1.23 \text{ Set } x = \text{sqrt}(z) + y.$$

but not

$$1.23 \; x = \text{sqrt}(z) + y.$$

The unconditional transfer is of the form

$$\text{To step } 1.35.$$

or

$$\text{To part } 4.$$

Built-in functions are invoked by writing the function name with the parameter(s) in parentheses. Conditions can be attached to the end of steps. Thus one can write

$$\text{Set } y = q + 4.2 \text{ if } a > b.$$

or

$$\text{To } 3.9 \text{ if } c = d + 1.$$

Numerical relations can be combined with the words and, or, e.g.,

$$\text{Set } y = a-b \text{ if } c>d \text{ or } d<e \text{ or } a<b<c.$$

It is possible to write loops by writing

$$\text{Do part } n \text{ for } x = a(b)c.$$

where a and c are the initial and terminal values and b is the increment. The range can either be a single line (using *step*) or many lines (using *part*). The parameters can also be specified as a sequence of values separated by commas after the equals sign, e.g.,

$$\text{Do step } 3.1 \text{ for } x=1, 2, 3, 100.$$

Expressions requiring computation can also be used, e.g.,

$$\text{Do step } 3.1 \text{ for } x=\sin(.5) \, (.0001), \sin(.9).$$

There are no direct error condition statements in the language, although typing errors can be corrected by strikeovers before releasing the input line.

The output command is *Type*, which can be followed with any number of expressions, a *format number*, literals, or even the phrase all values, etc. There are two types of fields provided for formatted numeric output: A string of underlines with an optional decimal point is used to indicate fixed point, while a string of periods specifies tabular form for floating point numbers. The number of digits typed is determined by the length of the field. The user can also specify page and line to control format and, in fact, can put conditionals after some of these; thus he can write

$$\text{Page if } \$ \text{ equals } 50.$$

where the $ carries a numerical value equal to the line number of the typewriter's current position on the page. He can also refer to a form with conditional, e.g.,

$$\text{Type t, a in form 6 if } a = b + 2 \cdot c.$$

Thus there is great output flexibility, considering the basic simplicity of the system.

The functions available to the user include the basic mathematical routines *sqrt, log, exp, sin*, and *cos*, as well as the following:

arg(a,b)	Argument of point (a, b)
ip(a)	Integer part
fp(a)	Fraction part
dp(a)	Digit part
xp(a)	Exponent part
sgn(a)	Sign
max(a,b,c,...)	
min(a,b,c,...)	

In addition, absolute value is denoted using vertical bars, e.g., $|x-3|$.

There is obviously considerable interaction with the system. There is a *Delete* command which can contain any number of variables or can delete all variables. Higher aggregates can be deleted by using expressions such as all steps, all parts, all forms, all values, and all. The system provides error messages of two kinds. The first involves language violations, and the message is fairly explicit. The second involves malformed expressions, steps, etc. In this case, the system responds by writing Eh? thus leaving it to the user to find the difficulty; this method was chosen as being preferable to one in which some attempt was made to pinpoint (perhaps erroneously) the error.

There is no doubt but that some of the language constraints are imposed by implementation considerations. Since JOSS is interpretive and was originally implemented on an old, small, slow machine, some user conveniences could not be provided. What is probably the most amazing aspect of the whole JOSS activity is that it provided such a useful tool to so many people at the RAND Corporation.

JOSS II has been implemented on the PDP-6. A summary of its commands is given as Figure IV-14. The discussion below is based on indicating the additions to JOSS I.

Data names can now have up to 10 subscripts instead of the previous 2.

Logical expressions were formerly allowed only after the *If*. In JOSS II they can appear in a number of other places, e.g., in a *Type* statement and in a *Let* statement (discussed below). In addition to the operators and and or, the word *not* can be used; parentheses may be used for grouping, thus permitting such statements as

To 3.2 if x = y + 1 and not (x < z or r > s).

A library function *tv* has a single argument consisting of a Boolean expression and converts it into the number 1 if true and 0 if false. This can be used in an expression, e.g.,

Set a = 5 + tv(x>3.1 or x<y).

Conditional expressions can be written as the following, where the p_i represent logical expressions and the e_i are functions:

$(p_1: e_1; p_2: e_2; \ldots p_n: e_n)$

(This is of course similar to LISP 1.5—see Section VI.5.) This is interpreted as *if p_1 then e_1 else if p_2 then e_2, etc.*, where the evaluation proceeds from left to right until the first true p_i is found. These conditional expressions can be used anyplace that arithmetic expressions are valid.

The *Do* statement has been extended to permit execution of a part or step *n* times by writing

Do part 3, 5 times.

It is now possible to nest *Do* statements to any depth by enclosing all but the outer one in parentheses. The variables of the *interrupted Do* are not changed.

One of the most significant additions in JOSS II is the provision for user-defined functions. These are defined using the *Let* statement and a single-

Appendix A--*JOSS COMMANDS AND FUNCTIONS*

L = letter
S = subscripted letter
P = proposition
F = formula

int = expression with integer value
num = expression with numerical value
rng = range of values of a letter,
    such as $1(.1)3, 3.64, 4$

- An expression, e.g., 3+sqrt(2), has a numerical value. A proposition, e.g., a≤b, has a truth value (true or false).
- An if P clause can be appended to any command except a short Set.
- A conditional expression may be used wherever an expression is allowed; e.g., define f(x) by a conditional expression: Let $f(x)=(x<0:0;0≤x<1:x*2;1)$ defines f(x) to be 0 for x<0, x*2 for 0≤x<1, and otherwise 1.

JOSS COMMAND LIST

TYPE COMMANDS
Type num.
Type S.
Type $S(int,int)$.
Type P.
Type "any text".
Type _.
Type form int.
Type step num.
Type part int.
Type formula F.
Type $F(num)$.
Type $F(P)$.
Type all steps.
Type all parts.
Type all formulas.
Type all forms.
Type all values.
Type all.
Type time.
Type timer.
Type size.
Type users.
Type item-list.

- Several individual Type commands may be combined in one, except for Type "any text". and Type item-list.
- in form int may be appended to Type commands that specify individual values.

SET COMMANDS
Set $L=num$.
Set $L=P$.
Set $S(int,int)=num$.
Set $S(int,int)=P$.

SHORT SET COMMANDS
$L=num$
$L=P$
$S(int,int)=num$
$S(int,int)=P$

DELETE COMMANDS
Delete L.
Delete S.
Delete $S(int,int)$.
Delete form int.
Delete step num.
Delete part int.
Delete formula F.
Delete all steps.
Delete all parts.
Delete all formulas.
Delete all forms.
Delete all values.
Delete all.

- Several individual Delete commands may be combined in one, such as
Delete L, form int, all parts.

LET COMMANDS
Let $F=num$.
Let $F=P$.
Let $F(L)=num$.
Let $F(L)=P$.

DO COMMANDS
Do step num.
Do step num, int times.
Do step num for $L=rng$.
Do part int.
Do part int, int times.
Do part int for $L=rng$.

TO COMMANDS
To step num.
To part int.

DEMAND COMMANDS
Demand L.
Demand $S(int,int)$.
Demand L as "any text".
Demand $S(int,int)$ as "any text".

FILE COMMANDS
Use file *int* (ident).
File ... as item *int* (code).
Recall item *int* (code).
Discard item *int* (code).

- *file int* and *ident* are assigned by the Computer Sciences Department.
- *item int* and *code* are assigned by the user at time of filing. 1≤*int*≤25. *code*, if used, ≤5 letters and numbers.
- "..." stands for any combination of elements that can be deleted by a Delete command.

SINGLE-WORD COMMANDS
Page.
Line.
Go.
Stop.
Done.
Quit.
Cancel.

PARENTHETICAL COMMANDS
Cancel or any Do command may be enclosed in parentheses.

SPECIAL COMMANDS
Reset timer.
Let S be sparse.

JOSS FUNCTION LIST

ip(x) = integer portion of x
fp(x) = fraction portion of x
dp(x) = digit portion of x
xp(x) = exponent portion of x
sgn(x) = -1 if x<0, 0 if x=0, +1 if x>0

sqrt(x) = square root of x for x≥0
log(x) = natural logarithm of x for x>0
exp(x) = e^x

sin(x) = sine of x, x in radians, |x|<100
cos(x) = cosine of x, x in radians, |x|<100
arg(x,y) = angle in radians formed by the positive x-axis and the line from the origin to (x,y)

sum(x,y,z) or sum[i=*rng:num*] = sum of values in argument list
prod(x,y,z) or prod[i=*rng:num*] = product of values in argument list

min(x,y,z) or min[i=*rng:num*] = minimum of values in argument list
max(x,y,z) or max[i=*rng:num*] = maximum of values in argument list

conj(p,q,r) or conj[i=*rng:P*] = true if and only if P is true for all i
disj(p,q,r) or disj[i=*rng:P*] = true if any P is true

first[i=*rng:P*] = first i for which P is true

tv(P) = the truth value of P; 0 if P is false, 1 if P is true

- *num* or P is evaluated for each value of *i* in the range *rng*; *i* usually appears in *num* or P.

Figure IV-14. JOSS II commands and functions.
Source: Marks and Amerding [ZL67], p. 37.

letter name. Up to 10 parameters may be used, and they are local to the definition, e.g.,

$$\text{Let } h(a,b,c) = (a+1) \cdot (b+2) \cdot (c+3).$$

The function can be invoked by writing its name and arguments in an expression such as

$$\text{Set } y = x + 3/h(3, 2, 4.1).$$

Two new operators, *sum* and *prod* are similar to the mathematical operators Σ and Π. Thus

$$\text{sum}[i=1(1)100 : i * 2] = \sum_{i=1}^{10} i^2$$

A new function called *first* gives the value of the first index which satisfies the proposition which is its argument. Its format is similar to that of *sum*, e.g.,

$$\text{first}[i=1(1)10:i>4] = 5$$

There are statements permitting the user to file (and retrieve) programs and/or data.

As stated in the introduction to this entire section, there have been a number of variations of JOSS developed. Some of these are available commercially to on-line users.

IV.6.3. QUIKTRAN

The work on QUIKTRAN was started in IBM in 1961 by a group under the direction of John Morrissey. Key individuals included T. Dunn, J. Keller, E. Strum, and G. Yang. The original objective was to improve user-debugging facilities. This objective eventually took the form of a *dedicated* system which was essentially FORTRAN, but with powerful debugging and terminal control facilities added. The two major constraints that the designers imposed upon themselves were to use only existing standard equipment configurations (which turned out to be the 7040/44 computers and 1050 terminal) and to use and to stay consistent with an existing language (which turned out to be USASI Basic FORTRAN). A first version was running in mid-1963. The best references for an overall view of the system and its objectives are Dunn and Morrissey [DM64] and Keller, Strum, and Yang [KR64]. The official manual for the system as it was made available for customer use is [IB66d].

The system was designed to handle most legitimate Basic FORTRAN

SAMPLE PROGRAM—QUIKTRAN†

```
101. —READY       PROGRAM ROOTS
102. +READY       ROOT1—
     CANCL PREV LINE
                  READ 0,A,B,C,
103. +READY       D=B**2—4.*A*C
104. +READY       ROOT1=(—B+SQRT(D)/(2.*A)
104. +RJECT UNPAIRED PAREN
104. +READY       ROOT1=(—B+SQRT(D))/(2.*A)
105. +READY       ROOT2=(—B—SQRT(D))/(2.*A)
106. +READY       START(0)
102. =1 00  1./—5./6.
105. =HALT END OF PROGRAM ENCOUNTERED DURING EXECUTION
106. +READY       PRINT 0, ROOT1, ROOT2
107. +READY       ALTER (106., 106.)
107. +NOTE MODE RESET — ALTER LINE NUMBER 106. HAS REACHED UPPER BOUND
106. +READY       PRINT 0 ROOT1, ROOT2
107. +READY       START(0)
102. =1 00  1./—5./6.
106. =0 00     0.30000000E 01    0.20000000E 01
106. =HALT END OF PROGRAM ENCOUNTERED DURING EXECUTION
107. +READY       START(0)
102. =1 00  1./1./1.
104. =XEQER THE SQUARE ROOT OF A NEGATIVE NUMBER IS IMAGINARY
107. +READY       ALTER(103.1)
103.1+ALTER      IF(D)20,30,30
103.2+ALTER      ALTERX
107. +READY 20   PAUSE
108. +READY      START(0)
102. =1 00  1./1./1.
107. =PAUSE
108. +READY      START(0)
102. =1 00  1./—5./6.
103.1=XEQER TRANSFER POINT 30 DOES NOT EXIST
108. +READY      ALTER(104.,104.)
104. +ALTER 30   ROOT1=(—B+SQRT(D))/(2.*A)
104.1+ALTER      ALTERX
108. +READY      START(0)
102. =1 00  1./—5./6.
106. =0 00 0.30000000E 01    0.20000000E 01
107. =PAUSE
108. +READY      START(0)
102. =1 00  1./1./1.
107. =PAUSE
108. +READY      SAVE
108. +READY      LIST
101. =        CF   PROGRAM ROOTS
102. =        CF   READ 0,A,B,C
103. =             D=B**2—4.*A*C
103.1=             IF(D)20,30,30
104. =          30 ROOT1=(—B+SQRT(D))/(2.*A)
105. =             ROOT2=(—B—SQRT(D))/(2.*A)
106. =        CF   PRINT 0,ROOT1,ROOT2
107. =          20 PAUSE
```

†[IB67e], extracts from pp. 56–57. Reprinted by permission from *Fundamentals of Using QUIKTRAN*. © 1967 by International Business Machines Corporation.

programs, as defined in [AA66a]. The main places in which QUIKTRAN does *not* accept legitimate Basic FORTRAN are the following: No arithmetic function statements; statement numbers must not exceed 199; some restrictions on the form of the EQUIVALENCE statement; library function names and built-in function names are reserved; some restrictions on functions, subroutines, and their arguments; and restrictions on the input/output.

The terminal can be in either the COMMAND or the PROGRAM mode; in the former case, each statement entered by the user is executed immediately upon entry and the result is printed at the terminal; this is the *desk calculator* mode and the statements are not retained by the system. In the PROGRAM mode, the statements are saved and executed only at the request of the user through *process codes*. They permit the user to do such things as execute and print the result of the statement on the same line, execute the last assignment or input/output statement that was entered, execute and print or execute all subsequent assignment and input/output statements, or execute and print all subsequent statements entered. In addition to the two modes just mentioned which control execution, there are a number of program control statements which assist in debugging. Several involve changes between command and program mode or handling entries and deletions from the users' library. The START initiates execution of the current active program, while the RESET cancels the effect of this execution. The XEQER statement specifies options for the system when an error is encountered. The AUXOP statement provides control over additional terminal devices. The DELTA allows differing incrementing rules for the system to use in generating line numbers. The EDIT statement specifies an output format for all variables not under control of a FORMAT statement.

The user can add, change, delete, and renumber program statements by using appropriately the statements ALTER, ALTER X, and NUMBER.

The user has five test statements available to him: SNAP, TRAP, GUARD, STEP, and TRAIL. An X appended to such a test statement will cancel it. The SNAP causes the printing of the value of the variable on the left of an assignment statement whenever that value changes during execution. The TRAP prints the origin and destination of every control transfer that takes place during execution of a region. (A region represents a portion of the program specified by parameters in the test statement.) The GUARD statement specifies a region or statement that is *protected* during execution so that the system does not execute it but prints a message and returns control to the user. The STEP statement permits the user to execute one statement at a time. The TRAIL statement prints the source and destination of subprogram linkages.

There are a number of display statements which permit the user to print selective information; these are all designed to permit the user to interrupt their execution. For example, the LIST statement causes a listing of the statements in the user's program. The COPY does the same thing without the line numbers. The *PDUMP* statement produces an alphabetical listing of each variable in a specified region, together with its current value or an indication that it does not have one. The *QDUMP* is like the *PDUMP* except that it prints only those variables which have changed in values since the beginning of the program or since the last execution of a *PDUMP* or *QDUMP* statement. The *INDEX* statement produces a cross-reference listing of the statement numbers and variables used in a program, together with the line numbers of the statements in which each appears. The *CHECK* statement produces a listing of variables and line numbers which have not been defined or referred to. The *AUDIT* statement produces a listing of all unexecuted regions and unreferenced variables.

QUIKTRAN was significant from several viewpoints. It was the first on-line system using standard equipment; it retained compatibility with an existing language and thus made it possible for the user to debug a program on line and then to use a regular FORTRAN compiler for batch production runs.

IV.6.4. BASIC

BASIC (standing for *B*eginner's *A*ll Purpose *S*ymbolic *I*nstruction *C*ode) is a system developed at Dartmouth College in 1965 under the direction of J. Kemeny and T. Kurtz [KM66]. It was implemented for the G.E. 225. It was meant to be a very simple language to learn and also one that would be easy to translate. Furthermore, the designers wished it to be a stepping-stone for students to learn one of the more powerful languages such as FORTRAN or ALGOL. The reader can understand the BASIC language constituents merely by looking at the list in Figure IV-15. There are only a few other significant points which are not fairly self-evident from that list. Data names are single letters possibly followed by a single digit. Single letter data names can have one or two subscripts and can be the same as unsubscripted data names, e.g., A and A(2) have no connection with each other. Subscripts can be expressions and can themselves be subscripted. A dimension statement is not needed if the value of a subscript does not exceed 10. Each statement must be numbered, and the program is executed in the numerical sequence of the statement identifiers.

The relational operators are: =, <, <=, >, >=, and <>. The last symbol represents *not equal*.

Declarations need not appear in any particular sequence. The program must terminate with an *END*.

The *LET* and *GOTO* are the assignment and control transfer statements, respectively. The *GOSUB* is a subroutine call. Functions are invoked by enclosing the parameter in parentheses following the function name. The

SAMPLE PROGRAM—BASIC

Problem: Find greatest common divisor using Euclid's algorithm.

Program:
```
100 PRINT " A", " B", " C", "GCD"
110 READ A, B, C
120 LET X = A
130 LET Y = B
140 GOSUB 500
150 LET X = G
160 LET Y = C
170 GOSUB 500
180 PRINT A, B, C, G
190 GO TO 110
200
300 DATA 60, 90, 120
310 DATA 38456, 64872, 98765
320 DATA 32, 384, 72
330
500 LET Q = INT(X/Y)
510 LET R = X — Q*Y
520 IF R = 0 THEN 560
530 LET X = Y
540 LET Y = R
550 GO TO 500
560 LET G = Y
570 RETURN
580
999 END
```

IF...THEN allows a single relational operator between expressions, and control passes to the statement number following the *THEN* on the true path. The *NEXT* is used to terminate the range of the *FOR* statement and must use exactly the same variable as in the *FOR* statement. If the third expression is omitted, it is assumed to be one.

The *READ* assigns to the listed variables the values obtained from a *DATA* statement. The latter is used to specify all the values needed for the variables. For output, the user can specify variable names or literals; the literals are enclosed within quotation marks. Thus the statement PRINT "THE SQUARE ROOT OF" X, "IS" SQR(X) might cause the following to be printed: THE SQUARE ROOT OF 625 IS 25. For normal printing purposes, the output line is divided into five zones of 15 spaces each. The user can change the width of these zones, however, through the use of commas and semicolons. A *PRINT* command without anything following it signals a new line. The program terminates with the *END* statement; the *STOP* acts like

```
LET ⟨variable⟩ = ⟨expression⟩
GOTO ⟨statement number⟩
GOSUB ⟨statement number⟩
RETURN
IF ⟨expression⟩ ⟨relation⟩ ⟨expression⟩ THEN ⟨statement number⟩
FOR ⟨unsubscripted variable⟩ = ⟨expression⟩ TO ⟨expression⟩ STEP ⟨expression⟩
NEXT ⟨unsubscripted variable⟩
READ ⟨variable⟩, ⟨variable⟩, . . . , ⟨variable⟩
PRINT ⟨literal or expression⟩, ⟨literal or expression⟩, . . .
STOP
END
DIM ⟨variable⟩ (⟨integer⟩ [, ⟨integer⟩])
DATA ⟨number⟩, ⟨number⟩, . . .
REM ⟨any string of characters⟩
DEF FN ⟨letter⟩ (⟨unsubscripted variable⟩) = ⟨expression⟩
```

Figure IV-15. Summary of BASIC statements. The angular brackets are used to define metalinguistic variables. Square brackets denote an optional element.

a GOTO where the statement number represents the END command. In addition to eight mathematical functions, the INT and RND functions are provided. The former is [x], i.e., the greatest integer not greater than x; the latter produces a random number between 0 and 1.

The DIM is used for subscripts whose value exceeds 10. The DATA specifies values. The REM is used for comments. Subroutines must terminate with a RETURN, but they have no special way of indicating their beginning. Functions are defined by the DEF statement; the function name consists of the letters FN followed by a letter. Any expression which fits on one line can be used to define a function, including another function. They cannot, however, be recursive.

In addition to the commands shown, there is also a set of 11 *matrix* commands. The user can write MAT C = A + B or MAT C = TRN(A) where the latter represents the transpose. Vectors can also be used, as long as the dimensions are correct.

The original compiler included a large number of error messages at both compile and object time.

A newer version on the 635 added the following features: (1) Functions can have any number of input parameters as long as the definition fits on one line. (2) A computed GOTO with the form ON var = expression GO TO $S_1, S_2, \ldots, S_n$ where as many S's as will fit on one line can be used and the value of var designates which S is used. (3) Additional formatting facilities can be used for output. (4) Multiple assignment statements are permitted, e.g., LET X = Y3 = Z(2, I) = 5 + A. (5) Strings of up to 60 alphanumeric characters are permitted and can be named, and a vector of strings is permitted; the name is followed by a $ and the string itself is enclosed in quotes (and cannot contain any quotes); the strings can be used in LET and IF... THEN statements as well as in input/output commands, e.g.,

```
10 LET Y$ = " YES "

20 IF Z7$ = " YES " THEN 200
```

and strings can be compared. (6) Additional flexibility is allowed in defining and using matrices.

IV.6.5. CPS

CPS (*C*onversational *P*rogramming *S*ystem) is a small PL/I-like on-line extended subset. It was developed jointly by the Allen-Babcock Corporation (primarily involving J. D. Babcock and P. R. DesJardins) and the IBM Corporation, under the overall direction of N. Rochester of IBM with significant participation by D. A. Schroeder (IBM). It was started early in 1965, and the initial version became operational in the fall of 1966. The philosophy of the system was to provide a language that was in the middle area between JOSS and QUIKTRAN. This meant that it should have as much simplicity as possible for the terminal user, but the language should have as much of the syntax of PL/I as possible. Considering the time scale, the major objective of simplicity, and the size of PL/I, the amount of PL/I which was included naturally had to be small. (This does not mean that CPS itself is small.)

A second major objective of the system was to investigate the effectiveness of microprogramming. The system was originally implemented on an IBM 360/50 with a special *Read-Only* store which was used for special machine instructions to make the language interpreter more efficient. However, subroutines were also written to replace the microprograms and thus avoid the necessity for special hardware.

A version of the system called RUSH (*R*emote *U*se of *S*hared *H*ardware) was the basis of a commercial time-sharing service offered by the Allen-Babcock Corporation, while the system in use by IBM was called CPS. Starting from the same base, the two systems added different facilities. CPS was released as a Type III system in September 1967. (A Type III program is one issued by the authors or developers as individuals and is not part of IBM's regularly delivered software.) A "one-page" summary of its commands is shown in Figure IV-16. (It requires only one physical page in the $8\frac{1}{2}''\times 11''$ source manual [IB67a].)

SAMPLE PROGRAM—CPS

```
#1,1    /* start automatic line numbering with 1 and */
   2.   /* with increments of 1 */
   3.   /* This is a CPS program to calculate mean and */
   4.   /* standard deviation of a table of numbers */
   5.   DECLARE p(10),psqr(10)
   6.   LET s(x)=sqrt(x/n-M**2)    /* define standard deviation*/
   7.   GET LIST(n,p)
   8.   psqr=p**2  /* square the entire array */
   9.   sum#=0
  10.   sum2=0
  11.   label: DO i=1 TO n
  12.   sum#=sum#+p(i)
  13.   sum2=sum2+psqr(i)
  14.   END label
  15.   PUT IMAGE(sum#/n,s(sum2)) (im1)
  16.   im1: IMAGE
Mean=...........       Standard Deviation=---.----------
  17.   /* ........indicates scientific notation*/
  18.   /*---.------indicates decimal conversion*/
  19.   GO TO start
  20.
p(10)=7/3  /* Set value requiring arithmetic operation*/
execute 1 thru ...  /* execute entire program */
n
10
p(1)
7.5,3.5,8.1,10,11,8.2,5.5,7,8.2,,
   15.        6911 VALUE OF M  IS NOT DEFINED
      *-*-*-* XEQ ERROR.
14.5 M=sum#/n /* Insert correction after statement 14*/
execute 9 thru ...  /* Run program from statement 9 to end */
Mean=.7133333E01     Standard Deviation= 2.5581243128
      *-*-*-*   19. "GOTO" OPERAND NOT LABEL.
7 start: GET LIST (n,p)
/* Correct statement number 7 */
p=p-M  /* As a test, calculate distance to mean.*/
xeq 9 thru ...  /* and do a rerun */
Mean=.444089E-15     Standard Deviation= 7.5781557416
      *-*-*-*   19. "GOTO" TARGET OUTSIDE XEQ RANGE.
```

The character set consists of the 26 upper- and lower-case letters, the 10 digits, the 3 special characters, $, @, and # (which together with the letters and digits are called *alphameric*), and the following 21 symbols:

```
     +  —  *  /  (  )  =  >  <  .  ,  '  :  ;  %
            ⌐  &  |  _  ?       blank
```

Some pairs of special characters are used to denote special operations. These

CPS - ONE PAGE SUMMARY

SIGN ON AND SIGN OFF

```
%LOGIN(acct,subacct);    %LOGIN(acct,subacct,CPS);                      01
%LOGOUT(OFF);            %LOGOUT(OFF,ACCT);                              02
%LOGOUT(RESUME);         %LOGOUT(RESUME,ACCT);
```

STATEMENTS THAT ARE DIRECT ONLY

```
%EXECUTE;   /*This may be abbreviated as %XEQ;*/                         03
%XEQ linenumber;  %XEQ linenumber THRU linenumber;
%XEQ THRU linenumber;   %XEQ label;
%LIST;   %LIST linenumber THRU linenumber;                               04
%RESEQ linenumber THRU linenumber FROM start BY STEP;                    05
%RESEQ THRU linenumber FROM start BY step;
%RESEQ linenumber;   /*RESEQ means RESEQUENCE*/
%ERASE linenumber;  %ERASE linenumber THRU linenumber;                   06
%ERASE THRU linenumber;   %ERASE label;
%LOAD(member);    %LOAD(member,key);                                     07
%SAVE(member);    %SAVE(member,longkey);                                 08
%LIB LIST;        /*LIB means LIBRARY*/                                  09
%LIB LIST,member;    %LIB LIST,member,longkey;                           10
%LIB SCRATCH,member;   %LIB SCRATCH,member,longkey;                      11
```

STATEMENTS: DIRECT WITH %; COLLECT WITH LINENUMBER

```
/*assignment*/                                                           12
Identifier=expression;/*as in the following examples*/
X=A**2+P/(5.63*W);  %D=IF-3;  ROOT=SQRT(ABS(A-B));
L,M,N=0;  ARRAY1=ARRAY2+ARRAY3;  RSLT=A*(X>0)+B*(X<=0);
B(1)=R;  B(2)=S;  B(3)=T;
```

```
¬A    +A   -A   A**B     A*B    A/B    A+B    A-B                        13
A=B   A>=B  A<=B  A¬=B   A¬>B   A¬<B   A>B    A<B    A¬B
A&B         A|B
```

```
GET LIST(variable);   %GET LIST(variable);                               14
PUT LIST(variable);      PUT IMAGE(variable)(label);                     15
%PUT LIST(variable);    %PUT IMAGE(variable)(label);
```

STATEMENTS THAT ARE COLLECT ONLY AND NEED LINENUMBER

```
GO TO label;                                                             16
IF expression THEN statement;   ELSE statement;                          17
IF expression THEN statement;
label:DO;         label:DO WHILE(expression);                            18
label:DO variable=specification,specification,...;
/*Follow one of the above with other statements*/
/*Then write*/    END label;
LET function(variable)=expression;                                       19
STOP;                                                                    20
label:IMAGE;                                                             21
Image specification;  /*for example, the following*/
TEMP:----.-DEG. PRESSURE:.....BARS. MEAS NO:----
DECLARE variable(dimension list) DEC(integer);                           22
/*examples below show some possibilities*/
DECLARE A(3);    DCL B(4,5);   DCL C(-1:2);
DCL D DEC(6);    DCL E(6,-3:4)  DEC(8);
DCL C(-1:2), D DEC(6), E(6,-3:4) DEC(8);
/*comments (like this) may replace blanks*/                              23
```

BUILT-IN FUNCTIONS

```
ATAN(e)       ATAND(e)                                                   24
ATAN(e,e)     ATAND(e,e)                                                 25
COS(e)        COSD(e)     SIN(e)     SIND(e)                             26
EXP(e)        LOG(e)                                                     27
ABS(e)        SQRT(e)                                                    28
MIN(e,e,...,e)           MAX(e,e,...,e)                                  29
MOD(e,e)      SIGN(e)    FLOOR(e)    CEIL(e)   TRUNC(e)                  30
```

STATEMENTS FOR REMOTE JOB ENTRY - DIRECT ONLY

```
%LOGIN(acct,subacct,RJE);                                                31
%IMAGE 029;   %IMAGE TEXT;   %IMAGE ?;                                   32
%SCHEDULE(member);   %SCHED(member||member);                             33
%FIND(jobname);                                                          34
%ERASE;  %RESEQ;  %LIST;  %SAVE;  %LOAD;  %LOGOUT;                       35
%LIB LIST;  %LIB SCRATCH;
```

234

NOTES FOR ONE PAGE SUMMARY

In the one page summary, lower case letters form names of things you may write while other characters stand for themselves. You may use upper and lower case letters, digits, $, @, and # for identifiers.

In CPS and RJE write exactly one statement per line.

You may omit the final semicolon on any line because CPS will provide it.

BACK SPACE kills a character. ATTN kills a line.

01 acct and subacct are identifiers assigned to you. After computer requests AUTO-SAVE parameter, type identifier or hit ATTN.

Identifiers must start with a letter or with $, @, or # and may have up to six alphanumeric characters.

03 A linenumber begins with 1 to 4 digits followed by an optional decimal point and up to 2 fraction digits. A label is an identifier that labels part of a program.

You may use ... Instead of a linenumber after THRU.

Anywhere in EXECUTE, LIST, RESEQ, and ERASE you may use a label in place of a linenumber.

05 The first linenumber after resequencing is start and the next linenumber is start+step. You may omit 'FROM start' and/or 'BY step'.

07 A member is an identifier used to name a member of a file; it is the name of a filed program.

A key is a one to four character identifier used to read a protected member of a file.

08 A longkey is a one to six character identifier used to protect a member; the longkey is needed to change this member. The first four characters of a longkey are the corresponding key.

12 Any ordinary algebraic expression may be used on the right side of an assignment statement. Bracket complex divisors. You may use builtin functions, functions you have defined, assign the same value to several variables, and use logical operators.

13 Operators grouped by order of evaluation. The logical operators", &, | are not, and, or. A variable is an identifier.

14 With GET and PUT you may replace the variable with a list of variables separated by commas. In PUT you may replace variable with expression. In PUT IMAGE label refers to an IMAGE statement.

15 An expression is considered true if its absolute value is 1.0 or greater and false otherwise. A logical expression has the value of 1.0 or 0.0. Nested IF statements require you to include all ELSE clauses.

17 e stands for expression.

18 A specification may be either of the following: 'e BY e TO e WHILE e' or 'e TO e BY e WHILE e'. Any TO, BY or WHILE phrase may be omitted. A DO statement must have a label.

19 In LET, function is an identifier that you use to name the function you are defining and its argument may be a single variable or a list of variables separated by commas.

21 Five or more dots – floating point. One or more dashes – integer. One or more dashes with a point – decimal. Other characters stand for themselves.

22 DCL and DEC stand for DECLARE and DECIMAL. A is a 3 element vector and B is a 4 by 5 array. C is a 4 element vector: C(-1), C(0), C(1), C(2). D is single precision; others are double. You may declare several arrays at once. All arrays in an expression have identical bounds. Array variables are OK in assignment, GET, PUT.

24 arctangent in radians and degrees
25 arctangent in radians and degrees of x,y coordinates
26 cosine, sine; of radians and degrees
27 exponential and natural logarithm
28 absolute value and square root
29 minimum and maximum of list of expressions
30 first expression modulo the second sign (+1, 0, -1)
largest integer not exceeding expression
smallest integer not exceeded by expression
integer part of expression, with sign

32 Follow IMAGE with specification as above using: 'card fold, | tab, - input character, # line number

33 You may use as many members as you wish.

34 CPS will tell you what jobname to use.

35 These are the same as described in 2-11.

235

Figure IV-16. Summary of CPS facilities (one-page summary with notes).
Source: [IB67a], pp. 131-132. Reprinted by permission from *Conversational Programming System.* © 1967 by International Business Machines Corporation.

are

	>=	Greater than or equal to
	<=	Less than or equal to
	¬=	Not equal to
	**	Exponentiation
	¬>	Not greater than
	¬<	Not less than

The ¬, &, and | represent *not*, *and*, and *or* respectively. Exponentiation is denoted by **. As seen by referring to Section VIII.4, this is exactly the PL/I character set, except that in CPS both upper- and lower-case letters are allowed.

Identifiers used for both data names and program unit labels consist of an alphabetic character followed by zero to five alphameric characters. Identifiers can be the same as key words. In addition, line numbers consisting of up to four digits followed optionally by a decimal point and one or two digits can also be written. (These are used for referencing and changing statements in the program.) Subscripts are shown separated by commas, contained in parentheses after the data name. Subscripts can be expressions and can themselves be subscripted. Punctuation is significant in a number of cases. Blanks are used as delimiters and can also follow punctuation symbols. (In general, the PL/I rules are followed.)

The input is free form, but a statement cannot extend over one line; only one statement may appear on one line.

The basic concept in writing a CPS program is that there are two forms of statements: *Collect* and *direct*. The former define the steps to be executed; while the latter specify control functions. A collect statement must be preceded by a line number and is automatically inserted in the correct sequence determined by the numerical value of the line number. It will replace any earlier statement with the same line number. A direct statement is immediately preceded by a percent sign, %. Some statements can be written in both modes. Single statements can be written and also grouped by a DO...END pair. Loops are written using either IF...THEN or a DO statement. Built-in functions are provided, and the user can also define his own. Both kinds can be used in expressions or in user-defined functions. The latter can have an arbitrary number of parameters but can contain only a single statement. Statements are terminated by either a carriage return or a semicolon followed by a carriage return. A statement label can immediately precede a collect statement and is itself followed by a colon.

Arithmetic floating point variables are the only type permitted. The net effect of Boolean variables can be obtained by appropriate combining of conditions, e.g., A>B & C<=D. Individual relations and their logical combinations are assigned the values 1 or 0 for truth or falsity, respectively. The relations permitted are >=, <=, =, ¬=, <, >. In addition, A|B evaluates to +1 if either |A| >= 1 or |B| >= 1 and to 0 otherwise. Both single- and double-precision floating point arithmetic is done; the system automatically retains 14 significant digits in all computations. If an expression contains a variable naming an array, the operations are performed on all corresponding elements of these arrays.

The assignment statement permits array variables on both sides of the equals sign and also permits multiple assignments, e.g.,

$$F(3*a - b), G(2 + B(3, 5)) = A + SQRT(x**2 + y**2)$$

The unconditional transfer is written as GOTO or GO TO *label*. Functions are invoked by writing them in expressions. Although subroutines as such are not allowed in the program, there is a convenient way of causing execution of groups of statements. This is done by the direct command *EXECUTE* (which can also be written as *XEQ*). This has the following format, where braces { } denote a choice and square brackets [ ] indicate something optional:

$$\begin{Bmatrix} EXECUTE \\ XEQ \end{Bmatrix} \begin{bmatrix} \begin{Bmatrix} \textit{line-number-1} \\ \textit{label-name-1} \end{Bmatrix} \begin{bmatrix} THRU & \begin{Bmatrix} \textit{line-number-2} \\ \textit{label-name-2} \\ \ldots \end{Bmatrix} \end{bmatrix} \end{bmatrix}$$

It is required that *line-number-1* be less than or equal to *line-number-2*. The option of three dots (. . .) represents the end of the program. If a *line-number* is not specified, execution resumes at the line following the last line executed by the previous *EXECUTE*. When an *EXECUTE* specifying only an initial label name is used and it is the label of a *DO* group, the entire *DO* group is executed.

The conditional statement is of the form IF *expression* THEN *statement-1;* [ELSE *statement-2;*]. The expression is true if its absolute value is greater than or equal to 1 and false otherwise. Both *statement-1* and *statement-2* can be any collect statements except *DO*, *END*, *LET*, *IMAGE*, or *DECLARE*. For example,

$$\text{IF } a=b \text{ \& } c=d \text{ THEN } x = 3.14/b; \text{ ELSE } x=c;$$

$$\text{IF } a\neg=b \mid SQRT(C**2) < d \text{ THEN GO TO end};$$

If nested *IF*s are used, the *ELSE* clause must be written explicitly.

The loop control is handled by means of a *DO* statement. This is of the form

$$\textit{label: } DO \begin{bmatrix} \begin{Bmatrix} \textit{var} = \textit{spec-1, spec-2, } \ldots, \textit{spec-n} \\ WHILE \text{ (expression)} \end{Bmatrix} \end{bmatrix};$$

. . .

END *label;*

where the range is denoted by the first succeeding *END* statement that has the same label as the *DO*. The *spec-i* can be any of the following forms:

> expr-1 [WHILE (expr-4)]
> expr-1 BY expr-3 [TO expr-2] [WHILE (expr-4)]
> expr-1 TO expr-2 [BY expr-3] [WHILE (expr-4)]

If *BY* is omitted, the value +1 is assumed; if *TO* is omitted, the largest positive value representable is assumed. The following examples show a few cases:

> A: DO X = 1 WHILE (Y<A), 5 BY 2;
>   . . .
> B: DO WHILE (K <= 4);
> C: DO I = −4 TO −19 BY −3 WHILE (J = 0)
>   . . .
>   END C;
>   END B;
>   END A;

Note that it would be illegal to reverse the position of the *END* statements.

The input/output statements are *GET*, *PUT*, and *LIST*. Their formats are as follows:

> GET LIST (var-1, var-2, ..., var-n);
>
> PUT $\begin{Bmatrix} \text{LIST} \\ \text{IMAGE} \end{Bmatrix}$ (e-1, e-2, ..., e-n) [label];
>
> LIST $\left[\begin{Bmatrix} \text{line-number-1} \\ \text{label-1} \end{Bmatrix} \left[\text{THRU} \begin{Bmatrix} \text{line-number-2} \\ \text{label-2} \\ \ldots \end{Bmatrix} \right] \right]$;

The *GET* causes the system to obtain (or prepare to obtain since *GET* can be both collect and direct) data from the terminal. Values can be assigned to both single variables and arrays. The *PUT* provides output to the terminal, with optional formatting and spacing, and allows expressions rather than just variable or array names. The *IMAGE* option in the *PUT* is a format statement providing various types of numeric conversion, specific number and placement of characters, or other text reproduced as it was input. The physical placement of the carriage while typing the *IMAGE* statement is used to specify the output positioning, so the user can easily see his format.

The *LIST* command causes printing of the specified portion of the program.

There are a number of built-in functions provided. In addition to the fairly standard mathematical ones, there are functions for determining *maximum, minimum, absolute value, floor, ceiling, truncation, sign,* and *remainder.* The routines for *sine, cosine,* and *arctangent* are available in both radians and degrees.

The ERASE command has the same format as the LIST command and causes the deletion of the specified statement(s).

The STOP terminates execution. It can be resumed with the statement following the STOP by using one of the EXECUTE forms in which the line number is omitted.

An interesting command is the RESEQ, which causes immediate resequencing of a program based on the information shown in the command; the format is

$$\text{RESEQ } [\textit{line-1}] \left[\text{THRU } \left\{\begin{matrix}\textit{line-2}\\ \ldots\end{matrix}\right\}\right] [\text{BY } \textit{incr}] [\text{FROM } \textit{line-3}];$$

where the line numbers identify the beginning and ending lines of the set to be resequenced and the increment to be used is shown. The *line-3* specifies the first line number to be assigned. Standard values are defined for each optional parameter. Thus the user can write

% RESEQ 2 THRU ...

% RESEQ 7. THRU 105 BY 10 FROM 1;

% RESEQ BY 100;

There are three declarations: DECLARE, IMAGE, and LET. The DECLARE (also written DCL) provides the dimension information for arrays, e.g., DCL A(2:15, −2:8, 7). The dimension ranges from the first number to and including the second number in steps of 1. The second number must be greater than or equal to the first. If a single number is shown, it is treated as the upper dimension with an implied first number of 1.

The IMAGE provides formatting information; this was mentioned earlier. The LET provides for function definitions by writing

$$\text{LET } \textit{function-name } (p_1[,p_2, \ldots, p_n]) = \textit{expression};$$

where the p_i are parameters, e.g.,

LET ROOT (A,B,C) = (−2*A + SQRT(B**2 − 4*A*C)

Other commands provided in CPS relate to its usage as an on-line system.

Of the statements which have been discussed here, the following can only be used as direct statements (which means they *must* be preceded by a % when used): *ERASE, EXECUTE, RESEQ, LIST*. All other executable statements can be either collect or direct. The three declarations can be used as collect statements only.

CPS provides a very small PL/I-like subset with some additional commands to make the terminal usage more efficient. It thus can provide a nontrivial introduction to a complex programming language to someone who has never programmed before. It is also a useful on-line system for experienced programmers with small problems, and has been successfully used by administrators for simple clerical, bookkeeping, and budgeting activities.

IV.6.6. MAP

MAP (*M*athematical *A*nalysis *P*rogram) is a system developed to run under M.I.T.'s Compatible Time-Sharing System (see Crisman [ZR65]) by R. Kaplow, S. Strong, and J. Brackett, all of M.I.T. It provides a higher level of required (and permitted) interaction and higher level mathematical commands than most of the other systems. The manual by Kaplow, Strong, and Brackett [KP66] states in the introduction that "The system is intended for the solution of mathematical problems. It should be usable by a person with no knowledge of computers or programming and little knowledge of numerical analysis." That manual, together with the paper by Kaplow, Brackett, and Strong [KP66a] constitute the primary documentation of the system. MAP has been used by Professors Averbach and Kaplow in several physics courses to permit the students to do computational problems related to the subject of the lectures or the experiments. A comparison of this system with AMTRAN (Section IV.6.11), Culler-Fried System (Section IV.6.9), and the Lincoln Reckoner (Section IV.6.7) is given by Ruyle, Brackett, and Kaplow [RU67].

The system is designed for use with the normal communication devices for CTSS, namely teletypes and IBM 1050's. There is also provision for using two types of display equipment: The M.I.T. Electronics Systems Laboratory terminal (see Stotz [UU63]) or storage oscilloscopes (see Stotz and Cheek [UU67]). These are not generally available, however, because of the cost.

The arithmetic operators are the five basic ones. Data names consist of one to six characters, but they must not end in the letter f and must not conflict with a few specified key words. Arrays play a very important role in MAP but the system makes a distinction between what would normally be considered a subscript and what is really the argument of a function. It is the latter which is used in MAP.

Although there are provisions for stored programs (as seen later), the primary usage of the system is through constant communication with the computer. Up to 72 characters per line can be used and each message by user and system is terminated by a carriage return; if the expression to be computed is too long for one line, the user gives a carriage return anywhere prior to the end (but not in the middle of a name). The computer will then respond by noting that the number of left and right parentheses do not match and the user has the option of continuing the expression or correcting a mistake. The system also has a very elaborate set of printouts which convey information to the user and/or tell him what input is needed; these messages can be cut short by the user who is familiar with the system.

There are two basic types of statements initiated by the user. The first requests the computation of an expression or a function; the second specifies one of the higher level mathematical operations to be performed. In the first case, the user would type something of the form

$$(v=3.295 ** (a * 3. * 10 ** (-5)))$$

where the outer parentheses are required to indicate that an assignment is involved. If no value has been assigned to a, then the system responds with

DECIMAL VALUE OF CONSTANT A PLEASE

The user then types in the value of the parameter and the system then responds

COMMAND PLEASE

This means that the computation has been completed; the results are not automatically printed out.

The basic element of data is a mathematical function of a single variable. If the user types an assignment statement involving functions which are undefined, e.g.,

$$(g(y) = h(y) * cosf(i(y)))$$

then the system will type back[19]

H(Y) IS NOT DEFINED. IF IT HAS A DIFFERENT NAME, TYPE THE NAME. IF YOU WANT TO TYPE IN NUMERICAL VALUES NOW, TYPE THE WORD INPUT. OTHERWISE, GIVE A CARRIAGE RETURN AND DEFINE THE FUNCTION BEFORE USING THE NAME AGAIN.

The user is thus given the opportunity either to define the function in terms of some known quantities or to define it by means of tabulated values which are typed into the computer. If he chooses the latter, the system asks him to

---

[19] Kaplow *et al.* [KP66], p. 13.

specify the minimum, maximum, and incremental values for the independent variable y. These are used to compute the subscripts. Thus if the user specifies −3.5, 1.5, and .5, the system will have the subscripts range from −7 to 3 in steps of 1. The user can specify a different set of values of the independent variable for each occurrence of a different function using that variable; the system will automatically use the old values if no new ones are given.

After the values of the independent variable are specified, the system gives a five-line message telling the user how to type in the values of the function. There is an editing command which permits corrections to be made and a provision for filing the data for later use. This process is repeated for each function as needed. The user can suppress the five-line message by typing the word no after the word input.

The specific mathematical functions of a single argument available for use in expressions include *sine, cosine, tangent, cotangent, arcsine, arccosine, arctangent, hyperbolic sine, cosine* and *tangent, exponential* and *logarithm* (base *e*), *absolute value* and *square root*. The code name for each of these ends in the letter f.

There are functions whose argument is an expression containing at least one function. These are[20]

    sumf(expr(y))   Sums over all values of expr(y) from the minimum to the maximum.
    intf(expr(y))   Computes the definite integral over the whole range of expr(y) using Simpson's rule.
    derif(expr(y))   Computes derivative of expr(x) at the same values of y for which expr(y) is tabulated.

In addition to these, there are other more complex mathematical commands, which the user invokes by writing the name of the command. Among those available are[21]

    integrate   (between fixed or variable limits.)
    basis   G(y)→G *(any function of y)*.
    transform   Fourier sine, cosine, or sine and cosine.
    convolv   Folding of two functions.
    least square   Least square analysis.
    minimax   Changes the range of definition of a function.
    select   Manipulation of a portion of a function.

There are also matrix operations similar to those in the Lincoln Reckoner (see Section IV.6.7).

---

[20] Kaplow *et al.* [KP66], p. 15.
[21] Kaplow *et al.* [KP66], p. 16.

least square
I CAN FIT EQUATIONS OF THE FORM
V(Y)=XA*FA(Y) + XB*FB(Y) + XC*FC(Y) + XD*FD(Y) + XE*FE(Y)
WITH A MAXIMUM OF 5 UNKNOWNS, XA, XB, ETC., AND 100 DATA POINTS.

WHAT IS THE NAME OF THE VARIABLE COMPARABLE TO V(Y).   data(x)
HOW MANY FUNCTIONS, FA(Y), FB(Y), ETC., WILL BE REQUIRED TO FIT THE DATA.   4
PLEASE PRINT ON THE NEXT LINE THE NAMES OF THE 4 FUNCTIONS REQUIRED.

        a(x)   b(x)   c(x)   d(x)

THE EQUATIONS RESULTING FROM THE LEAST SQUARE ANALYSIS ARE

   .3133E03 = .9455E02 *XA+   . . .   + .3137E02 *XD
   .5072E02 = .8074E01 *XA+   . . .   + .2993E01 *XD
   .9911E02 = .1806E02 *XA+   . . .   + .7587E01 *XD
   .1123E03 = .3137E02 *XA+   . . .   + .1231E02 *XD

LEAST SQUARE SOLUTION.   THE FOLLOWING ARE THE 4 UNKNOWNS CORRESPONDING TO XA, XB, ETC.

    .2162E 01   .1309E 01   .3399E 01   .1273E 01

ESTIMATES OF THE ERROR IN THESE VALUES ARE

    .1426E−04   .1091E−03   .6910E−04   .1344E−03

DO YOU WANT THE VALUES OF THE FITTED CURVE PRINTED.   THESE VALUES ARE AVAILABLE FOR FURTHER CALCULATIONS AS THE FUNCTION 'FITTED (X)'.   TYPE YES OR NO AND GIVE A CARRIAGE RETURN.   yes

THE FOLLOWING ARE THE VALUES OF THE FITTED CURVE AT VALUES OF X FOR WHICH DATA (X) IS TABULATED.

    .2583E01   . . .   .4656E01
    .5091E01   . . .   .6745E01
    .6926E01   . . .   .7160E01
    .7123E01   . . .   .6984E01
    .6870E01   . . .   .6999E01

DO YOU WANT THE DIFFERENCES BETWEEN THE FITTED AND ORIGINAL CURVES PRINTED.   yes

    −.2498E−01   . . .   .1683E−01
    .1164E−01   . . .   .1398E−01
    −.1469E−01   . . .   .1398E−01
    .1585E−01   . . .   −.8650E−02
    −.1400E−01   . . .   .9052E−02

COMMAND PLEASE

Figure IV-17. Example of the use of Least Square command in MAP. Words and symbols in lower case are input by the user. Upper-case characters and all of the numbers are typed out by the system. Dots are used in place of numerical results which were omitted in this figure, although they appeared in the actual example.
Source: Kaplow, Strong, and Brackett [KP66], pp. 22–23.

The most interesting thing about these commands is that the user does not have to know what information is to be supplied. For example, see Figure IV-17, where the upper-case characters are those printed by the system and the lower-case ones are user input. (The dots represent actual values which would be typed out.) On the other hand, if the user is familiar with the system, he can merely type in

$$\text{least} \quad \text{data}(x) \quad 4 \quad a(x) \quad b(x) \quad c(x) \quad d(x)$$

which accomplishes the same thing as the first eight lines in Figure IV-17.

Input is either typed in directly upon request from the system or it can be loaded off-line and then retrieved and referenced by the user.

The user obtains output by writing *print* and the name of the function; he can also specify the range desired. MAP can be used to generate graphs either on the ESL display console or on storage oscilloscopes, and the user can use the *plot* command with a choice of linear, log-log, linear-log, or log-linear scale. By using the *compare* command, a scaled graph is obtained by plotting the designated function(s) on the vertical axis and the values of another function on the horizontal axis.

Commands for examining and handling data are as follows:

data	System types list of all temporary files.
data restore	Types *data* list and then erases all data on the list.
delete	Erases files from disc storage.
data update	Moves information from temporary to permanent storage.
data delete	Equivalent to *data update* and *data restore*.

Although the facilities above do not provide stored programs, they can be formed. By writing *create*, the user can define a sequence of 18 or fewer lines of MAP statements to be defined and stored. There are no control sequence statements, however, e.g., IF, GOTO. When retrieved, each command is handled as if it had just been typed by the user. These command sequences can themselves contain references to other stored programs, with a restriction to a nesting of three such references.

The user is able to execute programs written in other languages by writing *execute prog*. Hence routines can be written in MAD, FAP, AED, or any other language available under CTSS. The MAP routines are available as subroutines and can be invoked from those languages.

Work on a newer version of MAP started around the middle of 1967 with participants from M.I.T. and Bell Telephone Laboratories (Whippany, N. J.). It is initially being implemented on the GE645 under MULTICS. Among the planned major differences or improvements are the following:

1. The command *help* is used to request detailed interrogation and explanation. Otherwise, the input must be correct or a brief (but mnemonically understandable) message will be typed out. This contrasts with the verbosity of the first system if any assistance is needed.
2. All entities have a single name. No information about the dimension or type of the data is contained in the name. The data type can be integer, real, or complex, and real numbers can be single- or double-precision.
3. Arrays of two and three dimensions can be handled.
4. The equation format is extended to include operations on matrices and functions of two variables. Matrix operations similar to those provided in the Lincoln Reckoner (see Section IV.6.7) are provided.
5. Control facilities of the type normally appearing in a complete programming language are provided, e.g., control transfers, logical operations, and parameter passage for subroutines.
6. The user is able to obtain time estimates for certain operations after supplying information about his data.
7. Expanded graphical capabilities are made available.
8. An attempt is being made to provide a framework and design to permit the addition of symbol manipulation operations (e.g., differentiation and substitution).

IV.6.7. LINCOLN RECKONER

The Lincoln Reckoner is an on-line system developed at Lincoln Laboratory under the direction of D. B. Yntema by R. A. Wiesen, A. N. Stowe, J. Forgie, and others. It runs on the experimental computer TX-2, under the APEX time-sharing system (see Forgie [FY65]). In the words of Stowe *et al.* [SW66], "The Reckoner is primarily a facility for making use of routines, not for writing them". Furthermore, rather than providing a general facility, "it offers a library of routines that concentrate on one particular application, numerical computations on arrays of data.".[22] A compiler called *Junior* for a small language with some of the facilities of FORTRAN or ALGOL is also available. The intended users of the system are engineers and scientists who wish to work with data from a laboratory experiment or explore small- or medium-sized computations. In the paper by Stowe *et al.* [SW66], three user case studies are described. A comparison of this system with AMTRAN (see Section IV.6.11), the Culler-Fried system (see Section IV.6.9), and MAP (see Section IV.6.6) is given by Ruyle, Brackett, and Kaplow [RU67].

A name can be 50 characters long and may contain any combination of digits, periods, and Roman capital letters, except that it must contain at least one letter. Because of the internal workings of the system, it is unwise

---

[22] Stowe *et al.* [SW66], p. 433.

to choose a name beginning with a period or a digit. There is a command for typing a synonym for a given name, thus permitting easy abbreviations for long names.

The basic data element is an array, and the basic operations of the system are those which operate on one or more than one array, e.g., transpose, multiply, invert, etc.

The statements available are relatively high level mathematical operations and can operate on arrays. For example, to multiply two matrices, the user might write

MATMUL YANS ZANS RESULT
INVERT RESULT NEWRESULT

which causes the name **RESULT** to be assigned to the product of the matrix **YANS** by **ZANS**, and this is inverted and given the name **NEWRESULT**. Arrays can be added, subtracted, multiplied, and divided. The functions of *reciprocal, square root, log* (base e or 10), *exponent, sine*, and *cosine* can be applied to a single array. Individual elements can be extracted and operated on. Other types of operations available include *array shuffling* (copy part of an array, join arrays), *matrix operations* (e.g., calculate determinant, eigenvectors), *control* (synonym, drop an array), and *input/output*. A lengthy list is given in Stowe *et al.* [SW66]. The format for invoking each of these usually consists of a key word followed by a sequence of variables relevant to the routine.

In order to create subroutines, the user writes

BUILD NAME

and the system responds

GO ON

whereupon the user types the body of the subroutine. When finished, he types

FINIS

and the system responds

OK

Statements are written one on a line and executed immediately. They can be numbered, with up to four digits before and after the decimal point. If the user does not type line numbers, the system will automatically supply integers, i.e., 1, 2, 3, etc. Lines are replaced and deleted by numbers, and new lines are inserted by giving intermediate line numbers. Conditional and unconditional jumps are identified by a pointing finger ☞ . Thus

7.3 BETA = GAMMA

means transfer control to line 7.3 if BETA equals GAMMA.

To define formal parameters, the user types their names preceded by a code *p* as

p VAR1 VAR2 VAR3

Temporary variable names in the subroutine are written with a period preceding the name and the system automatically assigns the name of the subroutine as a *local name*.

The subroutine is invoked merely by writing its name (and the parameters if there are any). While executed, a subroutine may be interrupted by the interrupt button, an error message, or something in the subroutine itself. The subroutine execution is suspended, control is returned to the keyboard, and the user can print results, do other calculations, or define and/or execute other subroutines. If desired, he can then resume the interrupted subroutine.

One of the main concepts in the Reckoner is that it is really a collection of programs which can work together; this concept is referred to by the developers as *coherence*, which means that the results of any program in the set can be used as inputs by any other program for which the results are meaningful.

IV.6.8. APL/360 AND PAT

APL (standing for *A Programming Language*) was originally developed by Iverson [IV62] and is discussed in Section X.4. It is pointed out there that the greatest handicap to any widespread use of the language is its notation, including a large character set and proliferation of unusual symbols. Some of this difficulty has been eliminated in the on-line subset denoted by APL/360 which runs as a stand-alone interpretive system on the IBM 360/50 at the IBM T. J. Watson Research Center; it is described in Falkoff and Iverson [FA67] and [FA67a]. The key people in developing this running system were K. E. Iverson, A. D. Falkoff, L. M. Breed, P. S. Abrams, R. D. Moore, and R. H. Lathwell. The full language was used by Rose [RJ66] to describe an assembly language, and a small portion was actually run on APL/360. The latter was also used to program and run algorithms to detect logic circuit failures (see Roth, Bouricius, and Schneider [RQ67]), to help solve mathematical computations arising in physics, and also to program COGO (described in Section IX.2.2.1) for inclusion in the system itself. In fact, it is somewhat misleading to place this discussion in a chapter on languages for numerical scientific problems since APL is actually useful in many other areas. Its inclusion here, however, makes it easier for the reader

SAMPLE PROGRAM—APL/360†

Problem: Matrix inversion by Gauss-Jordan elimination with pivoting.

Program:

```
        ∇ B←REC A;P;K;I;J;S
   [1]    →3×⍳(2=⍴⍴A)∧=/⍴A
   [2]    →0=⍴⎕←'NO INVERSE FOUND'
   [3]    P←⍳K←S←1⍴⍴A
   [4]    A←((S⍴1),0)\A
   [5]    A[;S+1]←S⍺1
   [6]    I←J⍳⌈/J←|A[⍳K;1]
   [7]    P[1,I]←P[I,1]
   [8]    A[1,I;⍳S]←A[I,1;⍳S]
   [9]    →2×⍳1E⁻30>|A[1;1]≠⌈/|,A
  [10]    A[1;]←A[1;]÷A[1;1]
  [11]    A←A-((~S⍺1)×A[;1])∘.×A[1;]
  [12]    A←1⌽[1]1⌽A
  [13]    P←1⌽P
  [14]    →5×⍳0<K←K-1
  [15]    B←A[;P⍳⍳S]
        ∇
```

†Falkoff and Iverson [FA67a]. By permission of Association for Computing Machinery, Inc.

to contrast it with other small on-line systems. Further references for the full language are given Section X.4.

The input device is an IBM 1050 or 2741, with a very special Selectric typewriter ball. The character set consists of the 26 upper-case letters, the 10 digits, and the following 52 characters:

```
+   -   ×   ÷   *   <   ≤   =   ≥   >   ≠   ⁻
.   :   ,   ;   ¨   (   )   [   ]   ?   '   _   ∘
∨   ∧   ←   ↓   →   ↑   \   /
⊤   ⊥   |   ∩   ∪   ⊃   ⊂
⌈   ⌊   ~   ∆   ∇   ○   ⎕
α   ω   ε   ρ   ⍳
```

The five arithmetic operations are denoted by the first five characters shown. The use of these and other operators (called *scalar functions* in APL) are shown in Figure IV-18. Most of the symbols involved are both unary and binary; in the latter usage the symbol is considered binary wherever it is preceded by an operand.

A data name or function name can be any string of letters and/or digits beginning with a letter. Literals are enclosed in quote marks. Mathematical expressions are evaluated from right to left within parentheses and therefore

Name	Symbol	Definition or Example
Multiplication	A × B	
Addition	A + B	
Division	A ÷ B	
Subtraction	A − B	
Negation	− B	
Exponential	A * B	
Natural exponential	* B	$e * B$, where $e = 2.71828...$
Minimum	A ⌊ B	
Maximum	A ⌈ B	
Floor	⌊ B	Integer part of B ⌊3.14 ≡ 3 ⌊3 ≡ 3 ⌊ − 3.14 ≡ −4
Ceiling	⌈ B	−⌈−B ⌈3.14 ≡ 4 ⌈3 ≡ 3 ⌈ − 3.14 ≡ −3
Residue	A ∣ B	B − A × ⌊B ÷ A 3∣4 ≡ 1 3∣3 ≡ 0 3∣ − 4 ≡ 2
Absolute value	∣ B	B⌈ − B
Less than	A < B	
Less than or equal	A ≤ B	Relations: result is 0 if the relation does not hold and is 1 if it does. Examples:
Equal	A = B	I = J is the Kronecker Delta function
Greater than or equal	A ≥ B	For logical arguments (0 or 1) A ≠ B is the exclusive-or
		3 < 4 ≡ 1 4 < 4 ≡ 0 1 ≠ 0 ≡ 1
Greater than	A > B	3 > 4 ≡ 0 3 = 3 ≡ 1 1 ≠ 1 ≡ 0
Not equal to	A ≠ B	
And	A ∧ B	A ∧ B ≡ A ⌊ B
Or	A ∨ B	A ∨ B ≡ A ⌈ B For arguments A and B restricted to the values 0 and 1
Not	∼B	∼B ≡ 1 ≠ B
Combinations†	A ! B	B things taken A at a time: 1!4 ≡ 3!4 ≡ 4; 2!5 ≡ 3!5 ≡ 10 5!4 ≡ 0
Factorial†	! B	!B ≡ B × !B − 1 if B is an integer. In general !B ≡ ⌈B + 1

†The symbols for most standard functions are single characters, but some (such as !) are formed by backspacing and overstriking.

Figure IV-18. Standard scalar operators in APL.
Source: Falkoff and Iverson [FA67], p. 25.

do not obey the normal arithmetic precedence rules; i.e., A × B + C is evaluated as A × (B + C). Parentheses should be used to ensure the proper grouping. Square brackets are used to enclose subscripts.

Expressions put into the system will be evaluated immediately and the result printed out. If a name is assigned, the result is not displayed but can be obtained upon request. Thus

$$3 \times 4$$

will cause 12 to be typed, whereas

$$A \leftarrow 3 \times 4$$

does not; the left-pointing arrow is used to denote an assignment statement. By typing the statement

$$A$$

the result 12 is printed out.[23]

The user types in a single statement and it is immediately executed, unless he defines a function. Hence, branching is meaningful only within functions, whose method of definition is described later. A branch is designated by →.

There are a number of operations on vectors and matrices, many of which can be extended to higher-rank arrays. *Catenation* of vectors is defined as the concatenation of the elements of the vectors; catenation is denoted by a comma; i.e., X, Y is the catenation of vectors X and Y. A vector is defined by using an assignment statement, and components can be obtained using subscripts. Thus

$$X \leftarrow 2,3,5,7$$

followed by X[3, 1, 4] yields the value 5, 2, 7. The dimension of a vector Z is denoted by ρZ.

The expression ιN yields a vector of the first N positive integers, while $N\rho X$ yields a vector of dimension N whose components are the successive elements of X, repeated cyclically if necessary. Thus $6\rho\iota 4$ is the vector 1, 2, 3, 4, 1, 2.

Component-by-component addition (or any other arithmetic operation) of two vectors with the same dimension is indicated with the plus sign (or the appropriate operator), e.g., X+Y.

The sum-reduction of a vector X is denoted by +/X and it is the sum of all the components of X. More generally, for any binary scalar function #, the expression #/X is equivalent to

$$X[1] \# X[2] \# \ldots \# X[\rho X]$$

Thus

$$+/2,4,5,9$$

equals 20.

The statement $M \leftarrow (A, B)\rho X$ defines M as a matrix of A rows and B columns whose elements in row-major order are the successive components of X replicated cyclically if necessary. Writing M[; J] yields the Jth column, while M[I ;] selects the Ith row. The expression ρM produces the vector (A, B). For example,

---

[23] The examples in this section are taken from Falkoff and Iverson [FA67].

$$M \leftarrow (3,4) \rho \iota 12$$

defines M as

$$\begin{array}{cccc} 1 & 2 & 3 & 4 \\ 5 & 6 & 7 & 8 \\ 9 & 10 & 11 & 12 \end{array}$$

Then M[3;] yields

$$9, 10, 11, 12$$

and +/M yields

$$10, 26, 42$$

and +/[1]M yields

$$15, 18, 21$$

A vector X can be *compressed* by writing Y←U/X where U is a *logical* vector, i.e., it contains only 1's and 0's; the result Y consists of the elements of X corresponding to the 1's in U, and the dimension of Y is obviously +/U. The converse operation of *expansion* is denoted by X←U\Y and defined by U/X ≡ Y and (~U)/X ≡ (+/~U), ρ0 (where ≡ means equivalence). If X is a matrix, then U/X compresses each row of X by the vector U. Similarly, U/[1]X denotes column compression. One of the uses of compression is in a conditional branch since → (X < Y)/5 branches to 5 if X is less than Y (but does not change the sequence otherwise).

A logical vector of dimension N with N⌊J leading 1's is called a *prefix vector* and is obtained from the expression N α J. Similarly, the suffix vector is obtained from N ω J using trailing 1's. Thus 7 ω 3 yields 0, 0, 0, 0, 1, 1, 1.

Cyclic left and right rotations of a vector X by K places are denoted by K↑X and K↓X, respectively. If A←1, 2, 3, 4, 5, then 2↑A yields 3, 4, 5, 1, 2.

The expression R⊥X denotes the value of the vector X evaluated in a number system with radices R[1], R[2], etc. Thus (R⊥X)≡+/W×X where W is the weighting vector determined as follows: W[ρW] ≡ 1 and W[I−1] ≡ R[I] × W[I]. The representation function RTN denotes the base representation of the scalar N; thus if Z ≡ RTN, then ((×/R)|N − R⊥Z) = 0. For example, writing (24, 60, 60)⊥1, 2, 3 yields 3723 seconds.

Operations of ranking, transposition, and reversal are also defined.

The regular matrix (i.e., inner) product is denoted by C ← A +.× B and defined by

$$C[I;J] \equiv +/A[I;] \times B[;J]$$

The outer product is denoted by

$$Z \leftarrow X \circ . \times Y$$

Other operators can be used in both the inner and outer product forms.

Set theoretic operations of membership, intersection, difference, and union are defined as follows: If X and Y are vectors, then $X \in Y$ produces a logical vector U of dimension ρX such that $U[I] \equiv 1$ if $X[I]$ is a member of Y. The intersection of X and Y is denoted by $X \cap Y$ and defined by

$$X \cap Y \equiv (X \in Y)/X$$

The difference is denoted by $X \sim Y$ and defined as

$$X \sim Y \equiv (\sim X \in Y)/X$$

The union of X and Y is denoted by $X \cup Y$ and defined as

$$X \cup Y \equiv X, Y \sim X$$

Function definitions are delimited by writing the character ∇ at the beginning and end. Statement numbers are typed automatically by the system, but they may be input by the user by typing $[n]$ where n is an integer and can be followed by a decimal point and up to four digits. Statements are ordered according to their statement numbers, and so they can be easily replaced, deleted, or inserted. The method of invoking the function is defined in the first line, e.g.,

$$\nabla D \leftarrow A \; B \; C$$

Functions can have 0, 1, or 2 arguments and 0 or 1 results. The use of the right-pointing arrow denotes a branch to the indicated statement number. The arrow can be followed by any expression, and its value determines the statement to which control is transferred. A variable name followed by a colon can precede a statement; in such a case, the variable is given the value of the statement number and is called a *label*.

A trace of a function P can be obtained by writing

$$\Delta P \; i, j, \ldots$$

and then statements $i, j, \ldots$ will have the value of the result variable printed out each time P is executed.

A much earlier attempt to implement a subset of the programming language defined by Iverson was the *P*ersonalized *A*rray *T*ranslator (PAT)

System described by Hellerman [HH64]. This was implemented on a 1620 which permitted a single user on-line interaction through typewriter-entered console commands as well as sense switches.

The data types permitted are Boolean, floating point, and alphanumeric. Each variable denotes one- or two-dimensional arrays. The executable commands are primarily designed for use on vectors and/or matrices rather than single elements. Blanks are critical. Each statement contains a single operator, and each of the common ones automatically applies to arrays of the same dimension on an element-by-element basis. Thus, writing $A=B+C$ causes the corresponding elements of B and C to be added, producing a corresponding element for A. To permit a large number of statements, the special character @ is used, followed by a specific code for each operator. The relations equal, not equal, less than, less than or equal, greater than, and greater than or equal are each designated by three bits. In addition to four arithmetic and three logical operators, the system includes some elementary mathematical functions, matrix transposition, left and right rotation, reduction, base value polynomial evaluation, residue, compression, input/output, and a few other miscellaneous ones. There is a statement comparing two scalars with a three-way branch and a statement which increments one of the variables by 1 and then compare-branches. There is no unconditional transfer; its effect is obtained from the conditional control transfer.

IV.6.9. CULLER-FRIED SYSTEM

As so often happens, the work originally done by G. Culler and B. Fried has appeared in several versions and under differing names. Quoting Ruyle, Brackett, and Kaplow

> The Culler-Fried system is one name for two physically separate but direct descendants of the system developed by Glen Culler and Burton Fried at Thompson Ramo Wooldridge, Canoga Park, California, beginning in 1961 [Culler, Fried [CU63]; [CU65]]. The first of these is the On-Line Computer System (OLC) at the University of California at Santa Barbara (UCSB) [Winiecki [WN66]] which operates on the same computer as the original system. The OLC system is used for research and teaching on the Santa Barbara campus as well as from remote terminals at UCLA and Harvard. The second, an expanded version of the original system, has been implemented at TRW Systems (formerly Space Technology Laboratories) in Redondo Beach, California for the use of scientists and engineers and has been operating since late 1964 [Fried, Farrington, Pope [FQ64], Fried [FQ66]]. The original system, which in some ways was more sophisticated than the present versions, was

widely used by a variety of scientists and numerical analysts [Culler and Huff [CU62]], and significantly influenced the initial development of AMTRAN.[24]

A comparison of the Culler-Fried system with AMTRAN (see Section IV.6.11), the Lincoln Reckoner (see Section IV.6.7), and MAP (see Section IV.6.6) is given in the paper cited above.

Only a brief description of the basic ideas will be given here, without concern to which system is being described, because the concepts are completely dependent upon very special equipment. For that reason, it completely violates the characteristic of relative machine independence which was cited in Chapter I as a defining characteristic for programming languages. Nevertheless, this work is of sufficient interest to be mentioned briefly.

The system is implemented on an RW-400 and has two special keyboards. One is used for inputting alphanumeric information (i.e., operands), while the other is a set of push buttons representing the operators; both are in the system on each of several "levels". The basic levels are

I —Real functions or vectors.
II —Matrices.
III —Display operations.
IV —Complex functions or vectors.
V & VI—Systems management and data transfer.

The operations on levels I and IV include, in addition to normal arithmetic, the elementary functions (exp, sin, log, square root, etc.) for functional operands. Thus, an addition on level I would involve $N + 1$ additions of real numbers, and the operation *EXPON* would yield the exponential of all $N + 1$ values constituting the argument function. There are also operations which apply only to functions, such as differentiation and integration. (Note that these are definitely numeric, not symbolic.) Facilities for displaying a curve in graphical form, showing the scale of a curve, obtaining numerical values of individual points, etc., are available on Level III. All these are shown on a display scope. "The total capability provided to the user is a fair representation of that part of mathematics known as classical analysis which forms the mathematical basis for most theoretical work in the physical sciences and engineering."[25]

As an illustration[26] of an elementary operation to be performed, the user might push the *INTERVAL* button on Level III and type in some num-

---

[24] Ruyle, Brackett, and Kaplow [RU67], pp. 151–52, but presenting the reference citations in the form that is being used in this book. By permission of Association for Computing Machinery, Inc.
[25] Culler [CU65], p. 70.
[26] Based on examples in Culler and Fried [CU65], pp. 71–72.

ber, say 99. Until this number is changed, operations will deal with functions represented by 100 points. To define a variable T with the range $0 \leq T \leq 1$, the following 10 keys are pushed:

$$\underline{IJ} + 1 \cdot 0.5 = T$$

which defines the variable T. To compute functions of T, the *LOAD BUTTON* and the subsequent appropriate keys are pressed. For example, to form

$$S = \sin 4e^{T^2}$$

the following keys are pushed:

$$\underline{I}\ LOAD\ T\ SQUARE\ EXPON \cdot 4\ SIN = S$$

The results can be displayed in either discrete or continuous form. One of the significant features of this system is its ability to create programs which can be quickly invoked by pushing a single button. This is done by using a key called *Program* and then specifying a particular button which will be defined by the program which immediately follows it. Each of these keys can in itself be used as components in a new program, thus making it very easy for the user to create complex programs and simply execute them by pushing a single button.

The emphasis in this system is on allowing the engineer or scientist with a complex problem to use an interactive system, with the minimum of physical effort on his part. The values and advantages of this system are most appreciated by people with mathematical problems which are referred to as *fixed point* problems.

A low level list processing facility has been added to this by Blackwell [QK67] and is discussed in Section VI.9.3.

IV.6.10. DIALOG

The DIALOG system is an interesting and unusual approach to the use of an on-line system. It has been in experimental operation at the I.I.T. (Illinois Institute of Technology) Research Institute since February, 1966. It runs under the UNIVAC 1105 time-sharing monitor, and a batch version was implemented on an IBM 7094. Descriptions of the system are given by Cameron, Ewing, and Liveright [CS66] and [CS67]; the former contains a (not very rigorous) syntactic definition of the language.

DIALOG is an algebraic language which uses a stylus and single push button for input; it displays programs and results on a display screen; it also has provisions for obtaining hard copy for the programs and results; and it uses paper tape input.

The user points with the stylus to one of the available characters displayed on the screen; these consist of the 26 upper-case letters, the 10 digits, and the following special characters:

$$
\begin{array}{cccccc}
+ & - & \times & / & \uparrow & \\
\leftarrow & < & > & = & \wedge & \vee \\
\neg & \downarrow & (&) & [&] \\
\sqcup & ? & \subset & . & , & ; & '
\end{array}
$$

The characters $\sqcup$ and $\subset$ represent space and backspace, respectively. The ? is used to remove all characters from the line being created.

The DIALOG interpreter accepts and acts on one character at a time. The character is shown on a line on the screen immediately above the display of the character set. Then the system determines what the legally allowable next characters are and displays *only those* on the screen for the user to choose from. The screen permits the display of portions of the program in addition to the line being created. A summary of the language elements is given in Figure IV-19.

There is a list of key words, and each is surrounded by '. Wherever logically possible, the system will automatically provide the whole word when the initial character is given. Thus, if 'P is entered, the system automatically supplies LOT' since this is the only legal sequence permitted after the 'P.

Variable names begin with a letter followed by up to five alphanumerics. One or two subscripts are permitted, where any expression can be used and the value of the expression is rounded to an integer. Subscripts are enclosed within square brackets and separated by a comma. Literals are delimited by square brackets; the context distinguishes between literals and subscripts.

Statements can be entered with a line number for later execution or without a number for immediate execution. The numbers can contain a decimal point. Sequence of execution is based on the value of the statement number. All statements are terminated by a semicolon. Comments are of the form

statement-number) [any character string] ;

Assignment is designated by ← and control transfer by GOTO. A conditional statement is of the form 'IF' *(conditional expression)* 'GOTO' *number;* where the statement is executed if the conditional expression is true or nonnegative, e.g., 'IF' ((A>B) V (A<C)) 'GOTO' 3.5;. The EXECUTE command causes execution of a single statement or a set of statements, with control returning to the line under the EXECUTE (unless there is a control transfer out or another EXECUTE).

A *PLOT* command is followed by two expressions separated by a comma. Their values are used as the horizontal and vertical coordinates, respectively,

DIALOG Elements	DIALOG Functions
E represents any expression.	'SIN' (E)
L represents a string of keyboard characters.	'COS' (E)
	'ARCTAN' (E)
N represents a line number.	'ARCTAN' (E_1, E_2)
S represents any statement.	'LN' (E)
V represents a variable.	'ROUND' (E)

DIALOG Operations	Type	Format	Execution
$+ \quad - \quad \times \quad / \quad \uparrow$	Console	S	Immediate
$> \quad < \quad \wedge \quad \vee$	Program	N S	Future

DIALOG Statement Formats

'CLEAR' $F_1, F_2, \ldots F_n$;
  where F_i may be N_i or N_i
  'THRU' N_j or 'ALL' or 'PLOT'
'DISPLAY' $F_1, F_2, \ldots F_n$;
  where F_i may be V or $[L]$ or 'PLOT'
'EXECUTE' N_i;
'EXECUTE' N_i 'THRU' N_j;
'FREE-MODE';
'GOTO' N_i;
'HALT';
'IF' (E) S;
  where S is executed if E is not negative or if E is true

'LOAD';
'PLOT' $F_1, F_2, \ldots F_n$;
  where F_i may be (E_i, E_j)
  or $[L]$ (E_i, E_j)
'REQUEST' $V_1, V_2, \ldots V_n$;
'SAVE';
'SENSE' $(V_1, V_2), (V_3, V_4), \ldots (V_n, V_m)$;
'STEP-MODE';
$V \leftarrow E$;
$V[F] \leftarrow E$
  where F may be E_i or E_i, E_j
'WRITE' $F_1, F_2, \ldots F_n$;
  where F_i may be V or $[L]$

Figure IV-19. Summary of DIALOG facilities.
Source: Cameron [CS67], p. 357. By permission of Association for Computing Machinery, Inc.

in an image automatically associated with the *PLOT*; i.e., the command is used to define points. The *SENSE* command senses the coordinates of the stylus and assigns the values to the variables named in the *SENSE* command. A *DISPLAY* command displays on the screen a cited list which can contain literals, variables, or a plot. In the latter case, the information previously prepared by the *PLOT* command is used.

Built-in functions are *SIN*, *COS*, *ARCTAN*, *LN*, and *ROUND*.

The *CLEAR* statement erases the designated portion of the program or the *PLOT*. The *HALT* returns control to the user from the stored program. A *STEPMODE* causes line-by-line execution; in this mode the result of each assignment statement is automatically displayed. The *FREEMODE* restores the system to normal operation after the *STEPMODE*.

A *REQUEST* command requires input from the console for the variables in the list, e.g., 'REQUEST' A, B[Y,2];.

A *LOAD* command permits input from a photoreader of a paper tape prepared off line, normally consisting of a stored program. The SAVE program causes punching out on paper tape of the stored program in a form suitable for reentry by the LOAD command. A WRITE command prepares a magnetic tape for off-line printing.

There are no declarations required, even for type or dimension information.

IV.6.11. AMTRAN

The AMTRAN (*A*utomatic *M*athematical *TRAN*slation) system was developed at NASA (National Aeronautics and Space Administration), Huntsville, Ala. The work was initiated by R. N. Seitz, and the key people involved were L. H. Wood, P. L. Clem, Jr., J. Reinfelds, and M. A. Sparks. All except Clem and Sparks were from NASA Marshall Space Flight Center.

The basic objective of the work was to provide an on-line system to facilitate the solution of mathematical problems by nonprofessional programmers. Although the developers state that one of the basic goals is "To use the natural language of mathematics as a programming language without any arbitrary restrictions whatsoever",[27] it is clear that this objective at least is not achieved.

The overall system was inspired by the work of Culler and Fried (see Section IV.6.9) and was influenced by JOSS (Section IV.6.2) and the work at the Hudson Laboratories of Columbia University by Klerer and May (Section IV.7.5). The documentation of the system appears in a number of NASA reports and in published articles by Clem *et al.* [CM66], Reinfelds *et al.* [RF66], Wood *et al.* [WD66], and Seitz *et al.* [UV67]. A comparison of this system with MAP, Lincoln Reckoner, and the Culler-Fried system is given by Ruyle, Brackett, and Kaplow [RU67].

Unfortunately, any discussion of AMTRAN is clouded by the difficulty of separating what has actually been accomplished from what is planned and from what is claimed; the published reports do not agree in their technical descriptions. While I have made an attempt to make this delineation, there is far less certainty of accuracy than in most of the other descriptions in this book.

An operational version of the system was implemented as an interpreter on the IBM 1620, which is rather surprising since this seems a rather complex system for such a small machine. There is one large specially designed keyboard with 224 push buttons that includes a typewriter keyboard with a special character set. The user can define operations besides those automatically provided by the system (approximately half the total number) and

---

[27] Reinfelds, *et al.* [RF66], p. 469.

SAMPLE PROGRAM—AMTRAN†

Problem: Adaptive Runge-Kutta-Simpson's rule integration.

Program:
1. N,A1,X1,X2,Y,ENTRY,H=(X2−X1)/10,X=X1,Y2=Y,I=0.
2. J=0,REPEAT N,Y3 SUB I=Y SUB J AND I=I+1 AND J=J+1.
3. IF X GT=X2, THEN GO TO 19.
4. Y1=Y.
5. RUN,A=H Z,X=X+H/2,Y=Y1+A/2,F=A.
6. RUN,B=H Z,Y=Y1+B/2.
7. RUN,C=H Z,X=X+H/2,Y=Y1+C.
8. RUN,D=H Z,Y2=Y=Y1+(A+2 B+2 C+D)/6.
9. RUN,A=H Z,X=X+H/2,Y=Y2+A/2,G=A.
10. RUN,B=H Z,Y=Y2+B/2.
11. RUN,C=H Z,X=X+H/2,Y=Y2+C.
12. RUN,D=H Z,Y=Y2+(A+2B+2C+D)/6,RUN,D=H Z.
13. E=ABS(Y−Y1−(F+4 G+D)/3)/(.000001 ABS Y+A1), F=LOG MAGNITUDE K+1.
14. IF E GT 1, THEN X=X−2H AND Y=Y1 AND IF E GT 2,H=H/(EXP(F LN 1.5848931))
14.C AND GO TO 3,OTHERWISE H=H((.5/E)**.2) AND GO TO 3.
15. IF E LT=.016,THEN H=2 H,OTHERWISE H=H((.5/E)**.2).
16. J=0,REPEAT N,Y3 SUB I=Y2 SUB J AND I=I+1 AND J=J+1.
17. J=0,REPEAT N,Y3 SUB I=Y SUB J AND I=I+1 AND J=J+1.
18. GO TO 3.
19. Y3.
20. NAME.THIS RUNGE1   104.

---

†Seitz, Wood, and Ely [UV67]. By permission of the Association for Computing Machinery, Inc.

---

assign these to a specific button. One version of the system called the *Sampler* uses only the regular console typewriter and card reader of the 1620; mnemonic labels are used to call the operations otherwise provided by the special keyboard. It is the (larger) combined system which is being described here. It includes 5-inch display scopes. The program (or other desired information) can be displayed on the scope and also printed as hard copy.

The typewriter has the following character set (in addition to the 26 upper-case letters and the 10 digits):

$$+ \quad - \quad \times \quad / \quad *$$
$$< \quad \leq \quad = \quad \geq \quad > \quad \equiv$$
$$[\quad] \quad (\quad) \quad \rightarrow \quad \leftarrow \quad \sqrt{}$$
$$' \quad '' \quad , \quad . \quad ? \quad | \quad \_ \quad \sim$$

Greek letters:

$$\alpha \quad \beta \quad \gamma \quad \delta \quad \epsilon \quad \zeta \quad \eta \quad \theta \quad \lambda$$
$$\mu \quad \nu \quad \xi \quad \pi \quad \rho \quad \sigma \quad \tau \quad \phi \quad \psi \quad \omega$$

Miscellaneous characters:

$$\Sigma \quad \infty \quad \int \quad \partial \quad \Delta \quad \nabla$$

The special keyboard contains at least the buttons shown in Figure IV-20. The meaning of some of these is discussed later.

The system is said to be approximately equivalent to FORTRAN II in overall capability (although definitely not in notation), plus of course having all the additional facilities from the push-button keyboard. Although provides "picturebook" integer and floating point formatting, it does not afford the extensive formatting capabilities of FORTRAN II nor does it allow for fixed point or variable-precision arithmetic.

A data name can consist of a single letter, one of six Greek letters, or a letter followed by up to three digits. (This permits implicit multiplication.) Subscripts are permitted and can themselves be subscripted. However, they are designated by following the variable name with the word SUB, e.g., Y SUB I. Any operand read into the system is dimensioned automatically, and array arithmetic is performed automatically in future operations. Thus, if B has been defined to be an *N*-dimensional vector, then multiplying a scalar A by B will automatically cause the result to be an *N*-dimensional vector.

The assignment statement in AMTRAN is denoted by an =. The conditional statement is of the form

IF condition THEN statement-1 OTHERWISE statement-2

where condition apparently involves two variables and a relation (e.g., GT or GT=); statement-1 is either an assignment statement or a GO TO and statement-2 can be an assignment followed by GO TO. A more generalized form is available to provide operations on dimensioned variables automatically. For example, the statement

IF |Y|>.5, THEN Y SUB IF.INDEX=0.

causes each subscripted value of X to be tested and to be set equal to 0 if the absolute value of Y is greater than .5.

The push-button keyboard has fixed buttons, some of which are shown in Figure IV-20. It has the facility to allow the user to assign either operands or sequences of existing commands (or operators) to the undefined buttons. Thus, for example, he could define a button for Runge-Kutta or for Legendre polynomials. Some of the less obvious operators have the following meanings. LABELS will list the names of all the user-defined procedures which are currently resident in the machine. CORE tells how much program space is still available. The → (mnemonic: *REPLACES*) indicates replacement: N+1→N. *MINIMAX* locates relative extrema by cubic interpolation. *MAGNITUDE* gives the power of 10 for a number. *LET* changes variables (numerically). *SUM.OF.SERIES* is used in series expansions. *REPRESENT* is part of the mathematical problem-solving package. *TYPE.OUT* is used for format control. *X.LOG*, *Y.LOG*, *POINT.PLOT*, *INCREMENTAL*, *LINEAR*, *POLAR*, *VS* and *DISPLAY.DATA* are used as modifiers in the *PLOT.ON.SCOPE* routine. *REPEAT* has the form *REPEAT (scalar expression), (compound statement)*; the result of the evaluation of the scalar expression is rounded to the nearest integer. *ENTRY* and *RUN* provide for the acceptance of code strings in procedures. *TRANSFER* is a crude symbol transfer operator

The user can either work in a direct execution mode, in which the statements are executed as entered, or he can merely enter statements for execution later. In either case, the user can request permanent retention by entering *NAME* this *label*, where *label* is the name he assigns. Various facilities for designating different types of scopes for variables are provided. Recursive calls are permitted but only to a certain level of depth. Effort has been made to extend subroutine definition to facilitate bootstrapping of the language to higher levels. These extensions consist of a crude facility to pick up code strings for subsequent execution (i.e., similar to *call by name* facilities), the ability to define dyadic operators (at only one level of hierarchy), and the ability to start executing a procedure called from the keyboard before all the parameters have been passed to the procedure. The latter examines parameters or code strings interactively to determine the number and kind of additional parameters to request from the user. Assuming compatible parameter strings, such subroutines can then be embedded in other subroutines without reprogramming.

Significant use of the scope is made, including the display of the program and instructions to the user. JOSS-like standard format operators such as *TYPE*, *SET*, *PUNCH*, and *SCOPE* are available.

Later versions of AMTRAN are said to be running on a Burroughs 5500 computer and an IBM 1130. The Burroughs 5500 implementation utilizes a reentrant time-sharing monitor written in ALGOL 60 that initially ran from teletypes. An experimental keyboard-typewriter terminal is also running. This uses a version of the typewriter with controls for subscripts and superscripts, similar to the machines for the languages described in Section IV.7. Differences in syntax exist, such as allowing *AND*, *OR*, ~, →, *NOT* in conditional clauses, an additional array *IF test*, and the use of square brackets for subscripts. ALGOL procedures may be attached to the system. Future plans call for the inclusion of some symbol manipulation capability, provision for graphic input and output, and other extensions and modifications.

The 1130 implementation, like the B5500 system, executes interpretively from pseudo object code. It was written in FORTRAN IV, and FORTRAN

Control Operators Specific to the Interactive Mode

SUPPRESS/EXECUTE	LIST	HALT
DELETE	LABELS	GENERAL INSTRUCTIONS
BACKSPACE	CORE	SPECIFIC INSTRUCTIONS
EDIT	TRACE	TURN PAGE
RESET	MOVE PROGRAM	
CLEAR	FULL TYPEOUT	

Mathematical Operators

$+$	Σ	STEP.FCT
$-$	Δf	CUBIC
$\times$	Δb	SUM.OF.SERIES
$/$	$\int$	ERF
$=$	d/dx	LAPLACE
$\rightarrow$	Array	SOLVE
$\sqrt{}$	Left (shift)	AVERAGE
SQ	Right (shift)	SIGMA
$*$ (exponentiation)	MINIMUM	MOMENTS
ABS	MAXIMUM	REGRESSION
SIN	MINIMAX	CORRELATION
COS	MAGNITUDE	LEAST.SQUARES
ARCTAN	INTERPOLATE	(all the TRIG and
TAN	REFLECT	HYPERBOLIC fcts.)
EXP	ZEROES	SUB
LN	INVERT	REPRESENT
LOG	LET	

Input-Output Operators

TYPE	TYPE.ON.SCOPE	X.LOG
PRINT	WRITE.SCOPE	Y.LOG
PUNCH	ERASE	POINT.PLOT
PLOT.ON.SCOPE	PUNCH.PROGRAM	LINEAR
TYPE.OUT	READ.CARDS	INCREMENTAL
PRINT.OUT	$*$	VS
PUNCH.PROGRAM	''	POLAR
	SET	DISPLAY.DATA

Programming and Logical Operators

IF	SWITCH	ENTRY
$<$	DIV.ZERO	RUN
$>$	EXIT	TRANSFER
THEN	GO	ACC
OTHERWISE/ELSE	TO	CALL
AND	REPEAT	NAME.THIS
IF.INDEX	INPUT	

Special Variables and Data Operators

α	ψ	REG
β	ξ	' (prime)
γ	η	INSERT
δ	X.MIN	CONCATENATE
ρ	X.MAX	SORT
ϕ	INTERVALS	ORDER
θ	ROW	

Basic Graphics Operators

LINE	AUTOSCALE	YZ.CURVE
ARC	SYMBOL	XZ.CURVE
ROTATE	THREE.D.MATRIX	XY.CIRCLE
TRANSLATE	THREE.D.AXES	YZ.CIRCLE
MAGNIFY	XYZ.CURVE	XZ.CIRCLE
AXES	XY.CURVE	

Graphic Circuit Elements

RESISTOR	GROUND	FILTER
CAPACITOR	BATTERY	
INDUCTOR	NODE	

Special Operators

PARALLEL	WYE.DELTA	DETERMINANT
SERIES	DELTA.WYE	QUADRATIC
		SIMPSON

Complex Routines

TIMES	SIN	EXP
OVER	COS	LN
POWER	ARCTAN	CONJUGATE
		POLAR.CONV

Figure IV-20. List of AMTRAN operations. Source: Seitz, Wood and Ely [UV67]. By permission of the Association for Computing Machinery, Inc.

IV subroutines may be added to the system. Input and output were initially limited to card, printer, and the console keyboard/printer.

This system seems to have a great many excellent features combined in a convenient way. The fact that it can be implemented on a small computer gives it a reasonable possibility of becoming a significant language (system.)

IV.7. LANGUAGES WITH FAIRLY NATURAL MATHEMATICAL NOTATION

IV.7.1. INTRODUCTORY REMARKS

As noted in all the languages described earlier in this chapter, the user is seriously handicapped, both in writing the language and looking at the form in which it is put into the computer, by the constraints of the hardware equipment that has been most commonly available, namely the key punch. The use of a typewriter has proved of some advantage because of a larger character set, but it still does not provide naturalness of notation. The primary disadvantage is the lack of two dimensions, which would permit subscripts and superscripts to be written where they belong. There have been small but significant efforts devoted to developing languages which have a format that is closer to normal mathematical notation. Because the inherent difficulty is one of hardware rather than software, each system has involved either the construction of a special typewriter or the major modification of existing equipment.[28]

In discussing languages with natural mathematical notation and nomenclature, there are three factors which should be kept in mind. The first are the notational problems which can be corrected or improved by special equipment. In this category fall such things as the ability to write exponents above and subscripts below the line; ease of displaying fractions; and existence of special symbols such as partial differentiation, integral, and sigma signs. The second factor involves ease of learning and writing programs in the language. This can be relatively independent of the hardware facilities. In particular, some work was done under my direction to see how natural a language could be defined using only key punch equipment (Boyer [QI65]). The third factor, namely ease of readability, is more readily achieved than ease of learning and writing. Reading and understanding a program which was written based on use of this special equipment is usually quite

---

[28] Since writing this, a manual for a system called MAC-360 has appeared ([MT67]). This system basically provides the user with *three* input lines for a single equation in a manner that can be keypunched, as shown by the following example (from page 1 of the manual) to represent the equation $R_i = A_i^2 e^{2\pi} + B_i^2 e^{-2\pi}$

```
E                 2   2 PI      2   —2 PI
M      R  =  A    E         + B     E
S            I    I              I
```

simple; whereas at least in the COLASL case (Section IV.7.2), the rules for writing it are extremely complicated. It is my contention that a mathematical language (for use on computers) which is truly natural in notation and nomenclature will convey to a prospective user about 85 percent of the information that he needs to write correct programs merely from a large series of well-chosen examples. This hypothesis assumes the user is not trying to trick the system and is not making any effort to obtain maximum efficiency. A partial proof of this contention is given by the two-page manual of Klerer and May (Section IV.7.5).

All the systems in this section involve special hardware, and they have usage limited to one or just a few places. They are included because they represent a significant potential direction for development of more natural languages. The primary interest, however, is due to the special input/output characteristics and not the new language concepts. With the exception of some facilities in MADCAP (Section IV.7.3) which are significant and have not appeared in any major way elsewhere, the languages in this section do not provide conceptual new ideas. However, the availability of special input/output equipment coupled with concepts that are impractical on standard equipment could eventually cause a major breakthrough in languages for scientific problems.

IV.7.2. COLASL

COLASL is one of two systems which were developed at the Los Alamos Scientific Laboratory of the University of California; the other is MADCAP (Section IV.7.3). They were developed by different groups, and COLASL tended to be used in a more production-oriented environment than MADCAP. Source materials available for COLASL are the paper by Balke and Carter [BQ62] and the unpublished programmer's manual by Carter, Balke, and Bacon [CA63]. COLASL runs on the IBM 7030 (STRETCH).

COLASL is moderately general and has very good notational facilities for mathematics. It is easy to read a written program, but the specifications on how to write programs are very complicated.

The application area is numerical scientific problems. The language is procedural and problem-solving. It is a hardware language for use with special equipment (described below) but it is also a good publication language. The intended user is a nonprogrammer.

The language is completely designed for use with special prototype hardware known as the IBM-9210 Scientific Descriptive Printer (4) which was available at Los Alamos in May, 1961. It contains 132 characters, many of which can be easily combined to provide additional symbols, and 15 carriage control functions, including the ability to turn the carriage up and down for superscripts and subscripts. Because of the special hardware and the localized usage, all other functional characteristics are insignificant except that the language was designed and implemented by the same people, and seems to have been useful.

The COLASL character set is shown in Figure IV-21. All the items in

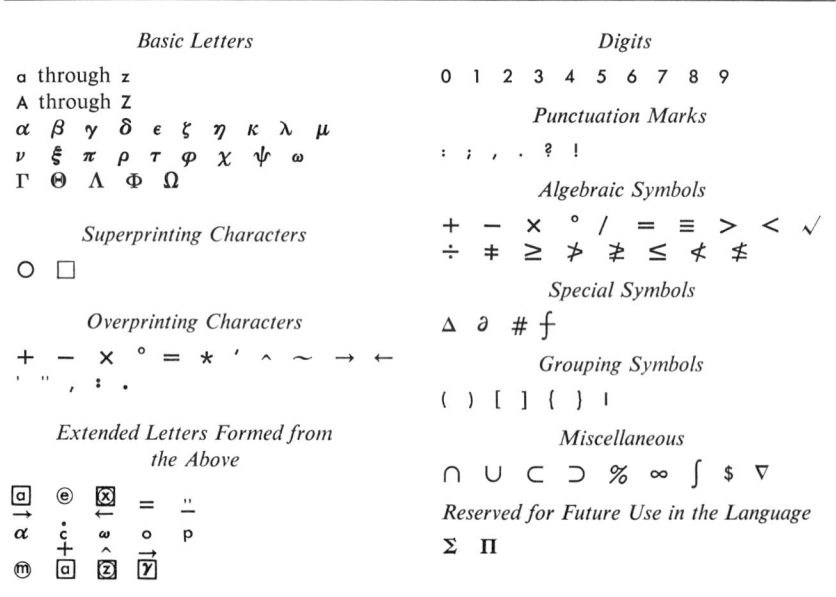

Figure IV-21. The COLASL alphabet.
Source: Balke and Carter [BQ62], p. 506.

the alphabet are either direct characters on the typewriter or can be formed from them, e.g.,

There are 16 key words, namely and, attach, end, exit, for, from, function, go, if, or, range, routine, stop, then, through, to. The graphic operators and punctuation symbols appear in Figure IV-21. The formation rules for data names and program unit labels are significantly different. Data names consist of a single letter (in the extended sense of Figure IV-21), together with descriptive subscripts, and this string may be prefixed by either a Δ or ∂. The total length must not exceed eight symbols. For example, the following are data names:

$$A \quad \Delta B \quad \partial \omega_{i24}$$

$$y_{nut} \quad \Delta_{\boxed{a}} \quad \zeta_{\boxed{b}3\boxed{c}4}$$

Statement labels contain eight or fewer characters and are of two different forms: (1) The # followed by one to seven digits; (2) an extended letter followed by extended letters or digits, with or without a numeric subscript. Examples are as follows:

$$\#012 \quad \#1234567 \quad \#3$$

$$\text{Temp} \quad \text{Heat}_{23} \quad yA5_{12}$$

The reserved words *can* be used for statement labels since the latter appear in a fixed place, but it is not advisable. Variables can have *descriptive* and *numeric* subscripts. The former are for mnemonic purposes only. The latter are variable or constant subscripts to denote elements in vectors or matrices, and they are separated by commas. There is no stated limit to the number of subscripts, and they can be subscripted but the expression must all fit on one line. Subscript variables can have constants added to them and can take on negative values.

The operators are not clearly defined as such, but they are the fairly obvious ones from the COLASL alphabet shown in Figure IV-21. Note particularly the square root symbol, which is an operator with the obvious meaning. The delimiters are the punctuation marks and some of the reserved words. The punctuation characters are those shown in Figure IV-21, and although a sentence can (and must) be terminated by a period, exclamation point, or question mark, there is no syntactic distinction among these. Commas, colons, and semicolons are also used for punctuation. Blanks may not be used in some places but are permitted in others. Use of noise words is permitted; however, see the discussion on comments in the next paragraph, and on p. 269. The only literals that are permitted are numeric constants.

The input form is an interesting combination of very natural free form but with three fixed fields prior to the main part of the statement. The first field cannot be used, the second one denotes class, and the third contains the statement name. The reason for choosing this format was to permit the programmer to interject statements from the STRETCH assembly program into a COLASL program. The only reason for the class column is to permit markers which cause the compiler to ignore the material between them. The conceptual form is maximum naturalness, partially caused by the free-form interspersion of comments with the meaningful program. The following device to accomplish this has been used: The programmer writes his program using a red pencil and a black one. Everything appearing in red is typed in this color on the IBM-9210 and is completely ignored by the COLASL translator. As an illustration, see Figure IV-22, where the material originally written in red has been set in *italics* to make it understandable in printing.

The only declaration is to specify the dimension, which is done by the use of the words *Range* or *Ranges*. The smallest executable unit is called a *phrase*, and phrases may be grouped through the use of commas as separators into clauses which in turn may be grouped into sentences through the use of colons, semicolons, or *if ... then* statements. Sentences end with a period.

We are given the coefficients of a set of quadratic equations of the form:
$$ax^2 + bx + c = 0 .$$

We wish to find those equations which have real roots, and print their coefficients and roots; and to print the coefficients of those equations having complex roots. Let there be a marker, m, for each equation such that $m_i = 0$ if the roots are real and $m_i = 1$ if the roots are complex. We read (*according to format* F1) *the number of sets of coefficients, I. If* ($I \leq 0$ or $I > 100$) then *the data are incorrect, and we* stop. *Otherwise we read* (*per format* F2) *the coefficient triples,* a(i), b(i), *and* c(i); *we also set the initial value of each* $m_i = 0$, *for* (i=1, 2, ..., I).

We now compute the discriminant of the current equation, which we will call d, from

#1 $$d = b_i^2 - 4a_i c_i.$$

If (d<0) then $m_i = 1$, *and we continue* from #2. *Otherwise, we compute the roots,* x_{pos} *and* x_{neg}, *from the relations:*

$$x_{pos_i} = \frac{-b_i + \sqrt{d}}{2a_i}.$$

$$x_{neg_i} = \frac{-b_i - \sqrt{d}}{2a_i}.$$

#2   To complete the calculation repeat the above from #1 *for* (i=1, I—1, ..., 1).

The subscripted variables have the ranges: 100 to 1 *for* a, b, c, m, x_{pos}, *and* x_{neg}. *When the calculation is complete,* print (*format* F3) *which provides headings; then print the coefficients and roots as indicated in the following.*

#3   If ($m_i = 0$) then print (*using* F4) *each of the variables*, a(i), b(i), c(i), x_{pos}(i), *and* x_{neg}(i); *then proceed* from #4. *Otherwise,* print (per F3), *the variables* a(i), b(i), *and* c(i).

#4   Do everything from #3 *for* (i=1, 2, ..., I); *then* stop.

F1   *is the* format *for reading* I. *It is* (E0*3).

F2   *is the* format *for reading the coefficients and printing them where the roots are complex. It has the form* (3E2.1.13.3).

F3   *is the heading* format, (S0, X27, H*COEFFICIENTS*, X50, H*ROOTS*, S2).

F4   *is the last* format, *which controls printing of coefficients which have real roots. It is* (3E2.1.13.3, X3, 2E2.1.13.3). *To make it accessible to the input-output package, we* attach i *to an index register; this is the* end *of the problem.*

Figure IV-22. Example of COLASL program including heavy commentary. Information which is normally typed in red by the user to indicate commentary has been set in *italics* here.
Source: Balke and Carter [BQ62], p. 520.

Loops can be written in three ways. Both functions and subroutines in the FORTRAN sense are permitted. COLASL has an extremely flexible method of specifying comments because any material which does not conform to defined syntactic units in the language is treated as noise and discarded. However the designers quite wisely caution the user to use some discretion in taking advantage of this facility, because it is easy to change the meaning of a program by inserting an extra key word or a punctuation mark in a critical place (i.e., the user might inadvertently create correct syntactic units even though he was only writing commentary). To avoid this difficulty, the use of the red and black pencils as described earlier is recommended.

Assembly language can be written in the midst of COLASL programs and will be appropriately handled.

The only types of data variables and constants are arithmetic. "An expression in the COLASL language is translated into a machine code which, when executed, will produce the same number for the value of the expression as a human computer might be expected to produce if he worked to the same significance as the machine."[29]

The most interesting aspect of COLASL is the flexibility of the expressions which can be created. Examples are given in Figure IV-23. There are complicated rules for creating these expressions so that the scope of underlines, radical signs, absolute value signs, etc., is correctly defined. Since variable names are only a single letter, juxtaposition can be used to denote multiplication.

There are two kinds of assignment statements, called *algebraic* and *parametric*. Algebraic is the standard type, but it also allows multiple naming on the left-hand side, e.g., X=Y=C+5. The parametric type is identified by the equivalence sign, e.g., $g \equiv 1 + 5$. Parametric variables must ultimately be defined in terms of constants; they are evaluated at compilation time.

Unconditional control transfers are written as either GO TO #S or FROM #S, which are equivalent. A GO TO statement appearing at the beginning of the sentence has its action deferred until the end of the sentence. Thus, in GO TO #3, X=5 the variable X will be assigned the value 5 before transferring control to statement #3. A facility for computing switch control is provided. Functions are invoked from within expressions. Subroutines are invoked by writing the name in a phrase under the same circumstances as a key word. The conditional statement is of the form IF B THEN S, where B has a truth value and S itself can be conditional; e.g.,

IF (x>y>z, or P≠3) then $Z=X^2+Y^2$.

IF (x>2) then IF (P=3) then X=Y+3.

---

[29] Balke and Carter [BQ62], pp. 511 and 513.

Division

$$\frac{c+d(e+f)}{ghi(k-1)} \quad \frac{b+c}{d} + \frac{e+f}{g}$$

$$4ac/3\pi\omega$$

Radication

$$\sqrt{b} \quad \sqrt[9]{b} \quad \sqrt[n]{b}$$

$$\sqrt[n]{\frac{b+c}{d-e}}$$

Exponentiation

$$y^3 \quad y^{z+3} \quad e^{z^2}$$

$$\lambda\mu^{\frac{x+y}{x-y}} \frac{y+z}{y-z} + 3$$

Function Calling

$$\sin(y) \quad \cos(y)$$

$$\text{EXP}\left[\frac{\sin(\boxed{x}_i)}{\cos\left(\frac{1}{\boxed{x}_i}\right)}\right]$$

Absolute Value Indication

$$|a|-|b| \quad \frac{\partial \tau}{\partial \sigma}\left|\frac{x_{i-1}-x_{i+1}}{2}\right|$$

$$|(|a|-|b|)|$$

Subscripting with Constants

$x_0 \quad\quad x_{25} \quad\quad P_{temp-3}$

$x_{1,5} \quad\quad P_{temp_{15,-1}} \quad\quad x_{-5,-10}$

$P_{temp_{9,4,2}} \quad x_{1,0,-1} \quad P_{temp_{-2,-3,-16}}$

Subscripting with Variables and Constants

$P_{temp_{i,2}} \quad\quad x_{2,i} \quad\quad x_{-2l,-i_{sub,j}}$

$x_{i_{sub}+2} \quad\quad P_{temp_{-i+4}} \quad\quad x_{j-5}$

$P_{temp_{-k-12}} \quad x_{i+2,j+4} \quad x_{-i-4,k_{sub}-3}$

$P_{temp_{i+1,j+2,k_{sub}+3}}$

Figure IV-23. Expressions in COLASL.
Source: Balke and Carter [BQ62], pp. 512-13.

Loops in COLASL are indicated in one of three ways. The first is by specifying the parameter and its values in the same sentence with the scope of the loop; e.g.,

$x_{i,j}=3\cos(y_i)$, $(i=0, 1, \ldots, 5)$; $z_j = j\sin(jw_i^{i+3})$, $(j=5, 4, \ldots, 0)$.

A constant increment or decrement can be implied from the notation; i.e., writing J=5,4,...,0 is the same as writing J=5,4,3,2,1,0. A second way of specifying loops is by the use of the word THROUGH, which operates the same as the DO statement in FORTRAN. The word FROM followed by the parameter list can also be used. These can all be nested.

There are no input/output statements specified in COLASL; symbolic assembly macros must be used. This was done to permit use of the existing input/output package on STRETCH.

There is a standard set of mathematical subroutines available in the library. The word EXIT is defined in functions and routines and has the same effect as the RETURN statement in FORTRAN. The command STOP terminates the execution of the program.

Dimension information can be written in a flexible format, called the *range sentence*. It must contain the word range or ranges, and all the rest of the clauses must be of the form

$$m_1 \text{ to } p_1, \ldots, m_n \text{ to } p_n \text{ for } v_1, v_2, \ldots, v_n$$

where the m_i and p_i represent constants or variable names and the v_i represent variables; the word and can precede v_n. For example, the following are legal:

A range is: 0 to 27 for x.

The ranges of the variables are -2 to 3, 4 to 8, and 3 to -7 for x, y_{23}, and z_n.

I have no knowledge about the actual effectiveness of this particular system. It appears on the surface at least to have the advantage that accrues from being able to write mathematical expressions in a natural notation; however, there appear to be many features in the language itself which do not entirely lend themselves to naturalness. In general, this seems to be a language that is much easier to read than to write. It is apparently no longer in use.

IV.7.3. MADCAP

Another project at Los Alamos (besides the COLASL system described in the preceding section) is MADCAP, which is used on the MANIAC II Computer. It tended to be developed and used in a more experimental environment than COLASL. The objectives of both projects are quite similar, although they vary in detail. MADCAP has undergone numerous additions; the first two versions did not use a typewriter permitting two-dimensional input. Since the main interest in the language is because of the natural notation, there is no need for a discussion of the early systems. The sources for the description of this work are the items listed in the references at the end of the chapter, and private communications from M. B. Wells.

MADCAP is moderately general and certainly more general than the other languages discussed in Section IV.7. It has very good notational facilities for mathematics. It is easy to read a written program, and not too difficult to write one. The application area is numerical scientific problems, plus some combinatorial and set theoretic problems. It is a hard-

SAMPLE PROGRAM—MADCAP

"Calculation of $Z*$-factor coefficients"
subscript range: r, R_0 to $_{69}$, $Z*_1$ to 10, 1 to 7
subscript range: Constant$_0$ to $_4$: 21.16, 2.37×10^{-2}, π, $\pi/4$, 19

#100 $a, b = 0$
       read: I_{MAX}, c
#1000   read: for $i = 0$ to I_{MAX}: R_i
       for $i = 0, 1, \ldots$; $j = 0, 3, \ldots$
          if $R_i < 0$, exit from loop
$$r_i = j\frac{R_{i+1} - R_i}{2}$$
       for $k = 1, i+1, \ldots I_{MAX}$: $r_k = \frac{3\pi}{k}R_k$
#2000   if $a = c$: $a+1 \to a$, go to #4000   "coeff. ng"
       for $i = 1$ to 10
          for $j = 1, 2, \ldots 7$
             $k = [(i/2)]$; $x = 7i+j$
$$Z*_{i,j} = \text{Constant}_k + \frac{\text{coef}(r_x, x)R_x^2}{(r_{x+1})^j - (r_x)^j}$$
#3000   (20 characters)Format = "$Z*_{X,X}$=XX.XXX"
       for $i = 1$ to 10
          for $j = 1$ to 7
             print by Format: i, j, $Z*_{i,j}$
       if $b = a$ or sense 3 is on: stop 1
       $b+1 \to b$, go to #1000
#4000   number type: (40 characters)Mess
       print by "M": Mess
       go to #1000
(...   coef(a,b)
       if b is even
          coef(a,b) = $a+b$
       otherwise
          coef(a,b) = $a-b$
...)

---

†[MP64], Appendix I, p. 1.

ware language for use with special equipment (described below) and also a good publication language. The user is meant to be a nonprogrammer.

The language is completely designed for use with a modified Frieden Flexowriter which records all the key strokes on a paper tape as well as on the typed page. It does not have nearly the number of characters that are available on the machine used with COLASL. Because of the special hardware and the localized usage, the other functional characteristics are insignificant, except that the system appears to have been developed in an experimental environment and undergone many levels of improvement.

The characters available from the modified Flexowriter may be printed in red as well as black and are as follows:

Alphabetic: a *through* z   A *through* Z
Numeric: 0 through 9 π
Punctuation: , . : ( ) " [ ]   blank
Arithmetic: + $-$ × / $\_$ $\sqrt{}$ = $\rightarrow$ | > <
Miscellaneous: Δ * ' #

There are also a number of control functions available as keys on the typewriter. In particular, subscripts and superscripts can be placed where they belong through use of control keys. As in COLASL, characters can be combined to make syntactically meaningful units. The following are specifically defined:

$$\not< \quad \not> \quad \neq \quad \leq \quad \geq \quad ;$$

The graphic operators and punctuation symbols are shown in the list above. There are a large number of key words in MADCAP.

Data names and program unit labels are defined quite differently. The former can be one of five forms, with the constraint that the name can contain only seven characters: (1) A single letter; (2) "letters" composed from three basic characters, e.g., θ; (3) a capital letter followed by lowercase letters, e.g., Count, Pb; (4) one of the preceding followed by digits, prime, double prime, or asterisk, e.g., a2*, B ", C291; (5) any allowable name prefixed by Δ or #. The data name can be terminated by a space, some form of punctuation, or arithmetic operators (including juxtaposition for multiplication). Statement labels consist of the number sign, #, followed by up to four digits; statement label variables can be subscripted. There is a large list of key words in MADCAP, and the formation rules for data names do not cause them to coincide. Where key words and products appear to coincide, the compiler distinguishes between them from context. The key words are usually written in lower case in programs but can be either initial or all capitals. The latter will be used here for ease of reading. Data names can have any number of subscripts (separated by a comma but no space intervening). A subscript can itself be subscripted by any one-line expression (i.e., no displayed division).

The operators are shown in the character set, although several of those listed under arithmetic are in the set of relational operators, namely $=, <, >, \leq, \geq, \not<, \not>, \neq$. Punctuation is used for delimiting and is quite significant. The characters for comma, colon, and semicolon; the words OR and AND; and the *blank* are all used for punctuation. A comma separates elements on a list. It is also used in specific statement formats. In certain "natural" cases, the words AND, OR can be used in place of the comma. The semicolon is a higher-order delimiter and usually separates items which might naturally contain commas. The colon is used when a list is to follow or to separate the independent part of a conditional or iterative statement. Blanks are significant; the rules are "natural" so that if something looks reasonable, then it should work.

Noise words are permitted; any sequence of five or more letters all in the same case, which is not a key word, the result of juxtaposing variables, or a subroutine name, is ignored by MADCAP.

In addition to numeric literals, character strings and bit patterns for use in set operations can be defined; the former are delimited by red quotation marks, and the latter by red parentheses.

The physical input form is quite natural. No special coding sheets are required. The statements are generally written one per line, with no particular rules about the horizontal spacing or the length of the statement. There can be more than one statement per line, however, and statements can continue on another line. The first statement on a line may have a label placed at the left margin. In certain cases, indentation of successive lines is either permitted or required to define the scope of control clauses. Tabs are critical in some situations.

The conceptual form is to permit the most normal mathematical terminology which could be devised within the limitations of the equipment and implementation technology.

There are declarations for variable type and array dimensions. There is no explicitly defined smallest executable unit, although it is similar to statements in other languages. There is no real concept of grouping the statements. Loops are allowed and are often controlled by indentation. If two or more statements are written per line, they must be separated. A semicolon is used to delimit assignment statements, and a colon delimits control statements (i.e., conditional and loop statements). It is also possible to use a comma followed by a key word such as GO, LOOP, THEN, LET instead of a colon. The word AND may follow a semicolon which is being used to separate statements, and it will have no logical significance.

A MADCAP function or procedure is essentially an independent subprogram, i.e., a block, although the procedure concept is not introduced into the main part of the program.

Comments can be inserted almost anywhere and are delimited by black quotation marks. The comment cannot contain a quotation mark, and a line cannot end within a comment.

There is no interaction with the operating system or environment, except through some powerful input/output facilities. There is a way of inserting machine language instructions at arbitrary points. A program consists of statements and procedures; the terminal marker for the program itself is punched on the Flexowriter.

There are a number of data types permitted in MADCAP. Arithmetic variables and constants are defined as *real*, with no distinction between floating point and integer. Hexadecimal constants are permitted. There are no Boolean variables as such, but there are character strings which can be alphanumeric or bit patterns. Complex variables and string variables are permitted. There are no formal, list, or hierarchical variables. There is a matrix data type (with real elements only). Arrays containing arithmetic variables or matrices are permitted. Statement label is considered a data type.

Through the use of set operations on the bit patterns, individual bits can be accessed. The rest of the variables or constants occupy one or more computer words. All the data types can be accessed by some command, but the string variables have only the operations of definition, equality, and concatenation (denoted by +). An entire array can be referred to as a single entity.

Arithmetic is done in single- or double-precision. Complex number arithmetic is performed. Boolean arithmetic computations are performed as set theoretic operations; those available are union, intersection, subtraction, symmetric subtraction, and complementation. (Since the customary symbols do not exist on the Flexowriter, red symbols for +, ×, −, Δ, _ and ' are used, respectively.) Comparisons of equality are carried out on strings. Matrix operations of addition, subtraction, multiplication, inversion, and transposition are denoted, respectively, by +, −, × or *juxtaposition*, / or *exponent* −1, and *red apostrophe*.

There is no distinction made between floating point and integer arithmetic. The *real* data type is a single-precision number with or without a fractional part. If a number with a nonzero fractional part is used as a subscript, then only the integer part is used. Expressions of mixed type are permitted and can contain real and complex or real and matrix elements. The *simpler* form is interpreted in an appropriate *higher* form; e.g., a real number R appearing in an expression with matrices is interpreted as a matrix with R along the main diagonal and zeros elsewhere. Obvious conversions are made for mixtures of double-precision and/or complex numbers.

Appropriate arithmetic is done for each parenthesized level; hence in the expression A×(B+C), if B and C are real and A is a matrix, the real sum will be computed and then converted to matrix form. While standard arithmetic operator precedence applies, the set theoretic operators have no precedence defined, so parentheses must be used.

Mathematical expressions are formed in the normal way, except that juxtaposition (with or without a blank) can be used to indicate multiplication; if there is any ambiguity because of coinciding with a key word, the compiler makes the determination from context. A number of special features are also available, e.g., displayed division, vertical lines for absolute value, a square root sign, sigma and pi symbols with the scope of the operation written as a subscript on the symbol, an integral sign, factorial using!, binomial coefficients shown in normal notation, and greatest integer using square brackets.

The assignment statement can be written either with an equals sign or a right-pointing arrow. Thus,

$$y = q; \; y = y+1$$

and

$$q \rightarrow y; \; y+1 \rightarrow y$$

have the same meaning. There is also an interchange statement written with two arrows; thus $x \leftarrow \rightarrow y$ causes the values of x and y to be interchanged. In the second example above, the positions of *left* and *right* side are obviously reversed. It is possible to have several variables on the *left* side. Unlike most other programming languages, if the modes of the variables on the two sides differ, the one on the left is assigned the type of the one on the right, unless there is a specific declaration to the contrary. Thus, writing Z = 14.2 would cause Z to be assigned the real value 14.2, while writing (COMPLEX)Z=14.2 would assign to a complex number Z the real part 14.2 and the imaginary part 0.

There are no direct character-handling statements, although characters can sometimes be handled by word-set operations (described below).

Unconditional control transfer is designated by go to, followed by either a statement number or a variable taking on such values, e.g., go to #17 or go to r. A procedure is invoked by giving the procedure title followed by the arguments enclosed in parentheses, with the arguments preceding a semicolon called by value and those following called by name. A function call can appear within a statement or as a statement by itself. Procedure calls are separate statements. Recursive procedures are *not* permitted.

There are many forms for the conditional control transfer. The one which is most similar to those in other languages is

IF condition : statement-1 ; OTHERWISE : statement-2 next-statement

This can also be written with indenting as

    IF condition
        statement-1
    OTHERWISE
        statement-2
    next-statement

The statements can also be conditionals. The form of conditional transfer without the alternative path given is

    IF condition
        statement

The words UNLESS, WHILE, or UNTIL can be used in place of IF. The condition following UNLESS is the negation of the one following IF. The WHILE causes recycling through the condition as long as it is true. The UNTIL has the combined effect of an UNLESS and a WHILE. Thus the following four examples are all equivalent:

```
#5    if A>2
          B = C
      go to #5
------------------------
      while A>2
          B = C
------------------------
#3    unless A≤2
          B = C
      go to #3
------------------------
      until A≤2
          B = C
```

The conditional expression (i.e., *condition*) is considered a statement by itself and must be terminated by the end of the line, a colon, or a comma followed by an appropriate key word. The *statement* follows either on the same line or on subsequent lines indented one tab stop to the right of the condition.

The *condition* can be one of many forms. In the simplest case, it consists of two expressions separated by one of the relations =, <, >, ≤, ≥, ≠, ≮, ≯. (For label, string, matrix, and complex variables, only the equality and nonequality relations apply.) Another condition is *congruence*, which can only be used between real numbers and is written expr-1=expr-2 (mod expr-3); this permits such statements as

$$\text{if } [k/2] = a_2(\text{mod } 5), \text{ go to } \#24$$

Another whole class of conditions are those involving the set theoretic operations of equivalence, contains, is contained in, is not equivalent, does not contain, and is not contained in (all between sets); and is a member of and is not a member of (between a real expression and a set). Other conditions allowed involve various machine status information, including sense lights and dials.

The simple conditions above may be compounded using AND and OR; commas or semicolons may be used in place of the words when three or more phrases are similarly connected, e.g., (A, B, OR C) AND D and [A OR (B, C, AND D)] AND E.

The looping facility appears to be more powerful and flexible than those available in other languages and it may be stated in many alternate ways. The range of the loop can be written on the same line if it is short enough; but since in general it is not, it is written on subsequent lines, each indented one tab from the loop-control statement itself. Since the range itself can contain loops, this will cause successive indenting to the right. In cases where the depth is variable or unspecified, a special notation is used. (Further details on the point of variably nested iteration can be found in Wells [WS64].) The loop statement is usually identified by the key word FOR and contains only one parameter. Its variations can be expressed in the following ways, where p is a variable, a and b are variables or constants, c is a variable, d is a constant, expr is an expression, cond is a condition, anything enclosed in square brackets is optional, a brace means to make a choice, and the three dots are actually written in the program:

```
for p = a, a+b, ... [expr]
for p = a, a−b, ... [expr]
for p = expr1 to expr2
for p = expr1 to infinity
for p = a+d1, a+d2, ... [expr]
for p = (expr1), (expr1)+a, ... [expr2]
for p = a, a+(expr1), ... [expr2]
for p = (expr1), (expr1)+(expr2), ... [expr3]
for p = d, d+a, ... [expr]
```

$$\text{for } p = a, a-b, \ldots, \begin{cases} \text{until} \\ \text{except when} \\ \text{while} \\ \text{such that} \end{cases} \text{cond}$$

In the case where the final expression is omitted, the termination condition must be handled from within the range of the loop. In the third and fourth cases, the increment is +1 unless *expr1* is a constant greater than *expr2*, which is also a constant. In the fifth case, the increment is *d2—d1*.

In addition to the cases above, the parameter variation can be defined in terms of sets; in particular,

$$\begin{aligned}&\text{for } i \text{ in } S \\ &\text{for } i \text{ in } S' \\ &\text{for } x \text{ in } \{0, 5, 23, 99\} \\ &\text{for } V \subset W \\ &\text{for } V \subset W'\end{aligned}$$

where V and W are sets and the ' denotes set complementation. Even a few more variations are possible, but they will not be listed here.

There can be more than one parameter per loop-control statement; they are separated by , as e.g.,

$$\text{for } i = 1 \text{ to } l, \text{ as } k \text{ in } S, \text{ as } j = 0, 3, \ldots$$

Each parameter is incremented simultaneously, so the loop terminates whenever a parameter reaches its terminal value. There is a significant difference between the form above, which expresses a single iteration, and the following, which expresses an iteration within an iteration:

$$\text{for } i = 1 \text{ to } l; \ k \text{ in } S: \ A_k = 2i$$

In the range of the loop there are no restrictions against entry or exit. Furthermore, there is a *loop back* statement which causes control to return to the incrementing and testing part of the loop control from a point other than the physical end. Similarly, there is an explicit *EXIT FROM LOOP* statement. For example,

$$\begin{aligned}&\text{for } i = 3, 5, \ldots, 21 \\ &\quad a = b + ci \\ &\quad \text{if } a = m, \text{ loop back} \\ &\quad \text{if light 3 is on, exit from loop}\end{aligned}$$

There is a very flexible input/output and format control system, with numerous commands provided to the user to cope with each of the various peripheral devices. Since the main interest of this language is in its two-dimensional notation and its flexibility in the control statements, there is nothing significant to be gained by a discussion of the specific input/output commands. However, the following is an example of a *print* statement:

$$\text{print by } x: a, b + c, \text{ for } n = 0 \text{ to } l: R_n, \text{ by } \text{"M"}: \text{Mess}$$

There is a library containing the basic mathematical functions, as well as others such as numerical integration.

There are no debugging statements. There are some statements by which the programmer can control memory allocation.

The program is stopped by use of the word **STOP** optionally followed by a label number which appears in one of the computer registers.

The most unusual facility in MADCAP is its handling of set theoretic terms and concepts, the most significant of which have already been described.

The only declarations are those involving arrays, data types, and the format. A *subscript range* statement defines the extent of arrays, e.g.,

$$\text{subscript range:} \ D_{0 \text{ to } 3, \ 0 \text{ to } 11}$$

Variables are assumed to be real, unless they are declared otherwise. These declarations can be *REAL, PRECISE* (=double-precision), *COMPLEX, MATRIX, LABEL,* and *STRING*. Sets are declared by *WORD—SET* and *SET*. The user can specify the size of matrices by writing m **COMPONENTS** and the length of the strings by writing n **CHARACTERS**. Each of these declarations is enclosed within parentheses and written immediately preceding the variable to which it applies.

Procedures and functions are declared through the use of the bracketing symbols

$$(\ldots \quad \text{and} \quad \ldots)$$

at the beginning and end, respectively. After the opening delimiter, the name of the procedure is given, followed by the names of the variables in parentheses. The variables are separated by commas, with the *call by value* variables first and that list terminated by a semicolon. If the procedure (or function) will fit on one line, then it can be written this way:

$$(\ldots \ \text{sum}(a,b,c) = a+b+c \ \ldots)$$

The **stop** Y statement terminates execution of the program.

The most interesting facet of the language relative to implementation is the necessity for linearizing the two-dimensional input. This is described carefully in Wells [WS61]. Essentially, an internal map of the external page is created and then the map is translated. One of the key problems is to reduce displayed division to a single line, which is done by inserting parentheses around each numerator and denominator and replacing the corresponding line of underscores by the slash symbol.

Although the language was originally designed for use with numerical scientific problems, the addition of the set theoretic notation and facilities has made it quite suitable for work with combinatorial problems. It is unfortunate that this feature, which is one of the significant technological contributions of MADCAP, has not been made available to a larger group of users.

It appears that even though the COLASL character set is larger, the language specifications for it are much more complicated. This indicates that sheer improvement in the hardware is no panacea nor any guarantee of good language design.

IV.7.4. MIRFAC

The MIRFAC system, developed in England by Gawlik [GK63], is another language which aims at ease of programming through improved hardware and has objectives similar to those of other languages in this section. Very few details are given in the cited reference, but MIRFAC appears to be the simplest of this class of languages. Newer, but unpublished, material given in Gawlik and Berry [GK67] shows more detail and power in the language, but arrived too late for inclusion here.

The machine is a Dura Mach 10 with a special sphere providing 88 symbols and various editing operations which include the facility for printing subscripts and superscripts and recording that information on paper tape. The available symbols can be typed in red and black and are as follows:

Alphabetical: a to z (lower case only), and 20 lower-case Greek letters (excluding π)
Numerical: 0 to 9, π, ∞
Mathematical: $+ \; - \; \times \; / \; \div \; = \; \neq \; > \; < \; \geq \; \leq \; ! \; (\;)$
  $\{ \; \} \; [\;] \; | \; \sqrt{} \; ^{-1} \; ^{2} \; . \; \sim \; d \; \int \; \Sigma$
Others: , $*$ ' ■

Note that $^{-1}$ is a single symbol and is treated as a letter by MIRFAC, with obvious simplifications of dealing with inverse trigonometric functions. The d is used with $\int$, deletion is denoted by ■, and $*$ can be used as a suffix to a variable. There are a number of key words, most notably the commands.

Variables are represented by English or Greek letters. Statement labels consist of integers because all MIRFAC statements must be numbered consecutively from 1 upwards; fractional numbers, e.g. 17.3 may be interpolated. The symbol . always means a decimal point and may not be used for punctuation. The comma is used only for punctuation. Noise words are very definitely permitted because the basic philosophy is to have sentences

SAMPLE PROGRAM—MIRFAC

Problem: Construct a subroutine with parameters A and B such that A and B are integers and $2 < A < B$. For every odd integer K with $A \leq K \leq B$, compute $f(K) = (3K + \sin(K))^{1/2}$ if K is a prime, and $f(K) = (4K + \cos(K))^{1/2}$ if K is not a prime. For each K, print K, the value of $f(K)$, and the word PRIME or NONPRIME as the case may be.

Assume there exists a subroutine or function PRIME(K) which determines whether or not K is a prime, and assume that library routines for square root, sine and cosine are available.

Program:

begin

1 *print title sample problem for miss jean sammet.*

2 *print 2 blank lines*

3 *put w_0=word prime, w_2=nonprime*

4 *set m=1, r=1, p_0=2*

5 *cycle for k from 3 to 100*

6 *put n=m and k*

7 *jump to line 18 if n=0*

rejects even k.

8 *enter section prime*

this section leaves s=-1 if k is prime and s=1 if k is composite.

9 *jump to line 14 if s=1*

10 *f=√{3k+sink}*

11 *p_r=k*

add k to the list of primes

12 *add 1 to r*

13 *jump to line 15*

14 *f=√{4k+cosk}*

15 *print k to 0 decs and f to 5 only*

16 *print word w_{s+1}*

17 *print 1 blank line*

18 *take next k*

19 *stop*

Comment: The material shown on unnumbered lines is actually typed in red on the input typewriter.

which begin with a verb and have all the necessary information as fixed words but in an arbitrary order. Thus find by newtons method the root near x=a of the equation is syntactically equivalent to find newtons method root near x=a or find in the equation the root near x=a by newtons method. Note that this is an example of a very high-level command.

The input has all statement labels starting in the same column, followed immediately by the statements, with a single statement per line.

The smallest executable unit is also the only one, namely a sentence which begins with a verb in imperative form and contains one and only one operand together with whatever qualifiers are needed to define the required operation uniquely. The verb must appear first, but the other information can be in any order. (See the example given above.) The sole exception to beginning everything with a verb is in the assignment statements which can be of the form *variable*=. Loops are handled by using *if* or *cycle* statements. Comments are shown by typing in red, as are the names of mathematical functions.

Data variables and constants can be numeric, alphabetic, or bit arrays. Juxtaposition denotes multiplication. The computer COSMOS on which this system runs has double-precision fixed point arithmetic. Computation is mixed number, with 40 bits on both sides of the radix point.

The assignment statement is written in the form θ = A + B, or set $\theta=2$, w=1/3. However, the technique of writing $\theta=\theta+3$ is not permitted, and is replaced by add 3 to θ. There are actually two kinds of unconditional control transfers, designated by the words jump and return. The former is used to go forward (i.e., to higher statement numbers), while the latter goes backward, thus permitting the compiler to check such statements. The most general form of return or jump instruction which MIRFAC will accept is of the form return [or jump] to line N if A R B where R is one of the comparison symbols =, $\neq$, >, <, $\geq$, $\leq$. One illustration of a loop control statement is given by the use of the word tabulate in the two programs shown in Figure IV-24. Note that the print

```
1   set α = 0
2   φ = ∫₀¹ θ tan⁻¹ αθ dθ
3   print α to 3 figs and φ to 8
4   add 0.1 to α
5   return to line 2 if α ≤ 50

1   tabulate for α = 0(0.1)50
2   φ = ∫₀¹ θ tan⁻¹ αθ dθ
3   print α to 3 figs and φ to 8
```

Figure IV-24. Two forms of the same MIRFAC program.
Source: Gawlik [GK63], p. 547. By permission of Association for Computing Machinery, Inc.

statement shows the number of figures to be used and also permits differing variables to be printed to differing degrees of accuracy.

All the mathematical function subroutines signal an error automatically if one occurs, e.g., if the user tries to take the square root of a negative number.

The ■ is used to correct punching errors; if an incorrect letter or symbol is punched, it may be followed at once by a ■ and then the correct symbol. Two black squares effectively delete the entire line.

The system is used for mathematical work of various kinds, data reduction, design automation (using a plotter), and information retrieval.

IV.7.5. KLERER-MAY SYSTEM

The system developed by M. Klerer and J. May at the Hudson Laboratories of Columbia University for use on the GE 225 or 235 is another system with special equipment to permit ease of programming for solution of scientific numerical problems. Various descriptions of the system are given in the references listed at the end of the chapter. The designers here seem to be quite interested in general exploration of software-hardware methods which seem promising for increasing automation in the problem-solving process. In particular, a certain amount of teaching of the user is done by having the system print out the way in which it interprets the user input. Perhaps the most unusual characteristic of this system is the *Reference Manual*, which consists of one $8\frac{1}{2}$ by 11-inch laminated card printed on both sides and a one-page addendum, the latter primarily for on-line usage information. See Figure IV-25a, b, and c. When the manual was issued, there was insufficient evidence to indicate to the designers whether the two pages would be adequate. However, Klerer has stated that after more than three years of extensive usage in a scientific research laboratory environment, he feels that a one-sheet (two pages) manual is quite satisfactory for normal production programming.[30] I feel that while this is probably true for the organization involved, it might not be true if the users were 3000 miles from the system designers and implementers and were lacking a more lengthy definitive language description.

As with the other languages in this category, the other functional characteristics are insignificant except that the language designers were also heavily involved with the implementation.

The original input device is similar to the one in MADCAP (see Section IV.7.3) in that it is a Friden Flexowriter modified so that subscript

---

[30] Private communication; April, 1967.

IV.7.5. KLERER-MAY SYSTEM

SAMPLE PROGRAM—KLERER-MAY SYSTEM†

Problem: Solution of n linear equations in n unknowns for $n \leq 20$.

Program:

MAXIMUM n=20.

READ n.

READ A_{ij} FROM j=1 TO n AND i=1 TO n.

READ C_i FROM i=1 TO n.

FROM j=1 TO n AND i=1 TO n   IF i>j THEN $a_{ij} = A_{ij} - \sum_{k=1}^{i-1} a_{ik}a_{kj}$ OTHERWISE $a_{ij} = \dfrac{A_{ij} - \sum_{k=1}^{i-1} a_{ik}a_{kj}}{a_{ii}}$.

FROM i=1 TO n   COMPUTE $\gamma_i = \dfrac{C_i - \sum_{k=1}^{i-1} a_{ik}\gamma_k}{a_{ii}}$.

FROM i=n BY -1 UNTIL i<1 COMPUTE $X_i = \gamma_i - \sum_{k=i+1}^{n} a_{ik}X_k$.

PRINT 1{2}, X_i FOR i=1, 2, ..., n.   FINISH.

†Klerer and May [KL65a], p. 66.

and superscript positioning can be done automatically under keyboard, paper tape reader, or computer control. There are 88 typable symbols (which can be printed in red or black), consisting of 26 capital letters, 10 digits, 14 lower-case letters, 18 Greek letters, and the following characters:

Arithmetic operators: + − × / |
Relational operators: > ≥ < ≤ =
                        (with overtyping permitted, e.g., ≠)
Punctuation: . , ( )   *blank*

and 6 special characters. These latter, shown at the lower right corner of Figure IV-25b, have been very carefully designed to permit combining and hence the creation of symbols of arbitrary size. Even large-sized integral signs can be created. Not only can the newly formed symbols be recognized by the compiler, but the strokes can be typed in any order and need not be combined neatly; e.g., in typing a summation sign, the lines that make up the physical sigma as well as the indication of the limit can be typed in any sequence.

REFERENCE MANUAL

Vocabulary List

ABS	CARD	END	LN	READ	TANGENT
ABSOLUTE	CARDS	EOF	LOG	RETURN	TANH
AND	COMPUTE	EQUALS	LOOP	REWIND	TAPE
ARC	CONTINUE	EXP	MAXIMUM	ROUND	THE
ARCCOS	COS	FILE	MESSAGE	SEC	THEN
ARCCOSH	COSECANT	FINISH	MINUS	SECANT	TIMES
ARCCOT	COSH	FOR	OF	SECH	TO
ARCCOTH	COSINE	FORMAT	OR	SIN	TOP
ARCCSC	COT	FORMULA	OTHERWISE	SINE	TRUNCATE
ARCCSCH	COTANGENT	FRACTIONAL	PART	SINH	TYPE
ARCSEC	COTH	FROM	PAUSE	SLEW	UNTIL
ARCSECH	CSC	GO	PERFORM	SPECIAL	UPPER
ARCSIN	CSCH	HEADING	PLOT	SQRT	VARIABLE
ARCSINH	CYCLE	IF	PLUS	STATEMENT	VARIABLES
ARCTAN	DIMENSION	INFINITY	PRINT	STOP	WITHIN
ARCTANH	DIVIDED	LABEL	PROCEDURE	SUBROUTINE	WRITE
BY	DO	LINE	PROGRAM	SWITCH	
CALL	ELSE	LINES	PUNCH	TAN	

A period denotes the end of a statement or the end of an implied loop.
Corrections can be made by overtyping or by pressing the control key ERASE when positioned over the error.
Each program must be terminated by the statement END OF PROGRAM. or FINISH.
More than one statement per typing line is acceptable.
To continue a statement beyond the maximum typing length for one line, press the carriage return as many times as desired.
Names of variables with more than one character should be defined by a SPECIAL VARIABLES statement before use.
A comma or the word AND may be used to separate computable statements.
FROM i = 1 TO 10 COMPUTE $A_i = B_i + C_{i+1}$, $C_i = A_{i+1} X$ AND $D = SIN \theta_i$.
Superscripts and subscripts must be in straight line form but forms such as $(A^x)^2$ are permissible.

Examples of Acceptable Forms

The letters E, F, G denote an arithmetic expression, e.g., E may denote the expression A + 2B + i, otherwise a single variable is meant. Braces { } denote a choice of forms. Square Brackets [ ] denote those forms that are optional.

Note: The horizontal extension of the lower limit equation and upper limit expression should not exceed the corresponding arms of the sum symbol. The operand of the sum should be outside the symbol.

A_1 A_E A_{1J} $A_{E,F}$ $\prod_{1,J=E}^{F}$ $\sqrt{E}$

DIMENSION A = (N, M).
This indicates that A is an (N + 1) by (M + 1) array
DIMENSION B = 40, Z = 30, Q = (10, 50).
SPECIAL VARIABLE [S] = DIMENSION
SPECIAL VARIABLES TEMPERATURE, HUMIDITY, PRESSURE, COUNT, LBJ = (14, 200), a_y = 10.
UPPER is used in the same manner as DIMENSION and SPECIAL VARIABLES except that the indicated arrays are stored in upper memory.
UPPER C, WEIGHT = 56, K = (20, 30).

Example

C = -1. D = 15. E = 3. F = 4. G = 2. M = 1.
FROM r = 1 UNTIL 4 COMPUTE B_r = 5 - r.
PERFORM H_r = r FOR r = 1 UNTIL r = 5.

$A = \sum_{i=1}^{4} \left[\frac{1}{C + \frac{D}{E + \frac{F}{G}}} \left(\sum_{J=1}^{4} (B_J \prod_k H_k) \right) + M \right]$

PRINT A.
FINISH.

MAXIMUM n = 20.
READ n.
READ A_i, B_i FROM i = 0 TO n.

$X = \sum_{i=0}^{n} \left[A_i \prod_{J=1}^{n} B_J A_J \right]$

PRINT X. FINISH.

Subscripted variables need not be dimensioned when used in forms such as:

(1) $A_{ik} = B_i Q_{i,k}$ FOR k = 0(2)20 AND i = 1 TO 5

or

(2) MAXIMUM n = 10, J = 15
$A_{ik} = B_i Q_{i,k}$ FOR k = 4, 5, . . . , J WITHIN i = 0 BY 3 UNTIL n.

or

(3) $A = \sum_{i=1}^{40} B_i$, $P = \prod_{1,J=0}^{30} C_{1,J}$

or

(4) READ TAPE C, 2, 2, 10.

FROM i = E [BY F] {TO / UNTIL} [$k \overset{=}{<}$ etc.] G

FROM i = E TO G (Unit steps assumed)
FROM i = N BY 2.34 UNTIL A + B
FROM A = B + 5 BY 2 UNTIL Q > 20
FROM i = E TO INFINITY

FOR i = 1, 2, . . . , 5
FOR j = 5(10)55
FOR i = 0, .5, . . . , 7.5

Note: Any number of dots permissible but no extra spaces before terminating comma. The difference between the first two numbers specifies the increment in the first FOR form.

FROM or FOR forms can be used either to begin or end a statement.
$C_i = A_i = iB_i$ FROM i = 1 TO 10.
FROM i = 1 TO 10 COMPUTE $A_i = iB_i$.

DO [TO / UNTIL] ≡ LOOP [ ] ≡ CYCLE [ ]

DO STATEMENT 5 FROM J = 1 TO 10.

This indicates that all statements up to but not including 5 will be executed. (No two LOOP statements should terminate at the same statement number. Otherwise, any number of LOOP procedures within or external to other LOOP procedures is permitted.)

FROM ≡ WITHIN ≡ AND
FOR ϕ = 0, 5, . . . , 90 WITHIN r = 1 TO 10 AND σ = 1 TO 5 LOOP TO FORMULA 6.

The loop to be performed most often is the first one; the least often is the last.

READ ≡ READ CARD ≡ READ CARDS
READ A_i FROM i = 1 TO $A_i > 15$.

Card Format is free field; number of data points may vary from card to card and may be in either fixed or floating point form.
READ X.
READ A_i, B_{i+1} FROM i = E UNTIL A_i = 93.643.

Data may be punched into cards in the following forms:
2 -2 1.596 $+3.213$ -4.60 $2.78T2[=2.78 \times 10^7]$
$2.78T-2[=2.78 \times 10^{-2}]$ $2.78E-3[=2.78 \times 10^{-3}]$
Each datum should be separated by at least one blank space and the value should be within $\pm 10^{\pm 76}$ and not exceed nine significant digits.

Three Alternate Formulations Of The Same Problem

DIMENSION x=20, y=20.	MAXIMUM w=20.
α=0, READ ω.	READ W. p=0.
FORMULA 1. READ x_α, y_α.	FROM X=0 TO W READ u_x, v_x.
$\alpha = \alpha+1$. IF $\alpha \leq \omega$ GO TO FORMULA 1.	DO FORMULA 3 FROM X=0 TO W.
S=α=0. STATEMENT 1. $\beta = \alpha$, P=1.	α=1.
STATEMENT 2. P=P$x_\beta y_\beta$, $\beta = \beta+1$.	FROM Y=X TO W COMPUTE $\sigma = \sigma u_y v_y$.
IF $\beta \leq \omega$ THEN GO TO STATEMENT 2.	p=p+$u_\alpha \sigma$.
S=S+Px_α AND $\alpha = \alpha+1$.	FORMULA 3. PRINT p.
IF $\omega \geq \alpha$ GO TO STATEMENT 1.	END OF PROGRAM.
PRINT S. END OF PROGRAM.	

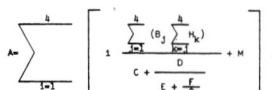

Figure IV-25a. Page 1 of the Klerer and May *Reference Manual*.
Source: Klerer and May [KL65b].

286

PRINT X, i |A|, Y$_i$|A.B|, Z$_i$ = SIN (θ_i+Y$_i^2$) FOR i = 1, 2, ..., N.
PRINT E, F |A.B|, X = G|A|.
PUNCH E, F |A.B|, X = G|A|.

A and B are integers between 0 and 9 but their sum may not exceed 9. F and Y$_i$ will be printed (or card punched) with A places to the left of the decimal point and B places to the right. The value of G and i will be printed (or card punched) as an integer of A places. G will be stored in X. E, and Z$_i$ will be printed (or card punched) in floating point form.

PRINT Y = E |A.B.C|.

Same as above except that E is first divided by 10^c to change its range.

In the print statement a maximum of 8 expressions (including a blank between commas) are allowed. Each is centered in a 15 position field.

PRINT LABEL A, COUNT, X-Y, SIGMA (J).
PRINT LABEL ≡ LABEL ≡ HEADING ≡ PRINT HEADING

Each label, separated by commas, in a PRINT LABEL statement may be up to 15 characters in length and will be printed in a 15 position field. A maximum of 8 labels per statement is permitted and should contain only those characters used on the high-speed printer.

The PRINT FORMAT statement may be used when it is desired to mix literals and answers or to have more than 8 answers per line.

PRINT FORMAT n, E, F, X = G.
FORMAT n LLL...L xxxx LLL...L x.xx y.

n is an integer of up to four places, LLL stands for any literals that are printable on the high-speed printer. Small x's are used to denote the actual position and number of digits of fixed point quantities while one small y is used for each floating point quantity. The first set of x's denotes the first expression/equation/variable mentioned in the PRINT FORMAT statement, the second set of x's denotes the second expression...etc. FORMAT statements may be located anywhere in a program:

PRINT FORMAT 12, θ_i, SIN θ_i, $\phi_i = \frac{180\theta_i}{\pi}$ FROM i = 1 TO N.
FORMAT 12 ANGLE (RADIANS) = y SIN THETA = x.xxxx AND THE
ANGLE IS xxx DEGREES.

If θ_i=3π/4 then the following would be printed on the high speed printer:

"ANGLE (RADIANS) = .23561945   1 SIN THETA = .7071 AND THE
ANGLE IS 135 DEGREES."

SLEW N (Printer paper spaced N lines)
SLEW [[TO]TOP] (Paper will advance to top of page)

Messages on the typewriter or printer are printed using the following forms:

TYPE NEGATIVE SQUARE ROOT.
PRINT MESSAGE (END OF PROGRAM) AND SLEW.

IF F = G THEN GO TO STATEMENT 1.
IF F = G GO TO STATEMENT 1.
IF F = G THEN B = C + E.
IF F = G THEN READ ... .
IF F = G THEN CONTINUE.

IF F = G THEN ... $\left[\begin{Bmatrix} \text{ELSE} \\ \text{OTHERWISE} \end{Bmatrix} \begin{Bmatrix} \text{E} \\ \text{GO TO} \\ \text{COMPUTE ...} \end{Bmatrix}\right]$

Examples of multiple conditions:

IF r = 5 OR G<H OR SIN θ_i>β^2 THEN $\begin{Bmatrix} \text{COMPUTE ...} \\ \text{READ } \alpha \\ \text{C = D} \\ \text{GO TO FORMULA 3} \\ \text{CONTINUE} \end{Bmatrix}$ OTHER-WISE

IF P = G AND H>ϵ/2 AND ...
IF U = O OR (G = r SIN θ AND H ≤ Cω) ...
IF E<F ≤ G THEN ...

1) COMPUTE A = B + 2, (IF i = j THEN (IF m = n THEN T = r SIN θ)
OTHERWISE   T = r COS θ) and PRINT T, A.
2) COMPUTE A = B + 2, (IF i = j THEN (IF m = n THEN T = r SIN θ)
OTHERWISE   T = r COS θ) and PRINT T, A.

In case 1 T = r sin θ if i = j and m = n
           T = r cos θ when i ≠ j
           T is not computed when i = j and m ≠ n.
In case 2 T = r sin θ when i = j and m = n
           T = r cos θ when i = j and m ≠ n
           T is not computed when i ≠ j.

GO ≡ GO[TO]
    GO TO STATEMENT 20

PAUSE will cause the object program to go into a loop. Exit out of the loop will occur if console switch No. 0 is toggled.

Comments (non-computable statements) are entered between | | symbols.

FROM i = 1 TO 10 READ X$_i$ |READ VALUES|.
Y|i,j|=i+12j.

Use of the next forms eliminates the necessity of using "DO" or "LOOP" statements. Computable sub-statements within an implied loop are separated by a comma or AND.

FOR i = 1(1)50 AND k = 0 BY 2 UNTIL Y > 2000 READ X$_{ik}$,
COMPUTE Y = 2X$_{i,k}$ AND PRINT Y.

FROM i = 1 TO INFINITY READ X$_i$, IF X$_i$ ≠ 10 COMPUTE Y = Y + X$_i$, n = n + 2 OTHERWISE GO TO STATEMENT 1.

Superscripts that are red are used to form new characters rather than being interpreted as exponents. The following is a short program to determine the maximum absolute value of a set of positive numbers X.

FROM i = 1 TO 100 IF |X$_i$| > X^{MAX} THEN X^{MAX} = |X$_i$|.   (red)

In the following magnetic tape commands L is the number of elements in the array V, T is the tape number and P is the controller (plug) number.

READ TAPE V$_0$, T, P, L. The first L elements of the tape record is
          read into locations V$_0$ to V$_{L-1}$.
WRITE TAPE V$_2$, T, P, 5. (Locations V$_2$ – V$_6$ are written on tape)
REWIND T, P.   RWD T, P.
WRITE END OF FILE T, P.   EOF T P.
IF END OF FILE P THEN ...   IF EOF P GO TO ...

In the following example Y is the variable to be plotted, X is the "independent index" (i.e. Y = f(X), A = the minimum value of Y and B = the maximum value of Y.

PLOT Y, X, A, B.   PLOT Z$_i$, i, 0, 1 FROM i = 1 TO 565π.

EXAMPLES

READ A$_1$, COMPUTE Y = $\frac{A_1}{A_{MAX}}$ AND PLOT Y, 1, -1, 1 FROM i=1
UNTIL Y>1.

IF α>k COMPUTE x = $\sqrt{(a-k)\Delta}$, Y=B$_{1j}$x+C$_0$T AND PRINT Y,
a, T, k, OTHERWISE COMPUTE x=2ak, Y=B$_{1j}$x+C$_0$TΔ AND PRINT Y,
a, T, k FROM α=1 TO n WITHIN T=2 BY .01 UNTIL 3 AND FOR
k=0(5)90.

FROM i=1 TO 10 AND j=1 TO 10 READ A$_{1j}$,
COMPUTE B$_{1j}$=A$_{1j}$+X$_1$+Y$_j$ AND PRINT A$_{1j}$, B$_{1j}$, X$_1$, Y$_j$, 1, J.

FOR r=1, 2, ..., 10 AND FOR θ=-π(.01)π COMPUTE S$_r$=rSIN$^2\theta$,

C$_r$= $\sqrt{r COS^{-1}\theta}$, A=T$_r$= $\sum_{q=1}^{30}$ TAN(.1$\pi q\theta$),

V$_r$= $\int_{q=1}^{25}$ $\frac{LOG_2 \theta}{A + \frac{r}{\theta C + \frac{DEF}{G}}}$ AND PRINT r, θ, V$_r$, A.

IF (X>Y AND γ>0) OR | 42-γ/ϵ | > (X-Y)^2 THEN COMPUTE
T$_{XY}$=γ(ϵ=$\frac{X}{\gamma}$)^2 AND ω=(YT$_{XY}$)$^{\gamma\epsilon}$ AND PRINT W, T$_{XY}$, X, γ FROM
γ=2k+3 BY .01τ UNTIL W>5800 AND FROM X=1 TO 100
OTHERWISE GO TO STATEMENT 2.

To define a procedure within a program:

... $\begin{Bmatrix} \text{SUBROUTINE} \\ \text{PROCEDURE} \end{Bmatrix}$ (Name). ....

........................RETURN ....

RETURN .... [END [(Name)] $\begin{bmatrix} \text{SUBROUTINE} \\ \text{PROCEDURE} \end{bmatrix}$].

The name of a subroutine can be an alphanumeric string of any length but must begin with an alphabetic character and cannot be identical to any item in the vocabulary list. As many RETURN's as desired may be inserted to branch out of the subroutine back to the main program. The END statement is optional. A STOP or GO TO should precede subroutines.

To call a procedure:

... CALL (Name) $\begin{bmatrix} \text{SUBROUTINE} \\ \text{PROCEDURE} \end{bmatrix}$ ...

Relative Positions of Special Characters

Suggested Reference Citation:
M. KLERER and J. MAY, REFERENCE MANUAL.
Columbia University, Hudson Laboratories
Dobbs Ferry, New York
Revised Edition July, 1965
This work has been supported by the Office of Naval Research and the Advanced Research Projects Agency under Contract Nonr-266(84)

Figure IV-25b. Page 2 of the Klerer and May *Reference Manual.*
Source: Klerer and May [KL65b].

ON-LINE FLEXOWRITER USE

To input, turn on the Flexowriter punch and set console switch 5 down. For on-line input/output at run time, set console switch 8 down. After the computer types "what mode" you may reply with "START.". Start typing your new program after the computer types "ready". (Anything typed between a period and a carriage return or between two carriage returns is defined as a "statement segment"; anything typed between two periods is defined as a "statement". Typing "START." anytime during program input will cancel the current program and permit a new program input.)

TAB

Pressing the "TAB" key omits or "erases" the current statement segment (which may not be complete).

OMIT

Typing "OMIT." omits the previous statement or all previous statement segments. "OMIT. OMIT." omits the previous two statements, and so on.
Example:
A=B. C=D, E=F. OMIT. G=H, (carriage return)
I=J "TAB" K=L. Q=R, (carriage return)
OMIT. M=N. FINISH.

The remaining program is:

A=B. G=H, K=L. M=N.

Computer error analysis and interpretation of ambiguous statements will be typed on the Flexowriter. You may retype the corrected statement after the computer message has been typed.

READ

When input data is expected by the program, the word "INPUT" is typed, and the data may be typed in any of the linear forms used for card input. Corrections may be made in the usual manner. Pressing the "TAB" key erases the entire line of data. The input data is terminated by pressing the carriage return key.

IMAGE

"IMAGE" and "FORMAT" are similar but for the fact that an "IMAGE" may be any two-dimensional construction. As in "FORMAT", output data is represented by small x's for fixed point and a small y for floating point format. The left-uppermost output data fields are given precedence.
Example.

PRINT IMAGE 1, A, B, $\frac{A}{B}$, i, $\sum_{r=0}^{i} r$, i, $\sqrt{1}$

IMAGE 1 $\frac{A}{B}$ = $\frac{xxx.xx}{xx.xxx}$ = y $\sum_{r=0}^{x}$ r=xxx THE SQUARE ROOT OF x EQUALS y.

Note: The "IMAGE" is not permanently stored with the object program and therefore the program must be run immediately after compilation.

All other output which usually goes to the printer or typewriter is also typed on the Flexowriter, i. e., PLOT, PRINT, PRINT FORMAT, PRINT LABEL, etc.

Pressing the "carriage return" key answers all toggle 0 requests and overrides the PAUSE instruction.

The program may be interrupted in two ways during execution.
(1) Press the tape feed button to restart the program.
(2) Press the stop-code key to stop execution and call the compiler.

LIST

A listing of the system's interpretation of your entire program or "old" program segment may be obtained by typing "LIST". anytime. Typing "LIST Axxxxxx.", where x may be 0-9 or blank, starts the listing at statement Axxxxxx. You may interrupt listing by pressing the "tape-feed" button. The listing will terminate upon completion of the typing of the current statement.

EDIT

Typing "EDIT." at anytime allows the editing of the "old" program, where "old" refers to any statements typed after the word START or the last program compiled. Statement numbers of the compiled program begin with A000010 and are stepped by 10. After editing and compiling, all statement numbers are readjusted so that the stepping increment in the statement number is again 10.

FINISH

Typing "FINISH." causes the edited or listed program to be compiled. The last compiled program is always available for editing or listing.

INSERT

To insert a statement or a set of statements after statement A000030, for example, type "INSERT A32." followed by the statement(s). To insert a statement(s) at the beginning of the program, i.e. before statement A000010, type "INSERT A4." followed by the statement(s).

REPLACE

To remove or replace statement A000050, for example, type "REPLACE A0050." followed by the statement(s) if any. Note that to remove statements 40, 50 and 60, for example, you may type either "REPLACE A40. REPLACE A50. REPLACE A60." or more simply "REPLACE A60. OMIT. OMIT.".

Example

what mode
START.
  ready
SLEW. FROM i=1 TO 3 PRINT i.
C=D. FINISH.
  .10000000 1
  .20000000 1
  .30000000 1
what mode
LIST.
  A000010 slew
  A000020 from i=1 to 3 print i
  A000030 c=d
  ready
EDIT.
  ready
REPLACE A20. FROM i=1 TO 6 PRINT i.
INSERT A31. E=F. G=H. K=L. LIST.
  A000010 slew
  A000020 from i=1 to 6 print i
  A000030 c=d
  A000040 e=f
  A000050 g=h
  A000060 k=l
  ready
FINISH.
  .10000000 1
  .20000000 1
  .30000000 1
  .40000000 1
  .50000000 1
  .60000000 1
what mode

OFF-LINE USAGE

To edit a program off-line by paper tape, set console switch 18 and read in the program tape. When the paper tape has been read in "toggle 19" is automatically typed on the console typewriter. Place new paper tape containing additions and corrections in the paper tape reader and toggle switch 19. After all corrections have been read in, lift switch 18 and toggle 19 to compile the program. Note that each time switch 19 is toggled a revised program listing is printed. All editing should refer to the last (most recent) revised listing.

To define a function:

FUNCTION $\begin{Bmatrix} \text{Name} \\ \text{Name } (X,Y,Z,\ldots) \end{Bmatrix}$ = E .

E may include previously defined functions. X, Y, Z,... are dummy variables local to the function definition. The arguments of the function name, if any, may include a replacement operator.

To call a function:

... $\begin{Bmatrix} \text{Name} \\ \text{Name } (F,G,\ldots) \end{Bmatrix}$ ...

Examples:

(Definition) FUNCTION Q(X) = X^2 + 1
              FUNCTION R(X,Y) = XY + Q(XY^2)
              FUNCTION H = A + B SINE θ

(Calls)   A = 3 Q(2) + $\sqrt{R(U+V, TAN H)}$

              B = Q(R(Q(π), H))

              C = $\sum_{X=1}^{N}$ XQ(X) ; D = $\int_{0}^{y}$ Q(X) d**X**

The range of FOR and FROM statements may be delimited by parentheses.

Suggested Reference Citation:
Working notes addendum (Jan. '66 edition) to:
M. KLERER and J. MAY, REFERENCE MANUAL
Columbia University, Hudson Laboratories
Dobbs Ferry, New York
Revised Edition July, 1965

Figure IV-25c. "Addendum" to the Klerer and May *Reference Manual.* Source: Klerer and May [KL66].

A data name consists of a single character, unless defined by a *SPECIAL VARIABLES* statement which permits any number of characters. Statement labels are optional and are defined by writing either **STATEMENT N** or **FORMULA N**, where **N** is an integer. There is a large list of key words but there is obviously no conflict.

Two subscripts are permitted and can be written with or without a comma between them, but the subscripts cannot be subscripted. An interesting situation occurs in writing something of the form A_{ij} because this can be interpreted as implied multiplication of the i and j or two different subscripts. The compiler makes a determination based on context and then notifies the user as to which choice is being made.

Superscripts are interpreted as an index if typed in red and as an exponent if typed in black. This permits extension of data names without the need of explicit definitions.

A period is used to denote the end of a statement, a sequence of statements, or an implied loop. Commas are used between numbers or items in a list and to separate statements causing computation. The words **AND** or **WITHIN** may also be used to separate computational statements. Blanks are critical in a number of places, including within a **FOR** statement. Wherever one blank is permitted, any number can be used.

The physical input form is free form with more than one statement per line. To continue a statement beyond the maximum length of a line, only a carriage return is needed. A new statement can start on the same line as the termination of another.

As indicated repeatedly in this entire section, the conceptual form for this language is naturalness with respect to mathematical notation, the language, and the connectives that are needed to combine expressions.

The smallest executable unit is a statement. A statement can be preceded or followed by **FROM** or **FOR** statements. Certain statements can be combined into a *larger statement*, which can conveniently be called a sentence. Loops can be stated and controlled in three ways.

Standard mathematical elementary functions are provided. Subroutines (=procedures) can be defined but there are no formal parameters; they are global and the values are transmitted from the program. Functions with formal parameters can be defined.

Comments are permitted anywhere in the text by enclosing them in braces.

Executable statements can be terminated by either a period, a comma, the word **AND**, parentheses, brackets, or a space. The end of the program is denoted by the word **FINISH** or the phrase **END OF PROGRAM**.

The only data variables are arithmetic and all computation is done in floating point. Juxtaposition or an explicit multiply sign is used for multiplication.

An assignment statement is designated either by a variable name to the left of the equals sign (e.g., A=B+3) or by preceding the statement with the word COMPUTE (e.g., COMPUTE A=B+3).

The unconditional control transfer is written as GO TO FORMULA N or GO TO STATEMENT N. Functions are invoked by writing them in expressions. Subroutines are invoked by writing CALL Subroutine-Name anywhere in the statement. The conditional statements are quite flexible. One form is represented by

$$\text{IF } A \text{ rel } B \text{ rel } C \text{ THEN } S_1 \text{ ELSE } S_2$$

where *rel* is one of the relations <, ≤, =, >, ≥, and S_1 and S_2 can be one of the statements Y=..., READ, COMPUTE, GO TO, CONTINUE, or IF; OTHERWISE can be used in place of ELSE. Parentheses may be used to define the scope of subsidary IF statements. More complicated logical expressions can be used, thus allowing something of the form IF *B-expression* THEN S_1 ELSE S_2 where *B-expression* is any combination of relations separated by AND and ORs. Parentheses (rather than precedence rules) are used in case of ambiguity, e.g., IF A=G OR (E<S AND C<D<F) THEN (IF Q=R $SIN^2\theta$ THEN P=Q) ELSE P=R $COS^2\theta$.

Loops are controlled in one of several ways, namely through the FROM, FOR, or DO statements. The words FROM, WITHIN, and AND are functionally equivalent and are used to define a loop whose range is exactly the sentence containing the loop specifications. The words DO, LOOP, and CYCLE are equivalent to each other and to conventional loop control statements. The parameter and terminating information can be controlled in several ways as follows:

FOR var = 1st-value, 1st-value + increment, ..., final-value

FOR var = 1st-value (increment) final-value

FROM var = 1st-value [BY increment] $\begin{Bmatrix} TO \\ UNTIL \end{Bmatrix}$ terminal-condition

The *terminal-condition* can be either equality with a variable or a constant, or a conditional expression such as Q>25. The values need not be integers. The implicit loop phrase can be "anywhere meaningful" within a statement. As illustrations, the following are all equivalent:

$C_n = A_n + 5$ FROM n = 4 BY 2 UNTIL 12

FOR n = 4, 6, ... , 12 COMPUTE $C_n = A_n + 5$

$C_n = A_n + 5$ FOR n = 4 (2) 12

A DO statement permits designation of a range, e.g., DO STATEMENT 5 FROM J = 1 to 10. means that all statements up to but not including 5 will be executed. There is a restriction that no two loop statements should terminate at the same statement number; otherwise, any number of loop procedures within or external to other loops is permitted, e.g., FOR A=0, 5, ..., 90 WITHIN R = 1 TO 10 AND B = 1 TO 5 LOOP TO FORMULA 6. The loop nesting is from left to right; i.e., the first one is performed most often.

The input statement is READ, and it can be used for cards or magnetic tape. One acceptable form permits specification of subscript ranges such as READ A_i, B_i FROM i=E UNTIL A_i=17.2. The data can be on cards in fixed or floating point without any predefined format except that each number should be followed by at least one blank space.

Output commands include PRINT, PUNCH, PRINT LABEL, SLEW, TYPE, PLOT. In some cases the formatting information may be provided within the PRINT statement; in other cases it can be specified by a format statement permitting two-dimensional output. The latter allows output which is an image of the format statement. The user can specify fixed or floating numbers, but most importantly vertical control can be specified. In addition, computed labels may be inserted within constructed *pictures* formed by the special type characters.

Computation and looping can be included in the PRINT statement, e.g., PRINT X, i(A), Y_i, (A.B), Z_i=SIN(θ_i+Y_i) FOR i=1, 2, ..., N. The SLEW command provides control of the printer paper. Specific messages can be typed using either the word TYPE or the word PRINT followed by the message. There is also a simple plotting command, namely PLOT Y, X, A, B where A and B specify the minimum and maximum of Y, respectively, as a function of X.

Standard mathematical functions are available.

The only declarations available are the DIMENSION, MAXIMUM, and SPECIAL VARIABLES. The former need only be given when the information is not otherwise deducible from the program. For example, if a statement in the program says READ X_i FROM i=1 TO 500 then there is no need for a DIMENSION statement. In special cases, a DIMENSION statement might be needed. Where explicit dimensioning is desirable for several arrays, all of which have the same dimension, this can be more easily handled through a declaration which is of the form MAXIMUM var_1, var_2, ..., var_n = *integer* where var_i is the array name. A discussion of cases which do not require explicit dimensioning and the implementation technique involved is given in Klerer and May [KL67]. Names of variables with more than one character can be defined by writing, for example, SPECIAL VARIABLES, HOT, AIR.

Subroutines are defined by writing the word SUBROUTINE or PROCEDURE, followed by the subroutine name (which cannot be one of the key words), followed by the body of the subroutine. The word **RETURN** is used to indicate each branch back to the calling program; there can be as many as needed. There are several options for indicating the end of the subroutine.

While there are no specific debugging statements in the language itself, there are some separate ones for use with batch and on-line programming. The system is designed to provide the user with a significant amount of information. In particular, because the user can write statements which have more than one interpretation, the system prints out in a modified language the particular interpretation that has been given. The user can then decide whether the system has chosen the correct meaning. For example, in Figure IV-26, there are a number of ambiguities. Not only is A_{ij} ambiguous, but the argument of SIN might be A or A COS B. Similarly, the argument of $CSCH^{-1}$ is either A or A−L but which is not obvious. In the summation statement, it is not clear whether M is within the summation or not. In this particular instance, the system interprets L and M as outside the argument of the $CSCH^{-1}$ and the sigma, respectively, but it does interpret A_{ij} as a two-dimensional array. In differing contexts, different decisions might be made by the compiler. Thus every attempt is made to provide the user with reasonable information and judgment.

$$X = A_{ij}.$$

$$Y = SIN\ A\ COS\ B.$$

$$Z = SEC\ TAN^{-1}\ \frac{A}{2B}.$$

$$A = CSCH^{-1}\ A - L.$$

$$B = \sum_{i=1}^{100} A_1 B_1 + M.$$

$$C = A/2B.$$

$$D = COS^{3n-2} t - 3.6\ e^x.$$

$$E = LOG_{2k+1}\ 4P^2 + \frac{\pi}{2}.$$

Figure IV-26. Example of expressions with ambiguities in Klerer-May system. Source: Klerer and May [KL65a], p. 70.

```
THIS IS THE WAY WE INTERPRET YOUR STATEMENTS.

IF ANY ARE INCORRECT PLEASE RETYPE THE STATEMENT CORRECTLY.

A00001      X=A SUB (I,J)

A00002      Y=SIN(A)*COS(B)

A00003      Z=SEC(ARCTAN((A)/(2*B)))

A00004      A=ARCCSCH(A)-L

A00005      B=SUM WITHIN (100,I=1) OF (A SUB (I)*B SUB (I))+M

A00006      C=A/2*B

A00007      D=(COS(T)) RAISED TO (3*N-2)-3.6*E RAISED TO (X)

A00008      E=LOG(4*P RAISED TO (2))/LOG(2*K+1)+((PI)/(2))

FINISH.
```

Figure IV-27. Interpretation of ambiguities in expressions shown in Figure IV-26.
Source: Klerer and May [KL65a], p. 71.

This system has been designed as a compromise between maximum user ease and reasonable efficiency in the compiler. Illustrations of some of the tradeoffs are as follows: The restriction on variable names to one character unless specially indicated; if a variable has not been predefined, then the system assumes that its initial value is zero; superscripts are distinguished from exponents by typing the superscript expression in red; comments are enclosed in braces. The authors state that these are not necessary restrictions and they could have been avoided but at the expense of a more elaborate translator than they wanted to construct. On the other hand, the flexibility in statement formats and sequencing allows the user great freedom in what he writes. Some of the restrictions in the current version will be eliminated in future implementations.

Among the significant contributions to the technology made by this system are (1) the generality of the two-dimensional input/output; (2) the analysis and handling of ambiguous source language statements; (3) elimination of the requirement for many of the normal dimension declarations; and (4) the minimal size of the *manual* relative to the power of the language.

In connection with the manual, the reader may wonder why a text description of this length was needed. A careful study will show that some of the syntactic information in this book is either not shown in the manual or is at best implied by examples. However, the requirement for user-learning by system interaction is a deliberate design philosophy. More specifically, the designers state

> Thus, except for what is outlined in . . . [the reference manual] we are asking the user to "play the game" without first telling him all the rules of the game. He learns whatever rules he needs depending on the type of game he plays, i.e. the type of problem he presents. In the first place, the immediate output of the system is a detailed presentation of how the system interprets his presentation of the problem. If the system interpretation is in disagreement with the intention of the user, then the output indicates that he should retype the questionable statement (to which the system assigns a code number) in a more explicit form.[31]

IV.8. MISCELLANEOUS

There are a few languages and/or concepts which seem worth mentioning very briefly, but they do not conveniently fit into one of the preceding sections. Hence they are lumped together here in a *miscellaneous* section.

IV.8.1. CORC

CORC is an experimental language developed at Cornell University to run on the Burroughs 220 and the CDC 1604. Although conceptually derived from FORTRAN and ALGOL, it really bears no relation to them since its objective is to be as simple as possible for students and inexperienced people. A key feature in the overall system is a very powerful error-correcting facility in the compiler.

In addition to the letters and digits, the following 11 characters are used.

$$+ \quad - \quad * \quad / \quad \$ \quad = \quad (\quad) \quad . \quad , \quad blank$$

There are 43 reserved words. Data names and statement labels each can consist of up to eight nonblank characters and are terminated by a blank or special character. All arithmetic is done in floating point, although

---

[31] Klerer and May [KL64], pp. 291–92. By permission of Association for Computing Machinery, Inc.

numbers can be input with the decimal point in any desired place. Variables can have one or two subscripts, and subscripts can be nested to any depth.

The relational operators allowed are EQL, NEQ, LSS, LEQ, GTR, and GEQ.

All executable statements start with a key word. The assignment statement starts with the word LET. In addition, the statements

        INCREASE *variable-name* BY *expression*

        DECREASE *variable-name* BY *expression*

can be used, and the abbreviations INC and DEC are permitted.

Control transfer is designated by GO TO *statement-name*. The conditional control transfer is one of the following forms, where R is one of the relations listed above:

IF *expr-1* R *expr-2* THEN GO TO *statement-1* ELSE GO TO *statement-2*

IF *expr-1a* R1 *expr-1b* AND *expr-2a* R2 *expr-2b* ... AND *expr-na* Rn *expr-nb* THEN GO TO *statement-1* ELSE GO TO *statement-2*

IF *expr-1a* R1 *expr-1b* OR *expr-2a* R2 *expr-2b* ... OR *expr-na* Rn *expr-nb* THEN GO TO *statement-1* ELSE GO TO *statement-2*

Loop control can be accomplished by using one of the following forms:

REPEAT *label* *expression* TIMES

REPEAT *label* UNTIL *expr-1a* R1 *expr-1b* AND *expr-2a* R2 *expr-2b* ... AND *expr-na* Rn *expr-nb*

REPEAT *label* UNTIL *expr-1a* R1 *expr-1b* OR *expr-2a* R2 *expr-2b* ... OR *expr-na* Rn *expr-nb*

REPEAT *label* FOR *variable* = *expr-1*, *expr-2*, ... , *expr-i*, *expr-j*, *expr-k*

where the latter is a triplet as in the ALGOL *for* statement. The *label* which appears in the *REPEAT* statement is used to define a closed subroutine, delimited by BEGIN and END using the same label with each. The user writes

        label   BEGIN
                          body of subroutine
        label   END

These subroutines can only be invoked by the *REPEAT* statement. There is no direct parameter passage, and each variable name has the same meaning throughout the program.

READ and WRITE statements provide for input and output of variables. Comments are denoted by NOTE, and the program ends with a STOP statement. A TITLE command causes the printout of the remainder of the card and continuation cards.

The user is required to declare *all* his variables at the beginning of the program and to specify the dimensions for the subscripted variables.

IV.8.2. OMNITAB

It is clearly a borderline case as to whether the OMNITAB system developed at the National Bureau of Standards on the IBM 7090/94 satisfies the criteria for a programming language. In its basic form, it definitely does not since its framework is to simulate the usage of a desk calculator. This is done by having the user write such statements as

ADD 3.257 TO 2 AND STORE IN COL 4

RAISE 3 TO 4., MULT BY −1.234567, ADD TO COL 3

where a number with a decimal point is considered a literal; without it, it is a column number; and the omission of a specified column number means the result is stored in a specific column inherent to the particular command. While this facet of the language is not significant, it does provide a large package of subroutines which can be addressed in a fairly natural way, and it even includes some control statements. The following sample statements should provide the reader with the flavor of the language, where the following notation is used:

    + + specifies that a column number must be used.
    * * specifies that a constant must be used.
    $ $ allows either a column number or a constant.
    ,, specifies that an integer must be used.

and all words in a statement but the first are optional. The meaning is presumed to be intuitively obvious unless specifically shown.

Matrix operations

TRACE OF (A) IN ,, ++ R=,, STORE IN COL ++

INVERT (A) IN ,, ++ R=,, STORE INVERSE STARTING IN ,, ++

SAMPLE PROGRAM—OMNITAB†

Problem: Compute tables of compressibility factors for hydrogen from the relations:

$$Z = 1 + B + [(\tfrac{1}{2})B + C]\rho^2 + [(\tfrac{1}{6})B^3 + BC]\rho^3$$
$$+ [(\tfrac{1}{24})B^4 + (\tfrac{1}{2})C^2 + (\tfrac{1}{2})B^2C]\rho^4.$$

where

$$B = 0.0055478T^{-1/4} - 0.036877T^{-3/4} - 0.22004T^{-5/4}$$

and

$$C = 0.004788T^{-3/2} - 0.04053T^{-2}$$

for

$$T = 210° (10°) 600°K$$

and

$$\rho = 100\,(100)\,500 \text{ Amagats.}$$

Program:
```
OMNITAB PROBLEM 9 - 2 COMPRESSIBILITY FACTORS FOR HYDROGEN
NOSUMMARY
GENERATE 210.(10.)600. IN 1
RAISE 1 TO -.25 .0055478 ,2
RAISE 1 -.75 -.036877 2
RAISE 1 -1.25 -0.22004 2
RAISE 1 -1.5 .004788 3
RAISE 1 -2. -0.04053 3
READ 11 12 13 14 15
100.,200.,300.,400.,500.
DUPLICATE 49 1 11 1 5   INTO 2 11
ADD 0. 1. 41
ADD 0. 2 42
ADD 3 0. 43
MULT 2 BY 2 BY .5       43
MULT 2 BY 3 44
RAISE 2 TO 3. POWER MULT BY .16666666 ADD TO 44
MULT .5 3 4
MULT 3 BY 4 45
RAISE 2 TO 4. POWER MULT BY .04166667 ADD 45
MULT 2 2 4 45
ADD 1. 0. 20
BEGIN
EXPAND COL 11 TO 4TH POWER IN STEPS OF 1 START STORING IN 21
INCREMENT 1 BY 1 0 0
AMULT 1 20 50 5 BY 1 41     1 26
ROWSUM COLS 26 27 28 29 30 STORE 31
INDEX    4 BY 1
FINISH
REPEAT 1 5 5
HEAD CCL 1/         T
HEAD 31/        100.
HEAD 32/        200.
HEAD 33/        300.
HEAD 34/        400.
HEAD 35/        500.
FIXED 5
PRINT 1 31 32 33 34 35 1
      STOP
```

---

†Hilsenrath *et al.* [HR66], p. 213.

LINEAR EQ COEF IN ,, ++ R=,, RHSIDE IN COL ++, STORE SOLUTION IN ++
(solves set of N linear equations in N unknowns)

EIGENVALUES OF (A) IN ,, ++ R=,, STORE ROOTS IN COL ++

Special functions

TSUB ,, OF ++   (Chebyshëv polynomial)

PSUB ,, OF ++   (Legendre polynomial)

Numerical and statistical analysis

POLYFIT COL ++ WEIGHTS IN ++ X IN ++ USE ,, DEGREE
(polynomial fitting)

GQUAD WITH ** POINTS A = ** B = ** STORE X IN ++ WTS IN ++
(Gaussian integration formula)

HARMONIC ANALYSIS OF COL ++ FOR ,, ORDINATES STORE COEF IN ++

Bessel functions

BEJZERO OF $$, STORE IN COL ++

BEKONE OF ++, STORE IN ++
(modified Bessel function of the second kind of the first order)

Special operators

EXPAND $$ TO ,, POWER IN INTERVALS OF ,, START STORING IN ++

MOLWT Z=,, AMOUNT=,, Z=,, AMOUNT=,, ... STORE SUM IN COL ++
(computes molecular weight of indicated molecule)

Control operations

BEGIN STORING INSTRUCTIONS
(system is normally interpretive)

COMPARE COL ++ AND ++ TO A TOLERANCE OF $$
(if comparison is not satisfied, a diagnostic message is given)

ITERATE X IN ++ Y IN ++ NEW Y IN ++ STORE IN ++

REPEAT INSTRUCTIONS ,, THRU ,, ,, TIMES

Input/output commands and a library of elementary functions are also available.

IV.8.3. MORE NONPROCEDURAL LANGUAGES

In Chapter I the main concept presented about nonprocedural languages was that nonprocedural is a relative term which changes as the state of the art changes. For that reason, it is impossible to predict what will be happening in this area as this book is being read. It is worth pointing out a few attempts, however, which have been made to provide languages which are more nonprocedural than the others specifically described in this book. These will only be mentioned briefly since their implementation status is unclear (as of this writing), and in at least one case there are no plans to implement the language.

The NAPSS (*N*umerical *A*nalysis *P*roblem *S*olving *S*ystem) described by Rice and Rosen [RI66] is a combination of more flexible language facilities and better numerical analysis techniques. The latter will not be discussed. As for the former, the basic concept is to allow the user to provide a number of statements needed in the solution of the problem and yet allow him to be less concerned with the details. For example, he can specify the accuracy he wants and he has a command to *SOLVE DIFF EQ*, and *PLOT CURVE*, and create a *TABLE*.

The POSE system described by Schlesinger and Sashkin [QL67] permits statements such as *INITIAL CONDITION* . . . , *PLOT, RANGE OF* . . . , *INTEGRAL,* and *EIGENVALUES* and considerable flexibility on sequencing of statements.

COMPROSL (*COM*pound *PRO*cedural *S*cientific *L*anguage) was a project started under my direction, with the objective of giving the scientist or engineer an intuitive and natural language with which to program his problems. It assumed the use of only the key punch and had obvious restrictions as a result. A complete syntax was worked out and is described by Boyer [QI65]. Sample legal sentences included the following:

CALCULATE THE SQUARE ROOT OF ALL MULTIPLES OF 7 FROM 14 TO 1400 TO 3 DECIMALS, AND PRINT THEM 4 TO A LINE.

FORM THE MATRIX PRODUCT OF A AND B THEN TRANSPOSE AND PRINT IT.

SET Y = LARGEST EIGENVALUE OF X USING JACOBI'S METHOD.

J = −F(0)/Z + (SUM OF B(M)/(A(M) * (A(M)−Z)) OVER M = 1 TO N) + F(Z)/Z.

IF R IS A MULTIPLE OF 5 AND F2 EXCEEDS 1.65, SET X=SQRT (1 + F2/R), ELSE REPEAT STATEMENT 14A WITH I=1,2,..., 20.

Note that in the fourth case the existence of input equipment which would permit a sigma symbol would make the statement an entirely natural mathematical one.

Obviously, in order to permit the user to provide less detailed information about his statements and their sequencing, algorithms are needed for examining the *program* and translating it. One attempt at this has been made by Homer [HM66]. In this system the user provides (1) arithmetic statements defining the problem area, (2) statements to define the available and desired data, and (3) control statements. Information from all these statements is entered into a matrix. For the arithmetic statements, the variables from the right side are put into rows and the left side variables are entered into the columns. Variables needed for input or output [category (2) above] are entered into rows. Each column is marked as to its nature, and each row is marked to show the availability of the variable it represents, e.g., in storage, input, and to be computed. The matrix then is scanned to determine what can be done.

REFERENCES

IV.2.1.1. *SHORT CODE*

[RR52] *UNIVAC SHORT CODE*, Remington Rand Inc., Philadelphia (Oct., 1952). (unpublished).

IV.2.1.2. *Speedcoding*

[BS54] Backus, J. W. and Herrick, H., "IBM 701 Speedcoding and Other Automatic Programming Systems", *Symposium on Automatic Programming for Digital Computers*, Office of Naval Research, Dept. of the Navy, Washington, D.C. (1954), pp. 106–45.

[BS54a] Backus, J. W., "The IBM 701 Speedcoding System", *J. ACM*, Vol. 1, No. 1 (Jan., 1954), pp. 4–6.

[IB53] *Speedcoding System for the Type 701 Electronic Data Processing Machines*, IBM Corp., 24-6059-0 (Sept., 1953).

IV.2.1.3. *Laning and Zierler System*

[AD54] Adams, C. W. and Laning, J. H., Jr., "The M.I.T. Systems of Automatic Coding: Comprehensive, Summer Session, and Algebraic",

Symposium on Automatic Programming For Digital Computers, Office of Naval Research, Dept. of the Navy, Washington, D.C. (1954), pp. 40–68.

[LA54] Laning, J. H. and Zierler, W., *A Program for Translation of Mathematical Equations for Whirlwind I*, M.I.T., Engineering Memorandum E-364, Instrumentation Lab., Cambridge, Mass. (Jan., 1954).

IV.2.1.4. *A-2 and A-3*

[RR55] *The A-2 Compiler System*, Remington Rand, Inc. (1955).

IV.2.1.5. *BACAIC*

[GR55] Grems, M. and Porter, R. E., *A Digest of the Boeing Airplane Company Algebraic Interpretive Coding System*, Boeing Airplane Co., Seattle, Wash. (July, 1955).

IV.2.1.6. *PRINT*

[IB56a] *PRINT 1* (Programmer's Reference Manual), IBM Corp., 32-7334-1 (1956).
[QF57] Bemer, R. W., "PRINT 1—An Automatic Coding System for the IBM 705", *Automatic Coding, Jour. Franklin Inst.*, Monograph No. 3, Philadelphia, Pa. (Apr., 1957), pp. 29–36.

IV.2.2.1. *MATH-MATIC (AT-3)*

[AS57] Ash, R. et al., *Preliminary Manual for MATH-MATIC and ARITH-MATIC Systems* (for Algebraic Translation and Compilation for UNIVAC I and II), Remington Rand Univac, Philadelphia (Apr., 1957).
[RR60] *MATH-MATIC* (Automatic Programming System), U-1568 Rev. 1, UNIVAC, © 1960, Sperry Rand Corporation.
[TB60] Taylor, A., "The FLOW-MATIC and MATH-MATIC Automatic Programming Systems", *Annual Review in Automatic Programming*, Vol. 1 (R. Goodman, ed.). Pergamon Press, New York, 1960, pp. 196–206.

IV.2.2.2. *UNICODE*

[RR59] *UNICODE—Automatic Coding for UNIVAC Scientific Data Automation System 1103 or 1105*, U-1451 Rev. 3, UNIVAC, © 1958, 1959, Sperry Rand Corporation.

IV.2.2.3. *IT, FORTRANSIT, GAT*

[GM00] Graham, R, and Arden, B., *Generalized Algebraic Translator*, U. of Michigan Statistical and Computing Lab., Ann Arbor, Mich. (unpublished).
[IB57a] *Programmer's Reference Manual: FOR TRANSIT, Automatic Coding System for the IBM 650*, IBM Corp., 32-7842 (1957).

[PR57]   Perlis, A. J., Smith, J. W., and van Zoeren, H. R., *Internal Translator (IT)—A Compiler for the 650*, U. of Michigan Statistical Research Lab., Ann Arbor, Mich. (Jan., 1957).

[PR57a] Perlis, A. J. and Smith, J. W., "A Mathematical Language Compiler", *Automatic Coding, Jour. Franklin Inst., Monograph No. 3*, Philadelphia, Pa. (Apr., 1957), pp. 87–102.

IV.3. FORTRAN

[AA66]   *USA Standard FORTRAN*, United States of America Standards Institute, USAS X3.9-1966, New York, Mar., 1966.

[AA66a] *USA Standard Basic FORTRAN*, United States of America Standards Institute, USAS X3.10-1966, New York, Mar., 1966.

[AX63]   Allen, J. J., Moore, D. P., and Rogoway, H. P., "SHARE Internal FORTRAN Translator", *Datamation*, Vol. 9, No. 3 (Mar., 1963), pp. 43–46.

[AY63]   Ayers, J. A., "Recursive Programming in FORTRAN II", *Comm. ACM*, Vol. 6, No. 11 (Nov., 1963), pp. 667–68.

[BS57]   Backus, J. W. *et al.*, "The FORTRAN Automatic Coding System", *Proc. WJCC*, Vol. 11 (1957), pp. 188–98. (Also in Rosen [RO67].)

[BS64]   Backus, J. W. and Heising, W. P., "FORTRAN", *IEEE Trans. Elec. Comp.*, Vol. EC-13, No. 4 (Aug., 1964), pp. 382–85.

[BU65]   Burkhardt, W. H., "Metalanguage and Syntax Specification", *Comm. ACM*, Vol. 8, No. 5 (May, 1965), pp. 304–305.

[CC64]   "FORTRAN vs. Basic FORTRAN—A Programming Language for Information Processing on Automatic Data Processing Systems", *Comm. ACM*, Vol. 7, No. 10 (Oct., 1964), pp. 591–625.

[CT64]   Carleton, J. T., Lego, P. E., and Suarez, R. S., "A FORTRAN Extension to Facilitate Proposal Preparation", *IEEE Trans. Elec. Comp.*, Vol. EC-13, No. 4 (Aug., 1964), pp. 456–62.

[FH64]   Fowler, M. E. and MacMasters, J. A., *A FORTRAN Program for Polynomial Manipulation*, IBM Corp., TR-24.012, Data Processing Division, Kingston, N.Y. (Mar., 1964).

[FP64]   Fimple, M. D., "FORTRAN vs. COBOL", *Datamation*, Vol. 10, No. 8 (Aug., 1964) pp. 34, 39–40.

[HE63]   Heising, W. P., "FORTRAN", *Comm. ACM*, Vol. 6, No. 3 (Mar., 1963), pp. 85–86.

[HE64]   Heising, W. P., "History and Summary of FORTRAN Standardization Development for the ASA", *Comm. ACM*, Vol. 7, No. 10 (Oct., 1964), p. 590.

[HE64a] Heising, W. P., "FORTRAN: Compatibility & Standardization", *Datamation*, Vol. 10, No. 8 (Aug., 1964), pp. 24–25.

[IB54]   *Preliminary Report: Specifications for the IBM Mathematical FORmula TRANslating System, FORTRAN*, IBM Corp., Programming Research Group, Applied Science Division (1954).

[IB56]   *The FORTRAN Automatic Coding System for the IBM 704 EDPM* (Programmer's Reference Manual), IBM Corp., 32-7026 (Oct., 1956).

[IB57] *Programmer's Primer for FORTRAN Automatic Coding System for the IBM 704*, IBM Corp., 32-0306-1 (1957).

[IB58] *FORTRAN II for the IBM 704 Data Processing System* (Reference Manual), IBM Corp., C28-6000 (1958).

[IB60] *IBM 709-7090 FORTRAN Monitor*, IBM Corp., C28-6065 (1960).

[IB61] *FORTRAN* (General Information Manual), IBM Corp., F28-8074, Data Processing Division, White Plains, N.Y. (1961).

[IB62] *IBM 1620 FORTRAN* (Reference Manual), IBM Corp., C26-5619-0 Data Processing Division, White Plains, N.Y. (1962).

[IB64a] *IBM Operating System/360: FORTRAN IV*, IBM Corp., C28-6515-2, Data Processing Division, White Plains, N.Y. (1964).

[IB66h] *IBM 7090/7094 IBSYS Operating System—Version 13: FORTRAN IV Language*, IBM Corp., C28-6390-3, Data Processing Division, White Plains, N.Y. (Apr., 1966).

[IC62a] "General Panel Discussion: Is a Unification ALGOL-COBOL, ALGOL-FORTRAN Possible? The Question of One or Several Languages", *Symbolic Languages in Data Processing*. Gordon and Breach, New York, 1962, pp. 833–49.

[JU65] Junker, J. P. and Boward, G. R., "COBOL vs. FORTRAN: A Sequel", *Datamation*, Vol. 11, No. 4 (Apr., 1965), pp. 65–67.

[MO67] Moulton, P. G. and Muller, M. E., "DITRAN—A Compiler Emphasizing Diagnostics", *Comm. ACM*, Vol. 10, No. 1 (Jan., 1967), pp. 45–52.

[MR65] McCracken, D. D., "How to Tell If It's FORTRAN IV", *Datamation*, Vol. 11, No. 10 (Oct., 1965), pp. 38–41.

[OL65] Olsen, T. M., "Philco/IBM Translation at Problem-Oriented, Symbolic and Binary Levels", *Comm. ACM*, Vol. 8, No. 12 (Dec., 1965), pp. 762–68.

[OS64] Oswald, H., "The Various FORTRANS", *Datamation*, Vol. 10, No. 8 (Aug., 1964), pp. 25–29.

[QH66] Budd, A. E., *A Method for the Evaluation of Software: Procedural Language Compilers—Particularly COBOL and FORTRAN*, Mitre Corp. (DDC) AD651142, Commerce Dept. Clearinghouse, Springfield, Va. (Apr., 1966).

[RM62] Robbins, D. K., "FORTRAN for Business Data Processing", *Comm. ACM*, Vol. 5, No. 7 (July, 1962), pp. 412–14.

[RN62] Rabinowitz, I. N., "Report on the Algorithmic Language FORTRAN II", *Comm. ACM*, Vol. 5, No. 6 (June, 1962), pp. 327–37.

[RO61] Rosen, S., "ALTAC, FORTRAN, and Compatibility", *Preprints, ACM 16th Nat'l Conf.*, 1961, pp. 2B-2(1)–(4).

[RO65] Rosen, S., Spurgeon, R. A., and Donnelly, J. K., "PUFFT—The Purdue University Fast FORTRAN Translator", *Comm. ACM*, Vol. 8, No. 11 (Nov. 1965), pp. 661–66. (Also in [RO67].)

[SA65] Sakoda, J. M., *DYSTAL Manual—Dynamic Storage Allocation Language in FORTRAN*, Brown U., Dept. of Sociology and Anthropology, Providence, R.I. (1965, revised).

[SY67] Shantz, P. W. *et al.*, "WATFOR—The University of Waterloo FORTRAN IV Compiler", *Comm. ACM*, Vol. 10, No. 1 (Jan., 1967), pp. 41–44.

[WR66] Wright, D. L., "A Comparison of the FORTRAN Language Implementation for Several Computers", *Comm. ACM*, Vol. 9, No. 2 (Feb., 1966), pp. 77–79.

[WZ63] Weizenbaum, J., "Symmetric List Processor", *Comm. ACM*, Vol. 6, No. 9 (Sept., 1963), pp. 524–44.

[YS62] McMahon, J. T., "ALGOL vs. FORTRAN", *Datamation*, Vol. 8, No. 4 (Apr., 1962), pp. 88–89.

IV.4. ALGOL

[AC66] "Collected Algorithms, from the *Comm. ACM*", Association for Computing Machinery, Inc., New York (1966ff.).

[AR61] Arden, B. W., Galler, B. A., and Graham, R. M., "Criticisms of ALGOL 60" (letter to editor), *Comm. ACM*, Vol. 4, No. 7 (July, 1961), p. 309.

[AU67] Auroux, A., Bellino, J., and Bolliet, L., "DIAMAG: A Multi-Access System for On-Line ALGOL Programming", *Proc. SJCC*, Vol. 30 (1967), pp. 547–52.

[BH62] Bottenbruch, H., "Structure and Use of ALGOL 60", *J. ACM*, Vol. 9, No. 2 (Apr., 1962), pp. 161–221.

[BN64] Baumann, R. *et al.*, *Introduction to ALGOL*. Prentice-Hall, Inc., Englewood Cliffs, N.J., 1964.

[BS60] Backus, J. W., "The Syntax and Semantics of the Proposed International Algebraic Language of the Zurich ACM - GAMM Conference", *Proc. Internat' l Conf. Information Processing, UNESCO, Paris, 1959*, R. Oldenbourg, Munich; Butterworths, London, 1960, pp. 125–32.

[CC59] "Recommendations of the SHARE ALGOL Committee", *Comm. ACM*, Vol. 2, No. 10 (Oct., 1959), pp. 25–26.

[CC61] "ACM's ALGOL Resolution", *Comm. ACM*, Vol. 4, No. 11 (Nov., 1961), p. 476.

[CC61a] "ALGOL References in *Communications of the ACM*, 1960–61", *Comm. ACM*, Vol. 4, No. 9 (Sept. 1961), p. 404.

[CC61b] "SMALGOL-61", *Comm. ACM*, Vol. 4, No. 11 (Nov., 1961), pp. 499–502.

[CC63a] "ALCOR Group Representation of ALGOL Symbols", *Comm. ACM*, Vol. 6, No. 10 (Oct., 1963), pp. 597–99.

[CC63b] "ECMA Subset of ALGOL 60", *Comm. ACM*, Vol. 6, No. 10 (Oct., 1963), pp. 595–97.

[CC64a] "A Proposal for Input-Output in ALGOL 60 (A Report of the Subcommittee on ALGOL of the ACM Programming Languages Committee)", *Comm. ACM*, Vol. 7, No. 5 (May, 1964), pp. 273–83.

[CC64b] "Report on Input-Output Procedures for ALGOL 60 (IFIP)", *Comm. ACM*, Vol. 7, No. 10 (Oct., 1964), pp. 628–30.

[CC64c] "Report on SUBSET ALGOL 60 (IFIP)", *Comm. ACM*, Vol. 7, No. 10 (Oct., 1964), pp. 626–28.

[CC64d] "CORRIGENDA: 'ALCOR Group Representations of ALGOL Symbols'", *Comm. ACM*, Vol. 7, No. 3 (Mar., 1964), p. 189.

[FS61]   Forest, B., "BALGOL at Stanford", *Datamation*, Vol. 7, No. 12 (Dec., 1961), pp. 24–26.

[GA67]   Galler, B. A. and Perlis, A. J., "A Proposal for Definitions in ALGOL", *Comm. ACM*, Vol. 10, No. 4 (Apr., 1967), pp. 204–19.

[GT59]   Green, J., "Possible Modifications to the International Algebraic Language", *Comm. ACM*, Vol. 2, No. 2 (Feb., 1959), pp. 6–8.

[GW00]   Garwick, J. V., Bell, J. R., and Krider, L. D., *The GPL Language*, Control Data Corp., TER-05, Palo Alto, Calif.

[GW64]   Garwick, J. V., "Remark on Further Generalization of ALGOL", *Comm. ACM*, Vol. 7, No. 7 (July, 1964), pp. 422–23.

[HN65]   Haynam, G. E., "An Extended ALGOL Based Language", *Proc. ACM 20th Nat'l Conf.*, 1965, pp. 449–54.

[HU61]   Huskey, H. D. and Wattenburg, W. H., "Compiling Techniques for Boolean Expressions and Conditional Statements in ALGOL 60", *Comm. ACM*, Vol. 4, No. 1 (Jan., 1961), pp. 70–75.

[IB66i] *IBM System/360 Operating System: ALGOL Language*, IBM Corp., C28-6615-0, Data Processing Division, White Plains, N.Y. (1966).

[IC62a] "General Panel Discussion: Is a Unification ALGOL-COBOL, ALGOL-FORTRAN Possible? The Question of One or Several Languages", *Symbolic Languages in Data Processing*. Gordon and Breach, New York 1962, pp. 833–49.

[IO65]   *ISO Draft Recommendation on the Programming Language ALGOL*, International Organization for Standardization, Technical Committee ISO/TC 97 Subcommittee 5, Programming Languages (Oct., 1965).

[IR59]   Irons, E. T. and Acton, F. S., "A Proposed Interpretation in ALGOL", *Comm. ACM*, Vol. 2, No. 12 (Dec., 1959), pp. 14–15.

[IR61]   Irons, E. T., "A Syntax Directed Compiler for ALGOL 60", *Comm. ACM*, Vol. 4, No. 1 (Jan., 1961), pp. 51–55. (Also in [RO67].)

[IV64]   Iverson, K. E., "A Method of Syntax Specification", *Comm. ACM*, Vol. 7, No. 10 (Oct., 1964), pp. 588–89.

[KF59]   Kanner, H., "Letter to Editor", *Comm. ACM*, Vol. 2, No. 6 (June, 1959), pp. 6–7.

[KN61]   Knuth, D. E. and Merner, J. N., "ALGOL 60 Confidential", *Comm. ACM*, Vol. 4, No. 6 (June, 1961), pp. 268–72.

[KN67]   Knuth, D. E., "The Remaining Trouble Spots in ALGOL 60", *Comm. ACM*, Vol. 10, No. 10 (Oct. 1967), pp. 611–17.

[LD65]   Landin, P. J., "A Correspondence Between ALGOL 60 and Church's Lambda-Notation: Part I", *Comm. ACM*, Vol. 8, No. 2 (Feb., 1965), pp. 89–101.

[LD65a] Landin, P. J., "A Correspondence Between ALGOL 60 and Church's Lambda-Notation: Part II", *Comm. ACM*, Vol. 8, No. 3 (Mar., 1965), pp. 158–65.

[LR67]   Leroy, H., "A System of Macro-Generation for ALGOL", *Proc. SJCC*, Vol. 30 (1967), pp. 663–69.

[MR61]   McCracken, D. D., "Basic ALGOL", *Datamation*, Vol. 7, No. 12 (Dec., 1961), p. 29.

[NA60]   Naur, P. (ed.), "Report on the Algorithmic Language ALGOL 60", *Comm. ACM*, Vol. 3, No. 5 (May, 1960), pp. 299–314.

[NA63]   Naur, P. (ed.), "Revised Report on the Algorithmic Language ALGOL 60", *Comm. ACM*, Vol. 6, No. 1 (Jan., 1963), pp. 1–17. (Also in [RO67].)

[NA63a] Naur, P., "Documentation Problems: ALGOL 60", *Comm. ACM*, Vol. 6, No. 3 (Mar., 1963), pp. 77–79.

[PK67]   Peck, J. E., "A List Processing Extension of ALGOL", *Symbol Manipulation Languages and Techniques, Proceedings of the IFIP Working Conference on Symbol Manipulation Languages* (D.G. Bobrow, ed.). North-Holland Publishing Co., Amsterdam (1968), pp. 254–59.

[PR58]   Perlis, A.J. and Samelson, K. (for the committee), "Preliminary Report—International Algebraic Language", *Comm. ACM*, Vol. 1, No. 12 (Dec., 1958), pp. 8–22.

[QG61]   *Burroughs Algebraic Compiler* (Reference Manual), Bulletin 220-21011-P, Equipment and Systems Marketing Division, Burroughs Corp., Detroit, Mich. (Jan., 1961).

[QG66]   *Burroughs B 5500 Information Processing Systems Extended ALGOL* (Language Manual), Equipment and Systems Marketing Division, Burroughs Corp., Detroit, Mich. (1966).

[QN62]   Schwarz, H. R., "An Introduction to ALGOL", *Comm. ACM*, Vol. 5, No. 2 (Feb., 1962), pp. 82–95.

[SM61]   Sammet, J. E., "A Method of Combining ALGOL and COBOL," *Proc. WJCC*, Vol. 19 (1961), pp. 379–87.

[SQ61]   Strachey, C. and Wilkes, M. V., "Some Proposals for Improving the Efficiency of ALGOL 60", *Comm. ACM*, Vol. 4, No. 11 (Nov., 1961), pp. 488–92.

[SS63]   Sanders, N. and Fitzpatrick, C., "FORTRAN and ALGOL Revisited", *Datamation*, Vol. 9, No. 1 (Jan., 1963), pp. 30–32.

[TA61]   Taylor, W., Turner, L., and Waychoff, R., "A Syntactical Chart of ALGOL 60", *Comm. ACM*, Vol. 4, No. 9 (Sept., 1961), p. 393.

[TS67]   Stone, H. S., "One-Pass Compilation of Arithmetic Expressions for a Parallel Processor", *Comm. ACM*, Vol. 10, No. 4 (Apr., 1967), pp. 220–23.

[UE67]   deBakker, J. W., *Formal Definition of Programming Languages with an Application to the Definition of ALGOL 60*. Mathematical Centre Tract 16, Mathematisch Centrum, Amsterdam (1967).

[VS67]   von Sydow, L., "Computer Typesetting of ALGOL", *Comm. ACM*, Vol. 10, No. 3 (Mar., 1967), pp. 172–74.

[VW63]   van Wijngaarden, A., "Generalized ALGOL", *Annual Review in Automatic Programming*, Vol. 3 (R. Goodman, ed.). Pergamon Press, New York, 1963, pp. 17–26.

[VW68]   van Wijngaarden, A. (ed.), *Draft Report on the Algorithmic Language ALGOL 68* (Supplement to ALGOL Bulletin 26), Mathematisch Centrum MR93, Amsterdam, (Jan., 1968).

[WG62]   Wegstein, J. H. and Youden, W. W., "A String Language for Symbol Manipulation Based on ALGOL 60", *Comm. ACM*, Vol. 5, No. 1 (Jan., 1962), pp. 54–61.

[WL65]   Weil, R. L., Jr., "Testing the Understanding of the Difference Between Call by Name and Call by Value in ALGOL 60", *Comm. ACM*, Vol. 8, No. 6, (June, 1965), p. 378.

[WO64] Woodger, M., "ALGOL", *IEEE Trans. Elec. Comp.*, Vol. EC-13, No. 4 (Aug., 1964), pp. 377-81.

[WT63] Wirth, N., "A Generalization of ALGOL", *Comm. ACM*, Vol. 6. No. 9 (Sept., 1963), pp. 547-54.

[WT66] Wirth, N. and Hoare, C. A. R., "A Contribution to the Development of ALGOL", *Comm. ACM*, Vol. 9, No. 6 (June, 1966), pp. 413-31.

[WT66a] Wirth, N. and Weber, H., "EULER: A Generalization of ALGOL, and its Formal Definition: Part I", *Comm. ACM*, Vol. 9, No. 1 (Jan., 1966), pp. 13-23.

[WT66b] Wirth, N. and Weber, H., "EULER: A Generalization of ALGOL, and its Formal Definition: Part II", *Comm. ACM*, Vol. 9, No. 2 (Feb., 1966), pp. 89-99.

[YE66] Yershov, A. P., ALPHA—An Automatic Programming System of High Efficiency", *J. ACM*, Vol. 13, No. 1 (Jan., 1966), pp. 17-24.

[YS62] McMahon, J. T., "ALGOL vs. FORTRAN", *Datamation*, Vol. 8, No. 4 (Apr., 1962), pp. 88-89.

IV.5.1. NELIAC

[HS62] Halstead, M. H., *Machine-Independent Computer Programming*. Spartan Books, Washington, D.C., 1962.

[HS63] Halstead, M. H., "NELIAC", *Comm. ACM*, Vol. 6, No. 3 (Mar., 1963), pp. 91-92.

[HS67] Halstead, M. H., Uber, G. T., and Gielow, K. R., "An Algorithmic Search Procedure for Program Generation", *Proc. SJCC*, Vol. 30 (1967), pp. 657-62.

[HS67a] Halstead, M. H., "Machine-Independence and Third-Generation Computers", *Proc. FJCC*, Vol. 31 (1967), pp. 587-92.

[HU60] Huskey, H. D., Halstead, M. H., and McArthur, R., "NELIAC—A Dialect of ALGOL", *Comm. ACM*, Vol. 3, No. 8 (Aug., 1960), pp. 463-68.

[HU63] Huskey, H. D., Love, R., and Wirth, N., "A Syntactic Description of BC NELIAC", *Comm. ACM*, Vol. 6, No. 7 (July, 1963), pp. 367-75.

[JO60] Johnsen, R. F., Jr., *Implementation of NELIAC for the IBM 704 and IBM 709 Computers*, U.S. Navy Electronics Lab., TM-428, San Diego, Calif. (Sept., 1960).

[LO66] *NELIAC Users Guide: UNIVAC 1107/1108 NELIAC*, Lockheed Missiles & Space Co., Sunnyvale, Calif. (Mar., 1966).

[MS60] Masterson, K. S., Jr., "Compilation for Two Computers with NELIAC", *Comm. ACM*, Vol. 3, No. 11 (Nov. 1960), pp. 607-11.

[SI65] Singman, D. *et al.*, "Computerized Blood Bank Control", *Jour. AMA*, Vol. 194 (Nov., 1965), pp. 583-86.

[SX66] Saxon, J. A. *et al.*, *Programming in NELIAC Mod 7 with Mod 7 Star Operating System*, U.S. Navy Electronics Lab., AD 635 179, San Diego, Calif. (Feb., 1966).

[WJ62] Watt, J. B. and Wattenburg, W. H., "A NELIAC-Generated 7090-1401 Compiler", *Comm. ACM*, Vol. 5, No. 2 (Feb., 1962), pp. 101-102.

IV.5.2. MAD

[AR61a] Arden, B. W., Galler, B. A., and Graham, R. M., "MAD at Michigan", *Datamation*, Vol. 7, No. 12 (Dec., 1961), pp. 27–28.

[UM60] *TheMichigan Algorithm Decoder* (The MAD Manual), U. of Michigan Computing Center, Ann Arbor, Mich. (Sept., 1960).

[UM66] *The Michigan Algorithm Decoder* (The MAD Manual), U. of Michigan Computing Center, Ann Arbor, Mich. (Aug., 1966).

IV.6.1. ON-LINE SYSTEMS: INTRODUCTORY REMARKS

[OU67] O'Sullivan, T. C., "Terminal Networks for Time-Sharing", *Datamation*, Vol. 13, No. 7 (July, 1967), pp. 34–43.

[RU67] Ruyle, A., Brackett, J. W., and Kaplow, R., "The Status of Systems for On-Line Mathematical Assistance", *Proc. ACM 22nd Nat'l Conf.*, 1967, pp. 151–67.

IV.6.2. JOSS (AND RELATED SYSTEMS)

[BK64] Baker, C. L., *JOSS: Scenario of a Filmed Report*, RAND Corp., RM-4162-PR, Santa Monica, Calif. (June, 1964).

[BK66] Baker, C. L., *JOSS: Introduction to a Helpful Assistant*, RAND Corp., Memorandum RM-50580-PR, Santa Monica, Calif. (July, 1966).

[DD66] *Preliminary Reference Manual for CAL*. Dial-Data, Inc. (Oct., 1966).

[JC64] Shaw, J. C., "JOSS: A Designer's View of an Experimental On-Line Computing System", *Proc. FJCC*, Vol. 26, pt. 1 (1964), pp. 455–64.

[JC65] Shaw, J. C., "JOSS: Experience with an Experimental Computing Service for Users at Remote Typewriter Consoles", RAND Corp., P-3149, Santa Monica, Calif. (May, 1965). Also in *Proceedings of the IBM Scientific Computing Symposium on Man-Machine Communication*, IBM Corp., 320-1941-0, Data Processing Division, White Plains, N.Y. (1966), pp. 23–32.

[MY66] Myer, T. H., *Manual for Users: TELCOMP Computation Service*, Bolt Beranek and Newman Inc., Cambridge, Mass. (Oct., 1966).

[UJ67] Bryan, G. E., "JOSS: 20,000 Hours at the Console: A Statistical Summary, *Proc. FJCC*, Vol. 31 (1967), pp. 769–77.

[UJ67a] Bryan, G. E. and Smith, J. W., *JOSS Language: Aperçu and Précis, Pocket Précis, Poster Précis*, RAND Corp., Memorandum RM-5377-PR, Santa Monica, Calif. (Aug., 1967).

[UJ67b] Bryan, G. E. and Paxson, E. W., *The JOSS Notebook*, RAND Corp., Memorandum RM 5367-PR, Santa Monica, Calif. (Aug., 1967).

[WK67] Waks, D. J., "Conversational Computing on a Small Machine", *Datamation*, Vol. 13, No. 4 (Apr., 1967), pp. 45–49.

[ZL67] Marks, S. L. and Armerding, G. W., *The JOSS Primer*, RAND Corp., Memorandum RM-5220-PR, Santa Monica, Calif. (Aug., 1967).

IV.6.3. QUIKTRAN

[AA66a] *American Standard Basic FORTRAN*, American Standards Association, ASA X3.10-1966, New York, Mar., 1966.

[DM64] Dunn, T. M. and Morrissey, J. H. "Remote Computing: An Experimental System Part 1: External Specifications", *Proc. SJCC*, Vol. 25 (1964), pp. 413–23.

[IB66d] *Information Marketing QUIKTRAN User's Guide*, IBM Corp., E-20-0240, Data Processing Division, White Plains, N.Y. (1966).

[IB67e] *Fundamentals of Using QUIKTRAN*, IBM Corp., J20-0002-0, Data Processing Division, White Plains, N.Y. (1967).

[KR64] Keller, J. M., Strum, E. C., and Yang, G. H., "Remote Computing: An Experimental System Part 2: Internal Design", *Proc. SJCC*, Vol. 25 (1964), pp. 425–43.

[MJ65] Morrissey, J. H., "The QUIKTRAN System", *Datamation*, Vol. 11, No. 2 (Feb., 1965), pp. 42–46.

IV.6.4. BASIC

[GZ66] *'BASIC' Language Reference Manual*, General Electric Information Systems Division (Sept., 1966, revised).

[KM66] Kemeny, J. G. and Kurtz, T. E., *BASIC* (User's Manual) (3rd ed.), Dartmouth College Computation Center, Hanover, N. H. (Jan., 1966).

[KM67] Kemeny, J.G. and Kurtz, T. E., *BASIC Programming*. John Wiley & Sons, Inc., New York, 1967.

IV.6.5. CPS (AND RUSH)

[AB66] *RUSH Terminal User's Manual*, Allen-Babcock Computing Inc. (Nov., 1966, plus updating material).

[AB67] "RUSH: An Interactive Dialect of PL/I", (presented at ACM sponsored PL/I Forum, Aug., 1967) (unpublished).

[IB67a] *Conversational Programming System*, IBM Corp., Contributed Program Library #360D 03. 4. 016, Program Information Dept., Hawthorne, N.Y. (Sept., 1967).

[IB67h] *CPS—Terminal User's Manual*, IBM Corp., Technical Report TM 48.67.006, Boston Programming Center, Cambridge, Mass. (Sept. 1967).

IV.6.6. MAP

[KP66] Kaplow, R., Strong, S., and Brackett, J., *A System for On-Line Mathematical Analysis*, M.I.T., MAC-TR-24, Project MAC, Cambridge, Mass. (Jan., 1966).

[KP66a] Kaplow, R., Brackett, J., and Strong, S., "Man-Machine Communication in On-Line Mathematical Analysis", *Proc. FJCC*, Vol. 29 (1966), pp. 465–77.

[UU63] Stotz, R. H., "Man-Machine Console Facilities for Computer-Aided Design", *Proc. SJCC*, Vol. 23 (1963), pp. 323–28.

[UU67] Stotz, R. H. and Cheek, T. B., *A Low-Cost Graphic Display for a Computer Time-Sharing Console*, M.I.T., ESL-TM-316, Electronic Systems Lab., Cambridge, Mass. (July, 1967).

IV.6.7. LINCOLN RECKONER

[FY65] Forgie, J. W., "A Time- and Memory-Sharing Executive Program for Quick-Response On-Line Application", *Proc. FJCC*, Vol. 27, pt. 1 (1965), pp. 599–609.

[SW66] Stowe, A. N. *et al.*, "The Lincoln Reckoner: An Operation-Oriented, On-Line Facility with Distributed Control", *Proc. FJCC*, Vol. 29 (1966), pp. 433–44.

[WU67] Wiesen, R. A. *et al.*, "Coherent Programming in the Lincoln Reckoner", *Proceedings of the Symposium on Interactive Systems for Experimental Applied Mathematics*, Washington, D. C., August 26–28, 1967, Academic Press, Inc., New York (1968).

IV.6.8. APL/360 AND PAT

[FA67] Falkoff, A. D. and Iverson, K. E., *The APL Terminal System: Instructions for Operation*, IBM Corp., T. J. Watson Research Center, Yorktown Heights, N.Y. (Mar., 1967).

[FA67a] Falkoff, A. D. and Iverson, K. E., "APL/360 Terminal System", *Proceedings of the Symposium on Interactive Systems for Experimental Applied Mathematics*, Washington, D.C., August 26–28, 1967, Academic Press, Inc., New York (1968).

[HH64] Hellerman, H., "Experimental Personalized Array Translator System", *Comm. ACM*, Vol. 7, No. 7 (July, 1964), pp. 433–38.

[IV62] Iverson, K. E., *A Programming Language*. John Wiley & Sons, New York, 1962.

[RJ66] Rose, A. J., *The Use of APL for Describing Programs at Many Levels of Detail*, IBM Corp., RC 1700, T. J. Watson Research Center, Yorktown Heights, N.Y. (Oct., 1966).

[RQ67] Roth, J. P., Bouricius, W. G., and Schneider, P. R., *Programmed Algorithms to Compute Tests to Detect and Distinguish Between Failures in Logic Circuits*, IBM Corp., RC 1764, T. J. Watson Research Center, Yorktown Heights, N.Y. (Feb., 1967).

[UD67] Bouricius, W. G. *et al.*, *On-Line Reliability Calculations to Achieve a Balanced Design of an Automatically Repaired Computer*, IBM Corp., RC 1800, T. J. Watson Research Center, Yorktown Heights, N.Y. (Apr., 1967).

IV.6.9. CULLER-FRIED

[CU62] Culler, G. J. and Huff, R. W., "Solution of Nonlinear Integral Equations Using On-Line Computer Control", *Proc. SJCC*, Vol. 21 (1962), pp. 129–38.

[CU63] Culler, G. J. and Fried, B. D., *An On-Line Computing Center for Scientific Problems*, Thompson Ramo Wooldridge Inc., MI9-3U3, TRW Computer Division, Canoga Park, Calif. (June, 1963).

[CU65] Culler, G. J. and Fried, B. D., "The TRW Two-Station On-Line Scientific Computer: General Description", *Computer Augmentation of Human Reasoning* (M. Sass and W. Wilkinson, eds.). Spartan Books, Washington, D.C., 1965, pp. 65–87.

[CU67] Culler, G. J., "User's Manual for an On-Line System", *On-Line Computing* (W. J. Karplus, ed.), McGraw-Hill, New York, 1967, pp. 303–24.

[FQ64] Fried, B. D., Farrington, and Pope, *STL On-Line Computer: General Description and User's Manual*, Thompson Ramo Wooldridge Inc., 9824-6001-RU-000, Redondo Beach, Calif. (1964).

[FQ66] Fried, B. D., *On-Line Problem Solving*, Thompson Ramo Wooldridge Inc., 9863-6001-R000, Redondo Beach, Calif. (1966).

[QK67] Blackwell, F. W., "An On-Line Symbol Manipulation System", *Proc. ACM 22nd Nat'l Conf.*, 1967, pp. 203–209.

[RU67] Ruyle, A., Brackett, J. W., and Kaplow, R., "The Status of Systems for On-Line Mathematical Assistance", *Proc. ACM 22nd Nat'l Conf.*, 1967, pp. 151–67.

[WN66] Winiecki, K. (ed.), *Culler On-Line System User Manual*, Harvard U. Computation Lab., Cambridge, Mass. (1966).

IV.6.10. DIALOG

[CS66] Cameron, S. H., Ewing, D., and Liveright, M., *DIALOG: A Conversational Programming System with a Graphical Orientation*, I.I.T. Research Inst., Tech. Note No. 109, Computer Sciences Division, Chicago (Sept., 1966).

[CS67] Cameron, S. H., Ewing, D., and Liveright, M., "DIALOG: A Conversational Programming System with a Graphical Orientation", *Comm. ACM*, Vol. 10, No. 6 (June, 1967), pp. 349–57.

IV.6.11. AMTRAN

[CM66] Clem, P. L., Jr. *et al.*, "AMTRAN—A Conversational-Mode Computer System for Scientists and Engineers", *Proc. IBM Scientific Computing Symposium on Computer-Aided Experimentation*, IBM Corp., Data Processing Division, White Plains, N.Y. (1966), pp. 115–50.

[RF66] Reinfelds, J. *et al.*, "AMTRAN, A Remote-Terminal, Conversational-Mode Computer System", *Proc. ACM 21st Nat'l Conf.*, 1966, pp. 469–77.

[UV67] Seitz, R. N., Wood, L. H., and Ely, C. A., "AMTRAN—Automatic Mathematical Translation", *Proceedings of the Symposium on Interactive Systems for Experimental Applied Mathematics*, Washington, D.C., August 26–28, 1967, Academic Press, Inc., New York (1968).

[WD66]   Wood, L. H. et al., "The AMTRAN System", *Datamation*, Vol. 12, No. 10 (Oct., 1966), pp. 22–27.

IV.7.1. LANGUAGES WITH FAIRLY NATURAL MATHEMATICAL NOTATION: INTRODUCTORY REMARKS

[QI65]   Boyer, M. C., *COMpound PROcedural Scientific Language*, IBM Corp., TR00.1242, Data Systems Division, Development Lab., Poughkeepsie, N.Y. (Feb., 1965).
[MT67]   *Users Guide to MAC-360*, M.I.T., Instrumentation Laboratory, Cambridge, Mass (Sept., 1967).

IV.7.2. COLASL

[BQ62]   Balke, K. G. and Carter, G. L., The COLASL Automatic Coding Language, *Symbolic Languages in Data Processing*. Gordon and Breach, New York, 1962, pp. 501–37.
[CA63]   Carter, G. L., Balke, K. G., and Bacon, B. A., *COLASL 1*. Los Alamos Scientific Lab., Los Alamos, N.M. (June, 1963) (unpublished).

IV.7.3. MADCAP

[BD61]   Bradford, D. H. and Wells, M. B., "MADCAP II", *Annual Review in Automatic Programming*, Vol. 2 (R. Goodman, ed.). Pergamon Press, New York, 1961, pp. 115–40.
[MP64]   *MADCAP Manual* (Dec., 1964) (unpublished).
[WS61]   Wells, M. B., "MADCAP: A Scientific Compiler for a Displayed Formula Textbook Language", *Comm. ACM*, Vol. 4, No. 1 (Jan., 1961), pp. 31–36.
[WS63]   Wells, M. B., "Recent Improvements in MADCAP", *Comm. ACM*, Vol. 6, No. 11 (Nov., 1963), pp. 674–78.
[WS64]   Wells, M. B., "Aspects of Language Design for Combinatorial Computing", *IEEE Trans. Elec. Comp.*, Vol. EC-13, No. 4 (Aug., 1964), pp. 431–38.

IV.7.4. MIRFAC

[DJ64]   Dijkstra, E. W., "Some Comments on the Aims of MIRFAC" (letter to the editor), *Comm. ACM*, Vol. 7, No. 3 (Mar. 1964), p. 190.
[GK63]   Gawlik, H. J., "MIRFAC: A Compiler Based on Standard Mathematical Notation and Plain English", *Comm. ACM*, Vol. 6, No. 9 (Sept., 1963), pp. 545–47.
[GK67]   Gawlik, H. J. and Berry, F. J., *Programming in MIRFAC*, Second Edition, (Feb., 1967) (unpublished).

[WS64a] Wells, M. B., "In Defense of MIRFAC" (letter to the editor), *Comm. ACM*, Vol. 7, No. 6 (June, 1964), p. 379.

IV.7.5. KLERER-MAY SYSTEM

[KL64] Klerer, M. and May, J., "An Experiment in a User-Oriented Computer System", *Comm. ACM*, Vol. 7, No. 5 (May, 1964), pp. 290–94.

[KL65] Klerer, M. and May, J. "A User-Oriented Programming Language", *Computer Jour.*, Vol. 8, No. 2 (July, 1965), pp. 103–109.

[KL65a] Klerer, M., and May, J., "Two-Dimensional Programming", *Proc. FJCC*, Vol. 27, pt. 1 (1965), pp. 63–75.

[KL65b] Klerer, M. and May, J., *Reference Manual*, Columbia U., Hudson Labs., Dobbs Ferry, N.Y. (revised edition, July, 1965).

[KL66] Klerer, M. and May, J., *Working Notes Addendum to Reference Manual*, Columbia U., Hudson Labs., Dobbs Ferry, N.Y. (Jan., 1966 edition).

[KL67] Klerer, M. and May, J., "Automatic Dimensioning", *Comm. ACM*, Vol. 10, No. 3 (Mar., 1967), pp. 165–66.

[KL67a] Klerer, M. and Grossman, F., "Further Advances in Two-Dimensional Input-Output by Typewriter Terminals", *Proc. FJCC*, Vol. 31 (1967), pp. 675–87.

IV.8.1. CORC

[CN63] Conway, R. W. and Maxwell, W. L., "CORC—The Cornell Computing Language", *Comm. ACM*, Vol. 6, No. 6 (June, 1963), pp. 317–21.

[FM64] Freeman, D. N., "Error Correction in CORC, the Cornell Computing Language", *Proc. FJCC*, Vol. 26, pt. 1 (1964), pp. 15–34.

IV.8.2. OMNITAB

[DT63] "OMNITAB on the 90", *Datamation*, Vol. 9, No. 3 (Mar., 1963), p. 54.

[HR66] Hilsenrath, J. *et al.*, *OMNITAB—A Computer Program For Statistical and Numerical Analysis*. National Bureau of Standards Handbook 101, Washington, D.C. (1966).

IV.8.3. MORE NONPROCEDURAL LANGUAGES

[HM66] Homer, E. D., "An Algorithm for Selecting and Sequencing Statements as a Basis for a Problem-Oriented Programming System", *Proc. ACM 21st Nat'l. Conf.* 1966, pp. 305–12.

[QI65] Boyer, M. C., *COMpound PROcedural SCientific Language*, IBM Corp., TR00.1242, Data Systems Division, Development Lab., Poughkeepsie, N.Y. (Feb., 1965).

[QL67] Schlesinger, S. and Sashkin, L., "POSE: A Language for Posing Problems to a Computer", *Comm. ACM*, Vol. 10, No. 5 (May, 1967), pp. 279–85.

[RI66] Rice, J. R. and Rosen, S., "NAPSS—A Numerical Analysis Problem Solving System", *Proc. ACM 21st Nat'l Conf.*, 1966, pp. 51–56.

V LANGUAGES FOR BUSINESS DATA PROCESSING PROBLEMS

V.1. SCOPE OF CHAPTER

The languages in this chapter are those whose primary intent is the effective solution of business data processing problems. The scope of this application area is assumed to be understood by the reader; briefly summarized, it is meant to include problems which have very large files on which straightforward operations must be performed. Some common illustrations of such problems are payroll, inventory control, and insurance files. Fortunately or unfortunately, there is only one language in major current use which falls into this category—namely COBOL. The forerunners to COBOL are FLOW-MATIC, AIMACO, Commercial Translator, and FACT; each of these is described very briefly to indicate its historical significance and technical contributions. GECOM, which largely paralleled some of the COBOL effort in time scale, is also discussed. Lest any reader wonder about the absence of SURGE ([NM00], Longo [LN62]), 9 PAC ([IB61b]), or report generators ([IB65d], Leslie [LS67]) from this list, it must be emphasized that none of those are *languages* by my definition of the term. Thus it is not the function of this chapter to debate the merits of languages versus fixed format forms for the solution of problems in business data processing. Only those systems which meet the characteristics of Chapter I are included.

The basic pattern for programming languages in business data processing problems was set—as shown Section V.2.1—by FLOW-MATIC, which established the concept of an English-like language with "natural" words for both the operations to be performed and the data on which they are to be acting. All major language developments in this area have followed this concept.

Other languages, or classes whose absence might be questioned are

DETAB X, DETAB 65, and decision tables in general. They are not considered languages within the framework of this book. Some references are given at the end of Chapter I for the reader wishing to pursue this area.

One of the unfortunate things about this particular application field is the fantastic variety of ways to say exactly the same thing. In Willey *et al.* [WY61a] there is a description of eight languages available (or at least defined) in 1960 and 1961. Four of these, or at least versions of them, are the ones mentioned in this chapter, and the other four are from England. It is shocking to look at this small reference booklet and find there are eight ways of trying to add three numbers and define the name of the result. Lest anyone doubt this, the actual specifications are given in Figure V-1.

	ADD $a + b + c = z$	ADD $a + b + c = c'$
FLOW-MATIC	ADD ⓐ TO b TO c; STORE THE SUM IN z	As Column 1
IBM COMMERCIAL TRANSLATOR	SET z = ⓐ + ⓑ + ⓒ	As Column 1, or ADD ⓐ TO c, ADD ⓑ TO c
COBOL	ADD ⓐ AND ⓑ AND ⓒ GIVING z	ADD ⓐ AND ⓑ TO c
FACT	SET z [EQUAL] TO ⓐ + ⓑ + ⓒ (PLUS ≡ +)	ADD ⓐ PLUS ⓑ TO c
CODEL	CALCULATE z = ⓐ + ⓑ + ⓒ	As Column 1
ELLIOTT'S	TAKE ⓐ ADD ⓑ AND ⓒ GIVING z	TAKE ⓐ ADD ⓑ {ADD INTO / INCREASE} c
NEBULA	COPY ⓐ + ⓑ + ⓒ {TO / → / INTO} z	ADD ⓐ + b TO c
SEAL	TAKE ⓐ ADD ⓑ [and] ⓒ MOVE RESULT [to] z	ADD ⓐ [to] c ADD ⓑ [to] c

Figure V-1: Eight different ways to add three numbers. The last four systems were developed in England. The use of a circle means that either a constant or piece of data can be used. Square brackets denote an option and braces indicate a choice to be made. Lower-case words are program supplied. Source: Willey *et al.* [WY61a], p. 10.

This gives added proof of the necessity for letting people define their own artificial languages (discussed in Chapter XII).

V.2. LANGUAGES OF PRIMARILY HISTORICAL INTEREST

It is not altogether surprising that the higher level languages for mathematical problems appeared before the development of languages suitable for business data processing problems, for the former field had a formalized and accepted notation which could be used as a common basis. There are far fewer systems to discuss here than there were in Chapter IV.

V.2.1. FLOW-MATIC (AND B-0)

As early as January, 1955, Dr. Grace Hopper and her staff at Remington Rand UNIVAC (among the key people were F. Delaney, L. Cousins, M. Harper, M. Hawes, T. Jones, M. Mulder, R. Rossheim, E. Somers, and D. Sullivan) had preliminary specifications of a language which would be suitable for doing business data processing on computers and still be easy to use. They made some early and unsuccessful attempts to provide abbreviations, on which numerous people could agree, for things like *GROSS PAY* or *COMPUTE* (as they could agree that *SIN X* is a reasonable abbreviation for *COMPUTE THE SINE OF X*). Then the idea of abbreviations was dropped and the designers introduced the concept of having a noun corresponding to a data description, rather than a symbol which would require lookup in a list to understand the meaning. A preliminary manual for the running system was marked *Company Confidential* and dated July, 1957; it was available to me at that time since I was an employee of the Sperry Rand Corporation. The first generally distributed version was available early in 1958. Its revision [RR59a] contained the first fairly complete list of commands; they are shown in Figure V-2.

Dr. Hopper and her group pioneered not only in the development and convincing that was necessary for the computer input of English-like notation but also in the general problem of getting users to accept programming languages of any kind. (See Chapter IV, for the early mathematical systems A-2 and MATHMATIC.) One of the first published user commentaries on this concept is given by Kinzler and Moskowitz [KB57].

The two most significant concepts introduced in FLOW-MATIC are (1) the use of understandable English words for both the operations to be performed and the data on which they are to operate, and (2) the realization that the data designs can and should be written completely independently of the procedures to be executed. Thus, it was possible to write

ADD

    $(h)\triangle$ADD$\triangle$field-name$\triangle(f_1)\triangle$TO$\triangle$field-name$(f_2)\triangle$[TO$\triangle$...etc.$\triangle$]
;$\triangle$STORE$\triangle$THE$\triangle$SUM$\triangle$IN$\triangle$field-name$\triangle(f_n)\triangle.\triangle$

CLOSE-OUT

    $(h)\triangle$CLOSE-OUT$\triangle$FILE$\triangle f_1 \triangle f_2 \triangle f_3 \triangle$...etc.$\triangle.\triangle$

COMPARE

  Option I:

    $(h)\triangle$COMPARE$\triangle$field-name$\triangle(f_1)\triangle$WITH$\triangle$field-name$\triangle(f_2)\triangle;\triangle$
    $\{$ IF$\triangle$EQUAL$\triangle$GO$\triangle$TO$\triangle$OPERATION$\triangle h_1\triangle;\triangle$ $\}$
      IF$\triangle$GREATER$\triangle$GO$\triangle$TO$\triangle$OPERATION$\triangle h_1\triangle;\triangle$
    OTHERWISE$\triangle$GO$\triangle$TO$\triangle$OPERATION$\triangle h_2\triangle;\triangle$

  Option II:

    $(h)\triangle$COMPARE$\triangle$field-name$\triangle(f_1)\triangle$WITH$\triangle$field-name$\triangle(f_2)\triangle;\triangle$
    IF$\triangle$ $\left\{ \begin{array}{l} \text{EQUAL}\triangle \\ \text{GREATER}\triangle \end{array} \right\}$ GO$\triangle$TO$\triangle$OPERATION$\triangle h_1\triangle;\triangle$
    IF$\triangle$ $\left\{ \begin{array}{l} \text{GREATER}\triangle \\ \text{EQUAL}\triangle \end{array} \right\}$ GO$\triangle$TO$\triangle$OPERATION$\triangle h_2\triangle;\triangle$
    OTHERWISE$\triangle$GO$\triangle$TO$\triangle$OPERATION$\triangle h_3\triangle.\triangle$

COUNT

  Option I:

    $(h)\triangle$COUNT$\triangle$field-name$\triangle(f_1)\triangle[,\triangle$PRESET$\triangle$VALUE$\triangle$IS$\triangle n_1\triangle][,\triangle$INCREMENT$\triangle$
IS$\triangle n_2\triangle$]
;$\triangle$WHEN$\triangle$COUNTER$\triangle$ $\left\{ \begin{array}{l} \text{EQUALS} \\ \text{EXCEEDS} \end{array} \right\}$ $\triangle n_3\triangle[,\triangle$RESET$\triangle$TO$\triangle n_4\triangle,\triangle$AND$\triangle$]GO$\triangle$TO$\triangle$
OPERATION$\triangle h_1\triangle[\triangle$OTHERWISE$\triangle$GO$\triangle$TO$\triangle$OPERATION$\triangle h_2\triangle].\triangle$

  Option II:

    $(h)\triangle$COUNT$\triangle$field-name$\triangle(f_1)\triangle[,\triangle$PRESET$\triangle$VALUE$\triangle$IS$\triangle n_1\triangle$]
[,$\triangle$INCREMENT$\triangle$IS$\triangle n_2\triangle$][,GO$\triangle$TO$\triangle$OPERATION$\triangle h_1\triangle].\triangle$

DIVIDE

    $(h)\triangle$DIVIDE$\triangle$field-name$\triangle(f_1)\triangle$BY$\triangle$field-name$\triangle(f_2)\triangle$
GIVING$\triangle$field-name$\triangle(f_3)\triangle.\triangle$

EXECUTE

    $(h)\triangle$EXECUTE$\triangle$OPERATION$\triangle h_1\triangle$[THROUGH$\triangle$OPERATION$\triangle h_2\triangle].\triangle$

FILL

    $(h)\triangle$FILL$\triangle$ $\left\{ \begin{array}{l} f_1\triangle \\ \text{sub-item-name}\triangle\text{IN}\triangle f_1\triangle \end{array} \right\}$
[,$\triangle f_2\triangle,\triangle f_3\triangle$...$\triangle$][sub-item-name$\triangle$IN$\triangle f_2\triangle,\triangle$
sub-item-name$\triangle$IN$\triangle f_3\triangle$...$\triangle$]WITH$\triangle$ $\left\{ \begin{array}{l} \text{SPACES} \\ x \\ \text{PERIODS} \end{array} \right\} \triangle.\triangle$ $x =$ *any character other than a period or space.*

Figure V-2. (cont. next page)

317

Figure V-2. (cont.)

HALT

    Option I:
        (h)△HALT△[Any-Descriptive-English]△.△

    Option II:
        (h)△HALT△BREAKPOINT△m△FORCE△TRANSFER△TO△GO△TO △OPERATION△h_1△[Any-Descriptive-English]△.△
        $m = 0, 2, 4$ through 9

IGNORE

    (h)△IGNORE△.△

INPUT

(h)△INPUT△name-of-file△FILE-f_1△$\begin{bmatrix} \text{SERVO}△s_1△ \\ \text{SERVOS}△s_1△,△s_2△ \end{bmatrix}$

name-of-file△FILE-f_2△$\begin{bmatrix} \text{SERVO}△s_1△ \\ \text{SERVOS}△s_1△,△s_2△ \end{bmatrix}$

    ⋮

;△OUTPUT△name-of-file△FILE-f_3△$\begin{bmatrix} \text{SERVO}△s_1△ \\ \text{SERVOS}△s_1△,△s_2△ \end{bmatrix}$

name-of-file△FILE-f_4△$\begin{bmatrix} \text{SERVO}△s_1△ \\ \text{SERVOS}△s_1△,△s_2△ \end{bmatrix}$

    ⋮

[;PRESELECTION][;△HSP△f_1△,△f_2△,△...△f_n△]

[;△T/C△f_1△,△f_2△,△...△f_n△][;△RERUN△$\begin{Bmatrix} \text{ON} \\ \text{WITH} \\ \text{FROM} \end{Bmatrix}$△OUTPUT△$f_n$△].△

INSERT

    Option I:
        (h)△INSERT△constant△INTO△field-name△(f_1)△[,△field-name△(f_2)△,△...△].△

    Option II:
        (h)△INSERT△constant△INTO△field-name△(f_1)△[,△field-name△(f_2)△,△...△]
        ;△WHERE△x△EQUALS△$\begin{Bmatrix} \text{SPACES} \\ \text{PERIODS} \end{Bmatrix}$△.△

    Option III:
        (h)△INSERT△constant△INTO△field-name△(f_1)△[,△field-name△(f_2)△,△...△]
        ;△WHERE△xEQUALS△SPACES△AND△y△EQUALS△PERIODS△.△

JUMP

    (h)△JUMP△TO△OPERATION△h_1△.△

MOVE

    (h)△MOVE△field-name△(f_1)△TO△field-name△(f_2)△[,△field-name△(f_3)△...etc.△]
    [;△field-name△(f_1')△TO△field-name△(f_2')△[,△field-name△(f_3')△...etc.△]].△

Figure V-2. (cont.)

MULTIPLY

$$(h)△MULTIPLY△\begin{Bmatrix} field\text{-}name△(f_1) \\ constant \end{Bmatrix}△BY\begin{Bmatrix} field\text{-}name△(f_2) \\ constant \end{Bmatrix}$$
$$△GIVING△field\text{-}name△(f_3)△.△$$

NUMERIC–TEST

Option I:

$(h)△NUMERIC\text{-}TEST△field\text{-}name△(f_1)△[,△field\text{-}name△(f_2)△,△...△$
$field\text{-}name△(f_n)△];△IF△\begin{Bmatrix} NUMERICAL \\ NUMERIC \end{Bmatrix}△GO△TO△OPERATION△h_1△$
$[;△OTHERWISE△GO△TO△OPERATION△h_2△].△$

Option II:

$(h)△NUMERIC\text{-}TEST△field\text{-}name△(f_1)△;△IF△\begin{Bmatrix} NUMERICAL \\ NUMERIC \end{Bmatrix}△MOVE△TO△$
$field\text{-}name△(f_2)△AND△GO△TO△OPERATION△h_1△[;△OTHERWISE△GO△TO$
$△OPERATION△h_2△].△$

OVERLAY

$(h)OVERLAY△FROM△OPERATION△h_1△.△$

PRINT–OUT

Option I:

$(h)△PRINT\text{-}OUT△field\text{-}name△(f_1)△[,△...etc.△].△$

Option II:

$(h)△PRINT\text{-}OUT△constant△,△constant△[,△...etc.△].△$

Option III:

$(h)△PRINT\text{-}OUT△field\text{-}name△(f_1)△,△constant△[,△...etc.△].△$

READ–ITEM

$(h)△READ\text{-}ITEM△f_1△[;△IF△END△OF△DATA△GO△TO△OPERATION△h_1△].△$

REPLACE

$$(h)△REPLACE△\begin{Bmatrix} LEADING△ZEROES \\ LEADING△0△ \\ PERIODS△ \\ SPACES \\ x△ \end{Bmatrix}△WITH△\begin{Bmatrix} PERIODS△ \\ SPACES△ \\ ZEROES△ \\ y△ \end{Bmatrix}$$
$$IN△field\text{-}name△(f_1)△[,△field\text{-}name△(f_2)△...,△field\text{-}name△(f_n)△].△$$

REWIND

$(h)△REWIND△f_1△[,△f_2△,△f_3△...,etc.△].△$

Figure V-2. (cont. next page)

Figure V-2. (cont.)

SELECT

$(h)\triangle$SELECT$\triangle\begin{Bmatrix}\text{INPUT}\triangle\\\text{OUTPUT}\triangle\end{Bmatrix}$name-of-file$\triangle$FILE-$f_1\triangle$
[name-of-file$\triangle$FILE-$f_2\triangle...\triangle$]
[;$\triangle\begin{Bmatrix}\text{OUTPUT}\triangle\\\text{INPUT}\triangle\end{Bmatrix}$name-of-file$\triangle$FILE-$f_n\triangle$
[name-of-file$\triangle$FILE-$f_m\triangle...\triangle$]]$\triangle.\triangle$

SELECT–LEAST

$(h)\triangle$SELECT–LEAST$\triangle$KEY$\triangle$;$\triangle$IF$\triangle f_1\triangle$GO$\triangle$TO$\triangle$OPERATION$\triangle h_1\triangle,\triangle$
IF$\triangle f_2\triangle$GO$\triangle$TO$\triangle$OPERATION$\triangle h_2\triangle,\triangle$[IF$\triangle f_3\triangle$GO$\triangle$TO$\triangle$OPERATION$\triangle h_3\triangle,\triangle$...
IF$\triangle f_3\triangle$GO$\triangle$TO$\triangle$OPERATION$\triangle h_3\triangle$]$.\triangle$

SET

$(h)\triangle$SET$\triangle$OPERATION$\triangle h_1\triangle$TO$\triangle$GO$\triangle$TO$\triangle$OPERATION$\triangle h_2\triangle$[,$\triangle$OPERATION$\triangle$
$h_3\triangle$TO$\triangle$GO$\triangle$TO$\triangle$OPERATION$\triangle h_4\triangle$...,$\triangle$OPERATION$\triangle h_5\triangle$TO$\triangle$GO$\triangle$TO$\triangle$
OPERATION$\triangle h_6\triangle$]$.\triangle$

STOP

$(h)\triangle$STOP$\triangle.\triangle$(END)

SUBTRACT

$(h)\triangle$SUBTRACT$\triangle$field-name$\triangle(f_1)\triangle$[AND$\triangle$field-name$\triangle(f_2)\triangle$AND$\triangle$etc.$\triangle$]
$\triangle$FROM$\triangle$field-name$\triangle(f_n)\triangle$;$\triangle$STORE$\triangle$THE$\triangle$REMAINDER$\triangle$IN$\triangle$field-name$\triangle(f_m)$
$\triangle.\triangle$

SUPPLEX

Option I:

$(h)\triangle$SUPPLEX$\triangle xxx\triangle$[BLK–RELATIVE$\triangle$][Any-Descriptive-English$\triangle$]$.\triangle$
where xxx is a section number

Option II:

$(h)\triangle$SUPPLEX$\triangle nnn\triangle$[BLK–RELATIVE$\triangle$]:$\triangle$
USE$\triangle v_1\triangle$[WHERE$\triangle x\triangle\begin{Bmatrix}\text{IS}\\-\end{Bmatrix}\triangle\begin{Bmatrix}\text{SPACE}\\\text{PERIOD}\end{Bmatrix}\triangle$][Any-Descriptive-English]
[AND$\triangle v_2\triangle$[WHERE$\triangle x\triangle\begin{Bmatrix}\text{IS}\\-\end{Bmatrix}\triangle\begin{Bmatrix}\text{SPACE}\\\text{PERIOD}\end{Bmatrix}\triangle$][Any-Descriptive-English]]
$\vdots$
[AND$\triangle v_n\triangle$...etc.$\triangle$]$.\triangle$

where nnn must be any alphabetic or alphanumeric designation assigned to this SUPPLEX routine

where each of the values $v_1, v_2, ... v_n$ may take any one of the following forms:

    a
    bbbb
    cccccc
    dddddddddd
    field-name$\triangle(f_1)$

Figure V-2. (cont.)

SWITCH

        (h)△SWITCH△[any-alphabetic-or-numeric-designation-for-the-switch]△.△

TEST

    Option I:

        (h)△TEST△field-name△(f_1)△AGAINST△test-value△;△

        IF△ { GREATER△ / EQUAL△ / LESS△ / UNEQUAL△ } GO△TO△OPERATION△h_1△;△

        OTHERWISE△GO△TO△OPERATION△h_2△.△

    Option II:

        (h)△TEST△field-name△(f_1)△AGAINST△test-value△;△

        IF△ { GREATER△ / EQUAL△ / LESS△ } GO△TO△OPERATION△h_1△;△

        IF△ { GREATER△ / EQUAL△ / LESS△ } GO△TO△OPERATION△h_2△;△

        OTHERWISE△GO△TO△OPERATION△h_3△.△

    Option III:

        (h)△TEST△field-name△(f_1)△AGAINST△test-value-1△;△

        IF△ { GREATER△ / EQUAL△ / LESS△ } GO△TO△OPERATION△h_1△;△

        AGAINST△test-value-2△;△

        IF△ { GREATER△ / EQUAL△ / LESS△ } GO△TO△OPERATION△h_2△;△

        IF△ { GREATER△ / EQUAL△ / LESS△ } GO△TO△OPERATION△h_3△;△

        ⋮

        [AGAINST△test-value-n△;△

        IF△ { GREATER△ / EQUAL△ / LESS△ / UNEQUAL△ } GO△TO△OPERATION△h_m△;△

        OTHERWISE△GO△TO△OPERATION△h_n△.△

TRANSFER

    Option I:

        (h)△TRANSFER△f_1△TO△f_2△.△

    Option II:

        (h)△TRANFER△sub-item-name△IN△f_1△TO△f_2△.△

Figure V-2. (cont. next page)

Figure V-2. (cont.)

    Option III:
        (h)△TRANSFER△f_1△TO△sub-item-name△IN△f_2△.△
    Option IV:
        (h)△TRANSFER△sub-item-name△IN△f_1△TO△sub-item-name△IN△f_2△.△

TYPE

    (h)△TYPE△[any-descriptive-words△]INTO△field-name△(f_1)△,△ [field-name△(f_2)△,△...etc.△].△

UNIVAC

    (h)△UNIVAC△xxx△[BLK–RELATIVE△][Any-Descriptive-English△].△
    where xxx is a section number

WRITE–ITEM

    (h)△WRITE–ITEM△f_1△.△

X–1

    (h)△X–1△[BLK–RELATIVE][Any-Descriptive-English△].△

Figure V-2. FLOW-MATIC verb formats. The △ represents a required blank character, the square bracket denotes an option, and braces indicate a choice. Source: [RR59a], extracts from pp. 87–92.

a complicated description of a file quite independently from a specific procedure to be executed on that file.

An example of an early (circa 1955–56) FLOW-MATIC program is shown in Figure V-3, and a list of the available commands is given in Figure V-2. Preprinted forms were used for describing the data design. Admittedly, the language is stilted and the English is not very natural; it is also designed with significant attention paid to the 12-character word available on UNIVAC. In spite of these shortcomings, FLOW-MATIC was a milestone in the significant concepts of programming languages.[1]

While the sort generator written by F. Holberton [HF54] was an earlier example of the development of the concept of code generators (they were also used in A-2), the implementation of FLOW-MATIC really brought this concept into compilers in a major way. In the more mathematically oriented compilers, an addition could take place between either two fixed point numbers or two floating point numbers, and each of those was of a fixed length corresponding to the word size of the machine. However, in a busi-

---

[1] Note that FLOW-MATIC was originally called B-0. All the early Remington Rand systems had a letter(s) followed by a digit to indicate which version it was, until some marketing people decided that names such as FLOW-MATIC and MATH-MATIC were more appealing than B-0 and AT-3.

(0) INPUT INVENTORY FILE–A PRICE FILE–B ; OUTPUT PRICED–INV FILE–C UNPRICED–INV FILE–D ; HSP D .
(1) COMPARE PRODUCT–NO (A) WITH PRODUCT–NO (B) ; IF GREATER GO TO OPERATION 10 ; IF EQUAL GO TO OPERATION 5 ; OTHERWISE GO TO OPERATION 2 .
(2) TRANSFER A TO D .
(3) WRITE–ITEM D .
(4) JUMP TO OPERATION 8 .
(5) TRANSFER A TO C .
(6) MOVE UNIT–PRICE (B) TO UNIT–PRICE (C) .
(7) WRITE–ITEM C .
(8) READ–ITEM A ; IF END OF DATA GO TO OPERATION 14 .
(9) JUMP TO OPERATION 1 .
(10) READ–ITEM B ; IF END OF DATA GO TO OPERATION 12 .
(11) JUMP TO OPERATION 1 .
(12) SET OPERATION 9 TO GO TO OPERATION 2 .
(13) JUMP TO OPERATION 2 .
(14) TEST PRODUCT–NO (B) AGAINST ZZZZZZZZZZZZ ; IF EQUAL GO TO OPERATION 16 ; OTHERWISE GO TO OPERATION 15 .
(15) REWIND B .
(16) CLOSE–OUT FILES C ; D .
(17) STOP . (END)

Figure V-3. FLOW-MATIC program.

ness data processing problem, the size of the field varied and alignment of decimal points was always required. Thus although the FLOW-MATIC generators were an extension of those developed in A-2, the generators for the business data processing compilers were far more essential. The generators referred to here are used by the compiler to create object code for each operation in which the possible choices are almost infinite. For example, instead of having at object time all the code needed for each possible type of addition, (only) the instructions needed for each case are ascertained from the data description and included in the object program. The major alternative is some type of interpretation at object time.

FLOW-MATIC, and its later modification known as AIMACO (see Section V.2.2), was a major input to the Short Range Committee developing COBOL (discussed in Section V.3.1). In actual fact, FLOW-MATIC was the only language with which there was any experience at that time. Perhaps the only disadvantage that accrued from the experience was what can be defined as bending over backwards to provide an English-like language. In other words, the Remington Rand people felt that no businessman or person concerned with business data processing problems was really interested in writing symbolic formulas and that if he actually wanted to compute $(A + B \times C)/D$ or $((BASE\text{-}PRICE) + INCREMENT \times (DISCOUNT\text{-}PERCENT)) \times NO\text{-}OF\text{-}UNITS$, he would prefer to write it as a series of individual statements. This concept, along with the corresponding one that any mathematical symbolism was not suitable for a business data processing language, did cause a fair amount of conflict within the committee; in fact, some of this conflict still rages. (The individuals involved at the time actually conveyed much stronger views along these lines than those stated here.) The question of whether people prefer to write *equals* or *equal to* instead of an = is one that requires solution by a psychologist rather than by programmers.

V.2.2. AIMACO

Using the concepts developed in FLOW-MATIC (see the previous section), the Air Force Air Materiel Command at the Wright-Patterson Air Force Base in Ohio developed a system called AIMACO. Direction and supervision were provided by Col. Alfred Asch and John L. Jones. The original system was implemented on the UNIVAC 1105. The verbs available were almost all those provided in FLOW-MATIC. Because of this, the first phase of the compilation could be made on the UNIVAC I or II, using part of the B-0 compiler; the intermediate files were then translated to 1105 code on the 1105. This was done to save calendar time in developing the system.

The most interesting aspect of the AIMACO development is that work was started to implement the language for the IBM 705. This is probably the first attempt at deliberately planning a language that could be used to run the same source program on two significantly different computers. The work on the 705 system was underway when the Short Range Committee started to define COBOL (See Section V.3.1); the 705 work was never finished so that the resources could be diverted to the development of COBOL.

AIMACO was used quite extensively for about two and a half years on the 1105 until a COBOL-60 compiler, developed by the Air Materiel Command for the 1105, became operational.

V.2.3. COMMERCIAL TRANSLATOR

As early as January, 1958 there were some preliminary specifications for a language to be used for business data processing (although not limited to that area) being developed in IBM under the technical leadership of R. Goldfinger. It was originally known as COMTRAN and eventually received the official title of Commercial Translator. Following the philosophy established in FLOW-MATIC, this was an English-like language. It introduced several significant concepts. One was the introduction of formulas which are of course standard in scientific languages but were new to the business data processing field. A second key feature was the introduction of the IF... THEN... facility. A third idea was the concept of allowing differing levels in the data description. The *Picture Clause* that eventually appeared in COBOL was a direct contribution from Commercial Translator, along with the concept of suffixing. Perhaps the weakest part of the Commercial Translator proposal was the avowed statement in the first manual that "... a data description is not directly transferable to a different machine system...."[1a]

The position of IBM relative to COBOL and Commercial Translator was one of oscillation for many years. The original Commercial Translator manual [IB59] was a significant input to the Short Range Committee developing COBOL (see Section V.3.1); a number of significant ideas from Commercial Translator were taken over directly, such as the *Picture Clause*, the IF... THEN... clause (which was however significantly extended), suffixing, and the use of formulas. The 1960 Commercial Translator manual [IB60a] with addenda [IB61b] shows certain concepts that were obtained from COBOL (and indeed even from FACT—see Section V.2.4). It is perhaps unfortunate that in some cases the Short Range Committee chose deliberately to do things in a different way simply to avoid accusation of domination by IBM. During much of 1960 and 1961, IBM took the position that COBOL was not a well-defined language and that Commercial Translator was really very much better even in the portions of COBOL that had been defined. IBM eventually found itself in the position of having to implement Commercial Translator for several different machines because of customer commitments, but it also had to implement COBOL because of pressure from government users and other customers. The very efficient implementation of Commercial Translator on the 709 [IB62a] caused the language to stay in use on that machine somewhat longer than it did on the 7070 and 7080. After the initial implementations on the latter two machines, IBM dropped Commercial Translator except for the 7090 customers who insisted on it and then dropped it completely when going to System/360.

From a technical point of view, there are several useful features or concepts in Commercial Translator which have not yet found their way into COBOL. These include the use of floating point numbers (although some implementations of COBOL have provided this facility), the truth operator, the ability to specify functions, parametric substitution in the Commercial

---

[1a] [IB59], p. 33.

Translator *DO* (which is not in the COBOL *PERFORM*), the absolute value operator, the *CALL* verb which permits alternate (and thus abbreviations for) data names, and the ability to assign a value to a condition name at object time. A list of the verbs and their formats is given in Figure V-4.

A comparison of COBOL and Commercial Translator as implemented for the IBM 7090/94 is given in [IB63].

ADD [CORRESPONDING] *data.name.1* TO *data.name.2, data.name.3, ... data.name.n*
BEGIN SECTION [USING *parameter.1, parameter.2, ... parameter.n*] [GIVING *function.1, function.2, ... function.n*]
CALL (*old.name.1*) *new.name.1,* (*old.name.2*) *new.name.2, ...* (*old.name.n*) *new.name.n*
CLOSE {*file.name.1, file.name.2, ... file.name.n* / ALL FILES}
DISPLAY {'any message' / *data.name* / *any.combination.of.the.above*}
DO *procedure.name* [EXACTLY *n* TIMES] [USING *data.name.1, data.name.2, ... data.name.n*] [GIVING *result.name.1, result.name.2, ... result.name.n*]
DO *procedure.name* FOR *index.name.1* = *p.1(q.1)r.1* [, *index.name.2* = *p.2(q.2)r.2, index.name.3* = *p.3(q.3)r.3*] [USING *data.name.1, data.name.2, ... data.name.n*] [GIVING *result.name.1, result.name.2, ... result.name.n*]
END *procedure.name*
ENTER *coding.language*
FILE *record.name* [IN *file.name*]
GET {RECORD FROM *file.name* / *record.name*} AT END *any imperative clause*
GO TO *procedure.name*
GO TO *procedure.name.1* WHEN *conditional.expression.1, procedure.name.2* WHEN *conditional.expression.2 ... procedure.name.n* WHEN *conditional.expression.n*
GO TO (*procedure.1, procedure.2, ... procedure.n*) ON *index.name*
INCLUDE [HERE] *library.procedure* [AS *procedure.name*] [WITH *new.name.1* FOR *old.name.1, new.name.2* FOR *old.name.2, ... new.name.n* FOR *old.name.n*]
LOAD *procedure.name*
MOVE [CORRESPONDING] *data.name.1* TO *data.name.2, data.name.3, ... data.name.n*
NOTE *any sentence.*
OPEN {*file.name.1, file.name.2, ... file.name.n* / ALL FILES}
OVERLAP *procedure.name.1 procedure.name.2, ... procedure.name.n*
SET *variable.1, variable.2, ... variable.n* = *arithmetic.expression* [TRUNCATED] [, ON OVERFLOW *any.imperative.clause*]
SET *condition.name*
STOP *n*

Figure V-4. List of Commercial Translator commands. The square bracket denotes an option and braces indicate a choice.
Source: [IB60a], pp. 108–109. Reprinted by permission from *General Information Manual: IBM Commercial Translator.* © 1960 by International Business Machines Corporation.

V.2.4. FACT

Early in 1959 a contract was given by the Minneapolis-Honeywell Regulator Company, Datamatic Division (currently called Honeywell) to the newly formed Computer Sciences Corporation (containing fewer than 10 people) to produce a business compiler for the Honeywell 800. The supervision from Honeywell's side was done by R. Clippinger; R. Nutt was the key technical man from CSC. The work on FACT (*Fully Automatic Compiling Technique*), as it was later called, was started prior to the work on COBOL and of course ran parallel with it. Fortunately or unfortunately, the Honeywell representative was remarkably silent about the technical work being done on FACT during most of the early deliberations of the Short Range Committee (see Section V.3.1). The first report issued early in the fall of 1959 as a preliminary description [HO59] came as a shock to the COBOL committee. Because of the obvious difficulties of developing a machine independent language, there were a number of more advanced features in the Honeywell-800 Business Compiler than in the preliminary specifications of COBOL, for the former did not have to worry about all the problems of machine independence. On the other hand, later versions of FACT [HO61] were influenced by COBOL and Commercial Translator, which in turn were affected by ideas in FACT.

The FACT Compiler, when produced (and it was significantly late), had several hundred thousand instructions and was probably the most complex system of its kind produced up to that time. Because of the existence of COBOL, accompanied by government pressure, Honeywell did not implement FACT for any machine beyond the 800. It is still in use by a few customers; but over the years, the major attention of that company in this area has been devoted to COBOL. Nevertheless, there are some interesting ideas in FACT which are worthy of mention. These are described more fully in the manual [HO61], and a description of FACT together with a comparison with Commercial Translator and COBOL are given by Clippinger [CP61].

FACT has a somewhat more flexible input/output system (including data description) than COBOL. The hierarchy present up to the record level in COBOL is actually extended up to the file level in FACT; in addition, the user is able to access by name any particular subrecord and it will be brought in automatically. The implicit assumption is that people using FACT will have their data on an input deck from which a tape file must be created by the FACT system. (A similar facility is available for paper tape file data.) The FACT system does quite a bit of error checking as this conversion is taking place. A distinction is made in FACT between what are called *primary* and *secondary* groups, where the former are present a specific number of times and can appear in memory all at once, while the latter are present an indefinite number of times and are therefore brought in one at a time. This makes it possible to keep information more compactly on tape and the user needs to do less bookkeeping in order to reach the particular fields that he wants. As seen from the list of verbs in FACT—Figure V-5—they appear very similar to those of COBOL. However, the input/output verbs are able to access any group and not just the record. A specific *UPDATE* verb is available in FACT. Sort and report writing facilities were included in Extended COBOL 61 (see Section V.3.3), but they appeared in the earlier specifications of FACT.

ADD	GO	REPLACE
ARE	IGNORE	REVERSE
CLOSE	IS	REWIND
CONTROL	LEAVE	SEE
DELETE	LOCK	SET
DIVIDE	MULTIPLY	SORT
DO	OPEN	SUBTRACT
EQUALS	PERFORM	UNWIND
FILE	PUT	UPDATE
FIND	RELEASE	USE
GET	REMOVE	WRITE

Figure V-5. List of FACT verbs.
Source: [HO61], extracts from pp. 172–173.

FACT provides certain facilities for validity checking of arithmetic. FACT also provides a number of synonyms for specific verbs, e.g., DO, SEE. An automatic search for the tabular value of an argument is available in FACT.

While there are a number of detailed differences in the way in which facilities or concepts are expressed, the items stated above indicate the most significant facility additions beyond those of COBOL. No attempt has been made here to show detailed differences for format.

V.2.5. GECOM

The GECOM system for the GE-225 was supposed to be based on COBOL-61, but there were enough changes to it, and elements deleted from and added to it, so that GECOM is worth describing very briefly. The additions are based primarily on ALGOL, either in syntax or in function. (It is assumed that the reader is familiar with both COBOL and ALGOL.) At the beginning of an article by Katz [KX62], who led the development of GECOM, he states "GECOM, the General Compiler for the GE-225 is not a new source language, but rather a compiling technique. It's [sic] source language is made up of four parts: ALGOL, COBOL, FRINGE, and TABSOL. The construction of the compiler is such that languages can be added, extended or removed."[2] (TABSOL is a language for decision tables and FRINGE is defined by Katz as "a problem oriented language for sorting and merging of data, writing reports, and file maintenance".[3]) The specifications given in the manual [GZ61] make no mention of the latter two languages, so their exact usage is unclear to me. They are shown in a diagram contained in an article by Schwalb [SB63] discussing GECOM usage, but apparently they were not actually available because he indicates that a report writer would be desirable in GECOM and in future compilers.

Some of the changes to COBOL are

1. Data names cannot exceed 12 characters and cannot contain all numerals and the letter E.
2. Data descriptions are given in a fixed format on a printed form.
3. Several verbs are omitted, both by name and function, e.g., *EXAMINE, USE, INCLUDE*.
4. Several verbs have functional capabilities changed or omitted or they incorporate those from others, e.g., *OPEN, CLOSE, READ, WRITE* (but no *ACCEPT* or *DISPLAY*), and *PERFORM* (see under extensions).
5. No *THEN* clause is allowed in the *IF* statement; the latter is only of the form IF . . . GO TO . . . .

Among the more significant extensions are

1. Subscripts can be arithmetic expressions and can be subscripted.
2. Floating point numbers and arithmetic.
3 Eight elementary mathematical functions.
4. Sections specify input and output formal parameters and use a *BEGIN* . . . *END* to identify the body of code.
5. The *PERFORM* verb specifies parameter passage for sections. The loop control function is performed by a new verb called *VARY*, but it is much weaker than the COBOL *PERFORM*.
6. Additional file description entries.
7. A verb to *EXCHANGE* the contents of two fields.

It should be clear, even from this very brief outline, that GECOM *syntactically* resembles no particular language. It is not enough like COBOL to be considered a dialect or even COBOL-like, and it is much further from ALGOL. The flavor and spirit resemble COBOL, but that is all that can be said for the resemblance. From the viewpoint of the functions it performs, the designers chose the features from both languages which they felt were

---

[2] Katz [KX62], p. 495.
[3] Katz [KX62], p. 495.

necessary. I wonder whether the ideas expressed and methods suggested in Sammet [SM61] for combining ALGOL and COBOL might have helped the GECOM designers come closer to both languages if my referenced work had been completed first. Both sets of ideas were being developed at about the same time but quite independently.

V.3. COBOL

V.3.1. HISTORY OF COBOL

On May 28 and 29, 1959, a meeting was called in the Pentagon by Charles A. Phillips of the Department of Defense. The suggestion that the Department of Defense call this conference was made by a small group representing users, manufacturers, and universities which had met at the University of Pennsylvania Computing Center on April 8, 1959 to discuss the problem of developing a common business language. The purpose of this May meeting was to consider both the desirability and the feasibility of establishing a common language for the adaptation of electronic computers to data processing. About forty representatives from users, government installations, computer manufacturers, and other interested parties were present.[4] There was almost unanimous agreement that the project was both desirable and feasible at this time. The concept of three committees was agreed upon. They were called the Short Range, Intermediate Range, and Long Range, with appropriate time scales.

One interesting point which is not widely known is that in spite of all the references to the CODASYL (*CO*nference on *DA*ta *SY*stems *L*anguages) Committee which have appeared for years, it was never a committee in the normal sense of the word. It was created—as a concept—at this May, 1959 meeting, but the group never met again and really consisted only of a mailing list. There was—and is—an Executive Committee for CODASYL, but the parent group never really functioned or existed as an organization.

The Short Range Committee was composed of six manufacturers (Burroughs, IBM, Minneapolis-Honeywell, RCA, Remington Rand Division of Sperry Rand, and Sylvania Electric Products) and two government agencies (Air Materiel Command, USAF, and David Taylor Model Basin, Department of the Navy), in addition to Chairman J. Wegstein of the National Bureau of Standards. This Committee held its first meeting on June 23, 1959 and working groups were established.

---

[4] Since I was present not only at this initial meeting (representing Sylvania Electric Products), but also was chairman of two different task groups of the Short Range Committee, the historical description of the early COBOL work is based on firsthand knowledge and participation, and is supported by appropriate documents (some of which I wrote at the time).

The assigned mission of the Short Range Committee was actually "to do a fact finding study of what's wrong and right with existing business compilers (such as FLOW-MATIC, AIMACO, COMTRAN, etc.) and the experience of users, thereof. This short range group is due to complete its work in three months, i.e., by September 1, 1959."[5] However, the Committee actually set itself the very ambitious goal of developing a language within three months. Thus this was the first attempt to have an intercompany committee, consisting primarily of competitive computer manufacturers, specify a complex machine-independent language on any time scale, let alone such a short one.

Working groups on data description and procedural statements prepared proposals for consideration by the full committee which met in August, 1959 for the purpose of preparing a report to the Executive Committee. The report, dated September 4, 1959, was presented and it stated that the Short Range Committee felt it had prepared a framework upon which an effective common business language could be built. It was recognized that the technical material contained rough spots and needed additions. The report requested that the Short Range Committee be authorized to complete and polish the system by December 1, 1959. It was also requested that the Short Range Committee continue beyond that date in order to monitor the implementation. Both these requests were granted.

The Committee held several meetings between September 18 and October 21, 1959 and proceeded steadily in its task of resolving problems and completing the language. The name COBOL, which suggests a *CO*mmon *B*usiness *O*riented *L*anguage, was adopted. From October 26 to November 7, 1959, H. Bromberg and N. Discount (RCA), V. Reeves and J. Sammet (Sylvania), and W. Selden and G. Tierney (IBM) worked continuously, integrating the rough specifications into a systematic language.

The COBOL System was reviewed and approved by the Short Range Committee during the week of November 16 to 20. Final editing by the people named above was done (with myself as chairman of the group), and initial distribution was accomplished December 17, 1959.

In January, 1960 the Executive Committee of CODASYL accepted and approved the report of the Short Range Committee. During the period from January to April, 1960, the report underwent editing for typographical and other minor errors, and it was published by the Government Printing Office in April 1960 [US60]. (After considerable debate, it was decided to list in that report only the names of the organizations involved and not the specific individuals representing them. Aside from the six people listed above (who were acknowledged by the Short Range Committee to have

---

[5] C.A. Phillips, "Summary of Discussions at Conference on Automatic Programming of ADPS for Business-Type Applications. The Pentagon, May 28–29, 1959", p. 3.

made the greatest contribution), a number of others participated. At the time of this writing, the draft of Appendix A of the USASI COBOL standard ([AA68] in preparation) contains the complete list of names, and this is the only place that they will be published.

The most significant aspect of this entire activity is that it was the first attempt (known to me) to have a group of competitors work together with the prime objective of developing a language that would be usable on computers from each of the manufacturers. That it succeeded is a tribute not only to the hard work of the individuals actually serving on the committee but more importantly to the management of the various companies involved who were able to recognize the value of subordinating their own individual plans and specialities to the broader overall benefit of the customers. This was particularly significant because many manufacturers were beginning or had already done considerable work on developing their own "commercial" languages and these developments naturally had to be eliminated or subordinated to the committee results.

The only companies which actually implemented the 1960 version of COBOL were Remington Rand and RCA, and they had compilers running in 1960. (Remington Rand actually used FLOW-MATIC to write a significant part of their COBOL compiler.) The two companies conducted an experiment in December, 1960 in which programs were interchanged; with only a minimum of modifications primarily due to differences in implementation, the programs were run on both machines—UNIVAC II and the 501. A description of this activity was given by Bromberg [BJ61].

During 1960, a Maintenance Committee existed for the purpose of initiating and reviewing recommended changes to keep COBOL up-to-date and clarifying points of confusion about the original specifications. The Maintenance Committee consisted of users' and manufacturers' groups which met both separately and jointly. A number of new (relative to the original Short Range Committee) organizations were represented on these committees, namely Bendix Computer Division, Control Data, DuPont, General Electric, National Cash Register, Philco, and U.S. Steel. In many cases the individuals representing the original organizations were new to the committee. As a result of both these factors, much attention was devoted to changes and improvements, with less attention paid to maintaining consistency with the 1960 specifications. (A description of the issues involved at that time as noted by me is given in Sammet [SM61b] or [SM61c].)

In order to devote concentrated attention to bringing out a revised and updated COBOL-1961, a Special Task Group was created by the Executive Committee. The sessions of this group were chaired by J. L. Jones of the Air Materiel Command and G. M. Dillon of the DuPont Company.

The net result of this activity was the report entitled *COBOL-1961: Revised Specifications for a Common Business Oriented Language,* issued by the Government Printing Office in June, 1961 [US61]. It differed significantly in some places from the 1960 specifications, but the basic concepts and principles remained the same. These 1961 specifications are the ones on which all ensuing work has been based.

It was recognized even by the Short Range Committee that there were certain major components of business data processing programming which needed to be put into COBOL, but there was just insufficient time in which to do it. After the issuance of the 1961 specifications, the work of the COBOL Committee was devoted primarily to developing some of these additional features. This resulted in the issuance by the Government Printing Office in November, 1962 of the *COBOL-1961 Extended: Extended Specifications for a Common Business Oriented Language* [US62] which contained major (Report Writer facilities and SORT verb), and a few minor, additions to COBOL-61 and a minimum of changes in specifications.

It was well recognized from the start that good facilities for table handling were an important feature in a language like COBOL. Since there were already ways of defining and operating on tables in COBOL-60, a separate verb was of much lower priority than a number of other problems which had to be solved; therefore nothing was done for several years. Then it was realized that although the emphasis in 1959–61 was on the use of magnetic tapes for large files, an increasing number of manufacturers were beginning to supply mass storage devices as a major component in their computing systems. It was clear that the input/output facilities suitable for magnetic tape could not make effective use of the mass storage equipment. For these reasons, the primary developmental effort of the COBOL Committee after the issuance of the *COBOL-61 Extended* manual was to prepare specifications for these two items. The designers naturally continued the work of cleaning up any ambiguities which were found.

In January, 1964 the COBOL Maintenance Committee was reorganized. The separate user and manufacturer groups were combined into the COBOL Committee with three subcommittees: Language, evaluation, and publication. The language subcommittee concerned itself with clarifications and additions, the evaluation subcommittee conducted surveys and evaluations of implementations and user activities, and the publication subcommittee was concerned with the preparation of publications and liaison with the USASI X3.4.4. Subcommittee, which was working on the development of a standard and the publication of the *COBOL Information Bulletin* (discussed later). In 1966 the *COBOL: Edition 1965* manual was issued through the Government Printing Office [US65]. An interesting feature of the work going on during that period is that significant contributions were made by the European Computer Manufacturer's Association (ECMA). The "new" version is still based on COBOL-61 and includes only "extensions, resolutions of ambiguities, deletion of redundancies, or removal of unused or poor language specifications."[6] It also contains a historical section which in my opinion is incorrect or misleading in a few small but significant comments. The COBOL-65 manual is a major reorganization and rewrite of the earlier ones.

The acknowledgment which the 1965 manual requests be used is as follows; it differs significantly from the earlier two versions, which listed the names of the companies involved.

> Any organization interested in reproducing the COBOL report and specifications in whole or in part, using ideas taken from this report as the basis for an instruction manual or for any other purpose is free to do so. However, all such organizations are requested to reproduce this section as part of the introduction to the document. Those using a short passage, as in a book review, are requested to mention "COBOL" in acknowledgment of the source, but need not quote this entire section.
>
> COBOL is an industry language and is not the property of any company or group of companies, or of any organization or group of organizations.
>
> No warranty, expressed or implied, is made by any contributor or by the COBOL Committee as to the accuracy and functioning of the programming system and language. Moreover, no responsibility is assumed by any contributor, or by the committee, in connection therewith.
>
> Procedures have been established for the maintenance of COBOL. Inquiries concerning the procedures for proposing changes should be directed to the Executive Committee of the Conference on Data Systems Languages.
>
> The authors and copyright holders of the copyrighted material used herein
>
>> FLOW-MATIC (Trademark of Sperry Rand Corporation), Programming for the Univac (R) I and II, Data Automation Systems copyrighted 1958, 1959, by Sperry Rand Corporation; IBM Commercial Translator Form No. F 28-8013, copyrighted 1959 by IBM; FACT, DSI 27A5260-2760, copyrighted 1960 by Minneapolis-Honeywell
>
> have specifically authorized the use of this material in whole or in part, in the COBOL specifications. Such authorization extends to the reproduction and use of COBOL specifications in programming manuals or similar publications.

V.3.2. Functional Characteristics of COBOL

In considering the general properties of languages, COBOL is not particularly general in the sense that it is aimed at that class of problems known as business data processing. COBOL is definitely *not* a succinct language; its objective was to be natural, where natural was defined as

---

[6] [US65], p. III-1-1.

being *English-like*. This led to the introduction of certain concepts in the language designed specifically to permit this type of naturalness. There is a certain amount of minor internal inconsistency in COBOL, particularly relative to the change of key and noise words in different divisions. Since one of the objectives was to make the statements easily understandable when read, it turned out that words which were logically necessary in one place were really only desirable noise words in another; this problem was resolved by permitting different noise and key words in different divisions. With regard to the efficiency, as indicated earlier, people have different views on what this may mean. COBOL does not permit minimal writing; on the contrary, it encourages a certain amount of verbosity. The benefit gained from this, however, is increased readability and understandability in looking at programs.

The purpose of COBOL was to provide a common business-oriented language. The word *common* was interpreted to mean that the source program language would be compatible among a significant group of computers. A realistic goal of achieving the maximum amount of compatibility on existent computers was the philosophy of the framework in which all the work was done. The application area is definitely defined as being that for business data processing, with no attempt to generalize the facilities. COBOL *has* been used however for some significant problems outside this area, e.g., for creating a differential equation writing system (Bennett [BE65]) and for writing a programming system (Callahan and Chapman [ZQ67]). In the latter case, a *COBOL program* was written to *translate* DETAB/65 (developed by SIGPLAN Working Group 2 of the ACM Los Angeles chapter) to *COBOL statements*. For further details, see the cover and page 125 of *Comm. ACM*, Vol. 9, No. 2 (Feb., 1966).

COBOL is definitely a language which specifies the problem solution by permitting the programmer to specify the algorithms. It is also very definitely a hardware language. In particular, close notice was made of the fact that the reference language and lack of input/output made it impossible to use ALGOL directly on a computer, and so COBOL was designed as a language which could be used as direct input to a computer.

The users for whom COBOL was designed were actually two subclasses of those people concerned with business data processing problems. One is the relatively inexperienced programmer for whom the naturalness of COBOL would be an asset, while the other type of user would be essentially anybody who had not written the program initially. In other words, the readability of COBOL programs would provide documentation to all who might wish to examine the programs, including supervisory or management personnel. Little attempt was made to cater to the professional programmer; in fact, people whose main interest is programming tend to be very unhappy with COBOL because so much writing is required. An attempt

SAMPLE PROGRAM—COBOL†

IDENTIFICATION DIVISION.
PROGRAM–ID. 'SORT360'.
REMARKS. THIS PROGRAM WAS WRITTEN TO DEMONSTRATE THE USE OF THE SORT FEATURE. THIS PROGRAM PERFORMS THE FOLLOWING TASKS –
1. SELECTS, FROM A FILE OF 1000–CHARACTER RECORDS, THOSE RECORDS HAVING FIELD–A NOT EQUAL TO FIELD–B.
2. EXTRACTS INFORMATION FROM THE SELECTED RECORDS.
3. SORTS THE SELECTED RECORDS INTO SEQUENCE, USING FIELD–AA, FIELD–BB, AND FIELD–CC AS SORT KEYS.
4. WRITES THOSE SORTED RECORDS HAVING FIELD–FF EQUAL TO FIELD–EE ON FILE–3 AND WRITES SELECTED DATA OF THE OTHER RECORDS ON FILE–2.

ENVIRONMENT DIVISION.
CONFIGURATION SECTION.
SOURCE–COMPUTER. IBM–360 F50.
OBJECT–COMPUTER. IBM–360 F50.
INPUT–OUTPUT SECTION.
FILE–CONTROL. SELECT INPUT–FILE–1 ASSIGN TO 'F401' UTILITY.
    SELECT SORT–FILE–1 ASSIGN 'SF1' UTILITY.
    SELECT FILE–2 ASSIGN 'F402' UTILITY. SELECT
    FILE–3 ASSIGN 'F403' UTILITY.

DATA DIVISION.
FILE SECTION.
FD INPUT–FILE–1 BLOCK CONTAINS 5 RECORDS
RECORDING MODE IS F
LABEL RECORDS ARE STANDARD
DATA RECORD IS INPUT–RECORD.

    01 INPUT–RECORD.
        02 FIELD–A   PICTURE X (20).
        02 FIELD–C   PICTURE 9 (10).
        02 FIELD–D   PICTURE X (15).
        02 FILLER   PICTURE X (900).
        02 FIELD–B   PICTURE X (20).
        02 FIELD–E   PICTURE 9 (5).
        02 FIELD–G   PICTURE X (25).
        02 FIELD–F   PICTURE 9 (5).
    SD SORT–FILE–1 DATA RECORD IS SORT–RECORD.
    01 SORT–RECORD.
        02 FIELD–AA   PICTURE X (20).
        02 FIELD–CC   PICTURE 9 (10).
        02 FIELD–BB   PICTURE X (20).
        02 FIELD–DD   PICTURE X (15).
        02 FIELD–EE   PICTURE 9 (5).
        02 FIELD–FF   PICTURE 9 (5).

Sample Program—COBOL (*cont. next page*)

Sample Program—*COBOL* (cont.)

```
FD  FILE-2 BLOCK CONTAINS 10 RECORDS
RECORDING MODE IS F
LABEL RECORDS ARE STANDARD
DATA RECORD IS FILE-2-RECORD

01  FILE-2-RECORD.
    02  FIELD-EEE   PICTURE $$$$$9.
    02  FILLER-A    PICTURE X (2).
    02  FIELD-FFF   PICTURE 9 (5).
    02  FILLER-B    PICTURE X (2).
    02  FIELD-AAA   PICTURE X (20).
    02  FIELD-BBB   PICTURE X (20).
FD  FILE-3 BLOCK CONTAINS 15 RECORDS
RECORDING MODE IS F
LABEL RECORDS ARE STANDARD
DATA RECORD IS FILE-3-RECORD

01  FILE-3-RECORD PICTURE X (75).

PROCEDURE DIVISION.

        OPEN INPUT INPUT-FILE-1, OUTPUT FILE-2, FILE-3.
        SORT SORT-FILE-1 ASCENDING FIELD-AA DESCENDING FIELD-BB,
        ASCENDING FIELD-CC INPUT PROCEDURE RECORD-SELECTION OUTPUT
        PROCEDURE PROCESS-SORTED-RECORDS. CLOSE INPUT-FILE-1, FILE-2,
        FILE-3. STOP RUN.

RECORD-SELECTION SECTION.
PARAGRAPH-1. READ INPUT-FILE-1 AT END GO TO PARAGRAPH-2.
    IF FIELD-A = FIELD-B GO TO PARAGRAPH-1 ELSE
        MOVE FIELD-A TO FIELD-AA MOVE FIELD-F TO FIELD-FF
        MOVE FIELD-C TO FIELD-CC MOVE FIELD-B TO FIELD-BB
        MOVE FIELD-D TO FIELD-DD MOVE FIELD-E TO FIELD-EE
        RELEASE SORT-RECORD. GO TO PARAGRAPH-1.
PARAGRAPH-2. EXIT.
PROCESS-SORTED-RECORDS SECTION.
PARAGRAPH-3. RETURN SORT-FILE-1 AT END GO TO PARAGRAPH-4.
    IF FIELD-FF = FIELD-EE WRITE FILE-3-RECORD FROM
        SORT-RECORD GO TO PARAGRAPH-3 ELSE
        MOVE FIELD-EE TO FIELD-EEE MOVE FIELD-FF TO FIELD-FFF
        MOVE FIELD-AA TO FIELD-AAA MOVE FIELD-BB TO FIELD-BBB
        MOVE SPACES TO FILLER-A, FILLER-B WRITE FILE-2-RECORD.
        GO TO PARAGRAPH-3.
PARAGRAPH-4. EXIT.
```

---

†Reprinted by permission from *IBM Operating System/360 COBOL Language*, pp. 142–43. © 1965 by International Business Machines Corporation, C28-6516-3, Data Processing Division, White Plains, N.Y. (1965).

to achieve the somewhat contradictory objectives of minimizing writing and obtaining good documentation is the Rapidwrite system developed in England (see Humby [HY62] and [HY63]), whereby people were able to write a very shorthand and formalistic version of COBOL and have the compiler turn out the actual legal official COBOL program. COBOL was definitely designed for use in a batch environment.

It is undoubtedly in the area of compatibility that the most misinformation and confusion has arisen concerning COBOL. First of all, it was recognized that "Differences in computers relating to size, types of peripheral equipment, and different order structure make *complete* compatibility impossible. Thus, the realistic goal of achieving the maximum amount of compatibility on present day computers was the philosophy or framework within which all work was done."[7] As noted later in the technical discussion, COBOL is divided into four divisions: IDENTIFICATION, PROCEDURE, DATA, and ENVIRONMENT. The IDENTIFICATION Division is trivial in size and clearly compatible across all computers and compilers, except that some implementers have imposed differing rules from those specified. The ENVIRONMENT Division (which defines the hardware to be used) by its very nature is completely dependent on the machine on which the source program is to be run and, in fact, even the machine on which the compiler is to be run. It is therefore completely machine dependent; it is probably compiler independent. In general, the PROCEDURE Division (which contains the executable operations) is machine and compiler independent, providing appropriate care is used in writing statements. The biggest difficulty in maintaining compatibility in the PROCEDURE Division is the lack of a standard collating sequence defined in connection with the language. Thus if letters test higher than numbers on one machine and lower on another, the *IF* statement that would be written to separate these might be different. The DATA Division (which describes the files and records to be processed) is the area in which there is the greatest difficulty in maintaining compatibility. Every attempt has been made to provide external descriptions of data, i.e., in terms of letters and numerals and types of usage rather than internal representation and format. This can be done to a very large extent, providing the user is less concerned about efficiency than about compatibility. In other words, if he wishes to establish a standard data description for a file that can be used on many machines, he may have to sacrifice certain specific features that would increase the efficiency on a particular computer. Problems have been run using the same program (except for the ENVIRONMENT Division, of course) with few or minimum modifications. (See, e.g., Fredericks and Warburton [FD65].) On the basis of hindsight and experience, some incorrect choices were made in placing certain features in certain divisions; a realignment would make the com-

---

[7] [US62], p. II-1.

patibility issues much clearer. On the other hand, the majority of incompatibilities arise from implementors who choose to deviate from clearly defined specifications. To deal with this problem, many organizations have written internal papers or manuals which tell users either how to hand-convert from one machine (or compiler) to another or how to write programs initially to avoid (or minimize) incompatibilities. Other groups have prepared lists of quirks which affect efficiency and/or compatibility (e.g., the *COBOL Programming Tips* of Westinghouse [WM67]).

Obviously, certain information pertaining to individual computers would never carry over to another machine. It was felt, however, that the advantage of having a common means of expression even for these features was sufficiently great to warrant the development of a standard form.

Dialects have not been a problem with COBOL, except for differences in interpretation and actual implementation, probably because there was a broadly based group which was defining the language. However, the problems of subsetting and extensions have been very significant from the beginning, and the subsetting is dealt with in an unusual way by the USASI standard (discussed later).

Although at the start of the Short Range Committee's activity it was tacitly assumed that COBOL was being developed only for large computers, an increasing number of manufacturers and users became interested in having this available on smaller machines. As a result, it was necessary to try to provide some official subsets that would be an adequate subsection of the language but still be more easily implemented on small machines than the entire language would be. In the COBOL-60 specifications, a subset called *Basic COBOL* was defined. In the 1961 specifications, a subset entitled *Required COBOL-1961* was defined to consist of "that group of features and options, within the complete COBOL specifications for the year 1961, which have been designated as comprising the minimum subset of the total language which must be implemented (to the extent of hardware capability) by any implementor claiming a "proper" COBOL-1961 compiler."[8] All other features and options were considered *elective*; but if they were implemented, they had to be done so in accordance with the specifications given in the manual. The manual for COBOL-1961 Extended kept this idea, but the concept was later dropped entirely. A subset known as *Compact COBOL* was defined by a COBOL Committee subcommittee but was not published because of possible confusion with the standards work.

As with all other languages, there have been some difficulties in compatibility arising from different interpretations of the specifications. As time progressed, these became minimized.

Converting COBOL programs from one machine (and/or compiler)

---

[8] [US61], p. I-3.

to another is relatively simple. As indicated earlier, various informal documents exist which provide directives or tips on how to do this. There has been no effort known to me to translate COBOL into another language either by sifting or by direct translation. Contrary to popular belief, a formal metalanguage was really used for much of the syntax definition. The connection between this and the notation used for ALGOL is discussed in Sammet [SM61a], and also in Section II.6.2.2.

The standardization of COBOL under ASA (USASI) was started in January, 1963 when Task Group X3.4.4 met, with H. Bromberg as chairman. At that time it was stated that "X3.4 recognizes [sic] CODASYL COBOL Maintenance Committee as the development and maintenance authority for COBOL."[9] This differs significantly from the creation of the FORTRAN standard, where there was no specified development and maintenance authority and, therefore, X3.4.3 became this authority by default. X3.4.4 undertook several activities, including (1) the establishment of a periodic *COBOL Information Bulletin* (CIB) which was to be widely distributed and permit rapid dissemination of information on the standardization activity; (2) a survey of features of existing or proposed COBOL processors; (3) the writing of test problems; (4) maintenance of close liaison with other standards bodies interested in COBOL, particularly ECMA (European Computer Manufacturer's Association) and ISO/TC 97/Subcommittee 5. Throughout the entire standardization activity, close liaison with international groups was considered important, and this was maintained. A similar statement applied to interaction with the CODASYL COBOL Committee, and this was particularly aided by a large overlapping membership on both that committee and X3.4.4 (or its working task groups).

An early view held that there should be one subset of COBOL defined in the standard. By March, 1964, however, it was felt that multiple levels should be defined. The criteria for allocation of elements to specific levels were

1. General usefulness, as determined by
   a. Degree of implementation.
   b. User acceptance.
   c. User desires.
   d. Experience.
2. Cost of implementation versus advantages of use.
3. Functional capability of element, considering redundancy.
4. Overall consistency of a defined level.
5. Upward compatibility.
6. Processing system capability.

The resulting proposal for the standard took the form of a nucleus and

---

[9] Shown as a note in Scope of X3.4.4 in *Minutes, Task Group X3.4.4*, Jan. 15–16, 1963.

eight functional processing modules. The proposed standard (known as pUSASI COBOL) was distributed as *COBOL Information Bulletin* (CIB) No. 9 and also as the April, 1967 issue of the *ACM SICPLAN Notices* [XB67]. The X3 committee balloted during the period July, 1967 through July, 1968. The final result was affirmative. However, during this period X3 voted to remove the Random Processing Module (discussed below) from the standard and to place it in an Appendix with suitable remarks to indicate that it was not an official part of the standard. Other suggestions for changes to the pUSASI COBOL were made and accepted. In August 1968 COBOL was officially approved as USA Standard X3.23–1968 although the final physical document was still being prepared and thus was unavailable at the time the approval was given. During the preparation of the U.S. Standard, a document containing a close logical subset was being prepared by an ISO COBOL editing committee. The preparation of that document was based on the continuing technical involvement of international standards organization representatives; the plan was to forward it directly upon completion to ISO/TC97 for distribution and ballotting as the ISO COBOL standard.

In order to understand the layout of the standard, it is necessary to assume familiarity with the technical specifications of COBOL. Those readers who do not have this knowledge are advised to read Section V.3.3 and then return to the next paragraphs.

The eight modules (besides the Nucleus) are Table Handling, Sequential Access, Random Access, Random Processing, Sort, Report Writer, Segmentation, and Library. Each module is divided into two or more levels; in some cases the lowest level of a module is null, meaning that none of its facilities is required in the minimum standard. A schematic description of the modules is shown in Figure V-6. The minimum which can be implemented is the level-1 Nucleus, which includes those elements necessary for internal processing on a small machine; e.g., the ADD, SUBTRACT, MULTIPLY, and DIVIDE verbs are provided in their simplest form, but the COMPUTE verb is not. The high level Nucleus contains the minimum package, plus additional facilities in the DATA Division and the remaining options of the internal processing verbs.

The low level Table Handling allows fixed tables, one level of subscripting or indexing, and manipulation of indices. The middle level permits three subscripts and more indexing operations. The highest level includes the SEARCH and SET verbs, the OCCURS, KEY, ASCENDING/DESCENDING, INDEXED BY, and USAGE IS INDEX in the DATA Division.

The Sequential Access low level module provides facilities for basic serial file processing. The high level includes extended capabilities in storage allocation and file assignment, file organization and labeling, and provisions for user-designed labels and error procedures. In addition, the more complex operations of the file processing verbs are permitted.

				Functional Processing Modules				
Nucleus	Table Handling	Sequential Access	Random Access	Random Processing*	Sort	Report Writer	Segmentation	Library
	3TBL 1, 3		2RAC 0, 2		2SRT 0, 2	2RPW 0, 2	2SEG 0, 2	2LIB 0, 2
2NUC 1, 2		2SEQ 1, 2		1RPR 0, 1				
	2TBL 1, 3		1RAC 0, 2		1SRT 0, 2	1RPW 0, 2	1SEG 0, 2	1LIB 0, 2
1NUC 1, 2	1TBL 1, 3	1SEQ 1, 2						
			null	null	null	null	null	null

*After the original issuance of the proposed standard it was decided to remove the Random Processing Module from the standard and place it in an Appendix for information purposes only. When the RPM is present, either 1RAC 0, 2 or 2RAC 0, 2 must also be present.

Figure V-6. Schematic diagram showing the structure of pUSASI COBOL standard. The first digit in all the codes represents the level's position in the hierarchy, and the last two digits indicate the minimum and maximum levels of the module to which the level belongs. For example, 2NUC 1, 2 denotes the second level of the Nucleus which is composed of two levels, neither one of which is a null set.
Source: [XB67], pp. 1–6a.

The Random Access module is similar to the Sequential Access module except that it provides for handling randomly ordered files on direct access devices and that its lowest level is a null set. If the Random Processing module is implemented, then one of the levels of the Random Access modules must also be implemented. The Random Processing allows the user to specify a number of asynchronous processing cycles and to specify *Saved Areas* to be associated with them. Facilities to start the asynchronous processing within the main program flow, control the cycle, and call a temporary halt to asynchronous processing to permit synchronization are all provided. The main facilities in this module are the verbs *HOLD*, *PROCESS*, and *USE FOR PROCESSING*, and the *FILE-CONTROL* paragraph statement *PROCESSING MODE IS RANDOM*.

The Sort module has a null lowest level, and its first meaningful level provides a single *sort* as a COBOL program, with provision for including first and last pass own-coding for the sort. The high level permits more than one sort per program and allows separate processing before and after each sort. (The main facilities are the *RELEASE*, *RETURN*, and *SORT* verbs.)

The Report Writer has a null lowest level, and its low level permits page formatting. The high level provides various controls for reporting. These include the *GENERATE*, *INITIATE*, *TERMINATE*, and *USE BEFORE REPORTING* verbs and, of course, the Report Group and Report Group Description entries from the DATA Division.

Segmentation also has a null lowest level, and its low level permits the user to overlay portions of the object program by assigning priority numbers. The high level also permits the assignment of segment limits in the ENVIRONMENT Division.

Finally, the Library has a null lowest level and its low level includes the *COPY* statement from all divisions, which permits the user to include elements in his source program at compile time. The high level adds the *REPLACING* option.

In considering the types and methods of language definition, much of the administrative information was supplied in the early history. An important thing to keep in mind is that the basic work from which all later developments arose was accomplished by a small group, who had the COBOL activity as only a part-time assignment from their employers, operating under the most tremendous time pressures.

The basic objective of COBOL was to supply an "English-based" common business-oriented language, independent of any make or model of computer, and open-ended.

The implementation of COBOL has been carried on by virtually every computer manufacturer and most of the independent software companies. One reason for this was based on interest in obtaining a commercial language for customers, and a more important reason was based on direct and indirect pressure from the government, which essentially said that a company which wanted to sell or rent computers to the federal government had to have a COBOL compiler unless they could clearly demonstrate that it was not needed for the particular class of problems involved.

The maintenance has been done with different organizational structures but always by a group of people under the official direction and sponsorship of the CODASYL Executive Committee. Both users and manufacturers have participated heavily in the maintenance. There have been two major problems in the maintenance: One is that many times implementers needed clarification on a particular point and could not obtain it from the maintenance committee fast enough to suit their schedules; this resulted in diverse interpretations of the language and a natural reluctance by each group to sacrifice its meaning at a later point in time. The second problem in maintenance was that in some cases the individuals serving on the committee changed quite frequently; the new person had to be educated rapidly, and in many cases he would bring up points that had been previously discussed at great length. Furthermore, most people had this activity as a low-priority assignment. It is very interesting to note that there are at least as many differing opinions on technical points within a company as across companies, and it was not at all uncommon to see one individual reversing the vote of another individual from his organization who had been on the committee somewhat earlier.

There was a great deal of talk in the earlier days about the poor definition of COBOL, climaxed by the often repeated comment that the language was not really defined. In actual fact, there was at least as much rigor in the definition of COBOL as in the definitions of FORTRAN. The complaints stemmed from two different problems. The first was that there were ambiguous statements in the descriptions (i.e., the semantics) which led to different interpretations by the implementers. This problem is still inherent in the state of the art of language definition, even for ALGOL, which was considered to have a more rigorous definition than either FORTRAN or COBOL. The second problem was that since the format of the COBOL manual was not the same as that of the ALGOL report, then the former was assumed to be vague. A paper (Sammet [SM61a]) showed this complaint to be invalid since the notation used for the COBOL report was actually a metalanguage which is roughly equivalent to the type of metalanguage used for the ALGOL report. More recent work in providing a formal definition has been done by ECMA [EC67]. Thus the syntax for COBOL is as well-defined as that of ALGOL. A set of syntactical charts for COBOL-61 was also produced (see Berman, Sharp, and Sturges [BF62]).

With regard to documentation, there has always been exactly one definitive manual, namely the one issued by the Government Printing Office under the auspices of the CODASYL Executive Committee. Other manuals which exist have been written by the manufacturers, sometimes as tutorial manuals and sometimes to indicate just what particular portions of the language they were implementing. Various descriptive articles and books have been published, e.g., Sammet [SM62] and Saxon [SX63]. The general status of COBOL up to 1963 is given by Cunningham [CG63]. The May, 1962 issue of the *Communications* of the *ACM* (Association for Computing Machinery) is devoted primarily to COBOL. Some comments on actual usage and/or company policies are given in Whitmore [WH62] and Cowan [CW64]. Various other articles are shown in the bibliography. A fairly good general discussion is given in *EDP Analyzer* [EP63]. Unfortunately, the majority of articles appear to have been written in 1961 and 1962, which was before the language was in widespread use and while it was still undergoing growing pains. A fairly complete list of articles (and books) is given in X3.4 *COBOL Information Bulletin* No. 8 (June, 1966). The CIB's [XB00] have been prepared by ASA (now USASI) Working Group X3.4.4 and distributed by BEMA (Business Equipment Manufacturer's Association). They have served as a good—although informal—means of disseminating information about COBOL from various sources, particularly the CODASYL COBOL Committee. The proposed standard was issued as CIB No. 9, and discussed earlier.

COBOL has proved its usefulness, and a number of organizations do all or most of their business data processing programming using it. Its strongest points (among the advantages normally expected from higher level languages) appear to be its convertibility from one machine to another and its ease of use for communication and documentation purposes; many of its disadvantages (although certainly not all) pertain to specific implementations. The early ones in particular were so unusable in many cases that people were reluctant to use, or prevented on practical grounds from using the language. Many professional programmers dislike the amount of writing which is required, and some people dislike the number of redundant features which make teaching or management more difficult. The lack of mass storage facilities until COBOL-65 was a handicap to installations with that equipment. As with any language, there continue to be other features which people would like to have included.

In total, COBOL seems to have met most of the objectives which were established for it, although perhaps not as completely as we had hoped. There is no doubt that it has achieved a successful and useful place in the computing community.

V.3.3. Technical Characteristics of COBOL

The language definition in this section is based on COBOL-65 [US65], plus those changes approved by the CODASYL COBOL Committee through

March, 1968. In a few cases this includes elements not contained in the standard, which is based on COBOL-65 and changes approved through January 1, 1967. Although the standard does not include all of COBOL-65, the latter was chosen as the basis for the description in the book because of the continuing changes in the standard as it was being developed.

The character set in COBOL consists of 51 characters. These include the 10 digits, 26 upper-case letters, and the following 15 symbols:

$$+ \quad - \quad * \quad / \quad = \quad \$ \quad < \quad > \quad , \quad . \quad ; \quad " \quad (\quad) \quad blank$$

There are a number of key words in the language. The graphic operators and punctuation are shown in the character set.

Identifiers are composed of a combination of not more than 30 characters chosen from the digits, letters, and the hyphen. An identifier may not begin or end with a hyphen; the hyphen is usually used for readability purposes, for example QUANTITY—ON—HAND. Data names and statement labels are formed this way.

Some other types of identifiers exist in COBOL. In particular, a *condition-name* is assigned to a specific value or set or range of values within the complete set that a data name may assume. For example, if there is a conditional variable called TITLE, then the condition names ANALYST, PROGRAMMER, and CODER can be used instead of their numeric equivalents, and a test may be made by writing IF CODER. Certain values exist which have been assigned fixed identifiers; these are called *figurative constants*, for example SPACES or ZEROS.

There are three types of reserved words, and these must not be used as identifiers. The types of reserved words are (1) connectives which denote the presence of qualifiers (e.g., OF, IN) or form compound conditionals (AND, OR, etc.); (2) optional words which have been defined to improve the readability of the language (e.g., IS, KEY); and (3) key words which are either verbs (e.g., ADD, READ) or required words. The latter are either required in the format of the verbs or they are words which are not in any format but which have a specific functional meaning such as NEGATIVE, SECTION, and TALLY.

Up to three subscripts can be associated with each data name, and these are shown in parentheses following the data name and are separated by commas, e.g., RATE (3, STATE, CITY). The subscript can be either an integer, a data name with an integer value, or the special register TALLY. Subscripts cannot be subscripted. An effect equivalent to subscripting can be achieved by defining and using an *index-name*. Data names can be qualified by indicating the names in the data hierarchy which are necessary to uniquely determine which one is meant. The qualifiers are considered part of the data name. A data name cannot be subscripted when being used as a qualifier; in such a case the subscripts are written with the lowest element in the hierarchy.

The operators are the four arithmetic ones and three relational ones, GREATER THAN, LESS THAN, EQUAL TO, each of which can be preceded by the word NOT.

Blanks and punctuation are considered significant in COBOL and, in particular, are used to delimit identifiers and reserved words. The punctuation characters are the following:

. , ; " ( )   blank

The rules were designed to make the formats as natural as possible. Thus sentences end with a period, statements end with a semicolon, and it is legal but not required that nouns written in a sequence are separated by commas.

Noise words were introduced into COBOL to improve the readability of the language. The programmer may either include or omit a noise word which has been specifically defined in the format; however, he is not allowed to replace the designated noise word with another one, nor to misspell. Literals are designated by the use of quote marks at the beginning and the end. The figurative constant QUOTE is used to designate the " symbol itself.

The type of physical input form used for COBOL is string-oriented, but it also has columnar restrictions if the Reference Format is used. A Reference Format has been specified to be one of the acceptable input forms to a compiler, but it is also the required format in which the output listing must be produced after a compilation. The Reference Format does impose certain requirements about where label names are placed and where certain subunits of the program can appear.

COBOL was definitely designed to provide an English-like language. Within its frame of reference, every attempt was made to create formats which were natural and which could be easily understood by somebody examining a program that he had not written. The introduction of the concept of noise words was definitely motivated by this objective.

The most striking feature of the organization of a COBOL program is the definitive separation of four aspects of a program into four separate and clearly defined divisions of the language. These are the IDENTIFICATION, ENVIRONMENT, DATA, and PROCEDURE Divisions. The IDENTIFICATION Division provides a standard way of specifying the name, author, and date of a program, together with other remarks. The ENVIRONMENT Division contains information about the hardware on which the program is to be compiled and run. This clearly differs from machine to machine and could differ from compiler to compiler. The DATA Division uses descriptions of files and records to describe the data which the object program is to manipulate or create. The PROCEDURE Division specifies the steps that the user wishes the computer to follow. These steps are expressed as statements, sentences, paragraphs, and sections.

The executable statements appear only in the PROCEDURE Division, while the declarations in a COBOL program occur in the IDENTIFICATION, ENVIRONMENT, and DATA Divisions.

The smallest executable unit in the COBOL program is called a *statement*. This can be either imperative, conditional, or compiler directing. An imperative statement consists of either a verb and its operands or a sequence of verbs and operands. A conditional statement is either of the form IF ... THEN ... ELSE or it is an imperative statement followed by a conditional statement. (This format is shown in more detail later.) A compiler-directing statement simply consists of a compiler-directing verb and its operands. There are several hierarchical levels of executable units. The one immediately above the statement is a *sentence* which consists of a sequence of one or more statements. Sentences can be grouped together to convey one idea, and such a grouping is called a *paragraph*. A paragraph is the smallest grouping which can be named. Paragraphs can be combined to form *sections*, which must be named.

Looping is controlled in two ways in COBOL. The first is the fairly standard IF ... THEN ... statement. The second (and main) looping facility in COBOL is provided by the PERFORM verb.

There are no functions provided in COBOL. Subroutines and procedures without parameter replacement can appear as either paragraphs or sections. Library routines can be included at compile time by using the COPY verb. At object time other programs can be invoked by the CALL verb.

Comments are designated by an asterisk in a fixed location on the line.

The primary way in which the program is able to interact with the operating system is through error returns that the user may specify in some verbs (e.g., ADD) but most significantly through the USE verb. This latter permits the user to specify procedures for hardware input/output error and label handling, which are in addition to the standard procedures supplied by the input/output system. The OPEN and CLOSE verbs also provide a form of interaction in the sense that they provide for label checking (on input) and creation (on output).

Other languages can be written following the ENTER verb which serves as a flag to the compiler.

A complete COBOL program is comprised of entries from the four divisions written in the order indicated above.

Statements do not have any required delimiter since the beginning of the next statement can always be determined when a verb is encountered.

Statements can be separated, however, by the optional symbols THEN or a semicolon; sentences must be ended by a period rather than by a statement terminator. A paragraph is delimited by following the requirements of the reference format, which provide a fixed columnar location for the paragraph name and fixed location for the beginning of each new line. A section is delimited by writing the name of the section followed by the key word SECTION and a period all on the same line, with the paragraph beginning on the next line. Paragraphs and sections are ended only by the appearance of a new section name or a new paragraph; there is no specific delimiter for the end.

Procedures cannot be recursive. Parameters are replaced at compile time by name and at object time by value.

The main type of embedding permitted in COBOL is within conditional statements, where the executable statements can themselves be conditional statements to any depth.

COBOL has arithmetic variables and also alphanumeric data. There are no Boolean variables as such, but the same effect is achieved without calling them that through the use of *condition names* and *conditions* (described later). There are no complex, formal, string, or list variable types. COBOL permits arrays of up to three dimensions and a very elaborate hierarchical structure. The latter is given for each program by the Record Description in the DATA Division.

The most basic subdivisions (those not further subdivided) of a logical data record are called *elementary items*; consequently, a record is said to consist of a sequence of elementary items. Often it is desirable to reference a set of elementary items—particularly with the MOVE verb. For this reason, elementary items may be combined into *groups*, each group consisting of a sequence of one or more elementary items. Groups, in turn, may be combined into groups of two or more groups, etc. The term *item* in future discussions denotes either an elementary item or a group. These can be accessed by various verbs.

A system of level numbers is employed in COBOL to show the organization of elementary items and groups. Level numbers start at 01 for records since records are the most inclusive groups possible. Less inclusive groups are assigned higher level (not necessarily successive) numbers not greater in value than 49. (Note: There are three *special* level numbers, 66, 77, and 88, which have specific meanings; hence they are exceptions to this rule.) Separate entries are written in the source program for each level.

Using a TIME—CARD as an example, a skeleton source program listing showing the use of level numbers (which were arbitrarily chosen nonconsecutive) to indicate the hierarchical structure of the data might appear as follows:

```
01  TIME—CARD
    04  NAME
        06  LAST—NAME
        06  FIRST—INITIAL
        06  MIDDLE—INITIAL
    04  EMPLOYEE—NUMBER
    04  DATE
        05  MONTH
        05  DAY
        05  YEAR
    04  HOURS
```

For the sake of simplicity, only the level number and data name of each entry have been given in the example above.

A group includes all groups and elementary items described under it until a level number less than or equal to the level number of that group is encountered. Thus, in the example above, HOURS is not a part of the group called DATE. MONTH, DAY, and YEAR are a part of the group called DATE because they are described immediately under it and have a higher level number.

The hardware data units which are accessible are controlled entirely by the data description which is given. The basic principle of describing data is that it is assumed to be stored contiguously from the beginning of the storage area. Since the basic description of the data refers to its external format, it is up to the compiler to interpret this in the most efficient way internally; three of the entries in the Record Description format which are useful for this purpose are *SYNCHRONIZE, JUSTIFIED*, and *FILLER*. The first indicates that data should be aligned with word boundaries; the second one indicates whether data should be left or right, justified within some storage unit; and the third term indicates that there is now to be some empty space with no meaningful data in it. Since all of this is controlled by the entries in the DATA Division, the commands have access to each of these subunits simply by referencing the data name. The variable types which are meaningful for the commands are the arithmetic, alphabetic, and of course the arrays and hierarchies.

The type of arithmetic done in COBOL is integer and mixed number. The system is responsible for aligning numeric data items so that the decimal points appear in appropriate places and then proceeding to do the correct arithmetic. There is no floating point or rational arithmetic nor of course complex numbers, although some implementers have provided floating point. With regard to precision, a rule exists limiting the maximum size of any numeric operand to 18 decimal digits; obviously the question of whether or not this is double or multiple precision depends on the word size of the computer. In a peculiar sense, arithmetic can be done on vectors and arrays through the use of a *CORRESPONDING* clause, which means that in two different data hierarchies corresponding elements are added (or subtracted). The only types of data that can be combined arithmetically are the arithmetic variables, but there are no restrictions on where the decimal points need to be relative to the numbers which are to be combined together. It is allowable to have data of different internal representations combined together (e.g., binary and *BCD*); the resulting field can be of any form: binary, *BCD*, or anything else which is capable of holding an arithmetic result. The system provides the conversion automatically. The arithmetic verbs permit the user to specify whether or not he wants the result rounded, and the language specifications require the implementer to carry enough digits to ensure that no loss of significance occurs.

Logical expressions are called *conditions* in COBOL. A simple condition is one of four types of tests—a *relation* test for comparing numeric and nonnumeric items, a *class* test to determine whether a variable is numeric or alphabetic, a *conditional variable* test, and a test for the *status* of a *switch* which is named in the ENVIRONMENT Division. Simple conditions can be combined with the logical operations AND, OR, NOT; these compound conditions have truth values associated with them and thus serve the same purpose as a Boolean variable. More specifically, it is possible to compare variables, literals, or formulas against each other, using the relations GREATER THAN, LESS THAN, and EQUAL TO, preceded by the word NOT if desired. The relations UNEQUAL TO, EQUALS, and EXCEEDS are also permitted. In addition to these, variables and formulas can be tested for POSITIVE, NEGATIVE, or ZERO, again with the word NOT permitted. Variables can be tested for being NUMERIC or ALPHABETIC, or NOT being either of these. Finally, it is possible to test the status of a conditional variable or a switch.

Each of the immediately preceding has been considered a *simple condition*. These can be combined with the connectives AND, OR, NOT, according to the normal rules of logic. In many cases, the operands can be implied rather than specifically repeated. For example, each of the following is a legitimate condition which can be tested for truth or falsity:

AGE IS GREATER THAN 21 AND LESS THAN MAXIMUM—AGE

AGE IS GREATER THAN 25 OR MARRIED

STOCK—ON—HAND IS LESS THAN DEMAND OR STOCK—SUPPLY IS GREATER THAN DEMAND PLUS INVENTORY

A IS EQUAL TO B AND C IS NOT EQUAL TO D OR E IS UNEQUAL TO F AND G IS POSITIVE OR H IS LESS THAN I + J

STOCK—ACCOUNT IS GREATER THAN 72 AND (STOCK—NUMBER IS LESS THAN 100 OR STOCK—NUMBER EQUALS 76290)

There is no scope problem in COBOL.

The actual format of the COBOL verbs is shown separately in Figure V-7. The formats use the metalanguage discussed in Section II.6.2.2 as pertaining to COBOL. The discussion given below assumes the reader will examine the formats.

There is no actual separate assignment statement provided in COBOL. The equivalent result is achieved by having many of the verbs contain a *receiving field*. This means that the verb contains not only the operands to be acted upon but also the name of the field into which the result is to be placed. The MOVE verb and the five arithmetic verbs (*ADD, SUBTRACT, MULTIPLY, DIVIDE, COMPUTE*) provide this facility. The actual conversion is done by requiring that the format of the resulting data be made the same as that specified for the receiving field. This applies particularly to the length of fields, as well as to their internal format. The latter is converted— if necessary—according to the specifications in the Record Description. Thus, binary numbers are converted to *BCD* if this is specified, or numeric results are converted to their equivalent alphanumeric form if desired. By controlling the size of the receiving field, unnecessary digits or characters can be lopped off or added at either end. Alignment of decimal points is done automatically by the compiler. There are no rules given to control the way in which arithmetic should be done; thus the implementer has a choice of whether he performs his arithmetic in binary, *BCD*, or some other mode. The implementer is merely required to make sure that the result is in the proper form.

It is natural that a large portion of COBOL should be devoted to providing alphanumeric data handling. The most important facility for doing this on a small scale is the MOVE verb. Among the editing operations it will perform, in addition to converting from one form of internal representation to another, are zero suppression, insertion of dollar signs or commas, and decimal point alignment. Some of the compilers implement the arithmetic verbs by performing the indicated operation and then calling the MOVE verb to carry out the conversion. On a much larger editing scale, there is the Report Generation facility. The primary controls for this are in the DATA Division, but the PROCEDURE Division contains three relevant verbs. The GENERATE verb causes specific action to be taken to create the necessary information for the report. Different actions are performed, depending upon whether it is the first or a subsequent time that GENERATE is called for a particular report group, since obviously totals must be created and maintained. The INITIATE verb essentially zeros out all variables and all page and line counters. The TERMINATE produces final totals and prepares final headings.

ACCEPT

        ACCEPT identifier [FROM mnemonic-name]

ADD

Format 1:

    ADD $\begin{Bmatrix} identifier\text{-}1 \\ literal\text{-}1 \end{Bmatrix}$ $\begin{bmatrix} , \ identifier\text{-}2 \\ , \ literal\text{-}2 \end{bmatrix}$ ... , identifier-n [ROUNDED]

    [; ON SIZE ERROR imperative-statement]

Format 2:

    ADD $\begin{Bmatrix} identifier\text{-}1 \\ literal\text{-}1 \end{Bmatrix}$ $\begin{bmatrix} , \ identifier\text{-}2 \\ , \ literal\text{-}2 \end{bmatrix}$ ... TO identifier-m [ROUNDED]

    [, identifier-n [ROUNDED]] ...

    [; ON SIZE ERROR imperative-statement]

Format 3:

    ADD $\begin{Bmatrix} identifier\text{-}1 \\ literal\text{-}1 \end{Bmatrix}$, $\begin{Bmatrix} identifier\text{-}2 \\ literal\text{-}2 \end{Bmatrix}$ $\begin{bmatrix} , \ identifier\text{-}3 \\ , \ literal\text{-}3 \end{bmatrix}$ ... GIVING

    identifier-m [ROUNDED] [, identifier-n [ROUNDED]] ...

    [; ON SIZE ERROR imperative-statement]

Format 4:

    ADD $\begin{Bmatrix} CORR \\ CORRESPONDING \end{Bmatrix}$ identifier-1 TO identifier-2 [ROUNDED]

    [; ON SIZE ERROR imperative-statement]

ALTER

    ALTER procedure-name-1 TO [PROCEED TO] procedure-name-2
    [, procedure-name-3 TO [PROCEED TO] procedure-name-4] ...

CALL

    CALL $\begin{Bmatrix} literal\text{-}1 \\ identifier\text{-}1 \end{Bmatrix}$ [USING identifier-2 [, identifier-3] ...]

CANCEL

    CANCEL $\begin{Bmatrix} literal\text{-}1 \\ identifier\text{-}2 \end{Bmatrix}$ $\begin{bmatrix} , \ literal\text{-}2 \\ , \ identifier\text{-}2 \end{bmatrix}$ ...

CLOSE

    CLOSE file-name-1 $\begin{bmatrix} REEL \\ UNIT \end{bmatrix}$ $\begin{bmatrix} WITH \begin{Bmatrix} NO\ REWIND \\ LOCK \end{Bmatrix} \end{bmatrix}$

    $\begin{bmatrix} , file\text{-}name\text{-}2 \ \begin{bmatrix} REEL \\ UNIT \end{bmatrix} \begin{bmatrix} WITH \begin{Bmatrix} NO\ REWIND \\ LOCK \end{Bmatrix} \end{bmatrix} \end{bmatrix}$ ...

Figure V-7. (cont. next page)

Figure V-7. (cont.)

COMPUTE

COMPUTE identifier-1 [ROUNDED][, identifier-2 [ROUNDED]] ...
$$\begin{Bmatrix} \text{FROM} \\ = \\ \text{EQUALS} \end{Bmatrix} \begin{Bmatrix} \text{identifier-n} \\ \text{literal-1} \\ \text{arithmetic-expression} \end{Bmatrix}$$
[; ON SIZE ERROR imperative-statement]

COPY

procedure-name . COPY library-name
$$\left[\text{REPLACING} \begin{Bmatrix} \text{word-1} \\ \text{identifier-1} \end{Bmatrix} \text{BY} \begin{Bmatrix} \text{word-2} \\ \text{identifier-2} \end{Bmatrix} \right.$$
$$\left. \left[\begin{Bmatrix} \text{word-3} \\ \text{identifier-3} \end{Bmatrix} \text{BY} \begin{Bmatrix} \text{word-4} \\ \text{identifier-4} \end{Bmatrix} \right] ... \right] .$$

DISPLAY

DISPLAY $\begin{Bmatrix} \text{literal-1} \\ \text{identifier-1} \end{Bmatrix} \left[, \begin{matrix} \text{literal-2} \\ \text{identifier-2} \end{matrix} \right]$ ... [UPON mnemonic-name]

DIVIDE

Format 1:

DIVIDE $\begin{Bmatrix} \text{identifier-1} \\ \text{literal-1} \end{Bmatrix}$ INTO identifier-2 ROUNDED
[, identifier-3 [ROUNDED]] ...
[; ON SIZE ERROR imperative-statement]

Format 2:

DIVIDE $\begin{Bmatrix} \text{identifier-1} \\ \text{literal-1} \end{Bmatrix}$ INTO $\begin{Bmatrix} \text{identifier-2} \\ \text{literal-2} \end{Bmatrix}$ GIVING identifier-3 [ROUNDED]
[, identifier-4 [ROUNDED]] ...
[; ON SIZE ERROR imperative-statement]

Format 3:

DIVIDE $\begin{Bmatrix} \text{identifier-1} \\ \text{literal-1} \end{Bmatrix}$ BY $\begin{Bmatrix} \text{identifier-2} \\ \text{literal-2} \end{Bmatrix}$ GIVING identifier-3 [ROUNDED]
[, identifier-4 [ROUNDED]] ...
[; ON SIZE ERROR imperative-statement]

Format 4:

DIVIDE $\begin{Bmatrix} \text{identifier-1} \\ \text{literal-1} \end{Bmatrix}$ INTO $\begin{Bmatrix} \text{identifier-2} \\ \text{literal-2} \end{Bmatrix}$ GIVING identifier-3 [ROUNDED]
REMAINDER identifier-4 [; ON SIZE ERROR imperative-statement]

Format 5:

DIVIDE $\begin{Bmatrix} \text{identifier-1} \\ \text{literal-1} \end{Bmatrix}$ BY $\begin{Bmatrix} \text{identifier-2} \\ \text{literal-2} \end{Bmatrix}$ GIVING identifier-3 ROUNDED
REMAINDER identifier-4 [; ON SIZE ERROR imperative-statement]

Figure V-7. (cont.)

ENTER

        <u>ENTER</u> *language-name* [*routine-name*] .

EXAMINE

$$\left\{ \begin{array}{l} \underline{\text{EXAMINE}}\ identifier \\ \underline{\text{TALLYING}} \left\{ \begin{array}{l} \underline{\text{UNTIL FIRST}} \\ \underline{\text{ALL}} \\ \underline{\text{LEADING}} \end{array} \right\} \left\{ \begin{array}{l} literal\text{-}1 \\ identifier\text{-}1 \end{array} \right\} \left[\underline{\text{REPLACING BY}} \left\{ \begin{array}{l} literal\text{-}2 \\ identifier\text{-}2 \end{array} \right\} \right] \\ \underline{\text{REPLACING}} \left\{ \begin{array}{l} \underline{\text{ALL}} \\ \underline{\text{LEADING}} \\ [\underline{\text{UNTIL}}]\ \underline{\text{FIRST}} \end{array} \right\} \left\{ \begin{array}{l} literal\text{-}3 \\ identifier\text{-}3 \end{array} \right\} \underline{\text{BY}} \left\{ \begin{array}{l} literal\text{-}4 \\ identifier\text{-}4 \end{array} \right\} \end{array} \right\}$$

EXIT

        <u>EXIT</u> [PROGRAM] .

GENERATE

        <u>GENERATE</u> *identifier*

GO TO

    Format 1:
        <u>GO</u> [<u>TO</u>] [*procedure-name-1*]

    Format 2:
        <u>GO</u> [<u>TO</u>] *procedure-name-1* [, *procedure-name-2*] ... *procedure-name-n*
        <u>DEPENDING</u> ON *identifier*

HOLD

        <u>HOLD</u> $\left\{ \begin{array}{l} \underline{\text{ALL}} \\ section\text{-}name\text{-}1\ [,\ section\text{-}name\text{-}2]\ \ldots \end{array} \right\}$

IF

        <u>IF</u> *condition* $\left[\begin{array}{l} \text{THEN} \\ ; \end{array} \right] \left\{ \begin{array}{l} statement\text{-}1 \\ \underline{\text{NEXT SENTENCE}} \end{array} \right\}$
                      $\left[\begin{array}{l} \text{THEN} \\ ; \end{array} \right] \left\{ \begin{array}{l} \underline{\text{OTHERWISE}} \\ \underline{\text{ELSE}} \end{array} \right\} \left\{ \begin{array}{l} statement\text{-}2 \\ \underline{\text{NEXT SENTENCE}} \end{array} \right\}$

INITIATE

        <u>INITIATE</u> $\left\{ \begin{array}{l} report\text{-}name\text{-}1\ [,\ report\text{-}name\text{-}2]\ \ldots \\ \underline{\text{ALL}} \end{array} \right\}$

MOVE

        <u>MOVE</u> $\left\{ \begin{array}{l} \left[\begin{array}{l} \underline{\text{CORRESPONDING}} \\ \underline{\text{CORR}} \end{array} \right] identifier\text{-}1 \\ literal \end{array} \right\}$ <u>TO</u> *identifier-2* [, *identifier-3*] ...

Figure V-7. (cont. next page)

Figure V-7. (cont.)

MULTIPLY

> *Format 1:*
>
> MULTIPLY $\begin{Bmatrix} \text{identifier-1} \\ \text{literal-1} \end{Bmatrix}$ BY identifier-2 [ROUNDED]
> [, identifier-3 [ROUNDED]] ...
> [; ON SIZE ERROR imperative-statement]
>
> *Format 2:*
>
> MULTIPLY $\begin{Bmatrix} \text{identifier-1} \\ \text{literal-1} \end{Bmatrix}$ BY $\begin{Bmatrix} \text{identifier-2} \\ \text{literal-2} \end{Bmatrix}$
> GIVING identifier-3 [ROUNDED]
> [, identifier-4 [ROUNDED]] ...
> [; ON SIZE ERROR imperative-statement]

OPEN

> OPEN $\left[\text{INPUT} \; \{ \text{file-name} \begin{bmatrix} \text{REVERSED} \\ \text{WITH NO REWIND} \end{bmatrix} \} \; ... \right]$ ...
> [OUTPUT {file-name [WITH NO REWIND]} ...] ...
> $\left[\begin{Bmatrix} \text{I–O} \\ \text{INPUT–OUTPUT} \end{Bmatrix} \{ \text{file-name} \} \; ... \right]$ ...

PERFORM

> *Format 1:*
>
> PERFORM procedure-name-1 [THRU procedure-name-2]
>
> *Format 2:*
>
> PERFORM procedure-name-1 [THRU procedure-name-2]
> $\begin{Bmatrix} \text{identifier-1} \\ \text{integer-1} \end{Bmatrix}$ TIMES
>
> *Format 3:*
>
> PERFORM [procedure-name-1 [THRU procedure-name-2] UNTIL condition-1
>
> *Format 4:*
>
> PERFORM procedure-name-1 [THRU procedure-name-2]
> VARYING $\begin{Bmatrix} \text{index-name-1} \\ \text{identifier-1} \end{Bmatrix}$ FROM $\begin{Bmatrix} \text{index-name-2} \\ \text{literal-2} \\ \text{identifier-2} \end{Bmatrix}$
> BY $\begin{Bmatrix} \text{literal-3} \\ \text{identifier-3} \end{Bmatrix}$ UNTIL condition-1
> $\left[\text{AFTER} \begin{Bmatrix} \text{index-name-4} \\ \text{identifier-4} \end{Bmatrix} \text{FROM} \begin{Bmatrix} \text{index-name-5} \\ \text{literal-5} \\ \text{identifier-5} \end{Bmatrix} \right.$
> BY $\begin{Bmatrix} \text{literal-6} \\ \text{identifier-6} \end{Bmatrix}$ UNTIL condition-2
> $\left[\text{AFTER} \begin{Bmatrix} \text{index-name-7} \\ \text{identifier-7} \end{Bmatrix} \text{FROM} \begin{Bmatrix} \text{literal-8} \\ \text{identifier-8} \\ \text{index-name-8} \end{Bmatrix} \right.$
> BY $\begin{Bmatrix} \text{literal-9} \\ \text{identifier-9} \end{Bmatrix}$ UNTIL condition-3 $\left. \right] \right]$

Figure V-7. (cont.)

PROCESS

        PROCESS section-name [FROM identifier] $\left[\text{USING} \begin{Bmatrix} \text{area-name} \\ \text{record-name} \end{Bmatrix} \right]$

READ

    Format 1:
        READ file-name RECORD [INTO identifier]
        ; AT END imperative-statement

    Format 2:
        READ file-name RECORD [INTO identifier]
        ; INVALID KEY imperative-statement

RELEASE

        RELEASE record-name [FROM identifier]

RETURN

        RETURN file-name RECORD [INTO identifier] ; AT END imperative-statement

SEARCH

    Format 1:
        SEARCH identifier-1 $\left[\text{VARYING} \begin{Bmatrix} \text{index-name-1} \\ \text{identifier-2} \end{Bmatrix} \right]$
        [; AT END imperative-statement-1]
        ; WHEN condition-1 $\begin{Bmatrix} \text{imperative-statement-2} \\ \text{NEXT SENTENCE} \end{Bmatrix}$
        $\left[\text{; WHEN condition-2} \begin{Bmatrix} \text{imperative-statement-3} \\ \text{NEXT SENTENCE} \end{Bmatrix} \right]$ ...

    Format 2:
        SEARCH ALL identifier-1 [; AT END imperative-statement-1]
        ; WHEN condition-1 $\begin{Bmatrix} \text{imperative-statement-2} \\ \text{NEXT SENTENCE} \end{Bmatrix}$

SEEK

        SEEK file-name RECORD [WITH KEY CONVERSION]

SET

    Format 1:
        SET $\begin{Bmatrix} \text{index-name-1 [, index-name-2] ...} \\ \text{identifier-1 [, identifier-2] ...} \end{Bmatrix}$ TO $\begin{Bmatrix} \text{index-name-3} \\ \text{identifier-3} \\ \text{literal-1} \end{Bmatrix}$

Figure V-7. (cont. next page)

Figure V-7. (cont.)

    Format 2:

$$\underline{SET}\ \text{index-name-4}\ [,\ \text{index-name-5}]\ \ldots\ \begin{Bmatrix}\underline{UP\ BY}\\\underline{DOWN\ BY}\end{Bmatrix}\begin{Bmatrix}\text{identifier-4}\\\text{literal-2}\end{Bmatrix}$$

SORT

$$\underline{SORT}\ \text{file-name-1 ON}\ \begin{Bmatrix}\underline{DESCENDING}\\\underline{ASCENDING}\end{Bmatrix}\ \underline{KEY}\ \text{data-name-1}\ [,\ \text{data-name-2}]\ \ldots$$

$$\left[\ ;\ \text{ON}\ \begin{Bmatrix}\underline{DESCENDING}\\\underline{ASCENDING}\end{Bmatrix}\ \underline{KEY}\ \text{data-name-3}\ [,\ \text{data-name-4}]\ \ldots\ \right]\ \ldots$$

$$\begin{Bmatrix}\underline{INPUT\ PROCEDURE}\ \text{IS section-name-1}\ [\underline{THRU}\ \text{section-name-2}]\\\underline{USING}\ \text{file-name-2}\end{Bmatrix}$$

$$\begin{Bmatrix}\underline{OUTPUT\ PROCEDURE}\ \text{IS section-name-3}\ [\underline{THRU}\ \text{section-name-4}]\\\underline{GIVING}\ \text{file-name-3}\end{Bmatrix}$$

STOP

$$\underline{STOP}\ \begin{Bmatrix}\text{literal}\\\underline{RUN}\end{Bmatrix}$$

SUBTRACT

    Format 1:

$$\underline{SUBTRACT}\ \begin{Bmatrix}\text{literal-1}\\\text{identifier-1}\end{Bmatrix}\ \begin{bmatrix},\ \text{literal-2}\\,\ \text{identifier-2}\end{bmatrix}\ \ldots$$
$$\underline{FROM}\ \text{identifier-m}\ [\underline{ROUNDED}]\ [,\ \text{identifier-n}\ [\underline{ROUNDED}]]\ \ldots$$
$$[;\ \underline{ON\ SIZE\ ERROR}\ \text{imperative-statement}]$$

    Format 2:

$$\underline{SUBTRACT}\ \begin{Bmatrix}\text{literal-1}\\\text{identifier-1}\end{Bmatrix}\ \begin{bmatrix},\ \text{literal-2}\\,\ \text{identifier-2}\end{bmatrix}\ \ldots$$
$$\underline{FROM}\ \begin{Bmatrix}\text{literal-m}\\\text{identifier-m}\end{Bmatrix}\ \underline{GIVING}\ \text{identifier-n}\ [\underline{ROUNDED}]$$
$$[,\ \text{identifier-o}\ [\underline{ROUNDED}]]\ \ldots$$
$$[;\ \underline{ON\ SIZE\ ERROR}\ \text{imperative-statement}]$$

    Format 3:

$$\underline{SUBTRACT}\ \begin{Bmatrix}\underline{CORR}\\\underline{CORRESPONDING}\end{Bmatrix}\ \text{identifier-1}\ \underline{FROM}\ \text{identifier-2}\ [\underline{ROUNDED}]$$
$$[;\ \underline{ON\ SIZE\ ERROR}\ \text{imperative-statement}]$$

SUSPEND

$$\underline{SUSPEND}\ \begin{Bmatrix}\text{file-name-1}\begin{bmatrix}\text{literal-1}\\\text{identifier-1}\end{bmatrix}\\\text{report-name-1}\begin{bmatrix}\text{literal-2}\\\text{identifier-2}\end{bmatrix}\end{Bmatrix}$$

$$\left[,\ \begin{Bmatrix}\text{file-name-2}\begin{bmatrix}\text{literal-3}\\\text{identifier-3}\end{bmatrix}\\\text{report-name-2}\begin{bmatrix}\text{literal-4}\\\text{identifier-4}\end{bmatrix}\end{Bmatrix}\ \ldots\ \right]$$

Figure V-7. (cont.)

TERMINATE

                TERMINATE $\left\{\begin{array}{l}\text{report-name-1 [, report-name-2] ...}\\ \text{ALL}\end{array}\right\}$

USE

    Format 1:

           USE AFTER STANDARD ERROR PROCEDURE ON

$$\left\{\begin{array}{l}\text{file-name-1 [, file-name-2] ...}\\ \text{INPUT}\\ \text{OUTPUT}\\ \text{INPUT–OUTPUT}\\ \text{I–O}\end{array}\right\}.$$

    Format 2:

           USE $\left\{\begin{array}{l}\text{BEFORE}\\ \text{AFTER}\end{array}\right\}$ STANDARD $\left[\begin{array}{l}\text{BEGINNING}\\ \text{ENDING}\end{array}\right]$ $\left[\begin{array}{l}\text{REEL}\\ \text{FILE}\\ \text{UNIT}\end{array}\right]$

           LABEL PROCEDURE ON $\left\{\begin{array}{l}\text{file-name-1 [, file-name-2] ...}\\ \text{INPUT}\\ \text{OUTPUT}\\ \text{INPUT–OUTPUT}\\ \text{I–O}\end{array}\right\}.$

    Format 3:

           USE BEFORE REPORTING identifier-1 [, identifier-2] ... .

    Format 4:

           USE FOR KEY CONVERSION ON $\left\{\begin{array}{l}\text{ALL}\\ \text{file-name-1 [, file-name-2 ...]}\end{array}\right\}.$

    Format 5:

           USE FOR RANDOM PROCESSING.

WRITE

    Format 1:

           WRITE record-name [FROM identifier-1]

$$\left[\left\{\begin{array}{l}\text{BEFORE}\\ \text{AFTER}\end{array}\right\} \text{ADVANCING} \left\{\begin{array}{l}\text{identifier-2 LINES}\\ \text{integer LINES}\\ \text{mnemonic-name}\end{array}\right\}\right]$$

$$\left[; \text{AT} \left\{\begin{array}{l}\text{END–OF–PAGE}\\ \text{EOP}\end{array}\right\} \text{imperative-statement}\right]$$

    Format 2:

           WRITE record-name [FROM identifier-1]
             ; INVALID KEY imperative-statement

Figure V-7. Formats of COBOL verbs. The general notation is the COBOL metalanguage described in Section II.6.2.
Source: [US65] plus CODASYL approved changes through March, 1968.

As indicated already, the conversion is controlled by the specifications in the DATA Division. Each verb which operates on one or more data fields obtains its information about the format from the Record Description in the DATA Division.

There is a *SORT* verb in COBOL. Although it may seem unnecessary to include such a verb in a higher level programming language, the real objective is to provide the facilities of the programming language with the sort routines; the best way to do this is to include a *SORT* verb in COBOL. The *SORT* verb specifies what input and output procedures are to be performed on a file before and after sorting, and specifies the keys and other necessary information for the sort to take place. The *RELEASE* and *RETURN* verbs are used to obtain the records to be acted on initially before the sort takes place and to put them back into the main portion of the executable program after the sort is executed.

The simplest type of control transfer is the *GO* verb, which exists in either the simple form of transferring control to the designated procedure or as a switch by allowing a sequence of names to be associated with it. The simple case can have the specified procedure name changed by the *ALTER* verb.

The conditional statements in COBOL are quite powerful. The basic conditional statement is of the following form, where underlined words are required:

$$\underline{\text{IF}} \text{ condition} \begin{bmatrix} \text{THEN} \\ ; \end{bmatrix} \begin{Bmatrix} \textit{statement-1} \\ \underline{\text{NEXT SENTENCE}} \end{Bmatrix} \begin{bmatrix} \text{THEN} \\ ; \end{bmatrix} \begin{Bmatrix} \underline{\text{OTHERWISE}} \\ \underline{\text{ELSE}} \end{Bmatrix} \begin{Bmatrix} \textit{statement-2} \\ \underline{\text{NEXT SENTENCE}} \end{Bmatrix}$$

It is also permitted to have an imperative statement followed by a conditional statement. The power of this form comes from two different features. The first is that nesting of conditional statements is permitted. More specifically, either *statement-1* or *statement-2* can themselves be conditional, and this nesting is permitted to any depth. Ambiguity is avoided through the rule which causes every *OTHERWISE* to be paired with the immediately preceding *IF*. The second feature providing power is the number and types of conditions which can be used; these were discussed earlier.

In several of the examples given earlier, both sides of the comparison were *not* repeated. This is particularly useful in trying to write something like IF X = 2, Y OR Z, which is equivalent to the much longer form IF X = 2 OR X = Y OR X = Z. This type of abbreviation is allowed to exist even across the key words in the sentence, thus permitting statements of the form

IF A = B MOVE X TO Y; OTHERWISE IF GREATER, MOVE M TO N ELSE MOVE P TO Q.

IF A EXCEEDS B OR EQUALS B AND X EQUALS Y THEN IF GREATER THAN B MOVE C TO D.

The former is an abbreviation for

...OTHERWISE IF A IS GREATER THAN B...

while the latter is an abbreviation for

IF A EXCEEDS B OR (A EQUALS B AND X EQUALS Y) THEN IF X IS GREATER THAN B...

Naturally, with this amount of flexibility, great care must be taken. This method of writing conditions is more general than in any other language in this book.

The nesting permits a different type of flexibility. For example, the single sentence IF C1 S1 IF C2 S2 OTHERWISE S3 OTHERWISE S4 IF C3 S5 OTHERWISE S6 represents the complicated flow chart shown in Figure V-8.

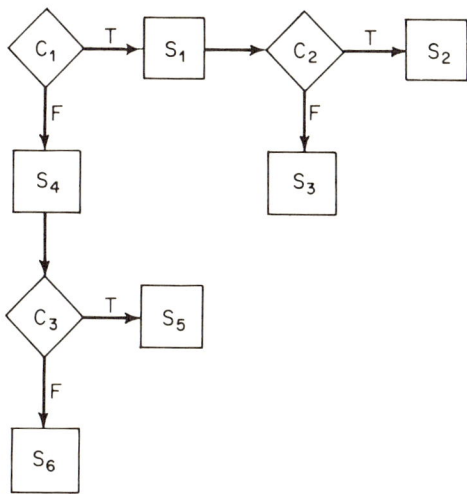

Figure V-8. Flowchart represented by a single COBOL sentence.

The main loop control statement in COBOL is the *PERFORM* verb. It permits the execution of named procedures which can be either open or closed subroutines and can consist of a single paragraph (this can be a single statement) or several sections. It is possible to specify the loop in a number of different ways. One is by specifying the number of times the range is to be performed. Another way is by executing the range until some condition is satisfied, with or without incrementing a variable. Finally, up to three variables can be varied sequentially until certain conditions are satisfied. (Sequentially means that there are really three nested loops.) The *EXIT* verb is used when there are two or more paths to the end of a loop.

A very specialized type of loop control statement was included in COBOL-65, namely the *SEARCH* verb. This is used to search a table for an element that satisfies a specified condition. This is used in conjunction with an index name which has been established in the Data Description and is used as a way of identifying items in a table. This index name can be preset by the *SET* command. The *SEARCH* command has two options which permit the user to specify whether he wishes a serial search made or some type of nonserial search which is specified by the implementer (but presumably would be a binary search).

The main error condition statement is the possibility of including with the arithmetic verbs a clause that says *ON SIZE ERROR any-imperative-statement*. Thus, if the result of an arithmetic calculation exceeds the size of the data field in which it is to be stored, a size error has occurred. When the error condition is specified, the single statement shown will be executed. A completely different type of error condition statement (namely the *USE* verb) is associated with input/output and is discussed in connection with those verbs.

COBOL provides a very simple and limited type of string handling through the *EXAMINE* verb which permits a data field to be examined for the occurrence of a particular literal. There are several options permitted, including the replacement of all the occurrences of one literal by another, just the leading occurrences, or until another specified literal is encountered. It is also possible to tally the number of occurrences in these various cases.

The main input/output verbs are the *READ* and *WRITE*; the former serves the purpose of putting the next logical record into the storage area for processing, whereas the latter places the specified record into an output processing area. The programmer need not worry about the problems of differences between logical and physical records, movement across tapes, etc.; all these are handled by the *READ* and *WRITE* commands automatically, with some work done through the *OPEN* and *CLOSE* verbs discussed later. The *READ* and *WRITE* commands provide the facility for specifying what is to be done when the end of file is reached. The *WRITE* command can also be used to control the vertical positioning of each record on the printed page. Furthermore, it permits the user to specify additional action to take place when the logical end of page is reached. Before the *READ* and *WRITE* verbs can operate, however, checking or creating of tape labels must be done; this is accomplished by the *OPEN* verb in which the programmer specifies whether the file is an input or an output file. The *OPEN* verb must be executed prior to the first *READ* or *WRITE* for a file. The *CLOSE* verb handles the final closing conventions for both input and output files. All the information which is needed to specify the format of files is given in the DATA Division.

Although the main input/output verbs are *READ* and *WRITE*, which deal with the normal large files, it is recognized that there may be low volume input and output; for these purposes, the *ACCEPT* and *DISPLAY* verbs are used. They would normally apply to something like an on-line typewriter. Late in 1967 a *SUSPEND* statement was added to stop the release of subsequent logical records to a graphic display device until re-initiated by the operator.

When the concept of handling mass storage was added in COBOL-65, certain additions were made to the *READ* and *WRITE* to handle the mass storage, and the *SEEK* verb was added to facilitate efficient programming. In particular, the programmer may give a *SEEK* command which will start searching the storage device for the indicated record and then perform other operations until he gives a *READ* or *WRITE*. This is only used for random access with sequential processing. If random access with random processing is desired, then a *USE FOR RANDOM PROCESSING* statement must be given and the execution of the related procedures is initiated by a *PROCESS* verb in the main program. This *PROCESS* can also be used in conjunction with a *HOLD* statement which provides a delay point that causes synchronous processing to be resumed when operating in an asynchronous environment.

The *USE* verb has a variety of functions. It specifies procedures for input/output labels and error handling which are in addition to the standard procedures provided by the input/output system. It is also used to specify PROCEDURE Division statements that are executed just before a report group named in the Report Section of the DATA Division is produced. Finally, it specifies out-of-line and key conversion procedural statements for processing mass storage files.

The library reference facility in COBOL is the *COPY* verb in the PROCEDURE Division and the *COPY* clauses in the DATA and ENVIRONMENT Divisions. *COPY* operates at compile time. Other programs can be obtained at object time by the *CALL* verb.

There are no direct debugging statements, but the *ON SIZE ERROR* and *USE*, which were both mentioned earlier, tend to help in that direction.

There are no executable statements for allocating storage or specifying segmentation. However, there are ways for the programmer to control the latter, and they are discussed later. A CANCEL statement releases the memory areas occupied by the named program.

The primary interface with the operating system is almost a default one, namely the USE verb which (as indicated earlier) allows the programmer to specify procedures which are in addition to those assumed supplied by the normal input/output system. The STOP verb has a RUN option in which the ending procedure established by the installation and/or compiler is instituted. Alternatively, the STOP verb can cause a particular literal to be displayed to the machine operator and from that he can perform actions depending on what the compiler and/or installation has specified.

There are no facilities for really using specific machine features in COBOL. However, the SPECIAL—NAMES paragraph and the APPLY clauses in the ENVIRONMENT Division allow some connection between general language facilities and specific machine or implementation facilities.

COBOL is the first major language to devote significant attention to the problem of describing wide varieties of data. There are three major categories of data to be concerned with: (1) Information on files coming to or from the computer, (2) data developed internally and placed in working storage (which might be used for report purposes), and (3) constants. The COBOL DATA Division has four sections: FILE, REPORT, WORKING—STORAGE, and CONSTANT. The most important is the FILE Section, which is itself subdivided into *File Descriptions* (including *Sort File Descriptions* as a special type), and *Record* (i.e., detailed data) *Descriptions*. These correspond to the *physical* and *conceptual* characteristics of the data. The *physical* aspects include items such as the mode in which the file is recorded, the grouping of logical records within the physical limitations of the file medium, and the means by which the file can be identified. The term *conceptual characteristics* means the explicit definition of each logical record within the file itself. A logical record is any consecutive set of information. In an Inventory Transaction File, for example, a logical record could be defined as a single transaction or as all consecutive transactions which pertain to the same stock item. Several logical records can occupy a block (i.e., physical record), or a single logical record can extend across physical records.

The sections in the DATA Division consist of a series of related and unrelated *entries* as will be shown in Figures V-9 to V-13. This is different from the *paragraph-sentence-statement* structure which characterizes the other three divisions. An entry consists of a *level indicator*, a *data name*, and a series of *independent clauses* which may be separated by the use of semicolons. The clauses may be written in any sequence, except when the entry format specifies otherwise. The entry itself is terminated by a period.

A detailed data description consists of a set of entries (shown in Figure V-9). In defining the lowest level or subdivision of data, the following infor-

mation may be required:
1. A *level-number* which shows the hierarchical relationship between this and other units of data.
2. A *data-name*.
3. The *SIZE* in terms of the number of standard data format characters (which are alphabetic or decimal numbers).
4. The dominant *USAGE* of the data (computational or display).
5. The number of consecutive occurrences (*OCCURS*) of elements to specify a table or list.
6. The location and type of *SIGN*.
7. Justification and/or synchronization (i.e., occupying single computer word) of the data (*JUSTIFIED, SYNCHRONIZED*).
8. Location of an actual or an assumed radix point.
9. The *CLASS* or type of data (alphabetic, numeric, or alphanumeric).
10. The *RANGE* of values which the data may assume.
11. Location of editing symbols such as dollar signs and commas.
12. Special editing requirements such as zero suppression and check protection.
13. Initial *VALUE* of a working storage item or fixed *VALUE* of a constant.

In some cases this information can be defined either through specific clauses or by the *PICTURE* clause.

The large (although not total) amount of machine independence, while still retaining some efficiency, comes through the use of features such as assuming only decimal numbers regardless of whether the computer is binary; controlling spacing within the word boundaries through such general clauses as *SYNCHRONIZED* or *JUSTIFIED*, which can be implemented appropriately on different computers; and indicating *USAGE* so that the compiler knows the best internal form in which to store the number.

To specify a particular format for input or output, the programmer obviously provides the appropriate data description. However, in order to provide the full flexibility of a report generator, the Report Section is included in the DATA Division. This consists of the Report Name (RD) and Report Group description entries, shown in Figures V-10 and V-11, respectively. The former contains the information pertaining to the overall format such as number of physical lines per page, limits for specified headings, footings, and details within a page structure. The latter allows the programmer to "fill in" the pictorial representation of a line or series of lines; the format is shown in Figure V-11.

The *WORKING—STORAGE* and *CONSTANT* Sections in the DATA Division have a format similar to that of the Record Description.

A File Description (*FD*) entry (shown in Figure V-12) generally includes the following: The manner in which the data is recorded on the file, the size of the logical and physical records, the names of the label and data records and reports contained in the file, and finally the keys on which the data

Format 1:

    01 data-name-1 ; <u>COPY</u> library-name

$$\left[\underline{\text{REPLACING}} \begin{Bmatrix} \text{word-1} \\ \text{identifier-1} \end{Bmatrix} \underline{\text{BY}} \begin{Bmatrix} \text{word-2} \\ \text{identifier-2} \end{Bmatrix} \left[, \begin{Bmatrix} \text{word-3} \\ \text{identifier-3} \end{Bmatrix} \underline{\text{BY}} \begin{Bmatrix} \text{word-4} \\ \text{identifier-4} \end{Bmatrix} \right] \ldots \right].$$

Format 2:

    level-number $\begin{Bmatrix} \text{data-name-1} \\ \text{FILLER} \end{Bmatrix}$

    [; <u>REDEFINES</u> data-name-2]

$$\left[\begin{array}{l} ; \underline{\text{SIZE}} \text{ IS integer-2} \begin{Bmatrix} \text{CHARACTERS} \\ \text{DIGITS} \end{Bmatrix} \\ ; \underline{\text{SIZE}} \text{ IS [integer-1 } \underline{\text{TO}}\text{] integer-2} \begin{Bmatrix} \text{CHARACTERS} \\ \text{DIGITS} \end{Bmatrix} \\ [\underline{\text{DEPENDING}} \text{ ON data-name-3}] \end{array} \right]$$

$$\left[; \begin{Bmatrix} \underline{\text{PICTURE}} \\ \underline{\text{PIC}} \end{Bmatrix} \text{ IS character-string } [\underline{\text{DEPENDING}} \text{ ON data-name-4}] \right]$$

$$\left[; [\underline{\text{USAGE}} \text{ IS}] \begin{Bmatrix} \text{COMPUTATIONAL} \\ \text{COMP} \\ \text{COMPUTATIONAL-}n \\ \text{COMP-}n \\ \text{DISPLAY} \\ \text{DISPLAY-}n \\ \text{INDEX} \\ \text{INDEX-}n \end{Bmatrix} \right]$$

$$\left[\begin{array}{l} ; \underline{\text{OCCURS}} \text{ integer-4 TIMES} \\ \quad \left[\begin{Bmatrix} \underline{\text{ASCENDING}} \\ \underline{\text{DESCENDING}} \end{Bmatrix} \text{ KEY IS data-name-5 [, data-name-6] } \ldots \right] \ldots \\ \quad [\underline{\text{INDEXED}} \text{ BY index-name-1 [, index-name-2] } \ldots] \\ ; \underline{\text{OCCURS}} \text{ integer-3 } \underline{\text{TO}} \text{ integer-4 TIMES } [\underline{\text{DEPENDING}} \text{ ON data-name-4}] \\ \quad \left[\begin{Bmatrix} \underline{\text{ASCENDING}} \\ \underline{\text{DESCENDING}} \end{Bmatrix} \text{ KEY IS data-name-5 [, data-name-6] } \ldots \right] \ldots \\ \quad [\underline{\text{INDEXED}} \text{ BY index-name-1 [, index-name-2] } \ldots] \end{array} \right]$$

$$\left[; \begin{Bmatrix} \underline{\text{SIGNED}} \\ \underline{\text{SIGN}} \text{ IS data-name-9} \end{Bmatrix} \right]$$

$$\left[; \begin{Bmatrix} \underline{\text{SYNCHRONIZED}} \\ \underline{\text{SYNC}} \end{Bmatrix} \begin{Bmatrix} \text{LEFT} \\ \text{RIGHT} \end{Bmatrix} \right]$$

$$\left[; \underline{\text{POINT}} \text{ LOCATION IS } \begin{Bmatrix} \text{LEFT} \\ \text{RIGHT} \end{Bmatrix} \text{ integer-5} \begin{Bmatrix} \text{PLACES} \\ \text{BITS} \end{Bmatrix} \right]$$

$$\left[; \underline{\text{CLASS}} \text{ IS } \begin{Bmatrix} \text{ALPHABETIC} \\ \underline{\text{NUMERIC}} \\ \underline{\text{ALPHANUMERIC}} \\ \underline{\text{AN}} \end{Bmatrix} \right]$$

$$\left[; \begin{Bmatrix} \underline{\text{JUSTIFIED}} \\ \underline{\text{JUST}} \end{Bmatrix} \underline{\text{RIGHT}} \right]$$

    [; <u>RANGE</u> IS literal-1 <u>THRU</u> literal-2]

Figure V-9., Format 2 (cont.)

$$\left[\;;\;\left\{\begin{matrix}\text{ZERO SUPPRESS}\\ \text{CHECK PROTECT}\\ \text{FLOAT DOLLAR SIGN}\\ \text{FLOAT CURRENCY SIGN}\end{matrix}\right\}\text{[LEAVING }\textit{integer-6}\text{ PLACES]}\right]$$

[; BLANK WHEN ZERO]
[; VALUE IS *literal-3*] .

Format 3:

    66 *data-name-1*; RENAMES *data-name-2* [THRU *data-name-3*] .

Format 4:

    88 *condition-name* ; $\left\{\begin{matrix}\underline{\text{VALUE IS}}\\ \underline{\text{VALUES ARE}}\end{matrix}\right\}$ *literal-1* [THRU *literal-2*]
    [, *literal-3* THRU *literal-4*]] ... .

Figure V-9. Data description (= Record description) skeleton in COBOL. This describes a logical record. The general notation is the COBOL metalanguage described in Section II.6.2.

Source: [US65] plus CODASYL approved changes through March, 1968.

Format 1:

    RD *report-name* ; COPY *library-name*

$$\left[\underline{\text{REPLACING}}\;\left\{\begin{matrix}\textit{word-1}\\ \textit{identifier-1}\end{matrix}\right\}\;\underline{\text{BY}}\;\left\{\begin{matrix}\textit{word-2}\\ \textit{identifier-2}\end{matrix}\right\}\right.$$
$$\left.\left[,\;\left\{\begin{matrix}\textit{word-3}\\ \textit{identifier-3}\end{matrix}\right\}\;\underline{\text{BY}}\;\left\{\begin{matrix}\textit{word-4}\\ \textit{identifier-4}\end{matrix}\right\}\right]\cdots\right]\;.$$

Format 2:

    RD *report-name*
    [; WITH CODE *mnemonic-name-1*]

$$\left[\;;\;\left\{\begin{matrix}\underline{\text{CONTROL IS}}\\ \underline{\text{CONTROLS ARE}}\end{matrix}\right\}\left\{\begin{matrix}\underline{\text{FINAL}}\\ \textit{identifier-1}\text{ [, }\textit{identifier-2}\text{] ...}\\ \underline{\text{FINAL}}\text{, }\textit{identifier-1}\text{ [, }\textit{identifier-2}\text{] ...}\end{matrix}\right\}\right]$$

$$\left[\;;\;\underline{\text{PAGE}}\left\{\begin{matrix}\underline{\text{LIMIT IS}}\\ \underline{\text{LIMITS ARE}}\end{matrix}\right\}\textit{integer-1}\left\{\begin{matrix}\underline{\text{LINE}}\\ \underline{\text{LINES}}\end{matrix}\right\}\text{[, }\underline{\text{HEADING}}\textit{ integer-2}\text{]}\right.$$
        [, FIRST DETAIL *integer-3*][, LAST DETAIL *integer-4*]
        $\left.\text{[, }\underline{\text{FOOTING}}\textit{ integer-5}\text{]}\right]$.

Figure V-10. Report description (**RD**) skeleton in COBOL. The general notation is the COBOL metalanguage described in Section II.6.2.

Source: [US65] plus CODASYL approved changes through March, 1968.

---

records have been sequenced, e.g., FD TRANSACT RECORDING MODE IS F; BLOCK CONTAINS 80 CHARACTERS; LABEL RECORDS ARE STANDARD; DATA RECORD IS TRANSACTION—RECORD; SEQUENCED ON ACCOUNT— NUMBER.

Format 1:

```
01 [data-name] ; COPY library-name
    [ REPLACING  {word-1     }  BY  {word-2     }
                 {identifier-1}     {identifier-2}
        [ , {word-3     }  BY  {word-4     } ] ... ] .
            {identifier-3}     {identifier-4}
```

Format 2:

```
01 [data-name-1]
    [ ; [CLASS IS]  {ALPHABETIC   } ]
                    {NUMERIC      }
                    {ALPHANUMERIC }
                    {AN           }

    [ ; LINE NUMBER IS  {integer-1       } ]
                        {PLUS integer-2  }
                        {NEXT PAGE       }

    [ ; NEXT GROUP IS  {integer-3       } ]
                       {PLUS integer-4  }
                       {NEXT PAGE       }

    [ ; SIZE IS integer-5  {CHARACTERS} ]
                           {DIGITS    }

    ; TYPE IS  { REPORT HEADING                                  }
               { RH                                              }
               { PAGE HEADING                                    }
               { PH                                              }
               { OVERFLOW HEADING                                }
               { OH                                              }
               { {CONTROL HEADING}  {identifier-2}               }
               { {CH             }  {FINAL       }               }
               { DETAIL                                          }
               { DE                                              }
               { {CONTROL FOOTING}  {identifier-3}               }
               { {CF             }  {FINAL       }               }
               { OVERFLOW FOOTING                                }
               { OV                                              }
               { PAGE FOOTING                                    }
               { PF                                              }
               { REPORT FOOTING                                  }
               { RF                                              }

    [ ; [USAGE IS]  {DISPLAY-n} ] .
                    {DISPLAY  }
```

Format 3:

```
level-number [data-name-1]
    [ ; [CLASS IS]  {ALPHABETIC   } ]
                    {NUMERIC      }
                    {ALPHANUMERIC }
                    {AN           }
    [; COLUMN NUMBER IS integer-1]
```

Figure V-11., Format 3 (cont.)

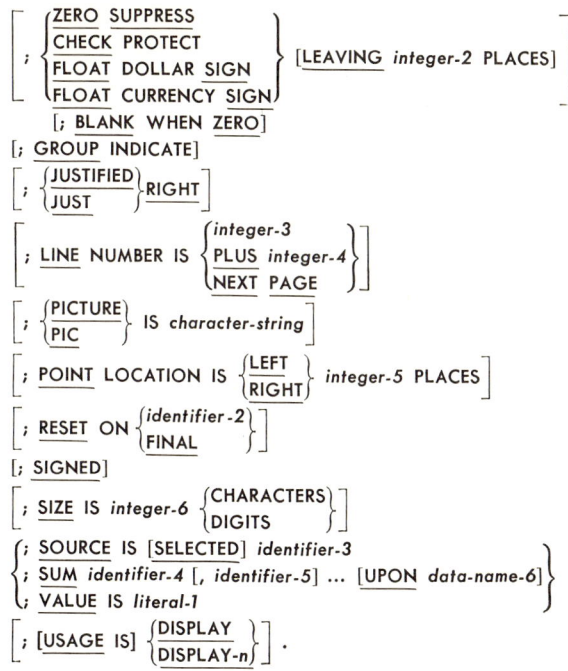

Figure V-11. Report group skeleton in COBOL. The general notation is the COBOL metalanguage described in Section II.6.2.
Source: [US65] plus CODASYL approved changes through March, 1968.

The listing of data and label record names in a File Description entry serves as a cross-reference between the file and the records in the file.

A *sort file* is a name for the set of records to be sorted by a SORT statement. The format of its description (SD) is shown in Figure V-13. There are no label procedures which the user can control, and the rules for blocking and internal storage are peculiar to the SORT verb. Each of the sorted records can be made available, in order, by being RETURNed from the sort file during the output procedures specified by the SORT statement. The sort file created by the execution of RELEASE statements can only be SORTed and its records can only be obtained by being RETURNed from the sort file during the OUTPUT PROCEDURE.

The storage allocation is handled automatically by the compiler. The prime unit for allocating executable code is a group of sections called

Format 1:

    FD file-name ; COPY library-name
    $\left[\underline{\text{REPLACING}} \begin{Bmatrix} \text{word-1} \\ \text{identifier-1} \end{Bmatrix} \underline{\text{BY}} \begin{Bmatrix} \text{word-2} \\ \text{identifier-2} \end{Bmatrix} \right.$
    $\left. \left[, \begin{Bmatrix} \text{word-3} \\ \text{identifier-3} \end{Bmatrix} \underline{\text{BY}} \begin{Bmatrix} \text{word-4} \\ \text{identifier-4} \end{Bmatrix} \right] \ldots \right]$.

Format 2:

    FD file-name
    [; RECORDING MODE IS mode]
    [; FILE CONTAINS ABOUT integer-1 RECORDS]
    $\left[; \underline{\text{BLOCK}} \text{ CONTAINS } [\text{integer-2 } \underline{\text{TO}}] \text{ integer-3 } \begin{Bmatrix} \underline{\text{RECORDS}} \\ \underline{\text{CHARACTERS}} \end{Bmatrix} \right]$
    [; RECORD CONTAINS [integer-4 TO] integer-5 CHARACTERS]
    $; \underline{\text{LABEL}} \begin{Bmatrix} \underline{\text{RECORDS}} \text{ ARE} \\ \underline{\text{RECORD}} \text{ IS} \end{Bmatrix} \begin{Bmatrix} \text{STANDARD} \\ \text{OMITTED} \\ \text{data-name-1 [, data-name-2] } \ldots \end{Bmatrix}$
    $\left[; \underline{\text{VALUE OF}} \text{ data-name-3 IS } \begin{Bmatrix} \text{data-name-4 [HASHED]} \\ \text{literal-1} \end{Bmatrix} \right.$
    $\left. \left[, \text{data-name-5 IS } \begin{Bmatrix} \text{data-name-6 [HASHED]} \\ \text{literal-2} \end{Bmatrix} \right] \ldots \right]$
    $\left[; \underline{\text{DATA}} \begin{Bmatrix} \underline{\text{RECORD}} \text{ IS} \\ \underline{\text{RECORDS}} \text{ ARE} \end{Bmatrix} \text{data-name-7 [, data-name-8] } \ldots \right]$
    $\left[; \begin{Bmatrix} \underline{\text{REPORT}} \text{ IS} \\ \underline{\text{REPORTS}} \text{ ARE} \end{Bmatrix} \text{report-name-1 [, report-name-2] } \ldots \right]$
    [; SEQUENCED ON data-name-9 [, data-name-10] ...]
    $\left[; \underline{\text{LINAGE}} \text{ IS } \begin{Bmatrix} \text{identifier-1 } \underline{\text{LINES}} \\ \text{integer-6 } \underline{\text{LINES}} \\ \text{mnemonic-name} \end{Bmatrix} \right]$.

Figure V-12. File description (*FD*) skeleton in COBOL. The general notation is the COBOL metalanguage described in Section II.6.2.
Source: [US65] plus CODASYL approved changes through March, 1968.

Format 1:

    SD file-name ; COPY library-name
    $\left[\underline{\text{REPLACING}} \begin{Bmatrix} \text{word-1} \\ \text{identifier-1} \end{Bmatrix} \underline{\text{BY}} \begin{Bmatrix} \text{word-2} \\ \text{identifier-2} \end{Bmatrix} \right.$
    $\left. \left[, \begin{Bmatrix} \text{word-3} \\ \text{identifier-3} \end{Bmatrix} \underline{\text{BY}} \begin{Bmatrix} \text{word-4} \\ \text{identifier-4} \end{Bmatrix} \right] \ldots \right]$.

Format 2:

    SD file-name
    [; FILE CONTAINS ABOUT integer-1 RECORDS]
    [; RECORD CONTAINS [integer-2 TO] integer-3 CHARACTERS]
    $\left[; \underline{\text{DATA}} \begin{Bmatrix} \underline{\text{RECORD}} \text{ IS} \\ \underline{\text{RECORDS}} \text{ ARE} \end{Bmatrix} \text{data-name-1 [, data-name-2] } \ldots \right]$.

Figure V-13. Sort file description (*SD*) skeleton in COBOL. The general notation is the COBOL metalanguage described in Section II.6.2.
Source: [US65] plus CODASYL approved changes through March, 1968.

a *segment*. The programmer combines sections by specifying a *priority number* with each section's name. Sections having low priority numbers up to a limit specified in the ENVIRONMENT Division are included in the fixed portion of the object program. Sections which are not included in the fixed portion are grouped into segments by their priority number. The compiler is required to see that the proper control transfers are provided so that control among segments which are not stored simultaneously can take place.

COBOL is the first language to provide a complete description of the physical environment in which a program is to be run. This is done through the ENVIRONMENT Division. This is clearly machine and compiler dependent; in fact, its primary purpose is to segregate into one place all the information which cannot possibly be made machine independent. The ENVIRONMENT Division is divided into two sections—*CONFIGURATION* and *INPUT—OUTPUT*. The former contains three paragraphs: (1) SOURCE—COMPUTER, which defines the computer on which the compiler is to be run; (2) OBJECT—COMPUTER, to define the computer on which the object program is to be run (see Figure V-14); and (3) an optional SPECIAL—NAMES, which shows

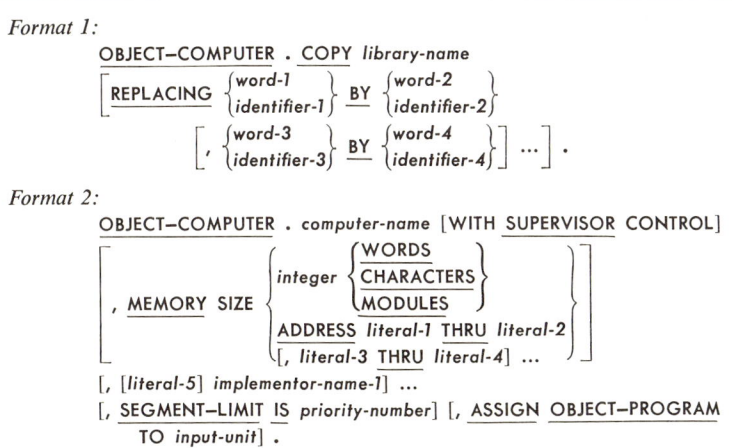

Figure V-14. *OBJECT—COMPUTER* format in COBOL. The general notation is the COBOL metalanguage described in Section II.6.2.
Source: [US65] plus CODASYL approved changes through March, 1968.

the relationship among certain hardware devices used by the COBOL compiler and the names they are referred to by in the program (see Figure V-15). Thus there is an easy mechanism for providing information about compiling on one machine for another; to the best of my knowledge, this "cross compilation" has not been done by anyone.

Format 1:

SPECIAL—NAMES. COPY library-name

$$\left[\underline{\text{REPLACING}} \begin{Bmatrix} \text{word-1} \\ \text{identifier-1} \end{Bmatrix} \underline{\text{BY}} \begin{Bmatrix} \text{word-2} \\ \text{identifier-2} \end{Bmatrix} \right.$$

$$\left. \left[, \begin{Bmatrix} \text{word-3} \\ \text{identifier-3} \end{Bmatrix} \underline{\text{BY}} \begin{Bmatrix} \text{word-4} \\ \text{identifier-4} \end{Bmatrix} \right] \cdots \right].$$

Format 2:

$$\underline{\text{SPECIAL—NAMES.}} \left[\text{implementor-name} \begin{Bmatrix} \underline{\text{IS}}\ \text{mnemonic-name }[,\ \underline{\text{ON}}\ \text{STATUS} \\ \underline{\text{IS}}\ \text{mnemonic-name }[,\ \underline{\text{OFF}}\ \text{STATUS} \\ \underline{\text{ON}}\ \text{STATUS}\ \underline{\text{IS}}\ \text{condition-name-1} \\ \underline{\text{OFF}}\ \text{STATUS}\ \underline{\text{IS}}\ \text{condition-name-2} \end{Bmatrix} \right.$$

$$\left. \begin{matrix} \underline{\text{IS}}\ \text{condition-name-1 }[,\ \underline{\text{OFF}}\ \text{STATUS}\ \underline{\text{IS}}\ \text{condition-name-2}]] \\ \underline{\text{IS}}\ \text{condition-name-2 }[,\ \underline{\text{ON}}\ \text{STATUS}\ \underline{\text{IS}}\ \text{condition-name-1}]] \\ [,\ \underline{\text{OFF}}\ \text{STATUS}\ \underline{\text{IS}}\ \text{condition-name-2}] \\ [,\ \underline{\text{ON}}\ \text{STATUS}\ \underline{\text{IS}}\ \text{condition-name-1}] \end{matrix} \right\} \cdots$$

[, $\underline{\text{CURRENCY}}$ SIGN $\underline{\text{IS}}$ *literal*] [, $\underline{\text{DECIMAL—POINT}}$ IS $\underline{\text{COMMA}}$] .

Figure V-15. *SPECIAL—NAMES* format in COBOL. The general notation is the COBOL metalanguage described in Section II.6.2.
Source: [US65] plus CODASYL approved changes through March, 1968.

The *INPUT—OUTPUT* Section provides the information needed to transmit the data on the external media to the object program. The *FILE—CONTROL* paragraph (see Figure V-16) names files and their media and specifies alternate input/output areas of file-control, e.g., SELECT UPDATED—MASTER ASSIGN 729—3 0210, 729—3 0220 RESERVE 3 ALTERNATE AREAS.. In COBOL-65 the File-Control paragraph has been extended to include information about mass storage—in particular, the user is allowed to specify whether the processing is to be random or sequential, the data names defining the end of a file, and the actual key to be used in accessing records. The *I—O—CONTROL* paragraph (see Figure V-17) specifies input/output techniques, points for rerun, shared memory areas, and location of files on a multiple file reel, e.g., I—O—CONTROL. APPLY SHORT—LENGTH—RECORD PROCEDURE ON MASTER—FILE; RERUN ON TAPE—WITH—LABEL—22 EVERY END OF REEL OF MASTER—FILE..

There is no provision in COBOL for modification of COBOL programs. With regard to self-extension of the language, this was permitted by the *DEFINE* verb which unfortunately was not implemented by anybody and so was finally removed in one of the COBOL Committee's modifications of the COBOL-65 manual. Essentially the *DEFINE* verb allowed the user to specify a new verb, show the format that it was supposed to have, and specify the subroutine that this new verb defined; it thus would have provided the first really powerful macro facility in a higher level language if it had ever been implemented.

Format 1:

FILE–CONTROL. COPY *library-name*

$$\left[\text{REPLACING} \begin{Bmatrix} word\text{-}1 \\ identifier\text{-}1 \end{Bmatrix} \underline{\text{BY}} \begin{Bmatrix} word\text{-}2 \\ identifier\text{-}2 \end{Bmatrix} \left[, \begin{Bmatrix} word\text{-}3 \\ identifier\text{-}3 \end{Bmatrix} \underline{\text{BY}} \begin{Bmatrix} word\text{-}4 \\ identifier\text{-}4 \end{Bmatrix} \right] \ldots \right] .$$

Format 2:

FILE–CONTROL. {SELECT [OPTIONAL] *file-name*

ASSIGN TO [*integer-1*] *implementor-name-1* [, *implementor-name-2*] ...

$$\left[\text{FOR MULTIPLE} \begin{Bmatrix} \underline{\text{REEL}} \\ \underline{\text{UNIT}} \end{Bmatrix} \right], \underline{\text{RESERVE}} \begin{Bmatrix} integer\text{-}2 \\ \underline{\text{NO}} \end{Bmatrix} \underline{\text{ALTERNATE}} \begin{bmatrix} \text{AREA} \\ \text{AREAS} \end{bmatrix} \right]$$

[, PRIORITY IS *implementor-name-3*]

$$\left[, \begin{Bmatrix} \underline{\text{FILE–LIMIT IS}} \\ \underline{\text{FILE–LIMITS ARE}} \end{Bmatrix} \begin{Bmatrix} data\text{-}name\text{-}1 \\ literal\text{-}1 \end{Bmatrix} \underline{\text{THRU}} \begin{Bmatrix} data\text{-}name\text{-}2 \\ literal\text{-}2 \end{Bmatrix} \left[, \begin{Bmatrix} data\text{-}name\text{-}3 \\ literal\text{-}3 \end{Bmatrix} \underline{\text{THRU}} \begin{Bmatrix} data\text{-}name\text{-}4 \\ literal\text{-}4 \end{Bmatrix} \right] \ldots \right]$$

$$\left[; \underline{\text{ACCESS MODE IS}} \begin{Bmatrix} \underline{\text{SEQUENTIAL}} \\ \underline{\text{RANDOM}} \end{Bmatrix} \right]$$

$$\left[, \underline{\text{PROCESSING MODE IS}} \begin{Bmatrix} \underline{\text{SEQUENTIAL}} \\ \underline{\text{RANDOM}} \end{Bmatrix} \text{FOR } integer\text{-}3 \underline{\text{RECORDS}} \right]$$

[, ACTUAL KEY IS *data-name-5*]

$$\left[, \underline{\text{SYMBOLIC}} \begin{Bmatrix} \underline{\text{KEY IS}} \\ \underline{\text{KEYS ARE}} \end{Bmatrix} data\text{-}name\text{-}6 \; [, data\text{-}name\text{-}7] \ldots \right] . \Big\} \ldots$$

Format 3:

FILE–CONTROL. {SELECT [OPTIONAL] *file-name*

ASSIGN TO *implementor-name-4*

[, *implementor-name-5*] ... OR *implementor-name-6*

[, *implementor-name-7*] ...

$$\left[\text{FOR MULTIPLE} \begin{Bmatrix} \underline{\text{REEL}} \\ \underline{\text{UNIT}} \end{Bmatrix} \right] \left[, \underline{\text{RESERVE}} \begin{Bmatrix} integer\text{-}4 \\ \underline{\text{NO}} \end{Bmatrix} \underline{\text{ALTERNATE}} \begin{bmatrix} \text{AREA} \\ \text{AREAS} \end{bmatrix} \right]$$

[, PRIORITY IS *implementor-name-8*] .} ...

Figure V-16. *FILE–CONTROL* format in COBOL. The general notation is the COBOL metalanguage described in Section II.6.2.
Source: [US65] plus CODASYL approved changes through March, 1968.

A distinction is made between *compiler-directing verbs* and *compiler-directing declaratives*. The former cause specific action to be taken at the point of appearance in the program, whereas the latter category provides information to the compiler which must be used sometime and which operates under the control of the main body of the PROCEDURE Division or the input/output system. The declaratives must appear at the beginning of the PROCEDURE Division. The compiler-directing verbs are: *ENTER*, which signals to the compiler that some other language is about to be used so that the compiler can invoke the necessary other translation system; *EXIT*, which provides a common exit point from the range of a *PERFORM*

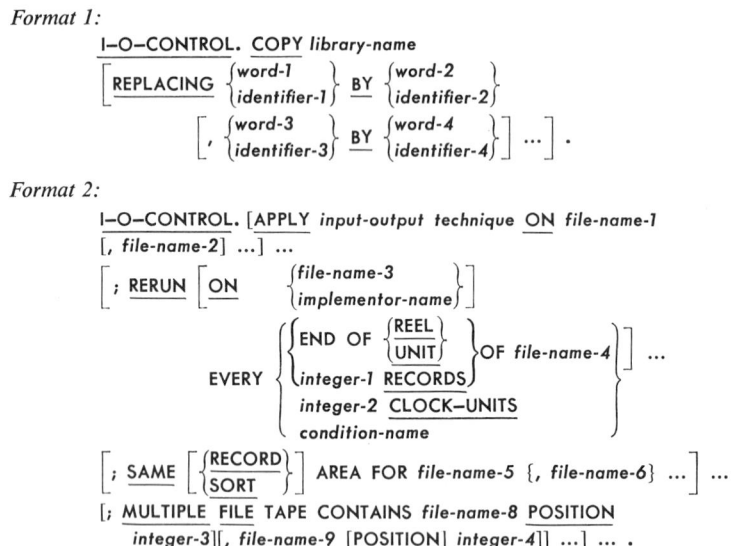

Figure V-17. *I–O–CONTROL* format in COBOL. The general notation is the COBOL metalanguage described in Section II.6.2.
Source: [US65] plus CODASYL approved changes through March, 1968.

verb; and **NOTE**, which signifies that the following sentence or paragraph is a comment and hence is not to be translated by the compiler.

The compiler-directing declaratives are **COPY** and **USE**, both of which were discussed earlier.

I am not aware of any serious attempts to write a COBOL compiler in COBOL, although I believe it probably could be done. The **EXAMINE** verb was included partially to aid in such a process. Some amount of concern was given in the design of COBOL to the ease of implementation, but it was not a major factor.

Some of the early implementations were carried on in parallel with the language development; this had the obviously valuable effect of providing instances in which the draft specifications were unworkable from an implementation point of view. Many of the early COBOL compilers were absolutely dreadful in terms of performance; in many cases this was considered a reflection on the language, whereas in reality it was merely an indication of the fact that techniques for efficient implementation of complex data processing languages like COBOL had not yet been developed. Current COBOL compilers are much more satisfactory from both the compile time and object time point of view. Some of the difficulties in the implementation are caused by the fact that it may be very hard to do something on a particular machine which is simple on another.

COBOL does not impose any *implied* storage allocation problems on the compiler since all the requirements are given in the source program. This obviously does not mean that some compilers will not be more effective in their storage handling than others, but such features as recursion and dynamic storage allocation are not required by the language. The most significant element in this area is the necessity of providing buffers for input and output data.

There are no separate requirements for debugging aids and error checking; these are left to the implementer. The language specifications point out a number of cases in which the results of a given error type are unpredictable.

V.3.4. Significant Contribution to Technology

COBOL has made a number of contributions to the technology of programming languages (and has forced development of new implementation techniques).

The first contribution was the development of a language in which a clear distinction was made among the actions to be performed, the description of the data on which it is to be done, and the physical environment in which the activity is being carried out. Some of these concepts appeared in FLOW-MATIC and Commercial Translator, but they were really solidified in COBOL. Experience has shown that the splitting of information between the DATA and ENVIRONMENT Divisions was not correct in every case, and some items are in one division which really belong in another (e.g., some of the File Description should be in the ENVIRONMENT Division).

A second major contribution has been the development of a description for data which is logically machine independent, although it does not *simultaneously* preserve efficiency and compatibility across machines.

A third contribution has been the creation of a programming language which is effective for handling problems with large files and simple processing.

A fourth contribution has been the development of a programming language which is very natural to read, permits really mnemonic names, and provides English-like executable statements. This has given us practical experience with a language which is simple relative to the scope of the natural English language; yet it is powerful and complex relative to its functional capabilities.

The conditional statements are more flexible than those of any other language in this book.

COBOL provided the first rudimentary indications of what types of information were required to interface with an operating system since many of the items in the ENVIRONMENT Division relate to an operating system more than to specific hardware.

Finally, although not a technical contribution per se, the COBOL activity has fostered the continued cooperation of competitive manufacturers, together with the users, in the development of a major language.

V.4. FILE HANDLING

Since a large part of the manipulations required in business data processing involve file manipulation, a number of systems have been developed for performing such operations. A few of the earlier systems were mentioned in Section V.1. This whole area is considered beyond the scope of this book, for the reasons indicated in Section V.4.2.

Since COBOL is the major language in use for business data processing, it would seem reasonable to suppose that individual groups who wished to provide more sophisticated file handling capabilities but within the framework of a powerful language would define additions to COBOL. The IDS system has done this, and is described very briefly below.

V.4.1. EXTENSIONS OF COBOL

1. *IDS*

The IDS (*I*ntegrated *D*ata *S*tore) system has been developed by C. Bachman and others at the General Electric Company. Although implemented as an extension of COBOL, the concepts are general and can be applied to any computer which has a mass memory device.

The basic concept is to place in a normal data record some additional fields called *chain fields* which contain the address of other IDS records. (This concept is closely related to the concept of *lists* developed in Chapter VI.) Additional entries are provided in the COBOL Record Description, e.g., RETRIEVAL VIA CALC CHAIN, PLACE NEAR *data-name* CHAIN, and PAGE-RANGE. Other additions to the DATA Division are provided. New verbs are also added. In particular, the verbs STORE (a new record into the file and link it into the chains as specified in the DATA Division), RETRIEVE (a record which is already in the IDS system), MODIFY (change the content of one or more fields with automatic relinking of the chains if necessary), DELETE (a record from its chains and the files) are provided.

V.4.2. GENERAL (*cross-reference only*)

File handling systems are beyond the scope of this book for several reasons. First, much of the emphasis is (quite rightly) on the file organization with the language elements being secondary and/or primitive. Second, many of the systems concentrate on the use of printed forms rather than languages. Third, the overlap and interconnnection between query languages (or systems) and file handling languages (or systems) is quite high. In some instances they are almost synonymous, i.e., a system might be considered a file handling or query system based primarily on the bias of the user.

References to some file handling systems are included with those listed in the bibliography for Section IX.3.2.1.

REFERENCES

V.1. SCOPE OF CHAPTER

[DG63] d'Agapeyeff, A., Baecker, H. D., and Gibbens, B. J., "Progress in Some Commercial Source Languages", *Annual Review in Automatic Programming*, Vol. 3 (R. Goodman, ed.). Pergamon Press, New York, 1963, pp. 277–98.

[IB61b] *IBM 7090 Programming Systems, SHARE 7090 9PAC Part 1: Introduction and General Principles*, IBM Corp., J28-6166, Data Processing Division, White Plains, N.Y. (1961).

[IB65d] *IBM System 360/Operating System Report Program Generator Specifications*, IBM Corp., C24-3337, Data Processing Division, White Plains, N.Y. (1965).

[LN62] Longo, L. F., "SURGE: A Recoding of the COBOL Merchandise Control Algorithm", *Comm. ACM*, Vol. 5, No. 2 (Feb., 1962), pp. 98–100.

[LS67] Leslie, H., "The Report Program Generator", *Datamation*, Vol. 13, No. 6 (June, 1967), pp. 26–28.

[MG59] McGee, W. C., "Generalization: Key to Successful Electronic Data Processing", *J. ACM*, Vol. 5, No. 1 (Jan., 1959), pp. 1–23.

[MG60] McGee, W. C. and Tellier, H., "A Re-Evaluation of Generalization", *Datamation*, Vol. 6, No. 4 (July–Aug., 1960), pp. 25–29.

[MG63] McGee, W. C., "The Formulation of Data Processing Problems for Computers", *Advances In Computers*, Vol. 4 (F. L. Alt and M. Rubinoff, eds.). Academic Press, New York, 1963, pp. 1–52.

[NM00] *SURGE: A Data Processing Compiler for the IBM 704*, North American Aviation, Inc., Columbus, Ohio.

[RR55a] *BIOR (Business Input-Output Rerun) Compiling System*, Remington Rand, Inc., ECD-2 (1955).

[SH65] Shaw, C. J., *Theory, Practice, and Trend in Business Programming*, System Development Corp., SP-2030/001/02, Santa Monica, Calif. (July, 1965).

[WY61a] Willey, E. L. *et al.*, "Some Commercial Autocodes—A Comparative Study," *A.P.I.C. Studies in Data Processing No. 1*. Academic Press, Inc. (London) Ltd., 1961.

V.2.1. FLOW-MATIC

[HF54] Holberton, F. E., "Application of Automatic Coding to Logical Processes", *Symposium on Automatic Programming for Digital Computers*, Office of Naval Research, Dept. of the Navy, Washington, D.C. (1954), pp. 34–39.

[KB57] Kinzler, H. M. and Moskowitz, P. M., "The Procedure Translator—A System of Automatic Programming", *Automatic Coding, Jour. Franklin Inst., Monograph No. 3*, Philadelphia, Pa. (Apr., 1957), pp. 39–49.

[RR59a] *FLOW-MATIC Programming*, U 1518 Rev. 1, UNIVAC. © 1958, 1959, Sperry Rand Corporation.

[TB60] Taylor, A., "The FLOW-MATIC and MATH-MATIC Automatic Programming Systems", *Annual Review in Automatic Programming*, Vol. 1 (R. Goodman, ed.). Pergamon Press, New York, 1960, pp. 196–206.

V.2.2. AIMACO

[AM58] *AIMACO Compiler*, The *AIMACO Compiling System Manual*, Air Materiel Command (Aug., 1958).

[YL59] Miller, E. R. and Jones, J. L., "The Air Force Breaks Through the Communications Barrier", *UNIVAC Rev.* (Winter, 1959), pp. 8–12.

V.2.3. COMMERCIAL TRANSLATOR

[IB59] *General Information Manual: IBM Commercial Translator*, IBM Corp., F28-8013 (1959).

[IB60a] *General Information Manual: IBM Commercial Translator*, IBM Corp., F28-8043, Data Processing Division, White Plains, N.Y. (1960).

[IB61c] *Addenda to the Commercial Translator General Information Manual*, IBM Corp., J28-8072, Data Processing Division, White Plains, N.Y. (1961).

[IB62a] *Preliminary Reference Manual: IBM 709/7090 Commercial Translator Processor*, IBM Corp., J28-6169-1, Data Processing Division, White Plains, N.Y. (1962).

[IB63] *COBOL and Commercial Translator: A Comparison*, IBM Corp., J28-6310, Data Processing Division, White Plains, N.Y. (1963).

V.2.4. FACT

[CP61] Clippinger, R. F., "FACT—A Business Compiler: Description and Comparison with COBOL and Commercial Translator", *Annual Review in Automatic Programming*, Vol. 2 (R. Goodman, ed.). Pergamon Press, New York, 1961, pp. 231–92.

[HO59] *The Honeywell-800 Business Compiler: A Preliminary Description*, Minneapolis-Honeywell Regulator Co., Datamatic Division, Newton Highlands, Mass. (1959).

[HO61]　　*FACT Manual (Interim Edition)*, DSI-27E 1161, Minneapolis-Honeywell Electronic Data Processing Division, Wellesley Hills, Mass. (1961).

V.2.5. GECOM

[GZ61]　　*GE 225: GECOM Language Specifications*, General Electric, Computer Dept., Phoenix, Ariz. (Dec., 1961).
[KX62]　　Katz, C., "GECOM: The General Compiler", *Symbolic Languages in Data Processing*. Gordon and Breach, New York, 1962, pp. 495–500.
[SB63]　　Schwalb, J., "Compiling in English", *Datamation*, Vol. 9, No. 7 (July, 1963), pp. 28–30.
[SM61]　　Sammet, J. E., "A Method of Combining ALGOL and COBOL", *Proc. WJCC*, Vol. 19 (1961), pp. 379–87.

V.3. COBOL

[AA68]　　*USA Standard X3.23-1968 COBOL*. (Final copy in preparation.)
[BE65]　　Bennett, N. W., *DEMON—A Programme Generator for Problems Involving Ordinary Differential Equations*, Australian Atomic Energy Commission Research Establishment, AAEC/E142, Sydney, Australia (Aug., 1965).
[BF62]　　Berman, R., Sharp., J., and Sturges, L., "Syntactical Charts of COBOL 61", *Comm. ACM*, Vol. 5, No. 5 (May, 1962), p. 260 plus insert.
[BJ61]　　Bromberg, H., "COBOL and Compatibility", *Datamation*, Vol. 7, No. 2 (Feb., 1961), pp. 30–34.
[BJ67]　　Bromberg, H., "The COBOL Conclusion", *Datamation*, Vol. 13, No. 3 (Mar., 1967), pp. 45–50.
[CG63]　　Cunningham, J. F., "COBOL", *Comm. ACM*, Vol. 6, No. 3 (Mar., 1963), pp. 79–82.
[CW64]　　Cowan, R. A., "Is COBOL Getting Cheaper?", *Datamation*, Vol. 10, No. 6 (June, 1964), pp. 46–50.
[DO62]　　Donally, W. L., "A Report Writer for COBOL", *Comm. ACM*, Vol. 5, No. 5 (May, 1962), p. 261.
[EC67]　　ECMA (European Computer Manufacturers Association), *Formal Definition of the Syntax of COBOL* (preliminary edition), Geneva, Switzerland (Aug., 1967).
[EM62]　　Emery, J. C., "Modular Data Processing Systems Written in COBOL", *Comm. ACM*, Vol. 5, No. 5 (May, 1962), pp. 263–68.
[EP63]　　"Time to Switch to COBOL?", *EDP Analyzer*, Vol. 1, No. 11 (Dec., 1963), pp. 1–11.
[FD65]　　Fredericks, D. S. and Warburton, C. R., "Across Machine Lines in COBOL", *Comm. ACM*, Vol. 8, No. 12 (Dec., 1965), pp. 731–35.
[FP64]　　Fimple, M. D., "FORTRAN vs. COBOL", *Datamation*, Vol. 10, No. 8 (Aug., 1964), pp. 34, 39–40.
[GN63]　　Gordon, R. M., "COBOL and Compatibility", *Datamation*, Vol. 9, No. 7 (July, 1963), pp. 47–48.

[GV62] Greene, I., "Guides to Teaching COBOL", *Comm. ACM*, Vol. 5, No. 5 (May, 1962), pp. 272–73.

[HI62] Hicks, W., "The COBOL Librarian—A Key to Object Program Efficiency", *Comm. ACM*, Vol. 5, No. 5 (May, 1962), p. 262.

[HY62] Humby, E., "Rapidwrite—COBOL Without Tears", *Symbolic Languages in Data Processing*. Gordon and Breach, New York, 1962, pp. 573–83.

[HY63] Humby, E., "Rapidwrite", *Annual Review in Automatic Programming*, Vol. 3 (R. Goodman, ed.). Pergamon Press, New York, 1963, pp. 299–310.

[IB63] *COBOL and Commercial Translator: A Comparison*, IBM Corp., J28-6310, Data Processing Division, White Plains, N.Y. (1963).

[IC62a] "General Panel Discussion: Is a Unification ALGOL-COBOL, ALGOL-FORTRAN Possible? The Question of One or Several Languages", *Symbolic Languages in Data Processing*. Gordon and Breach, New York, 1962, pp. 833–49.

[JU65] Junker, J. P. and Boward, G. R., "COBOL vs. FORTRAN: A Sequel", *Datamation*, Vol. 11, No. 4 (Apr., 1965), pp. 65–67.

[KS62] Kesner, O., "Floating-Point Arithmetic in COBOL", *Comm. ACM*, Vol. 5, No. 5 (May, 1962), pp. 269–271.

[LN62] Longo, L. F., "SURGE: A Recoding of the COBOL Merchandise Control Algorithm", *Comm. ACM*, Vol. 5, No. 2 (Feb., 1962), pp. 98–100.

[LP62] Lippitt, A., "COBOL and Compatibility", *Comm. ACM*, Vol. 5, No. 5 (May, 1962), pp. 254–55.

[MN61] Makinson, T. N., "COBOL: A Sample Problem", *Comm. ACM*, Vol. 4, No. 8 (Aug., 1961), p. 340.

[MU62] Mullin, J. P., "An Introduction to a Machine-Independent Data Division", *Comm. ACM*, Vol. 5, No. 5 (May, 1962), pp. 277–78.

[NF64] Naftaly, S. M., "Compiling a COBOL Questionnaire", *Datamation*, Vol. 10, No. 8 (Aug., 1964), pp. 30–33.

[QH66] Budd, A. E., *A Method for the Evaluation of Software: Procedural Language Compilers—Particularly COBOL and FORTRAN*, Mitre Corp., (DDC) AD 651142, Commerce Dept. Clearinghouse, Springfield, Va. (Apr., 1966).

[SG62] Siegel, M. and Smith, A. E., "Interim Report on Bureau of Ships COBOL Evaluation Program", *Comm. ACM*, Vol. 5, No. 5 (May, 1962), pp. 256–59.

[SM61] Sammet, J. E., "A Method of Combining ALGOL and COBOL", *Proc. WJCC*, Vol. 19 (1961), pp. 379–87.

[SM61a] Sammet, J. E., "A Definition of the COBOL 61 Procedure Division Using ALGOL 60 Metalinguistics", *Summary in Preprints of 16th Nat'l Meeting of the ACM*, Sept., 1961, pp. 5B-1 (1)–(4).

[SM61b] Sammet, J. E., "General Views on COBOL", *Annual Review in Automatic Programming*, Vol. 2 (R. Goodman, ed.). Pergamon Press, New York, 1961, pp. 345–49. (Same article as [SM61c].)

[SM61c] Sammet, J. E., "More Comments on COBOL", *Datamation*, Vol. 7, No. 3 (Mar., 1961), pp. 33–34. (Same article as [SM61b].)

[SM62]   Sammet, J. E., "Basic Elements of COBOL 61", *Comm. ACM*, Vol. 5, No. 5 (May, 1962), pp. 237–253. (Also in [RO67].)

[SX63]   Saxon, J. A., *COBOL: A Self-Instructional Manual*. Prentice-Hall, Inc., Englewood Cliffs, N.J., 1963.

[US60]   *COBOL: Initial Specifications for a Common Business Oriented Language*, Dept. of Defense, U.S. Govt. Printing Office, Washington, D.C. (Apr., 1960).

[US61]   *COBOL-1961: Revised Specifications for a Common Business Oriented Language*, Dept. of Defense, U.S. Govt. Printing Office, Washington, D.C. (1961).

[US62]   *COBOL-1961 Extended: Extended Specifications for a Common Business Oriented Language*, Dept. of Defense, U.S. Govt. Printing Office, Washington, D.C. (1962).

[US65]   *COBOL: Edition 1965*. Dept. of Defense, U.S. Govt. Printing Office, Washington, D.C. (Nov., 1965).

[WH62]   Whitmore, A. J., "COBOL At Westinghouse", *Datamation*, Vol. 8, No. 4 (Apr., 1962), pp. 31–32.

[WM67]   *COBOL Programming Tips*, Westinghouse Electric Corp., Management Systems Dept. (Apr., 1967).

[WY61]   Willey, E. L. *et al.*, "A Critical Discussion of COBOL", *Annual Review in Automatic Programming*, Vol. 2 (R. Goodman, ed.). Pergamon Press, New York, 1961, pp. 293–304.

[XB00]   X3.4 COBOL Information Bulletins, BEMA/DPG, Numbers 1–11 (Apr., 1963–present) (continuing publication, distributed irregularly).

[XB67]   X3.4 COBOL Information Bulletin No. 9, "Proposed USA Standard COBOL", *ACM SICPLAN Notices*, Vol. 2, No. 4 (Apr., 1967).

[ZQ67]   Callahan, M. D. and Chapman, A. E., "Description of Basic Algorithm in DETAB/65 Preprocessor", *Comm. ACM*, Vol. 10, No. 7 (July, 1967), pp. 441–46.

V.4.1.1. *IDS*

[GZ65]   *Introduction To Integrated Data Store*, General Electric, Computer Dept., Phoenix, Ariz. (Apr., 1965).

[QJ64]   Bachman, C. W. and Williams, S. B., "A General Purpose Programming System for Random Access Memories", *Proc. FJCC*, Vol. 26, pt. 1 (1964), pp. 411–22.

[QJ65]   Bachman, C. W., "Software for Random Access Processing", *Datamation*, Vol. 11, No. 4 (Apr., 1965), pp. 36–41.

VI STRING AND LIST PROCESSING LANGUAGES

VI.1. SCOPE OF CHAPTER

For several reasons, this chapter is probably a more difficult one to comprehend than most of the others. The main reason is that relatively few people have had experience with, or exposure to, the types of problems for which the languages in this chapter are needed. In the other areas, far less background and/or direct experience is required to appreciate the language facilities described. Thus, this chapter may well have the unfortunate characteristic that the people who are not familiar with the languages involved may not understand the descriptions, and those who are familiar with them will not need the information contained herein. In spite of this dilemma, it is obvious that no book on programming languages would be complete without a description of the languages used for list processing, string manipulation, and pattern matching. A further difficulty arises because the outline of Chapter III is difficult to follow for this class of languages; hence the technical descriptions in this chapter do not particularly match the sequence or emphasis of material in that chapter.

In order to have any understanding at all of the language facilities provided in differing ways, it is necessary to comprehend the basic concept of a *list*. This concept was first introduced by Newell, Simon, and Shaw [NW56] to help solve the particular problem that they were working on, namely the proof of theorems in the propositional calculus. The list concept has since become one of the cornerstones of a great deal of work in the computer field. Its need arises in the many applications which require dynamic storage allocation of a very large and complex kind. One specific type of problem in which this arises occurs when there is a need to maintain tables for several different kinds of data, but the user has no way of telling ahead of time how much space should be allocated to each table; in other words, he cannot tell (in fact, it changes from problem to problem) how many elements will be in each table. In addition to that difficulty, there are many applications in which the major activity consists of inserting and deleting elements from tables. If the information is stored in sequential memory locations, this requires constant work (by both the programmer, and the computer at object time) to move the data elements.

Some of the specific areas using this type of activity are compiler writing, theorem proving, manipulation of formal algebraic expressions, picture processing, some types of linguistic data processing, and most aspects of work in artificial intelligence. These applications have tended to become lumped together under the general title *symbol manipulation*.

The basic concept of a list is quite simple. Instead of storing data sequentially in memory, each item contains not only the data element but the address of the next data element in logical sequence. (See Figure VI-1.) It is immediately obvious that this may (but does not always) waste storage. If the data item and the address fit together in some logical storage unit in the machine (e.g., a word), then perhaps not much space is wasted. On the other hand, it may be necessary to use a second memory location for the

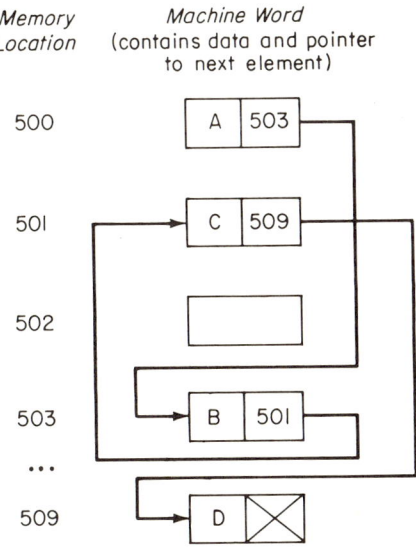

Figure VI-1. Illustration of list. Assume there is a sequence of data items *A, B, C,* and *D* which are to be represented in the form of a list; i.e., each element contains a pointer to the memory location containing the next one.

address (commonly called a *pointer*) to the next element. This is one of the penalties accruing from the use of list processing. On the other hand, to insert or delete an element is extremely simple. Thus, to insert an element between B and C in Figure VI-1, it is only necessary to change one address rather than moving all the elements. (See Figure VI-2.) Beyond this basic concept, one of the significant elements in list processing is whether the element of a list can itself be a list, as illustrated in Figure VI-3. Another situation occurs when each element of a list can consist of many data elements. This is sometimes called an *n-component element* or *plex*.

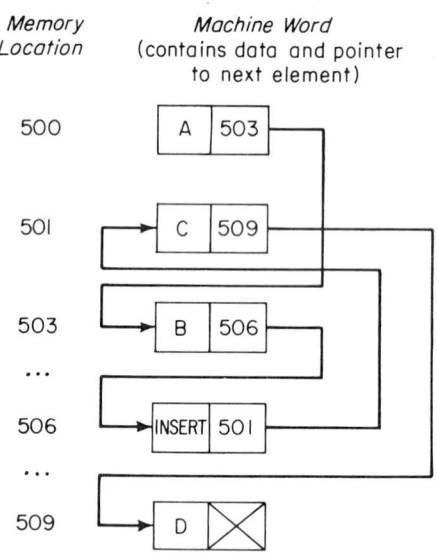

Figure VI-2. Inserting element in a list. Contrasting this with Figure VI-1 shows that only one address, namely the one in position *503* (associated with B) has been changed to point to the *INSERT* item in *506*. The pointer from *INSERT* goes to C in *501*. Nothing else had to change.

One of the major problems is that of returning currently unused data to the so-called *free list*. In Figure VI-2, if we wish to delete item C, all that is necessary is to change the pointer associated with B from C to D and the list is now in correct form. However, the memory location that was occupied by the item C really is no longer needed. It is characteristic of list processing problems that this type of thing happens quite frequently, and there is a need to "clean up" memory so that positions no longer being actively used are now available for new data. This becomes one of the major problems in the design of a list processing system.

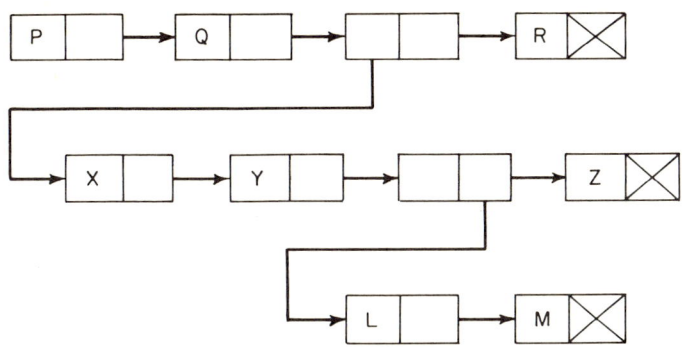

Figure VI-3. List structure, i.e., list containing list as element. This shows several levels of sublists. The element *Q* points to an element which contains no data but instead points both to a single element *R*, and to a sublist whose first element is *X*. *X* points to *Y* which in turn points to an element containing no data but points to a single element *Z*, and to a sublist containing the elements *L* and *M*.

From these basic concepts, an entire area of development has appeared, involving both languages and techniques. The techniques include such words or phrases as *threaded lists, backward pointers, reference counts, garbage collection,* and *free list*. It is beyond the scope of this book either to deal with these ideas or even to provide an adequate bibliography for them. For the reader who is interested in pursuing this matter in detail, however, an annotated bibliography is given in Sammet [SM66] and its updated version [SM67a].

Some introductory papers on symbol manipulation which discuss the types of applications involved are Raphael [RA66a] and [RA66b] and other references cited throughout this chapter. A good source for a number of papers discussing artificial intelligence is Feigenbaum and Feldman [FG63]. A few other relevant articles are shown in the reference list (given at the end of this chapter) for this section or for the individual languages.

A second—and tangentially related—concept of importance is that of string processing. A *string* is generally considered to be a variable length sequence of characters. It is important to understand immediately that strings and lists are neither opposites nor synonyms. A list is a particular way of representing information in a computer, and a string is one of the types of information that can be so represented. The significance of strings occurs to a large extent in the handling of text material. The text material can be either natural language of some kind (e.g., this sentence), a string composed of a program in any language, or any arbitrary sequence of characters from some particular data area.

From a language point of view, the types of operations that people wish to perform differ significantly between list processing and string processing. In list processing, the main concern is to put information into a list or to delete it, to combine lists in different ways, and to deal with the *free list* problem.

In considering string processing, the major types of operations to be performed include searching for patterns and transforming them into different patterns. This may also involve deletions or insertions in the string itself. If the string has been represented as a list, some of the list processing facilities are needed.

From the point of view of programming languages as discussed in this book, it would be quite logical to have two separate chapters, one dealing purely with the list processing languages and the other with the string processing languages. They have been combined into one chapter, primarily because the areas in which they are used tend to overlap somewhat and there is sometimes a need for these facilities to be intermingled in one way or another. Although I question the validity of the title, the article by Bobrow and Raphael [BB64a] provides a useful overview of some of the similarities and differences along these lines.

The oldest of the list processing languages is the IPL family culminating in IPL-V, which was considered a higher level language during the course of much of its development but it is now viewed largely as a conceptual assembly language for a mythical machine capable of doing list processing. A more recent development, namely L^6, also operates at a very low level from the point of view of the facilities provided, but it offers the user much more flexibility (and requires more work) in the creation of his lists than IPL-V does. In that sense, it is even lower level than IPL-V.

The LISP systems (really LISP i for $i < 2$) represent an entirely different approach to the entire subject. LISP has its fundamental basis in certain aspects of mathematics and recursive function theory. It represents an approach to language design that is significantly different from virtually all other languages in this book because LISP is to a large extent a functional language rather than a statement-oriented one. That is to say, the user tends to write functions in LISP, and his program consists of instructions to evaluate the functions to produce his desired result.

The first of the string processing languages was COMIT. This was developed to assist linguists doing work in natural language translation. COMIT has served as a pattern for other developments, most notably the more recent SNOBOL. In some cases, the capabilities represented by COMIT have been added to list processing languages.

While a discussion of essential differences between COMIT and SNOBOL is premature in an introductory section if the reader is not familiar with either language, it seems more logical to include the remarks here and refer the reader to this paragraph after he has studied Sections VI.6 and VI.7. COMIT seems to be a more complex language for professional programmers than SNOBOL since some of the facilities in the latter are closer to those of other programming languages. On the other hand, linguists and programmers have certainly used both languages successfully. I am unaware of any carefully written detailed comparisons; the following comments are of necessity superficial. COMIT permits a more complex data structure through the use of its logical and numeric subscripts than SNOBOL does; SNOBOL provides only strings. Furthermore, the ability in COMIT to define individual elements (i.e., constituents) which are either single characters or groups of characters (e.g., English words) provides more flexibility in programming and scanning. The list rule feature in COMIT to provide fast table lookup is a distinct advantage. On the other side, the ability of SNOBOL to assign names to specific strings and to name unknown strings provides an ease of programming far above equivalent actions in COMIT. The ability to define functions and the elaborate types of pattern searching which are available in SNOBOL provide that language with definite advantages in certain cases. In conclusion, as in deciding between any two languages, the potential user must weigh carefully all the technical features, the specific implementations he is dealing with, and the factors discussed in Chapter II.

TRAC[1] is a still more recent development which attempts to combine some of the basic facilities from the list and string processing languages and to do this in an on-line environment.

Finally, there are several langugaes which have been developed and implemented but are not in any way widely used. These include AMBIT, TREET, CLP, CORAL, SPRINT, and LOLITA.

A general description of many *symbol manipulation* languages is given in Raphael *et al.* [RA67]. This provides an excellent bird's-eye view of most of the languages described in this chapter and in Chapter VII. However, some of the items in that paper are not languages by the criteria established in this book.

Any reader who has had previous exposure to list processing or who has even absorbed some of the atmosphere in this field may wonder at the absence of SLIP and DYSTAL from the list of languages discussed in this chapter. The reason is simply that they are not languages but merely packages of subroutines which are embedded in a language, e.g., FORTRAN. To quote from the SLIP designer, "It [SLIP] is a language system designed to be imbedded in a higher order language capable of calling machine language subroutines",[2] and his personal comment to me "of course SLIP is not a

---

[1] TRAC is the trademark and service mark of Rockford Research Institute Incorporated in connection with their standard computer controlling languages.

[2] Weizenbaum [WZ63], p. 524.

language". Thus from a purely language point of view there is no conceptual difference between the subroutines associated with SLIP (described in Weizenbaum [WZ63]) and a package of routines to do matrix manipulation, integration, or anything else that is invoked by a CALL statement. The same remark applies to the more recent development DYSTAL (Sakoda [SA65]). It also applies to a significantly different type of package, namely the addition of pattern-matching facilities to LISP in the CONVERT system (see Guzman [ZH66]). This comment, and the exclusion of these systems, in no way represent a negative value judgment on their importance or significance. It is merely a reaffirmation of the fact that adding a package of subroutines to a language does not change the language per se. This point was discussed earlier in Section IV.3 (FORTRAN). The reader who is still skeptical or unhappy about this viewpoint might wish to contrast this situation with that of FORMAC which is discussed in Section VII.3. In that case, while there was definitely a package of subroutines added to FORTRAN, there was also a language placed on top of FORTRAN. Thus the user invoked the new facilities by means of new statements in FORTRAN rather than by merely writing a CALL to the appropriate subroutine. The paper by Bobrow and Weizenbaum [BB64b] discusses this problem, but in my opinion it does not meet the issue directly. The authors do not really consider whether one has a new language if one adds a set of packages which are able to be used without changing the host language or the compiler. There does not seem to be any illustration (in the United States) of an *implemented* basic language extension to FORTRAN or ALGOL to just provide list or string processing facilities, although there have been numerous suggestions for doing so. Some of these are listed under the ALGOL references at the end of Chapter IV.

Since FLIP and DYSTAL are not considered languages, the earlier FLPL (*F*ORTRAN Compiled *L*ist *P*rocessing *L*anguage) (see Gelernter, Hansen, and Gerberich [GE60]) is not a language for the same reason. It was a much earlier development along the same line, namely adding subroutines to FORTRAN.

VI.2. LANGUAGES OF HISTORICAL INTEREST ONLY

There are no list or string processing languages which fall in this category.

VI.3. IPL-V

VI.3.1. HISTORY

One of the most significant events that has ever occurred in programming was the development of the concept of list processing by Allen Newell,

J. C. Shaw, and Herbert Simon. The classic paper in which these ideas were presented is by Newell and Simon [NW56]. Because of the historical significance of this work, a list of most of the principal primitive instructions from that paper is included as Figure VI-4; however, this language was never implemented. From a conceptual point of view, the early language was fairly application-oriented, where the applications involved proving theorems in the propositional calculus and playing chess. The first language (which was really a collection) was called IPL-I (*Information Processing Language*-I). The first implemented version was IPL-II, which is described in Newell and Shaw [NW57a]. This was implemented on the JOHNNIAC at the RAND Corporation. The first major use of the language was for proving theorems in the propositional calculus and was reported in Newell, Shaw, and Simon [NW57]. The next version, IPL-III, was abandoned shortly after it became operational because it required too much space. IPL-IV was used for a number of significant programs in the field of artificial intelligence; IPL-IV strongly resembles IPL-V[3] but has never actually been documented. Work on IPL-V actually started in late 1957 at Carnegie Institute of Technology on the IBM 650 as a modification of IPL-IV. A running version of IPL-V was available in early 1958. A newer (and essentially final) version of the language became operational on the 704 at the end of the summer in 1959. The first published description of IPL-V appeared in 1960 (see Newell and Tonge [NW60]).

The implementation and development of this line of language stopped with IPL-V because the people most vitally concerned were more interested in the problems they were trying to solve than in further language developments. This is a refreshing contrast to other areas in which the language development began to overshadow the problem-solving. However, an IPL-VI was proposed as an order code for a computer (see Shaw *et al.* [JC58]).

IPL-V has been one of the most widely implemented languages, and versions exist on at least the following machines: IBM 650, 704, 709/7090, 1620, UNIVAC 1105 and 1107, Control Data 1604 and G20, Burroughs 220, Philco 2000, and AN/FSQ-32.

VI.3.2. FUNCTIONAL CHARACTERISTICS OF IPL-V

The most significant property of IPL-V is that it has a closer notational resemblance to assembly language than any other language in this book (except perhaps L[6]). IPL-V is really the machine language for a hypothetical computer which does list processing. Newell notes[4] that it is curious that

---

[3] Newell *et al.* [NW65], p. xxi.
[4] Private communication, June, 1967.

Numerical

NAG	x		Add one to G(x).
NAGG	x	y	Add G(x) to G(y); result in G(y).
NAH	x		Add one to H(x).
NAK	x		Add one to K(x).
NAJ	x		Add one to J(x).
NAW			Add one to W (work done).
NSG	x		Subtract one from G(x).
NSGG	x	y	Subtract G(x) from G(y).

Assign

AA	x	Assign an unused list to A(x).
AN	x	Assign an unused name to E(x).

Compare

CC	x	y	b	If $C(x) = C(y)$, $\longrightarrow b$.
CN	x	y	b	If $N(x) = N(y)$, $\longrightarrow b$.
CGG	x	y	b	If $G(x) > G(y)$, $\longrightarrow b$.
CKG	x	y	b	If $K(x) > K(y)$, $\longrightarrow b$.
CWG			b	If W (work done) $>$ limit, $\longrightarrow b$.
CPS	x	y	b	If E(x) is subelement of E(y), $\longrightarrow b$.

Find

FEF	x	y	b	Find the first E in A(x) and put in y; if none, $\longrightarrow b$.
FEN	x	y	b	Find the E in A(x) next after E(y), put in y; then $\longrightarrow b$. If none (end of list), $\rightarrow$ next.

Store

S	x		Store E(x) back in A(x); if not there, store E(x) at end of A(x).
SEN	x	y	Store E(x) as next E in A(y); E(x) now last item in A(y).
†SX	x	y	Store a copy of X(x) at (new) A(y).
†SXE	x	y	Store X(x) in A(y) in place of E(y).
†SXM	x	y	Store X(x) at (new) A(y) as main expression.

Put

PE	x	y	Put E(x) in E(y); E(x) remains.
PK	x	y	Put K(x) in K(y).

Test

TC	x		b	If $C(x) = \longrightarrow$ (implies), $\longrightarrow b$.
TB	x		b	If E(x) is blank, $\longrightarrow b$.
TU	x		b	If E(x) is a unit, $\longrightarrow b$.
TGG	x		b	If $G(x) > 0$, $\longrightarrow b$.
TF	x		b	If E(x) is free, $\longrightarrow b$.

Figure VI-4. (cont.)

Branch

B	b	Branch to *b*.
BHB		In higher instruction, ⟶ *b*.
BHN		In higher instruction, ⟶ next.

Figure VI-4. Fairly complete list of basic processes in first Information Processing Language. In the figure, *x, y* and *b* denote memory locations and ⟶ denotes *branch to*. *X* is a logic expression and *E* a set of elements. *A* is the location of the whole expression in memory. *C* is the connective in the expression. *G* is the number of negation signs before the expression. *H, J*, and *K* represent respectively the number of variable places, distinct variables, and number of levels in the expression. Processes preceded by a dagger can be defined in terms of simpler operations.
Source: Newell and Simon [NW56], extracts from pp. 63–78.

IPL has come to be described as "like assembly code" when its description in the early 1960's was as a *problem-oriented language*. The IPL application area is "for problems that are sufficiently complex, ill-structured, and difficult to require intelligence for their solution by humans."[5] It was certainly meant to be used only by a professional programmer. It was definitely developed for use as a batch system, but it has been successfully used in general time-sharing systems or in those with remote job entry facilities.

By its very nature, it is not hard for IPL-V to be fairly machine independent, providing only that the data items will fit into the portion of the machine word allocated to data. Since most of the implementations are interpretive, there is slightly less of a problem about compiler dependence and the systems are essentially (although not 100 percent) interchangeable. As far as dialects are concerned, because the structure of the language consists of a long list of available routines, it is very easy for somebody to add other routines. The problem of subsettings and extensions has not really arisen, except in the sense of having these new routines added by individual groups.

There has been no consideration of official USASI standardization.

In addition to A. Newell, F. M. Tonge, E. A. Feigenbaum, B. F. Green, Jr., and G. H. Mealy, who are the editors of the manual [NW65], numerous other people were involved in the implementation for their individual computers. A list is given at the beginning of the IPL-V manual. There is no official maintenance. However, in an effort to provide some coordination

---

[5] Newell *et al.* [NW65], p. xv.

SAMPLE PROGRAMS—IPL-V

Problem: Reverse the symbols after the head cell of the list named in H0.

Program:

Name	PQ	Symb	Link	Comment
R2		J60		Step pointer to cell containing first symbol
	70	J8		Clean up and stop if list was empty
	40	H0		
	40	H0		Push down two copies of pointer to first symbol (needed as arguments for J65 and J68 below)
		R2		Recursively reverse symbols after the first
	12	H0		Get the first symbol
		J65	J68	Insert it at the end and delete it from the beginning

Problem: Test whether two list structures are equal, where "equal" means they have the same hierarchical structure and the same elementary symbols in corresponding positions.

Program:

Name	PQ	Symb	Link	Comment
R1		J51		Spread arguments in working storage
9–3	12	W0		Get next symbols
	12	W1		
		J2		Compare symbols
	70	9–2		Branch if not identical
9–4	11	W0		
		J60		Step first list
	20	W0		
	70	9–1		Branch if first list ended
	11	W1		
		J60		Step second list
	20	W1		
	70	J31	9–3	Clean up and quit, or loop
9–2	12	W0		Symbols not identical
		J132		Test if local, i.e., name of a sublist
	70	J31		Quit if either not local
	12	W1		
		J132		
	70	J31		
	12	W0		Both local. Input substructure names
	12	W1		
		R1		Recursively test sublists for equality
	70	J31	9–4	Quit if substructures differ
9–1	11	W1		
		J60		Are main lists the same length?
	30	H0		
		J5	J31	Set H5 to + if R1's two inputs are names of "equal" structures, and set H5 to – otherwise. Cleanup and quit.

of the activity, an IPL secretary was established at the RAND Corporation. This was solely for communication exchange and not for maintenance of a subroutine library.

The language definition was given using English and illustrations. The latter are essential since in many cases the description is not sufficient to determine either syntax or semantics.

The main documentation of IPL-V is the manual published in 1961, with a second edition in 1964 (Newell [NW65]). It contains a historical and conceptual introduction, a primer, and the official description of the language. In many cases the individual implementations have manuals. One philosophical aspect of the general documentation problem led to significant developments in the language itself. Newell says "Originally, as in most languages, only a very small, almost minimal, set of primitives were specified. All the rest, it was argued, could be built up from these. On the other hand, a different argument went, since list processing was relatively new, people should not have to discover for themselves how to put the little pieces together into useful medium-sized processes. Hence, a substantial array of basic processes should be provided in the language as a means of (automatically) communicating to the user the things he could do with the language. This argument, which derives from a concern with documentation in the broad sense, was accepted and led to the 200 odd primitives that now exist in IPL."[6]

Two related documents which are of interest are the demonstration problem consisting of the IPL-V coding of the Logic Theorist (see Stefferud [SF63]) and another teaching aid called TIPL (*Teach IPL*) (see Dupchak [DU63]). This examines IPL code written by students and tests it against various inputs.

Some of the problems for which IPL-V has been used are described in Newell and Simon [NW61], Feigenbaum [FG61], and Tonge [TN60], and other references at the end of this chapter. The work by Tonge actually used IPL-IV.

VI.3.3. TECHNICAL FEATURES OF IPL-V

IPL-V is really an assembly language level set of commands for doing list processing. As such, it is extremely difficult to describe in terms of the general outline which was given in Chapter III, and so this will be done only where feasible.

It is convenient to think of an IPL machine which consists of a set of cells (i.e., storage locations) which hold expressions, a set of symbols used to form expressions (the symbols are actually all addresses), and a set of primitive processes (i.e., instructions). (A CDC 3600 was actually modified

---

[6] Newell [NW63], p. 88.

to respond directly to IPL-V commands and used at least for playing checkers. This is described in Hodges [HX64] and Cowell and Reed [ZN65].) From a very general overview, IPL can be considered a postfix system with a specific cell, H0, containing the operand stack and the cell H1 containing the operator stack.

IPL (used as an abbreviation for IPL-V) allows two kinds of expressions: *data list structures*, which contain the information to be processed, and *routines*, which define the processes.

IPL distinguishes three types of symbols: *regional*, *internal*, and *local*. A regional symbol is a letter or punctuation mark followed by one to four decimal digits, e.g., A2, B1234. An internal symbol consists of a positive integer and is used primarily by the computer. Both regional and internal symbols always stand for the same object throughout the program. Local symbols connect lists and list structures and are defined only through a specific list structure; a local symbol is a 9 followed by a hyphen – followed by one to four decimal digits. All IPL expressions are written in terms of what is called an IPL *word*. A standard word consists of four parts: *P*, *Q*, *SYMB*, and *LINK*. *P* and *Q* are called the *prefixes* of the word and consist of an octal digit; *Q* is the designation prefix; and *P* is the operation prefix (for routines) or data type prefix (for data list structures).

The standard IPL word mentioned above does not suffice to represent numbers, alphabetic characters, etc., so data terms are defined and have three parts: *P*, *Q*, and *DATA*. Again, *P* and *Q* are prefixes, and *DATA* contains the information.

The input form is fixed with specific columns assigned to the cell *NAME*, a *SIGN*, *P*, *Q*, *SYMB* and *LINK*. For data, the information is placed in the combined *SYMB* and *LINK* fields. A version of IPL called LIPL (*Linear IPL*) permits a more horizontal format. See Dupchak [DU65].

On the coding sheet, the appearance of a regional or internal symbol in the *NAME* area defines the start of a list structure, and all local symbols that occur after this line belong to the same list structure until another regional or internal *NAME* occurs.

NAME	PQ	SYMB	LINK
L1		S2	
		9–3	
		A4	0
9–2		A3	
		9–4	0
9–3		A4	
		L2	0

A storage cell holds symbols; it is created by giving the cell a regional name and putting the termination symbol 0 for *LINK*. *SYMB* is the symbol contained in the cell, and it may be placed there initially or during execution.

Associated with each storage cell is a data list called a *pushdown list*. The storage cell is the head of the list and the cells used in the storage system are list cells. By convention, W0, ..., W9 are used for temporary storage, although any cell can be a storage cell.

Since the basic purpose of IPL-V is to operate on lists, it is essential to know what types are defined. IPL permits *data lists*, *data list structures*, and *list structures*.

A *data list* is a set of cells whose sequence is defined by the rule that the *LINK* part of the cell contains the name of the next cell in the list. (When nothing is written, there is an automatic link to the next word.) The link of the last cell in the list is a termination symbol, denoted by 0. A data list is created by writing a symbol in the *NAME* field of some line, and this symbol is the name of both the list and the symbol at the head of the list. The termination symbol 0 is written for the link of the last cell. The system will assign an internal name to any cell not explicitly named by the programmer; however, the programmer may give names to list cells by using local symbols (since using regional symbols usually starts a new list). This can be done to avoid writing the data list information in sequential order on the coding sheet. For example,

NAME	PQ	SYMB	LINK
L1		0	9–1
9–2		S2	9–3
9–1		S1	9–2
9–3		S3	0

It is possible to associate with a list a description list which contains information about the list in the form of values for specified attributes. The *SYMB* of the head contains the name of the description list. The value of an attribute can be the name of a list of values. A description list contains alternately the symbols for attributes and their values. The same symbol cannot occur more than once as an attribute on the description list. The following example shows a list with the name L2 described by the attributes A1 and A2 with values V1 and V2, respectively:[7]

NAME	PQ	SYMB	LINK
L2		9–0	
		S1	
		S2	
		S3	0
9–0		0	
		A1	
		V1	
		A2	
		V2	0

---

[7] Newell *et al.* [NW65], p. 150.

A *list structure* is a set of lists connected by the names of the list occurring on other lists in the set. A *data list structure* is a fairly simple form of the list structure since there are a number of restrictions on what a data list structure can contain. A data list structure is made up of data lists and can be more complicated than a tree since the name of a sublist can appear in the structure.

The smallest executable unit is a single instruction which always has the standard form: *P Q SYMB LINK*, where the *LINK* defines the location of the next instruction and the rest shows the operation to be performed. Single instructions can be combined to form a program list which can be reentrant. A program list can have regional symbols as *LINK*s and names of list cells as *SYMB*. Programmed lists can be combined into a higher executable unit called a *routine*; this is a list structure with one programmed list (called the *main list*) that has a regional or internal name while all the other lists are called *sublists* and have local names. The program list cannot have a description list associated with it. A program is a set of routines. Every routine may be a subroutine or a main program; no special linkage conventions are needed.

There are three types of cells defined relative to a routine. A storage cell is called *safe* over a routine if the routine leaves the symbol in the cell (and the pushdown list) the same as it was prior to beginning the routine, except where the latter calls for modification. A routine can have input and output symbols. By convention, inputs for a routine are placed in a special storage cell called the *communication cell* which is designated H0. If there are multiple inputs, they are in the pushdown list of H0 in the sequence determined by the definition of the routine. Similarly, multiple outputs are placed in the pushdown list. The communication cell must be safe over all routines, which means that a routine must remove all the input symbols from the communication pushdown list before the routine terminates. The outputs of course are to be left in H0. A cell called the *test cell* and designated H5 is used to record *yes* or *no* information, where the symbols + and − are used, respectively; they are denoted internally by J4 and J3.

The *Q* prefix is called the *designation operation* in an instruction. The operand of the *Q* prefix is *SYMB* and the result is a new symbol *S*. *Q* can have eight values, including the provision for one or two levels of indirect addressing (denoted by 1 and 2) and tracing provisions (designated by 3 and 4).

The *P* prefix actually specifies the operation in an instruction. The eight *P* operations on *S* permit the execution of subroutines ($P=0$), the transferral of symbols to and from the communication cell H0 ($P=1, 2, 5$, and 6), push or pop the pushdown list of *S* to *S* ($P=3, 4$), and transfer control to *S* if H5 has a minus sign ($P=7$); the last is thus a conditional transfer.

The heart of the system is the set of *basic processes* (designated by J).

Some of these are primitive, and some are elementary routines built up from the primitives (i.e., coded in IPL-V). (In the definition of the J processes, we let (0) represent the symbol in H0, (1) represents the first symbol beneath the top of the pushdown list, (2) represents the second symbol from the top of the pushdown list, etc.) There are system regions and system cells which are used by these basic processes. The regions H, J, and W cannot have new symbols defined by the programmer. There are a number of specified cells which have special functions, e.g., H0, ..., H5, W0, ..., W43. As illustrations of how these are used, H3 is a tally of the interpretation cycles executed, W11 is the remainder of integer division, W21 holds the name of the integer determining the print column, W28 holds the symbol indicating the cause of the trap, and W38 holds the name of the routine which tests whether slow auxiliary storage should be compacted at this time and compacts it if *yes*.

While not all the basic titles of the J routines are self-explanatory, the general categories are described here in order to give a feel for what is available. The complete list (by title) is shown in Figure VI-5.

General processes J0–J9. These are shown completely to give the general idea.

- J0 *NO OPERATION*. Proceed to the next instruction.
- J1 *EXECUTE* (0). The process (0) is removed from H0, H0 is restored (this positions the process's inputs correctly), and the process is executed (as if its name occurred in the instruction instead of J1).
- J2 *TEST IF* (0) = (1). (The identity test is on the *SYMB* part only; P and Q are ignored.) H5 is set + or − for true or false, respectively.
- J3 *SET* H5−. The symbol in H5 is replaced by the symbol J3.
- J4 *SET* H5+. The symbol in H5 is replaced by the symbol J4.
- J5 *REVERSE* H5. If H5 is +, it is set −; if H5 is −, it is set +.
- J6 *REVERSE* (0) and (1). Permutes the symbol in H0 with the first symbol down in the H0 pushdown list.
- J7 *HALT, PROCEED ON GO*. The computer stops; if started again, it interprets the next instruction in sequence.
- J8 *RESTORE* H0. Identical to 30H0, but can be executed as *LINK*.
- J9 *ERASE CELL* (0). The cell whose name is (0) is returned to the available space list, without regard to the contents of the cell.

Description processes (J10–J16). The name of the describable list, rather than the name of the description list, is the input. The name of the

```
* Indicates processes which set H5
General Processes (§ 5.0)
  J0    No operation
  J1    Execute (0) after restoring H0
 *J2    TEST (0) = (1)
 *J3    Set H5-
 *J4    Set H5+
 *J5    Reverse sense of H5
  J6    Reverse (0) and (1)
  J7    Halt, proceed on GO
  J8    Restore H0
  J9    ERASE cell (0)
Description Processes (§ 6.0)
 *J10   FIND value of attribute (0) of (1)
  J11   Assign (1) as value of attribute (0) of (2)
  J12   Add (1) at front of value list of attribute
        (0) of (2)
  J13   Add (1) at end of value list of attribute
        (0) of (2)
  J14   ERASE attribute (0) of (1)
  J15   ERASE all attributes of (0)
 *J16   FIND attribute of (0) randomly
Generator Housekeeping Processes (§ 7.1)
  J17   Gen set up: context (0), subprocess (1)
 *J18   Execute subprocess of Gen
 *J19   Gen clean up
Working Storage Processes (§ 8.0)
  J2n   MOVE (0)-(n) into W0-Wn
  J3n   Restore W0-Wn
  J4n   Preserve W0-Wn
  J5n   Preserve W0-Wn; MOVE (0)-(n) into W0-Wn
List Processes (§ 9.8)
 *J60   LOCATE next symbol after cell (0)
 *J61   LOCATE last symbol on list (0)
 *J62   LOCATE (0) on list (1) (1st occurrence)
  J63   INSERT (0) before symbol in cell (1)
  J64   INSERT (0) after symbol in cell (1)
  J65   INSERT (0) at end of list (1)
  J66   INSERT (0) at end if not on list (1)
  J67   Replace (1) by (0) on list (2) (1st occur.)
 *J68   DELETE symbol in cell (0)
 *J69   DELETE (0) from list (1) (1st occurrence)
 *J70   DELETE last symbol from list (0)
  J71   ERASE list (0)
  J72   ERASE list structure (0)
  J73   COPY list (0)
  J74   COPY list structure (0)
  J75   Divide list after location (0); name of
        remainder is output (0)
 *J76   INSERT list (0) after (1), locate last symbol
 *J77   TEST if (0) is on list (1)
 *J78   TEST if list (0) is not empty
 *J79   TEST if cell (0) is not empty
 *J8n   FIND the nth symbol on list (0)
  J9n   Create list of n symbols, (n-1) to (0)
 *J100  Gen symbols on list (1) for (0)
 *J101  Gen cells of list structure (1) for (0)
 *J102  Gen cells of tree (1) for (0)
 *J103  Gen cells of block (1) for (0)
  J104
Auxiliary Storage Processes (§ 10.1)
 *J105  MOVE list structure (0) in from auxiliary
  J106  File list structure (0) in fast-auxiliary
  J107  File list structure (0) in slow-auxiliary
 *J108  TEST if list structure (0) is on auxiliary
  J109  Compact auxiliary data storage system (0)
Arithmetic Processes (§ 11.0)
  J110  (1) + (2) → (0), leave (0)
  J111  (1) - (2) → (0), leave (0)
  J112  (1) x (2) → (0), leave (0)
  J113  (1) / (2) → (0), leave (0)
 *J114  TEST if (0) = (1)
 *J115  TEST if (0) > (1)
 *J116  TEST if (0) < (1)
 *J117  TEST if (0) = 0
 *J118  TEST if (0) > 0
 *J119  TEST if (0) < 0
  J120  COPY (0)
  J121  Set (0) identical to (1), leave (0)
  J122  Take absolute value of (0), leave (0)
  J123  Take negative of (0), leave (0)
  J124  Clear (0), leave (0)
  J125  Tally 1 in (0), leave (0)
  J126  Count list (0)
 *J127  TEST if data type (0) = data type (1)
  J128  Translate (0) to be data type of (1),
        leave (0)
  J129  Produce random number between 0 and (0)

Data Prefix Processes (§ 12.2)
 *J130  TEST if (0) is regional symbol
 *J131  TEST if (0) names data term
 *J132  TEST if (0) is local symbol
 *J133  TEST if list (0) has been marked processed
 *J134  TEST if (0) is internal symbol
  J135
  J136  Make (0) local, leave (0)
  J137  Mark list (0) processed, leave (0)
  J138  Make (0) internal, leave (0)
  J139
Read and Write Processes (§ 14.0)
 *J140  Read list structure
 *J141  Read symbol from console
  J142  Write list structure (0)
  J143  Rewind tape (0)
  J144  Skip to next tape file
  J145  Write end-of-file
  J146  Write end-of-set
Monitor System (§ 15.6)
  J147  Mark routine (0) to trace
  J148  Mark routine (0) to propagate trace
  J149  Mark routine (0) to not trace
Print Processes (§ 16.1, 16.2)
  J150  Print list structure (0)
  J151  Print list (0)
  J152  Print symbol (0)
  J153  Print data term (0) w/o name or type
  J154  Clear print line
  J155  Print line
 *J156  Enter symbol (0) left-justified
 *J157  Enter data term (0) left-justified
 *J158  Enter symbol (0) right-justified
 *J159  Enter data term (0) right-justified
  J160  Tab to column (0)
  J161  Increment column by (0)
 *J162  Enter (0) according to format W43
  J163
  J164
In-process Loading (§ 19.0)
  J165  Load routines and data
Save for Restart (§ 20.0)
 *J166  Save on unit (0) for restart
 *J167  Skip list structure
  J168
  J169
Error Trap (§ 21.0)
  J170  Trap on (0)
Block Handling Processes (§ 17.0)
  J171  Return unused regionals to H2
  J172  Make block (0) into a list
 *J173  Read into block (0)
 *J174  Write block (0)
 *J175  FIND region control word of regional symbol
        (0)
  J176  Space (0) blocks on unit 1W19
  J177
  J178
  J179
Line Read Processes (§ 22.0)
 *J180  Read line
 *J181  Input line symbol
 *J182  Input line data term (0)
 *J183  Set (0) to next blank, leave (0)
 *J184  Set (0) to next non-blank, leave (0)
 *J185  Set (1) to next occurrence of character
        (0), leave (0)
 *J186  Input line character
  J187
  J188
 *J189  Transfer field to line (0)
Partial Word Processes (§ 23.0)
  J190  Input P of cell (0)
  J191  Input Q of cell (0)
  J192  Input SYMB of cell (0)
  J193  Input LINK of cell (0)
  J194  Set (1) to be P of cell (0)
  J195  Set (1) to be Q of cell (0)
  J196  Set (1) to be SYMB of cell (0)
  J197  Set (1) to be LINK of cell (0)
  J198
  J199
Miscellaneous Processes (§ 24.0)
 *J200  LOCATE (0)th symbol on list (1)
  J201  ERASE routine (0)
  J202  Print post mortem and continue
```

Figure VI-5. List of IPL-V basic processes.
Source: Newell *et al.* (eds.) [NW65], p. 263.

description list is found in the head of the describable list and is a local symbol whenever it is created by these processes.

Generator housekeeping processes (J17–J19). Generators permit repetitive operations by producing a sequence of outputs and applying to each a specified process. The process that the generator applies is called the *subprocess* and is an input. Generators are different from the other processes because the generator and its sub- or superprocesses must coexist at the same time. Hence the normal hierarchy of routines and subroutines is violated and care must be taken to see that each routine is always working in its appropriate context.

Working storage processes (J2n–J5n). The programmer can create storage cells for either permanent or temporary use. The advantage in using the W's lies in having 40 processes to manipulate them.

List processes (J60–J104). In these commands, the locate operation produces an output which is the name of the cell containing the desired symbol, and it sets H5 to + if the symbol is located and to − if it is not located; in the latter case the output is the name of the last list cell. In the case of insertion, there are no outputs. For deletion, the symbol is removed and H5 is set negative if the symbol did not exist, otherwise it is set at plus. When a structure of any kind is erased, all the cells on it are returned to available space and there is no output.

Auxiliary storage processes (J105–J109). The system provides both fast and slow auxiliary storage, and it also separates auxiliary storage for data structures and for routines. A list structure can be filed in an auxiliary storage by the programmer; but when he does this, the information is no longer available in major storage and cannot be accessed directly by the J processes. A filed list structure may be moved back into main storage.

Arithmetic processes (J110–J129). All the input and output symbols in this category are the names of data terms. In the arithmetic operations, if both elements are integers, the result is integer; otherwise, the result is in floating point.

Data prefix processes (J130–J139). Since the basic technique in processing list structures is recursion, there must be a way of avoiding multiple processing of the same list (when this is not desired). These processes provide relevant facilities.

Read and write processes (J140–J146). Only data list structures can be input or output by these processes.

Monitor system (= *program tracing*) (J147–J149). These processes provide quite powerful tracing facilities.

Print processes (J150–J162). The printing processes provided are for units of data and for a line of information. There are actually quite flexible facilities for controlling the format of these, as well as the form of arithmetic.

There are a number of other processes available but they become heavily involved with actually loading the program or other very specialized areas. There are also partial word processes, however, which read individual lines, partial words, etc.

VI.3.4. SIGNIFICANT CONTRIBUTION TO TECHNOLOGY

IPL has made major contributions to the technology of programming in general, as well as to programming languages. In both categories there is obviously the concept of a list, its implementation in a practical system, and an initial solution to a number of problems. The development of the pushdown and popup operations has contributed not only to general programming techniques but also to the language terminology. In some ways, the subroutine hierarchy implementation and usage in IPL is easier than in languages such as FORTRAN or ALGOL since the user pays less in setup and execution inefficiencies from subroutine calls; on the other hand, somewhat more cleanup is required in IPL. Attribute value lists have turned out to be significant data structures in their own right. As secondary benefits, the availability of this system has permitted the attack of problems in artificial intelligence that could not be approached any other way.

In a more general framework, IPL introduced a concept and provided a consistent means for dynamically expanding the data areas needed during the running of a program.

VI.4. L⁶

L^6 (standing for Bell Telephone Laboratories *Low-Level Linked List Language*) was developed in 1965 by Kenneth C. Knowlton (Bell Telephone Laboratories). The original implementation was on the IBM 7094; since then, versions have been developed for at least the 7040, 360, MOBIDIC B, PDP 6, and SDS 940. Implementations on a number of other machines have been planned. The most complete description is by Knowlton [KO66].

L^6 is a borderline case in being considered a language within the scope of this book. It is really used to build a system (language) to solve problems. For that reason, relatively little detail will be given. L^6 is similar in format to an assembly program but in a very different way than IPL-V. It provides the

SAMPLE PROGRAMS—L⁶

Problem: Reverse the list pointed to by bug W, and connected by A-fields.

Program:
```
     THEN (S,FC,X)(S,FC,Y)(Y,E,0)
LOOP IF   (W,N,0)THEN(X,P,W)(W,P,WA)(XA,P,X)(Y,P,X)LOOP
     THEN (W,P,Y)(R,FC,Y)(R,FC,X)
```

Explanation: In first line, save on field-contents pushdown bugs X and Y so they can be used temporarily; set Y to 0. Second line: if W not zero (i.e., there is one more block on list) let bug X point to first block by (X,P,W), then W bypasses first block (W,P,WA), X's block's A field points to Y's current list and Y points to X's block; then try again. Third line: W finally points to Y's reversed list and Y and X are restored from field-contents pushdown.

Problem: Test whether two list structures are equal, where "equal" means they have the same hierarchical structure and the same elementary symbols in corresponding positions. Each node is assumed to have three fields: R holds pointer to right subtree or zero, L holds pointer to left subtree or zero, S contains a symbol. List structures to be compared are pointed to by bugs X and Y. Exit is to the place just beyond the call if structures are equal; otherwise to the designated failure exit.

Program:
```
TEST IFALL (X,E,0)(Y,E,0)DONE
     IFALL (X,E,0)(Y,N,0)FAIL
     IFALL (X,N,0)(Y,E,0)FAIL
     IF    (XS,N,YS)FAIL
     THEN  (S,FC,X)(S,FC,Y)(X,P,XL)(Y,P,YL)(FL,DO,TEST)(R,FC,Y)(R,FC,X)
     THEN  (S,FC,X)(S,FC,Y)(X,P,XR)(Y,P,YR)(FL,DO,TEST)(R,FC,Y)(R,FC,X)DONE
FL   THEN  (R,FC,Y)(R,FC,X)FAIL
```

Explanation: Line 1: if both lists are null, test succeeds. Lines 2 & 3: If one list is null but not the other, test fails. Line 4: if symbols of top blocks are different, test fails. Line 5: after saving pointers to top of lists, X and Y go to their respective left subtrees and perform the entire subroutine recursively, going to FL if subtrees are not equal, otherwise X and Y are restored. Line 6: likewise with right subtrees. Line FL: if subtrees were not equal, restore X and Y but use failure exit from this level.

---

user with very primitive facilities for defining storage allocation, and structures for lists and their pointers. The general structure of a program is a sequence of individual commands, each of which appears on a separate line and is highly symbolic. For example,

IF(XA,E,0)THEN(R,FC,X)DONE

provides a test and a resultant operation (described later). Actions and/or tests are performed from left to right.

The basic data element is a memory block, which can be defined by the user to contain 1, 2, 4, 8, ..., 128 words. The programmer can define up to 36 fields in blocks; the fields can vary in size up to a complete machine word, and they can overlap each other. A field which is large enough may contain a pointer to another block. The names of the fields can be single digits or single letters. The contents of a field can be alphanumeric, nonnegative integers, bit patterns (for use in logical operations), or a pointer to another block; in the latter case the field must be large enough to contain a machine address. L^6 also contains a set of 26 base fields, each 36 bits wide and named by a single letter; these are referred to as *bugs*. Access to all blocks and their data is achieved by pointers in the bugs. Remote fields are named by concatenating the bug name, names of fields holding successive pointers to blocks, and the name of the field in the block. Thus UVWXY means that base register U is pointing to a block with a field named V which points to a block with a field named W which points to a block with a field named X. X points to a block whose Y field contains the data being referenced.

There are two major classes of instructions in L^6: Tests and operations. The latter are further subdivided into: Setup storage and get and free blocks; copy blocks and fields; arithmetic operations; logical operations; shifts and bit counts; input/output and conversion; and pushdown and popup. The general format for all instructions is a sequence of arguments (most commonly three) separated by commas and enclosed in parentheses, e.g., (a,A,cd) means add to the contents of the field a either the contents of an indicated field c or a decimal number specified d. As another illustration, a field is defined by writing (cd_1,Df,cd_2,cd_3), where f is the name of the field being defined, cd_1 is the word of the block in which the field falls (0, 1, ..., 127), cd_2 is the leftmost bit (0, 1, ..., 35), and cd_3 the rightmost bit. For example,

THEN(0,DA,18,35)(1,DB,21,35)(1,DC,6,20)(0,DD,0,17)

defines a field named A in word 0 and occupies bits 18 to 35; a field B in word 1, occupying bits 21 through 35; etc. The summary of all instructions is shown in Figure VI-6.

Individual instructions and tests can be combined into conditional statements whose general form is

[*label*] *if-statement test-statements* THEN [*operation-statements*] [*label*]

The limitation on the number of individual statements permitted is that they must all fit on one card. The *if-statements* permitted are IFNONE, IFANY, IFALL, IFNALL, which refer to the truth of none, any, all, or not all the indicated tests. If a label is given at the end, control is transferred there, otherwise to the next line. If the operations are missing, the label is required.

I. Mnemonic Notation Used for Describing Elementary Tests and Operations—with Permissible Ranges of Arguments for 7094 L^6

Field Designators
- c "*c*ontents", i.e., designation of a field whose contents are used in a test or operation: either a bug, A, B, ..., Z, or a remote field, A0, A1, ..., ZZZZZZ.
- a "*a*ffected field", i.e., designation of a field whose contents are affected by an operation: a bug A, B, ..., Z, or remote $f$ield, A0, A1, ..., ZZZZZZ.

Names
- f name of a definable $f$ield: 0, 1, ..., 9, A, B, ..., Z
- s a program *s*ymbol (i.e., name of a program location)

Literals
- o an *o*ctal number specified directly: 0, 1, ..., 777777777777
- d a *d*ecimal number specified directly:
  0, 1, ..., 34359738267 ($2^{35} - 1$).
- h a *H*ollerith literal: 0, 1, ..., ZZZZZZ. Permissible characters are the ten digits, 26 letters and period. Other characters must be specified in terms of their octal equivalences.

Alternatives
- cd "*c*ontents or *d*ecimal"—i.e., either c or d as explained above.
- co "*c*ontents or *o*ctal"—i.e., either c or o as explained above.

II. Resumé of Tests and Operations of L^6—Lower-Case Mnemonics Are Explained in I.

A. Tests

Equality	*Inequality*	*Greater than*	*Less than*
(c, E, cd)	(c, N, cd)	(c, G, cd)	(c, L, cd)
(c, EO, o)	(c, NO, o)	(c, GO, o)	(c, LO, o)
(c, EH, h)	(c, NH, h)	(c, GH, h)	(c, LH, h)
Pointers to same block		*One-bits of*	*Zero-bits of*
(c_1, P, c_2)		(c, O, co)	(c, Z, co)
		(c, OD, d)	(c, ZD, d)
		(c, OH, h)	(c, ZH, h)

B. Operations

Setup Storage, Get and Free Blocks

Setup storage	*Define field*	*Get block*	*Free block*
(s_1, SS, d, s_2)	(cd_1, Df, cd_2, cd_3)	(a, GT, cd)	(a, FR, 0)
↑ size ↗ ↑	↑ word ↑ ↗	(a, GT, cd, a_2)	(a, FR, c)
└ first word ╲	first bit ╯		
last word ╯	last bit ╯		

Copy Blocks and Fields

Copy field	*Duplicate block*	*Interchange field contents*	*Point to same block as*
(a, E, cd)	(a, DP, c)	(a_1, IC, a_2)	(a, P, c)
(a, EO, o)			(a, Δ) ≡ abbrev.
(a, EH, h)			for (a, P, aΔ)

Figure VI-6. (cont. next page)

Figure VI-6. (cont.)

	Arithmetic Operations		
Add	*Subtract*	*Multiply*	*Divide*
(a, A, cd)	(a, S, cd)	(a, M, cd)	(a, V, cd)
(a, AO, o)	(a, SO, o)	(a, MO, o)	(a, VO, o)
(a, AH, h)	(a, SH, h)	(a, MH, h)	(a, VH, h)

	Logical Operations		
Logical or	*Logical and*	*Exclusive or*	*Complement*
(a, O, co)	(a, N, co)	(a, X, co)	(a, C, co)
(a, OD, d)	(a, ND, d)	(a, XD, d)	(a, CD, d)
(a, OH, h)	(a, NH, h)	(a, XH, h)	(a, CH, h)

	Shifts and Bit Counts		
Shift left	*Shift right*	*Locate bits*	*Count bits*
(a, L, cd)	(a, R, cd)	(a, LO, c)	(a, OS, c)
(a, L, cd, co)	(a, R, cd, co)	(a, LZ, c)	(a, ZS, c)
(a, LD, cd, d)	(a, RD, cd, d)	(a, RO, c)	
(a, LH, cd, h)	(a, RH, cd, h)	(a, RZ, c)	

	Input/Output and Conversion		
Input	*Print*	*Convert*	*Microfilm*
(a, IN, cd)	(cd, PR, co)	(a, BZ, c)	(cd_{xmin}, XR, cd_{xmax})
	(cd, PRH, h)	(a, ZB, c)	
Print List		(a, BD, c)	(cd_{ymin}, YR, cd_{ymax})
(c, PL, f)	*Punch*	(a, BO, c)	(cd_{x_0}, TvH, cd_{y_0}, co, cd)
(c, PL, f, cd)	(cd, PU, co)	(a, DB, c)	(cd_{x_0}, TvHH, cd_{y_0}, h, cd)
	(cd, PUH, h)	(a, OB, c)	(DO, ADVANC)

	Pushdown and Pop-Up Operations		
Save and restore field contents	*Save and restore field definition*	*Do subroutine* (DO, s) (s_2, DO, s)	*Go-To's for exiting from subroutine*
(S, FC, c)	(S, FD, f)	(DO, STATE)	DONE
(R, FC, c)	(R, FD, f)	(DO, DUMP)	FAIL

Figure VI-6. Summary of all L⁶ instructions.
Source: Knowlton [KO66], p. 619. By permission of Association for Computing Machinery, Inc.

The word **IF** can be used for **IFALL** and **NOT** for **IFNONE**. There is also an unconditional **THEN** which is followed only by operations. Operations can be omitted, in which case the label is required.

As an illustration,

LOOP2 IFNONE (XD,E,Y)(XA,E,0) THEN (XD,E,1) (X,P,XA) LOOP2

says that *if none* of the following is true

that the contents of XD (i.e., the D field of the block that bug X points to) *E*quals the contents of Y or that the contents of XA *E*quals 0

then perform the following operations:

set the contents of XD *e*qual to 1, make X *P*oint where current contents of XA point, then go to the instruction labeled LOOP2 (the same instruction in this case).

Otherwise no operations are to be performed and control goes to the next line of coding.

A subroutine can be invoked by using (DO,*label*) as an operation. The form (*failexit*,DO,*label*) specifies the place to which control should be transferred if the subroutine is exited by a special *go-to* FAIL. Subroutines can be recursive.

L^6 permits redefinition of the position of fields in blocks at object time.

Because L^6 provides the programmer with such detailed control over the arrangement of the size and arrangement of both lists and fields within them, it is very suitable for problems requiring maximum efficiency at object time. It may replace IPL-V in some applications.

VI.5. LISP 1.5

VI.5.1. HISTORY OF LISP 1.5

Work was started in 1959 by the Artificial Intelligence Group at M.I.T. under the direction of Professor John McCarthy to develop a programming system

> ... designed to facilitate experiments with a proposed system called the Advice Taker, whereby a machine could be instructed to handle declarative as well as imperative sentences and could exhibit "common sense" in carrying out its instructions. ... The main requirement was a programming system for manipulating expressions representing formalized declarative and imperative sentences so that the Advice Taker system could make deductions.
>
> In the course of its development the LISP system went through several stages of simplification and eventually came to be based on a scheme for representing the partial recursive functions of a class of symbolic expressions.[8]

This description of motivation is taken from the first published paper on LISP (McCarthy [MC60a]). The first manual (McCarthy *et al.* [MC60])

---

[8] McCarthy [MC60a], p. 184.

was published in March, 1960 with the following authors: J. McCarthy, R. Brayton, D. Edwards, P. Fox, L. Hodes, D. Luckham, K. Maling, D. Park, and S. Russell.

The work was originally done for the 704, whose word layout played a monumental although indirect role in the naming of two of the fundamental concepts in the LISP language, namely the now famous car (contents of the address register) and cdr (contents of the decrement register). (These acronyms had actually been used earlier in FLPL (see Gelernter, Hansen, and Gerberich [GE60]), but they did not have the significant role there that they had in LISP.) When the IBM 709 became available, a new version was prepared for it. Then, as with all other languages, ways of making significant improvements were discovered, and this led to the development of LISP 1.5 (see McCarthy *et al.* [MC62]). (The reason for this numbering scheme presumably was the realization by the developers that they had not reached a sufficiently major change in the language to warrant calling it LISP 2. However, as a result of this policy, later contemplated individual versions of the language became known as LISP 1.75, LISP 1.9, etc.) The development of LISP 2, which is discussed in Section VIII.6, is such a monumental change over the earlier versions that its placement in a completely different chapter is justified. As will be seen, many of the difficulties which existed with LISP 1.5 stemmed from its unusual and rather difficult notation; for several years this problem was recognized and small attempts at correcting this situation were made, such as providing small routines on top of the normal LISP system to do more reasonable input/output or allowing algebraic expressions to be written in a more natural form. The major change in LISP 2 is that the source language is based on ALGOL 60, with some LISP concepts added. (For further details, see Section VIII.6.)

LISP proved sufficiently popular among a small group of people so that it became implemented on a number of machines, although in most cases a subset of LISP 1.5 was implemented. For example, versions exist on at least the IBM 1620, the PDP-1, and the SDS 940. Several versions were made available on the 709/90/94, and one was put under M.I.T.'s CTSS (see Crisman [ZR65]); a version was also put under the SDC time-sharing system. LISP was released to SHARE.

LISP is unusual, in the sense that it clearly deviates from every other type of programming language that has ever been developed. Even IPL-V (see Section VI. 3) bore a reasonable similarity to the concept of an assembly program whose main instructions were really those of a mythical machine to do list processing. The theoretical concepts and implications of LISP far transcend its practical usage. However, there has been considerable practical usage of LISP by a certain group of devotees, who have done interesting work using it for symbolic integration and other forms of algebraic manipulation, theorem proving, solving geometry analogy problems, and contract bridge bidding, etc. (See the list of references at the end of the chapter.)

VI.5.2. FUNCTIONAL CHARACTERISTICS OF LISP 1.5

LISP is not a very general language, in the sense that it is ill-suited for anything except general symbol manipulation and list processing. Its hardware representation is succinct, formal, and highly unnatural. It is reasonably consistent but there are a number of particular cases which involve special rules. It is extremely difficult to read and write because of the existence of large numbers of parentheses and the problem of matching them; as a result, the use is extremely error prone. Depending on the mental set and receptivity of the audience, it is either easy or hard to learn.

The application area is primarily problems which have the following characteristics, although some may be more significant than others: (1) They require list processing; (2) they require significant amounts of recursion in the operations they want to perform; and (3) they are dealing with some type of symbol manipulation.

LISP is sufficiently unique in its concept and notation so that it does not really fit at all well into the classifications of programming languages given in Chapter I. About the only statement that can be made which is both true and relevant is that there are two forms of language, a hardware

SAMPLE PROGRAMS—LISP 1.5

Problem: Reverse a list

Program:
```
DEFINE ((
(REVERSE (LAMBDA (X) (COND
   ((NULL X) NIL)
   (T (APPEND (REVERSE (CDR X)) (LIST (CAR X)) )) ))) ))
```

Problem: Test whether two list structures are equal, where "equal" means they have the same hierarchical structure and the same elementary symbols in corresponding positions.

Program:
```
DEFINE ((
(EQUAL (LAMBDA (X Y) (COND
   ((ATOM X) (COND
        ((ATOM Y) (EQ X Y)) (T NIL)))
   ((ATOM Y) NIL)
   ((EQUAL (CAR X) (CAR Y)) (EQUAL (CDR X) (CDR Y)))
   (T NIL) ))) ))
```

language and another form which is really a reference language. Publications tend to be written in the hardware language, although this is certainly not necessary. The main elements of the *hardware language* are referred to as *S-expressions*, while those of the *reference language* are called *M-expressions*, standing for *symbolic* and *meta expressions*, respectively.

In my opinion, LISP is definitely *not* intended for use by either a novice programmer or a nonprogrammer; it requires a certain amount of sophisticated background to appreciate and use the language effectively.

LISP was definitely designed for use in a batch mode, although on-line versions have been developed for M.I.T.'s CTSS and SDC's Q32 time-sharing system, as well as for some other small systems.

The language is actually machine independent, even though its original implementation and conception were based on a machine order code which had two addresses in a single word. LISP is implementation dependent to a certain extent, although this can be minimized by using the various versions of LISP which have been written in their own language. The problem of dialects exists for LISP because virtually every group using it has made some modifications to the language and the system. The problem of extensions really disappears because LISP has self-extension capabilities built in, which will be seen later. The method for the language definition has not caused severe compatibility problems. Programs run on one system can usually be converted fairly easily to run on another.

There has certainly been no consideration of USASI standardization for LISP. There has not been any other kind of standardization control either, because almost every group that uses LISP has made some modifications to the language and to the system.

The designers of the language were essentially those people listed in the historical section. The objectives were those cited there also. The language designers and implementers have tended to be the same throughout the history of LISP, and there is no official maintenance of the language as such.

The language was originally defined simply through English language descriptions. However, in the LISP 1.5 manual, McCarthy *et al.* [MC62] (pp. 8 and 9) formally define the LISP syntax. However, that is such a small part of the system and its concepts that the authors who state "all parts of the LISP language have been explained. That which follows is a complete syntactic definition of the LISP language, together with semantic comments." do themselves quite an injustice with that comment. The LISP concepts are far more complex and powerful than those included on less than one page of material given in Backus Normal Form. For the interested reader, the syntactic definitions of the data and publication languages (which the authors refer to as the symbolic and metalanguages) are given in Figure VI-7. However, a more significant factor in the language definition is that the heart of the system, namely the interpreter (APPLY and EVAL),

Data Language

<LETTER> ::= A, B, C, ...
<number> ::= 0, 1, 2, ...
<atomic-symbol> ::= <LETTER> <atom part>
<atom part> ::= <empty> | <LETTER> <atom part> | <number> <atom part>
<S-expression> ::= <atomic symbol> | (<S-expression> . <S-expression>) |
                   (<S-expression> ... <S-expression>)

Meta-Language

<letter> ::= a, b, c, ...
<identifier> ::= <letter> <id part>
<id part> ::= <empty> | <letter> <id part> | <number> <id part>

The names of functions and variables are formed in the same manner as atomic symbols but with lower-case letters.

<form> ::= <constant> |
           <variable> |
           <function> [<argument> ; ... ; <argument>] |
           [<form> → <form> ; ... ; <form> → <form>]
<constant> ::= <S-expression>
<variable> ::= <identifier>
<argument> ::= <form>
<function> ::= <identifier> |
               λ [<var list> ; <form>] |
               label [<identifier> ; <function>]
<var list> ::= [<variable> ; ... ; <variable>]

Figure VI-7. Syntactic definition of LISP *S*- and *M*-expressions. Source: McCarthy *et al.* [MC62], pp. 8–9.

can be expressed in LISP; a good understanding of what the language is doing can be obtained by studying the short LISP program for these functions which appears in the manual.

The types of documentation have occurred in several forms. First, there have been two manuals put out by the design group, namely the LISP 1 and the LISP 1.5 manuals. Secondly, a series of informal and unpublished documents were created by members of the Artificial Intelligence Project at M.I.T. and then later by a similar group at Stanford University. Some of these memos provide significant information relative to the meaning, use, and implementation of LISP. Thirdly, there have been various tutorial papers whose ostensible purpose was to describe list processing in general but whose authors found it convenient to discuss the concepts of LISP primarily, e.g., Wilkes [WI64a] and Woodward [WX66]. There have also been informal technical papers distributed among the university groups using LISP. A few users' meetings were held, and material was distributed there also. A primer by Weissman [WE67] exists. The last and single most useful document about LISP is the one edited by Berkeley and Bobrow [BY66].

The evaluation of LISP as a language is difficult to separate from an evaluation of its implementations. In particular, the early implementations were all interpretive, and when this proved to be too slow, then compilers were written. People now use both and are tending more toward compilers. (However, several significant programs were written using the interpreter, e.g., Slagle's integration program [SL61] and Evans' geometry analogy program [EV64].) The great advantage of LISP is its ability to express in a meaningful way solutions to problems which people cannot handle any other way and to express them in a form which is natural to that class of problems. The greatest disadvantage to LISP, and probably the one that prevented its widespread use, is the notation that is used. This has been and is being corrected in LISP 2, which would seem to have a much more promising future for widespread use than any of the LISP $i(i < 2)$.

VI.5.3. TECHNICAL CHARACTERISTICS OF LISP 1.5

The character set for the hardware representation of LISP consists of the 26 letters, the 10 digits, and the following characters:

$$+ \quad - \quad * \quad / \quad (\quad) \quad = \quad , \quad . \quad \$ \quad \text{blank}$$

The arithmetic operators are *not* used in their normal way for arithmetic operations.

Identifiers consist of a letter followed by letters and/or digits; some implementations place a limit on the number of characters allowed. Identifiers are called atomic symbols. They cannot be subscripted but there are other ways of dealing with arrays. No blanks are permitted within an atom.

The concepts of LISP are so different from other languages that it is better to try to present the fundamental ideas rather than to try to follow the general outline of Chapter III.

The whole language is built on the concept of operators or functions, and operands. In most cases the program is a collection of function definitions, which relate to each other as a set of subroutines do. A program is executed by invoking one function (which in turn invokes the others). While this is an oversimplification, it serves to convey the flavor of the language. The motivation for this concept, which is the lambda calculus, and its implications for computability theory, are beyond the scope of this book.

The fundamental data element in LISP is the *atom*, which is either an atomic symbol (i.e., identifier) or a number. All data and programs in LISP are written as *S*-expressions, which are binary tree structures built up from atoms. An *S*-expression can be empty, in which case it is written as

$$(\ )$$

or more commonly, as NIL. For example,

A
(A)
(A B C)
(A (B C) D)

are *S*-expressions.

Lists may have sublists; e.g., (A B C) is a list with three elements, whereas (A (B C)) is a list of two elements, A and (B C). The second element itself consists of two elements, namely B and C. Each expression in list notation has an internal computer representation.

The language for expressing operations on the *S*-expressions is called the metalanguage, and the legitimate strings which can be written are called *M*-expressions. There are five elementary functions which can be applied to *S*-expressions: *car*, *cdr*, *cons*, *eq*, and *atom*. In the metalanguage (i.e., in *M*-expressions), the arguments of these functions are enclosed in square brackets and separated by semicolons, and the functions are written in small letters. The *car* and *cdr* are functions to define, respectively, the first element of a list (i.e., the head), and the rest of the list after the first element (i.e., the tail). Thus

$$car[(A\ B)] = A$$
$$car[((A\ B)\ C)] = (A\ B)$$
$$cdr[((A\ B)\ C)] = (C)$$
$$cdr[(A\ B\ C)] = (B\ C)$$

The action of *car* and *cdr* on atomic elements is undefined. Because it is often useful to compose the operation of these functions, e.g., *car*[*car*[*cdr*(*S*-*expr*)]], abbreviations are defined. The single letter a or d can be used between the c and the r for each functional operation of this type to be performed. The application of the operators is from right to left. For example,

$$cadr[(A\ B\ C)] = car[cdr[(A\ B\ C)]] = B$$
$$caddr[A\ B\ C)] = C$$
$$cadadr[(A\ (B\ C)\ D)] = C$$

The third elementary function is *cons*, which is a *join* operation and is used to add an element to a list. It has two arguments. For example,

$$\text{cons}[(A;\text{NIL})] = (A)$$
$$\text{cons}[A;(B\ C)] = (A\ B\ C)$$
$$\text{cons}[\text{cons}[A;\text{NIL}];(B)] = ((A)B)$$

Note that cons[A;B] cannot be expressed as a list and is expressed using *dot notation* as (A . B). The use of the dot is a fundamental means of forming *S*-expressions. The *S*-expressions are used to represent data and are defined as follows: An *S*-expression is either an *atomic symbol*, or if e1 and e2 are *S*-expressions, then (e1 . e2) is an *S*-expression. (See Figure VI-7 for syntactic definition.)

LISP defines a predicate as a function whose value is *true* or *false*, denoted by T and NIL (or F), respectively. The predicate eq is a test for equality on atomic symbols and is undefined for equal nonatomic arguments. Thus

$$\text{eq}[A;A] = T$$
$$\text{eq}[A;B] = F$$
$$\text{eq}[A;(A)] = F$$
$$\text{eq}[(A);(B)] = F$$
$$\text{eq}[(A);(A)] = \text{undefined}$$

The predicate *atom* is true if the argument is atomic but false otherwise. Thus

$$\text{atom}[\text{ABCDEFGHIJKL}] = T$$
$$\text{atom}[\text{NIL}] = T$$
$$\text{atom}[(A\ B\ C)] = \text{NIL}$$
$$\text{atom}[\text{car}(A\ B\ C)] = T$$
$$\text{atom}[\text{cdr}(A\ B\ C)] = \text{NIL}$$

One of the useful concepts and notational devices introduced in LISP is the conditional expression, which provides branches in function definitions. A conditional expression is written as

$$[p_1 \rightarrow e_1;\ p_2 \rightarrow e_2;\ \ldots;\ p_n \rightarrow e_n]$$

where p_i has the value T or NIL, and e_i is any expression. The value of the entire conditional expression is the value of the e_i for the first true p_i. (This is of course equivalent to the more normal way of writing a conditional statement, namely IF p_1 THEN e_1 ELSE IF p_2 THEN e_2 ELSE ... IF p_n THEN e_n.) If none of the p_i are true, the value of the expression is undefined. The T or NIL can be used to designate truth or falsity. For example, if we have

$$\text{eq}[\text{car}[y];A] \rightarrow \text{cons}[B;\text{cdr}[y]]; T \rightarrow y]$$

then if y = (A B) the value is (B B). If y is of the form (A *anything*), the value is (B *anything*); whereas if y is of the form (*not* A, *anything*), the value is the value of y.

An important application of conditional expressions is in defining functions recursively. Thus, to define the function ff which selects the first atomic symbol of any given expression, we can write

$$ff[x] = [atom[x] \to x; T \to ff[car[x]]]$$

which is interpreted by noting first that if x is atomic, it is the answer; if it is not, then ff is to be applied to the first element in x. This process must terminate and will eventually produce the desired answer, e.g.,

$$ff[((A\ B)\ C)]$$

yields A as the result. However, this is really quite unrigorous, because the = has no meaning and the ff on the right side is not defined as being the same as the one on the left. This problem is dealt with later.

One of the fundamental conceptual inputs to LISP is Church's lambda calculus (Church [ZP41]). The need for this arises when we try to determine the meaning of something like

$$y^2 + 3x\ (3, 4)$$

It is not clear which number is associated with which variable. In Church's lambda notation, the expression $y^2 + 3x$ is called a *form*, and the lambda(λ) is introduced to provide binding (i.e., appropriate pairing) of the variables. Thus, if one writes

$$\lambda((x, y); y^2 + 3x)(3, 4)$$

the result will be 25. In a case like this, the function does not have a name and does not need one. However, if the function is to be recursive, there must be some way of providing its name. Recursive functions are achieved by the LISP functions *define* and *label*; the former assigns a name to a definition and the latter permits the name to be written in the definition.

We have now described two types of expressions: S and M, where the former have a specific internal computer representation. Since LISP functions have S-expressions as arguments and also since we wish to be able to represent M-expressions in the computer in a similar way, there is a method of converting M-expressions to S-expressions. Stated another way, S-notation is for data and M-notation is for programs and we wish to be able to represent programs as data. The rules can be summarized briefly as follows: Function names are changed to upper case, the word LAMBDA represents λ, COND is used with the p_i, e_i pairs, and constants (i.e., literals) are preceded by the word **QUOTE** (except for **T** and **NIL**, which need not be **QUOTE**d). Thus, for example,

$$[atom[x] \rightarrow x; T \rightarrow ff[car[x]]]$$

becomes

(COND ((ATOM X) X) ((T) (FF (CAR X))))

The execution of a LISP program is the evaluation of the function(s) given. A function of a list has as its value another list or an atom, i.e., the evaluation of an *S*-expression gives an *S*-expression. In order to know when to cause the evaluation to take place, the LISP operating system contains an operator known as *evalquote*. It is given a list of doublets, each consisting of a function and a list of arguments for that function. Since one of these functions can be *define*, programs may be brought in.

Looping can be performed through recursion and also through the program feature described later.

Numbers are represented as atoms and can be fixed or floating point. Hence

(1 2 A 5)
(ALPHA 327)

are legal *S*-expressions. There are a set of arithmetic functions, e.g., *plus, minus, difference, times, add1, sub1, max, min, recip, quotient, remainder, divide,* and *expt*. Unfortunately, the format for these is "Polish" (i.e., parenthesis free) notation. Thus, to express the computation A∗B+C, it is necessary to write

(PLUS (TIMES A B) C)

Although more than two operands for multiplication and addition are permitted, the inconvenience of this notation is obvious. The arithmetic functions can be used recursively, just as other functions can, as shown in the following definition of factorial:

DEFINE ((
  (FACTORIAL (LAMBDA (N) (COND ((ZEROP) N 1)
  (T (TIMES N (FACTORIAL (SUB1 N))))))
))

which is the machine input form for

$$n! = \lambda[[n][n=0 \rightarrow 1; T \rightarrow n \cdot [n-1]!]]$$

There are arithmetic predicates (which are truth-valued functions), e.g., *lessp, greaterp, zerop, onep, minusp, numberp, fixp, floatp* and additional functions, e.g., *logor, logand, logxor,* and *leftshift,* with the fairly obvious meanings.

LISP has a class of functions called *pseudo-functions*. These are executed for their effect on the system in memory, as well as for the value. The most important of these are *define,* which acts similarly to the idea of naming a function or subroutine in other programming languages; *input/output;* and *array,* which permits the user to allocate blocks of storage for data (i.e., *S*-expressions).

A number of other functions are available in the system, including *sub*stitute (an *S*-expression for occurrences of an atomic symbol) and various more general list operations. Other predicates are also available.

Machine language subroutines can be used.

Although one structure of a LISP program is the evaluation of a set of nested functions, there is also a *program* feature available. This is of the very general form

(PROG, *list of variables, sequence of statements and atomic symbols*)

Within this feature, there is an assignment statement, designated by SET or SETQ, and a control transfer GO which can be used only on the top level of PROG or immediately inside a COND at the top level. If none of the propositions in the conditional expression are true, then there is an automatic drop through to the next statement in sequence. A RETURN is written to designate the end of the program. The PROG function can be used recursively, so that a program can contain a program.

By using the DEFINE, the user is able to create new functions which have exactly the same system characteristics as those provided automatically.

The method of handling storage allocation in LISP is primarily done by implementation, but the very absence of a language facility is significant. One of the biggest problems of using a list processing language is the constant need to return unused memory locations to a *free list* which the system can then use again for new data. Unlike IPL and L^6, where the user must perform this task himself, LISP provides a *garbage collection* facility. This means that at periodic intervals the system automatically examines the memory to determine which locations are not currently in use and returns these to the free list. (It must be pointed out that this technique is a controversial one and has profound effect on the efficiency of the running system; however, from a pure language viewpoint, the most significant factor is that the user is freed from the responsibility of doing this.)

Although all the early versions were interpretive, there were some compilers built, and the trend seems to be shifting more that way. The most successful compilers have been written in LISP and have compiled themselves using an interpreter.

The greatest advantage to LISP is its recursive power and list processing capability for those who are able to appreciate these and use them properly. It is also easy to write interpreters for string and pattern-matching type facilities, e.g., CONVERT (see Guzman [ZH66]). The greatest disadvantages are the inconvenience of the notation in general (most particularly the arithmetic) and the enormous difficulty in keeping parentheses balanced. More recent implementations have added features which help to minimize this problem.

VI.5.4. Significant Contribution to Technology

The most significant contribution made by LISP seems to be its elegance and potential introduction to, and use in, developing a general theory of computability. The importance of this is very great, but beyond the scope of this book. LISP is unique among programming languages in every possible way, but its use has been limited to a relatively few diligent people who can cope with the notational inconveniences. A good description of many of the nuances of LISP is given in the review of LISP 2 by Weizenbaum [WZ67].

On a more practical level, the particular formalism for the conditional expression was new. The emphasis on functional notation and the provision for expressing recursive processes easily provided new ideas, although they have not actually been used elsewhere. LISP appears to have been the first language to allow Polish + and × operators with more than two operands—sometimes called *Cambridge Polish*. Although LISP can manipulate list structures easily, its essence is that it is a functional language.

LISP is the only higher level language (in this book) in which the internal representation of the program is defined to be exactly the same as that of the data; this makes it possible for a program X to create and execute a program Y or to operate upon itself.

One of LISP's most lasting contributions is a nonlanguage feature—namely the term and technique *garbage collection*, which refers to the system's method of automatically dealing with storage.

VI.6. COMIT

VI.6.1. History of COMIT

Work on COMIT for the IBM 704 started in 1957, and a brief and general description of the language appeared as early as December, 1957 (Yngve [YN57]). The COMIT system was designed and programmed at

M.I.T. as a joint project of the Mechanical Translation Group of the Research Laboratory of Electronics and the Computation Center. The people who worked on the development of COMIT include J. Bennett, S. Best, C. Bosche, W. Cooper, M. Greber, K. Hansen, F. Helwig, M. Kannel, K. Knowlton, G. Matthews, A. Siegel, M. Weinstein, and Dr. Victor Yngve, under the direction of the latter. The system was designed to provide the professional linguist with a programming system in which he could easily write the programs that he needed for his research. It will be shown that this motivation caused a certain number of terms and concepts to be brought over from linguistics into COMIT itself.

A shift was made to the 709/90 before the check-out of the 704 version was complete. The system was distributed through SHARE in September, 1961 and later made available on the IBM 7040/44. Complete control of the language remained with the design group, and so no other *complete* or *official* implementations exist. After some period of usage, the designers naturally found things that could be improved, and internal memoranda relating to COMIT II existed by late 1963. A partial version of COMIT II was operational under M.I.T.'s Compatible Time Sharing System (see Crisman [ZR65]) during the summer of 1965. Work continued on the development of COMIT II, with the base of operations shifting from M.I.T. to the University of Chicago when Yngve went there in 1965. Early in 1968, the COMIT II system on the 7090/94 was only lacking in some documentation before it could be released to SHARE. A version was being prepared for the IBM System/360.

Since COMIT II has only a few (although significant) improvements over the original system, the description will not distinguish between them except where the addition is a major one.

VI.6.2. FUNCTIONAL CHARACTERISTICS OF COMIT

Since COMIT is a language whose format is unlike most others, it is desirable to state the fundamental idea and show an example of a COMIT *rule* before trying to make any comments about the properties of the language. COMIT is a string processing language which introduced the concept of a *rewrite* rule to transform the string in a manner determined by specifying a pattern. The following is an illustration of a COMIT rule:

FIND BOY + $ + , + $ = 2 + 3 + GIRL // *WSMI 2, NEXT

This means that the label FIND is assigned to a statement which examines a string to see if it consists of the word BOY followed by any number of words, then followed by a comma followed by any number of words; if so, the word BOY is removed and the word GIRL is placed after the comma and the rest of the string is deleted; the material now preceding the comma, and the comma, are written out on tape and control is transferred to the statement labeled **NEXT**. At first glance this language appears highly formal and difficult to read and write. In actual practice, it is fairly simple in notation, compact, and both succinct and quite natural within its class of applications. Key design objectives were naturalness, ease of learning, and ease of use, all within the framework of doing language data processing, and these were definitely achieved.

SAMPLE PROGRAM—COMIT II

```
            COMIT II    EDITOR
        (THIS PROGRAM READS IN TEXT AND PRINTS N CHARACTERS PER LINE. EXTRA)
        (BLANKS ARE INSERTED BETWEEN WORDS TO FILL OUT LINES.)
        (READ IN N)
*           $=                          //*RCK1                              *
*           $+*. = 1                    //*K1                              READ
R1          $1+$+*. = 1+2+—+2+—        //*Q1 4 5 (COPY ON SHELF 1)  *
READ        $= 1+$0                    //*RCK2                              R1
*           $1 = N/*CSN1 + —*. —PRINT—THE—FOLLOWING—STRING— + 1
                    + — —CHARACTERS—PER—LINE.*.—*.   //*WAM2 3 4  *
        (PRINT INPUT TEXT)
PRINT       $1+$80 = 1+2+*.            //*WAM2 3                         PRINT
*           $1+$ = 1+2+*.—*.—*.       //*WAM2 3,*A1 2                      *
COPYN       N = 1+1                    //*S2 1 (SAVE N ON SHELF 2)  *
TEST        N/.GO+$1 = 1/.D1+2         //*Q3 2 (CHECK FOR N CHAR.)   TEST
*           N/.GO =                    //*A3 1                         LASTLINE
.*          $0+ N+—+$ = 4              //*S1 1,*A3 1 (CHAR N+1 BLANK)  WRITE
*           N+$=                       //*S1 2,*A3 1                       *
*           $0 = K/.0                                                      *
FINDK       K+$+$——+$0 = 1/.I1+2+3    //*S1 3                           FINDK
*           K+$+—+$0 = 1/.I1+2                                      INSERT+WRITE
*                                                                          ERR
WRITE       $= —+1+*.                  //*WAM1 2 3,*A2 1,*A1 2           COPYN
        (SUBROUTINE TO INTERSPERSE K BLANKS IN A LINE)
INSERT      $0+K/.0 = 0                (GO IF K=0)                          +
*           — =                        (AT LEAST 1 BLANK)                 LOOP
*           $1 = 0                     (NO BLANK)                           +
LOOP        $1+$+— = 1+2+3+—          //*Q4 2 3 4                        TESTB
*           $ = $0+1                   //*A4 1                            LOOP
TESTB       $0+K+— =                   //*Q4 3 (REMOVE ALL LEADING BLANKS)  TESTB
*           $0+K/.GO = 2/.D1                                              LOOP
*           K+$ =                      //*A4 1                              +
LASTLINE $ = —+1+*.                    //*WAM1 2 3                         END
ERR         $ = —*.—****—EDITOR—CANNOT—PRINT—LINE—BECAUSE—N—IS—
*.                  TOO—SMALL.—        //*WAM1
END                                                                        *
            END
```

While COMIT was originally designed as a tool for use by linguists in their research, it has actually received usage outside this area. It has been used effectively for experimental or small problems in symbol manipulation, information retrieval, and other areas which will be discussed later. It is a procedure-oriented, problem-oriented, problem-solving language. However, relative to most other programming languages, its fundamental executable unit—namely the *rule*—is fairly nonprocedural. It is simultaneously the reference, publication, and hardware language. The user was primarily intended to be a nonprogrammer and, in particular, a linguist.

The basic language design is for use in a batch environment. However, the version of COMIT II released to SHARE was put under M.I.T.'s Compatible Time Sharing System. Nevertheless, COMIT is no more of an on-line language than any of the other systems which were implemented under CTSS.

From a completely external view, COMIT is quite machine independent, although its usage is highly implementation dependent. Since the system is partly interpretive, it is more appropriate to use the term *implementation dependent* than *compiler dependent*. There are no dialects of COMIT because control was kept within the hands of the development group. Similarly, there are no significant subsets and extensions, except in the sense that COMIT II is an extension of COMIT I. Any programs which are legal in the latter are legal in the former. Both systems have been implemented on the 7090/94 and the 7040/44, and programs have been successfully interchanged. Aside from this, there has been no real attempt at conversion because the development group has not put this on any other machine.

Detailed flowcharts for COMIT were made available to some groups contemplating implementation on another computer so that new implementations could maintain source language compatibility. As far as known, however, no compatible implementation was made except the COMIT I 7040/44 version, which was done by revising the 709/90/94 code. Several partial and noncompatible versions were developed on other machines, e.g., LECOM on the GE 225 (Hilton and Hillman [HT66]). A compatible implementation for the IBM System/360 is reported to be underway at Washington State University.

Source language compatibility between different 709/90/94/40/44 installations was maintained at a high level by only distributing COMIT as an assembled binary program. Symbolic code and flowcharts were not made available to local systems programmers. These measures succeeded relatively well in maintaining compatibility standards but retarded the addition of COMIT to the various operating systems that were in use and undoubtedly alienated some systems programmers. For COMIT II the matter has been solved by providing a separate interface program that can be changed to suit the local machine configuration and operating system.

The compiler-interpreter is just a subroutine for this interface. Complete interface documentation and assembly language code is being provided for COMIT II.

There has been no consideration of official USASI standardization of COMIT; this would really be either trivial and/or unnecessary because of the control kept by the development group.

The people who designed the language and first system were listed under the description of the history. The primary objective was to provide a tool for research linguists. The language designers implemented and maintained the language; this does not mean that there has not been a turnover in personnel, but Yngve has remained as the supervisor throughout the effort.

The language definition itself is written in English, with no attempt at rigorous formalized notation, and there is considerable ambiguity in some places. It is, however, in the documentation available to the user that COMIT really shines. (A general discussion is given in Yngve [YN63b].) The two main sources originally were the *Introduction* [MT61a] and *Reference Manual* [MT61]. These were originally issued in 1961 and 1960, respectively, with a few minor updates since then. The reason that this type of documentation is significant is that the introductory manual is really quite informative; it gives people an understanding of the basic concepts of the language; the reference manual is valuable for those who understand the language but need a detailed description of the operations. It does happen however that a few concepts which are covered in the introductory manual are either not in the reference manual or are actually given in somewhat less detail. There is also an operator's manual containing just the information needed by the person at the computer console. The introductory manual gives numerous problems, with answers in the back of the book, and other problems for which a grader program is available. The manuals have been successfully used in a number of courses. Although I do not agree, Yngve states "We consider the very complete diagnostics provided by COMIT as actually part of the documentation".[9] It is certainly true that the diagnostics provided by the system when a program is actually run are good; however, I cannot agree with the conclusion that the diagnostics are really part of the documentation, even though the notation of the diagnostic comments is consistent with the notation used in the manuals. (As a counterstatement to my disagreement, Yngve indicated[10] that the diagnostics assist the novice who might otherwise jump to the wrong conclusion from an incomplete description in the introductory manual which avoids bogging him down with complete explanations.) Yngve also states "The reference manual is quite complete and explicit on syntax, but it is not complete enough on semantics to be

---

[9] Yngve [YN63b], p. 84.
[10] Private communication, June, 1967.

used as a basis for implementing the language on other machines. For such purposes the detailed flowcharts will also have to be studied."[11]

In this instance, the language and the method of translating and executing it are so intricately linked together that a separate evaluation is not reasonable.

In my opinion, the primary errors of omission are the lack of reasonable arithmetic facilities and the inability to conveniently and naturally assign names to strings. Certain other features, such as providing indirect naming for variables and the ability to access substrings to any reasonable depth, are also wanting; they have been provided in SNOBOL, which is discussed in Section VI.7.

VI.6.3. Technical Characteristics of COMIT

The character set in COMIT consists of the 26 capital letters, the 10 digits, and the following characters:

$$* \quad / \quad \$ \quad + \quad - \quad = \quad (\quad) \quad . \quad , \quad \text{blank}$$

In COMIT II, a special feature allows the use of the apostrophe, which replaces the hyphen on some key punches. It is essential to note at this early stage in the description that virtually none of these characters has the usual arithmetic meaning.

Identifiers can be up to 12 characters in length and can consist of letters and numbers and periods or hyphens, although the latter two cannot be at the beginning or end. Up to 10 periods and/or hyphens can be used.

The operations of addition, subtraction, multiplication, and division of integers are available, but they are denoted by symbols other than the normal arithmetic ones (even though the latter are in the character set). It was deemed more important by the designers to use the arithmetic symbols for purposes more central to the basically nonnumeric character of the language. In particular, since the + sign was normally used by linguists to denote concatenation, it was not used for addition. There are also comparison operators for integers, namely greater than, less than, and equals.

There are no commands in the usual sense of the word. COMIT is basically a language for handling strings and patterns. Everything to be done is embodied within a *rule* which depends primarily upon notation and position within the rule for specifying the action to be taken. (In this sense, COMIT is slightly more nonprocedural than most of the numerically oriented programming languages.) There are no data names as such because COMIT is designed to deal only with strings of text; they are in either a fixed storage

---

[11] Yngve [YN63b], p. 84.

area called the *workspace* or they can be stored in some temporary storage areas called *shelves*. Since the basic input to a COMIT program is a string of text and operations are performed on the string in the workspace, there is less need to provide data names. (However, such a facility does turn out to be very useful.) COMIT uses the terms *numerical* and *logical subscripts*, but these bear little relation to subscripts as normally used; their meaning will be described later. Although the COMIT subscripts can be used to denote arrays, this is seldom done. There is no hierarchical data structure and hence no qualification. There are statement labels and a certain number of key words; but since there are no variable names, there is no conflict. In order to represent as literals either the 10 digits or the 8 special characters, + − = * $ / ( ), the user must precede them by an asterisk. The reason for this is to allow the data strings read in or put out to be completely arbitrary and, simultaneously, to permit the specification of unambiguous rules and patterns. Spaces are critical in certain places. Punctuation characters are not used in the usual way, although they are certainly sprinkled throughout the language. There are no noise words.

The COMIT program is written with some fixed requirements, and the notation is certainly highly symbolic. Normally COMIT rules are punched one to a card. Each rule must always start at the beginning of the card with a nonblank character in column 1. A rule can extend onto more than one card if necessary; in such a case there must be a hyphen in the last nonblank character before column 73 on the card.

A COMIT rule has five sections and can be represented schematically as follows:

$$\text{name} \quad \text{left-half} \quad = \quad \text{right-half} \quad // \quad \text{routing} \quad \text{go-to}$$

The fundamental purpose of the *left-half* part of the rule is to specify a pattern which is to be searched for in the data in the workspace. If this pattern is found, then something is shown in the *right-half* part of the rule to indicate what is to be done with it, the *right-half* is executed, and control then depends on the contents of the *go-to* section. This "rewriting" of a string is the central concept in COMIT. If the pattern is not found, then control is automatically transferred to the next rule in sequence.

The *name* section, which is really a statement label, begins in column 1 and must be followed by one or more spaces. (An asterisk can be used instead of a name.) The *go-to* is defined as the last string of nonblank characters on the card and is preceded by one or more blanks. The other sections are optional except that some combinations are required together. The *left-half* is usually terminated by an equals sign. The *routing* section is always preceded by the two slashes.

There are no declarations in COMIT, nor is there any logical need for them. The smallest executable unit is the rule, whose format has just been shown above. There are two kinds of rules: Grammar rules, which may involve program branches under the control of information in the dispatcher (described later), and list rules, which may be used for dictionary lookup operations. Both kinds of rules can include a variety of instructions for operating on the data and affecting the flow of control and both may have a number of subrules.

Loops consisting of any number of rules can be written, and there are several ways of exiting from them, including presence or absence of specific data in the workspace, comparison tests on integers, and choice of subrule.

Closed subroutines can be written and used in COMIT. There are no functions. Comments enclosed in parentheses may be inserted in the rule format wherever spaces are allowed.

The program consists of a sequence of rules preceded by a title card and followed by an end card. In COMIT II, there may be one or more *COMSET* cards after the title card. These have various functions such as setting the limits on the length of the run, changing margin settings for printed output, and establishing options where the standard values may be inappropriate.

From a language point of view, the only interaction with the operating system and the environment is through the *COMSET* cards.

The rules are delimited by the format described earlier. They are recursive only in the sense that the same rule can be repeated many times before the pattern in the *left-half* fails to be found and control proceeds to the next rule. The user can easily program recursive subroutines because there is a built-in pushdown list for the returns and the *workspace* and *shelves* (discussed later) can be used for data pushdown lists. The only way of passing parameters to a subroutine is by leaving them in the workspace or on specified shelves.

There is a provision for binary-coded subroutines to be used with COMIT II programs. (This is one of the significant additions to COMIT I.)

There are no arithmetic variables directly permitted in COMIT, although a very awkward equivalent is available through the use of subscripts; only integers are allowed. The only type of variable allowed is a string of characters. The characters are themselves organized into groups called *constituents*. Constituents consist of either individual characters or groups of characters and may have subscripts. For example, in COMIT notation, where the boundaries between constituents are represented by plus signs, one type of input data might be represented in the following way:[12]

$$T + H + I + S + - + I + S + - + D + A + T + A + . +$$

(Ex. 1)

where the hyphens represent spaces.

---

[12] This example, and most of the others, are taken from Yngve [YN66] or from [MT61] and [MT61a].

If the data were in the form of one constituent per word, it could be written as

$$\text{THIS} + \text{IS} + \text{DATA} + . \qquad \text{(Ex. 2)}$$

or with spaces left in,

$$\text{THIS} + - + \text{IS} + - + \text{DATA} + . \qquad \text{(Ex. 3)}$$

The only arithmetic performed is the integer arithmetic on the integer constants. Thus there are no mode problems. Also, since COMIT does not have any significant type of program structure, there is no problem about scope of data.

There are no assignment statements, except for the transfer of strings between shelves and the workspace. There are no logical variables defined as such, but the logical subscripts (which are discussed later) allow the user to perform logical operations.

With regard to control transfers, rules are normally executed in the sequence in which they are written. The statement name (shown in the *go-to* section of the format given earlier) causes the named rule to be executed next. If the next rule in sequence is in fact the next one which is supposed to be executed, then an asterisk can be used in place of a specific statement label. Similarly, an asterisk can be used on the left side as a statement label if there is no need to reference the rule. Subroutines are called by writing in the *go-to* portion of the rule the name of the subroutine, followed by a plus sign, and then the name of the return point. Thus to execute the subroutine that starts at rule ALPHA and upon exit returns to the rule BETA, the user writes ALPHA+BETA.

The fundamental operation in COMIT is really the selection of a substring from a string and then the execution of transformations on the string and substring. This of course contrasts markedly with the numerical programming languages in which the fundamental operation tends to be an assignment statement. That portion of the rule format designated as the *left-half* directs the computer to search in the data for a matching expression. The *left-half* matches a workspace substring only if each *left-half constituent* matches a corresponding workspace constituent or substring, where the workspace constituents are consecutive and all in the same sequence as in the *left-half*. In the earlier examples of forms of data, if the beginning of a rule were written as * T = then in Ex. 1 this letter would be found, whereas in Exs. 2 and 3 it would not because the single letter is not a constituent.

The following items can appear in the *go-to* section of a rule:

name	Transfer control to the rule with the specified label.
*	Transfer control to the next rule.
**	Transfer control to the rule following the next one (i.e., skip a rule).
/	Repeat the same rule.
name1+name2	Go to the rule with name1 and store name2 on the pushdown list (i.e., subroutine call to name1 with return to name2).
name1++name2	Go to the rule with name1 and store name2 on the pushdown list on the level below the item that is next.
+	Go to the rule name which is the next item on the pushdown list and delete the name from the pushdown list.

Conditional statements are of several forms. The first is each rule itself, in which the *left-half* provides a test. The second form of conditional statement performs comparisons on numerical subscripts and is itself a form of *left-half* test. A third kind is grammar subrule selection which is dependent on the dispatcher setting (discussed later). A fourth type of conditional statement involves list subrule selection based on table lookup. The last two of these are really types of switch control.

COMIT permits a rule to have several subrules. The first subrule has the rule name starting in column 1, and succeeding subrules have one or more blanks starting in column 1. Each subrule also has a subrule name. The subrules permit a choice of alternate computations at a given point in the program. The choice is made by the dispatcher, which is a switch control mechanism in COMIT (discussed later).

Loop control can be specified in several ways. The simplest type is through the kind of test indicated in the basic rule itself, where essentially the only thing that controls the loop is the presence or absence in the workspace of the string indicated in the *left-half* section of the rule. Another way to control looping is by specifying the number of times to execute a loop. This is handled by using a numerical subscript as a counter and will be described when the notation is shown in somewhat more detail. Exit from a loop can also be determined by choice of a subrule.

The user can call a built-in dump facility at any point in his program by putting a comma in column 1. He can use a period to request a dump if the *left-half* fails.

COMIT is basically a language for handling strings and patterns. The *left-half* section of a rule indicates a pattern to be searched for in the workspace, which is the storage area containing the data, and the right-hand side (which includes the *right-half*, *routing*, and *go-to* sections) indicates what is to be done if that pattern is found. Let us consider the simple case where the workspace contains

$$A + B + C + A + B$$

and the following rule is executed:

$$\star \ \$ + C + A = 2 + 1 + 3 \ \star$$

The workspace now contains

$$C + A + B + A + B$$

The numbers on the right represent the sequence numbers of the constituents in the *left-half*. Thus, C is the second constituent in the *left-half*, and A is the third constituent. The $ indicates that an arbitrary number of constituents can be allowed when the test is made. Having numbered the constituents in the left-hand side of the rule, their sequence can be redefined in the right-hand side as shown. If the *left-half* had been B + $2 + A, then the test would have failed because the $ immediately preceding a specific integer requires that exactly that number of constituents must appear, and neither of the A's in the workspace follows a B with exactly two intervening constituents. Any constituent shown in the *left-half* which is not mentioned in the *right-half* is automatically deleted (if the rule succeeds thus causing the *right-half* to be executed). An interesting illustration of what can be done is the following rule which doubles any letter before the sequence of letters E and D:

$$\star \ \$1 + E + D = 1 + 1 + 2 + 3 \ \star$$

Note that this rule doubles *any* letter before the E + D, not just consonants; it would be possible to write the rule to exclude doubling of a vowel immediately preceding the ED if vowels and consonants had been named by different subscripts.

The following is a list of notations used for search patterns which can be expressed in the *left-half* section of a rule:

$1	A single but unknown constituent.
symbol	A single constituent which is *symbol*.
$–symbol	A single constituent which is not *symbol*.
$n	*n* arbitrary constituents ($n > 1$).
$	Any number of constituents or none.
$0	A null constituent.
n	A constituent matching the constituent already found and given the relative constituent number n ($n > 0$).

The *right-half* provides facilities for carrying out operations on the data found in the workspace. This is usually done by reference to the constituent shown in the *left-half* and can consist of such things as replacement, rearrangement, insertion, deletion, a null operation (which would be used for a test on the *left-half*), and duplication. As an illustration suppose we consider the following:

$$* A + B + C = 2 + D + 1 + F + 2$$

This means search in the workspace for an A followed by a B followed by a C; if found, replace that triplet in the workspace by B followed by D followed by A followed by F followed by B. Note that in this case the C is deleted and the B is duplicated.

The *routing* section is used to specify certain operations which cannot be specified in the *right-half*. There can be any number of routing operations, separated by commas and optional spaces. The operations shown in the *routing* are carried out from left to right in the order written. The types of operations permitted are special workspace operations, dispatcher entries, shelf operations, input/output, and free-storage count operations.

The most important special workspace operations are *expand* and *compress*, denoted by En and Kn, where n refers to the constituent number. The first causes each character of a constituent to become a separate constituent, while the Kn reverses the process by compressing several constituents into one. For example, if the workspace contained

$$\text{THE} + \text{DOG}$$

writing

$$* \$1 + \$1 = \quad // * E1\ 2\ *$$

causes the workspace to become

$$T + H + E + D + O + G$$

while the rule

$$* \$3 + \$ = \quad // *K1,\ *K2\ *$$

causes the first three constituents to be combined into one constituent, namely THE (from the *K1), and the remainder of the workspace to be combined into one constituent (from the *K2) since the numbers in the routing refer to the constituent numbers in the *left-half*.

Constituents in the workspace can have one numerical subscript and any number of logical subscripts. It is essential to realize that for their most common use in COMIT, subscripts do not bear any relationship to the term subscript as it is used in most programming languages. Subscripts are shown to the right of a slash which is itself immediately to the right of the workspace constituent. Numerical subscripts are shown preceded by a period, and logical subscripts are shown without a period but separated by commas. As an example of an expression with both numerical and logical subscripts, the following can be written:

GEORGE/.26, SEX MALE, OCCUPATION CLERK, HOBBIES GOLF CHESS

The numerical subscript is simply a way of associating a number with an element in the workspace. It is used primarily for counting purposes. Associated with numerical subscripts are the following operations:

In the *left-half* section, search for constituents with the specified property:

- .n   Numerical subscript with the value n.
- .*j Numerical subscript with a value equal to the numerical subscript on the constituent in the workspace corresponding to the jth sequence number in the *left-half*.
- .Gn Numerical subscript with value greater than n.
- .G.*j Numerical subscript with a value greater than the numerical subscript on the constituent in the workspace corresponding to the jth sequence number in the *left-half*.
- .Ln Similar to .Gn, except means less than.
- .L.*j Similar to .G.*j, except means less than.
- —.   No numerical subscript.

In the *right-half* section, for the referenced constituent,

- .n   Set the numerical subscript to n.
- .*j Set the numerical subscript equal to the numerical subscript on the jth constituent in the *left-half*.
- .Dn Decrease the numerical subscript by n.
- .D.*j Decrease the numerical subscript by the numerical subscript on the jth constituent in the *left-half*.
- .In Similar to .Dn, except increase.
- .I.*j Similar to .D.*j, except increase.

The two most frequent uses of these operations are either for setting counters in loops or for testing actual numerical values that a constituent might have. In the first case, a program to do something 50 times could look like the following:

```
  * $0 = C/.0              *   (insert counter constituent with sub-
                                script)
  A       do something     *
  *       do something     *
  * $0 + C/.L50 = 2/.l1 A      (test subscript on counter and
                                add 1)
  * C = 0                  *   (remove constituent with subscript)
```

A test for a subscript value between 25 and 35 would be written in the *left-half* as C/.G25, .L35.

Logical subscripts can be inserted, deleted, complemented, replaced, or merged (which is a combination of both the replacing and the *and* operation). Subscripts can be used as part of the *left-half* search specification in order to specify classes and subclasses of constituents. For example, suppose the workspace contained a number of constituents of the kind indicated above, representing data on people, and it is desired to search for somebody who is male, unmarried, a professional radio operator, and an amateur mountain climber. The following *left-half* will perform this search:

```
  * $1/SEX MALE, MARRIED NO, OCCUPATION RADIO-OPRTR,
        HOBBIES MOUNTAINEER =
```

However, if only a radio operator is needed and the other qualities are irrelevant, then all that needs to be written is the following:

```
  * $1/OCCUPATION RADIO-OPRTR =
```

If the workspace contains the GEORGE constituent defined above, the COMIT rule

```
  GEORGE = 1/OCCUPATION PROGRAMMER
```

changes the value of the logical subscript OCCUPATION from CLERK to PROGRAMMER (presumably a promotion). The COMIT rule

```
  GEORGE = 1/HOBBIES GOLF POKER
```

results in GEORGE having the HOBBIES of GOLF since only subscript values appearing in both the workspace and the rule are left on the constituent when there is an overlap in values. However, by writing

```
  GEORGE = 1/HOBBIES POKER
```

then GOLF and CHESS are dropped and replaced by POKER. In essence, when there is an overlap, then only the intersection is used; when there is no overlap, the values in the *right-half* of the rule replace the values in the workspace. Various other rules and facilities also exist for manipulating these logical subscripts. The subscripts are a powerful feature of COMIT which permit the efficient handling of some problems which do not involve a simple string data structure.

The entries in the dispatcher are in the form of *logical subscripts;* the subscript names designate the program branches, and there is a dispatcher entry for every subscript name (or rule name) in the program. In certain cases where the dispatcher is not itself actually set to a given value, the branch is chosen by using a random number generator. Information in the dispatcher can be altered by an entry in the routing or by an entry from the workspace. In the former case, it is written directly in the routing instruction in the same form as a *right-half* logical subscript and is merged into the dispatcher in the same way that subscripts are merged. Information is entered into the dispatcher from the workspace by writing * Dn as a routing instruction, where n is a single constituent number. When this instruction is executed, the logical subscripts attached to the indicated workspace constituent are merged into the dispatcher.

The general form of an output statement (which is written in the routing) is * WFDni, where the F can be either A or S (these represent two different types of formats); the D can be either M (monitor output tape), I (output tape for punching), L (on-line printer), P (on-line card punch), or other characters for other devices defined in the interface; the ni refers to *right-half* relative constituent numbers, up to a maximum of 128. Format A is used for normal output and essentially removes the plus signs between the constituents in the workspace; in addition, subscripts will also be lost. If the user asks to have less than a single line of material printed, it will be kept in an output buffer until enough characters are accumulated through write commands to cause the contents of the buffer to be written out; however, a specific end-of-line character can be used to cause the printine of a short line. Conversely, if the output requires more than one line, the COMIT system will automatically end the line for the user by means of bell and margin settings analogous to those on a typewriter. There are standard values for these, but they can be changed by the use of *COMSET* cards. The user is able to insert into the workspace specific characters to control the on-line printer paper-feeding mechanism.

Format S is used to print the contents of the workspace in detail, including full constituent structure; this is primarily useful for debugging.

Format B is for fast input and output for temporary storage of workspace data on tape or disc.

The general format for the read command is * RFDn, where F can be C (cards), A (single character per constituent), T (text), S (with subscripts), or B (binary); the *n* refers to a *right-half* relative constituent number and specifies where in the workspace the material that has been read is to be put. The D can be either K (regular input tape), R (on-line card reader), or some other character for some other input device. When format C is used, each data card is read into the workspace, with each character going into the workspace as a single constituent. Format A input is exactly like format C, except that the READ command brings in only one character (instead of a card or record) each time it is executed. The format T is especially useful for large quantities of text because it is fast. It reads a card or record just as format C does, but instead of bringing in each character as a separate constituent, it brings in words as separate constituents. Format S is used to read material that has been written or punched in format S.

COMIT will read a record of any length up to and including 3072 characters. In formats C, A, and T, various operations involving terminal blanks are performed before the characters are grouped into constituents and put into the workspace.

Normally COMIT is set to read physical records into the workspace, but it is possible to cause the reading of logical records.

The options that are available for input and output in COMIT are actually quite flexible and are not all being described here.

There are no library references or built-in functions in COMIT.

The main debugging statement in COMIT is the one called COMDUMP. This provides a copy of the contents of the workspace, shelves, and dispatcher. This dump is automatically provided at each error stop and at the normal end of each run, but it can also be called by the programmer at any point in his program. An unusual feature of COMDUMP is that the first time it is called it gives the complete content of workspace and shelves; however, on each succeeding call of the dump it reports the contents of the workspace or a shelf only in case the contents have changed since the last dump.

In order to discuss some of the storage allocation statements, it is necessary to describe the storage areas in COMIT called *shelves*. There are 127 of these, referred to by the numbers 1 through 127. They are used as temporary storage areas to avoid cluttering the workspace with material which is not needed at a given point in the program. The shelving instructions operate very rapidly and permit the transfer of data between the shelves and the workspace.

The general form of the commands is

$$*\#sn$$

where *s* is the shelf number and *n* designates a sequence of constituent numbers from the *right-half* section. The # can be one of the letters

Q, S, A, N, or X, which have the following meanings:

Q   Queues onto the right end of shelf *s* the workspace expressions numbered *n*.
S   Stores onto the left end of shelf *s* the workspace expressions numbered *n*.
A   Replaces the workspace expression numbered *n* with the contents of shelf *s*.
N   Replaces workspace expression *n* with the leftmost constituent of shelf *s*.
X   Exchanges all the data in the workspace for that on shelf *s*. (In this case, no *n* is given.)

As an example of the use of shelving to achieve more rapid execution, suppose that the problem is to change every B in the workspace to a C. The simplest way of doing this is simply to write

$$\star \ B \ = \ C \ //$$

However execution will be very slow because the rule replaces only one B at a time and then goes back to the beginning of the workspace again to search until it finds the next B to be replaced by a C. A faster routine to do this would be the following:

$$\star \ \$ \ + \ B \ = \ 1 \ + \ C \ // \ \star \ Q23 \ 1 \ 2 \ /$$
$$\star \ \$0 \ = \ // \ \star A23 \ 1 \ \star$$

The *left-half* of the first rule finds everything up to and including the first B and changes it to a C. The instruction following the double slash serves to move all constituents up to and including the C onto the right end of shelf 23, keeping them in the original workspace order. Thus when the rule is executed again, the string that has already been examined is no longer in the workspace. After all the B's have been changed to C's, the rule fails and the next rule returns everything from shelf 23 back to the beginning of the workspace again. (Note that if shelf 23 were not initially empty, it would be necessary to clear it before starting this routine. This avoids putting extraneous material into the workspace when the second rule is executed.)

The number of memory locations required for the various types of information in the workspace and shelves is shown in the manual. Storage is automatically reallocated to the various uses as needed.

The user is able to test the amount of free storage available and to allow certain corrective action to take place if there is insufficient free storage for further operations; such action might consist of deleting material from the workspace or the shelves or writing it out. The commands available for this testing are written in the routing section and consist of the following:

.Gn If the number of free storage registers is greater than *n*, continue with the rest of the *routing* and *go-to* sections; otherwise transfer control to the next rule.

.Ln Same as .Gn except for *less than* n.

.G.*n If the number of free storage registers is greater than the value of the numerical subscript on the first constituent in the workspace subexpression currently having the number *n*, continue with the rest of the *routing* and *go-to* sections; otherwise transfer control to the next rule.

.L.*n Same as .G.*n except for *less than* n.

There are actually quite a few statements that permit the user to have some effect on the overall operating environment; these are handled by means of *COMSET* cards. The facilities available to the user include items such as specifying that the minus sign and the hyphen are to be considered different characters, setting margins for output, eliminating the automatic *COMDUMP* at the end of the program, counting the number of constituents on indicated shelves, specifying ways of marking the end of output records, and providing controls over the way in which various formats are handled.

One of the special features in COMIT is the availability of list rules which facilitate rapid table lookup. A list rule consists of two or more list subrules, and it may be used like a dictionary with its entries. To illustrate the need for, and use of, this facility, the following example is given.[13]

Consider the following program for word-for-word substitution. We assume that the workspace contains a German sentence. Each letter is a constituent and words are separated by −'s, representing spaces. There is no punctuation, but − at the beginning and at the end. It is desired to look up the words in a list, substitute English equivalents, and print them out.

```
         A  −  +  $  +  −  +  $  =  //  *WAM1, *K2, *Q5  3  4  DICT
  *      STOP
  DICT  IST    = IS         B
  *     ALT    = OLD        B
  *     MANN   = MAN        B
  *     DER    = THE        B
                   .
                   .
                   .
  B  $  //  *WAM1, *A5  1  A
  STOP *
  END
```

[13] It is taken directly from pp. 41–42 of [MT61a].

What we are really doing in this program is searching through a number or rules for a *left-half* that matches the workspace. The trouble with the program is that it will run quite slowly if the list is very long. In COMIT, there is a facility for writing an equivalent program that runs quite rapidly. It is written as follows:

```
A  − + $ + −  // *WAM1, *L2 DICT
* STOP
−DICT IST    = IS          B
      ALT    = OLD         B
      MANN   = MAN         B
      DER    = THE         B
              .
              .
              .
B $ + −  // *WAM1  A
STOP *
END
```

In this program, the first rule locates a word as a string of characters between spaces. The first space is written out. The routing instruction *L2 says to compress these characters temporarily into one constituent without subscripts and look it up in the list mentioned in the *go-to* (DICT). The list rule −DICT is distinguished by a hyphen in column 1. The *lrft-halves* of the various list subrules are single constituents. Each list subrule can have a normal *right-half*, *routing*, and *go-to*. If the temporarily compressed constituent cannot be found in the list, the original uncompressed constituents are restored to the workspace complete with their subscripts, and control goes to the rule after the list rule B.

The reason why this program written with a list rule will run so much faster than the other program is because the COMIT compiler automatically sorts the *left-halves* of the list subrules into alphabetic order so that whenever a constituent is looked up in a list, COMIT can use a fast search procedure.

There are no declarations as such in COMIT. The closest to these would be the logical subscripts, which were discussed earlier (together with the operations on them).

COMIT does not permit self-modification of programs nor self-extension of the language. There are no real compiler directives. Attempts at writing parts of COMIT in itself have been tried, but there is no completed system prepared this way known to me.

Some of the language features, most notably the shelves and list rules, have been provided to improve the execution speed. Since COMIT is partly interpretive, in the sense that a translation is made to an intermediate pseudocode which is then interpreted, the issue of object time efficiency is not too significant.

The COMIT systems have a large number of error checks built into them both at compile and execution time.

COMIT has proved to be useful, at least on an experimental basis, in a large number of areas outside its primary application area. For example, it has been used to do information retrieval (Yngve [YN62]), symbol manipulation (e.g., differentiation), theorem proving (Darlington [DA65]), checkerboard puzzles, bridge bidding, and comparison of student programs (Chanon [ZS66]. The General Inquirer (Stone and Hunt [SJ63]) was initially programmed in COMIT.

VI.6.4. SIGNIFICANT CONTRIBUTION TO TECHNOLOGY

COMIT was the first language to provide an effective means of searching for a particular string pattern and then performing transformations when it was found. There are facilities in COMIT which have not yet come into the more multipurpose programming languages, although the latter sometimes wish or claim that they have. COMIT has served as a model for the type of facilities needed for string manipulation, and virtually every language which *has* included significant features of this kind has based them at least in spirit—and often in notation—on COMIT. Thus, the introduction of the CONVERT functions into LISP (Guzman [ZH66]) and many facilities in Formula ALGOL (see Section VIII.5) owe their ideas to COMIT. Finally, the creation of a major language (namely SNOBOL, see Section VI.7) which performs many of the same functions but also corrects some of the deficiencies, shows the value of the ideas brought forth in COMIT. In some cases, SNOBOL omitted features (e.g., subscripts) which are significant for efficiency in certain classes of problems.

One of the most astounding things about COMIT is the discrepancy between its apparent and surface difficult notation and the actual ease of writing and using the language. This contrasts sharply—at least in my opinion—to LISP, whose notation is inherently simpler than COMIT but which seems to be much harder to learn and to use. (This view is even shared by proponents of LISP.)

Another contribution to the technology is not unique to COMIT, but it is nevertheless worth mentioning. Several problems, or classes of problems, have been solved—or at least worked on—that would have been impractical to attack without COMIT. While this is true of almost any programming language, it is more significant for COMIT because COMIT provided a type of facility that did not exist anywhere else (at least until its imitators appeared).

Some comments on COMIT versus SNOBOL are included in Section VI.1.

VI.7. SNOBOL

VI.7.1. History of SNOBOL

After several years of exposure to, and experience with, COMIT (see Section VI.6), some of its significant deficiencies became extremely clear; the most notable of these were the inability to name strings or to do arithmetic conveniently. Motivated by a desire to improve this and by the need for a more effective language to do symbol manipulation, in 1962 at Bell Telephone Laboratories, D. J. Farber, R. E. Griswold, and I. P. Polonsky with assistance from M. D. McIlroy developed a system called SNOBOL which they defined as a string manipulation language. The first publication of information on this system was in Farber, Griswold, and Polonsky [FB64]. The system became popular among certain groups outside Bell Laboratories (primarily universities). After some period of time, certain improvements to the system were found desirable, and these were incorporated gradually, eventually leading to the creation of SNOBOL3, which is defined by the authors above in [FB66]. Still further improvements were made during 1967 and SNOBOL4 was created. The timing of the latter work made it impossible to include it in this book, so it is SNOBOL3 which is described here. A brief indication of the major improvements in SNOBOL4 is given at the end of Section VI.7.3.

SNOBOL3 has been implemented on at least the following machines: IBM 7090/94, 7044, 1620, System/360, CDC 3100, 3600, RCA 601/604, and SDS 930/940.

VI.7.2. Functional Characteristics of SNOBOL

Since SNOBOL is a language with a format unlike any other (except COMIT), it is desirable to show an example of a SNOBOL statement before discussing the general characteristics of the language:

HERE TEXT "THE" *WHOKNOWS* "IS" = WHOKNOWS /S(NEXT)

This means that the label HERE is assigned to a statement which examines a string named TEXT to see if it contains the words THE and IS with any string between them; if so, the string named TEXT is replaced by the string found between THE and IS and control is transferred to the statement named NEXT.

SAMPLE PROGRAM—SNOBOL 3†

```
*SNOBOL3 EDITOR
*          THIS PROGRAM READS IN TEXT AND PRINTS N CHARACTERS
*          PER LINE. EXTRA BLANKS ARE INSERTED BETWEEN WORDS
*          TO FILL OUT LINES.

           MODE("ANCHOR")
           DEFINE("INSERT(K,LINE)","IN","BLANK,WORD")
*          READ IN N
           TRIM(SYSPIT)                    *N*
READ       TEXT = TEXT TRIM(SYSPIT) " "                    /S(READ)

           SYSPOT =
           SYSPOT = "PRINT THE FOLLOWING STRING " N
                    " CHARACTERS PER LINE:"

           SYSPOT =

*          PRINT INPUT TEXT
           TEXT1 = TEXT
PRINT      TEXT1 *SYSPOT/"80"* =                           /S(PRINT)
           SYSPOT    = TEXT1
           SYSPOT =
           SYSPOT =

TEST       .GT(SIZE(TEXT),N)                               /F(LASTLINE)
           K       = "0"
SCAN       TEXT   *LINE/(N — K)* " " =                     /S(WRITE)
BUMPK      K      =    .LT(K,N) K + "1"                    /S(SCAN)F(ERR)
WRITE      SYSPOT = INSERT(K,LINE)                         /(TEST)

*          FUNCTION TO INTERSPERSE K BLANKS IN A LINE
IN         INSERT  = .EQ(K,"0") LINE                       /S(RETURN)
           LINE    ** " "                                  /S(BLINK)
           INSERT  = LINE                                  /(RETURN)
BLINK      BLANK   = BLANK " "
LOOP       LINE    *WORD* BLANK =                          /F(MORE)
           INSERT  = INSERT WORD BLANK " "
           K       = .GT(K,"1") K — "1"                    /S(LOOP)
           INSERT  = INSERT LINE                           /(RETURN)
MORE       LINE    = INSERT LINE
           INSERT  =                                       /(BLINK)
LASTLINE   SYSPOT  = TEXT                                  /(END)
ERR        SYSPOT  =
           SYSPOT  = "** EDITOR CANNOT PRINT LINE BECAUSE N IS TOO SMALL."
END
```

†Farber, Griswold, & Polonsky [FB66], p. 943 (with slight modifications). Copyright, 1966, The American Telephone and Telegraph Co., reprinted by permission.

At first glance this language appears highly formal and difficult to read and write. In actual practice, it is fairly simple in notation and is compact, succinct, and reasonably natural within its class of applications. It is easy to learn, to read, and to write, although it is somewhat more complex (and more powerful) than COMIT.

The primary application area is anything in which string naming and manipulations are important component elements, with at most simple integer arithmetic being needed. It is a problem-oriented language which is simultaneously a hardware, reference, and publication language. In the same sense as COMIT, it is slightly (and intuitively to me) more nonprocedural than most of the more numerically oriented programming languages. In my opinion, the user will tend to be at least a novice programmer, i.e., somebody who has some programming background but is not necessarily a professional programmer. It is definitely designed for use in a batch system, although both an early version and SNOBOL3 run under M.I.T.'s CTSS (see Crisman [ZR65]). The system is not machine dependent, but its description in the previously cited reference makes it more compiler dependent than most languages, particularly with regard to the input/output facilities. No particular concern has been expressed about the creation of dialects or subsetting or extensions; some of these variations exist on different implementations. Although never explicitly stated, there are several places in which the language is implicitly defined with respect to a particular method of implementation.

The whole problem of conversion has not arisen in any significant way. There has been no consideration of USASI standardization.

The designers of the first version of the language were primarily the authors of the two cited papers. Their avowed objective was to develop a language for use on problems involving the manipulation of character strings. Any maintenance of the language is being done by the designers. A *SNOBOL Bulletin* exists for informal interchange of information among users and any other interested people.

In addition to the two main published papers cited, some significant unpublished papers are included in the list of references at the end of this chapter. The language definition has been given in English, with heavy dependence upon examples for clarification of the definition.

SNOBOL3 is too new to be able to give it any type of reasonable evaluation based on usage, although a number of groups seem to find it worthwhile. It has been used in a number of courses.

VI.7.3. Technical Characteristics of SNOBOL

The character set for the language is never explicitly defined, but it appears to consist of the 26 capital letters, the 10 digits, and the following special characters:

$$+ \quad - \quad * \quad / \quad : \quad , \quad . \quad (\quad) \quad " \quad \$ \quad = \quad \textit{blank}$$

Data names can consist of any combination of letters, numbers, periods, and colons, e.g., 1:X.A, A.B.::1C, and .:A.3. (There is no stated requirement to begin a data name with a letter or number.) Implicit data names are constructed by indirect referencing and consist of any data name preceded by a dollar sign, $. Thus if the data name MONTH has the value MARCH, then writing $MONTH is equivalent to writing MARCH. Since the only data type in SNOBOL is a character string (which cannot be split into constituents as in COMIT), data names apply only to a string. An arbitrary string variable is designated by a name bounded by asterisks, e.g., *ARB*. Statement labels must begin with a letter or digit and can be followed by any other character except a blank. Data names and statement labels can be the same, and labels can be the same as some other structure, such as a function call. Thus S(S) is a legitimate label. There is no subscripting of any kind nor any qualification.

The word QUOTE has a preassigned value of a quotation mark at the beginning of the program; however, it can be given another value later in the program.

The operators fall into two groups: The five basic arithmetic operations denoted by +, −, *, /, and ** (for exponentiation); and six comparison operators with the obvious meanings: .EQ, .NE, .LT, .LE, .GT, and .GE. (These are actually considered functions in the SNOBOL terminology; the period preceding the two letters is part of the operator name.)

Punctuation has no meaning in the normal sense. Blanks are significant in a number of places since their presence denotes concatenation of strings, and blanks must be used to separate arithmetic operands and operators. They are also used to separate the differing parts of a statement. There are no noise words.

Literals are defined by the use of quotation marks at the beginning and at the end of the literal. The literal can consist of any set of characters except the quotation mark; the latter is indicated by the use of the word QUOTE.

In order to understand the format used, the general form of a statement must be shown; it is as follows:

label   string-reference   pattern = *replacement-expression   /go-to*

It is important to note that the main difference in statement format between SNOBOL and COMIT is the facility in the former to name the string being examined (shown above as the *string-reference*) and furthermore to name a substring.

A statement can contain some or all these fields. Labels must begin in column 1; a statement with no label must have a blank in column 1. Statements can be continued over one card, and a period in column 1 is used to indicate continuation of the preceding statement. Statements may be split across cards wherever a blank is permissible in the text. However, literals cannot be broken over a card, and a long literal which would not fit on one card must be represented as a concatenation of literals. A card with an asterisk in column 1 is treated as a comment.

The conceptual form is highly symbolic.

There are no declarations as such in SNOBOL. Some things which would normally be handled as declarations (e.g., arithmetic mode) are actually specified through function calls.

The smallest executable unit is a statement, and there is no way of grouping these units. Loops can be written in either of two ways. In the first instance, a rule can repeat itself, providing the transfer in the *go-to* section is to the statement itself; in other cases, the looping can be controlled by arithmetic counters and tests on them.

Functions of three kinds are permitted in SNOBOL. The first are those defined by the system; the second are defined in a particular SNOBOL program; and the third are machine language routines which can be invoked as functions in SNOBOL. There are no other procedures or subroutines.

Comments can be written anywhere in the program and are designated by an asterisk in the first column of the card.

A complete SNOBOL program consists of statements terminated by an **END** card. A function must be defined before it is invoked. There are a number of compiler-directing control cards which can be used in a program before the **END** card.

The end of a statement is determined by the beginning of a statement on the next card. Function calls must not have a blank between the function name and the left parenthesis; thus, **SIZE(A)** is a function call, while SIZE (A is a concatenation of a name and a parenthetical grouping.

Statements are recursive in the sense that control may be transferred back into the statement itself, depending upon the failure or success of the left-hand pattern matching. Most SNOBOL programs are recursive. Defined functions can be recursive.

Function calls can occur anywhere in a statement where a string value is appropriate. An argument of a function can be a string expression, no matter how complicated.

No other languages can be written in SNOBOL programs; however it is possible to include machine language functions which have been compiled or assembled separately. There is a set of conventions specifying how to do this.

Although the only variable type in SNOBOL is a string, there are certain strings which have the property of being integers. Thus, unlike COMIT, there are conceptually integer variables and constants in SNOBOL but no actual arithmetic data type. While there are no logical variables as such, the six comparison functions mentioned earlier return a null value if the condition is satisfied, and they fail otherwise.

Indirect addressing or naming can be used to any level desired. Function calls, parenthetical grouping, and names may be indirectly referenced. Parentheses are required between successive levels of indirect references. For example,

$$\$(\$(\$ROUTE))$$
$$\$SIZE(END)$$
$$\$("END" \ \$(I))$$

are all examples of legally written indirect references.

Only integer arithmetic is permitted, and there are various rules pertaining to what happens when the result of division or exponentiation is not an integer. It is possible to combine integer constants and names; thus the expression

$$N + "3"$$

is legal assuming that N has an integer value. Since each of the arithmetic operations is considered purely binary, it is necessary to group terms to create more complicated expressions. Thus, the user must write

$$N + ("3" * "2")$$

to achieve

$$N + "6"$$

Expressions can contain both concatenation and arithmetic operations, and the latter have precedence over the former. Hence the value of

$$"N" \ "5" + "7"$$

is N12 since juxtaposition of two strings with a space between them (i.e., "N" "5") denotes concatenation. Also, the value of

$$"3" * "2" \ "10" / "2"$$

is 65. However, the value of

$$"3" * ("2" \ "10" / "2")$$

is 75.

The most important "arithmetic" performed in SNOBOL is actually string concatenation. This is indicated by writing the items (which can be constants or data names) to be concatenated successively, separated by blanks. For example,

"RED" "ORANGE"
X "3.5" Y

The only problem relative to scope of data occurs in the creation of user-defined functions, and in that case it is possible to use names which are used in functions but whose values are not to be destroyed by a function call.

A subset of the complete statement format provides specifically for an assignment statement, where the right-hand side will be either a string or a literal and the left-hand side will be a name. For example, the user can write

Y = ABC + "5"
Y = "Y + SIN(Z)"
Y = ","
ABC = $("N" I)

If there is a need for any arithmetic shown on the right-hand side, it must follow the rules of precedence already described. The result in every case is a string.

There is no alphanumeric data handling as such.

As an illustration of the effect of the sequence of assignment statements, consider the following sequence of statements:

ALPHA = "ABCD"
KY = SIZE(ALPHA) + "1"
Z = KY + "3"
M = ("−" KY Z) + "2"

SIZE is a SNOBOL primitive whose value is the number of characters in the string indicated. Thus, the sequence of statements above would cause KY to have the value 5, Z to have the value 8, and M to have the value −56.

Statements are normally executed in sequence unless the statement contains a *go-to* field; transfers are always enclosed within parentheses and preceded by a / which is itself preceded by a space. The control transfers can themselves be conditional by prefixing them with either an S or an F for success or failure, respectively. For example, A = B /(K) will automatically cause control to transfer to the statement with the label K. On the other hand, the statement ALPHA "," /S(XYZ) will cause the statement XYZ to be executed next if there is a comma in the string named ALPHA. As in COMIT, failure of a statement is very significant in SNOBOL. The failure can occur in several ways; e.g., functions may specify failure explicitly (most normally this occurs with the comparison operators), a pattern match may fail, or an attempt to read a record may fail due to an end-of-file mark. In all cases, a failure signal stops execution of the statement and causes the appropriate control action to take place.

The order of execution of operations within a statement is extremely significant; this overall order of execution is as follows: (1) The string reference (if any) is evaluated; (2) the elements of the pattern (if any) are evaluated from left to right; (3) the pattern match (if any) is performed; (4) any naming (discussed below) as the result of the successful pattern match is performed; (5) if a string expression is specified as a replacement, that string expression is evaluated; (6) reformation (if specified) of the string reference is made; (7) the *go-to* (if any) corresponding to the success or failure of the statement is evaluated and transfer is made to the next statement accordingly. Note that all arguments of a function are evaluated, left to right, before the function is called. A function is invoked by writing it any place where a string variable is permitted; the name of the function is given, followed by the parameters, and a string value is returned. The relationship of the function evaluation to the general scan is quite tricky.

Conditional statements are specified either by the success of the pattern search or through the use of the comparison operators which can indicate either success or failure. There are no separate loop control statements in SNOBOL; they are programmed either implicitly through the test of the pattern appearance or nonappearance in the indicated string or through the use of the comparison operator in counting.

An example of the use of the comparison operators to do necessary counting is the following, which assigns the cubes of the first 50 positive integers to the names CU1 through CU50, respectively:

```
         N = "1"
COMPUTE $("CU" N) = N * (N * N)
         N = .LT(N, "50") N + "1"  /S(COMPUTE)
```

There are no error condition statements in SNOBOL, although error conditions in arithmetic operations can be used to change the flow of control through the use of a conditional *go-to*.

There are no algebraic or list statements in SNOBOL, but work has been done to provide the facilities to manipulate trees and certain link lists. These are described in unpublished memos shown in the reference list at the end of this chapter.

The heart of SNOBOL is its string-handling and pattern-matching statements. In the general form of the statements, the key items for the pattern matching and string handling are the *string-reference*, the *pattern*, and the *replacement-expression*. The string reference is searched for occurrences of the pattern, which itself can consist of a sequence of string variables and/or constants, and then the occurrence in the string reference is replaced by the designated replacement expression. For example, if we had the statement

>TEXT = THIS IS THE 5TH TIME

then

>TEXT "IS" *VAR* "5" "TH"

would cause the search to be successful and would assign to the variable named VAR the value THE. In addition to arbitrary string variables, there are two other types, namely fixed length and balanced. The former can be used to match a string with a specified number of characters; the name of the string variable is followed by a slash; this is followed by the string specifying the length, and the whole thing is delimited on both sides by the asterisks. Thus, for example, if we had the statement

>"+—*" *PLUS/"1"* *MINUS/"1"* *STAR/"1"*

the pattern successfully matches the string, and PLUS, MINUS, and STAR are assigned the values of the symbols indicated.

A balanced string variable can only match strings that have the same number of left and right parentheses in the usual order; i.e.,) (is not balanced. The notation consists of a name enclosed within parentheses and surrounded by a pair of asterisks. For example, if EXPRESSION has the value SIN(A * (B + C)), then the pattern match in the statement

>EXPRESSION "SIN(" *(ARG)* ")"

is successful and ARG is given the value A * (B + C).

Replacement can take place by naming substrings. For example, the statements

>SUM = "A1 + A2"
>SUM *X* "+" *Y* = "+(" X "," Y ")"

change the value of SUM to +(A1 , A2).

One of the interesting pattern-matching facilities in SNOBOL is that referred to as *back referencing*. This means that if a constant in the pattern has the same name as a variable named to the left of it in the pattern, th value of the constant is taken to be the substring currently matched by the variable. Thus if we have A = "C" and B *C* "," $A then this is not only back referencing, but has the indirect naming facility along with it; $A which is C will match the substring denoted by *C* i.e., the substring to the left of the , in the string named B. Another interesting illustration is the following:

"ABCDEFGHFGH" *A/"3"* A

this match succeeds with A = "FGH" because the pattern that is being looked for is three characters which are juxtaposed. For example, if the pattern had been written as *A/"2"* A then the pattern would not have matched.

The specifications in the basic reference for SNOBOL3 indicate very clearly that the input/output facilities may vary significantly on different machines. The following is a very brief description of the capability available on the 7090/94 system.

Input and output are accomplished by associating string names with logical files. There are three standard file names which are automatically assumed at the beginning of every program. The user can associate the value of a name with the value of a particular file, including the three named files. Thus, PRINT("X", "OUT") associates the name X with the logical file OUT. After execution of this function, copies of all values assigned to X will be placed in the file OUT. Similarily, PUNCH(NAME, FILE) and READ(NAME, FILE) associate the value of NAME with the value of FILE in the punch and read sense, respectively. Output occurs whenever an output-associated name is given a value. Printing and punching file associations are different in their carriage control. Various facilities for short and long lines and appropriate fillers are given. In reading input, all strings read from the standard input are 84 characters long, and blanks are used to fill out shorter records; records read from other files are not extended. In using a read-associated name, the loss of the previous value of the name occurs. There is a failure if the read operation fails as the result of an end-of-file record.

There are various other functions existing for performing standard input/output and file operations, including name deletion, rewinding, backspacing, and opening a file.

There are no library references as such, but there are a number of built-in functions. They are as follows:

SIZE(N)
  Specifies the number of characters in the string whose name is N.
EQUALS(X,Y)
  Returns a null value if the value of X is identical to the value of Y and fails otherwise.

UNEQL(X,Y)
>Returns a null value if the values of X and Y are not identical and fails otherwise.

TRIM(X)
>Returns the value of X with trailing blanks removed.

TIME()
>A function of no arguments which returns as a value the millisecond time from the beginning of the program to completion.

DATE()
>A function of no arguments which returns the value of the current date as a six-character number.

There are a certain number of debugging statements, some of which are as follows:

TRACE(LIST)
>The value of LIST is a list of function names whose trace is desired.

STOPTR(LIST)
>Stops the tracing of functions specified in the LIST.

STRACE(X,Y)
>Associates the value of X with the value of Y. Every time a value is assigned to X, a trace message will be printed on an associated file name indicating the names, its new value, and the statement value where the new number was assigned.

MODE("DUMPERR")
>Causes a string dump if execution is terminated by an error during execution.

MODE("DUMP")
>Causes a string dump following either normal or error termination.

There are no storage-allocating statements, and the only statements involving the operating system are the use of a statement with the label END whose result is dependent upon the monitor system. (This contrasts with the storage allocation facility of the *shelves* in COMIT.)

There are no data, file, format, or storage allocation descriptions.

Functions are defined by writing a heading of the form

$$\text{DEFINE}(form, label, names)$$

where *form* gives the function name and the formal arguments, *label* denotes the entry point, and the value of *names* is the list of local names separated by commas. The termination of the function is marked by either *RETURN* or *FRETURN*, the latter providing a failure signal. These functions can be recursive.

There are a number of control cards which provide directives to the compiler, e.g., stop listing source program and rewind specified file.

There are several other types of compiler directives. The most interesting are those which are associated with **MODE(X)**. If **X** is **ANCHOR**, then this means that the first pattern element must match a substring in the string reference beginning with the first character of the string reference. This is in contrast to the normal method which is **UNANCHOR**, which says that the first element of the pattern can start matching anywhere. Thus, if the MODE ("ANCHOR") has been given in the program, the pattern match

"ABC" "B"

will fail; whereas if that statement had not occurred or if it were followed by a **MODE("UNANCHOR")**, then the match would be successful.

A mode affecting results of a division is available.

SNOBOL programs cannot modify themselves nor is there any provision for self-extension of the language.

The implementation of SNOBOL on the CDC 3200 was done by writing the compiler in SNOBOL; the one developed for the IBM System/360 was also written in SNOBOL. At least one group is attempting to develop a version which is fast enough to be practical for compiler writing.

As with COMIT, the translation system consists of a compiler, which translates source programs into an intermediate language, and then an interpreter, which controls the execution. The language specification and the implementation of some of the pattern-scanning mechanism seem to be intimately related, i.e., the rules for the language match the algorithms which have been designed for scanning and vice versa.

The biggest problem with respect to the language and its implementation is the need for dynamic storage allocation.

SNOBOL4 (described in Griswold, Poage, and Polonsky [GF67] and [GF68]) is not a true extension of SNOBOL3, although it is certainly very similar. Two syntactic changes involve the following requirements in SNOBOL4: Identifiers begin with a letter and may not contain colons, and a colon **:** instead of a slash **/** is used to begin the *go-to* field. On the other hand, there are a number of significant improvements, including at least the following:

1. Parentheses are no longer required in arithmetic expressions; the normal precedence is used.
2. Unary operators can be written consecutively without parentheses.
3. Integers do not have to be surrounded by quotation marks.

4. Functional expressions can appear in more places than previously.
5. Facilities for pattern matching have been greatly extended; this is the most significant improvement. A pattern is considered a data object and hence can be named with the name then used in place of the actual pattern. A pattern can be formed by concatenation. There are new functions which have patterns as values, and new patterns have been added. It is possible to specify alternative patterns to be searched for.
6. Arrays can be created.
7. The programmer can define new data types.

SNOBOL4 has been implemented on at least the IBM System/360.

VI.7.4. Significant Contribution to Technology

SNOBOL's main contribution to the technology has been to provide significant improvements over COMIT, based on experience with the latter. This comment is not meant to negate the additional features which have been added but merely to emphasize the fact that SNOBOL built heavily (and effectively) on previous work.

More specifically, the ability to name strings and substrings is probably the most important new feature. The inclusion of facilities for doing arithmetic in a reasonable way, particularly in SNOBOL4 (where parentheses are not required to establish proper precedence), is important since even string- and pattern-handling problems usually require some arithmetic computations. The convenience of allowing user-defined functions of a general type is worth mentioning. Finally, the development of a large number of differing kinds of pattern-matching capabilities is an important contribution. Some of these bear a resemblance to facilities existing in the much earlier SHADOW system (see Barnett and Futrelle [BI62]).

SNOBOL appears to be a good tool for programmers to use in a wide variety of problems involving string handling and pattern matching. Its notation is probably as similar to languages discussed in earlier chapters as can be without violating the fundamental operations that the language is trying to provide effectively.

Some comments comparing SNOBOL with COMIT were included in Section VI.1.

VI.8. TRAC

The TRAC[14] (*T*ext *R*eckoning *A*nd *C*ompiling) language system (meaning somewhat more than just a language) was created by Calvin Mooers (Rockford Research Institute). Development started in 1960 and it was first imple-

---

[14] TRAC is the trademark and service mark of Rockford Research Institute Incorporated in connection with their standard computer controlling languages.

mented by Peter Deutsch on the PDP-1 in 1964. Experimental versions have since run on at least the following machines: The PDP-5, 8, and 8S; the GE Datanet-30; the IBM 360/67; and the European computers SAAB D-21 and ICT 1202. All the implementations use the basic concepts and specifications of TRAC; most have added their own primitives to accomplish specific objectives such as control of displays and code conversion.

TRAC is a language designed specifically to handle unstructured text in an interactive mode, i.e., by a person typing directly into a computer. An excellent discussion of the objectives for the system and an indication of how they were achieved is given in Mooers [ME65]. The specifications of the language are contained in Mooers [ME66] and [ME67], although not stated explicitly as such.

Although Mooers states that he considered IPL-V, LISP, and COMIT carefully, he found that none of them were suitable for achieving his desired goals. He said that the prime impetus for the TRAC language came from

SAMPLE PROGRAM—TRAC

Problem: Define a FIX program whose purpose is to make small changes in TRAC forms, and use the FIX program.

Program	*Comments*
↑(DS,FIX,(	
↑(SS,NAME,↑(PS,(	This is the complete input program
DELETE- -))↑↑(RS))	to define FIX.
↑(DS,NAME,↑↑(CL,NAME,↑(PS,(	
INSERT- -))↑↑(RS)))	
))	
↑(SS,FIX,NAME)'	
↑(SS,1@,↑(PS,(	Representation of the FIX macro.
DELETE- -))↑↑(RS))	"1@" represents a formal variable,
↑(DS,1@,↑↑(CL,1@,↑(PS,(	and the ordinal one. This takes the
INSERT- -))↑↑(RS)))	name of the form to be operated on.
↑(DS,A1,THIS IS THE TEXT)'	Define a text object A1.
↑(CL,FIX,A1)'	Call FIX.
DELETE- -THIS'	Interaction with user, to make
INSERT- -THAT'	changes.
↑(PS,↑(CL,A1))'THAT IS THE TEXT	Call and print out A1.
↑(CL,FIX,A1)'	Begin another example.
DELETE- -THE'	
INSERT- -THE VERY LONG'	
↑(PS,↑(CL,A1))'THAT IS THE VERY LONG TEXT	Result.

two early unpublished papers on macro assemblers by Eastwood and McIlroy.

The basic concept in TRAC is that a program consists of strings containing sequences of functions which can be nested indefinitely deeply. Evaluations of these proceed from the innermost level outward, and from left to right within a level, to cause the execution of the program. Furthermore, since the executable statements are treated in the same way as a general character string, a procedure can act upon itself as well as upon other executable statements, thus giving completely general self-referencing capabilities.

An independently developed system, GPM (*General Purpose Macro-generator*) by C. Strachey [SQ65] is similar in many respects to TRAC in concept, although not in format. GPM is not being discussed here because this book is limited primarily to systems developed in the United States.

Both the data strings and the strings specifying executable operations are intermingled in TRAC. *Active* and *neutral* functions, corresponding approximately to *use* and *mention* of commands, are denoted by #(...) and ##(...), respectively. The ellipsis, ..., represents an arbitrary string, normally divided into substrings separated by commas to specify the arguments of the function. The functions themselves can be enclosed within parentheses; this has the effect of completely preventing their execution.

All strings in TRAC are really literals rather than names, although strings for names are permitted in certain specified places. A TRAC function has a two-letter mnemonic to denote the specific action being invoked, followed by the appropriate number of arguments separated by commas, and all enclosed within parentheses; e.g., #(DS,AA,CAT) causes the data string CAT to be stored, and the name string AA to be stored in a table with a pointer to the string CAT. This data string can be brought back from storage by means of a *call* function written #(CL,AA). Upon execution, the expression #(CL,AA) is deleted and replaced by the string CAT. In this case the call function actually creates a value; whereas the function which stored the text did not and is considered to be a TRAC function with a null value. If the user writes

#(CL,#(DS,CAT,DOG)#(DS,AA,CAT)#(CL,AA))

then DOG is the final value of the function. This is obtained because the processor generates first

#(CL,#(CL,AA))

and then

#(CL,CAT)

and eventually DOG is obtained.

The overall control of the TRAC program comes from what is called the *idling* procedure, #(PS,#(RS)). The PS and RS stand for *print* and *read*, respectively, and *read* has the value consisting of the string of characters typed at the keyboard up to the first occurrence of a specified metacharacter or terminator. The latter is normally designated by an apostrophe, ', but there is a TRAC function which can be used to change it. Since both teletypes and typewriters can be used, upper- and lower-case letters are permitted and are handled appropriately, depending on the form of input device. In particular, if there is a shift character on the input device, then it will be retained. Carriage returns or line feeds are ignored except within strings so the user can format his functions conveniently.

The distinction between an active and a neutral function (mentioned earlier) is that the value from the active function is rescanned, while in the neutral function it is not. Scanning and execution of functions are done from left to right but from inside out. These concepts are illustrated in the following example:[15] Suppose that both #(DS,AA,CAT)' and #(DS,BB(#(CL,AA)))' have been presented to the processor. Then,

#(PS,(#(CL,BB)))' prints out #(CL,BB)
#(PS,# #(CL,BB))' prints out #(CL,AA)
#(PS,#(CL,BB))' prints out CAT

Now that the basic concepts in TRAC have been seen, it is possible to discuss the primitive functions that are available. A function defined to have a null value causes some action to occur, but it does not create a value which could then be used in other functions.

There are four arithmetic functions and they operate only on integers. The subtraction function is designated as #(SU,D1,D2,Z), which subtracts D2 from D1; if the result overflows the capacity of the arithmetic process, then the overflow value Z of the function is taken. A similar format is used for the other three arithmetic operations. Bit strings are permitted and the functions of Boolean union, intersection, complement, shift, and rotate are permitted; these are designated by the function mnemonics BU, BI, BC, BS, and BR, respectively. There are two conditional transfer functions, namely #(EQ,X1,X2,X3,X4), which has the value X3 if string X1 is equal to string X2 and the value X4 otherwise. The function #(GR,D1,D2,X1,X2) tests the decimal number at the tail of string D1 against the number at the tail of string D2; if the first is greater than the second, the value of the function is X1, otherwise it is X2.

There are a number of functions that relate to defining and retrieving strings. They are as follows:

---

[15] Mooers [ME66], p. 219.

#(DS,N,X)
> Stands for *define string* and has a null value. The string shown as the argument X is placed in storage and given a name shown as N. The string shown as X is placed in a list of *forms*, which are in storage; a *form* is merely a named string. If there is already a form in storage with the name N, then the first form is erased.

#(SS,N,X1,X2, . . .)
> Represents *segment string*, where there can be three or more arguments; this function has a null value. The form named N is scanned to see if it has a substring matching X1; if so, the location is marked and the matching substring is excluded from further scanning and creates what is called a *segment gap*. The rest of the form is then scanned with respect to X1 to create any additional segment gaps. These are all given an internal code marking of 1. Anything not in a segment gap is now scanned with respect to X2 and segment gaps of code marking 2 are created. This is continued until all the Xi are used; then the marked form, along with the identifiers for the gaps, is put back into storage (still with the name N). The untouched portions of the string are called *segments*. This function essentially creates a macro in which the X1, X2, etc., indicate the dummy variables.

The following *call* functions make use of an internal *form pointer* which initially points to the first character of the form and is moved as required by the functions used. These *call* functions only act upon the string to the right of the location of the form pointer. The calls CS, CC, CN, and IN each advance the form pointer to the next untouched character.

#(CL,N,X1,X2, . . .)
> Means *call segmented form* and has two or more arguments. The form named N is retrieved from storage and its segment gaps of code marking 1 are filled with string X1, the gaps of marking 2 are filled with string X2, etc.

#(CS,N,Z)
> Means *call segment* and it obtains the string up to the next segment gap of the form named N; if the form is empty, its value is Z.

#(CC,N,Z)
> Calls exactly one character.

#(CM,N,D,Z)
> Calls M characters from the form named N and obtains as many more as are designated by an integer at the end of the string D.

#(IN,N,X,Z)
> Searches the form named N for the first location where the string X produces a match. The value is the string up to the character just before the matching string.

#(CR,N)
> Restores the pointer of the form named N to the initial character.

#(DD,N1,N2, . . .)
> Deletes the forms N1, N2, etc., and removes their names from the list of names.

#(DA)
> Deletes all forms and names.

On the assumption that there is a large-scale secondary storage unit, there are two functions which cause forms to be stored in a block on this unit and to be retrieved from there.

There are four diagnostic functions, designated as (LN,X), (PF,N), (TN), and (TF). The LN lists the names of all the forms in storage. The PF prints the form named N with an indication of the location and ordinal number of the segment gaps. The TN initiates a trace mode in which the neutral strings for each function are typed out as the computation proceeds. The TF terminates the trace mode.

As an interesting illustration of what can be accomplished, the procedure #(DS,N,# #(CL,N)#(DA)) deletes all forms but the one named N. This is done because the ##(CL,N) reads the form N into the processor and it is held as a neutral string while all the forms in memory are erased; then it is redefined with the original name. In this particular method, the segment gaps are lost.

There are very nice ways of using the TRAC functions to define programs which can then be called at any time desired, and which can also modify themselves after each execution of themselves. Examples of this are given in Mooers [ME67].

TRAC has combined concepts of LISP, COMIT, and macro facilities in a very unusual way. It is still too new for its long-range significance to be determined. However, to allow compatible exchange of TRAC language programs, Mooers strongly urges adherence to the latest standards of the language (as set forth by Rockford Research Institute Incorporated). Mooers also states that he is working on extensions which promise to make the language more powerful. One contemplated direction is to extend the string storage apparatus to strings of machine code; this would permit operating systems to be built in TRAC.

VI.9. LANGUAGES NOT WIDELY USED

VI.9.1. AMBIT

AMBIT is a language developed by C. Christensen (Massachusetts Computer Associates) in 1964. It has been implemented on only one machine in the United States and one in England, and I am not aware of any usage outside the author's organization. However, the implementation made by Christensen was written in ALGOL, and this makes it relatively machine independent.

In the developer's own words, "The AMBIT programming language arose from the decision to base the design of a programming language for mathematical symbol manipulation on the conventional notation for the identity."[16] In spirit it is very similar to COMIT and SNOBOL, which are discussed in Sections VI.6 and VI.7, respectively. In overall framework and appearance, there is some usage of concepts from ALGOL. The actual definition of AMBIT is definitely given in terms of a reference language containing about 150 "characters".

An identifier is a letter followed by any number of letters and/or digits. It is used for both data names and program unit labels. Blanks are definitely considered significant; unlike most other languages, strings will be considered equivalent if they differ only in the number and positioning of blanks according to a specifically defined set of rules. Certain other punctuation is also significant; in particular, the use of a semicolon, ; , for ending subunits of the program, and the use of a colon, : , following a statement label.

The input form can be a continuous string, but it is most commonly written with significant elements on different lines. From a conceptual point of view, the language is based on the use of a replacement rule which has a left-hand side denoting a string, together with at least one pointer that is expected to occur in the data stream. The left-hand side is followed by a right-hand arrow, then followed by the right-hand side which is the transformed string together with a pointer. For example,

$$(A\ /p\Delta\ B) = C \quad \rightarrow \quad A\ p\Delta = (B \times C)$$

Another way of stating the basic concept is that the replacement rule

---

[16] Christensen [CQ64], p. 1.

SAMPLE PROGRAM—AMBIT†

Problem: Clear fractions: Given an equation in suitable notation, transform the equation into an algebraically equivalent equation which contains no division operators or which contains only division operators which cannot be cleared by general methods. Both the given equation and the result equation will be fully parenthesized.

Program:

1.		**begin string dummy** S;	**phrase dummy** A, B, C;	
2.		sign **dummy** sign;	**segment dummy** seg;	
3.	ENTER:	GivenΔ(S)	$\rightarrow$	GivenΔ(S pΔ);
4.	LOOP:	**if ∃ / pΔ**		
5.		**then** (A /pΔ B)=C	$\rightarrow$	A pΔ =(B×C)
6.		**or** (A /pΔ B)sign C	$\rightarrow$	(A sign(B×C)) /pΔ B
7.		**or** (A /pΔ B)×C	$\rightarrow$	(A×C) /pΔ B
8.		**or** (A /pΔ B)/C	$\rightarrow$	A /pΔ (B×C)
9.		**or** (A /pΔ B)↑C	$\rightarrow$	(A↑C) /pΔ (B↑C)
10.		**or** A=(B /pΔ C)	$\rightarrow$	(A×C)=B pΔ
11.		**or** A sign(B /pΔ C)	$\rightarrow$	((A×C)sign B) /pΔ C
12.		**or** A×(B /pΔ C)	$\rightarrow$	(A×B) /pΔ C
13.		**or** A/(B /pΔ C)	$\rightarrow$	(A×C)/B pΔ
14.		**or** sign(B /pΔ C)	$\rightarrow$	(sign B) /pΔ C
15.		**or** / pΔ $\rightarrow$ pΔ /		
16.		**else** seg pΔ $\rightarrow$ pΔ seg ;		
17.		GivenΔ(pΔ S) $\rightarrow$ ResultΔ(S) **or**		
18.		**go to** LOOP; **end**		

Explanation: The program incorporates ten identities for the clearing of fractions (Lines 5-14). Each of these identities either eliminates a division operator from the equation or moves a division operator one level outward in the parenthesis structure of the equation. The application of the identities is controlled by a scanning pointer, 'pΔ', which moves from right to left through the equation.

The letters A, B, and C designate an arbitrary phrase, S denotes an arbitrary string, and sign and seg represent arbitrary signs and segments respectively.

---

†Christensen [CQ65], p. 255. By permission of Association for Computing Machinery, Inc.

---

consists of a *citation* followed by the special symbol $\rightarrow$ followed by a *replacement*. The major differences between AMBIT, and SNOBOL or COMIT are (1) the existence of the AMBIT pointer which is moved through the string according to some well-defined rules and which permits access to a specified position in the data; (2) a program structure based on ALGOL, which includes the existence of data types (e.g., numbers and identifiers) which can be handled naturally instead of just as character strings.

There are several declarations. The smallest executable unit is either the replacement rule just discussed, a control transfer, or an *existence Boolean*. The latter consists of the special symbol, ∃, followed by a citation. Each of the three executable units is defined as a simple imperative which can be grouped to form a compound imperative. When the latter is followed by semicolon, it is called an *imperative statement*. A compound imperative can also consist of compound imperatives combined with the use of the logical operators **AND**, **OR**, **NOT**, or something of the form **IF** $C1$ **THEN** $C2$ [**ELSE** $C3$] where the Ci are compound imperatives. The existence Boolean does not modify the data string but can be used to control the flow of the program by testing the format of the data string and causing appropriate branching; hence it may be considered a form of conditional statement. The overall program structure is derived from ALGOL and consists of **begin**, followed by the imperative statements, and terminated by **end**.

There is only a single data object in AMBIT, namely the data string. The data string is a sequence of legal characters and is considered to be divided into subsequences called *segments*. Typical segments are an identifier, a number, a sign, or a parenthesis. Parentheses must appear in matched pairs. From the users' point of view, the data string is a linear notation for a tree structure, i.e., a sequence of characters in which the structure is indicated by parenthesis pairs; in its internal representation, the data string is a symmetric linked list in which there are no common sublists.

A particular kind of segment called a *pointer* is central to the whole concept of AMBIT. A pointer is an identifier followed by the Greek capital letter delta, Δ.

A string description (which appears on the left side of the replacement rule) is composed of literals, dummy variables, and names. A dummy variable is an identifier which has been declared to represent an arbitrary member of a set of subsequences specified by the declaration. A dummy variable may be declared to represent one of a number of things: An element (a word which is a sequence of one or more alphanumerics or an arithmetic, relational, or logical symbol); a phrase (an element or (followed by a string followed by)); a segment (an element, (or)); a string (a sequence of none or any number of phrases and blanks); a sign (plus or minus); etc.

As indicated earlier, the only executable statements in AMBIT are the replacement rule, an unconditional control transfer, and conditional statements (including the existence Boolean). Of these, the replacement rule is the heart of the language. The replacement rule has a simple but severe restriction: Its left-hand side must contain at least one pointer, and sufficient parentheses to be unambiguous; i.e., the citation must be such that at any given time it can be matched with the data string in no more than one way. This has the practical effect of requiring fully parenthesized data, although one does not input it or output it in that form. The replacement rule is executed by finding, if possible, a subsequence of the current data string which is equivalent to the symbol sequence designated by the citation, and replacing that subsequence of the data string by a symbol sequence which is equivalent to the symbol sequence designated by the replacement.

If this action is possible, then it is performed and the replacement rule has the execution value *true*; if the action is not possible, the current data string is left unchanged and the replacement rule has the value *false*.

As an example,[17] consider the following replacement rule:

$$p\Delta \ (A \times A) \rightarrow p\Delta \ (A \uparrow 2)$$

and assume that A is declared **phrase dummy**. If this replacement rule is executed for the data string

$$EQ\Delta \ ((ALPHA+p\Delta((M+1) \times (M+1))) = 6.0)$$

then the replacement rule succeeds and the modified data string is

$$EQ\Delta \ ((ALPHA+p\Delta((M+1) \uparrow 2)) = 6.0)$$

However if the data string is

$$EQ\Delta \ ((ALPHA+p\Delta((M+1) \times (M+5))) = 6.0)$$

then the replacement rule fails and the data string is not changed.

There are no input/output commands within AMBIT itself. The program is inserted, followed by a data string; after execution, the resultant data string is then put out automatically.

A number of declarations exist to permit the association of one of the data types with either a dummy name, multiple name, or literal.

The design of the language has been made with the overall implementation plan (see Christensen [CQ66]) very much in mind so as to permit certain types of maximal efficiency to be obtained.

This system does not seem particularly suited for practical usage even in the areas for which it provides appropriate facilities, and it does not seem to have a solid enough theoretical foundation to provide new insight in any particular direction. Perhaps further usage and development will make either or both of these statements less true.

VI.9.2. TREET

TREET is a list processing language that was originally developed by E. C. Haines [HA65] as a Master's thesis at M.I.T. in 1964 and improved since then. It is based on LISP, in the sense that it uses many of the same concepts and philosophies. As an avowed objective, however, "The author has adopted what he considers to be the most useful and important aspects

---

[17] Christensen [CQ64], pp. 41–42.

of LISP as a programming language"[18] and put them into a more convenient notation, with emphasis on reducing the parenthesis problem. The reader may be interested in contrasting TREET with LISP 2 (see Section VIII.6).

TREET was originally designed to do natural language processing and is well suited for manipulating tree structures. It can also be used on-line for some query work, as well as for displaying trees. (See the reference list at the end of this chapter for applications using TREET.) It is considered a research tool and hence undergoes frequent changes. It was implemented on the IBM 7030 (STRETCH) and is being implemented on the IBM System/360. The description here is based on the IBM 7030 version.

SAMPLE PROGRAMS—TREET

Problem: Reverse a list

Program:
```
DEF( REVERSE (A) (B)
    WHILE(A ADL(CHOP(A) B));
    RETURN(B);
    )
```

Problem: Test whether two list structures are equal, where "equal" means they have the same hierarchical structure and the same elementary symbols in corresponding positions.

Program:
```
DEF( EQUAL (A B) ()
    IF (A EQ B) THEN RETURN('TRUE');
    IF (ATOM(A) OR ATOM(B)) THEN RETURN(NIL);
    RETURN((EQUAL(MEM1(A) MEM1(B)) AND
           EQUAL(REM1(A) REM1(B)) ));
    )
```

There is both a reference language and a hardware representation. The character set for the former consists of the 10 digits, the 26 upper-case letters, and the following special characters:

+ − × / < = > ∧ ∨ ⌐ ( ) . : ; * '   blank

There are a large number of key words. An identifier in the hardware representation is a letter followed by up to nine letters and/or digits; it is used for both data names and statement labels. (Apparently it can be the same as one of the key words, since there is no restriction against this.)

The arithmetic operators are the first four graphics shown above; the

---

[18] Haines [HA65], p. 3.

relational operators are the next three graphics, and the logical operators are the following three. Identifiers must be delimited on each side by one of the following characters:

: ; * ( )   blank

Any number of blanks can be used where one is permitted. Literals are delimited by a prime, ' , on either side.

The input is free format.

A TREET program consists of one or more user-defined functions. Even things normally considered statements, e.g., IF, DO, are considered functions, although their format is not necessarily of that type. Almost any function can serve as the main program, invoking other functions. Comments can be inserted into any function definition; they must be preceded by an asterisk, *, and must end either with an * or at the end of a unit record (e.g., column 72 of a card). The programs themselves are stored as list structures; they will normally be compiled after they are debugged. The macro assembler TAP is part of the implementation and can be used to define functions; operations include both list processing and machine code.

Recursion is definitely permitted, but it is not considered as important as in LISP. Statements in the reference language are terminated with a semicolon except in a few special cases in which it can be omitted.

The data type allowed in TREET is (only) the LISP S-expression (defined in Section VI.5); all lists are terminated by a right parenthesis. However, an additional type of atom is a card, which contains 80 characters and is dynamically allocated and reclaimed just as single cells are. List cells have a third address field which may be used for backward pointers or additional data. Integers and floating point numbers are permitted. All calculation is done in floating point and converted to integer if the result is integral. There are no logical values defined as data types, since TREET adopts the convention that NIL (representing the empty list) means false and all other S-expressions are true.

There are certain system-defined variables which can be used by all functions.

The assignment operation is defined as a function called *replacement* and is written with := in the reference language, e.g., X := X + 5.

The unconditional transfer is written GOTO (*label*). The *RETURN* causes the current function to finish and have the value of the argument of *RETURN*. The conditional statement is of the form IF *function-1* THEN *function-2* [ELSE *function-3*];. *function-1* is evaluated and if it is not NIL (i.e., false), then the value of *function-2* is the value of the IF expression. The predicate (i.e., *function-1*) always has a value, assigned to the *free variable* P, which can then be used by other functions. For example, the statement IF (Y=6) THEN

RETURN ((IF Z THEN 5 ELSE 7)); assigns to P the value **TRUE** if Y = 6 and Z is true, and the value **NIL** if Y=6 and Z is false. If Y ≠ 6, P is **NIL**.

In cases where only a single argument to a function is normally allowed, the effect of executing several statements can be achieved by writing **DO** (*statements*).

The loop control is a *WHILE* function with two arguments. The second (which is usually a *DO*) is evaluated as long as the first is true, e.g.,

$$\text{WHILE } ((X < 5) \text{ DO } (\\ Y := X + Y;\\ X := X + 1;\\ ));$$

The following is a set of list processing functions which are either equivalent to or identical with certain LISP statements:

MEM1(*list*)	First member of a list (same as LISP CAR).
MEM2(*list*)	Second member of a list (same as LISP CADR).
MEM3(*list*)	Third member of a list (same as LISP CADDR).
MEM12(*list*)	First member of the second member of a list (same as LISP CAADR).
REM1(*list*)	Remainder list after first element is removed (same as LISP CDR).
REM2(*list*)	Remainder list after first two elements are removed (same as LISP CDDR).
REM3(*list*)	Remainder list after first three elements are removed (same as LISP CDDDR).
CHOP(*var*)	Value is the first member of the list that is the value of the variable and that variable is set to REM1 of the list.

The following LISP functions are also included in TREET: *LIST, NULL, CONS, ATOM, MEMBER,* = (same as LISP *EQUAL), RPLACA, RPLACD,* and *COPY.*

There are a number of output functions. One is *PRINT,* followed by any number of arguments which are printed with a single space between them. Writing

$$\text{PRINT ('Y =' Y 'AND ITS SQUARE IS' (Y × Y));}$$

causes

$$\text{Y = 7 AND ITS SQUARE IS 49}$$

to be printed if Y had the value 7.

Also available are commands to TYPE on the on-line printer and to display various information (including trees and lists of atoms) and arbitrary text on the on-line display. The light pen may be used on all these. Various input functions also exist. TREET is part of AESOP (see Bennett, Haines, and Summers [QA65]) and has available to it a different set of displays and operations from that system.

User-created functions are specified by writing DEF followed by the function name, the argument list, the variable list, and statements, all enclosed within parentheses, e.g., in

DEF (HELLO(ABC YES) (GOLD) IF ATOM (ABC) THEN RETURN (ABC) ELSE RETURN (GOLD);)

GOLD will have the value NIL.

TREET is intertwined with LISP in several ways. Its original objective was to provide almost the identical capability except in an easier notation. However this compatibility has not continued to be stressed. TREET functions are translated into the LISP 1.5 syntax.

There are various types of error-checking facilities; e.g., if a double asterisk, **, is found anywhere other than between functions, it is considered a read context error; this avoids the problem of matching parentheses count. Most routines check the legality of the structure with which they are dealing; when an error is detected, control is transferred to the ERR routine which does a number of things. A function EVALL dumps the values of all system variables.

VI.9.3. Others

A few other systems of limited usage are worth mentioning because they provide somewhat different approaches to some of the problems in the list and string processing area.

1. *CLP*

The *Cornell List Processor* (CLP) is a definite language addition to CORC (see Section IV.8.1). The basic concept is to add a new type of data element called *entity*, which is a particular kind of data record; the total number of entities can vary at object time. Entities can be subdivided into *lists* and can have attributes, which are simply variables with the entity actually being used as a subscript. Thus the user can write LENGTH OF CAR and this may be used as a variable. An ERASE statement eliminates such a variable. Elements can be inserted and deleted from lists by writing INSERT A AFTER B ON C or INSERT A LAST ON B. An element can be inserted in a sequenced list by writing INSERT A RANKED ON B. It is also possible to use the REPEAT statement over all the entities that exist at a given time, e.g., REPEAT A FOR ALL B ON C.

There are strong similarities between CLP and SIMSCRIPT (see Section IX.3.1.4).

2. CORAL

CORAL (*C*lass *O*riented *R*ing *A*ssociative *L*anguage) was developed by L. G. Roberts [RB65] at the M.I.T. Lincoln Laboratory for the TX-2 computer during 1964. It is used on-line for graphical display and systems work. CORAL used a type of list structure called a *ring* (based on earlier work of I. Sutherland on Sketchpad [QW63]). The basic principle involved having every element (except the first) point either to the start of the list or back to the immediately preceding element. From a language point of view, the notation and symbolism are highly dependent on the specific peculiarities of the TX-2 computer and the character set of the Lincoln writer keyboards. The instructions available to the user are at a very low level, as they are in IPL-V and L^6. Some of the facilities are: go up *N* memory locations from point *P*, go up or down, and go left or right from *P* in a ring *N*; create and rearrange list structure elements; test whether the pointer points to the start of the ring or to certain other specific types of elements; obtain information in the accumulator from commands such as find the length of the block and find a list length; and perform an action for each element in a ring, such as go right or left around a ring doing a specified action. Figure VI-8 shows two statement lines which are typical of CORAL programs, where the reader must realize that a very special keyboard is being used.

3. SPRINT

SPRINT is an attempt by C. A. Kapps [KC67] to develop a list processing language which is as machine independent as possible. It has been implemented on the IBM 7094 and 7040.

A word consists of a variable length string of alphanumeric characters (with an implementation-imposed upper limit of 179 characters). Lists contain either programs or data, and they cannot be distinguished except by context. Lists are named and stored in an associative type of memory area and are located by means of their names. Programs are executed in an area consisting of two pushdown stacks called the *instruction* and *operand* stacks. Basic SPRINT instructions refer to the top several levels of the operand stacks for their arguments and results.

Operations are denoted by a three-letter mnemonic followed by an appropriate number of operands and include things such as the four arithmet-

(⊛BLK1→NEWBLK)⊚((OLDBLK↓3)⦃(OLDBLK K̄5))

((𝟋)𝟋NAMRING)Ē▷((OLDBLK↓1)⊳SUBR1⊸LABEL1)

Figure VI-8. Example of CORAL statements. The main operator in the first line is ⊚ (put right). The element to be inserted is the first tie register of a newly created block of type BLK1. BLK1 is a symbolic name for some integer type number. The pointer named NEWBLK is assigned the address of the inserted ring element. The new tie element will be inserted next to an element determined by what follows the ⊚ operator. To find the reference element, the program will move the pointer OLDBLK down three registers and then left (backwards) around that ring. The number of steps moved is determined by the contents of the data structure register found by moving the pointer OLDBLK down five. The accumulator's contents are identical with those of NEWBLK after the statement has been executed.

The second line takes the accumulator pointer to the newly inserted tie element, moves to its ring start, then to the top of the block containing the ring start, and down a distance equal to NAMRING. If the resulting tie register is not empty, control goes on to the next line. If the register is empty, then the subroutine SUBR1 is performed for each member of the ring found by moving the pointer OLDBLK down one. After all the members of the ring have been treated, control transfers to LABEL1.

Source: Sutherland [SU66], p. 29.

ic operators (*ADD, SUB, MPY, DIV*); read and write data (*RDT, WDT*); reverse two elements (*REV*); repeat (*REP*); delete (*DEL*); concatenate (*CON*) and deconcatenate (*DCN*); bring (*BNG*) a named list; call (*CLL*), which puts a named list into the instruction stack; transfer (*TRA*); store (*STO*); find (*FND*), which affects the flow of control; and test for zero (*TZE*).

To evaluate the algebraic expression $((A+B)C+D)(E-10)$, one would execute the following program:

A B ADD C MPY D ADD E ⊕ 10 SUB MPY

where the ⊕ indicates data and all other marks are assumed to be instructions.

Elements of a matrix can be named independently of the size of the matrix; there is completely dynamic storage allocation which never requires the use of declarations for array size.

It is possible not only to write subroutines but to execute them after they have been written or modified at object time.

SPRINT is admitted by the developer to be inefficient because it is an experimental attempt and is designed to be efficient on machines of an entirely different logical structure.

4. LOLITA

LOLITA (*L*anguage for the *O*n *L*ine *I*nvestigation and *T*ransformation of *A*bstractions) developed by Blackwell [QK67] is a part of the extended Culler-Fried on-line system (see Section IV.6.9). The basic objective was simplicity in concept, implementation, and usage. LOLITA is essentially a language for processing strings of symbols, but they are really two-way lists (although without any sublist structure). The user has available to him nine lists in core storage and many more using the drum. The system uses the Culler-Fried console which consists of a display oscilloscope and two special keyboards. There is a specific memory location called the *symbol accumulator* (*SA*), which is fundamental to the use of the elementary operators and is similar to an arithmetic accumulator. The elementary operators which are provided in the system, *all of which are used by pushing a button*, include the following: Define a list of symbols (**DLIST**); create a single symbol (**DSYM**); load symbol into *SA* (**LS**); store symbol before (or after) (**SSB, SSA**), which puts the symbol in *SA* on a designated list immediately before (or after) the place pointed to by the list index; remove the symbol pointed to by the index list (**ES**); concatenate the indexed symbol and its predecessor to form one symbol (**CON**); set index, increase index (**SI, II**); unconditional and conditional branching (**BRANCH**); **DISPLAY** currently designated list on the scope; control the exact format of a display (**DC**); place the console in the symbol manipulation mode (**ISM**); **STORE, LOAD**; the **SUB** key causes the typed line to be half-spaced down so that subsequent characters appearing on the display scope show as subscripts; the superscript key is similar; and **LIST** the program on the scope and have the system remember it.

The main idea to realize is that the operations provided here are very elementary, but the console and the Culler-Fried system within which this is embedded permit the user to build up much more complex operations.

This system has been used to program some simple operations in algebra, e.g.,

Rule to be applied: $\quad D_T E^U = (D_T U) * E^U$
Input expression: $\quad \pi + D_X(E^{SIN\ X})$
Transformed expression: $\quad \pi + (D_X\ SIN\ X) E^{SIN\ X}$

REFERENCES

VI.1. SCOPE OF CHAPTER

[AH67] Abrahams, P. W., "List-processing Languages", *Digital Computer User's Handbook* (M. Klerer and G. A. Korn, eds.). McGraw-Hill, New York, 1967, pp. 1-239–1-257.

[BB64a]	Bobrow, D. G. and Raphael, B., "A Comparison of List Processing Languages", *Comm. ACM*, Vol. 7, No. 4 (Apr., 1964), pp. 231–40. (Also in [RO67].)
[BB64b]	Bobrow, D. G. and Weizenbaum, J., "List Processing and Extension of Language Facility by Embedding", *IEEE Trans. Elec. Comp.*, Vol. EC-13, No. 4 (Aug., 1964), pp. 395–400.
[BB67a]	Bobrow, D. G. (ed.), *Symbol Manipulation Languages and Techniques, Proceedings of the IFIP Working Conference on Symbol Manipulation Languages*. North-Holland Publishing Co., Amsterdam, 1968.
[DQ66]	Dodd, G. G., "APL—A Language for Associative Data Handling in PL/I", *Proc. FJCC*, Vol. 29 (1966), pp. 677–84.
[GB61a]	Green, B. F., "Computer Languages for Symbol Manipulation", *IRE Trans. Elec. Comp.*, Vol. EC-10, No. 4 (Dec., 1961), pp. 729–735.
[GE60]	Gelernter, H., Hansen, J. R., and Gerberich, C. L., "A FORTRAN-Compiled List-Processing Language", *J. ACM*, Vol. 7, No. 2 (Apr., 1960), pp. 87–101.
[JG67]	Gray, J. C., "Compound Data Structure for Computer Aided Design; A Survey", *Proc. ACM 22nd Nat'l Conf.* 1967, pp. 355–65.
[NW56]	Newell, A. and Simon, H. A. "The Logic Theory Machine—A Complex Information Processing System", *IRE Trans. Information Theory*, Vol. IT-2, No. 3 (Sept. 1956), pp. 61–79.
[RA66a]	Raphael, B., *Symbol Manipulation by Digital Computer*, Stanford Research Inst., Menlo Park, Calif. (Apr., 1966) (to be published).
[RA66b]	Raphael, B., "Aspects and Applications of Symbol Manipulation", *Proc. ACM 21st Nat'l Conf.*, 1966, pp. 69–74.
[RA67]	Raphael, B. *et al.*, "A Brief Survey of Computer Languages for Symbolic and Algebraic Manipulation", *Symbol Manipulation Languages and Techniques, Proceedings of the IFIP Working Conference on Symbol Manipulation Languages* (D. G. Bobrow, ed.). North-Holland Publishing Co., Amsterdam, 1968, pp. 1–54.
[SA65]	Sakoda, J. M., *DYSTAL Manual—Dynamic Storage Allocation Language in Fortran*, Brown U., Dept. of Sociology and Anthropology, Providence, R.I. (1965, revised).
[SM66]	Sammet, J. E., "An Annotated Descriptor Based Bibliography on the Use of Computers for Non-Numerical Mathematics", *Computing Rev.*, Vol. 7, No. 4 (July–Aug. 1966), pp. B1-B31
[SM67a]	Sammet, J.E., "Revised Annotated Descriptor Based Bibliography on the Use of Computers for Non-Numerical Mathematics", *Symbol Manipulation Languages and Techniques, Proceedings of the IFIP Working Conference on Symbol Manipulation Languages* (D. G. Bobrow, ed.). North-Holland Publishing Co., Amsterdam, 1968, pp. 358–484.
[SR66]	Satterthwait, A. C., "Programming Languages for Computational Linguistics", *Advances in Computers*, Vol. 7 (F. L. Alt and M. Rubinoff, eds.). Academic Press, New York, 1966, pp. 209–238.
[VB67]	Madnick, S. E., "String Processing Techniques", *Comm. ACM*, Vol. 10, No. 7 (July, 1967), pp. 420–24.
[WI64a]	Wilkes, M. V., "Lists and Why They Are Useful", *Proc. ACM 19th Nat'l Conf.*, 1964, pp. F1-1–F1-5.

[WZ63] Weizenbaum, J., "Symmetric List Processor", *Comm. ACM*, Vol. 6, No. 9 (Sept., 1963), pp. 524–44.

[ZH66] Guzman, A. and McIntosh, H. V., "CONVERT", *Comm. ACM*, Vol. 9, No. 8 (Aug., 1966), pp. 604–15.

VI.3. IPL-V

[CZ64] Chapin, N., "An Implementation of IPL-V on a Small Computer", *Proc. ACM 19th Nat'l Conf.*, 1964, pp. D1.2-1–D1.2-6.

[DU63] Dupchak, R., *TIPL: Teach Information Processing Language*, RAND Corp., RM-3879-PR, Santa Monica, Calif. (Oct., 1963).

[DU65] Dupchak, R., *LIPL: Linear Information Processing Language*, RAND Corp., RM-4320-PR, Santa Monica, Calif. (Feb., 1965).

[FG61] Feigenbaum, E. A., "The Simulation of Verbal Learning Behavior", *Computers and Thought* (E. Feigenbaum and J. Feldman, eds.). McGraw-Hill, New York, 1963, pp. 297–309.

[FN62] Feldman, J., "TALL—A List Processor for the Philco 2000 Computer", *Comm. ACM*, Vol. 5, No. 9 (Sept., 1962), pp. 484–85.

[GB61] Green, B. F. *et al.*, "Baseball: An Automatic Question-Answerer", *Proc. WJCC*, Vol. 19 (1961), pp. 219–24. (Also in [FG63].)

[HB63] Hunt, E. B. and Hovland, C. I., "Programming a Model of Human Concept Formulation", *Computers and Thought* (E. A. Feigenbaum and J. Feldman, eds.). McGraw-Hill, New York, 1963, pp. 310–25.

[HX64] Hodges, D., *IPL-VC, A Computer System Having the IPL-V Instruction Set*, Argonne Natl. Lab., ANL-6888, Applied Mathematics Division, Argonne, Ill. (May, 1964).

[JC58] Shaw, J. C. *et al.*, "A Command Structure for Complex Information Processing", *Proc. WJCC* (May, 1958), pp. 119–28.

[NW56] Newell, A. and Simon, H. A., "The Logic Theory Machine—A Complex Information Processing System", *IRE Trans. Information Theory*, Vol. IT-2, No. 3 (Sept., 1956), pp. 61–79.

[NW57] Newell, A., Shaw, J. C., and Simon, H. A., "Empirical Explorations of the Logic Theory Machine: A Case Study in Heuristic", *Proc. WJCC* (Feb., 1957), pp. 218–30.

[NW57a] Newell, A. and Shaw, J. C., "Programming the Logic Theory Machine", *Proc. WJCC* (Feb., 1957), pp. 230–40.

[NW60] Newell, A. and Tonge, F. M., "An Introduction to Information Processing Language-V", *Comm. ACM*, Vol. 3, No. 4 (Apr., 1960), pp. 205–11.

[NW61] Newell, A. and Simon, H. A., "GPS, A Program That Simulates Human Thought", *Computers and Thought* (E. Feigenbaum and J. Feldman, eds.). McGraw-Hill, New York, 1963, pp. 279–93.

[NW63] Newell, A., "Documentation of IPL-V", *Comm. ACM*, Vol. 6, No. 3 (Mar., 1963), pp. 86–89.

[NW65] Newell, A. *et al.* (eds.), *Information Processing Language-V Manual*, 2nd ed. Prentice-Hall, Inc., Englewood Cliffs, N.J., 1965.

[SF63] Stefferud, E., The Logic Theory Machine: A Model Heuristic Program, RAND Corp., Memorandum RM-3731-CC, Santa Monica, Calif. (June, 1963).

[TN60] Tonge, F. M., "Summary of a Heuristic Line Balancing Procedure", *Computers and Thought* (E. Feigenbaum and J. Feldman, eds.). McGraw-Hill, New York, 1963, pp. 168–90.

[ZK63] Gullahorn, J. T. and Gullahorn, J. E., "A Computer Model of Elementary Social Behavior", *Computers and Thought* (E. A. Feigenbaum and J. Feldman, eds.). McGraw-Hill, New York, 1963, pp. 375–85.

[ZN65] Cowell, W. R. and Reed, M. C., *A Checker-Playing Program for the IPL-VC Computer*, Argonne Natl. Lab., ANL-7109, Applied Mathematics Division, Argonne, Ill. (Oct., 1965).

VI.4. L^6

[KO65] Knowlton, K. C., "A Fast Storage Allocator", *Comm. ACM*, Vol. 8, No. 10 (Oct., 1965), pp. 623–25.

[KO66] Knowlton, K. C., "A Programmer's Description of L^6", *Comm. ACM*, Vol. 9, No. 8 (Aug., 1966), pp. 616–25.

VI.5. LISP 1.5

[AH63] Abrahams, P. W., *Machine Verification of Mathematical Proof*, International Electric Corp., Report No. P-AA-TR-(0045), ITT, Paramus, N.J. (May, 1963).

[BB64c] Bobrow, D. G., "A Question-Answering System for High School Algebra Word Problems", *Proc. FJCC*, Vol. 26, pt. 1 (1964), pp. 591–614.

[BB67] Bobrow, D. G. et al., *The BBN 940 LISP System*, Contract No. AF19(628)-5065, Project No. 8668, Bolt, Beranek and Newman, Scientific Report No. 9, Cambridge, Mass. (July, 1967).

[BY66] Berkeley, E. C. and Bobrow, D. G. (eds.), *The Programming Language LISP—Its Operation and Applications*. M.I.T. Press, Cambridge, Mass., 1966.

[EN65] Engelman, C., "MATHLAB—A Program for On-Line Machine Assistance in Symbolic Computations", *Proc. FJCC*, Vol. 27 (Nov., 1965), pp. 413–22.

[EV64] Evans, T. G., "A Heuristic Program to Solve Geometric-Analogy Problems", *Proc. SJCC*, Vol. 25 (Apr., 1964), pp. 327–38.

[GE60] Gelernter, H., Hansen, J. R., and Gerberich, C. L., "A FORTRAN-Compiled List-Processing Language", *J. ACM*, Vol. 7, No. 2 (Apr., 1960), pp. 87–101.

[MC60] McCarthy, J. et al., *LISP 1 Programmer's Manual*, M.I.T. Computation Center and Research Lab. of Electronics, Cambridge, Mass. (Mar., 1960).

[MC60a] McCarthy, J., "Recursive Functions of Symbolic Expressions and Their Computation by Machine, pt. 1", *Comm. ACM*, Vol. 3, No. 4 (Apr., 1960), pp. 184–95. (Also in [RO67].)

[MC62] McCarthy, J. et al., *LISP 1.5 Programmer's Manual*, M.I.T. Computation Center and Research Lab. of Electronics, Cambridge, Mass. (Aug., 1962).

[MT60] *LISP I* (Programmer's Manual), M.I.T. Computation Center and Research Lab. of Electronics, Cambridge, Mass. (Mar., 1960).

[SL61] Slagle, J. R., "A Heuristic Program that Solves Symbolic Integration Problems in Freshman Calculus", *J. ACM*, Vol. 10, No. 4 (Oct., 1963), pp. 507-20. (Also in [FG63].)

[UB62] Bastian, A. L., Foley, J. P., and Petrick, S. R., "On the Implementation and Usage of a Language for Contract Bridge Bidding", *Symbolic Languages in Data Processing*. Gordon and Breach, New York, 1962, pp. 741-58.

[UQ65] Brody, T. A., *A LISP Processor for the IBM 1620*, Centro Nacional de Calculo, Instituto Politecnico Nacional, Unidad Zacatenco, Mexico, D. F. (1965).

[WE67] Weissman, C., *LISP 1.5 Primer*. Dickenson Publishing Co., Belmont, Calif., 1967.

[WI64a] Wilkes, M. V., "Lists and Why They are Useful", *Proc. ACM 19th Nat'l Conf.*, 1964, pp. F1-1-F1-5.

[WX66] Woodward, P. M., "List Programming", *Advances in Programming & Non-Numerical Computation* (L. Fox, ed.). Pergamon Press, New York, 1966, pp. 29-48.

[WZ67] Weizenbaum, J., "Review R67-22 (of *The LISP 2 Programming Language and System*)", *IEEE Trans. Elec. Comp.*, Vol. EC-16, No. 2 (Apr., 1967), pp. 236-38.

[ZH66] Guzman, A. and McIntosh, H. V., "CONVERT", *Comm. ACM*, Vol. 9, No. 8 (Aug., 1966), pp. 604-15.

[ZP41] Church, A., *The Calculi of Lambda-Conversion*. Princeton U. Press, Princeton, N.J., 1941.

VI.6. COMIT

[DA65] Darlington, J. L., "Machine Methods for Proving Logical Arguments Expressed in English", *Mechanical Translation*, Vol. 8, Nos. 3 and 4 (June, Oct., 1965), pp. 41-67.

[HT66] Hilton, W. R. and Hillman, D. J., *The Structure of LECOM*, Lehigh U., Center for the Information Sciences, Bethlehem, Pa. (June, 1966).

[MT61] *COMIT Programmers' Reference Manual*, M.I.T. Research Lab. of Electronics and the Computation Center, Cambridge, Mass. (Nov., 1961).

[MT61a] *An Introduction to COMIT Programming*, M.I.T. Research Lab. of Electronics and the Computation Center, Cambridge, Mass. (Nov., 1961).

[SJ63] Stone, P. J. and Hunt, E. B., "A Computer Approach to Content Analysis: Studies Using the General Inquirer System", *Proc. SJCC*, Vol. 23 (1963), pp. 241-56.

[YN57] Yngve, V. H., "A Framework for Syntactic Translation", *Mechanical Translation*, Vol. 4, No. 3 (Dec., 1957), pp. 59-65.

[YN58]   Yngve, V. H., "A Programming Language for Mechanical Translation", *Mechanical Translation*, Vol. 5, No. 1 (July, 1958), pp. 25–41.
[YN62]   Yngve, V. H., "COMIT as an IR Language", *Comm. ACM*, Vol. 5, No. 1 (Jan., 1962), pp. 19–28. (Also in [RO67].)
[YN63b] Yngve, V. H., "COMIT", *Comm. ACM*, Vol. 6, No. 3 (Mar., 1963), pp. 83–84.
[YN66]   Yngve, V. H., *COMIT Programming* (course notes), U. of Chicago, Chicago, Ill. (Autumn quarter, 1966).
[ZH66]   Guzman, A. and McIntosh, H. V., "CONVERT", *Comm. ACM*, Vol. 9, No. 8 (Aug., 1966), pp. 604–15.
[ZS66]   Chanon, R. N., "Almost Alike Programs", *Proc. ACM 21st Nat'l Conf.*, 1966, pp. 215–22.

VI.7. SNOBOL

[DS67]   Desautels, E. J. and Smith, D. K., "An Introduction to the String Manipulation Language SNOBOL", *Programming Systems and Languages* (S. Rosen, ed.). McGraw-Hill, New York. 1967, pp. 419–54.
[FB64]   Farber, D. J., Griswold, R. E., and Polonsky, I. P., "SNOBOL, A String Manipulation Language", *J. ACM*, Vol. 11, No. 1 (Jan., 1964), pp. 21–30.
[FB66]   Farber, D. J., Griswold, R. E., and Polonsky, I. P., "The SNOBOL3 Programming Language", *Bell System Tech. Jour.*, Vol. XLV, No. 6 (July–Aug., 1966), pp. 895–944.
[FT67]   Forte, A., "The Programming Language SNOBOL3: An Introduction", *Computers and the Humanities*, Vol. 1, No. 5 (May, 1967), pp. 157–63.
[FT67a] Forte, A., *SNOBOL3 Primer: An Introduction to the Computer Programming Language*. M.I.T. Press, Cambridge, Mass., 1967.
[GF65]   Griswold, R. E. and Polonsky, I. P., *Tree Functions for SNOBOL3*, Bell Telephone Lab., Holmdel, N.J. (Feb., 1965) (unpublished).
[GF65a] Griswold, R. E., *Linked-List Functions for SNOBOL3*, Bell Telephone Lab., Holmdel, N.J. (June, 1965) (unpublished).
[GF67]   Griswold, R. E., Poage, J. F., and Polonsky, I. P., *Preliminary Description of the SNOBOL4 Programming Language*, Bell Telephone Lab., Holmdel, N.J. (July, 1967) (unpublished).
[GF68]   Griswold, R.E., Poage, J. F., and Polonsky, I. P., *Preliminary Report on the SNOBOL4 Programming Language, II*, Bell Telephone Lab., S4D4b, Holmdel, N.J. (March 1968) (unpublished).

VI.8. TRAC

[ME65]   Mooers, C. N. and Deutsch, L. P., "TRAC, a Text Handling Language", *Proc. ACM 20th Nat'l Conf.*, 1965, pp. 229–46.
[ME66]   Mooers, C. N., "TRAC, a Procedure-Describing Language for the Reactive Typewriter", *Comm. ACM*, Vol. 9, No. 8 (Mar., 1966), pp. 215–19.

[ME67] Mooers, C. N., "How Some Fundamental Problems Are Treated in the Design of the TRAC Language", *Symbol Manipulation Languages and Techniques, Proceedings of the IFIP Working Conference on Symbol Manipulation Languages* (D. G. Bobrow, ed.). North-Holland Publishing Co., Amsterdam, 1968, pp. 178–88.

[SQ65] Strachey, C., "A General Purpose Macrogenerator", *Computer Jour.*, Vol. 8, No. 3 (Oct., 1965), pp. 225–41.

VI.9.1. AMBIT

[CQ64] Christensen, C., *AMBIT: A Programming Language for Algebraic Symbol Manipulation*, Computer Associates, Inc., Report No. CA-64-4-R, Wakefield, Mass. (Oct., 1964).

[CQ65] Christensen, C., "Examples of Symbol Manipulation in the AMBIT Programming Language", *Proc. ACM 20th Nat'l Conf.*, 1965, pp.247–61.

[CQ66] Christensen, C., "On the Implementation of AMBIT, A Language for Symbol Manipulation", *Comm. ACM*, Vol. 9, No. 8 (Aug., 1966), pp. 570–73.

VI.9.2. TREET

[GQ67] Gross, L. N., *On-Line Programming System User's Manual*, MITRE Corp., MTP-59, Bedford, Mass. (Mar., 1967).

[HA65] Haines, E. C., Jr., *The TREET List Processing Language*, MITRE Corp., Information System Language Studies Number Eight SR-133, Bedford, Mass. (M.S. thesis, M.I.T.) (Apr., 1965).

[HA67] Haines, E. C., Jr., *The TREET Programming System: IBM 7030 Implementation*, MITRE Corp., MTP-58, Bedford, Mass. (Mar., 1967).

[QA65] Bennett, E., Haines, E. C., Jr., and Summers, J. K., "AESOP: A Prototype for On-Line User Control of Organizational Data Storage, Retrieval and Processing", *Proc. FJCC*, Vol. 27, pt. 1 (1965), pp. 435–55.

[ZT67] Chapin, P. G. et al., *SAFARI, An On-Line Text-Processing System User's Manual*, MITRE Corp., MTP-60, Bedford, Mass. (Mar., 1967).

VI.9.3. Others

[CN65] Conway, R. W. et al., "CLP—The Cornell List Processor", *Comm. ACM*, Vol. 8, No. 4 (Apr., 1965), pp. 215–16.

[KC67] Kapps, C. A., "SPRINT: A Direct Approach to List-Processing Languages", *Proc. SJCC*, Vol. 30 (1967), pp. 677–83.

[QK67] Blackwell, F. W., "An On-Line Symbol Manipulation System", *Proc. ACM 22nd Nat'l Conf.*, 1967, pp. 203–209.

[QW63] Sutherland, I. E., *Sketchpad: A Man-Machine Graphical Communication System*, M.I.T. Lincoln Lab., Tech. Report No. 296, Lexington, Mass. (Jan., 1963).

[RB65] Roberts, L. G., "Graphical Communication and Control Languages", *Information System Sciences: Proceedings of the Second Congress*. Spartan Books, Washington, D.C., 1965, pp. 211–17.

[SU66] Sutherland, W. R., *On-Line Graphical Specification of Computer Procedures*, M.I.T. Lincoln Lab., Tech. Report No. 405, Lexington, Mass. (May, 1966).

VII FORMAL ALGEBRAIC MANIPULATION LANGUAGES

VII.1. SCOPE OF CHAPTER

This chapter deals with languages which provide formal algebraic manipulation facilities. The phrase *formal algebraic manipulation* applies to the computer processing of formal mathematical expressions without any particular concern for their numeric values. The main criterion for being included in this area is one of intent rather than methodology; it will be seen later that in at least one case the implementation techniques are in fact numeric (rather than symbolic), although the motivation is to operate on algebraic expressions. The phrase (*formal*) *algebraic manipulation* is being used in preference to a slightly more common term, namely *formula manipulation*, because the latter is somewhat misleading. It has often been used by me, as well as others, to apply to formulas *and/or* expressions. Since there is often a need to distinguish between formulas and expressions, the more general phrase, *algebraic manipulation*, is being used. It must be emphasized that this term in no way excludes the use of trigonometric or other types of functions. Examples of the type of operations involved are differentiation, substitution, integration, and deletion of a portion of an expression. In order to appreciate the significance of this field, a little historical background is necessary. Detailed discussions have been given in several papers by the author (e.g., Sammet [SM66a] and [SM67]).

The primary motivation for the initial development of digital computers was to solve numeric problems. This is generally conceded to be true, in spite of the fact that the first UNIVAC was delivered to the United States Census Bureau, whose problems by more modern standards would be considered inherently data processing. Because of this tendency for using a computer to solve numeric problems, there has been a major shift in emphasis in the technology of applied mathematics. Prior to the existence of digital computers, numerical analysis was a very lightly taught subject. Many people who had problems in applied mathematics to solve either used analytic techniques or simply did not solve the problems. (A distinction is being made here between solving problems and merely doing large amounts of computation with a desk calculator, e.g., computing firing tables.) With the advent of computers came such a major development in numerical analysis that analytic techniques tended to be ignored or forgotten. Languages like the ones in this chapter are attempting to reverse the trend.

The first known uses of a computer to do what can reasonably be called formal mathematics or formal algebraic manipulation, were the differentiation programs written quite independently by Kahrimanian [KD54] and Nolan [NO53] in 1953. There seems to have been no real appreciation of the very significant step that was taken at that time. No further development occurred for many years, and the area of numerical analysis continued to flourish. However, starting around 1959, the tide began to turn and there has been an ever increasing amount of work either described in the public literature, in internal reports, or simply under development. The majority of the work was in the development of facilities via subroutines or specific packages, however, rather than through languages. Although LISP (see Section VI.5) provided a language of a potentially useful kind, it was too general and most people found it hard to learn and use. The first attempt at a practical language of any kind was ALGY (see Section VII.2), and the first one to provide wide flexibility and achieve significant usage was FORMAC (see Section VII.3).

It may not be clear to the reader why a computer is needed for doing work in this area. A partial answer is that for many mathematical applications, the amount of mechanical algebra that is necessary is extremely tedious and is as susceptible to human error as numerical calculations are. There are cases in which problems involving weeks or months of manipulation of mathematical expressions have been done by hand, checked two or three times, and then found to be wrong when the work was redone on a computer.

It is important to realize that significant *language* developments, and not significant systems, are being described in this chapter. The systems are far more numerous; descriptions of them can be found in Sammet [SM66a] and [SM67] and, in fact, in the whole August, 1966 *Comm. ACM* as well as other sources shown in Sammet [SM66] and its updated version [SM67a].

The languages discussed in this chapter have major similarities and dissimilarities of approach. FORMAC, in both the FORTRAN- and PL/I-based versions, adds a general capability to an existing language. ALTRAN is also based on FORTRAN, but it is limited to handling rational functions. MATHLAB is an independent language for use in an on-line situation, but it involves primarily typewriter (or teletype) input. FLAP is a LISP-based system which permits the user to specify which of several different mathematical *algebras* he wishes to use. The Symbolic Mathematical Laboratory and Magic Paper are independent languages whose implementations involve either completely special equipment or unusual modifications of standard computers.

In spite of the claims of some people, the ability to write recursive procedures is not of major significance in this area, and while most of these languages do not permit it, they are useful nevertheless.

Formula ALGOL is discussed in Section VIII.5 because in spite of its name it provides enough other facilities to warrant placing it in the multipurpose language category. It is an indication of both the relative newness and the rapid growth of this field that two of the languages (FORMAC and MATHLAB) are definitely in a second generation which differs significantly from the first, and the Magic Paper situation is somewhat similar.

VII.2. LANGUAGES OF HISTORICAL INTEREST ONLY

VII.2.1. ALGY

The earliest attempt at a reasonably general system found by me was ALGY, developed by Bernick, Callender, and Sanford [BM61] prior to 1961. It was an interpretive routine for the Philco 2000. It allowed expressions written in a notation similar to FORTRAN as input, except that the $ was used instead of the ** for exponentiation.

The following commands were available:

EQUAT	Record on tape an expression and its name.
INQT	Rename an expression already on tape.
BUGG	Search the tape for name that is to be "bugged" and delete it.
OPEN	Remove parentheses from an algebraic expression; it performs the necessary algebraic multiplication, groups identical terms, and sorts them in quasi-alphabetical manner.
SBST	Substitutes one or more expressions in a given expression.
FCTR	Factor a given expression with respect to a single variable or powers of a single variable.
TRGA	Expand a product of sine and cosine functions to a sum of sine and cosine functions of multiple angles.
DONE	Control word for permitting several independent problems to be processed during the same run.

Notice that there is no arithmetic defined, nor is there any facility for loop control or control transfers.

Although ALGY apparently never received too much usage or publicity, I consider it a major contribution to the field because it was the first system to try to provide multiple capabilities on a general class of expressions all in one system. In fact, for some ideas, ALGY was a conceptual forerunner to FORMAC.

VII.3. FORMAC[1]

VII.3.1. History of FORMAC

The basic concepts of FORMAC (*FOR*mula *MA*nipulation *C*ompiler) were developed by me (assisted by Robert G. Tobey) at IBM's Boston Advanced Programming Department in July, 1962. At that time I recognized that what was really needed was a formal algebraic capability associated with an already existing numerical mathematical language, and FORTRAN was the obvious choice. A first complete draft of language specifications was prepared in December, 1962; implementation design started shortly thereafter. The basic objective of the work was to develop a practical system for doing formal mathematical manipulation running under IBSYS/IBJOB on the IBM 7090/94. At that time, and for more than 18 months following, FORMAC was intended only as an experiment, with no plan to make it available outside of IBM.

In April, 1964 the first complete version was successfully running after extensive testing. (It is interesting to note that one of the most significant problems run using FORMAC was done by R. G. Tobey as a part of the systems test; it is described in Sconzo, LeSchack, and Tobey [SO65].) Experimentation with the system started in order to learn its strong and weak points. As a result of pressure from numerous people who were interested in trying it, and also as a way of obtaining feedback from users that would lead to developing better systems in the future, FORMAC was released as a Type III program in November, 1964. (This means that it was made available to users by individuals (i.e. the authors) in IBM, who were solely responsible for the maintenance if any was done; such programs are not considered part of normal IBM delivered software.) The 7090/94 version of FORMAC received extensive usage, in spite of its avowed experimental nature and lack of official status.

Consideration of language and implementation for the IBM System/360 started in the fall of 1964, with the intent to provide a better capability, and to

---

[1] It is hoped that the reader will attribute what might seem like an overly large amount of space devoted to this language to my sincere belief in the long range importance of this idea rather than to merely personal involvement.

associate it with PL/I rather than FORTRAN. The PL/I-FORMAC system was released as a Type III, i.e., contributed library program, in November, 1967.

VII.3.2. Functional Characteristics of FORMAC

The first FORMAC was defined to be an extension of FORTRAN IV on the 7090/94, so that all the characteristics associated with FORTRAN IV apply to FORMAC. (The standard FORTRAN described in Section IV.3 is sufficiently close to FORTRAN IV so that the same comments apply.) The following description applies to the "pure FORMAC" part, i.e., to the additions.

FORMAC was designed to look as much like FORTRAN as possible. It was aimed specifically (and only) at the class of mathematical problems which require large amounts of tedious algebraic manipulation by hand. The commands given to the user were relevant only to that problem area and, in particular, no attempt was made to provide general list- or string-handling capabilities. The anticipated user was not only the existing FORTRAN user but also those people with mathematical problems to solve who had only slight computer experience to date and for whom FORMAC would provide a significant new facility. It was definitely designed for use as a batch system, although a subset of FORMAC was made available under M.I.T.'s CTSS (see Crisman [ZR65]). A "desk calculator" version of FORMAC is described in Bleiweiss *et al.* [JL66].

It was originally felt that FORMAC should be designed to run on both the 7040/44 and 7090/94 computers. However, the combination of efficiency loss on the 7090 by restricting the instruction set to that of the 7040 and the implementation problems inherent in using a particular version of IBSYS, made it necessary to drop the idea of 7040 compatibility. The specific FORMAC implementation is thus dependent on the 7090, IBSYS/IBJOB, and the FORTRAN IV compiler. Because control remained completely in the hands of the development group, there were no problems of subsetting or extensions.

With regard to standardization, it is worth noting that I took specific pains to *avoid* this because since FORMAC was the first of its kind and was being developed at IBM, it could easily have perpetuated a *de facto* standard on the computing community prematurely.

The initial language design was done by S. F. Grisoff, R. J. Evey, R. G. Tobey, and the author, under the direction of the author. Implementation and further language design was done by a group working under my general direction, with detailed supervision performed by E. R. Bond. The other people most involved in the initial implementation were R. G. Tobey, R. J. Evey, M. Auslander, R. Kenney, M. Myszewski, and S. Zilles. The main objectives were (1) to develop a practical system running under the

7090/94 IBSYS and (2) to get the system running as rapidly as possible. Maintenance (in the sense of correcting the relatively few bugs which were found) was carried on by the development group on an informal unofficial basis.

The official and complete definition of the language is given in the manual [IB65c]. Numerous papers have been written about FORMAC; the language is described in increasing amounts of detail in Sammet and Bond [SM64], Bond et al. [BZ64], and the manual.

As well as can be determined, the *language* seems to have met its objectives; while some minor improvements were needed, the commands chosen seemed to fulfill the major needs of the users. Papers and reports describing its use are shown in the reference list at the end of the chapter. The biggest fault with the 7090/94 FORMAC was the storage or *free list* problem, although this is inherent in the problem and not specifically the fault of FORMAC. It is characteristic of this type of mathematical problem that expressions require an enormous amount of space, and the most common complaint has been that the user ran out of free list. This trouble presumably can be corrected by increasing the amount of main storage available. Unfortunately, the space required for problems tends to go up at an exponential rather than linear rate, so doubling the available core does not double the overall effectiveness. In fact, I have coined a new version of Parkinson's law, namely "expressions grow to exceed the space available." Some of the problems for which FORMAC has been used are described in articles listed in the references at the end of the chapter.

The original plan in developing a version for the IBM System/360 was to (1) define a specific extension of PL/I with a new data type and other necessary additions and (2) incorporate the improvements which were shown to be needed from experience with the 7090/94 system. A proposal for (1) is given in Bond and Cundall [BZ67]. Unfortunately, it became impossible to carry out the approach defined there because of time and funding constraints and so a somewhat "less clean" language design was developed. It should be emphasized that what is involved here is a subtle view of the language definition rather than any fundamental lack of capability. The language for the running PL/I-FORMAC system is defined in the manual [IB67c].

VII.3.3. Technical Characteristics of FORMAC[2]

The first FORMAC was designed as an extension to FORTRAN IV and, therefore, retains all the essential characteristics of the FORTRAN language form. Because of the method of implementation (a preprocessor

---

[2] Both the 7090/94 (FORTRAN) and System/360 (PL/I) versions are being described. Because of the greater familiarity of readers with FORTRAN, more emphasis is given to that system. The language in the PL/I version is given with much less explanation.

SAMPLE PROGRAM—PL/I-FORMAC[†]

Problem: Solve the system of differential equations
$$\frac{dy_1}{dx} = f_1(x, y_1, \ldots, y_n), \quad y_1(0) = y_{10}$$
$$\cdots$$
$$\frac{dy_n}{dx} = f_n(x, y_1, \ldots, y_n), \quad y_n(0) = y_{n0}$$
where $f_1, \ldots, f_n$ are polynomials. The variable data read in are
1. N—the number of equations
2. M—the number of iterations requested
3. Y0(1), ..., Y0(N), the initial values $y_{10}, \ldots, y_{n0}$
4. F(1), ..., F(N) as polynomials in the placeholders $X, $Y(1), ..., $Y(N)

Program:

```
GMPP3: PROC OPTIONS(MAIN);
FORMAC_OPTIONS;
OPTSET(LINELENGTH=72);
DCL AA CHAR(158);
CONVERT FIXED (I,J,K,L);
ON ENDFILE (SYSIN) GO TO END;
START: PUT PAGE; PUT LIST('TIME AT START'); PUT LIST(TIME);
       PUT SKIP(3);
       GET DATA (N);   /* # OF EQUATIONS */
       GET DATA (M);   /* # OF ITERATIONS DESIRED */
       GET LIST (AA);  /* TITLE OF EXAMPLE */
       PUT LIST (AA);
       PUT SKIP(2);
       DO I=1 TO N;
           GET LIST (AA); FORM (AA); /* INITIAL CONDITIONS: AT X=0 Y(I)=
                                        EXPRESSION = YO(I)    */
           GET LIST (AA); FORM (AA); /* RIGHT SIDE OF D.E.: F(I) IS A
                                        FUNCTION OF $X,$Y(1),...   */
           PRINT_OUT(F("I");YO("I"));
           END;
       CALL IDE; /* COMPUTE APPROXIMATE SOLUTION */
       PUT SKIP (2);
           DO I=1 TO N; PRINT_OUT(Y("I")); END;
           PUT SKIP(2); PUT LIST('TIME AT END'); PUT LIST(TIME);
           GO TO START;
END: PUT LIST ('END OF JOB');
/*                                                       */
IDE: PROC;
     DO I=1 TO N; LET (Y("I")=YO("I")); END;
DO KK=1 TO M;
     DO K=1 TO N; LET (K="K");
           /* COMPUTE THE INTEGRAND */
           LET (CHN  =(F(K),$X,X) );
           DO I=1 TO N; LET (CHN  =(CHN  ,$Y("I"),Y("I")) ); END;
           LET (Y=EVAL(CHN ));
               /* THE "TAIL" OF THE INTEGRAL OF Y IS THE SAME    */
               /* AS Y(K), SO THE TAIL OF Y IS THE SAME AS Y(K)' */
           LET (YY=EXPAND( X*(Y-DERIV(Y(K),X)) ) );
           LET (L=LOWPOW(YY,X) );
           LET (Y(K)=Y(K)+COEFF(YY,X**L)/L*X** L);
           END;
END;
END IDE;
END GMPP3;
```

[†][IB67c], pp. 88, 103 (slightly modified). Reprinted by permission from *PL/I FORMAC Interpreter.* © 1967 by International Business Machines Corporation.

to the FORTRAN IV compiler), a restriction was imposed that variable names could not be the same as, or begin with, a list of reserved words; these words also could not be used as subroutine or function names.

In considering the program structure, a deliberate decision was made to keep FORMAC at the same technical level as FORTRAN. This means that no attempt was made to provide features in FORMAC that could just as well be (or should have been) provided in FORTRAN; thus, no recursion was provided (nor has its absence been any significant handicap). Transmittal of FORMAC variables to subroutines is done with an explicit declaration (*SYMARG*) beyond the normal *CALL* statements. FORMAC and FORTRAN statements can be intermingled in the program, and FORTRAN variables can appear in FORMAC statements.

FORMAC was added to various versions of the 7090/94 IBSYS; Version 13 was the last for which this was done. The net result to the user was that if he had made no changes to IBSYS, then by merely using the proper FORMAC control card he could run FORMAC problems and all other normal IBSYS programs within the same system.

The most significant single element in FORMAC is the addition of the *FORMAC variable* concept to FORTRAN. A FORMAC variable is a formal (i.e., algebraic) data variable. This is best illustrated through an example:

Consider the equation $y = x^2 + 3xz$. A typical FORTRAN program would assign values to x and z and achieve a numerical value for y. Thus, if $x = 3$ and $z = 5$, the FORTRAN program

```
X = 3
Z = 5
Y = X**2 + 3*X*Z
```

would assign the value 54 to Y. However, we might wish to allow the variable x to assume a nonnumeric value, say the expression $a + b$. Then a FORMAC program would be written as

```
LET X = A + B
    Z = 5
LET Y = X**2 + 3*X*Z
```

and the variable Y would be assigned as the name of the expression (A+B)**2 + 15*(A + B). Note that the key word *LET* is used to identify FORMAC statements and that FORTRAN variables can appear in FORMAC statements.

For each FORTRAN operator there is a corresponding FORMAC one; thus for the FORTRAN *EXP*, there is *FMCEXP*; for *SIN*, there is added *FMCSIN*, where of course the arguments of the FORMAC operators are FORMAC expressions.

In addition to these, four new operators were added to FORMAC. Three of these, namely FMCFAC, FMCDFC, and FMCOMB, represent the numeric values for factorial, double factorial, and combinatorial, respectively; the symbolic form is retained until the user specifies that it should be evaluated. Thus, FMCFAC(4) would be carried along in that form until a request was made in the program to obtain the value (namely 24). The fourth operator added to FORMAC is the differentiation operator, written FMCDIF(Y, X, N), where Y can be either an expression or the name of an expression, X is the variable of differentiation, and N is the degree of differentiation. It is possible to have several variables and degrees of differentiation in one set of arguments, thus permitting successive partials to be obtained in one statement.

With regard to arithmetic, FORMAC permits mixed mode, floating point, and rational arithmetic but not complex or double-precision arithmetic. (The latter two may still be used in FORTRAN statements, of course, but no double-precision or complex variable can appear in a FORMAC statement.) In my opinion, one of the most notable features of FORMAC is its rational arithmetic facility. In FORTRAN compilers, if one adds $\frac{1}{3} + \frac{1}{3}$ in fixed point arithmetic, the result is zero because truncation is performed on each fraction in turn. On the other hand, if $\frac{1}{3} + \frac{1}{3}$ is calculated in floating point, the result is naturally 0.666 ... 7. Unfortunately, for the vast majority of problems that are done in an algebraic manipulation system, neither of these alternatives is tolerable. The former gives the wrong answers and the latter gives answers which may be correct but are certainly not understandable. Thus, for example, if one has coefficients that take on the form $\frac{5}{7}$, $\frac{7}{9}$, $\frac{9}{11}$, etc., the information in this form is essential to the correct solution of the problem and providing the floating point equivalent of these will not convey the appropriate information. (If one wished to evaluate $1 + x + x^2/2! + x^3/3! + x^4/4! + \cdots$, one would obtain the following result by using floating point: $1 + x + 0.5x^2 + 0.16666667x^3 + 0.04166667x^4 + \cdots$. This is correct but meaningless.) For that reason, if all the quantities in an expression—both variables and constants—are fixed point, the calculation performed in FORMAC (by the automatic simplification routine) will be done in precise arithmetic, using rational numbers where necessary. Thus $\frac{3}{4} + \frac{1}{2}$ will yield $\frac{5}{4}$ as a result. Some people have actually used FORMAC primarily to obtain the capability of doing the rational arithmetic.

Within a FORMAC statement, variables (both FORMAC and FORTRAN) can be either fixed or floating point. If all the variables and constants on the right side are in fixed point, then any necessary calculation is done in rational arithmetic; if at least one variable or number on the right side is floating point, then the entire computation is done in floating point, on the grounds that the absolute accuracy has been lost. Any numbers created on the right-hand side (which will occur as a result of automatic simplification) will be converted to the mode of the variable named on the left.

A complete list of executable statements, showing their format and an example, is shown in Figure VII-1. Hence only a brief description will be given here.

The basic assignment statement in FORMAC is obtained by merely writing the key word LET before the variable on the left-hand side, e.g., LET Y = A + B*C. The actual execution is to assign Y as the name of the expression obtained by carrying out the indicated operations on the right. The system knows that it is to perform the formal algebraic manipulation by the existence of the word LET on the left-hand side; the variables on the right-hand side are known to be FORMAC variables through either explicit declaration as ATOMIC (meaning the variable stands only for itself), or as the variable on the left-hand side of a previous FORMAC statement. All other variables are assumed to be FORTRAN variables which must, of course, have a numeric value currently assigned. The rules for conversion were indicated above.

The sequence control statements are exactly—and only—those of FORTRAN; no others were added.

There are three classes of executable commands: Those yielding FORMAC variables as results, those yielding FORTRAN variables, and miscellaneous commands. In the first category are

LET	Constructs specified expressions; it is the algebraic equivalent to the normal assignment statement in FORTRAN.
SUBST	Replaces a variable with an expression.
EXPAND	Removes all parentheses by applying the multinomial and distributive laws.
COEFF	Obtains the coefficient of a variable or a variable to a power.
PART	Separates expressions into terms, factors, exponents, etc.

The following statements yield FORTRAN variables; the first two create FORTRAN numeric variables, and the second two create FORTRAN logical variables.

EVAL	Evaluates an expression for specific numerical values.
CENSUS	Counts words, terms, or factors.
MATCH	Compares two expressions for equivalence or identity.
FIND	Determines dependence relations and/or whether a particular variable appears in an expression.

ALGCON

    algcon-statement := LET *let-var* = ALGCON *ftn-num-var, ftn-fxd-var*

    Example:

        If ARRAY is

(1)	A + B * I —
(2)	F M C D I F
(3)	(X , X , 1
(4)	) $

        then ATOMIC A, B
            J = 0
            LET I = EVAL (A + B), (A, 3), (B, 1)
       10 LET Y = ALGCON ARRAY (1), J
             results in Y ⟶ A + B*4. —1.
                      J = non-zero value

    If a second expression began after the $, then returning to execute statement 10 would convert the next expression since J has a non-zero value.

AUTSIM

    autsim-statement := AUTSIM $\begin{Bmatrix} \text{QINT} \\ \text{QNUM} \\ \text{QNINT} \\ \text{ON} \end{Bmatrix}$

    Example:

        ATOMIC X, Y, B
        I = 3
        LET A = I + 4 + I**I + FMCSIN(Y)*FMCFAC(I) + FMCDFC(B)**I

    If an *autsim-statement* has not been executed prior to the execution of the above *let-statement*, or if the last *autsim-statement* executed was
        AUTSIM QNINT
then the result of the above *let-statement* is
            A ⟶ 34. + FMCSIN(Y)*6. + FMCDFC(B)*3.
    If the last *autsim-statement* executed was
        AUTSIM ON
the above *let-statement* results in
            A ⟶ 7. + 3.**3. + FMCSIN(Y)*FMCFAC(3.) + FMCDFC(B)**3.

BCDCON

    bcdcon-statement := LET *ftn-num-var* = BCDCON *fmc-exp, ftn-fxd-var, fxd-num*
    fxd-num := $\begin{Bmatrix} \textit{ftn-fxd-var} \\ \textit{fxd-cons} \end{Bmatrix}$

    Example:

        ATOMIC AB, BC, X
        DIMENSION LIST (4)
        LET R = AB*BC + FMCSIN(X)*FMCCOS(X) + 4.3
        Q = 0.
        LET Q = BCDCON R, LIST, 4

                                Figure. VII-1. (cont. next page)

Figure VII-1. (cont.)

would result in LIST (1) | binary 16 | FORTRAN fixed point value
(2) | A B * B C + | in binary as opposed to BCD
(3) | F MC S I N |
(4) | (X)* ƀ ƀ |

Q ≠ 0. (there is more to do).

Upon re-entry to BCDCON with Q ≠ 0. the translation continues; and now

LIST (1) | 14
(2) | F M C C O S
(3) | (X) + 4.
(4) | 3 $ ƀ ƀ ƀ ƀ

Q = 0. (translation complete).

Note that $ is inserted as the last character of every expression which is translated by BCDCON. This is necessary in case the expression is to be read by the ALGCON routine.

CENSUS

$$\text{census-statement} := \text{LET } \textit{ftn-fxd-var} = \text{CENSUS } \textit{let-var}, \begin{Bmatrix} \text{WORD} \\ \text{TERM} \\ \text{FACT} \end{Bmatrix}$$

Example:

```
ATOMIC A, B, C
LET Z = A + B
LET M = CENSUS Z, TERM
    results in M = 2
LET M = CENSUS Z, FACT
    results in M = 1
LET M = CENSUS Z, WORD
    results in M = 4
```

COEFF

$$\text{coeff-statement} := \text{LET } \textit{let-var} = \text{COEFF } \textit{fmc-exp}, \textit{seek-var} \left[\begin{Bmatrix} \textit{ftn-flt-var}_1 \\ ,\ \textit{ftn-flt-var}_1,\ \textit{ftn-flt-var}_2 \end{Bmatrix} \right]$$

Example:

```
LET B = A*X**2 + C*X**4.2 + B*X**3.1
LET Z = COEFF B, X**2, R
    results in Z ⟶ A
              R = 3.1
```

ERASE

erase-statement := ERASE *let-var* {, *let-var*} ...

Example:

```
ERASE MAS, EOJ, A, X(4), M(3)
    where MAS, EOJ, A, X, M are all let variables
    results in MAS, EOJ, A, X(4), and M(3) all being erased
```

EVAL

eval-statement :=

$$\text{LET } \textit{ftn-num-var} = \text{EVAL } \textit{fmc-exp}_0 \left[,\ \begin{Bmatrix} \textit{param-label} \text{ max 9 } \{,\ \textit{param-label}\} \\ \textit{param-list} \end{Bmatrix} \right]$$

param-list := (*seek-var*, *fmc-exp₁*) {, (*seek-var*, *fmc-exp₂*)} ...
    (i.e., exactly the same as in the *subst-statement*)

Example:

```
RI = −.5
LET A = B(1)*X + B(1)*(Y**−1.)
LET R = X−4.
LET IANS(5) = EVAL A + R, (B(1),2), (X,ABSF(RI)), (Y,3)
```

After substitution A + R is 2.*(.5) + 2.*3.**−1. + .5−4. the floating point arithmetic yields 1. + 2.*3.**−1. + .5 − 4. = −1.8333, etc., but IANB (5) is fixed point mode and therefore IANS (5) = −1

EXPAND

expand-statement := LET *let-var* = EXPAND *fmc-exp* [, CODEM]

Example:

```
ATOMIC A,B,C,D
LET R = EXPAND(A + B)**2 + A*(C−D)
    results in R ⟶ A**2. + A*B*2. + B**2. + A*C−A*D
```

FIND

find-statement :=
    LET *ftn-log-var* = FIND *fmc-exp*, $\begin{Bmatrix} APP \\ DEP \end{Bmatrix}$, $\begin{Bmatrix} ALL \\ ONE \end{Bmatrix}$, (*seek-var* {, *seek-var*} ...)

Example:

```
ATOMIC B, C, D, F, X, Y, R(10)
LET A = B + C + D−F
LET Q = FIND A, APP, ALL, (B,C,D,F,G)
    results in Q ⟶ .FALSE.
whereas
LET Q = FIND A, APP, ONE, (B,C,D,F,G)
    results in Q ⟶ .TRUE.
```

LET

let-statement := LET *let-var* *fmc-exp*

Example:

```
ATOMIC Z
N = 4
LET C = (M + Z)**(1/2)
    results in C ⟶ (4. + Z)**(.5)
```

MATCH

match-statement :=
    LET *ftn-log-var* = MATCH $\begin{Bmatrix} ID \\ EQ, \textit{ftn-flt-num} \end{Bmatrix}$, *fmc-exp1*, *fac-exp2*

Figure VII-1. (cont. next page)

Figure VII-1. (cont.)

> *Example:*
>
>> ATOMIC A, B
>> LOGICAL Q
>> LET X = (A + B)**2
>> LET Y = A**2 + 2*A*B + B**2
>> LET Q = MATCH ID, X, Y
>>> results in Q = .FALSE.
>>
>> LET Q = MATCH EQ, .001, X, Y
>>> results in Q = .TRUE.

ORDER

> *order-statement* =
>> LET *let-var* = ORDER *fmc-exp*, $\begin{Bmatrix} \text{INC} \\ \text{DEC} \end{Bmatrix}$, $\begin{Bmatrix} \text{FUL} \\ \text{PRT} \end{Bmatrix}$
>> $\left[\begin{Bmatrix} , (seek\text{-}var_i \{, seek\text{-}var_j\}), (seek\text{-}var_i \{, seek\text{-}var_j\}) \\ , (seek\text{-}var_i \{, seek\text{-}var_j\}) \\ ,, (seek\text{-}var_i \{, seek\text{-}var_j\}) \end{Bmatrix} \right]$
>
> *Example:*
>
>> LET Z = ORDER X*3 + X**2*Z*2 + Y*X**3, INC, FUL, (X)
>>> results in Z ⟶ 3*X + 2*X**2*Z + X**3*Y

PART

> *part-statement* := LET *let-var$_1$* = PART *let-var$_2$*, *ftn-fxd-var*
>
> *Example:*
>
>> LET X = (A + B) + C
>> LET ANS = PART X, M
>>> results in ANS ⟶ A
>>> X ⟶ B + C
>>> M = 4

SUBST

> *subst-statement* :=
>> LET *let-var* = SUBST *fmc-exp$_0$*, $\begin{Bmatrix} param\text{-}label \ max \ 9 \ \{, param\text{-}label\} \\ param\text{-}list \end{Bmatrix}$
>
> *param-list* := (*seek-var*, *fmc-exp$_1$*) {, (*seek-var*, *fmc-exp$_2$*)} ...
>
> *Example:*
>
>> LET A = B + 3*C
>> M = 4
>> LET Z = SUBST A, (B, M), (C, FMCDIF(X**2, X, 1))
>>> results in Z ⟶ X*6. + 4.

Figure VII-1. 7090/94 FORMAC verb formats with examples. The general notation is the COBOL metalanguage described in Section II.6.2.
Source: [IB65c], extracts from pp. 141–184. Reprinted by permission from *FORMAC (Operating and User's Preliminary Reference Manual)*. © 1964 by International Business Machines Corporation.

The following commands are miscellaneous.

BCDCON Convert to BCD form from the internal representation.
ALGCON Convert to the internal representation from a BCD form.
AUTSIM Control type and amount of arithmetic done during automatic simplification.
ERASE   Eliminate expressions no longer needed.
ORDER   Specify sequencing of variables within expressions (for output editing purposes).
FMCDMP Symbolic dump.

The actual format for writing most of these is with a LET on the left-hand side and the command on the right, e.g.,

$$\text{LET Y = SUBST X**2 + Z, (X, A+B)}$$

which would yield Y as the name of the expression (A+B)**2 + Z.

Note that there are no separate input/output statements in FORMAC. The FORTRAN *READ* and *WRITE* are used with the FORMAC *ALGCON* and *BCDCON*, respectively, to convert to and from the standard BCD format and the FORMAC internal representation.

Although a discussion of algebraic simplification is not within the scope of this chapter, the apparent absence of a command for this might cause confusion. FORMAC's approach to the problem is to have a routine which is invoked automatically by the system after each executable command. This routine eliminates zeros and ones, combines like terms and products, and performs a few other transformations. Control of parentheses removal and limited factoring is left to the user with the *EXPAND* and *COEFF* commands, respectively. The types and amount of arithmetic done during the automatic simplification are controlled by the *AUTSIM* command.

Library references are made through regular *CALL* statements. A command called *FMCDMP* permits dumps to be taken either at the point when it is invoked or automatically whenever an error occurs (if the programmer so requests). The only other debugging facilities are inherent in the error checking done by the individual commands themselves.

Most of the storage allocation is done automatically by the system, by putting expressions on and removing them from the free list. If there is insufficient memory to execute a particular command, the system will automatically put out on tape any expressions not needed for that particular execution and will automatically see that they are brought back. The user has no concern with, nor any control over, this. The *ERASE* command permits the user to remove from storage any expressions which he no longer needs, thus making more space available.

There are four declarations in FORMAC.

ATOMIC   Declare those variables which represent only themselves.
DEPEND   Declare an implicit dependence relationship between atomic variables.
PARAM   Specify pairs to be used with *SUBST* or *EVAL*.
SYMARG   Declare subroutine arguments as FORMAC variables and flag program beginning.

Since FORMAC was designed for a specific purpose, with an objective of rapid implementation, it is not surprising that it provides no capabilities for self-extension, self-modification, nor any real use outside its primary application area. Two trivial places in which another use has been demonstrated are proving formulas by mathematical induction and verifying steps in a mathematical derivation. The former was illustrated in Sammet and Bond [SM64] and the latter was suggested by N. Rochester.[3]

The language was designed to make it easily translatable by a preprocessor to FORTRAN IV. In particular, the key word *LET* and certain other restrictions were placed on the language to make it possible to translate FORMAC programs to legal FORTRAN programs without redoing a major portion of the FORTRAN compilation.

The primary debugging aid is the *FMCDMP* command, and there are also a large number of diagnostic messages both at compile and object time.

In PL/I-FORMAC on System/360, many of the same principles were continued, with changes made both to accommodate the structure and flavor of PL/I and to incorporate improvements suggested by the first system. It is assumed that the reader is familiar with PL/I (see Section VIII.4). The main language limitation of the PL/I version over the FORTRAN version is the inability to use PL/I variables in FORMAC statements without putting quote marks around them. However, some of the FORMAC macros can be used in PL/I statements which is an added convenience.

FORMAC statements can of course be included in a sequence of PL/I statements. Most of the statements in the FORTRAN version are expressed as functions in the PL/I version.

FORMAC variables have the same rules for naming as PL/I variables do except that the former may not contain more than eight characters and may not contain the underscore character ＿. There are many reserved words in FORMAC, and they cannot be used for data names. There can be up to four subscripts; a subscript may be any FORMAC variable or expression which is or evaluates to an integer. No dimension information is required. In addition to fixed and floating point constants, rational number and systems constants are permitted. The latter are designated by #E, #P, and #I representing, respectively, the natural logarithm base, π, and $\sqrt{-1}$.

FORMAC variables are considered *atomic* or *assigned*, but there is no

---

[3] Private communication.

specific declaration required. The user may—if he wishes—change an assigned variable to an atomic one at object time by using *ATOMIZE*.

A notation for specifying functional forms is defined, using an array of atomic elements. This notation can be used to define the form of partial derivatives and to evaluate functions.

The basic assignment statement is denoted by the key word *LET*, followed by any number of FORMAC statements, all enclosed in parentheses. Each assigned variable on the right-hand side is replaced by its value, thus causing the evaluation of the given expression. The statement

$$\text{LET } (A = B + C; B = X * Y; B = A + B;)$$

causes the variable B to be assigned the value B+C+X*Y.

FORMAC expressions are written in essentially the same way as PL/I expressions, and they can contain a number of specific FORMAC functions.

The major FORMAC capabilities are functions, which can have an arbitrary degree of nesting. They are divided into the following categories.

User control of simplification

    MULT (*expr*)   Expands all sums in *expr* to positive integral powers, using the multinomial law. The combinatorials can be evaluated or left in symbolic form at the choice of the user.

    DIST (*expr*)   Applies the distributive law to all products of sums in *expr*. *DIST* does not apply the multinomial law, so integral powers of sums are left unchanged.

    EXPAND (*expr*)   Applies both the multinomial and distributive laws to *expr*.

    CODEM (*expr*)   Rewrites *expr* to place everything over a common denominator. Subexpressions are also placed over a common denominator.

    FRACTN (*expr*)   Same as *CODEM* except that only the outermost level is affected.

Substitution

    EVAL (*expr*, a_1, b_1, a_2, b_2, ..., a_n, b_n)
        Each occurrence of a_i in *expr* is replaced by b_i. This replacement is made in parallel for each a_i.

    REPLACE (*expr*, a_1, b_1, a_2, b_2, ..., a_n, b_n)
        Each occurrence of a_1 in *expr* is replaced by b_1. Then each occurrence of a_2 is replaced by b_2, etc. Thus the replacement is sequential rather than simultaneous.

Differentiation

DERIV (*expr*, v_1, n_1, v_2, n_2, ..., v_m, n_m)
The result is the partial derivative of *expr* as follows:

$$\frac{\partial^n}{\partial v_1^{n_1} \partial v_2^{n_2} \ldots \partial v_r^{n_r}} \textit{expr} \quad \text{where} \quad n = \sum_{i=1}^{r} n_i$$

DRV (*fncn-var*).(arg_1, ..., arg_m), $(1), n_1, $(2), n_2, ..., $(r), n_r)
Causes differentiation of unspecified functions with respect to arguments.

DIFF (*fncn-var*) = **CHAIN** (*expr-1*, *expr-2*, ..., *expr-n*)
DIFF is a pseudovariable which assumes *fncn-var* is an unspecified function of *n* variables. The *expr-i* are assigned as the values of the first partial derivatives of *fncn-var*. (This latter concept is described on page 489.)

Expression analysis

COEFF (*expr-1*, *expr-2*)	Returns the coefficient of *expr-1* in *expr-2*, considering only the top level of *expr-1*.
HIGHPOW (*expr-1*, *expr-2*)	Returns the highest power of *expr-1* in *expr-2*.
LOWPOW (*expr-1*, *expr-2*)	Returns the lowest power of *expr-1* in *expr-2*.
NUM (*expr*)	Returns the numerator if *expr* is a fraction, and *expr* if it is not.
DENOM (*expr*)	Returns the denominator if *expr* is a fraction and 1 if it is not.
LOP (*var*) **NARGS** (*var*) **ARG** (*n*, *expr*)	LOP, NARGS, and ARG provide information about the expression named by *var*, based on the internal representation.

Storage allocation

SAVE (var_1; var_2; ... ; var_n)
The expressions named by var_i are moved from main storage to secondary storage. If referenced in the program, they will be brought back automatically and then must be reSAVEd.

ATOMIZE (var_1; var_2; ... ; var_n)
The var_i become atomic and the space used by the expressions they name is released.

Output

PRINT_OUT (var_1; var_2; ... ; var_n)
  The indicated variables are printed out, each on a separate line. In addition to the format shown, an assignment statement without the *LET* can be used, e.g.,

PRINT_OUT (A = B + C; D = SIN (X); ANS)

The functions provided in FORMAC fall into the following categories.

Mathematical

SIN, COS, SINH, COSH, ATAN ATANH, LOG, LOG10, LOG2, ERF, EXP, SQRT, SIND, COSD, TAN, TAND, TANH, ATAND, ERFC

Integer-valued

FAC (n)	Factorial.
COMB (n, n_1 n_2 ... n_m)	Combinatorial.
STEP (expr-1, expr-2, expr-3)	Has the value 1 if expr-1 $\leq$ expr-2 $<$ expr-3 and 0 otherwise.

User-defined

FNC(F) = expr ($(1), $(2), ..., $(n))
  The $(i) represent the formal parameters in the expression.

Function variables

By using a dot after the function name (i.e., *fncn-var*), an undefined function of the given variables can be used, e.g., FNCNAM.(A+B,A−B).

FORMAC variables may be assigned a sequence (or chain) of FORMAC expressions by writing

CHAIN (var_1, var_2, ..., var_n)

This is useful in creating argument lists for FORMAC functions and routines.

A number of options can be specified by the user, using an *OPSET* statement. These provide control over certain aspects of automatic simplification and output editing.

FORMAC and PL/I variables can each be transformed (and used in the other part of the program) in several ways. A PL/I variable enclosed within double quotation marks can be used in *LET*, *PRINT_OUT*, and *SAVE*, and expressions. A PL/I variable can be used on the left side of a *LET* statement if it is a character string variable; executing the statement assigns the resulting FORMAC expression as the value of the PL/I variable.

Writing **CHAREX** (*var = variable*), where *var* must be a PL/I varying character string, assigns to *var* the character string value of the FORMAC variable.

Writing **CONVERT FIXED** $(x_1; \ldots; x_n)$ **FLOAT** $(y_1; \ldots; y_m)$ where the x_i and y_i are unsubscripted PL/I variables, causes assignment of the value of FORMAC expression (which must be numeric) to the specified variable whenever the latter appears (within quotes) on the left-hand side of a FORMAC assignment statement. If the variable appears within quotes on the right-hand side, the conversion is made from PL/I to FORMAC. The words *FIXED* and *FLOAT* actually refer to the conversions which will take place.

Using the functions *INTEGER* and *ARITH*, each followed by a FORMAC variable name enclosed in parentheses, causes the numeric evaluation of the expression with that name and its return in fixed point binary or double-precision floating point, respectively. This provides the mechanism for evaluating an expression and using the numeric result in a PL/I statement.

There are several macros which can appear in PL/I statements. In particular, it is possible to write

IF IDENT(A;B) THEN LET (C = D + E;)

The other macros permitted are *ARITH*, *INTEGER*, *LOP*, and *NARGS*, each of which requires a single FORMAC variable as an argument.

VII.3.4. Significant Contribution to Technology

There appear to be five main contributions of FORMAC to the technology. The first is that it introduced the concept of adding this type of facility *as a language* to an existing language used for numerical scientific problems. In my opinion, this is actually the most significant contribution. Second is the fact that it showed that a practical system could be developed to do formal algebraic manipulation on the computer, and could be easily learned and used to solve specific analytic problems arising in the course of engineering and mathematical work. The third contribution is that it demonstrated quite clearly that the amount of storage is a limiting factor in the problems which can be solved and that this is much more of a limitation than slow running time. Very few FORMAC problems had to be stopped because they were taking too long to run, but a number could not be completed because of lack of storage. A fourth contribution is the development of a reasonable algorithm for doing automatic simplification.

A fifth and more subtle contribution to the technology is that FORMAC will help steer people away from numerical analysis and back to analytic solution of problems.

VII.4. MATHLAB

VII.4.1. History of MATHLAB

MATHLAB is a system started by C. Engelman (MITRE) in 1964 and developed by him and a few summer employees. There has been very little attempt to do anything other than experiment with the system within the company and make improvements as time and manpower permit. It is continually undergoing development, and some of the information about the new system is indicated after the language description of the first one.

MATHLAB has run under the AN/FSQ-32 time-sharing system at SDC, on the IBM 7030 (STRETCH) at MITRE and on CTSS (see Crisman [ZR65]) at Project MAC.

The first version of MATHLAB has been replaced by a new system referred to as MATHLAB 68 when a distinction is necessary. Because the new system did not become operational until the fall of 1967, most of the following description pertains to the first version.

VII.4.2. Functional Characteristics of MATHLAB

The first language was fairly simple, with an unusual mixture of succinctness and naturalness as can be seen in Figure VII-2. For example, one command was pleasesimplify (x y) which is fairly clear, whereas the command factor ((x y...)) is not intuitively obvious because of certain precedence relations. Thus the language tended to be slightly inconsistent in style, although certainly not in content. There was a strong tendency toward English-like language in the output, where the system typed such messages as THANKS FOR THE EXPRESSION or ARE YOU FAMILIAR WITH THE FORMAT FOR TEACHING ME?.

The application area is mathematics, and the stated objective is to provide on-line facilities for mechanical operations such as addition of expressions and equations, substitution, differentiation, integration, and Laplace transforms, etc. Only a typewriter was used for input/output.

The language was definitely a problem-solving, hardware language, with the user a novice or perhaps nonprogrammer. Although the user of MATHLAB can be completely unaware of it, the system was coded in LISP (see Section VI.5) and actually invoked LISP for the execution of all its facilities.

Since the system was running on three machines, there was a compati-

```
DIFFERENTIATE(Z T DZ)
 (I DO NOT KNOW THE DERIVATIVE OF ARCSIN . DO YOU)
YES
 (ARE YOU FAMILIAR WITH THE FORMAT FOR TEACHING ME)
NO
 (PLEASE IMITATE THE FOLLOWING EXAMPLE)

 ARCTAN ,
 X ,
 1/(1 + X↑2) ,

 GO
ARCSIN,
U,
1/SQRT(1-U↑2),
 (THANK YOU)
!USERS

$ 7 USERS.
 " DZ  =  PI*SQRT (ARCSIN (PI*T)) + 1/2*PI↑2*T/(SQRT (ARCSIN (PI*T))*SQRT
 (1 - (PI*T).2))

RAISE(Z 2  Z2)
 Z2  =  (PI*T*SQRT (ARCSIN (PI*T)))↑2

EXPAND((T))
 (SIMPLIFY)
PLEASESIMPLIFY(Z2 Z2)
 Z2  =  PI↑2*T↑2*ARCSIN (PI*T)

NEWNAME(DIFFERENTIATE DIFF)
 (SORRY THE NAME DIFF IS ALREADY TAKEN)

 PRINT NEWNAME *EVQ *EVALQT *TEVALQT
NEWNAME(DIFFERENTIATE D)
 OK
D(Z2 T DZ2)
 DZ2  =  2*PI↑2*T*ARCSIN (PI*T) + PI↑3*T↑2/SQRT (1 - PI↑2*T↑2)

DENOTE NIL
CUBE = (A + B)↑3,
'
 (THANKS FOR THE EXPRESSION CUBE)
NEWNAME(PLEASESIMPLIFY PS)
 OK
PS"PS(CUBE CUBE)
 CUBE   =   (A + B)↑3

EXPAND((A))
 (SIMPLIFY)
PS(CUBE NEWCUBE)-
 NEWCUBE   =   A↑3 + 3*A↑2*B + 3*A*B↑2 + B↑3
```

Figure VII-2. Example of use of MATHLAB on the AN/FSQ-32.
Source: Part of actual console session.

bility problem, which really reduced to the compatibility of the LISP systems which were available on the three different machines, and to the differences in input hardware.

Since this was basically a system under continuing development and tight control, with no users, there was no problem of subsets or extensions. Dialects existed because of the compatibility problems.

The language was designed by a single person, assisted by a few others, who maintained tight control over it. The definition is simply given in English. The primary documentation of the first system is the first two papers listed in the references at the end of this chapter.

There has not been enough use to evaluate the language in either version.

VII.4.3. Technical Characteristics of MATHLAB

The character set consisted of the 26 letters, 10 digits, and the following symbols:

$$+ \quad - \quad * \quad / \quad \uparrow \quad (\quad) \quad , \quad = \quad \text{blank}$$

(Where the input keyboard did not contain the up arrow, $\uparrow$, then ** was used instead.)

Names of symbolic variables consist of a letter followed by letters and/or digits; no limit on length was stated. No subscripting or qualification was allowed. Variable names could be the same as the fixed commands of the system. Numeric literals were used without any special delimiter. Blanks were significant. The comma was the main punctuation character and was used to separate functional arguments and also to delimit an individual data item, e.g., an expression (which could itself be quite complicated). There were no noise words permitted as input, but the system typed out a number of polite but logically unnecessary words, e.g., THANKS ....

The input was from an on-line keyboard, with a number of commands defined as English words followed by parenthetical information to provide the parameters.

The structure of the program was quite primitive. There were a few declarations (which really had the effect of deferred commands; i.e., they caused something to happen but not until sometime later in the execution). There were a number of single executable commands, but they could not be combined. There were no loops or subroutines or functions in the normal sense. It was possible to name objects and have the system remember the name and retrieve the object when the name was given. The primary interaction with the environment was through the normal commands of whichever time-sharing system was being used. In essence, the system operated on a line-by-line basis rather than as a stored program, with all the attendant advantages and disadvantages.

The main delimiters for the commands and declarations were commas and carriage returns. Declarations were intermingled with the executable commands as needed.

The only individual data variable type permitted was formal (i.e., algebraic). However, these could be combined to form *expressions, equations,* or *functions*. Each of these could be named and could generally be transformed from one type into the other. This is one of the most interesting facilities in the system. There were no arithmetic or Boolean (or any other kind of) variables. The arithmetic done was only that available in the LISP system and in the automatic simplification routine being used. The former permits (only) rational numbers.

The assignment statement was written in the form

$$\text{DENOTE NIL}$$
$$A = B + C,,$$

which assigns to the formal expression B + C the name A; conversely the variable A has the *value* B + C. If we now write

$$\text{DENOTE NIL}$$
$$B = D + 2,,$$

then B has the value D + 2 but A still has the value B + C. (The first comma delimits the equation while the second indicates that there are no more equations.)

There was no alphanumeric data handling. Since this system was designed to operate essentially on a line-by-line basis, there were no control transfers, no conditional or loop-control statements, and no error condition statements.

The available algebraic expression manipulation statements were as follows:

An equation was named and defined by writing two equals signs. For example

$$S2 == T\uparrow2 = A\uparrow2 + B\uparrow2,,$$

defines an equation $T^2 = A^2 + B^2$ whose name is S.

Arithmetic commands on algebraic items

ADD ((Q1 Q2 ... Qn) NAME)
    The Qi could be equations, functions, expressions, or numbers. Equations were added by adding left and right sides independently. If expressions, functions, and/or numbers were to be added to equations, the first three were added to both sides of the equation. The result was given the NAME. Using the A and S from the example above, ADD ((A S) ANS) would produce (ANS) B + C + T↑2 = B + C + A↑2 + B↑2

MULTIPLY ((Q1 ... Qn) NAME)
:   Similar to addition

SUBTRACT (X Y NAME)
:   The X and Y could be equations, functions, expressions, or numbers.

DIVISION (X Y NAME)
:   Similar to subtraction.

RAISE (X Y NAME)
:   Defined by $NAME = X^Y$

NEGATIVE (X Y)
:   Defined by $Y = -X$

INVERT (X Y)
:   Defined by $Y = 1/X$

Commands involving simplification

PLEASESIMPLIFY (X Y)
:   This command simplified (using a particular routine developed at Stanford University) the expression X and named it Y. The action of this command could be affected by the next two commands.

EXPAND ((X1 X2 ... Xn))
:   This produced no immediate result, but in succeeding simplifications whenever one of the X_i occurred in a product of sums, that product was multiplied out.

FACTOR ((X1 X2 ... Xn))
:   This produced no immediate result, but in succeeding simplifications it caused the collection of all terms which contain X1 or a power of X1 as a factor, then collection of terms involving X2 or a power of X2, etc. Thus X1∗X2 + X1↑2 + 2∗X2 + X2∗C would be simplified to X1∗(X2+X1) + X2∗(2+C).

Commands involving changing of algebraic items

FLIP (X Y)
:   Y becomes the name of the equation X with the left and right sides interchanged.

MAKEEQUATION (X Y)
   An equation is formed with the name Y; the left side is X and the right side is the value of X. For example,

$$\text{DENOTE NIL}$$
$$A = B + C,,$$
$$\text{MAKEEQUATION (A R)}$$

caused the computer to respond

$$(R) \; A = B + C$$

where there is now an equation named R.

MAKEEXPRESSION (E)
   E must be an equation whose left side is a single word. This command created an expression whose name is the left side of E and whose value is the right side. Thus, using the information above,

$$\text{MAKEEXPRESSION (R)}$$

would result in

$$A = B + C$$

which is where we started.

MAKEFUNCTION (E)
   This applied to an equation E whose left side was of the form $F(X1, X2, \ldots, Xn)$ where each Xi was a single word. A new function F was formed whose dummy variable list was $(X1 \; X2 \ldots Xn)$ and whose value was the right side of E.

Higher level commands

DIFFERENTIATE (Y X Z)
   This differentiated Y with respect to X and called the result Z.

LEARNDERIVATIVE
   The user defined for the system the derivative of a new function.

INTEGRATE (X Y Z)

This integrated X over Y and called the result Z. X must be a rational function of Y, with rational numeric coefficients.

SOLVE (E X)

This command required that the equation E be rational (but it could have symbolic coefficients) in X and be really (although not necessarily explicitly) quadratic or linear in X.

System commands and miscellaneous

RENAME (X Y)

The name Y was assigned to the item which had the name X.

NEWNAME (A B)

This allowed the use of the command name B in place of the command name A. For example, instead of writing INTEGRATE, the user could say NEWNAME (INTEGRATE I), which would cause I (A B C) to have the same meaning as INTEGRATE (A B C).

REPEAT (X)

This command printed out X. (It is useful when the programmer has forgotten what he typed or created earlier.) For example, using the earlier information, REPEAT A causes A=B+C to be printed out.

FORGET (X)

This deleted the item X from storage.

SUBSTITUTE ((X_1 X_2 ... X_n) V W)

The X_i must be a list of names of expressions and/or functions. The value of each X_i was substituted in V at each occurrence of the name V_i. The new equation, expression, or function was named W. For example,

$$\text{DENOTE NIL}$$
$$A = B + C,$$
$$B = D + 2,,$$
$$\text{SUBSTITUTE ((B) A M)}$$

results in

$$M = D + 2 + C$$

498 FORMAL ALGEBRAIC MANIPULATION LANGUAGES

There is considerable dialogue with the system as can be seen from Figure VII-2. Much of the information typed by the system is not required logically.

The system just described is considered "dead" by its developer. It has been replaced by a new system, referred to as MATHLAB 68 when a distinction is necessary, which started operating in the fall of 1967 on a PDP-6 with a 256K core memory.[4] Input and output are by means of a teletype-like keyboard with a fixed character display scope. It still employs linear FORTRAN-like typewriter input for expressions.

Among the features of the old system which have been deleted are the excess verbiage, the restriction of input format caused by the old system's direct conversation with LISP, the restriction of stored data to those stored by name, and the restriction of references to data to references by name. References by name and value can now be freely mixed within any instruction.

The character set consists of the 26 upper-case letters, 10 digits, and the following symbols:

$$+ \quad - \quad * \quad / \quad (\quad) \quad , \quad = \quad : \quad ' \quad \$ \qquad blank$$

The variables and function names of MATHLAB are LISP atoms, and the rules for constructing them are precisely the rules for constructing LISP atoms in whichever LISP system MATHLAB is embedded.

The only numbers permitted for input/output are integers, but the internal arithmetic is rational.

The $ is used to terminate all input statements.

The : is used to store an expression or equation by name, e.g., V : X + Y $ and to specify the definition of a function, e.g., F(Z) : Z**2$. Blanks are significant.

The ' is used to evaluate symbols which are the names of stored data and to evaluate functions. Using the assignments of V and F above,

F(V)	evaluates to	F(V)
F('V)	evaluates to	F(X+Y)
'F(V)	evaluates to	V^2
'F('V)	evaluates to	$(X+Y)^2$

Another use of the evaluation or unquote symbol, ', is to distinguish between formal and active references to the commands. This distinction is illustrated in later examples.

Again assuming V as above and assuming, furthermore, that the formal

---

[4] The description of this new system is based on a private communication from C. Engelman and is included with his permission.

SAMPLE PROGRAM—MATHLAB 68

Problem: Given a random variable u with probability density $f(u)$ defined by:

$$f(u) = \begin{cases} \dfrac{3}{4sy}\left[1-\left(\dfrac{u-y}{sy}\right)^2\right] & y-sy \leq u \leq y+sy \\ 0 & \text{elsewhere} \end{cases}$$

We define a random variable p by:

$$p = \begin{cases} \dfrac{u-(1-r)x}{r} & u < x \\ b(u-x)+x+ax & u > x \end{cases}$$

The problem is to find the value of a such that the expectation of p is maximized when $x = y$.

Program:

```
P1:(U−(1−R)*X)/R+A*X$

    U − (1 − R)X
    − − − − − −  +  A*X
         R

P2:B*(U−X)+X+A*X$

    B(U − X) + X + A*X

F(U):3/(4*S*Y)*(1−((U−Y)/(S*Y))**2$

              3          U − Y  2
    WS (U) : − − − (1 − (− − −) )
            4S*Y          S*Y

          ALIAS SUBSTITUTE SUB$

DEFINT(EXPR,X,A,B):''('SUB(B,X,INT:'INTEGRATE(EXPR,X))−'SUB(A,X,'INT))$

WS(EXPR,X,A,B) :    'SUBSTITUTE(B,X,INT :    'INTEGRATE(EXPR,X))
−'SUBSTITUTE(A,X,'INT)

'SOLVE('SUB(Y,X,'DERIV('DEFINT('P1*'F(U),U,Y−S*Y,X)+'DEFINT('P2*'F(U),U,X,Y+S*Y),
X))=0,A)$

THE ROOTS ARE

              (B − 2)R + 1
    ML1 :     − − − − − −
                   2R

ONCE

FINISHED
```

Comments:

The command "integrate" in MATHLAB computes indefinite integrals. This was used, after introducing the initial data, to define the definite integral as a MATHLAB function. After that, the problem can be posed in one instruction. In the program, the lines typed by the user end with $ and the others are the computer response.

variables V and Y have not been declared dependent upon X,

DERIV(V,X)	evaluates to	$\dfrac{DV}{DX}$
DERIV('V,X)	evaluates to	$\dfrac{D}{DX}(Y+X)$
'DERIV(V,X)	evaluates to	0
'DERIV('V,X)	evaluates to	1

The result of any computation, whether assigned a name by the user or not, resides in an area named *workspace*. It is also temporarily assigned the name WS and may be referred to as such in the next instruction.

The command *SUBSTITUTE* is designed to imitate the evaluation concept above and, in effect, allows one to supply the evaluation signs as an afterthought. In its simplest utilization, 'SUBSTITUTE(X,Y,3∗Y) evaluates to 3∗X. Using the definition of F above, however, and assuming the expression in workspace is F(F(2)), then 'SUBSTITUTE('F,F,'WS) will evaluate to 16. *SUBSTITUTE* can be further employed to change formal references to commands to active references. Thus, if the value of the workspace is INTEGRATE(X^2,X), then the value of 'SUBSTITUTE ('INTEGRATE, INTEGRATE, 'WS) is $X^3/3$.

Arithmetic commands

The arithmetic commands of the old system have disappeared. Instead of typing ADD((E1 E2) *name*) we now type NAME : 'E1 + 'E2 $ if storage by name is desired, and 'E1 + 'E2 $ if no name is needed. (In this case, the expression is referred to as WS.) The real point is that input like X∗('E1 + 'E2)$ is now allowed. It would have taken three commands to achieve the same result in the old system.

Simplification

The user controls the usage of a simplification routine by means of on and off switches, e.g., SIMP ON $ causes simplification to occur after the execution of each statement. The *SIMP* $ switch reverses the status of the simplification invocation. In addition, there is a command RATSIMP, which uses the rational function routines to perform the required divisions. Thus 'RATSIMP(*DATUM,EXPRESSION1*, . . . , *EXPRESSIONn*) evaluates to the value of *DATUM*, expressed as a rational function with a single numerator and a single denominator, with all possible greatest common divisor cancellations having been made. The expressions *EXPRESSION1*, . . . , *EXPRESSIONn*, if present, are taken as the most important variables in decreasing order.

Commands for changing data types

The three fundamental elements of the formal algebraic data type are expression, equation, and function. The old system contained three MAKE commands which were removed from the current system since they are all special cases of the more flexible input syntax. Thus, to accomplish MAKEEQUATION(A R) of the old system, merely type R : A = 'A$ since the 'A really denotes the value of A.

Output

Output expressions are presented two-dimensionally on the scope or, when hard copy is desired, on the on-line printer. The display program is described in Millen [ZJ67].

Miscellaneous

Writing ALIAS A B permits A to be used in place of B.

VII.4.4. SIGNIFICANT CONTRIBUTION TO TECHNOLOGY

MATHLAB was the first complete on-line system with formal algebraic manipulation facilities. However, its effectiveness for man-machine interaction has been very limited because of the inadequacies of typed input and output. It was the first language to include higher level operations such as integrate, solve, etc. While these are of interest, they are not of major significance from a language point of view. The major contribution of MATHLAB is actually in an area which is beyond the scope of this book, namely the routines which were created to implement these higher level commands. Some useful work was also done so that the typewriter could be used to print expressions and equations in a two-dimensional form, using standard equipment.

The new version of the system (discussed briefly earlier) differs from the first one above in a number of ways. The notation and framework are more ALGOL-like, and they permit a stored program as well as the line-by-line mode. Probably the most significant part of the new system will be a method for defining and identifying mathematical subexpressions which have been typed; this is planned to be done by using input editing routines and by attempting to provide a typewriter facility equivalent to the concept of pointing at a scope with a light pen.

Somewhat different from the question of language or syntax, but quite important to the potential usefulness of MATHLAB, are improvements (both internal and external) in its rational function routines which allow them to operate in any number of variables and, in particular, to factor polynomials over the integers in any number of variables. This results in commands available to the user which provide such facilities as the computation of direct and inverse Laplace transforms; the inversion of matrices; and a spectrum of *solves*, including the solution of one equation rational in the desired unknown, several equations linear in the unknowns, one linear differential equation with constant coefficients, as well as the solution of several simultaneous implicit equations for several derivatives.

VII.5. ALTRAN

ALTRAN is a language developed at the Bell Telephone Laboratories in Murray Hill, N. J. by W. S. Brown, M. D. McIlroy, D. C. Leagues, and G. S. Stoller. Its design and approach were heavily influenced by the ALPAK system on which it is based; ALPAK is a set of subroutines for handling polynomials and rational functions and is used with FAP. A version of the system called *Early ALTRAN* was running in late 1964 on the IBM 7090/7094. It also runs on the 7040/44.

SAMPLE PROGRAM—ALTRAN†

Problem: Evaluate D in terms of the r's and the c's, where

$$D = \begin{vmatrix} a_1 & a_{12} & a_{31} \\ a_{12} & a_2 & a_{23} \\ a_{31} & a_{32} & a_3 \end{vmatrix}$$

$$a_i = (r_0 + r_i)^2$$

$$a_{ij} = r_0^2 + r_0 r_i + r_0 r_j - r_i r_j$$

$$r_i = \frac{1}{c_i}$$

Program:
```
C DECLARATIONS.
    LAYOUT (L1) C0 9, C1 9, C2 9, C3 9
    LAYOUT (L2) R0 9, R1 9, R2 9, R3 9
    POLYNOMIAL A1, A2, A3, A12, A23, A31, D
C ESTABLISH THE A'S.
    A1  = (R0+R1)**2
    A2  = (R0+R2)**2
    A3  = (R0+R3)**2
    A12 = R0**2+R0*R1+R0*R2−R1*R2
    A23 = R0**2+R0*R2+R0*R3−R2*R3
    A31 = R0**2+R0*R3+R0*R1−R3*R1
C COMPUTE AND PRINT THE CONDITION ON THE RADII
C AND THE CORRESPONDING CONDITION ON THE CURVATURES.
    D=A1*A2*A3*−A12**2*A3−A23**2*A1−A31**2*A2+2*A12*A23*A31
    PRINT  D, D(R0=1/C0,R1=1/C1,R2=1/C2,R3=1/C3)
    STOP
    END
```

†Brown [BW66a], pp. 6a, 6c–6e.

ALTRAN is an extension of FORTRAN, in a sense similar to that of FORMAC. Since the compiler was constructed from the beginning by using the TMG system (see Section IX.2.5.3 and McClure [MZ65a]), ALTRAN is not based on any particular version of FORTRAN; in fact, the language is primarily FORTRAN II with some elements from FORTRAN IV. Several new data types are added, most notably *numeric rational*, *polynomial*, and *algebraic*; the latter two are formal expressions, where *algebraic* really means *rational function*, i.e., the quotient of polynomials. Expressions are created as they are in FORTRAN, except that the constituents can be any of the three new data types and/or integers. Floating point numbers can be used but not with the new data types, i.e., neither as coefficients of polynomials or rational functions nor in expressions containing them.

Data names are defined as they are in FORTRAN. Arrays of the new data types are conceptually allowed, but they were not actually implemented. Arrays of polynomials were implemented.

As assignment is written the same way as in FORTRAN. The system knows whether it is to perform numeric computation or manipulation of rational functions by examining the data types involved. Much of this information is provided by declarations available at compile time.

The unconditional control transfers are the standard form GO TO and also the computed GO TO. The conditional control transfer can be of two forms. The first is

$$\text{IF } (expr) \ n_1, n_2, n_3$$

where *expr* is purely numeric and control is transferred to n_1, n_2, n_3, depending on whether *expr* is negative, zero, or positive, respectively. A second conditional statement has only two statement numbers associated with it and allows *expr* to be any of the five data types; control is transferred to n_1 if the value is not 0 and to n_2 if it is 0. The DO statement from FORTRAN II is permitted.

Arithmetic is done on rational numbers to produce exact results, just as in FORMAC. When computation is done on rational functions, the greatest common divisor is calculated automatically and used to simplify the rational functions.

The main algebraic statement directly permitted in ALTRAN is substitution. There are two kinds, namely complete and partial. The complete substitution has the general form $F(A_1, \ldots, A_n)$ where $A_1, \ldots, A_n$ are any of the five variable types discussed above. The variables $X_1, \ldots, X_n$ which appear in the expressions represented by F are replaced by the values of $A_1, \ldots, A_n$, respectively. The substitute expression is distinguished from a function reference because the latter has a function declaration. A partial substitution has the general form $F(Y_1=B_1, \ldots, Y_k=B_k)$, where the Y_i are some subset of the X_i and the B_i are expressions of type integer, rational, polynomial, or rational function. The Y_i are replaced simultaneously by the B_i in the expression represented by F. Note that every partial substitution could be written as a complete substitution by merely replacing each of the variables which is not to be changed by itself.

The only input statement is of the form

$$\text{READ } (L) \; B_1, \ldots, B_n$$

where L is a layout (discussed later) and the B_i are variables. When this statement is executed, the next n data items are read and converted appropriately. The items must have the same type as B_j. If none of the B_i are polynomial or rational functions, L is unnecessary and may be omitted.

The output statements are PRINT $E_1, \ldots, E_n$ and PUNCH $E_1, \ldots, E_n$ where the E_i are either literals or one of the five data types discussed earlier. The expression name followed by = will be printed or punched and then followed on one or more lines or cards by the value of the expression.

Subroutines may be invoked by writing

$$\text{CALL } S(A_1, \ldots, A_n)$$

where the A_i can be one of the five data types or arrays of variables.

The CONTINUE, RETURN, STOP, and END statements have the same meaning as in FORTRAN.

A new statement with the form MODE INTEGER OVERFLOW CHECK causes every integer arithmetic operation from then on to include a check for overflow.

Because it is desirable to obtain maximum efficiency by appropriate internal representation for polynomials, it is necessary for the user to supply certain information. This is done by means of the LAYOUT declaration. This has the form

$$\text{LAYOUT } (L) \; X_1 \; W_1, \ldots, X_n \; W_n$$

where L is the layout name, the X_i are the variables, and W_i are the corresponding field widths. The W_i refers to the amount of space to be allocated internally for exponents. All variables must be declared as either INTEGER, RATIONAL REAL, POLYNOMIAL, or ALGEBRAIC. The words FUNCTION and SUBROUTINE are used as in FORTRAN to declare functions and subroutines. Functions can also be declared to be any of the first five types. It is possible for the user to include FAP coded subprograms in his program. In particular, all the subroutines and facilities from ALPAK are made available to the user. ALTRAN is actually implemented by translating the expression manipulation statements to calls to ALPAK subroutines. Thus, while from a language point of view ALTRAN has very little beyond FORTRAN and the new data types in it, its ability to easily use all its ALPAK facilities makes it more powerful than it appears on the surface.

The following built-in functions are available in ALTRAN,[5] using this notation:

Arguments are denoted by the letters I, Q, R, P, A, and L representing the types integer, rational, real (= floating point), polynomials, algebraic (rational functions), and layout, respectively. Array arguments are underscored. Subscripts are used to distinguish different expressions of the same type. Data variables, which will be transmitted as expressions of type polynomial, are denoted by X. Finally, expressions of arbitrary type are denoted by E, and arrays of data variables or of language variables of arbitrary type are denoted by AR.

Integer functions

INT(E)	Equals the value that would be obtained by an integer variable I as a result of the assignment statement I = E.
IP(Q)	Equals integer part of Q (truncation toward 0).
INTTST(A)	Equals 1 if A is an integer and 0 if A is not an integer.
RATTST(A)	Equals 1 if A is a rational and 0 if A is not a rational.
POLTST(A)	Equals 1 if A is a polynomial and 0 if A is not a polynomial.
EQ (A_1, A_2)	Equals 1 if $A_1 = A_2$ and 0 if $A_1 \neq A_2$.
QUO(I_1, I_2)	Equals IP(I_1/I_2).
REM(I_1, I_2)	Equals $I_1 - I_2 * IP(I_1/I_2)$

Rational functions

RAT(E)	Equals the value that would be obtained by a rational variable Q as a result of the assignment statement Q = E.

Real functions

REAL(E)	Equals the value that would be obtained by a real variable R as a result of the assignment statement R = E.

---

[5] This list is taken from pp. A1-2–A1-4 of McIlroy and Brown [ML66].

Polynomial functions

POL(E)   Equals the value that would be obtained by a polynomial variable P as a result of the assignment statement P = E.

NUM(A)   Equals numerator of A.

DEN(A)   Equals denominator of A.

GCD(P_1, P_2)   Equals greatest common divisor of P_1 and P_2.

Algebraic functions

ALG(E)   Equals the value that would be obtained by an algebraic variable A as a result of the assignment statement A = E.

DIFF(A,I)   Equals $\partial A / \partial X(I)$.

This system has had very limited usage outside Bell Laboratories.

In conclusion, while admitting that my view may be biased, I feel that ALTRAN itself has not contributed anything to the technology. The significant part of the effort was the work done on ALPAK, which of course is not within the scope of this book.

VII.6. FLAP

FLAP is an experimental language that was written in LISP 1.5 (see Section VI.5) by A. H. Morris, Jr., at the U. S. Naval Weapons Laboratory at Dahlgren, Va.; it runs on the IBM 7090. FLAP permits the user to handle different types of symbolic mathematical data, e.g., differential equations, vectors and matrices, etc. Much of the nomenclature, although not the notation, comes from LISP.

An atom is defined in FLAP to be a letter followed by up to 29 numbers or letters. Atoms are used as variables, constants, or function names. There are two special atoms defined, namely TRUE and NIL, which have the values true and false, respectively.

A number of mathematical systems (which are really modes of arithmetic) are provided. The simplest is the *NUMBER, which assumes that only rational numbers are available. The systems available to the user are called *algebras* since they have been defined to be that from a mathematical point of view. The *SIMPLIFY is the second simplest algebra in FLAP. It assumes that the basic data are rational numbers and constant atoms such as A and PI, which are independent of each other. These constants can

SAMPLE PROGRAM—FLAP[†]

Problem: Given the two series $\sum_{n=0}^{\infty} a_n x^n$ and $\sum_{n=0}^{\infty} b_n x^n$ where $b_0 \neq 0$, evaluate the first $r + 1$ terms $c_0, c_1, \ldots, c_r$ of the series $\sum_{n=0}^{\infty} c_n x^n = \sum_{n=0}^{\infty} a_n x^n / \sum_{n=0}^{\infty} b_n x^n$. The variables A, B, N, R have the respective values a_n, b_n, n, r.

Program:

```
          DIVIDESERIES,DOTPROD
DEF       DIVIDESERIES
INPUT     A,B,N,R
DUMMY     K,L,V,BO,NUM
          NUM = 0
          MODE *SIMPLIFY
          BO = COMPUTE(SUBST(0,N,B))**(-1)
          L = LIST(COMPUTE(SUBST(0,N,A))*BO)
          V = L
LOOP      IF NUM=R THEN RETURN(CAR(V))
          FLAPPRINT CAR(V)
          NUM = NUM+1
          K = CONS(COMPUTE(SUBST(NUM,N,B)),K)
          RPLACD V,LIST(BO*(COMPUTE(SUBST(NUM,N,A))-DOTPROD(K,L)))
          V = CDR(V)
          GO LOOP
          END
DEF       DOTPROD
INPUT     X,Y
DUMMY     K,L,M
          K = X
          L = Y
          M = 0
A         M = M+CAR(K)*CAR(L)
          K = CDR(K)
          L = CDR(L)
          IF NULL(K) THEN RETURN(M)
          GO A
          END
          DIVIDESERIES(SIN(A*(N+1))*FACTORIAL(N)**(-1),FACTORIAL(B+N)**(-1),
          N,2)
```

[†]Morris [MB67], pp. 22–23.

---

be combined under the five normal arithmetic operators, and Morris applies his definition of simplification to expressions; this means expansion to eliminate parentheses, followed by combination of like terms. The algebra *NUMBER is imbedded in *SIMPLIFY and the latter in turn is included in every other algebra.

For arbitrary values, FLAP assumes that the variables are listed at the beginning of the program so that the system can determine whether an atom is a variable or a constant. A particular atom may be the name of a function, where assumptions may or may not have been made regarding its nature.

There are five types of expressions in FLAP: Atoms; lists of items, each of which is an expression; mathematical expressions; expressions of the form A=B, where A and B are themselves expressions (and this composite expression then has the value true or false); and functional expressions of the form $f(x_1, \ldots, x_n)$, where f is the name of a function and the x_i are expressions which are its arguments. Any expression which has only the values TRUE or NIL is called a *conditional* expression and its corresponding function is called a *predicate*. Thus the expressions A=B and EQUAL (A, B) have the same values and can be used interchangeably.

The assignment statement is of the form A = B, where B is an expression.

Blanks are critical in many places.

The unconditional statement is of the form GO *statement-name*. The conditional statement is of the form IF A THEN B, where A is a conditional expression and B is an arbitrary expression; the latter can be of the form GO (*statement-name*).

The main declaration is the MODE, which defines whatever algebra is needed. Its general format is MODE A, K, L, ..., M, where A is the name of the algebra and the sequence K, L, ..., M is an ordered sequence of parameters associated with A. An arbitrary number of mode statements can be used in any program and the evaluation of any expression is controlled by the most recent MODE declaration given prior to the evaluation of the expression. As an example, the *MULT algebra is used to manipulate expressions containing constant atoms that fall into either a *vector* or *scalar* category. Thus, after writing MODE *MULT(X), an expression such as (3*X+6*X**2)*(A+B*X+X**2) will have as its value the polynomial 3*A*X+(6*A+3*B)*X**2+(3+6*B)*X**3+6*X**4. In this case, A and B are the scalar coefficients of the polynomial in the vector X.

A COMPUTE function has a single argument S which is used after some algebraic mode has been set. The purpose of COMPUTE is to assign to S the value of the argument in this particular mode.

Axioms about specifically defined functions can be made. Suppose that we wish the function S to have the values $e_i(x_1, \ldots, x_n)$ whenever the conditional expression $p_i(x_1, \ldots, x_n)$ has the value TRUE. Then the axiom mapping h for f would be the function with the arguments $x_1, \ldots, x_n$ having the value $e_i(x_1, \ldots, x_n)$ whenever $p_i(x_1, \ldots, x_n)$ is true and the value $f(x_1, \ldots, x_n)$ otherwise. By means such as these and other facilities, it is possible to use a function, e.g., COS(A), to have different meanings in differing algebras.

A partial list of the algebras available is as follows:

*MULTILINEAR
: This has two modifier lists, K and L. The items of L are considered to be the mathematical data and each is assumed independent of the others. The *constants* of this mode are elements of the *MULT arithmetic based on the modifier list shown in K. The differentiation operator DIFF(A,X) is also defined. Using this operator, arbitrary symbolic differential expressions can be manipulated.

*EXTERIOR
: There are three *EXTERIOR algebras available, differing only by the number of modifier lists used. If there is a single list B, then B must be a basis for a vector space having as its ring of scalars the *SIMPLIFY algebra. For two or three modifier lists, then either the *MULT or the *MULTILINEAR algebras are used.

*MATRIX
: There are three *MATRIX modes available, depending on the number of modifier lists used. If no modifier list is used, then every expression in the mode is assumed to be a matrix whose components are values in the *SIMPLIFY arithmetic. If there are one or two modifier lists, then the values of the matrix elements are assumed to be in *MULT and *MULTILINEAR modes, respectively.

A great many LISP functions are made available to the FLAP programmer. These include the following:

AND, APPEND, ATOM, CAR, CDR, CAAR, CADR, ..., CDDDR, CONC, CONS, COPY, EQUAL, GET, LENGTH, LIST, MEMBER, NCONC, NOT, NULL, OR, REVERSE, RPLACA, RPLACD, SELECT, SETQ, SUBST.

There are functions which have arbitrary expressions as arguments. These include the following:

FLAPPRINTS(S)
: S is printed and NIL is returned as the value of the function.

INTEGERP(S)
: If S is an integer, the value of the function is TRUE, otherwise it is NIL.

LARGER(A,B)
: If A is greater than B, the value of the function is TRUE, otherwise it is NIL.

ONEPE(S)
: If $S=1$, the value of the function is TRUE, otherwise it is NIL.

ZEROPE(S)
: If $S=0$, the value of the function is TRUE, otherwise it is NIL.

The following functions have lists of mathematical expressions as arguments:

FACTORSINV(S)   If $S = (a_1, \ldots, a_r)$, the value will be $a_1 * \cdots * a_r$. If $S = $ NIL, the value is 1; if $S = (a)$, the value is a.

SUMMANDSINV(S)   Similar to above, except the results are $a_1 + \cdots + a_r$.

DIFF (S,X)   Involves differentiation.

The functions PLUSP(S), POWERP(S), SUMMANDS(S), and TIMESP(S) are used for testing the form of the expression S.

Functions having matrices as arguments include ADJOINTDATA(A), COEFFICIENTDATA(A), MATRIXMAP(A,X), SWITCH(A), and TRANSPOSE(A). Each has a specific definition, depending upon the form of A.

Mr. Morris states, "The system has been employed in the development of techniques for the manipulation of transformations of families of matrices and partial differential equations. It is currently being used to study the algebraic structures of the indefinite integral and related operators such as the Laplace and Fourier transforms."[6] FLAP is apparently only being used internally in the developing organization.

VII.7. SYSTEMS REQUIRING SPECIAL EQUIPMENT

As in the case of the numerical scientific languages, anybody attempting to develop an efficient system for doing formal algebraic manipulation is hampered by the constraints of the normal input/output equipment. This section discusses two quite different systems which are using either special equipment or special hookups of standard equipment; they are wholly or partially dependent on a light pen. Since this is a book on *languages*, only enough of the physical and system aspects will be mentioned to make the language capabilities meaningful to the reader.

VII.7.1. MAGIC PAPER

The first attempt at an on-line system was the Magic Paper described by Clapp and Kain [CL63]; it was never fully implemented but, nevertheless, it presented some interesting ideas which are worthy of note.

The system used the PDP-1, with a typewriter, display scope, paper tape reader and punch, magnetic tape, and light pen for input/output. Some

---

[6] Private communication, October, 1967.

of the control characters available to the user are shown in Figure VII-3. The main input device was considered to be the typewriter, but this is obviously not the most convenient means of typing mathematical expressions. For that reason, a number of symbols were designated as format control characters; these are shown in Figure VII-4.

The display scope was the principal on-line output device. Equations and other data could be displayed directly in fairly standard mathematical notation. The results could be saved by photographing the scope or by using punched paper or magnetic tape.

Numerical evaluation could be done; in addition to the five standard operators, there were functions indicating summation or product over an index between limits, and conditional expressions.

A new and independent version that retains the same name has been developed at Computer Research Corporation under the direction of L. Clapp, who was a developer of the first version. The newer version used a nonstandard PDP-1 and special visual display equipment. The system was *not* being used at the time of this writing because it needed to be integrated with other packages. For this reason, only the highlights of the language

Symbol	Control significance
I	Enter input mode
,	Leave input mode
P	Display pointer on scope
L	Label equation
→	Substitute
U	Underline equation
R	Replace underlined symbols
F	Factor out underlined terms
G	Get equation from storage and add to bottom of display
S	Enter sketch input mode (to construct a figure)
D	Display figure
T	Transpose underlined strings
N	Name underlined string
K	Expunge equation
W_n	Change window size to n
M	Move window (u-up, d-down, b-page back, f-page front)
EVAL	Evaluate function
GEN	Generate graph
$c <\ldots> i, j, \ldots, n$	Create new control function

Figure VII-3. Typical executive control characters in first Magic Paper. Source: Clapp and Kain [CL63], p. 511.

Symbol		Function
□	(Space)	Delimiter (except after nonspacing characters)
⇒	(Tab)	Delimiter
ϙ	(Carriage return)	Delimiter (end of line)
·	(Center dot)	Dismiss the symbol which follows from its control significance
↑		Denotes exponent
∧		Denotes superscript
∨		Denotes subscript
|	(Vertical bar)	Restore to main line
?		Drop exponent level one line
→		Equation label follows
⇐	(Backspace)	Ignore last character
⊃		Treat the string just typed as a defined symbol

Figure VII-4. Format control characters in first Magic Paper. Source: Clapp and Kain [CL63], p. 513.

capability will be given. The main report on this system is Clapp *et al.* [CL66].

The system contains a special keyboard, a flicker-free display, a light pen, a push-button panel, and two foot switches. The panel is the primary control for the system. Every button on it corresponds to a system command which is invoked by pressing the button. These commands control the display and also perform manipulations. The arguments either are selected by the light pen or are entered through the keyboard. Both input equations and calculated results are organized into a system of *scrolls* which are conceptually like pages of a book. The scrolls are used for displaying current results and also expressions being entered. When the scroll is full it is automatically rotated, but the user can also control the movement forward and back. The light pen permits expressions on the screen to be brightened or underlined, and it is used to select arguments for use with commands.

The special keyboard used contains the 52 upper- and lower-case letters; the 10 digits; Greek letters; keys for *Greek mode* and *English mode*; a *panic key*; control keys for space, carriage return, and backspace; and keys for subscript and superscript. It also contains the following characters:

Variables can only be single letters since juxtaposition is used to denote multiplication. Variables and function names can have subscripts and/or superscripts; the latter can themselves contain subscripts and superscripts. Multiple subscripts are permitted, with commas as separators; the same applies for superscripts. Each of the "scripting" (abbreviation for subscripting and superscripting) operations must be terminated by a question mark,

?, which has the meaning *return to the previous line*, or terminate last exponentiation or scripting operation. Superscripting and exponentiation are logically different but are displayed similarly. Denoting sub- and superscript operations by $\smile$ and $\frown$, respectively,

$$x^{(y_i^2)}$$

would be input as x ↑ y$\smile$i$\frown$2???. Rules for interpretation in potentially ambiguous or special cases are defined.

Spaces are generally ignored, but they cannot appear in the middle of a function name and must appear between a whole number and a fraction.

Numbers can be decimal, fractional (e.g., 1/2), or mixed (2 1/2); 1/2a is interpreted as (1/2)a, whereas 1/ab is taken to mean 1/(ab).

Nine elementary mathematical functions are available and typed in their normal form; arguments need not be in parentheses; e.g., sin 2x is interpreted as sin(2x).

The categories of commands available are scroll manipulation, equation input, equation editing, arithmetic and mathematical (applied to expressions or equations) simplification, application of system-defined transformations to expressions, definition and execution of user-created procedures, and system status changes. These commands are invoked by using the correct pushbutton.

The scroll manipulation commands are *rewind* (return to the starting equations), *backup* (shift the display back one equation), *advance* (shift forward one equation), *unwind* (place display window at end of scroll), *save* (*name*) (file the current scroll and assign the user-specified name), *display* (*name*) (retrieve the scroll name by the user), *start* (delete current scroll), and *delete* (*expr*) (delete the indicated expression from the current scroll).

An equation or expression is used as the argument of an *enter* command and is placed into the current scroll. A *label equation* command assigns the next label number to the current equation. An existing item can be *edited* by inserting and deleting characters.

The *apply* command has as arguments a mathematical operator (e.g., +, −) and a sequence of expressions or equations. The designated action is performed (on both sides of the equations if they are used) and the result is put into the scroll. Subtraction, division, and exponentiation allow only two arguments, while multiplication and addition allow any number. No simplification is performed automatically; this is done by the user with the commands *combine terms*, *combine fractions*, and *simplify products and fractions*.

Substitution is accomplished by *substitute*, with the arguments consisting of an equation of the form a = *expression* b and an expression. All occurrences of a in the designated expression are replaced by b.

An expression which is a summand on one side of an equation can be moved to the other side by the *transpose* command (and the sign will be changed automatically).

The system transformations include such things as changing $(a+b)^2$ to $a^2+2ab+b^2$. The transformations are stored in tables and can be displayed to the user by pressing a foot pedal. A particular one is activated by the *apply transformation* command in which the user specifies the transformation and the expression on which it is to operate. If the expression does not match, a *no match found* message is displayed. For example,

$$((c + 2d/(e+1)) + 2)^2$$

would match the transformation mentioned above, but $(c+1)^3$ would not. New transformations can be put into the table by the *insert information* command, which specifies the transformation; similarly, one can be deleted by the *remove transformation*. Transformations can be used in both directions by appropriate use of the → and ←.

The most recent result can be deleted and the system returned to its earlier condition by the *restart* command. A *cancel* command deletes the last step the user took in setting up the arguments for an operation.

The user has a *procedure define* which allows him to assign a procedure to a particular button, and he has ways to specify the operands.

The most interesting language feature of this system is its provision for user-defined new operators. The user specifies (or responds to questions from the system) whether the operator is prefix, infix, or suffix; whether its arguments must be parenthesized; how many arguments it must have; whether it is associative and commutative; and what its precedence value is relative to the other operators.

From a general point of view, most of the commands in this system relate to providing an appropriate means of using the hardware. It does not seem to have any significantly new or general concepts, although some of the details seem both interesting and novel. The one exception is the concept of the *scroll*, whereby previous information is available in a readable form and can be displayed upon proper request. This seems to be a valuable solution to one of the major problems in this type of system.

VII.7.2. SYMBOLIC MATHEMATICAL LABORATORY

The Symbolic Mathematical Laboratory, developed by W. A. Martin [ZB67] as his Ph.D. thesis at M.I.T., was running early in 1967. While the equipment used is standard, the interconnections are not, and Martin himself states that the system is inefficient for these and other reasons. The user

SAMPLE PROGRAM—Symbolic Mathematical Laboratory

Problem: Application of the Poincaré-Lighthill procedure to $\ddot{x} + \omega^2 x = \epsilon x^3$.

Program:

```
#'E1←'(DRV(:T,2,X(:T)) + OMEGA↑2*X(:T) = EP*(X(:T))↑3)#
#EDISPLAY('E1)#
#'E2←'(X(TAU) = SUM(I,0,INF,EP↑I*X[I](TAU)))#
#EDISPLAY('E2)#
#'E3←'(:T(TAU) = TAU + SUM(J,1,INF,EP↑J*:T[J](TAU)))#
#EDISPLAY('E3)#
#'E4←DRVFACTOR(E1,':T,1)#
#EDISPLAY('E4)#
#'E5←SUBSTITUTE(DELSUBST(E4,'(DEL(:T)), '(DEL(TAU))/DRV('TAU,1,RIGHT(E3))),
    '(X(TAU)),'(X(:T)))#
#EDISPLAY('E5)#
#'E6←DRVDO(SUBSTITUTE(E5, RIGHT(E2),'(X(TAU))),'TAU)#
#EDISPLAY('E6)#
#'E7←TRUNCATE(E6,'EP,1)#
#EDISPLAY('E7)#
#'E8←!E7,6=;!E7,88#
#EDISPLAY('E8)#
#'E9←SIMPLIFY(DRVDO(SUBSTITUTE(SOLVE(E8,'(X[1](TAU))),'(A*COS(OMEGA*TAU)),
    '(X[0](TAU))),'TAU))#
#EDISPLAY('E9)#
#'E10←COLLECT(EXPAND(SUBSTITUTE(E9,'((COS(3*OMEGA*TAU)+3*COS(OMEGA*
    TAU))/4),'((COS(OMEGA*TAU))↑3))),'((SIN(OMEGA*TAU),COS(OMEGA*TAU))))#
#EDISPLAY('E10)#
#'E11←SIMPLIFY(SOLVE(SUBSTITUTE(!E10,44,'C,'(DRV(TAU,1,:T[1](TAU))))=0,'C))#
#EDISPLAY('E11)#
```

Comment: An example of the first three displayed equations as plotted by the CALCOMP plotter is shown below.

(E1) $\qquad \dfrac{d^2}{dt^2} X(t) + \omega^2 \cdot X(t) = \epsilon \cdot X(t)^3$

(E2) $\qquad X(\tau) = \displaystyle\sum_{I=0}^{\infty} \epsilon^I \cdot X_I(\tau)$

(E3) $\qquad t(\tau) = \tau + \displaystyle\sum_{J=1}^{\infty} \epsilon^J \cdot t_J(\tau)$

†Martin [ZB67], extracts from pp. 27–30.

inputs information to a PDP-6 from teletype, or from a light pen acting on a display scope. Output is on a plotter, the display scope, or the teletype. Information is sent from the PDP-6 via a transmission device (the 7750) to a 7094. The calculations are done in the 7094, using a disc for auxiliary memory, and then transmitted back to the PDP-6 for output. This rather unusual configuration was required because it was the only way to provide minimum requirements of a time-shared computer with a large memory and a display with fast light pen response.

One of the most interesting aspects of this system is its effective use of the display scope, both for showing expressions and for providing commands to manipulate expressions based on pointing with the light pen. Expressions are displayed on the scope in essentially normal mathematical notation, including, e.g., Σ. The list of commands given below contains several commands which permit the user to institute action based on pointing with the light pen. While this particular system is not a practical one for actual usage, it nevertheless has demonstrated a number of internal techniques which are useful and necessary in developing future systems in this area.

Commands are typed and executed one at a time. The system is coded in LISP 1.5.

The five operations of addition, subtraction, multiplication, division, and exponentiation for algebraic expressions are denoted respectively by +, −, *, /, and ↑.

Variable names are composed of up to 120 letters. They can be subscripted with expressions to any number of levels, but only the first level is actually handled by the system. Subscripts are enclosed within square brackets and are separated by commas. Functional notation is shown by enclosing arguments in parentheses after the subscripts; e.g., A[I,J](C,D,E) is a function of $A_{I,J}$ with arguments C, D, and E. Sets are denoted by enclosing the elements in parentheses; e.g., (A,B,C) is a set with three elements.

Assignment of a name to an expression is denoted by ←; i.e., A ← B gives the name A to the expression B. There are other ways of identifying expressions through the use of the light pen (discussed later).

Either an expression or a variable name can be quoted, i.e., preceded by a '. Quoting a function name means that its arguments will be evaluated but the function will not be evaluated.

The following is a typical command in this system:[7]

#'E1→'(X + Y) * 'DRV(':T,1,DRV('U,2,E1)) + !E2,20 + '(F[I,J](X,Y)) ↑ 2#

This means that the name E1 is assigned to the expression which is the sum of three terms. The first term is the product of (X + Y) with the unevaluated first derivative with respect to lower-case T of the second derivative

---

[7] Martin [ZB67], p. 20.

with respect to U of the expression currently named E1. The second term is the twentieth subexpression of a displayed expression currently named E2; this subexpression has been indicated with the light pen. The final term is the square of a subscripted function of X and Y.

There are no control statements, either conditional or unconditional, except for a few which are inherent in some of the algebraic manipulating commands. The complete list of these is as follows, where the description of each command has been taken directly from Martin [ZB67],[8] but the subhead grouping was done by me. The presence or absence of blanks is *not* significant.

Differentiation

DRV (X_1, N_1, ..., X_n, N_n, Y)
  Differentiate Y N_i times with respect to X_i, for each *i*.
DRVDO (EXP, X)
  All indicated derivatives with respect to X in EXP are carried out as far as posible.
DRVFACTOR (EXP, X, N)
$$\frac{d^{N+M}f}{dx^{N+M}} \longrightarrow \frac{d^M}{dx^M}\left(\frac{d^N f}{dx^N}\right)$$
  for each such subexpression in EXP.
DRVZERO (EXP, X)
  All derivatives with respect to X in EXP are set equal to zero.
DELSUBST (EXP, OLDDEL, NEWDEL)
$$\frac{dx}{d\ OLDDEL} \longrightarrow \frac{dx}{d\ NEWDEL}$$
  for each such subexpression in EXP.

Types of substitution

REPLACE (E, X, Y)
  Expression X replaces Y in the expression named E. Y is a term indicated with the light pen or a group of terms indicated with GROUP.
SUBSTITUTE (EXP, X, Y)
  Substitue X for each occurrence of Y in EXP.
EVALUATE (EXP, SET)
  SET is a set of equations; whenever the left side of one of these equations can be matched to a subexpression in EXP, the right-hand side is substituted. The left sides must be variables or functions. A match occurs whenever a binding of the function variables and subscripts can be made.

---

[8] Martin [ZB67], pp. 21, 23–26

Commands involving simplification

SIMPLIFY (EXP)
　　Simplifies expression *EXP*.

EXPAND (EXP)
　　Multiplies out all expressions of the form a*(b+c) in *EXP*. In addition,
$$\frac{d}{dx}(a+b) \rightarrow \frac{da}{dx} + \frac{db}{dx}$$

COLLECT (EXP, SET)
　　Top level terms in *EXP* are collected on powers of the expressions in *SET*.

FACTOROUT (EXP, FACTOR, Y)
　　The factor *FACTOR* is factored from each term of *EXP*. The third argument *Y* is optional. If *Y* is present, the factor *FACTOR* is renamed *Y*.

NORMPOLY (EXP, X)
　　Every sum in *EXP* is treated as a polynomial in *X* and a power of *X* is factored out so that the lowest power of *X* in the polynomial will be zero.

SUMEACH (EXP)
$$\Sigma(a+b) \rightarrow \Sigma a + \Sigma b$$

SUMEXPAND (EXP)
　　Expands the finite summation *EXP*.

ALLSUMEXPAND (EXP)
　　Applies *SUMEXPAND* to every summation in expression *EXP*.

TRUNCATE (EXP, VAR, N)
　　Expands *EXP* up to power *N* in variable *VAR*.

Rearrangement of expressions

BRINGOVER (EXP, X)
　　Subexpression *X*, which has been indicated with the light pen, is brought to the other side of equation *EXP*.

EXCHANGE (EXP)
　　If the top level connective of *EXP* is binary, its arguments are exchanged from right to left.

GROUP (SET)
　　The set *SET* of terms which has been indicated by the light pen in *EXP* is grouped within the associated sum or product. The value of *GROUP* is the grouped set of terms.

MULTIPLYTHROUGH (EXP, X)
　　Multiplies each top level term of *EXP* by *X*.

VII.7.2. SYMBOLIC MATHEMATICAL LABORATORY

Obtaining or manipulating subexpressions

RIGHT (*EXP*)
> Returns the right argument of the main binary connective of *EXP*.

LEFT (*EXP*)
> Returns the left argument of the main binary connective of *EXP*.

SPLIT (*EXP*)
> Subparts of *EXP* are named and replaced by their names in *EXP* so that *EXP* will contain less than 100 subexpressions.

TERM (*EXP, N*)
> Returns the *N*th argument of the top level connective of *EXP*, or *NIL* if there is no *N*th argument.

!*A, N*
> The *N*th subexpression of *A* is the value of !. Intensify the desired subexpression of *A* by pointing to its main connective with the light pen. Then type !*A* and the computer will type *N*. If the expression has no main connective, point to one of its arguments and type ;!*A* instead of !*A*.

Higher level commands

SOLVE (*EXP, X*)
> Solves equation *EXP* for variable *X* as far as possible.

ITG (*X, L1, L2, Y*)
> Integrate *Y* with respect to *X* between limits *L1* and *L2*.

LIMIT (*EXP, X, N*)
> Determines the limiting value of *EXP* as *X* approaches *N*.

SUM (*I, N1, N2, Y*)
> Sum expression *Y* for values of *I* from *N1* to *N2*.

Input/output

EDISPLAY (*E*)
> Displays the expression named *E* on the PDP-6 scope.

EPRINT (*E*)
> Prints out the internal form of the expression named *E* with *PLS, PRD, EQN,* and *PWR* in infix form; the other operators are in prefix form.

EDELETE (*E*)
> Deletes expression named *E* from the disc.

Miscellaneous

NEWNAME ()
> Creates a name of the form *Fn*, where *n* is an integer.

DEPENDENCE (EXP)
Returns a set of the variable and function names in *EXP*.

There has been no real usage of the system except by the developer. The most significant elements are the methods for displaying and manipulating expressions and some of the internal techniques used for simplification.

REFERENCES

VII.1. SCOPE OF CHAPTER

[BB67a] Bobrow, D. G. (ed.), *Symbol Manipulation Languages and Techniques, Proceedings of the IFIP Working Conference on Symbol Manipulation Languages*. North-Holland Publishing Co., Amsterdam, 1968.

[KD54] Kahrimanian, H. G., "Analytical Differentiation by a Digital Computer", *Symposium on Automatic Programming for Digital Computers*, Office of Naval Research, Dept. of the Navy, Washington, D.C. (1954), pp. 6–14.

[NO53] Nolan, J., *Analytical Differentiation on a Digital Computer* (M. A. thesis), M.I.T., Cambridge, Mass. (May, 1953).

[RA67] Raphael, B. *et al.*, "A Brief Survey of Computer Languages for Symbolic and Algebraic Manipulation", *Symbol Manipulation Languages and Techniques, Proceedings of the IFIP Working Conference on Symbol Manipulation Languages* (D. G. Bobrow, ed.). North-Holland Publishing Co., Amsterdam, 1968, pp. 1–54.

[SM66] Sammet, J. E., "An Annotated Descriptor Based Bibliography on the Use of Computers for Non-Numerical Mathematics", *Computing Rev.*, Vol. 7, No. 4 (July–Aug., 1966), pp. B1–B31.

[SM66a] Sammet, J. E., "Survey of Formula Manipulation", *Comm. ACM*, Vol. 9, No. 8 (Aug., 1966), pp. 555–69.

[SM67] Sammet, J. E., "Formula Manipulation by Computer", *Advances in Computers*, Vol. 8 (F. L. Alt and M. Rubinoff, eds.). Academic Press, New York, 1967, pp. 47–102.

[SM67a] Sammet, J. E., "Revised Annotated Descriptor Based Bibliography on the Use of Computers for Non-Numerical Mathematics", *Symbol Manipulation Languages and Techniques, Proceedings of the IFIP Working Conference on Symbol Manipulation Languages* (D. G. Bobrow, ed.). North-Holland Publishing Co., Amsterdam, 1968, pp. 358–484.

VII.2. LANGUAGES OF HISTORICAL INTEREST ONLY

[BM61] Bernick, M. D., Callender, E. D., and Sanford, J. R., "ALGY—An Algebraic Manipulation Program", *Proc. WJCC*, Vol. 19 (1961), pp. 389–92.

VII.3. FORMAC

[BZ64] Bond, E. R. *et al.*, "FORMAC—An Experimental FORmula MAnipulation Compiler", *Proc. ACM 19th Nat'l Conf.*, 1964, pp. K2.1-1–K2.1-11.

[BZ67]   Bond, E. R. and Cundall, P. A., "A Possible PL/I Extension for Mathematical Symbol Manipulation", *Symbol Manipulation Languages and Techniques, Proceedings of the IFIP Working Conference on Symbol Manipulation Languages* (D. G. Bobrow, ed.). North-Holland Publishing Co., Amsterdam, 1968, pp. 116–132.

[CX65]   Cuthill, E., Voigt, S., and Ullom, S., *Use of Computers in the Solution of Boundary Value and Initial Value Problems*, Annual Progress Report, SR011-01-01 Task 0401 AML Problem 821-911, David Taylor Model Basin, Washington, D.C. (June, 1965).

[DB67]   Duby, J. J., "Sophisticated Algebra on a Computer—Derivatives of Witt Vectors", *Symbol Manipulation Languages and Techniques, Proceedings of the IFIP Working Conference on Symbol Manipulation Languages* (D. G. Bobrow, ed.). North-Holland Publishing Co., Amsterdam, 1968, pp. 71–85.

[HQ67]   Howard, J. C., "Computer Formulation of the Equations of Motion Using Tensor Notation", *Comm. ACM*, Vol. 10, No. 9 (Sept., 1967), pp. 543–48.

[IB65c]   *FORMAC (Operating and User's Preliminary Reference Manual)*, IBM Corp., No. 7090 R2IBM 0016, IBM Program Information Dept., Hawthorne, N.Y. (Aug., 1965).

[IB67c]   *PL/I-FORMAC Interpreter*, IBM Corp., Contributed Program Library, 360D 03.3.004, Program Information Dept., Hawthorne, N.Y. (Oct., 1967).

[JL66]   Bleiweiss, L. et al., *A Time-Shared Algebraic Desk Calculator Version of FORMAC*, IBM Corp., TR00.1415, Systems Development Division, Poughkeepsie, N.Y. (Mar., 1966).

[ND67]   Neidleman, L. D., "An Application of FORMAC", *Comm. ACM*, Vol. 10, No. 3 (Mar., 1967), pp. 167–68.

[SM64]   Sammet, J. E. and Bond, E., "Introduction to FORMAC", *IEEE Trans. Elec. Comp.*, Vol. EC-13, No. 4 (Aug., 1964), pp. 386–94.

[SM67]   Sammet, J. E., "Formula Manipulation by Computer", *Advances in Computers*, Vol. 8 (F. L. Alt and M. Rubinoff, eds.). Academic Press, New York 1967, pp. 47–102.

[SO65]   Sconzo, P., LeSchack, A. R., and Tobey, R. G., "Symbolic Computation of f and g Series by Computer", *Astronomical Jour.*, Vol. 70, No. 4 (May, 1965), pp. 269–71.

[SV66]   Swigert, P., *Computer Generation of Quadrature Coefficients Utilizing the Symbolic Manipulation Language FORMAC*, NASA TN D-3472, Lewis Research Center, Cleveland, Ohio (Aug., 1966).

[TO66]   Tobey, R. G., "Eliminating Monotonous Mathematics with FORMAC", *Comm. ACM*, Vol. 9, No. 10 (Oct., 1966), pp. 742–51.

[WB67]   Walton, J. J., "Tensor Calculations on Computer: Appendix", *Comm. ACM*, Vol. 10, No. 3 (Mar., 1967), pp. 183–86.

VII.4. MATHLAB

[EN65]   Engelman, C., "MATHLAB—A Program for On-Line Machine Assistance in Symbolic Computations", *Proc. FJCC*, Vol. 27, pt. 2 (Nov., 1965), pp. 117–126.

[MV67] Manove, M., Bloom, S., and Engelman, C., "Rational Functions in MATHLAB", *Symbol Manipulation Languages and Techniques, Proceedings of the IFIP Working Conference on Symbol Manipulation Languages* (D. G. Bobrow, ed.). North-Holland Publishing Co., Amsterdam, 1968, pp. 86–97.

[ZJ67] Millen, J. K., "CHARYBDIS: A LISP Program to Display Mathematical Expressions on Typewriter-Like Devices", *Proceedings of the Symposium on Interactive Systems for Experimental Applied Mathematics,* Washington, D.C., August 26–28, 1967, Academic Press, Inc., New York, 1968.

VII.5. ALTRAN

[BW63] Brown, W. S., Hyde, J. P. and Tague, B. A., "The ALPAK System for Nonnumerical Algebra on a Digital Computer", *Bell System Technical Jour.*, Vol. 42, No. 5, (Sept., 1963), pp. 2081–119; Vol. 43, No. 2 (Mar., 1964), pp. 785–804; Vol. 43, No. 4, pt. 2 (July, 1964), pp. 1547–62.

[BW66] Brown, W. S., "A Language and System for Symbolic Algebra on a Digital Computer", *Proc. of the IBM Scientific Computing Symposium on Computer-Aided Experimentation,* IBM Corp., 320-0936-0 Data Processing Division, White Plains, N.Y. (1966), pp. 77–114.

[BW66a] Brown, W. S., *The ALTRAN Language and the ALPAK System for Symbolic Algebra on a Digital Computer* (presented at the Princeton U. Summer Conference in Computer Sciences, Princeton, N.J., Aug., 1966) (unpublished).

[ML66] McIlroy, M. D. and Brown, W. S., *The ALTRAN Language for Symbolic Algebra on a Digital Computer*, Bell Telephone Lab., Murray Hill, N. J. (May, 1966) (unpublished).

[MZ65a] McClure, R. M., "TMG—A Syntax Directed Compiler", *Proc. ACM 20th Nat'l Conf.*, 1965, pp. 262–74.

VII.6. FLAP

[MB67] Morris, A. H., Jr., *The FLAP Language—A Programmer's Guide*, U.S. Naval Weapons Lab., K-8/67, Dahlgren, Va. (Jan., 1967).

VII.7.1. Magic Paper

[CL63] Clapp, L. C. and Kain, R. Y., "A Computer Aid for Symbolic Mathematics", *Proc. FJCC*, Vol. 24 (Nov., 1963), pp. 509–17.

[CL66] Clapp, L. C. et al., *Magic Paper: An On-Line System for the Manipulation of Symbolic Mathematics*, Computer Research Corp., Report No. R 105-1, Newton, Mass. (Apr., 1966).

VII.7.2. Symbolic Mathematical Laboratory

[ZB67] Martin, W. A., *Symbolic Mathematical Laboratory*, M.I.T., MAC-TR-36 (Ph.D. thesis), Project MAC, Cambridge, Mass. (Jan., 1967).

VIII MULTIPURPOSE LANGUAGES

VIII.1. SCOPE OF CHAPTER

Chapters IV, V, VI, and VII contained descriptions of languages which were designed primarily for use in solving, respectively, numerical scientific, business data processing, string and list processing, and formal algebraic manipulation problems. In some cases a specific language might have been used for more than its intended purpose, but this was a by-product and not a deliberate intent.

The term *general purpose* as applied to programming languages has been in vogue for many years. However, after careful deliberation, I have reached the conclusion that there is no programming language in this book which is truly general purpose by any reasonable definition of the term. Each language in this chapter provides effective capabilities for dealing with at least two of the areas in the preceding four chapters, but none is really good for all of them. Hence, the term *multipurpose* is used to provide a more accurate description of the languages in this chapter. This term also implies that a reasonable balance among the various areas is maintained in the definition of the language in order to qualify as multipurpose. In other words, a language which provides excellent facilities in one area and primitive ones in another is not considered multipurpose.

The first language developed which made a serious attempt to cover several areas was JOVIAL. Its primary objective was to solve command and control problems, and these involve both scientific calculation and a certain amount of data processing. Hence JOVIAL itself provided fairly balanced capabilities in both areas.

The major language in this chapter is, of course, PL/I, which is really a culmination of the line of languages discussed in Chapters IV and V, with some facilities of the types discussed in Chapter VI. (A discussion of the connection of PL/I with the concepts of formal algebraic manipulation is given in Section VII.3.) In addition to the numerical scientific and data processing facilities, PL/I also allows the user significant interaction with the compiler and/or operating system, and it has a macro facility to allow some user-defined extensions.

Both Formula ALGOL and LISP 2 are based on ALGOL. The former is a strict extension which provides facilities for formal algebraic manipulation and for string and list handling. LISP 2 is an ALGOL-like language based on LISP 1.5; it provides list processing capabilities as well as most of the regular ALGOL facilities.

VIII.2. LANGUAGES OF HISTORICAL INTEREST ONLY

There are no languages in this category.

VIII.3. JOVIAL

VIII.3.1. HISTORY OF JOVIAL

In June, 1958, the System Development Corporation initiated a research project to investigate the problems of automatic coding. This project resulted in the development of CLIP (*C*ompiler *L*anguage for *I*nformation *P*rocessing). Much of this work was similar in spirit to that being done for IAL (i.e., ALGOL 58); the specifications for IAL were published in December, 1958, and SDC adopted the notation of IAL but made certain additions and modifications to permit greater convenience in expressing data manipulations. Certain other facilities were omitted. Thus, CLIP was really a derivative of ALGOL 58. (A further discussion of CLIP, including some of the statements about its initial purpose and the list of references, is in Section IX.2.5.2.) Early in 1959, SDC's SACCS (Strategic Air Command Control System) Division in New Jersey decided to develop a similar language (eventually called JOVIAL) that would be useful for programming SACCS. Since CLIP was a research effort involving compiler writing and was being developed 3000 miles away, it was not possible to keep the languages the same. Whether right or not, JOVIAL has survived as the major SDC language.

SDC actually began work on JOVIAL in February, 1959. A year later, a JOVIAL interpreter was running on the 709, and by December, 1960 a compiler for the 709 was running on the 709.[1]

---

[1] Shaw [SH63b], p. 89.

The story of the name JOVIAL has been told many times in differing versions. According to the man who should know, namely Jules Schwartz, the key points are as follows: He wrote a draft proposal for a programming language to be used for the Air Force 465L program. The language was based on IAL, but it had additional facilities. The title he put on the draft was OVIAL (*O*ur *V*ersion of the *I*nternational *A*lgebraic *L*anguage). The language concept was accepted, but the word OVIAL was not considered satisfactory for several reasons. A suggestion for a new name was JOVIAL, and this was accepted. However, it was necessary to decide what JOVIAL stood for, and one joking suggestion was that it should be *J*ules' *O*wn *V*ersion of the *I*nternational *A*lgebraic *L*anguage. According to Schwartz, he felt this was a joke but did not become too concerned about it. During a business trip shortly thereafter, a statement of work was formally submitted to the contracting officer which listed among the deliverable items a *JOVIAL (Jules' Own Version of the International Algebraic Language) compiler*. Nobody would agree to changing the contract, and so the name remained, providing both more fame and more blame to Schwartz than he deserved.[2]

The original version of JOVIAL was implemented on both the IBM 709 and the Military Computer used for SACCS (AN/FSQ-31). By no later than May, 1961 (and possibly earlier), JOVIAL 1 became obsolete, was replaced by JOVIAL 2 (running on the 709, 7090, and the SACCS computer), and then eventually by JOVIAL 3. JOVIAL 3 (hereafter referred to as JOVIAL) was implemented at least for the CDC 1604 and 3600, the Philco 2000, the SAGE AN/FSQ-7, and the AN/FSQ-32. JOVIAL 2 was implemented on the IBM 9020.

As usually happens, even before the compilers were fully operational, work began on improvements to them and the language as well. In May 1960, a decision was made to standardize on JOVIAL as a common programming language for SDC.[3]

JOVIAL has been used not only by SDC for Air Force projects but it has been adopted by the Navy for use in its NAVCOSSACT and by the Army in some of its programming efforts. In June, 1967 the Air Force issued its own specifications to establish (a version of) JOVIAL 3 as the standard programming language for Air Force command and control applications (see [AF67]).

Although JOVIAL has been developed and used primarily by SDC, the history of its development, its modifications, and the implementations on various machines is almost as complicated as that of FORTRAN. Those people interested in further details should see Shaw [SH63b], Steel [ST66], and the references listed in [SH63b]. A list of usage in the Department of Defense and a list of compilers are given in Figure VIII-1.

---

[2] Private communication, August, 1967.
[3] Shaw [SH00].

VIII.3.2. Functional Characteristics of JOVIAL

JOVIAL tends definitely toward generality and falls somewhere in the middle relative to succinctness and naturalness; this *middle ground* characteristic occurs because the language is based on ALGOL 58, but it has added so many other features that there is no longer any resemblance. It is consistent and fairly easy to read or write. Since it has great power, it is not particularly easy to learn. The application area includes the whole command and control field, which itself subsumes much of the need for numerical scientific as well as some data-handling calculations, and a particular need for manipulation of logical entities. A significant objective was the necessity of handling system data described in a *COM*munication *POOL* (COMPOOL). While this concept is independent of, and precedes, JOVIAL (e.g., see FAST [MI62]), the ability to use a COMPOOL was a design criterion in the language. The COMPOOL serves as a central source of data description, and it is particularly valuable in the large command and control systems for which JOVIAL was designed and has been used. JOVIAL is a procedure-oriented, problem-oriented, and problem-defining language.

	JOVIAL Usage in USAF as of August 1, 1967	
Dialect	*User systems*	*Computer*
J3	SAGE 416L	FSQ-7
	BUIC III 416M	GSA-51
	RADC	CDC 1604B
	NORAD	Philco 2000/212
	AFSC (Sat. Cntrl)	CDC 3600
J2	SAC Planning	IBM 7090
	SACCS 465L	FSQ-31
	FTD	IBM 7090
JS	ADC(STP)	IBM S/360
	JOVIAL Usage in DOD other than USAF and other DOD related Federal Government Usage	
J3	NMCS/SC	CDC 1604
	NAVCOSSACT	CDC 1604A
	RADC	CDC 1604B
	NASA (GEMINI)	Philco 2000/212
	JOB CORPS	CDC 3600
J2	ARMY (ADSF)	IBM 7090
	NAVSPASUR	IBM 7090
	FAA–NAFEC	IBM 7090
	FAA–NAS	IBM 9020
JS	OASD (Manpower)	IBM 7090
	NAVCOSSACT	IBM 7090

Figure VIII-1. (cont.)

JB		NMCS/SC	IBM S/360/50
		COMPILERS AS OF AUGUST, 1967	
	Machine	*Developer*	*Use*
J3	CDC 3870	SDC	NRL
	Burroughs D825[1]	Burroughs	TWA
	Burroughs D825	SDC	BUIC III
	(IBM) FSQ-7	SDC	SAGE
	Philco 2000/210	SDC	IR
	Philco 2000/212	SDC	425L
	(IBM) FSQ-32	SDC	ARPA/TSS
	Univac M1218	UNIVAC	RADC
	(Univac) C1667	PRC	BuShips
	Hughes H3118	PRC	IPG
	CDC 3600	SDC	SSD
	CDC 1604	SDC	NMCS
	CDC 1604A	SDC	NAVCOSSACT, RADC
	GE 635	PRC	RADC
	CDC 1604	Teledyne	RADC (Exper.)
	IBM 360	CSC	IBM
J2	(IBM) FSQ-31	SDC	465L
	(IBM) FSQ-32	SDC	ARPA
	(IBM) 9020[1]	CUC/IBM	FAA
	IBM 7090	SDC	SAACS, SAC, SPASUR,
	IBM 7094	SDC	FTD, FAA, etc.
	RCA VIC[4]	RCA	Exper.
JS	(IBM) FSQ-32[3]	SDC	ARPA/TSS
	Univac 1107[2]	CSC	CSC Proprietary
	IBM S/360/50/65/67/75	SDC	NMCS/SC, Multiple
			SDC Cmptr. Cntr.
	IBM 7090/7094	SDC	CPSS, ACIS
	IBM 7094	SDC	NAVCOSSACT
JX2	Philco 2000/210	SDC	IR
	(IBM) FSQ-32	SDC	IR
	IBM 7090 7094	SDC	IR

[1] Coded in Machine Language
[2] Coded in FORTRAN IV
[3] JTS (JOVIAL Time Sharing)
[4] Variable Instruction Computer; micro programmed to simulate IBM 7090

Figure VIII-1. JOVIAL usage and compilers.
Source: [AA67]

SAMPLE PROGRAM—JOVIAL

Problem: Construct a subroutine with parameters A and B such that A and B are integers and $2 < A < B$. For every odd integer K with $A \leqslant K \leqslant B$, compute $f(K) = (3K + \sin(K))^{\frac{1}{2}}$ if K is a prime, and $f(K) = (4K + \cos(K))^{\frac{1}{2}}$ if K is not a prime. For each K, print K, the value of $f(K)$, and the word PRIME or NONPRIME as the case may be.

Assume there exists a subroutine or function PRIME (K) which determines whether or not K is a prime, and assume that library routines for square root, sine and cosine are available. This program also assumes the existence of three output routines. [Note: JOVIAL has ODD as a primitive.]

Program:
```
       PROC    SPEC (A1, B1)$   ITEM A1 I U 47 $
                                ITEM B1 I U 47 $
       BEGIN   WRITE (0) $
               IF NOT ODD (A1)$
               A1 = A1 + 1$
               FOR K = A1, 2, B1$
       BEGIN           WRITEN (15, K, 0)$
         IFEITH        PRIME (K) $
       BEGIN WRITEN (30, SQR, (3*K + SIN (K)), 5)$
             WRITEH (45, 5H(PRIME))$
             END    ORIF 1 $
       BEGIN WRITEN (30, SQR (4*K + COS (K)), 5) $
             WRITEH (45, 8H(NONPRIME))$
       END
         END           WRITE (1) $
       END             WRITE (4) $
       END
```

The language serves simultaneously as a reference, publication, and hardware language. JOVIAL was designed for the professional programmer and definitely to be used in a batch environment. However, a much later and much simpler version called JTS (see Sandin and Foote [SN65]) was installed under SDC's time-sharing system, and an interpretive extended subset version called TINT was specifically designed and implemented for on-line use. (See Kennedy [KE 65].)

JOVIAL has had the misfortune to suffer throughout its history from all the problems that could possibly arise from an attempt to have wide usage, maintain compiler independence, avoid dialects, and control subsetting and extensions. The proliferation of documents and systems on differing machines did not help the situation, although there were continuous attempts in SDC to control this problem. The earliest description seems to be the one by Schwartz, Petersen, and Olson [SC60]. The reader interested in pursuing which versions existed on which machines should see the papers by Shaw [SH63b] and Steel [ST66], but even these are not complete. (See also Figure

VIII-1.) There have been several versions of the *official specifications* with the latest one by Perstein [PE66a], based on earlier ones by Shaw. On the more positive side, in August, 1965 certain internal management decisions were made in SDC with the resulting policy, stated in Perstein [PE66], that any new JOVIAL compiler must implement Basic JOVIAL as defined in that manual. If the new compiler provides capabilities included in J3 (JOVIAL 3), it must implement them in accordance with the official description given in Perstein [PE66a]. A new compiler may implement features that are not included in the specifications of JOVIAL and are not incompatible with the specifications of JOVIAL; of necessity, some features will be machine dependent, e.g., precision of the arithmetic. Since most of the language definitions have been given in a fairly formal way, there has been relatively little problem of incompatibility caused by misunderstanding.

Until 1967 there was no significant attention paid to JOVIAL from the viewpoint of American standardization (i.e., through USASI), although obviously there has been tremendous attention paid to this within SDC itself. As a result of interest by the Air Force and SDC, there is a possibility that a USASI standard might be developed.

The original CLIP language work was started by J. Schwartz and E. Book, and the former supervised the development of the first JOVIAL system.[4] Since then, numerous people within SDC have contributed to the further development of JOVIAL; in not all cases were the language designers directly involved with the implementation effort since SDC set up various groups to control the maintenance.

The basic objective of the language was to create a language for use by professional programmers in solving large complex information processing problems. In the various documents on JOVIAL, several different notations for defining syntax have been used, ranging from reasonable to arbitrary notation that in my opinion did not seem to have any justification whatsoever, the latter appearing, for example, in Shaw [SH63a].

One complaint which never could be leveled against the JOVIAL activity is a shortage of documentation. Shaw states "My collection of documents of JOVIAL weighs almost 50 pounds and stands almost two feet high."[5] Since that was written in 1962, it seems reasonable that the material has not decreased in quantity. Naturally some of those documents are working papers of interest to limited groups only, but on the other hand at least some of it is of widespread interest. The material ranges from primers (e.g., Kennedy [KE62]) to detailed syntactic descriptions (e.g., Perstein [PE66a]) to general description and tutorial articles (e.g., Shaw [SH63a] and [SH61]). Other references are listed at the end of the chapter, and still more are

---

[4] Shaw [SH00].
[5] Shaw [SH63b], p. 90.

given in Shaw [SH63b]. Some technical comparisons with other languages are given by Coffman [CO61] and Shaw [SH64]. A JOVIAL Bulletin has been started and issued irregularly as part of SICPLAN notices [AC00].

JOVIAL appears to have fulfilled its objectives, even though it is perhaps ready to be replaced by PL/I, at least in the view of Steel [ST66]. It seems surprising, however, that although JOVIAL has a potential application area that was wider than any other language until PL/I, and in spite of the fact that it has been implemented on a large number of machines, it does not appear to have been used much outside the military command and control applications (and for writing its own compilers). In that area it has been used more heavily for writing utility and support programs than for the operational programs themselves. Three instances in which JOVIAL has been used in other areas are writing a program to simplify JOVIAL source programs (described by Clark [CE67]), creating a teaching program (see Marsh [MD64]), and automatic essay paraphrasing (see Klein [KK65]). One reason for JOVIAL's lack of major acceptance seems to be the lack of direct support by computer manufacturers and the natural reluctance by customers to produce a compiler themselves. In some other cases, an NIH (Not Invented Here) factor seems to have played a major role. However, it is not really clear to me why it has not received wider usage outside SDC.

VIII.3.3. Technical Characteristics of JOVIAL

The character set in JOVIAL consists of the 26 capital letters, the 10 digits, and the following 12 characters:

$$+ \quad - \quad * \quad / \quad . \quad , \quad = \quad (\quad) \quad ' \quad \$ \quad blank$$

An identifier is a letter followed by at least one letter, a numeral, or an apostrophe, ' (called a *prime* in JOVIAL), except that the identifier cannot end with the prime character nor contain two consecutive primes. Hence an identifier cannot consist of a single letter. The five standard graphic arithmetic operators are available.

Data names and program unit labels are formed as the identifiers defined above, except that they cannot be the same as any of the reserved words in the language. Since statement labels and data names can be distinguished from context, the same name can be used in each of those categories, although this is certainly not recommended because of potential confusion to the user. Data names can have any number of subscripts, separated by commas and delimited by the dollar sign, e.g., ALPHA ($ 3+A*B, B/C $). There is no qualification because there is no data hierarchy except for a limited facility in creating tables. However, it is possible to refer to bits and bytes specifically (which are of course largely or completely machine dependent). There are a number of reserved words in JOVIAL, and they cannot be used as names.

The relational operators are *EQ*, *GR*, *GQ*, *LQ*, *LS*, and *NQ* with the meanings of equal, greater than, greater than or equal, less than or equal, less than, and not equal, respectively. The logical operators are *AND*, *OR*, and *NOT*. Punctuation is used in JOVIAL, but it is not particularly significant except for the commas used to separate items in lists. Much of the delimiting normally done by punctuation (e.g., end of statement) is done through the use of the dollar sign. Blanks are quite significant, and the rules about when they can and cannot appear are complicated; however, the obvious cases where blanks can appear between operators and operands and between key words and names are allowed. Whenever one blank is permitted, any number are allowed. There are no noise words.

Two types of literals are permitted, namely alphanumeric (denoted as Hollerith) and transmission code; the latter is used to specify the exact form of the machine language representation of the literal. In both cases, the number of characters in the literal precedes the identifying letter, e.g., 4H(NUTS). It is also possible to denote literals by octal constants.

The format is quite free form, with the programmer being allowed to start any JOVIAL statement in columns 1 to 72; more than one statement per line may be written, and one statement may extend over several lines. The conceptual form seems to fall on rather neutral ground; it is not particularly symbolic or succinct, but on the other hand it is not as English-like as COBOL.

The following is a partial list of the declarations in JOVIAL: *ITEM*, *MODE*, *ARRAY*, *TABLE*, *OVERLAY*, *DEFINE*, *PROCEDURE*, *SWITCH*, *FILE*. Many of these in turn have further declarations associated with them, and all are discussed later.

The smallest executable unit is a statement containing one of the executable commands; it is terminated by $. Any such statement can be named, and the name is followed by a period. Groups of statements can be combined into a larger form which is composed of a *BEGIN* and *END* bracket, with statements (and possibly declarations) between them. These compound statements can be nested to any depth desired. Loops can be controlled by an *IF* statement or through the use of the *FOR* command. JOVIAL permits functions, procedures, and closed subroutines. The closed subroutine is a special kind of procedure that has no parameters. There are different rules involving the handling of all these with regard to loops.

Comments are delimited by two primes (i.e., '') at the beginning and the end; the intervening string can consist of any characters except the dollar sign and, of course, it cannot contain two primes prior to the end nor a prime immediately preceding the ending pair. A complete program consists of a list of declarations and statements preceded by the key word *START* and optionally preceded by the word **CLOSE**. If the latter is used, it indicates the program is a closed subroutine. The program ends with **TERM** $. It is possible to specify the name of the first statement to be executed immediately after the **TERM** delimiter.

The only interaction with the environment is from the input/output statements.

The main delimiter is the dollar sign, which is used to end an executable statement, and of course the **BEGIN**...**END** pair for compound statements. Call by name and call by value parameter passage are provided. As indicated above, compound statements themselves can be contained within compound statements. A procedure can invoke other procedures (but no recursion is permitted).

The executable units and the declarations can be intermingled in the program. It is possible to include machine code statements in a JOVIAL program by preceding the code with the key word **DIRECT** and following it with the key word **JOVIAL**. It is also possible to make a definite connection between the variables in the JOVIAL program and a machine register since the programmer is allowed to write **ASSIGN A** = *variablename* $ or **ASSIGN** *variablename* = **A** $. The A is considered an undefined machine register called the *accumulator*.

JOVIAL introduces the concept of a functional modifier which provides built-in operations on certain types of expressions or conceptually provides answers to questions about various program items. From a structural viewpoint, functional modifiers seem to fall between direct executable statements and declarations, although they actually appear in statements.

The types of data variables and constant permitted in JOVIAL are arithmetic, Boolean, status, literal, and dual. Status variables are essentially mnemonic names which can be associated with integer values of a data variable and can be tested against these values. (This is the same concept as the conditional variable in COBOL.) The dual constants represent an ordered pair which is *not* a rational number but is useful for calculations involving two-dimensional coordinate systems; it is only implemented on machines with dual arithmetic and was motivated by the early existence of such hardware. Arithmetic variables can be fixed, integer, or floating point. There are no complex, formal, string, or list variables. Arithmetic expressions are created in the normal way. Absolute value is designated by either

$$(/ \quad /)$$

or

$$ABS (\quad)$$

There are both *tables* and *arrays* in JOVIAL. An array is an *n*-dimensional collection of values or variables of the same type, all identified by a single name (with the necessary subscripts). A table on the other hand is a one-dimensional array, each element of which contains a fixed but arbitrary number of data items of (possibly) different types. For implementation reasons, tables are handled more efficiently. JOVIAL permits access to bits and bytes through functional modifiers BIT and BYTE by writing, e.g., BIT ($ 0, 6 $) (COLUMN) or BYTE ($ 3 $) (STATE). In the first case, the information designated is an unsigned integral variable, and it can be assigned to another variable name (which can itself be designated through the BIT modifier). The BYTE modifier designates a literal variable. The first bit or byte of the data item and the number of bits or bytes are specified by the two numbers associated with the modifier, and the number (of bits or bytes) can be omitted if it is 1.

All the data variables, constants, and aggregates of constants are accessible by the commands in JOVIAL.

There are three types of arithmetic in JOVIAL: Floating point, fixed point, and dual fixed point. In the dual mode, operations are done in parallel with the left component of one operand combined with the left component of the other to yield the left component of the result. (This is a special case of more general array operations.) Operands of different modes can be combined, and automatic conversions from fixed to floating, floating to fixed, and fixed (or floating) to dual are implied. In the last case, the single-valued operand is duplicated before doing the arithmetic. The precision of the arithmetic result is computer dependent because of the word size. Standard precedence and sequencing rules apply except that the unary minus takes precedence over exponentiation, e.g., $-3**2=9$.

Boolean expressions are constructed in the standard way from the three operators AND, OR, and NOT. Parentheses are used to indicate grouping when necessary. The status variables can be included within a Boolean formula.

Names defined within a procedure are defined only for that procedure, and they can be used outside it with a different meaning. Names defined within a program apply only to that program when it is used in conjunction with others. Names in a COMPOOL (described briefly on page 536) are defined for an entire system of programs, excluding programs or procedures that define identically spelled names. Names of data elements not defined in a COMPOOL should be defined before they are used. If they are not, a preset (compiler dependent) mode is assigned; this can be changed by the MODE directive.

JOVIAL has a basic assignment statement; in addition, it permits the interchange of values of a pair of variables; this is designated by two equals signs. In both cases, the numeric quantities on the right are converted to the appropriate form on the left. Thus, writing the sequence

$$A = 5 \;\$$$
$$B = 7 + A \;\$$$
$$A == B \;\$$$

causes A to have the value 12 and B to have the value 5.

There are no specific alphanumeric data-handling commands, although the easy access to bits and bytes make their programming fairly simple.

The unconditional control transfer is the (single word) GOTO. It may have statement label information with it, to provide both a *computed* and an *assigned* GOTO. This command can be used to invoke a subroutine with no parameters permitted; after execution of the subroutine, control is returned to the statement immediately following the GOTO unless there has been a transfer of control out of the subroutine. Procedures with formal parameters are invoked by specifying the procedure name and a list of the parameters enclosed in parentheses and separated by commas; an equals sign is written between the input and the output parameter lists to distinguish between them.

A function can be invoked from within an arithmetic or Boolean expression.

Conditional statements take one of two forms. The first is itself called a conditional statement by JOVIAL writers and is of the form IF *Boolean-expression* $ *statement* where the *statement* cannot contain an IF except within a BEGIN... END bracket. Thus, IF AR EQ BJ $ CR = DK + EM $ and IF ALPHA − BETA LS GAMMA + DELTA $ BEGIN IF HE EQ SHE $ GOTO BLAZES $ END are correct, but IF ALPHA − BETA LS GAMMA + DELTA $ IF HE EQ SHE $ GOTO BLAZES $ is not correct because the second IF is not within a BEGIN... END bracket.

Another form of conditional statement provides alternatives, using the following format:

    IFEITH *Boolean-expression* $ *statement*
    [*label* .] ORIF *Boolean-expression* $ *statement*
    [*label* .] ORIF *Boolean-expression* $ *statement*
    ...
    END

where [ ] indicate optional elements. Any number of alternatives can be listed, and the Boolean expressions are tested in sequence from the top; the statement associated with the first true *Boolean-expression* is executed and control is then passed to the statement following the END. For example,

```
           IFEITH AB GR BB $ AB = CD $
NAME . ORIF AB + CD LS CD $
                  BEGIN EF = (AB + CD) ** 2 $
                  LL. AB = BB + 2 $
                      CD = ABC $
                  END
                  ORIF 1 $ GOTO WARM $
                  END
        NEXTST . AB = CD + 1 $
```

In this example, if AB > BB, then AB = CD is executed and control passes to NEXTST. If AB ≤ BB, then control passes to NAME and the AB + CD < CD test is made. Depending upon its truth value, either the three statements following the BEGIN are executed or else control passes to the next ORIF; in that case, since the 1 is defined as the Boolean constant *true*, control is passed to WARM.

The loop-control statement is the FOR statement, which has the following general format;

```
FOR paramlist1 $ [name .] FOR paramlist2 $ ... $
    [name .] FOR paramlistn $
BEGIN statements END
```

the range of the FOR is the set of statements contained within the BEGIN ... END pair. The paramlist is of the form

```
        parameter = initial-value, increment, final-value $
```
or
```
        parameter = ALL (name) $
```

where the parameter is a single letter defined only within the range of the FOR, any of the three elements can be expressions, and the final value can be omitted (in which case the loop must terminate from within the range). The parameters following each separate FOR statement are varied in parallel, i.e., the initial value of each is set, then each is incremented, etc; nesting is accomplished by including a FOR statement within the BEGIN ... END pair. Only one *paramlist* can contain a *final-value* for its *parameter*. In the use of the ALL case above, the *name* refers to a table or an item in a table, and the looping is executed for each element in the table. The value of a parameter can be used in an expression to determine the value of a parameter that appears in a succeeding FOR paramlist.

A rather unusual TEST [*parameter*] statement exists. It permits control to transfer to the compiled code which causes incrementing and testing of a parameter in a loop.

There are no error condition statements nor any symbolic data-handling statements.

With regard to input/output, it is worth noting that neither JOVIAL 1 nor JOVIAL 2 on the 709/7090 contained any input/output commands because the executive program that they were using did all the input/output. The input/output commands that were finally put into JOVIAL look much like COBOL, although they are not necessarily based on it. The first time a file is activated, the user must give an OPEN INPUT filename [recordinformation] $. Subsequently the user writes INPUT filename recordinformation $. The recordinformation can either be a variable name, an array name, a table name, sequences of table entries, or an individual table entry. The file is shut by means of a SHUT INPUT filename [recordinformation] $ statement. The actual READ operation transfers the information from the physical file into internal storage. Similarly, for output, the user would write OPEN OUTPUT filename [recordinformation] $ the first time, OUTPUT filename [recordinformation] $ the rest of the time, and then SHUT OUTPUT filename [recordinformation] to close the file and deactivate it. For each logical record read or written, the file position, accessed by the functional modifier POS (filename) is incremented by one.

There is only one built-in function in JOVIAL, namely REM, and there is a built-in procedure REMQUO. They produce, respectively, the remainder and the remainder and quotient of division of two integers. Procedures and functions in the library are made part of the object program by just writing the names in the source program. The COMPOOL contains descriptions or definitions of both data declarations and programs. Any reference to data, procedures, or functions that are listed in the COMPOOL will be used as if they were a part of the program, unless the program itself defines such a name explicitly prior to its usage.

There are no specific executable debugging or storage allocation statements. The only type of interface with an operating system is through the COMPOOL. The closest to machine feature statements are the availability of the bit and byte modifiers and the provision for a mythical accumulator.

The basic units of data are called *items*, and the most basic data declaration is the ITEM, which must be written for each variable. A MODE declaration describes the item and provides an implicit declaration for all following and otherwise undefined simple items until another MODE declaration occurs.

The ARRAY declaration provides the dimensions along with the rest of the ITEM declarations. Associated with each item are a number of characteristics. Some are either self-explanatory or have already been described, namely *floating, fixed, integer, dual, signed, unsigned, Hollerith, transmission, status,* and *Boolean*. Many of these are written using only the initial letter, e.g., F specifies floating point.

A *rounded* attribute declares that any value assigned to the item should be rounded rather than truncated. It is also possible to show for each variable an estimated minimum and maximum absolute value of the item. For each status item, the list of status constants is given; these are encoded in order by the integers 0, 1, 2, etc. The number of bits can be specified but if it is omitted, then the compiler determines the maximum number of bits required.

It is often desirable to assign specific initial values; this can be done by inserting a preset declaration immediately after the name and description. As an illustration, the declaration

ITEM ALPHA I 15 U 5 P 97.18 $

describes a variable named ALPHA as an Unsigned 15 bit Integer with 5 fractional bits which is Preset to the value 97.18.

There are TABLE and ARRAY declarations. For the table, it is possible to specify *variable* or *rigid* (=fixed) lengths; *serial* or *parallel* entry structure; and either *no* item packing, *medium* item packing, or *dense* item packing. A variable table declaration indicates that the number of entries can vary during program execution, whereas for a rigid table length this cannot be done. The difference between serial and parallel entry tables is the way in which their data is stored. In the serial table, words needed for an entry are allocated consecutive memory positions; whereas in the parallel structure the nth word of each entry resides in a contiguous block of memory. The density of packing indicates how the items are to be stored within a particular entry. No packing means that each item is allocated a full word; medium packing means that storage is allocated in subword units, where these subunits can be of different sizes within the word and are of course completely machine dependent; finally, dense packing means that the storage is allocated primarily in consecutive bit positions. (Note that the language is heavily oriented toward a binary machine; it does not seem to have been implemented on a decimal machine.) If packing is not specified, then each compiler assumes some normal packing mode. (More recent specifications provide rules for this.)

It is possible to define a table structure completely through the use of the structured item declaration and the STRING item declaration (which declares more than one occurrence of the item per entry of a table). This plus some other facilities allow the user to specifically indicate what item appears in what part of a word. Obviously this plays havoc with machine independence.

The NENT (*number-of-entries*) is a parameter indicating the number of entries in a table. Thus one can write

NENT (PAYROLL) = NENT (PAYROLL) + 1 $

if one wants to add somebody to the payroll. For fixed length tables this is an integer constant, while for variable length tables it is a counter that the program itself must maintain. Another parameter for table processing is NWDSEN (*number-of-words-per-entry*) which is an integer constant established at compile time. The ALL modifier was already seen in the FOR clause to permit more effective processing of tables. The ENTRY modifier allows a table entry to be treated as a single value. The MANTissa and CHARacteristic modifiers permit the user to extract the indicated information from a floating point variable and assign it to another variable. The ODD modifier can be used to determine if the value of the least significant bit of a quantity is 0 or 1; it can be assigned to a Boolean variable which will then be true if it is odd and false if it is even.

The file declaration is of the following form:

FILE *filename filestructure statuslist devicename* $

where the *filestructure* specifies Hollerith or binary, the estimated maximum number of records, fixed or variable record size, and estimated maximum number of bits or bytes in a record. The list of possible *file statuses*, e.g., busy, ready, and error, are associated with integer values as status constants. The *devicename* is defined by the implementer, but it is assigned by the programmer.

The overlay declaration is the word OVERLAY followed by a sequence of overlay lists; storage is assigned sequentially to the data elements in each list and each separate list can occupy the same space as another.

Finally, JOVIAL contains a DEFINE declaration which permits a name to be substituted for a string. It is written DEFINE *name* " *definition* " $. It can be used, for example, to abbreviate lengthy expressions, introduce noise words, or make key words more readable, e.g.,

DEFINE AREA "3.1416 * RADIUS ** 2" $
DEFINE NUTS "   " $
DEFINE GREATER "GR" $

Procedure declarations usually have the following form:

PROC *name* (*inputparamlist* = *outputparamlist*) $
[*declaration list*] BEGIN *statements* END

The items in the *paramlists* are separated by commas. A function declaration also uses PROC, and the name is used as the output parameter.

The STOP statement terminates execution; however, a statement label can be written after the STOP statement, and if execution is resumed, then the indicated statement will be the next one executed.

Switch declarations also exist. The word SWITCH is followed by an identifier and then a sequence of identifiers in parentheses. The choice is determined as a *computed* GOTO (see Section III.5.3.1).

There is no self-modification of programs. There is also no self-extension of the language, except in the trivial sense permitted by the DEFINE which allows the user to substitute a name for a string. There are no specific compiler directives.

JOVIAL is the best documented and practical example of a language which can be used to write its own compiler. Almost all the recent JOVIAL compilers have been written that way, and even JOVIAL 1 was bootstrapped using itself. Most of the JOVIAL compilers are structured with a *generator* which accepts the source code, translates into an intermediate language and then a *translator* which converts the intermediate language into executable machine code. (In my opinion, these people are using the terms *generator* and *translator* backwards. However, they picked up this terminology from the early work on the UNCOL concept (see Section X.2).) Not only does this procedure make it easier to construct new compilers, but it also helps significantly with the documentation and somewhat with compatibility problems because both the generator and the translator themselves are written in JOVIAL.

It is clear that JOVIAL has been designed to provide efficiency at object time. The facilities for structuring the data internally all lead to efficient storage allocation and execution, with a minimum amount of incompatibility. Although the language is fairly general, there are annoying restrictions in it. Since the programmer is able to specify and control the storage allocation considerably through the declarations, the compiler needs to obey only the specifications given for each program. There are no particular debugging aids or error checking in the language, although most of the compilers provide them.

VIII.3.4. Significant Contribution to Technology

JOVIAL appears to have made several contributions to the technology. It was the first language (and until PL/I, the only one) to provide good facilities for simultaneously doing scientific numerical computation and nontrivial data handling, while at the same time it could also be used in general information handling areas. A second contribution, although not really introduced by JOVIAL, is the use of the COMPOOL (see Section 4.2 in Perstein [PE66a]) in connection with a compiler. A third contribution to the technology is its practical usage as its own compiler. Finally, it has made a very significant contribution in terms of allowing the programmer great flexibility for controlling storage allocation when he needs to but not requiring him to do so otherwise.

VIII.4. PL/I

VIII.4.1. History of PL/I

Considering the potential impact of PL/I upon the computing industry, it is not altogether surprising that it developed a significant history in a short time span after its inception.

It had been recognized by numerous people for many years that FORTRAN—while effective for solving numerical scientific problems—was sadly lacking in two main areas, namely in character and alphanumeric data handling and in having provisions for good interaction with more modern equipment and operating systems. Even the improvements which produced FORTRAN IV, while extremely significant and useful, did not really address themselves to these problems. In order to remedy this situation, SHARE and IBM agreed in September, 1963 to form a joint Advanced Language Development Committee under the SHARE FORTRAN project. The committee goal was to provide a language which would encompass more users while still remaining as a useful tool to the engineer. The members and their affiliations at that time were G. Radin, C. W. Medlock, and B. Weitzenhoffer of IBM (Radin being chairman of the IBM delegation) while the SHARE representatives were H. S. Berg (Lockheed) J. Cox (Union Carbide), and B. Rosenblatt (Standard Oil of California) (Rosenblatt serving as official chairman). Extensive help was also provided to the committee by T. Martin (Westinghouse), H. P. Rogoway (Hughes Dynamics, and also chairman of the SHARE FORTRAN project), and L. Brown and R. Larner of IBM. The committee and its advisors worked many long hours and weekends, particularly since several of the participants had many other responsibilities. Furthermore, in many cases there were specific (and somewhat unreasonable) deadlines imposed.

While there was no explicit or implied commitment on IBM's part to implement the final product resulting from the committee, it was certainly assumed that this would probably happen. The original definition of purpose was a 6-month project to specify a "major advance in FORTRAN". Although compatibility with FORTRAN was considered very desirable, it was not defined as a requirement. For a long time there was a feeling that the committee would merely extend FORTRAN in significant places and remove some of the restrictions that existed, e.g., in subscripts, lack of mixed mode expressions, and limited input/output. The implied commitment to extending FORTRAN was evidenced by the double fact that the committee was under the auspices of the FORTRAN project and that for quite a while it was informally called the FORTRAN VI Committee within IBM. (Some preliminary work had been done at IBM the preceding year, involving the addition of character-handling facilities into FORTRAN;

this accounted for the missing FORTRAN V.) It soon became clear to this committee that they could not maintain compatibility with FORTRAN and still develop a language which they felt was needed to meet modern programming techniques and equipment. For example, in the minds of many people, such issues as the use of blanks, card-oriented format, and methods of handling declarations prevented a direct extension from being desirable. However, even while the first report was being prepared, there was considerable internal discussion about this issue until it was finally and irrevocably decided that the new language would *not* be a compatible extension of FORTRAN. Obviously, the committee did not make arbitrary changes just for the sake of being different, but the designers were in no way bound by the provisions of FORTRAN.

During its deliberations, the committee heard presentations by numerous people from both inside and outside IBM about features that the speakers thought should be in the language. The committee also attempted to study languages (e.g., ALGOL, COBOL, and JOVIAL) which contained features or capabilities which should go into a new language even if in a different syntactical form. Thus, every reasonable attempt was made to consider concepts from as wide a source as possible.

A document dated March 1, 1964 and entitled "Report of the SHARE ADVANCED LANGUAGE DEVELOPMENT COMMITTEE" [XY64] was presented to SHARE in March, 1964. It was defined as "a status report, in the form of language specifications from the Advanced Language Development Committee". It was received with mixed reactions; some people felt that this was a new powerful language that would be well suited to their needs, while others felt that it was a hodgepodge and not suitable for actual usage.

It is of some historical interest to note a few of the items which were included in that report and which have subsequently been either dropped or significantly changed, e.g., the use of a dollar sign to terminate statements, restriction of an identifier to 8 characters, restriction of the number of subscripts to less than or equal to 15, and references to precision of 32 bytes (thus causing accusation of being machine dependent). Some of these items (most notably the first) were due to the planned use of only a 48-character set. A major omission was the block structure which first appeared in the April, 1964 version [XY64a] and the compile time (i.e., macro) facility which first appeared in the IBM December, 1964 report [IB64].

Following the initial presentation to SHARE in March, 1964, a number of comments, suggestions, and criticisms were received and an April revision [XY64a] was prepared. The original committee, augmented by M. D. McIlroy (Bell Telephone Labs) and R. C. Sheppard (Procter and Gamble), the latter representing GUIDE, worked diligently and produced a drastically modified version in June [XY64b]. Somewhat later that year, the language received the (temporarily) official IBM nomenclature of NPL for *New Programming Language*; in December, 1964 another drastically revised version of the language appeared as [IB64]. However, conflict between the letters NPL and the National Physical Laboratory in England caused IBM to drop that name. The letters MPL and MPPL were considered, and IBM finally adopted PL/I, which is officially considered the name and not an acronym. While the computing industry watched with great interest, implementation started at the IBM Laboratories in Hursley, England; language work continued there and in the United States. A PL/I department was formed within IBM to centralize all the PL/I activities. The first official manual [IB65e] was issued early in 1965, and subsequent manuals containing minor and major changes over previous versions were issued. The manual on which this description is based was issued late in 1966 [IB66b]. In August, 1966 the first compiler for System/360 was released; this F-level compiler implemented most (but not all) of the language as then defined. Subsequent releases included other features. The D-level compiler was based on a subset (see [IB65]). [IB66a] and [IB66g] give descriptions of these specific compilers.

While all this activity was going on within IBM, other groups also exhibited interest and concern. Several manufacturers set up their own internal teams to study the language and advise its management on the potential desirability of implementing PL/I. A small, limited, and modified subset called NICOL I was implemented by Massachusetts Computer Associates in the fall of 1965 on the IBM 7094 [CQ65a]. A version called EPL was used by Bell Laboratories and M.I.T. in the development of their MULTICS system. A few other groups implemented small versions. An ACM-sponsored 1-day forum on PL/I, held in August, 1967, heard presentations about an interpretive scientific subset on the UNIVAC 1108 (Glass [GX67]), and a PL/I-like extended subset for compiler writing developed at Stanford University on the IBM 360/67 under OS/360 (McKeeman *et al.* [ZG67]). Other versions have been under development but the exact status is unclear, and reminiscent of the early days of FORTRAN, ALGOL, and COBOL.

Within 3 years from the start of the project, PL/I had already had a profound impact upon the computing industry and caused many debates. Its place in the future is unknown, but there is a good chance that eventually it will replace the major existing languages, most specifically FORTRAN, COBOL, ALGOL, and JOVIAL.

VIII.4.2. FUNCTIONAL CHARACTERISTICS OF PL/I

The reference used for the description of PL/I is [IB66b] and not any specific implementation of it.

PL/I is very general with the widest scope of any language in this book. PL/I notation is succinct and semiformal rather than English-like; it follows in the FORTRAN and ALGOL traditions rather than the COBOL line. It is not particularly consistent in the sense that there are a large number of special cases and exceptions. It is not particularly efficient to implement because of its size and the fact that good implementation techniques have lagged behind the development of languages that are this complex. It is fairly easy to read and write, with no great tendency toward error-prone usage except for errors due to its power and complexity. In considering its ease of learning, one must specify the amount to be taught and the experience of the learner. A person without any experience in a programming language would find it very difficult to learn all of PL/I. On the other hand, a FORTRAN programmer would find it easy to learn the PL/I subset equivalent to the capability of FORTRAN. PL/I is an excellent illustration of the point made in Chapter I that an extremely complex and powerful programming language might be harder to teach and to learn than a very simple machine code. Thus if a computer had no more than a dozen machine instructions, a person could undoubtedly be taught machine code more rapidly and easily than he could be taught the whole of PL/I.

PL/I is definitely aimed at having an extremely wide application area. It was meant to be used in the fields for which FORTRAN and COBOL were individually designed and, similarly, with regard to JOVIAL, whose prime characteristic was the combining of such facilities into a single language. PL/I is also effective for systems programming.

PL/I is a procedure-oriented language and, furthermore, it is really the culmination of the whole line of procedure-oriented languages. There are relatively few concepts that are relevant to this conceptual line of languages which are not already in PL/I or are not possible or obvious potential extentions of it. PL/I is problem-oriented but only in the very broad sense. It is simultaneously a reference, publication, and hardware language; however, the 60-character set (defined in Section VIII.4.3) can be considered a reference language with a 48-character set representation. It has very definitely been aimed at the entire cross section of users; i.e., PL/I has been designed so that a novice programmer can write simple correct PL/I programs without too much difficulty, while on the other hand the professional programmer has more control over his machine and environment than he has ever had before. This environment, while primarily batch, has the facility of being used in a reasonable fashion in an on-line situation. The format was designed with that need in mind. In addition, there are facilities for use in a multiprogramming environment.

PL/I as a language is quite machine independent except for certain provisions for doing arithmetic; if it has any faults along these lines, PL/I is more likely to be operating system dependent than actually hardware dependent. There is insufficient experience with which to judge the difficulty of writing completely compatible compilers. However, allowing the implementer to specify the largest precision he will handle makes the language compiler dependent as well as machine dependent, but at least this happens in a well-defined way. It is assumed that a reasonable connection between the machine word size and implementer choice normally exists. The dialect problem does exist, primarily for implementation reasons but also for the normal problem of personal taste. However, relative to its size, there will probably be only a minor problem along these lines because the people who are using PL/I are quite likely to be doing so because it has a combination of features that they want. If they need only some subset or modification and do not care about doing all their programming in a single language, then they can probably fall back on FORTRAN, COBOL, or even ALGOL. Subsetting and extensions on the other hand present somewhat more of a problem. There are three major reasons for subsets of PL/I to be in existence. The first is based on a stated objective of modularity, i.e., to cater to or provide facilities for, previous users of FORTRAN, ALGOL, or COBOL who need only comparable capabilities provided to them but still wish to write in PL/I. Thus, it is both meaningful and even desirable to talk about subsets

SAMPLE PROGRAMS—PL/I

Problem: Construct a subroutine with parameters A and B such that A and B are integers and $2 < A < B$. For every odd integer K with $A \leqslant K \leqslant B$, compute $f(K) = (3K + \sin(K))^{\frac{1}{2}}$ if K is a prime, and $f(K) = (4K + \cos(K))^{\frac{1}{2}}$ if K is not a prime. For each K, print K, the value of $f(K)$, and the word PRIME or NONPRIME as the case may be.

Assume there exists a subroutine or function PRIME (K) which determines whether or not K is a prime, and assume that library routines for square root, sine and cosine are available.

Program:
```
PROBLEM: PROCEDURE(A,B);
         DECLARE (A,B,K) FIXED(15,0), Q(0:1)
              CHAR(8) INITIAL('NONPRIME','PRIME');
         DCL PRIME ENTRY(FIXED,FIXED) RETURNS(FIXED);
         DO K=2*(A/2)+1 TO B BY 2;
         E=PRIME(K)*SQRT(3*K+SIN(K))
              +¬PRIME(K)*SQRT(4*K+COS(K));
         PUT LIST(K,E,Q(PRIME(K)));
         END PROBLEM;
```

Sample Programs — PL/I (cont.)

Problem: Inventory control program for company with 20,000 stock items.†

Program:
```
INVCTL: PROCEDURE;
    DECLARE (OLDMAST INPUT, NEWMAST OUTPUT) BLOCK (FIXED,432,8),
        PFILE OUTPUT,
        1 WORK,
            2 PARTNO CHARACTER (7),
            2 DESCR CHAR(12),
            2 (QOH, QOO, RP, RQ) FIXED (5),
            2 UP FIXED (6),
            2 YTDSALE FIXED (8),
            2 CODE FIXED,
        1 TRANS,
            2 TNUMBER CHARACTER (7),
            2 TCODE FIXED,
            2 TQ FIXED (5),
        CODEIS (4) LABEL;
    ON ENDFILE (STANDIN) BEGIN; TNUMBER = '9999999';GO TO WRITNM; END;
    ON ENDFILE (OLDMAST) BEGIN; IF TNUMBER = '9999999' THEN DO; CLOSE
                                    OLDMAST DISCARD, (PFILE, NEWMAST)
                                    STORE, DISPLAY ('JOB FINISHED');
                                    END;
                                ELSE ERROR: DISPLAY ('FILE OR DATA
                                    ERROR'); EXIT; END;
    ON SUBSCRIPTRANGE BEGIN; DISPLAY ('BAD CLASS CODE JOB HALTED');
        EXIT; END;
    READ (TRANS)(A);
READM: READ FILE (OLDMAST), (WORK) (A);
TESTM: IF PARTNO < TNUMBER THEN WRITNM: DO; WRITE FILE (NEWMAST),
                                (WORK) (A); GO TO READM; END;
    IF PARTNO > TNUMBER THEN GO TO ERROR;
        /*THEN PARTNO = TNUMBER*/
    GO TO CODEIS (TCODE);
CODEIS(1): QOH = TQ; GO TO JOIN;
CODEIS(2): QOH = QOH + TQ; QOO = QOO - TQ; GO TO JOIN;
CODEIS(3): QOO = QOO + TQ; GO TO JOIN;
CODEIS(4): IF QOH < TQ THEN DO; WRITE ('ONLY', PARTNO, 'AVAILABLE',
        QOH, ' REQUESTED') (3A, F(5), A); TQ=QOH; END; QOH =
        QOH-TQ; IF CODE = 1 THEN YTDSALE = YTDSALE + TQ*UP;
JOIN: IF QOH + QOO =RP THEN WRITE FILE(PFILE), (PARTNO, CODE, RQ) (3 A);
    READ (TRANS) (A); GO TO TESTM; END INVCTL;
```

†[IB00a], p. 41. Reprinted by permission from *Introduction to PL/I (Student Text)*. © 1967 by International Business Machines Corporation.

for scientific or commercial programming. Similarly, the needs of the systems programmer can be met by appropriate subsetting. A second reason for subsetting is to reduce the size of the compilers; since this is a complex language requiring powerful compilation techniques, there will be some cases in which it is considered more desirable to have a smaller compiler and handle only a subset of the language. A third reason, which is related to the second but not identical to it, involves the use of small computer configurations. A computer (and operating system) configuration which is either small in memory or relatively low in speed or which does not provide facilities that would call for multitasking and other operating system interactions makes it appropriate to implement a subset of PL/I that can provide better performance on the smaller machines. In most cases, the subsets contain only those features which are common to many programming languages. Hence the subsets tend to have the syntax of PL/I but not its new or significant features.

PL/I appears to have sufficient facilities to permit some bootstrapping.

Because (and although) PL/I is such a powerful language, there is a great tendency for everybody to wish to incorporate in it facilities for areas that were not included initially. Among these proposed facilities are graphics, formal algebraic manipulation (see Bond and Cundall [BZ67]), associative data handling (see Dodd [DQ66]), and possibly even simulation. Since PL/I is being used for some experimental technical development in formal language definition, particularly in the semantic area (references are cited later), it seems likely that there will probably be less incompatibility due to problems in language definition than would otherwise be expected. (This judgment is relative to the size of the language; on an absolute basis, the number and complexity of the rules make this type of incompatibility inevitable.)

While some consideration has been given to sifting other languages to PL/I, there has been no work done (known to me) in going from PL/I to some other language. It would be almost a contradiction in objectives and timing if this were done.

It is indicative of the importance of PL/I that consideration of its official standardization was taking place even while new language manuals were under development and long before the first compiler was in existence. Because of its potentially large impact on programming in future years, it is natural that both manufacturers and users were concerned about the direction in which the language would develop. Early in 1965, IBM suggested that X3.4 consider the standardization of PL/I. After debate and consideration for several months, the offer was rejected in August, 1965 as being premature. Late in 1965 and early in 1966 an *ad hoc* working group under BEMA met several times and reached the conclusion that there was a reasonable basis for a prestandardization activity; it was agreed that the most desirable place for such an activity to take place was under the auspices of X3. As a result, X3.4 initiated a task group called X3.4.2.C, which held its first meeting in April, 1966. This group considered the problem of subsets, language development, character sets, and form of definition. Late in 1967, X3.4.2.C recommended that standardization of PL/I *not* be started at that time. In the spring of 1968 they reversed their position and issued an affirmative recommendation. Thus there will probably be a USASI PL/I standard at some point in time.

As indicated under the discussion of history, the first preliminary specifications were designed by a group of six people, three from IBM and three users from SHARE. That group, plus two more, also designed the second version of the language which appeared 3 months after the first. As time went on, an increasingly large number of people became involved, and some of the original participants were no longer associated with the effort. In particular, the joint IBM-SHARE effort was terminated as such, although a PL/I project was set up under SHARE.

Since the language is powerful and complex, its objectives cannot be stated succinctly. The following statements of objectives are extracted from the March 1, 1964 report [XY64]: "... to recommend a successor language for currently available FORTRANs to be used on unannounced IBM equipment. ... to provide a language which would encompass more users while still remaining a useful tool to the engineer. ... a deadline consistent with IBM scheduling ... in order to allow consideration of the language for the new equipment. ... to redefine those parts of the present FORTRAN language which appeared to be in conflict with, and to augment those parts which were inadequate for, the 'state of the art' today. The changes would, where practical, follow current syntax, but ... not restricted to this form. ... the need to keep the language simple, both to teach and to use. ... to allow subsetting of the language."[6]

The following statements of objectives are extracted from the Introduction in the original official IBM manual [IB66b]: "... None of the traditional high level languages, however, can be used with efficiency across the entire range of ability of these new computers. That is the reason for PL/I, a multipurpose programming language for use not only by commercial and scientific programmers but by the real-time programmer and the systems programmer as well. It is a language designed for efficiency, a language that enables the programmer to use virtually all the power of his computer. ... any programmer ... can use it easily at his own level. ... One of the primary aims ... was *modularity*, that is, providing different levels of the language for different applications and different degrees of complexity. ... every ... description of a variable, every option, and every specification has been given a 'default' interpretation. ... a 'default' interpretation ... is made by the compiler

---

[6] [XY64], p. 1.

if no choice is stated by the programmer. ... The 'modularity' and the 'default' aspects are the bases upon which the simplicity of PL/I has been built. They are also part of its power."[7]

The first compiler for the language was implemented in IBM by the laboratory in Hursley, England and released in August 1966 (see [IB66a]). The Hursley group was also responsible for setting up a language control board which met to resolve any ambiguities found in the manuals and to do language maintenance. Various definitions of the IBM formal syntax and semiformal translator and interpreter have been issued (see for example Beech *et al.* [BC66], [BC66a], and [BC67]). Significant language definition work for PL/I has been done by the IBM Hursley group (for the semiformal definition) and the IBM Vienna Laboratory; the latter developed formal semantic definitions described in Walk *et al.* [VK67], Bandat [BA67], [IB66], and other more recent reports not included in the list of references. [IB66] was the first significant attempt to provide a formal definition of the semantics of any real and implemented programming language, let alone one as complex as PL/I.

There are four types of documentation which existed for PL/I during its first few years, three issued by IBM and the fourth from other sources. The first set of documents were the reports of the Advanced Language Development Committee and its successors; these included the following: [XY64], [XY64a], [XY64b], and [IB64]. The specific and official reports put out by IBM were the language specification [IB66b], the subset definition [IB65], specific implementation guides [IB66a] and [IB66g], the subroutine descriptions [IB66e], and a reference data card [IB67]. The third type of documentation from IBM was student texts, e.g., [IB65a], [IB65b], and [IB66f]. The fourth type of documentation was not issued by IBM. It included papers prepared by individuals either critiquing the language or writing general descriptions of it, e.g., McCracken [MR64], Radin and Rogoway [RG65], Burkhardt [BU66], books by Bates and Douglas [QB67], and Weinberg [WC66]. In addition to these, an informal PL/I bulletin was issued irregularly by the Los Angeles ACM SICPLAN; this was issued later as part of SICPLAN notices [AC00].

As of this writing, there has been insufficient use of PL/I for a really valid evaluation.

VIII.4.3. Technical Characteristics of PL/I

The technical description of PL/I is being given in more detail than most other languages because of the relative newness, its potentially major significance, and also because of PL/I's power and complexity. The description is based on [IB66b] except where stated otherwise; however, not

---

[7] [IB66b], p. 9. Reprinted by permission from *IBM System/360 Operating System: PL/I Language Specifications*. © 1966 by International Business Machines Corporation.

all these features have been implemented. For this reason and because some of the specifications are being reconsidered, there may well be changes in the information given here.

PL/I very carefully (and intelligently) defined two character sets, one consisting of 60 characters and the other of 48. The 60-character set is composed of the 10 digits, 26 upper-case letters, 3 characters defined as alphabetic, namely

currency symbol	$
commercial "At" sign	@
number sign	#

and 21 special characters. The 48-character set does not have the commercial "At" sign or the number sign nor does it substitute graphics for them. The following list shows the graphics for the special characters in both sets, where the substitutions in the 48-character set are shown beneath the characters they replace:

| 60-character set | = | + | − | * | / | (|) | , | . | ' | % | ; | : |
| 48-character set | | | | | | | | | | | // | ,. | .. |

| 60-character set (cont.) | ¬ | & | \| | > | < | _ | ? | blank |
| 48-character set (cont.) | NOT | AND | OR | GT | LT | | | |

Certain special rules apply to the usage of the 48-character set equivalents of **:** ; and %. In addition, certain character combinations in the 60-character set are replaced by alphabetic equivalents, as follows:

| 60-character set | ¬> | >= | ¬= | <= | ¬< | \|\| | → |
| 48-character set | NG | GE | NE | LE | NL | CAT | PT |

An alphanumeric character is either a digit or one of the 29 alphabetic characters (26 letters, 3 special characters).

There is a long list of key words in PL/I and many of them can be abbreviated, e.g., **SUBSCRIPTRANGE**, which can be written as **SUBRG**, and **FIXEDOVERFLOW** as **FOFL**. The operators with graphics and punctuation are shown above.

An identifier starts with an alphabetic character and is followed by a string containing no more than 30 alphanumeric and/or break characters; an identifier is preceded and followed by a delimiter. There are no reserved words which cannot be used as identifiers in the 60-character set (although a few must be specifically declared as identifiers) but in the 48-character set the alphabetic equivalents used for the physical graphics are considered reserved. Identifiers can be used for either data names or program unit labels; there is no distinction made in the formation.

Both data names and statement labels can be subscripted. The subscripts are separated by commas and enclosed within parentheses, and they can be subscripted themselves. There is no limit on the number of subscripts nor on their formation; any expression that can be evaluated and converted to an integer may be used as a subscript. The lower limit can be specified, and it can be negative. The subscript notation is extended to include the concept of cross section; this is denoted by replacing one or more subscripts by asterisks, thus obtaining all the elements in that particular subscript position. Thus, A(3,*) denotes the third row of the array A. Similarly, if MATRIX is the array

$$\begin{matrix} 1 & 2 & 3 \\ 4 & 5 & 6 \\ 7 & 8 & 9 \end{matrix}$$

then MATRIX (*,2) is the (vertical) vector

$$\begin{matrix} 2 \\ 5 \\ 8 \end{matrix}$$

Hierarchies are permitted in PL/I (which uses the word *structure* rather than *hierarchy*). This allows and requires qualification, with the rule that only enough names to resolve any ambiguity are needed. The qualifiers are written before the data name from left to right in increasing level-number order; they are separated by periods. Thus **REGULAR.HOURS** and **OVERTIME.HOURS** represent different variables.

Structures can contain arrays and arrays can contain structures, so subscripted qualified names are permitted. The basic variable name is preceded by the qualifiers as just defined, and each qualifier itself can be subscripted. Subscripts can actually be attached to names at a lower or higher level than the one to which they belong. Thus, for an array A of structures of the following description (where the notation is defined later),

DECLARE 1 A (10, 12), 2 B (5), 3 C (7), 3 D;

the following all represent the same item:

A (10, 12) . B(5) . C(7)
A (10) . B(12, 5) . C (7)
A (10) . B . C (12, 5, 7)
A . B (10, 12) . C (5, 7)
A (10, 12, 5, 7) . B . C

and there are even more possible ways of writing this.[8]

---

[8] This example, and most of the others in this section, are extracted and/or modified from [IB66b]. Reprinted by permission from *IBM System/360 Operating System: PL/I Language Specifications*. © 1966 by International Business Machines Corporation.

The operators are divided into four types: *Arithmetic, comparison, bit string,* and *string*. The arithmetic operators are the standard five. The comparison operators are >, ¬>, >=, ¬=, <=, <, and ¬<, where ¬ denotes *not*. The bit string operators are *not, and,* and *or,* written with the graphics shown earlier. The string operator is ||, denoting concatenation.

The delimiters are operators, parentheses, and separators and other delimiters. The latter group includes the following:

$$= \quad : \quad ; \quad , \quad ' \quad . \quad \% \quad \rightarrow \quad \text{blank}$$

Various punctuation characters have both specific and general uses. In particular, the parentheses, colon, semicolon, comma, and period can be used to separate identifiers, constants, or picture specifications. Each of these also has specific separate usages which will be described in the appropriate places.

Identifiers, constants, or picture specifications may not be immediately adjacent. They must be separated by either an operator, assignment symbol, parenthesis, colon, semicolon, comma, period, blank, or comment. Blanks are optional between certain key words of a command, e.g., GO TO, but they cannot be used between composite operators, e.g., ¬=. At least one blank must appear between a level number and its following identifier. Blanks can be used to separate identifiers, constants, or picture specifications. Wherever one blank is used, any number of blanks or comments can be used. There are no noise words permitted.

The following types of constant are allowed: Decimal and binary, fixed and floating, imaginary, sterling fixed point, and strings. The strings are enclosed within quote marks and if it is desired to represent a quote mark, this must appear as two immediately adjacent quote marks. The actual string can be preceded by an integer in parentheses to specify repetition. Thus, (3)'TOM' is equivalent to 'TOMTOMTOM', and (5)'1'B is equivalent to '11111'B.

The input form is definitely free, and string-oriented; i.e., there are absolutely no restrictions or rules about card columns or equivalent concepts. The entire program can be written as one continuous string from beginning to end, naturally subject to all other rules of the language. From a conceptual point of view, the language is close in spirit to both ALGOL and FORTRAN, in the sense of providing reasonable but not complete formal symbolism.

From a conceptual point of view, PL/I has three types of declarations, namely explicit, contextual, and implicit.

The smallest executable unit is called a *simple statement*, which usually consists of a verb and its associated parameters and format. (PL/I uses the phrase *statement identifier* to mean a command or verb, and *statement label* for the name given to a particular statement written by the programmer.) There are two ways of grouping these smallest executable units. One is called a *compound statement*, of which there are two types, namely the IF and the ON; they are called compound because they precede a simple statement. The second type of grouping is actually called a DO group; this is used for control purposes. It has a simple and a looping form, and the looping form is discussed later. The simple form has an optional label, then the word DO and a semicolon, then the set of executable statements, and the key word END, which may have a label associated with it if the original DO statement itself did. For example,

ALPHA: DO; A = B*C; IF A > 0 THEN DO; B = 1; C = 0; END; END ALPHA;

Note that in this example any of the single statements except the DO or END statement is an example of a simple statement.

Statements can be combined to form a *block* whose primary purpose is to establish the scope of an identifier but this can also be used for control purposes. A block also contains declarations. There are two kinds of blocks, namely *begin* blocks and *procedure* blocks. The format of the begin block is the same as that of a simple DO group except for the use of the word BEGIN instead of DO. A procedure block must have a label (which is simply an identifier followed by a colon); it can have more than one label; and it has the word PROCEDURE at the beginning. The begin block and the procedure block are very similar in syntax and role relative to delimiting the scope of names. However, the begin block is executed in line whereas a procedure can be activated only by a CALL statement or by invocation from within an expression. A procedure block can also have more than one entry point.

Loops are written using a DO statement or using an IF ... THEN sequence.

PL/I distinguishes between two types of procedures, namely function procedures and subroutine procedures, and each type can have multiple entry points. A procedure is considered a function if there is a specific result obtained when it is invoked. This result is indicated as part of the RETURN statement by which control is returned to the calling location. Functional procedures are normally used as operands in expressions. A subroutine procedure does not provide a value as part of the RETURN statement but specifies results by setting some of its parameters. A subroutine procedure may only be invoked by a CALL statement or by a statement with a CALL option. Because a procedure may contain more than one RETURN statement, it is possible to use it both as a function and as a subroutine procedure, although this is seldom done.

A procedure that is not included in any other block is called an *external* procedure; a procedure included in some other block is called an *internal* procedure. Every begin block must be included in some other block; hence the only external blocks are external procedures. This distinction affects the scope of names, which is discussed later.

A comment is defined as any string preceded by the two characters /* and terminated by */ except that the comment cannot contain the two-character string */. A comment can be used wherever a blank is permitted (except of course within a literal).

PL/I has many facilities which can be considered as interaction with the operating system or the environment. The first could be considered an implementation problem in providing the necessary interface but because of the complexity, it has wider implications; the interaction referred to involves the various language features that the programmer can use to control storage allocation for the data variables. The second facility allows the programmer to create tasks, to synchronize them, to test whether or not tasks are complete, and to change the priority of tasks; thus he may perform operations asynchronously. The third class is a set of interrupt conditions, many of which relate to input/output conditions, both hardware and software. The final way of communicating with the system is the compile-time macro facility.

There is no provision for inclusion of other languages.

A program is composed of one or more external procedures. Thus, a program is a set of procedure blocks, each of which is completely nested and separate from the others. Statements and declarations can appear in any order except for the restrictions due to the placement of declarations to control the scope of variables.

The concept of *prologue* is used to refer to computations that must be done at object time before executing the statements within a block. Thus, the prologue must allocate storage or automatic variables and may need to evaluate expressions which define lengths, bounds, and iteration factors, or to supply initial values to variables. For this reason and others, there are various rules which require that the allocation or initialization of differing items not be circular; i.e., variable *A* cannot require information from variable *B* and simultaneously have the converse be true.

The primary method of delimiting is the semicolon, which indicates the end of all executable units and declarations. Labels are followed by colons, and the colon has other uses as well. Macro statements are preceded by a percent sign. A DO group is ended by reaching an END statement with the same label or by the first END without a label. It is this definition rather than a specific rule against it which prevents overlapping DO groups.

Procedures can be used recursively if they are so declared.

PL/I allows both call by location and call by value, although neither phrase is used in the manual. The call by value occurs whenever the argument to be passed is either a constant, an expression involving operators, an expression in parentheses, or an expression whose data attributes disagree with those of the parameter.

Statements can be embedded within others as indicated above. Arithmetic expressions can contain functions, Boolean expressions, and concatenated strings. An expression can be used wherever its value can syntactically be used.

There are arithmetic variables and constants. There are no separately defined Boolean variables, although the use of a bit string of length 1 can be used to accomplish the same effect. There are no separately defined character variables as such, but both bit and character strings are permitted; they can be either fixed or varying lengths (although the latter always has a maximum specified). Complex variables are permitted. There are no formal (i.e., algebraic) variables.

There are no lists as such provided in PL/I, but there are the concepts of *pointer variable* and *based variable* which permit the creation of list structures. Hierarchies are a legitimate data type.

Several other data types are identified in PL/I, namely *statement label*, *task variable*, *event variable*, and *cell*. A statement label constant is simply an identifier that appears in the program as a statement label. A statement label variable is simply a variable having statement label constants as its values. These can be grouped into arrays or structures. A task variable is the name of a task, and an event variable is the name of an event. Both of these can be elements of arrays or structures: they are described in more detail later. However, an event variable has an associated completion status noted by '0'B or '1'B for not completed and completed, respectively.

All the above variable types can be put into arrays. Furthermore, each of them can themselves be elements within a structure. Structures of arrays and arrays of structures are permitted. Expressions can consist of combinations of single variables of the types arithmetic, complex, and string; an array or structure expression is evaluated as a sequence of scalar expressions, using corresponding elements of the aggregates involved.

The only hardware data type accessible is a bit. Each of the variable types indicated can be accessed by many of the commands, although some combinations are illegal, e.g., structures in a GOTO. A command which is allowed to access a single variable can usually also access a structure or array. A single command can access a cross section of an array through the use of the asterisk (described earlier).

The arithmetic in PL/I includes decimal and binary, as well as fixed (including integer as a special case) and floating point. There is no rational arithmetic,[9] but complex arithmetic is performed. Arithmetic variables can

---

[9] This capability exists in the PL/I-FORMAC interpreter described in Section VII.3.3 and can be used in that system.

have any of the characteristics of base, scale, mode, and precision; these can be represented by a numeric picture. Formats for these are shown in the description of the declarations, but the concepts will be described briefly here. Base refers to binary or decimal, and scale refers to either fixed or floating point form. Mode refers to real or complex. There is no specific double or multiple precision in the language. Instead, the user specifies the precision he wants by defining the scale of the data and the total number of binary or decimal digits to be maintained, for both fixed and floating point data. The implementer specifies the largest precision that he will handle. Finally, it is possible to specify by means of a *PICTURE* the formats of the variables. This is described in some detail later. The higher level data units (i.e., arrays and structures) are handled by dealing with the individual subparts.

There are a number of default conditions specified for arithmetic variables; i.e., if the user does not provide a specific declaration, certain characteristics are automatically assumed. Some examples are as follows: if the first letter of a name is I, J, K, L, M, N, then **FIXED REAL BINARY** is assumed; otherwise **FLOAT REAL DECIMAL** is assumed. If some but not all the characteristics are specified, then it is assumed that the base is decimal, the scale is float, and the mode is real.

Although there are no Boolean variables defined as such, the net effect of Boolean arithmetic is permitted because arithmetic expressions can contain the eight relational operators; furthermore, binary arithmetic is permitted and the operations of *and*, *or*, and *not* are defined on binary strings.

Concatenation of bit and character strings is permitted and the result is a character string.

Arithmetic can be done on structures and arrays. The operands of a structure expression are structures or a combination of structures and variable expressions. The result is a structure; arrays are not allowed as operands in structure expressions. To permit this arithmetic to be done, all the structures in an expression must have the same number of contained scalars and arrays, their relative positioning must be the same, and similarly positioned arrays must have identical dimensions and bounds. The data types need not be the same.

Because of the numerous data types which are allowed in PL/I, the rules for intermingling, converting, and precision are extremely complex. No attempt will be made here to supply all the details but merely to give the general principles involved. The term *scalar expression* is introduced rather than *arithmetic expression* because the expressions involved may contain nonarithmetic quantities. Scalar expressions can contain any variable types except statement labels, area variables, task variables, and event variables. Only the comparison operators = and ¬= can appear with pointer data. However, as indicated earlier, structure expressions and/or array expressions

556   MULTIPURPOSE LANGUAGES

can be formed and their evaluation is done by evaluating the respective scalars. The most significant conversions are exactly what would be expected; namely decimals are converted to binary if both appear, fixed point operands are converted to floating point if both appear, and real numbers are converted to complex if both appear as operands of the same operator. Bit string operations can be performed on arithmetic data by converting them to bit strings. If comparisons are to be made among arithmetic, character strings, and bit strings, then the operand of lowest type is converted to the operand of highest type, where the priority is decreasing in the order just stated; i.e., bit string is the lowest priority. Only the operations of = and ¬= are defined if one of the operands is complex.

Comparisons are actually of three types: Algebraic, character (left to right pair-by-pair comparison of characters according to a given collating sequence), and bit (left to right comparison of the binary digits). The result of a comparison is a bit string of length one; the value is '1'B if true and '0'B if false.

The precedence of operations is as follows from the highest to the lowest:

```
    ¬       **      prefix +     prefix −
    *       /
    infix +  infix −
    ||
    >=      >       ¬>    ¬=    <    ¬<    <=    =
    &
    |
```

(This differs slightly from the manual [IB66b] since an approved language change has occurred since that publication.)

The rules of precision are based on a complicated formula. The aim is to give a large enough result field for a fixed point operation (or the maximum size in the case of division) or a result field for floating point with the greater of the precisions of the two operands.

All the text of a begin block except the labels preceding the heading statement of the block is said to be contained in the block. All the text of a procedure except the entry names of the procedure is said to be contained in the procedure. That part of the text of a block *B* that is contained in block *B* but not contained in any other block contained in *B* is said to be *internal* to block *B*. The *scope* of a declaration of an identifier is defined as that block *B* to which the declaration is internal, but it excludes from block *B* all contained blocks to which another declaration of the same identifier is internal. This definition of scope applies to all identifier declarations except those for entry names of external procedures; these have slightly different rules.

It is possible to use the same name for different declarations of the same identifier through the *EXTERNAL* attribute. All external declarations for the same identifier are considered to be related to the same name; the scope of the name is the union of the scopes of all the external declarations for the identifier. The declarations must not be contradictory, of course.

The specific formats for the executable statements are shown in Figure VIII-2. They use the metalanguage discussed in Section II.6.2.2 as pertaining to COBOL. The discussion given below assumes the reader will examine the formats. Because of differences in terminology, some items which are called executable statements in PL/I are not considered so within the framework of this book and so are not included in Figure VIII-2. Those omitted are: *BEGIN, DECLARE, END, FORMAT, PROCEDURE*. They are discussed in the body of the text and the format is described whenever it is appropriate to do so.

The assignment statement can have multiple variables to the left of the equals sign; furthermore, as noted from the specific format, they can be arrays, structures, or label or pointer variables. The pseudo-variables are discussed below. By writing the assignment statement, e.g., A,B = C+D; the right-hand side is computed and assigned to the variables on the left-hand side. When necessary, the expression for values on the right-hand side is converted to the characteristics of the variables on the left, according to the standard rules mentioned earlier. It is possible to have both complete arrays and single elements from an array on the right-hand side, and there is a well-defined rule for computation of arrays on the left-hand side. If the variable on the left of the equals sign is a variable character string, the assignment is performed from left to right starting with the leftmost position. Specific rules are defined for the various cases that can arise.

Three examples from p. 110 of [IB66b] are given below; the second is not self-evident without knowing the form of the *DECLARE* declaration which is shown in examples on page 578.

Example. Given the arrays A and B, respectively, as

```
2 4      1 5
3 6      7 8
1 7      3 4
4 8      6 3
```

then the assignment statement

$$A = (A+B)**2 - A(1, 1)$$

produces the following matrix for A (where the new A(1,1) is computed and

ALLOCATE

    Option 1:
        ALLOCATE [*level*] *identifier* [*dimension*] [*attribute*] ...
            [, [*level*] *identifier* [*dimension*] [*attribute*] ...] ... ;

    Option 2:
        ALLOCATE *based-variable-identifier* SET (*pointer-variable*)
                [IN (*area-variable*)]
        [, *based-variable-identifier* SET (*pointer-variable*)
                [IN (*area-variable*)]]
        ... ;

Assignment

    Option 1: (*Scalar assignment*)
$$\begin{Bmatrix} \text{scalar-variable} \\ \text{pseudo-variable} \end{Bmatrix} \begin{bmatrix} , \text{ scalar-variable} \\ , \text{ pseudo-variable} \end{bmatrix} \ldots = \text{scalar-expression} ;$$

    Option 2: (*Array assignment*)
$$\begin{Bmatrix} \text{array} \\ \text{pseudo-array} \end{Bmatrix} \begin{bmatrix} , \text{ array} \\ , \text{ pseudo-array} \end{bmatrix} \ldots = \begin{Bmatrix} \text{array-expression } [, \text{BY NAME}] ; \\ \text{scalar-expression ;} \end{Bmatrix}$$

    Option 3: (*Structure assignment*)
$$\begin{Bmatrix} \text{structure} \\ \text{pseudo-structure} \end{Bmatrix} \begin{bmatrix} , \text{ structure} \\ , \text{ pseudo-structure} \end{bmatrix} \ldots =$$
        structure-expression [, BY NAME] ;

    Option 4: (*Statement label assignment*)
$$\text{scalar-label-variable } [, \text{ scalar-label-variable}] \ldots = \begin{Bmatrix} \text{label-constant ;} \\ \text{scalar-label-variable ;} \end{Bmatrix}$$

$$\text{array-label-variable } [, \text{ array-label-variable}] \ldots = \begin{Bmatrix} \text{label-constant ;} \\ \text{scalar-label-variable ;} \\ \text{array-label-variable ;} \end{Bmatrix}$$

    Option 5: (*Pointer assignment*)
        pointer-variable [, *pointer-variable*] ... = *pointer-expression* ;
        array-pointer-variable [, *array-pointer-variable*] ... =
$$\begin{Bmatrix} \text{pointer-expression} \\ \text{array-pointer-variable} \end{Bmatrix} ;$$

CALL

    CALL *entry-name* [(*argument* [, *argument*] ...)] [TASK [(*scalar-task-name*)]]
        [EVENT (*scalar-event-name*)] [PRIORITY (*expression*)] ;

CLOSE

    CLOSE FILE (*filename*) [IDENT (*argument*)]
        [, FILE (*filename*) [IDENT (*argument*)]] ... ;

DELAY

    DELAY (*scalar-expression*) ;

DELETE

    DELETE FILE (*filename*) KEY (*expression*) [EVENT (*event-variable*)] ;

Figure VIII-2. (cont.)

DISPLAY

> *Option 1:*
> > DISPLAY (*scalar-expression*) ;
>
> *Option 2:*
> > DISPLAY (*scalar-expression*) REPLY (*character-variable*)
> > [EVENT (*event-variable*)] ;

DO

> *Option 1:*
> > DO ;
>
> *Option 2:*
> > DO WHILE (*scalar-expression*) ;
>
> *Option 3:*
> > DO $\begin{Bmatrix} \textit{pseudo-variable} \\ \textit{variable} \end{Bmatrix}$ = *specification* [, *specification*] ... ;
>
> A *specification* has the following format:
> > *expression1* $\begin{bmatrix} \text{TO } \textit{expression2} \text{ [BY } \textit{expression3}\text{]} \\ \text{BY } \textit{expression3} \text{ [TO } \textit{expression2}\text{]} \end{bmatrix}$ [WHILE (*expression4*)]

ENTRY

> > *entry-name*: [*entry-name*:] ... ENTRY [(*parameter* [, *parameter*] ...)]
> > [*data-attributes*] ;

EXIT

> > EXIT ;

FREE

> *Option 1:*
> > FREE *identifier* [, *identifier*] ... ;
>
> *Option 2:*
> > FREE [*pointer-variable* –>] *based-variable-identifier*
> > [, *pointer-variable* –>] *based-variable-identifier*] ... ;

GET

> > GET [FILE (*filename*) | STRING (*character-string-name*)]
> > *data-specification* [COPY] ;

GO TO,

> > $\begin{Bmatrix} \text{GO TO} \\ \text{GOTO} \end{Bmatrix}$ $\begin{Bmatrix} \textit{label-constant} \text{ ;} \\ \textit{scalar-label-variable} \text{ ;} \end{Bmatrix}$

IF

> > IF *scalar-expression* THEN *unit-1* [ELSE *unit-2*]

Figure VIII-2. (cont. next page)

Figure VIII-2. (cont.)

LOCATE

        LOCATE *variable* FILE *(filename)* SET *(pointer-variable)*
        [KEYFROM *(expression)*] ;

Null

        [*label* :] ... ;

    Example:

        .
        .
        .
        ON OVERFLOW;
        .
        .
        .

The *on-unit* (see ON) is a *null* statement.

ON

    Option 1:
        ON *condition* [SNAP] *on-unit*
    Option 2:
        ON *condition* [SNAP] SYSTEM ;

OPEN

        OPEN *options-group* [, *options-group*] ... ;
    Following is the format of *options-group*:
        FILE *(filename)* [IDENT *(argument)*] [TITLE *(expression)*]
        [INPUT | OUTPUT | UPDATE] [STREAM | RECORD]
        [DIRECT | SEQUENTIAL] [BUFFERED | UNBUFFERED] [EXCLUSIVE]
        [KEYED] [BACKWARDS] [PRINT] [LINESIZE *(expression)*]
        [PAGESIZE *(expression)*]

PUT

        PUT [FILE *(filename)* | STRING *(character-string-name)*] [*data-specification*]
        [PAGE] [SKIP [*(expression)*]] [LINE *(expression)*] ;

READ

        READ FILE *(filename)* { INTO *(variable)* | SET *(pointer-variable)* | IGNORE *(expression)* } [KEY *(expression)*]
        [KEYTO *(character-string-variable)*] [EVENT *(event-variable)*]
        [NOLOCK] ;

Figure VIII-2. (cont.)

RETURN

> *Option 1:*
>> RETURN ;
>
> *Option 2:*
>> RETURN (expression) ;

REVERT

>> REVERT ON-condition ;

REWRITE

>> REWRITE FILE (filename) [KEY (expression)] [FROM (variable)]
>> [EVENT (event-variable)] ;

SIGNAL

>> SIGNAL ON-condition ;

STOP

>> STOP ;

UNLOCK

>> UNLOCK FILE (filename) KEY (expression) ;

WAIT

> *General format:*
>> WAIT (event-name [, event-name] ...) [(scalar-expression)] ;

WRITE

>> WRITE FILE (filename) FROM (variable) [KEYFROM (expression)]
>> [EVENT (event-variable)] ;

Figure VIII-2. Executable PL/I statement formats. Note that because of differences in terminology, some things which are called executable statements in PL/I are not considered so within the framework of this book and so are not included on this list. Those omitted are: *BEGIN, DECLARE, END, FORMAT,* and *PROCEDURE*. They are discussed in the body of the text, as are some of the terms used in the format, e.g., *scalar-expression*. The general notation is the COBOL metalanguage described in Section II.6.2.

Source: [IB66b], extracts from pp. 104–132. Reprinted by permission from *IBM System/360 Operating System: PL/I Language Specifications*. © 1966 by International Business Machines Corporation.

then used to compute the remaining elements):

$$\begin{array}{cc} 7 & 74 \\ 93 & 189 \\ 9 & 114 \\ 93 & 114 \end{array}$$

Example. Given the structure defined by

DECLARE 1 X, 2 Y, 2 Z, 2 R, 3 S, 3 P,
1 A, 2 B, 2 C, 2 D, 3 E, 3 Q;

then the assignment statement

X = X*A;

has the same result as the following set of statements:

Y = Y*B;
Z = Z*C;
S = S*E;
P = P*Q;

Example. Given

A is a fixed-length string whose value is 'XZ/BQ'
B is a varying-length string of maximum length 8 whose value is 'MAFY'
C is a fixed-length string of length 3
D is a varying-length string of maximum length 5

Then in the statement

C = A;   The value of C is 'XZ/'.
C = 'X'   The value of C is 'X♭♭', where ♭ designates blank.
D = B;   The value of D is 'MAFY'.

When the BY NAME option is used, the basic principle of the rules is that only those names which are common to all the variables on both sides of the assignment statement are used to create appropriate assignments.

PL/I introduces the concept of using certain built-in functions as *pseudo-variables*. (In addition to being used in an assignment statement, pseudo-variables can also appear in a DO statement or in a data list in a GET statement.)

As an example,

COMPLEX (A, B) = COMPLEX (U,V) + REAL (Q);

is the same as writing

```
C = COMPLEX (U,V) + REAL (Q);
A = REAL (C); B = IMAG (C);
```

There are no specific separate statements for handling alphanumeric data; conversion of character strings was discussed earlier. A *SORT* statement appeared in earlier versions of the language, but it is not defined in the reference used for this description.

The unconditional control transfer is the *GOTO*. Switch control is accomplished by assigning values to a label variable from an assignment statement; there is no separate switch statement. Because of the block structure and the provisions for tasks, there are some specific rules indicating when it is illegal to use a *GOTO* statement or indicating what effect a *GOTO* has on the execution of blocks and procedures.

The *CALL* statement is used to invoke a procedure, and of course it causes control to be transferred to a specified entry point in the procedure. The *TASK*, *EVENT*, and *PRIORITY* options can be used alone or in any combination, and specify that both the invoked and invoking procedures are to be executed asynchronously. When the *EVENT* option is used, the *EVENT* name is associated with the completion of the task created by the *CALL* statement. Another task can then wait for completion of this created task by specifying the event name in a *WAIT* statement. If the *PRIORITY* option is used, the priority of the named task is made relative to the task in which the *CALL* is executed, based on the value of the expression.

The *RETURN* statement returns control to the invoking procedure; the expression must be used if and only if the procedure was invoked as a function procedure. The *END* statement also returns control to the point logically following the invocation.

The conditional statement is the *IF* whose format is shown in Fig. VIII-2. Each *unit* is either a group or a begin block; each of these is terminated by a semicolon. The *scalar-expression* is evaluated and if necessary, converted to a bit string with the understanding that all zero's mean *false* and a single occurrence of a one means *true*. In the true case, *unit-1* is executed and the control then goes to the next statement (assuming that *unit-1* does not contain a control transfer out). In the false case, *unit-2* is executed if it is present and, if not, then control is immediately transferred to the next statement. Both *unit-1* and *unit-2* can have labels and can also be *IF* statements themselves. Each *ELSE* clause is always associated with the innermost preceding *IF* which does not yet have an *ELSE* clause. An *ELSE* can be used with a null statement (i.e., just a semicolon) to provide proper pairing. In the following example if $X \neq Y$, then X is assigned the value 4 and S is given the value 5. If $X = Y$, $S = R$, and $W \not< P$ then P is given the value of Q and S is given the value 5. If $X = Y$, $S = R$, and $W < P$, then Y is assigned the value 1 and S is given the value 5.

```
A:  IF X=Y THEN
        IF S = R THEN
        IF W < P THEN Y = 1;
        ELSE P = Q;
        ELSE;
        ELSE X = 4;
J:  S = 5;
```

There is one loop-control statement, namely the DO, whose format is shown in Fig. VIII-2. The range of the loop is the DO block containing the given DO as its heading. In the DO WHILE option, there is no parameter but merely a termination condition indicating when to stop executing the range. In the other option, the variable can be not only arithmetic but also a label, string, or complex variable, providing the last three items produce valid PL/I programs when used with the appropriate expansion of the iteration. As is common, if the BY clause is omitted, then *expression3* is assumed to be 1. If the TO expression is omitted, then the iteration is performed until terminated by either the WHILE clause or by some other statement within the range of the DO. If both of these are omitted, there is a single execution of the DO group with the parameter having the value of *expression1*, which is the initial value. If the variable in *option3* is a label variable, then (only) the WHILE clause must be used. When the range is terminated, control is transferred to the statement immediately following the END which terminates the DO group. The most complex case actually occurs when the TO, BY, and WHILE clauses are all used; in that situation, the definition of the DO statement is defined in the manual in terms of other PL/I statements. The significant result of this is that all the unusual cases are specifically defined, unlike most other languages.

Some examples of DO statements are the following:

```
DO I = 1 TO K - 1, K + 1 TO N BY J;
DO COMPLEX (X,Y) = 0 BY 1 + 2I WHILE (X LESS THAN 5);
DO J = 1 TO 9, 22 TO 30;
DO INDEX = Z WHILE (A GREATER THAN B), 5 TO 9 WHILE
    (A = B), 815;
```

There are a number of conditions involving error or program checking which cause an automatic interrupt by the system unless something is done to prevent them. However, the user can override standard action by using what are known as ON-*conditions*. These are used in an ON statement and can be classified into computational, input/output, program check-out, list processing, programmer named, and system action. The specific list is given in Figure VIII-3. If an interrupt takes place before an ON statement for

Computational	
	CONVERSION
	FIXEDOVERFLOW
	OVERFLOW
	SIZE
	UNDERFLOW
	ZERODIVIDE
Input/Output	
	ENDFILE (filename)
	ENDPAGE (filename)
	TRANSMIT (filename)
	UNDEFINEDFILE (filename)
	NAME (filename)
	KEY (filename)
	RECORD (filename)
Program checkout	
	SUBSCRIPTRANGE
	CHECK (identifier-list)
List Processing	
	AREA
Programmer-Named	
	CONDITION (identifier)
System Action	
	FINISH
	ERROR

Figure VIII-3. List of ON-*conditions* in PL/I.
Source: [IB66b], extracts from pp. 162–166. Reprinted by permission from *IBM System/360 Operating System: PL/I Language Specifications.* © 1966 by International Business Machines Corporation.

a specific condition has been executed, standard system action is taken. However, the programmer may specify some other action to take place and in that case it is considered as a procedure internal to the block in which it appears. Thus one can write ON OVERFLOW GOTO ALPHA; or ON CONVERSION Y = X+1;. The *on-unit* to be performed when the condition occurs is either an unlabeled simple statement (other than BEGIN, DO, END, RETURN, FORMAT, PROCEDURE, or DECLARE) or an unlabeled begin block. If SNAP is specified, a calling trace is listed when the given condition occurs.

There are specific detailed rules about the scope of the ON statement. A condition raised during execution results in an interrupt if and only if the condition is *enabled* at the point where it is raised. Most conditions are enabled by default, and the remainder are *disabled* by default. For several, the enabling or disabling may be controlled by the use of condition prefixes.

As implied earlier, an interruption for most error conditions of a general type will occur whether or not an ON statement has been executed; the ON statement merely determines the action to be taken when the condition arises, but it has nothing to do with allowing or preventing an interruption to occur. However, the programmer can actually control certain interruptions through the use of condition prefixes. An *enabling condition prefix* is a list of condition names, enclosed in parentheses, and prefixed to a statement with a colon that precedes the label. Thus the user can write (SIZE, SUBSCRIPTRANGE): LABEL: *executable statements;*. By preceding certain condition names with the letters NO, the condition is disabled from causing an interrupt. If the condition name is prefixed to any statement other than a PROCEDURE or BEGIN statement, the condition is enabled (or disabled) only through the execution of that single statement. If it is prefixed to an IF statement, its scope is only through the evaluation of the expression in the IF clause. If a condition name is prefixed to a DO statement, its scope is only through the DO statement itself. If the condition name is prefixed to a PROCEDURE or BEGIN statement, its scope is through the entire block including all nested blocks except for any statements that lie within the scope of another condition prefix with a different specification for the same condition. Unlike the scope in an ON statement, the scope of a condition prefix does not extend to a block that is invoked remotely.

The REVERT command has essentially the effect of canceling an ON statement once the latter has been actually executed, assuming that they were internal to the same block. It also reactivates the most recent ON statement in the containing block.

It is possible to simulate the existence of one of the interrupts through the use of the SIGNAL statement, which causes the same action as if the specified condition had actually occurred.

There are no algebraic expression manipulating statements. (However, see the PL/I-FORMAC interpreter discussed in Section VII.3.)

The basic idea of the PL/I list processing facilities is to allow the user to determine the detailed structure of his lists. In other words, instead of specifying a particular form of list structure like in previous list processing languages such as IPL-V and LISP (see Sections VI.3 and VI.5), the user is given a variable type called *pointer* and a form of controlled variable called a *based variable*. (The newer list processing language L^6 (see Section VI.4) also permits the user to define his own structure.) The term *based* means that the actual storage position accessed in a reference to the variable is determined by the value of a pointer to it. For example, writing

DECLARE P POINTER, ALPHA FLOAT BASED (P);

defines P as a pointer variable and ALPHA as a floating point based variable, whose location is identified by P when reference is made to ALPHA. The value of P itself can be set during the program by assignment, by SET, or through the use of the built-in *ADDR* function, which returns a pointer value which "points to" its argument. (The word **BASED** replaced the word **CONTROLLED** which was used in [IB66b].)

In some list processing applications there is a need for more than one pointer to identify a given item of data. In these cases, other pointers may be used to refer to a based variable. The symbol —> (called a pointer qualifier) is used for this. To identify the last item in a list, the *NULL* built-in function is usually used to provide a *null* pointer.

A fairly thorough discussion of this subject is given by Lawson [LH67].

It is true but misleading to say that there are no specific string-handling statements in PL/I. Since a character string is a very legitimate data type, strings are handled in the same manner as most other data variables, with appropriate rules for conversion, etc. These rules will be discussed briefly here, but the reader is cautioned to realize that the strings are handled through the assignment or some other statement.

Both bit and character strings are permitted. Most of the relevant information has been provided earlier, but it is being summarized here for reference. Bit strings can be combined using **AND, OR, NOT**. Strings of any kind can be concatenated; if both are bit strings, then no conversion is done and the result is a bit string; but in all other cases the operands are converted to character strings. If the variable on the left of an assignment statement is arithmetic, then a string to be converted must be either an arithmetic constant with or without a sign or a real constant with a sign followed by either a plus or a minus sign and an imaginary constant. In such an instance, the arithmetic value of the constant is converted according to specified rules. If the variable on the left side of the equals sign is a bit or character string, the assignment is performed according to the rules of list-directed output which is described later. There are specific rules about truncating and filling out strings when needed.

There are a number of built-in string functions, and they are shown in Figure VIII-4.

There are no pattern-handling statements in PL/I. A primitive facility in this area is provided by the *INDEX* built-in function.

PL/I allows for two different kinds of data transmission, namely *stream-* and *record-*oriented. The verbs **GET** and **PUT** are used for input and output of data items in the stream, while the statements **READ** and **WRITE** do similar things for the record-oriented data. In the stream-oriented case, the data is considered to be a continuous stream of data items in character form, and an assignment must be made from the stream to the variables or vice versa. With record-oriented transmission, the data set is considered to consist of a collection of physically separate records, each of which consists of one or more data items in an encoded form; each record is transmitted as an entity directly without any conversion.

There are three types of stream-oriented transmission, namely list-

Arithmetic generic
    ABS
    MAX
    MIN
    MOD
    SIGN
    FIXED
    FLOAT
    FLOOR
    CEIL
    TRUNC
    BINARY
    DECIMAL
    PRECISION
    ADD
    MULTIPLY
    DIVIDE
    COMPLEX
    REAL
    IMAG
    CONJG

Float arithmetic generic
    EXP
    LOG
    LOG10†
    LOG2†
    ATAND†
    ATAN†
    TAND†
    TAN
    SIND†
    SIN
    COSD†
    COS
    TANH
    ERF†
    SQRT
    ERFC†
    COSH
    SINH
    ATANH
    ATAN
    ATAND

String generic
    BIT
    CHAR
    SUBSTR
    INDEX
    LENGTH
    HIGH
    LOW
    REPEAT
    UNSPEC
    BOOL

Generic functions for manipulation of arrays
    SUM
    PROD
    ALL
    ANY
    POLY
    LBOUND
    HBOUND
    DIM

Condition
    ONFILE
    ONLOC
    ONSOURCE
    ONCHAR
    ONKEY
    ONCODE
    DATAFIELD

List processing
    ADDR
    NULL

Others
    DATE
    TIME
    ALLOCATION
    LINENO (*filename*)
    COUNT (*filename*)
    ROUND (*expression, decimal-integer-constant*)
    STRING (*structure-name*)
    EVENT (*scalar-event-name*)
    PRIORITY (*scalar-task-name*)

†Defined only for real arguments.

Figure VIII-4. List of *Built-in* functions in PL/I.
Source: [IB66b], extracts from pp. 152–158. Reprinted by permission from *IBM System/360 Operating System: PL/I Language Specifications.* © 1966 by International Business Machines Corporation.

directed, data-directed, and edit-directed. In each case, the user usually supplies the file name and the list of variable names involved (which is called a *data list*). For edit-directed data, the format of each data item must be given. In several cases there are default conditions so the user need not always supply this information explicitly.

In the list-directed transmission, the data items in the stream are always written as arithmetic or string constants. The user provides in the GET or PUT statements a list of variables to which the data items are to be assigned or to be output from in sequence; the variables in the data list are separated either by commas or blanks when used as input, while on output the blanks are supplied automatically between items. The PUT statement also allows the user to write an expression in his list and the output is the value obtained by evaluating the expression. Some examples are as follows:

> GET LIST (A,B,C);
>> specifies the input transmission of values to be assigned to the variables A, B, and C from the default condition standard file.
>
> GET FILE (BETA) LIST (CITY, MONTH, MIN_TEM, MAX_TEM);
>> If the input file contains the following data:
>> 'NEW YORK', 'JANUARY', −6.5, 72.6 then the first two variables would be assigned as character strings and the second two would be assigned the numerical values; this of course assumes that the appropriate data declarations for the four variables have been given.
>
> PUT FILE (JUNK) LIST (NAME, 3.5*RATE, NUMBER − 10);
>> would cause the output of the value represented by NAME and the values resulting from evaluating the expressions 3.5*RATE and NUMBER − 10.

In the data-directed form of transmission, the data list need not appear in the GET statement because the stream is in the form of a series of assignment statements that specifies each variable name and the value assigned to it. Thus, the data in the input stream might look like the following:

> A = 7.3 B = 'ABC' C (4,2) = 1234;

Note that the last data item is followed by a semicolon which is used to delimit the number of items obtained by a single GET statement. On output, the data list must be written to specify which data items are to be written into the stream. The PUT statement referring to the data items just given could be

> PUT FILE (OUT) DATA (A,B,C(4,2));

On input, the assignments can be separated by commas or by blanks. On output, blanks are supplied and the semicolon is written after the last item specified in the data list.

In both the data-directed and list-directed cases, there are various rules about what types of data may appear in the data list, when and how subscripts can be used, and when subscripts are evaluated.

The edit-directed transmission allows the user to control the format by providing information about things such as precision, strings, conversion, etc. In addition, the user can control pagination and lines through the use of the *PUT* statement. The *LINE* option causes the data to be written on a new line; the *PAGE* option allows the user to start a new page; the expression in the *LINE* option essentially controls which line is used, i.e., which new line the data will start on. The *SKIP* option also allows the user to control the start of a new line and to indicate how many lines are to be skipped; however, it also permits the user to overprint a particular line.

It is possible to use the input/output statements for an internal character string; this is done by using the *STRING* option, which allows the user to obtain information from an internal character string and place data there. Although the string option can be used with any of the three types of stream-oriented transmission, it is usually most practical in association with a format list since individual items in the string need not be separated by commas or blanks.

Each *READ* and *WRITE* statement transmits a single logical record between the external medium and the variable specified, without any conversions. The variable specified must be a "level-1" item and normally contains several data items or arrays. If the file specified in a *READ* or *WRITE* statement is not open when the command is given, it is opened automatically. The variable associated with the words *INTO* and *FROM* in the *READ* and *WRITE*, respectively, specifies the variable in internal storage into which or from which the record is to be read or written. In the *READ* statement, the *SET* option places a record in a buffer and assigns a pointer variable as its identification so that a based variable can be subsequently referred to via the pointer value. The *IGNORE* option may be specified for **SEQUENTIAL INPUT** and **SEQUENTIAL UPDATE FILES**. It controls the number of records that are skipped. The *KEY* option must appear if the file is *DIRECT*; the expression is converted to a character string but it determines which record is read. The *KEYTO* option can be given only if the file is *SEQUENTIAL* and keyed; it specifies that the key of the record is to be copied onto the string variable. The *EVENT* option allows processing to continue while the record is being read or ignored. The *NOLOCK* option prevents the statement from causing a record on an *EXCLUSIVE* file from being locked against access by other tasks.

In the *WRITE* command, the *KEYFROM* option is converted to a character string and attached to the record as a key. As examples, the statement

READ FILE (BETA) KEY (VALUE) INTO (WORK); causes the record identified by the key VALUE to be transmitted from the data set associated with BETA into the variable WORK. The statement WRITE FILE (BETA) FROM (UPDATE) KEYFROM (UKEY); specifies that the record UPDATE is written as the next record in the data set associated with the file BETA, and the key identifying the record in the data set is taken from UKEY.

As noted later in the file description, a file can be INPUT, OUTPUT, or UPDATE. The REWRITE statement can be used for the UPDATE and serves the purpose of replacing an exiting record in the data set in the file involved. The other key words provide options similar to those in the WRITE command.

The OPEN statement causes the opening of a file and provides a number of additional file characteristics beyond those shown in the file description. The IDENT option associates the identifying user label on an input file with the variable given as the argument; for output, the argument is an expression which is evaluated and converted to a character string which is placed as a header label. If an input file is a BACKWARDS file, the label will be a trailer label; otherwise it will be a header label. The TITLE option causes the conversion of the specified expression to a character string which identifies the data set associated with the file; if this option does not appear, the file name is taken as the identifier. The other options are either self-explanatory or are discussed under the file description.

The CLOSE statement dissociates the named file from the data set with which it was associated by opening, and it also dissociates all the attributes declared for it in the original opening of the file. However, attributes for that file which are explicitly given in a DECLARE statement remain in effect. The argument in the IDENT option essentially serves as a trailer label.

The DELETE statement removes a record from a DIRECT UPDATE file. The expression associated with the KEY identifies the record to be deleted. The DELETE statement can cause implicit opening of a file.

The UNLOCK statement makes accessible a record which would otherwise be inaccessible as a result of the READ statement accessing it from an EXCLUSIVE file.

The LOCATE statement applies to BUFFERED OUTPUT files and allows a record to be created in buffer storage and later written out. The SET option specifies a POINTER variable which is to be set to identify the variable in the buffer.

The DISPLAY statement causes a message to be displayed to the machine operator; a response may be requested by using the REPLY option. If the EVENT option is used, execution of subsequent statements will continue before the reply is completed.

The library references which are specifically available to the user are the built-in functions, of which there are a large number; they are listed in

Figure VIII-4. Many of these are generic; this means that the same name can be used for differing types of arguments. For example, *EXP* is used for the exponential function, regardless of whether the argument is of *REAL* or *COMPLEX* mode, regardless of the precision, etc; the system automatically supplies the function of the *EXP* family that fits the requirements. Almost every built-in function, whether or not it is generic, has a specified number of arguments given. In some cases the actual number of arguments is optional beyond a specified minimum, while in others a maximum is specified. If a built-in function which is not generic is used, then any argument whose characteristics do not match the specified ones are converted to the appropriate form before the function is invoked; the return values are determined by the function. There are also many built-in functions that can return array or structure values. Although the names of the functions are fixed as far as the language is concerned, they can also be used for identifiers. However, in such a case, the name must be explicitly declared.

The functions can be divided into the following classifications: *Arithmetic generic* (e.g., *ABS*, *MAX*, and *TRUNC*); *float arithmetic*, which converts all input arguments to floating point before the function is invoked and produces floating point numbers, as results (e.g., *LOG2*, *SIND*, which are defined only for real arguments, and *SIN*, *LOG*, and *SQRT*, which are defined for real and complex arguments); *string generic* (e.g., *BIT*, *SUBSTR*, and *BOOL*); *array manipulation*, which has array expressions with scalar values as arguments but no arrays of structures (e.g., *SUM*, *POLY*, and *LBOUND*). All the built-in functions in the arithmetic and string generic categories may have *array* or *structure* expressions as arguments except where integer decimal constants are required, and they yield arrays or structures as results; *condition* (e.g., *ONFILE*, *ONCHAR*, *ONKEY*), *list processing* (*ADDR* and *NULL*), and the miscellaneous category (*DATE*, *TIME*, *LINENO*, *EVENT*, and *PRIORITY*).

The category of debugging statements in some cases overlaps with the error condition statements which were described earlier. There are currently none which can be truly classed as pure debugging statements, although some of the options under some of the commands have this effect, in particular, the *SNAP* in the *ON* statement.

Although there are only two actual commands dealing directly with storage allocation, namely *ALLOCATE* and *FREE*, there are actually four classes of storage which are controlled by the declarations. In order to understand the meaning of the statement, it is necessary to discuss the storage categories themselves here. The four storage classes are *static*, *automatic*, *controlled*, and *based*. (The latter was added after the issuance of [IB66d], and is described in later editions of [IB67d]. The formats of *ALLOCATE* and *FREE* shown in Figure VIII-2 do not reflect these changes which are minor syntactically.) Static storage is assigned before first entry to the program and remains in effect throughout the life of the program.

The other three classes define dynamic storage allocation. Automatic storage is assigned at object time upon entry to the block in which it is declared and released upon exit from that block. The dimensions can of course be variables or expressions.

Controlled and based storage are under programmer control. Variables declared as *CONTROLLED* and *BASED* can and must have storage assigned and released by the *ALLOCATE* and *FREE* statements, respectively. The *ALLOCATE* command essentially serves the purpose of placing something on top of a pushdown stack, while the *FREE* command performs the popup function. References to a stacked controlled variable always refer to the *most recent* allocation, whereas *all* current allocations for a based variable can be obtained by a pointer value. In Option 1 of the *ALLOCATE* statement, the *attribute* indicates a *BIT*, *CHARACTER*, or *INITIAL* attribute, where the first two can only appear with identifiers of that type. Since bounds or lengths can be specified in the *ALLOCATE* statement, they override any similar information which might be included in a *DECLARE* statement; if no bound or length is specified in the *ALLOCATE* statement, it must be specified in a *DECLARE* statement. When an identifier is allocated, the initial values will be assigned if the identifier has that attribute. To ascertain whether or not storage has been allocated for a particular identifier, the built-in function *ALLOCATION* may be used.

Both controlled and based variables can be specified in the same *ALLOCATE* and *FREE* statements. In Option 2 of the *ALLOCATE* statement, there is no pushdown list, and any generation of the based variable can be referenced through a pointer variable. The *SET* clause indicates the pointer variable that is to receive the pointer value identifying the particular value of the variable for which storage is to be allocated. If the *IN* clause appears in the *ALLOCATE* statement, storage will be allocated in the named area for the based variable; if that clause is omitted, space will be allocated in systems storage. For based variables, all characteristics must be specified in the declaration and cannot be included in the *ALLOCATE* statement. In the use of the *FREE* statement, if a specific pointer qualification is not given for the based variable, then the pointer declared with the based variable will be used. In Option 1, a *CONTROLLED* variable must be used.

There are no specific segmenting instructions in PL/I.

Aside from the normal input/output, the primary way in which PL/I has interaction with the operating system is through the creation and execution of tasks. As indicated under the *CALL* statement, the *TASK*, *EVENT*, and *PRIORITY* options specify that the called and calling procedures are to be executed asynchronously. The *task* is not a set of instructions but rather the execution of a set of instructions. There is always one major task and optionally available subtasks; each of the latter category can be named and the name can be used to refer to and set the priority of the task. A task can be suspended by the programmer until some particular point in the execution of another task has been reached; this specified point is known as an *event* and it can in fact be associated with the completion of a particular task. The WAIT statement causes the task in which it is executed to be suspended until the condition EVENT (*event-name*) equals '1'B is satisfied. The methods of defining this are described later. All the event names listed, or a number equal to the value of the optional expression (if used), must satisfy the condition in order for the task issuing the WAIT statement to be allowed to resume.

The DELAY statement causes execution of the controlling task to be suspended for *n* milliseconds, where *n* is the value of the expression shown in the format. Execution resumes after *n* milliseconds only if the controlling task is of sufficiently high priority to be selected in preference to all other ready tasks.

The STOP statement causes immediate termination of the major task and all subtasks. The EXIT statement terminates the task containing the EXIT statement and all tasks attached by this task; hence if the EXIT statement occurs in a major task, it is equivalent to a STOP statement.

The *Null* statement causes no action and has no effect on sequential operation. It can be used with an ELSE to obtain the desired pairing in an IF statement.

PL/I has a number of extremely significant (and some new) features connected with its data description. The first is the very large number of different data types which were discussed earlier. Because of that, the data declarations tend to be somewhat voluminous. However, two features make this easier for the programmer than might be expected on the surface. The first is the requirement that all characteristics of a data item (which is a term being used loosely here to indicate individual variables, arrays, structures, and all the different data types) must be shown together in a single DECLARE statement. (There are a few minor exceptions to this requirement.) This makes it clear to the reader of any program just what type of data is being dealt with at a particular point in the program. A second very useful feature is the ability to *factor* characteristics by writing a declaration only once if it applies to many variables. A third feature whose value is more controversial is the occurrence of *default* conditions; this means that in a number of specific instances if no characteristic is given for a data item, the compiler will automatically assume one; e.g., an identifier starting with one of the letters I though N is assumed to be FIXED BINARY REAL if no attributes are given. (PL/I uses the term *attributes* for *characteristics* of data items.) A fourth concept is the existence of *deduced* declarations, which means that certain characteristics are assumed when other information is given. For example, the use of a GET automatically assigns the attributes STREAM and INPUT to the file.

Related to this is the concept of *contextual* declarations, which means that identifiers appearing in certain specific contexts are recognized without an explicit declaration. For example, writing GET FILE (MASTER) DATA; declares MASTER to be a file without need for an explicit declaration. Finally, PL/I permits an identifier to be used in a block without any explicit or contextual declaration. In this case the identifier is said to be implicitly declared in the containing external procedure.

There are a number of attributes which can appear in a DECLARE statement, and they can be subdivided into major and minor categories. Those items marked with a dagger are described more fully below. The purpose of the others is shown briefly in the list.

Data description

data† (*arithmetic*, PICTURE, *string*, LABEL, TASK, EVENT).
INITIAL
> Specifies values to be assigned when storage is allocated, or names a procedure to be invoked to perform the initialization.

structure (LIKE)
> Data being declared has the same structure as the name following LIKE.

File description (Some of these are assumed when others are used or when the variable name is used in an appropriate statement).

FILE
> Identifies variable as a file.

file usage (STREAM, RECORD)
> Specifies type of data in file.

function (INPUT, OUTPUT, UPDATE)
> Specifies function of file; UPDATE means both input and output.

PRINT
> Specifies that data is eventually to be printed.

access (SEQUENTIAL, DIRECT)
> In SEQUENTIAL the next record is the next one physically available; in DIRECT a key must be specified.

buffering (BUFFERED, UNBUFFERED)
> Applies to SEQUENTIAL RECORD files only; specifies whether there is intermediate storage.

BACKWARDS
> A SEQUENTIAL INPUT file is to be accessed from back to front.

EXCLUSIVE
A *DIRECT UPDATE* file cannot be used by two tasks simultaneously.

ENVIRONMENT
Implementation defined in order to cover any missing aspects.

Format descriptions

PICTURE
Is used as part of data description.

Storage allocation

DIMENSION
Specifies upper and lower bounds for each dimension, separated by colons; missing lower bound is assumed to be 1; bounds can be expressions.

SECONDARY
Data does not require efficient storage.

storage class **(STATIC, AUTOMATIC, CONTROLLED, BASED)**
Concepts were defined earlier.

ALIGNED or **PACKED**
PACKED data is stored contiguous to the fields surrounding it; *ALIGNED* data may have each string data element start at a storage boundary defined by each implementer.

DEFINED
Data declared is to occupy the same storage area as that assigned to other data.

CELL
Alternative declarations share the same storage.

list processing **(AREA, POINTER, OFFSET)**
AREA defines storage for based data items; *POINTER* data identifies values in any storage class; *OFFSET* data identifies data relative to the start of an area. (This last item was added after the issuance of [IB66d].)

Procedure declarations

ENTRY
Declares entry names.

entry name† **(SETS, USES, GENERIC, BUILTIN, RETURNS, REDUCIBLE, IRREDUCIBLE)**.

scope **(INTERNAL, EXTERNAL)**
Specifies scope of data.

Miscellaneous

ABNORMAL or **NORMAL**
  ABNORMAL data may be accessed at an unpredictable time during execution.
scope (**INTERNAL, EXTERNAL**)
  Specifies scope of identifiers.

In considering the *data description*, the basic data attribute for a numerical variable is the *arithmetic*, which consists of base (**BINARY** or **DECIMAL**), scale (**FIXED** or **FLOAT**), mode (**REAL** or **COMPLEX**), and precision (specifies number of significant digits to be maintained and scale of the data). The *PICTURE* declaration shows the formats of numeric and character-string data fields and specifies editing. The *string* characteristic has the format

$$\left\{ \begin{matrix} \left\{ \begin{matrix} \text{BIT} \\ \text{CHARACTER} \end{matrix} \right\} \text{(length)} \quad [\text{VARYING}] \\ \text{PICTURE } \textit{specifications} \end{matrix} \right\}$$

LABEL defines the variable as a statement label. *TASK* specifies that the variable is used as a task name. *EVENT* specifies the use as an event name. These last two are really part of the interaction with the operating system.

ENTRY declares entry names referred to in a procedure. The attributes *SETS* and *USES* for entry name specify, respectively, that the invoked procedure reassigns, allocates, or frees that item, or that it is accessed but not reassigned unless it also has a *SETS* attribute. *GENERIC* defines the name as a family of entry names; the proper one is selected, based on the characteristics of the input arguments. *BUILTIN* specifies the reference is to a built-in function or pseudo-variable. *RETURNS* defines the attributes of the value to be returned by that entry. *IRREDUCIBLE* specifies that invocations of the specified entry may not be reduced to a smaller number of invocations.

The default conditions for arithmetic data are basically dependent on the first letter of the name; if it is **I** through **N**, **FIXED REAL BINARY** is assumed; otherwise, **FLOAT REAL DECIMAL**. If some but not all attributes are specified, then **DECIMAL**, **FLOAT**, and **REAL** are assumed for the missing ones. Other default conditions are **AUTOMATIC** (for storage unless scope is **EXTERNAL**, in which case **STATIC** is default), **PACKED** (for structures) and **ALIGNED** (for arrays not in structures), and **INTERNAL** (for scope).

In certain contexts, identifiers are implicitly defined, e.g., **FILE, EVENT**, entry name, and pointer.

Some illustrations of declarations and their meanings (taken from [IB66b]) are as follows:

DECLARE A FLOAT (3), B REAL (10) FLOAT, X FIXED (5,2);
: means A is (real) floating point with 3 significant digits; B is real, floating point with 10 significant digits, and X is fixed with 5 significant digits and 2 decimal places.

DECLARE (A(7), J BINARY (32)) FLOAT, C CHARACTER (5);
: means A and J are floating point but A is a decimal array with 7 elements and J is binary with 32 bits; C is a 5-character string. This illustrates the *factoring* of the attribute FLOAT.

DECLARE A BIT (10), B PICTURE 'XAA9AA', D BIT(*) VARYING CONTROLLED;
: means A is a field of 10 bits, B is a field consisting of any character followed by 2 alphabetic characters (or blanks) followed by a decimal digit (or blank) followed by 2 alphabetic characters (or blanks). D is a field of bits with a maximum length to be taken from a previous allocation or to be specified in a subsequent ALLOCATE statement.

DECLARE 1 A (10) PACKED, 2 B BIT (200), 2 C BIT (500), 2 D BIT(300), E (10, 15) ALIGNED BIT (15);
: means that the array of structures called A will occupy a continuous area of storage. Each element of the array E will start at a storage boundary defined by the implementer.

DECLARE MAX_VALUE INITIAL (99), MIN_VALUE INITIAL (_99), TABLE (20,20) INITIAL CALL INITIALIZE (X,Y);
: means that the variables MAX_VALUE and MIN_VALUE receive initial values of 99 and −99, respectively, and the array TABLE is initialized by a procedure called INITIALIZE.

DECLARE ((A FIXED, B FLOAT) STATIC, C CONTROLLED) EXTERNAL;
: is equivalent to DECLARE A FIXED STATIC EXTERNAL, B FLOAT STATIC EXTERNAL, C CONTROLLED EXTERNAL;.

DECLARE 1 A AUTOMATIC, 2 (B FIXED, C FLOAT, D CHAR (10));
: is an illustration of factoring of the level number and is equivalent to DECLARE 1 A AUTOMATIC, 2 B FIXED, 2 C FLOAT, 2 D CHAR (10);.

DECLARE A(0:99, −2:7) FIXED, B(M:A+1, N:Y+1);
: means that variable A is a two-dimensional array with fixed point, decimal elements whose first subscript ranges between 0 and 99 and whose second subscript goes between −2 and 7. B is a floating, real, decimal array with two subscripts whose initial and final values are evaluated when storage is allocated for it.

Procedure declarations are defined by the following format:

> entry-name **:** [entry-name **:**] ... PROCEDURE
> [(parameter [, parameter] ...)] [OPTIONS (option-list)]
> [RECURSIVE] [data-attributes];
> > body of procedure
>
> END;

The *parameters* are names that are associated with the entry point and of course are put into one-to-one correspondence with the arguments used with a CALL statement. The OPTIONS attribute is implementation dependent. The RECURSIVE attribute specifies that this procedure may be invoked recursively. It applies to all the entry points for the given procedure.

The data attributes that are permitted in the PROCEDURE declaration (which is actually defined as a statement in the PL/I manual) are the arithmetic, string, picture, and pointer attributes. They apply to the value returned by the procedure when it is invoked as a function. Default attributes are supplied if necessary. The value specified in the RETURN statement of the invoked procedure is converted to the specified data attributes.

There are no additional facilities that are specifically and only defined as compile-time directives in PL/I, except for the macro facility to be described later. However, there are a number of such language facets which have been discussed under other headings. In particular, a number of the declarations are really compiler directives.

There is no self-modification of programs at object time. However, the macro facility provides this facility at compile time.

PL/I has a macro facility which can be used either to make it easier to write some types of PL/I programs or alternatively to allow the user to write a program with new command names and then have the compiler automatically translate these into appropriate PL/I coding. There is no direct facility to extend the data types, although the system can be judiciously used (and tricked) to provide primitive capability in this direction (see, e.g., PL/I-FORMAC in Section VII.3.3).

The macro (i.e., compile-time) facility is defined at the language level to be handled by a preprocessor to the regular compiler. The preprocessor interprets compile-time statements and acts upon the source program accordingly. In most cases, the compile time-statements must be preceded by a percent sign. The facilities of PL/I which are available in the macro language are the following:

1. The assignment statement evaluates expressions. The expressions can contain decimal integer, bit string, or character string constants; compile-time variables; compile-time procedure references; or references to the SUBSTR built-in function. No exponentiation is permitted. For arithmetic operators only decimal integer arithmetic is performed.

2. The *GOTO* statement can of course only transfer control to a compile-time label.
3. The *IF* statement is available and has the following general form:

$$\% \;[label:] \ldots \text{IF } \textit{compile-time-expression}$$
$$\% \;\text{THEN } \textit{compile-time-group-1}$$
$$[\% \;\text{ELSE } \textit{compile-time-group-2}]$$

The compile-time group is either a single executable compile-statement, or a compile-time *DO* group.

4. The *DO* group has the following format:

$$\% \;[label:] \ldots \text{DO}\left[i=m1\begin{bmatrix}\text{TO } m2\;[\text{BY } m3]\\ \text{BY } m3\;[\text{TO } m2]\end{bmatrix}\right];$$
$$\% \;[label:] \ldots \text{END } [label];$$

where the *i* is a compile-time variable and the *mi* are compile-time expressions.

A compile-time procedure is an internal procedure that is executed only at the preprocessor stage. The only difference in its syntax is that the *PROCEDURE* and *END* statements must have a leading percent sign and the only types of parameters that are allowed are those which are either *CHARACTER* or *FIXED*. A compile-time procedure can contain the assignment, *GO TO*, *IF* and *RETURN* statements, and the *DO* group. In this situation, the individual statements are not preceded by the percent sign; it is used only at the very beginning and at the end. A compile-time *PROCEDURE* can also contain a null statement and a *DECLARE*.

In addition to the normal PL/I statements above, an additional statement called *INCLUDE* is permitted outside procedures. This incorporates a string from some external text into the program text which is being formed. Finally there are two other statements which are *executed* at the macro processor time, namely *ACTIVATE* and *DEACTIVATE*, which refer to compile-time variables and compile-time procedures. When one of those is activated, the specified identifier is replaced by its current (i.e., compile-time) value. When it is deactivated, this replacement no longer takes place.

Only the *CHARACTER* (with no length specification), *FIXED*, *ENTRY*, and *RETURNS* attributes are allowed in the compile-time *DECLARE* statement. Factoring is permitted.

As an example (from [IB66b]) of the use of the macro facility, suppose that the programmer wished to execute at object time the following loop in an expanded form for greater efficiency:

```
DO I = 1 TO 10;
Z(I) = X(I) + Y(I);
END;
```

This could be accomplished by writing the following:

```
% DECLARE I FIXED;
% I = 1;
% LAB:;
    Z(I) = X(I) + Y(I);
% I = I + 1;
% IF I <= 10 % THEN % GO TO LAB;
% DEACTIVATE I;
```

As a result of the macro program above, the following PL/I statements would be put into the program text and eventually compiled into the object program:

```
Z(1) = X(1) + Y(1);
Z(2) = X(2) + Y(2);
    ...
Z(10) = X(10) + Y(10);
```

While no PL/I compiler has yet been written in PL/I, it is believed that this is definitely possible. Some of the facilities, in particular the pointer and based variables, are provided very specifically to help with that problem. Furthermore, because of the many features in PL/I which provide interaction with the operating system, the use of PL/I itself might not introduce too many further inefficiencies in writing the compilers.

PL/I provides many more facilities for producing object-time efficiency than for improving compilation. Many of the features in the language have been chosen to aid the user at the expense of compilation time. (In my opinion, that is a desirable choice.) There are a number of features in PL/I which are there specifically to permit the compiler to generate optimized code, e.g., the *ABNORMAL, NORMAL, IRREDUCIBLE, REDUCIBLE,* and *USES* and *SETS* attributes. The *RECURSIVE* attribute avoids the need to make all object-time procedures recursive.

PL/I provides great flexibility to the user and, in particular, provides generality rather than restrictions. (One of the major design goals of PL/I was to provide this generality.) Among the specific features that have significant (but harmful) effect on compilation efficiency are the ability to use the key words as identifiers, the freedom to state many options in any order that is desired, and the ability to factor attributes. Some of the features that help to improve object-time efficiency have already been mentioned.

As noted earlier, the general problem of storage allocation has been given considerable attention at the language level in PL/I. A large amount of information is given to the compiler either directly or indirectly to assist it in doing reasonably good storage allocation. Of course, if the user writes inefficient programs, then he has only himself to blame if he obtains inefficient object code; thus, if he declares all variables to be *CONTROLLED,* he should certainly expect less efficient object code than if he deals more intelligently with that portion of the problem.

As noted earlier, there are a number of facilities in the language for both compile and object time debugging and error checking. There are a large number of *consistency rules*, i.e., requirements that items either must or must not appear together, and the compiler can and should certainly check for these. Since PL/I was designed to have an almost universal application area, it is not too surprising that it can be used for a wide variety of problems. There are some things, however, that are unavailable in the language as defined in [IB66b], e.g., graphics, simulation, formula manipulation, and complicated pattern matching. However, PL/I provides many powerful facilities in the areas heretofore covered by languages such as FORTRAN, ALGOL, COBOL, and JOVIAL.

VIII.4.4. Significant Contribution to Technology

PL/I has made a large number of significant contributions to the technology. In my opinion, it seems valid to say that PL/I is the culmination of the so-called standard procedural line exemplified by ALGOL, COBOL, FORTRAN, and JOVIAL. It has included virtually all the good features from each of those languages, although often in a different syntax; this is to be expected since it is not possible to extract both concept and syntax from four or five languages and expect them all to fit together. In addition to amalgamating the features from those languages, PL/I is the first language to address itself seriously to the problems arising from interacting with an operating system. It also has provided more facilities for dealing with storage allocation, tasking, and interrupt handling than any other language. Although the concept of a *default condition* for a variable has really been around for a long time, PL/I has carried this out to a very large extent. However, although the PL/I philosophy—that what the user does not know should not hinder or hamper him in writing a program—seems an interesting one, it is not completely obvious that it is really desirable. The user may in fact obtain very unexpected results because he did not even know that there was something in a particular area to be concerned about. Another significant contribution to the technology is the macro facility; it seems fair to assume that it will prove valuable and lead to further practical developments to match the numerous conceptual developments currently underway in this area.

In summary, PL/I seems to have made its major contribution to the technology by synthesizing in a very reasonable way almost all that has been known about languages that are useful for solving numerical scientific and business data processing problems, and combinations of them.

VIII.5. FORMULA ALGOL

The first version of the Formula ALGOL system was developed on the CDC G-20 at Carnegie Institute of Technology (now Carnegie-Mellon) in 1963 by Professor Alan Perlis and R. Iturriaga. The version described here was developed over the next few years by R. Iturriaga and T. Standish, with assistance from R. Krutar and J. Earley, under the direction of Professor Perlis. The initial development and motivation for this system was to add a formula-manipulating capability to ALGOL. (See Perlis and Iturriaga [PR64].) If that had been the sole extent of the work, then this language would appropriately belong in Chapter VII, where formal algebraic manipulation languages are discussed. However, over a period of several years, the original system evolved into a much more general language, containing not only the simple formula manipulation facilities but also string and list processing operations. It is because of the inclusion of these three somewhat different types of facilities within an ALGOL framework that the language appropriately belongs in a chapter on multipurpose languages.

Since the language was developed at a university as an experimental and educational tool (and on a fairly obsolete computer), it has been used only at Carnegie. Furthermore, because of the environment, there were numerous versions, and many of them underwent frequent change to better suit the needs of the designers and users. The version described here is reported in Perlis, Iturriaga, and Standish [PR66].

SAMPLE PROGRAM—Formula ALGOL†

Problem: Clear fractions in arithmetic expressions.

Program:
```
begin form F, X, A, B, C ; symbol S, P, T ;
    A ← A: any; B ← B: any; C ← C: any;
    P ← / [operator: +][comm: true]; T ← / [operator: ×][comm: true];
S ← [A ↑ (−B) → 1 / .A ↑ .B,
    A |P| (B / C) → (.A × .C + .B) / .C,
    A |T| (B / C) → (.A × .B) / .C,
    A − B / C → (.A × .C − .B) / .C,
    B / C − A → (.B − .A × .C) / .C,
    A / (B / C) → (.A × .C) / .B,
    (B / C) / A → .B / (.C / .A),
    (B / A) ↑ C → [.B ↑ .C / .A ↑ .C];
F ← (X + 3 / X) ↑ 2 / (X − 1 / X);
PRINT (F ↓ S) end
```

† Perlis, Iturriaga, and Standish [PR66], p. 44.

An interesting characteristic of Formula ALGOL relates to its implementation, much of which was done using various *compiler-compiler* techniques based on the work of Feldman [FJ66]; see Iturriaga *et al.* [IT66] and [IT66a]. An illustration of the use of the system is given by Iturriaga [IT67].

The power and complexity of the concepts and notation that are added to ALGOL make it impossible to describe them adequately in any reasonably short amount of space. Hence this description will attempt to concentrate more on the concepts that are being added than on any type of formal definition. Almost all the specific examples have been copied from the basic manual by Perlis, Iturriaga, and Standish [PR66].

A long list of additions to the basic ALGOL character set has been made (remember that any sequence of letters **in this** or *this* **type** is considered a single symbol; thus the word **element** is a single symbol and not a word composed of seven letters). The character set additions can be grouped in the following ways:

Related to formula manipulation:

form   eval   subs   replace   operator   comm   index

Related to list processing and retrieval:

symbol   nil   before   after   between   all after   all before   sublist   text   let   elements   attributes   parallel

Related to editing:

insert   delete   is   is not   is also

Miscellaneous:

of   the   st   th   nd   rd   first   last   all   atom   any

The most significant concepts introduced into Formula ALGOL relate to formal algebraic manipulation, symbols, and lists. Two new data types, **form** and **symbol**, corresponding to formulas and list structures, respectively, can be used to declare identifiers, arrays, or procedures. When any of these declarations is used for simple variables, not only is storage reserved but the value of each variable is initialized to the name of the variable; this is referred to as the *atomic name*. Furthermore, any identifier which is of type **form** or **symbol** can have a description list associated with it into which attributes and values can be entered and retrieved; variables of type **symbol** name a pushdown stack into which can be stored list structures and their degenerate cases, namely symbols and individual data terms.

Formulas are written in the normal way, except that they can also contain Boolean expressions. A *dot* (i.e., period) has several new meanings. When used with the name of an expression, it means the actual name and not the current value. A dot can also be used to prefix operators, identifiers, subscript lists, parameter lists, and the assignment arrow in an expression; when this is done, it causes a postponement of the indicated action. To institute the indicated action, the user must explicitly write the **eval** operator; this causes numeric evaluation of formulas and permits replacement by numbers, Boolean variables, formulas, etc. It also causes trivial simplifications for both numeric quantities and Boolean variables.

The assignment statement is indicated by a dot and a left arrow; e.g., F .← F + G which means actually to create the data structure representing the expression F + G as the (new) value of F. The trivial simplifications of **eval** primarily include appropriate handling and disposal of ones and zeros, such as $A + 0 \rightarrow A$ and $A * 0 \rightarrow 0$. In addition, certain transformations relating to negative signs are made, e.g., $A - (-n) \rightarrow A + n$, $(-n) - A \rightarrow -(n+A)$. The Boolean simplification carries out four commutative operations: $A \vee$ **true** $\rightarrow$ **true**; $A \wedge$ **false** $\rightarrow$ **false**. For example, executing the assignment statement

$$F \leftarrow G .\leftarrow A + B$$

where F is of type **form** and G, A, and B are type **real**, causes the construction of the formula $G \leftarrow K$, where $K =$ **value** $(A + B)$, and causes this formula to be assigned to F as a value. Execution of **eval** F causes **eval** (K) to be stored as the value of G, and the value of the expression **eval** F becomes **eval** (K).

There are two commands directly relating to formula manipulation, namely **subs** and **eval**; there are two pattern-matching commands which apply to formal expressions. A **replace** function is also available. The **subs** command is written

$$D \leftarrow \textbf{subs} \ (X_1, X_2, \ldots, X_{n1}) \ F(Y_1, Y_2, \ldots, Y_{n2})$$

in which X_i is replaced by Y_i (assuming a few other conditions are true). The **eval** command has a similar format and somewhat similar results, except that in addition to the action of the **subs** operation some simplification is done and one actually obtains the effect of **eval** F where F is an assignment statement, a procedure, or an array access. Writing **replace** (F) causes replacement of every atomic variable in F by its value and causes an **eval** of the result.

It should be clear to the readers of Chapter VII from the discussion above that the philosophy in Formula ALGOL is to provide to the user an absolute minimum in the way of built-in commands for formula manipulation. The advantage to this approach is that the user is not hampered by the existence of commands which are similar to what he wants and yet do not exactly suit his purpose. The best analogy—and one which is *not* meant to be derogatory—is that the facilities for manipulating formal algebraic expressions are similar in spirit to an assembly program. The user has the ability—and simultaneously the need—to write and hand-tailor the procedures which are provided automatically by many of the systems described in Chapter VII. In some cases this leads to great inefficiency for some very basic operations. This actually happened in the first several versions of Formula ALGOL because differentiation was not a built-in operation. It could be programmed very easily in the language but the result was quite inefficient. For that reason, it was eventually put into the language as a specific function.

It is possible for the user to specify transformation rules as productions, e.g.,

$$A: expr \times (B: expr + C: expr) \to .A \times .B + .A \times .C$$

which, using two given sequencing rules, can be applied to a given formula. If $F \leftarrow D \uparrow 2 \times (Y + sin(Z))$ is operated on by the production, then the subexpressions $D \uparrow 2$, Y, and $sin(Z)$ are extracted into the variables A, B, and C, respectively, and cause the replacement of the atomic names A, B, and C on the right-hand side of the production. This results in the transformation of the value of F into the formula $D \uparrow 2 \times Y + D \uparrow 2 \times sin(Z)$. It is clear that this type of facility is extremely powerful in terms of writing the types of operations that a user would want in doing formula manipulation.

A symbolic expression is a rule for computing either a single symbol or a list as a value. The list is a string of symbols separated by commas and contained within square brackets. The symbols can be any kind of expressions or patterns, including description lists. Sublists are permitted. A description list is of the form **/** *attribute* **:** [*list*]. Description lists are created and attached by assignment statements and can be assigned to variables of type **form**. Examples are as follows:

T ← /[properties: continuous, differentiable, integrable]
R ← /[color: blue, green, pink] [processed: **true**] [shape: round]

Specific values in description lists are retrieved by essentially writing the attribute, an atomic symbol, or the position in the list structure having the description list. More specifically, the user can write *attrib* (.*name*), *attrib* (*i***th of** *name*), or **the** *attrib* **of** .*name*. The first and third of these are essentially the same, except that in the third the attribute can be calculated by using any symbolic or formula expression. From the examples shown above, shape (.R) has the value round and 3rd of color (R) has the value pink.

There are a number of operations that can be performed on symbolic expressions. First, there are pushdown and popup operators indicated by ↓ and ↑, respectively. The actual contents of any variable of type **symbol** are considered a pushdown stack; by applying these operators as many times as needed, the desired information can be manipulated. It is also possible to insert, delete, and alter symbolic expressions, and the user can also add or delete values on the description lists.

Among the most interesting features in Formula ALGOL are the various ways of handling patterns. It is possible to determine in both formulas and lists whether one expression is an exact instance of another or whether it contains an exact instance of another; furthermore, it is possible to name these various subpatterns. More specifically, writing F = = P is a Boolean expression whose value is determined by whether or not F is an exact instance of P. The notation F>>P is a Boolean expression whose value is determined by whether F contains an instance of P; more specifically, this means that F>>P is *true* if F contains a subexpression W (which might equal F) such that W==P is *true*. To illustrate these, consider an expression and three cases (this is technically illegal because the Boolean and arithmetic expressions cannot be combined, but it is a useful illustration nevertheless):

P = **integer** + **form** × **Boolean**
Case 1:   F = 3+ (A − B) × (C < D)
Case 2:   F = 3.2 + (A − B) × (C < D)
Case 3:   F = Y + 3 + (A − B) × (C < D)

For *Case 1*, statement F==P is true; for *Case 2*, it is false. For *Case 3*, the statement F==P is false but F>>P is true.

A more complex illustration of the use of the search for a subpattern and how it can be used is illustrated by considering the execution of the statements

F ← 2 × (sin (X↑2 + Y↑2) + cos(X↑2 − Y↑2)) / 5;
G ← sin (**form**) + cos(**form**);

where all variables used are of type **form**. Then A : E>>T : G is a pattern with value **true**. The value of T will replace the first instance of G in F, i.e., the expression sin(X↑2+Y↑2) + cos(X↑2−Y↑2) (this being the first subexpression matching the pattern G). A is assigned the expression 2 × T / 5. Thus A is the same as F with the first subexpression of F matching G replaced by the value of T.

It is also possible to define operator classes and use them in a formula pattern. A definition of an operator class is accomplished by assigning to a variable of type **symbol** a description list of the form

/ [operator: operator-list] operator-attribute-list

It is then possible to write

R ← / [**operator**: +, −, /] [**comm**: **true, false, false**] [**index**: J]

where J is a variable declared of type *integer* and the words **operator**, **comm**, and **index** are reserved for special attributes. If P is a formula pattern structure such as A |R| B, then F==P is true if and only if F is of the form C *operator* D and one of the two following conditions holds: (1) C == A, D == B, and *operator* is a member of the operator value list on the description list of R which is [+, −, /]; (2) C==B, D==A, and *operator* is on the operator value list and has the value **true** following the attribute **comm** (i.e., in this case it must be +). If F==P is true, then the variable name which is the attribute of **index** will be set to an integer denoting the position of the operator in the value list.

List patterns are permitted to test whether a list is an instance of a certain linear pattern. The notation used in COMIT (see Section VI.6) is carried over here, where the symbols $ and $n represent an arbitrary number of, and n arbitrary, consecutive elements, respectively. For example, suppose that the statement S ← [A, B, C, D] has been executed, where all variables involved have been declared of type *symbol* and where the values of A, B, C, and D are their respective names. Consider the statement

if S == [$1, B, $] **then** T ← [T, B] **else** T ← [T, **last of** S];

Since the contents of S, which is the list [A, B, C, D], are an instance of the pattern [$1, B, $], the list pattern S == [$1, B, $] is **true**. Therefore, T ← [T, B] is executed, which has the effect of appending B to the end of the list stored as the value of T. If, however, the first part of the statement were written

if S == [$1, C, $]

then it would be false and D would be appended to the list stored as the value of T.

List-editing statements include *insert*, *delete*, and *alter*, with placement and selector locators of **before** *position* **of**, **after** *position* **of**, and **the** *symbolic-expression* **of** *symbolic-expression* **is** *expression* (where **is** can be replaced by **is not** or **is also**). As an illustration, suppose S ← [X, A, A, X] has been executed. Then the statement

insert [[Y, Z]] (**after** 1 **st of**, **before last of**) S;

changes the value of S to look like [X, [Y, Z], A, A, [Y, Z], X]. If the user wrote S ← / [class: freshman, sophomore], then the statement **the class of** S **is also** junior would append the value junior to the value list following the attribute class, while **the class of** S **is not** sophomore; would change the original description list to be of the form / [class: freshman].

It is possible to define classes and test for membership in a class.

Additions have been made to the **for** statement, both in terms of the variables which can be listed and the way in which they are processed. In the first instance, the user can write **elements of** or **attributes of** symbolic expressions or just show a symbolic expression. The purpose of this is to permit the user to assign to the control variable in a **for** statement the elements of a list or the attributes of a description list, one by one. This can be done either in series or in parallel.

There are about 10 special built-in functions available to the user; they include DERV(F, X), which takes the derivative of a formula F with respect to the variable X, and REPLACE, which was discussed earlier. Most of the others are specialized; they relate to modifying counters, erasing lists, or testing Boolean expressions.

It is difficult to evaluate Formula ALGOL because its use has been limited to one place and it continues to be changed as the need arises. It is certainly the first language to include the three components of algebraic formula manipulation (although to a very limited extent), list processing, and string manipulation and pattern matching. Many of these facilities have been included in a rather primitive way; the net overall effect is powerful (although awkward).

VIII.6. LISP 2

LISP 2 is too new to have much history except that which relates to its creation; even its future is uncertain at the time of writing. Since the motivation for LISP 2, as well as many of its technical developments, comes from the earlier versions of LISP (primarily LISP 1.5), it is assumed that the reader is familiar with LISP 1.5. (See Section VI.5.) Since LISP 2 is also based heavily on ALGOL, familiarity with that language is also assumed. (See Section IV.4.)

It was clear for a long time even to the staunchest advocates of LISP 1.5 that it had certain significant disadvantages. For one thing, the arithmetic was very slow, making it impractical to use LISP for any problem involving significant numerical computations. Secondly, the interpretation (which was for a long time the major method of running LISP programs) was also slow; that problem has been improved by the development of LISP compilers. The biggest disadvantage of LISP was its notation, which was bad from several points of view. One was the tremendous dependence upon parentheses with the resultant difficulty in reading and writing LISP programs. Another problem was the requirement for writing all mathematical expressions in Polish notation and receiving them in that form as output. Thus it became increasingly clear that while the fundamental concepts of LISP were extremely valuable, the general framework within which they existed was much more awkward to use than it really needed to be. For these reasons, starting late in 1963, several people at M.I.T. began to concern themselves with potential remedies for some of these defects. One of the earliest documents relative to this situation was the memorandum by Levin [LE64]. This paper described additions and changes to be made in ALGOL 60 to incorporate the LISP concepts. Work has subsequently been done jointly by the Systems Development Corporation and Information International Inc. of Cambridge, Massachusetts, with funds from ARPA (Advanced Research Projects Agency). The primary people participating were, from I.I.I., P. W. Abrahams, L. Hawkinson, M. I. Levin, and R. A. Saunders; from SDC, J. A. Barnett, E. Book, D. Firth, S. L. Kameny, and C. Weissman. Support and contributions were given by Professors Marvin Minsky of M.I.T. and John McCarthy of Stanford and their associates; significant assistance was given by Dr. Daniel Bobrow (Bolt, Beranek, and Newman).

The only published paper is by Abrahams *et al.* [AH66a], although numerous internal SDC documents have been accessible. Quoting from [AH66a] "Typical application areas for LISP 2 include heuristic programming, algebraic manipulation, linguistic analysis and machine translation of natural and artificial languages, analysis of particle reactions in high-energy physics, artificial intelligence, pattern recognition, mathematical logic and automata theory, automatic theorem proving, gameplaying, information retrieval, numerical computation, and exploration of new programming technology."[10] This list certainly makes LISP 2 the most ambitious language to date, at least in terms of application areas that are claimed to be covered. The careful reader will note that most of these areas are those in which earlier versions of LISP proved useful. It seems to me highly unlikely that someone wishing to do straight numerical computation, or for that matter linguistic analysis, would deliberately choose to use LISP 2. About the only area for which LISP 2 does not claim to be useful is business data processing.

Since LISP 2 is a frequently changing item, any description must be based on a particular version. The primary sources for the following material are Abrahams *et al.* [AH66a], Book [BO66], Firth and Kameny [FI66], and Kameny [KA66].

LISP 2 is designed for a professional programmer. The general LISP 2

---

[10] Abrahams *et al.* [AH66a], p. 661.

programming system description calls for not only a compiler but also object-time facilities, including capabilities for on-line interaction and communication with the computer monitor system, whatever it may be. LISP has been implemented on the AN/FSQ-32V with intentions for bootstrapping it onto other machines. Since this has not yet been accomplished,

SAMPLE PROGRAM—LISP 2†

```
SYMBOL SECTION EXAMPLES, LISP;
% LCS FIND S THE LONGEST COMMON SEGMENT
% OF TWO LISTS L1 AND L2
FUNCTION LCS(L1,L2); SYMBOL L1, L2;
  BEGIN SYMBOL X, Y, BEST ← NIL; INTEGER
      K←0, N, LX←LENGTH(L1);
    FOR X ON L1 WHILE LX > K DO
    BEGIN INTEGER LY ← LENGTH (L2);
      FOR Y ON L2 WHILE LY > K DO
      BEGIN N ← COMSEGL (X,Y);
        IF N <=K THEN GO A;
        K ← N;
        BEST ← COMSEG (X,Y);
      A: LY ← LY − 1
      END;
      LX ← LX − 1
    END;
    RETURN BEST;
  END;
% COMSEGL FINDS THE LENGTH OF THE
% LONGEST INITIAL COMMON SEGMENT
% OF
% TWO LISTS X AND Y.
  INTEGER FUNCTION COMSEGL (X,Y);
    IF NULL X OR NULL Y OR CAR X /=CAR Y
      THEN 0 ELSE COMSEGL (CDR X, CDR Y) + 1;
% COMSEG FINDS THE LONGEST INITIAL
% COMMON SEGMENT OF TWO LISTS X AND Y
  SYMBOL FUNCTION COMSEG (X,Y);
    IF NULL X OR NULL Y OR CAR X /= CAR Y
      THEN NIL ELSE CAR X. COMSEG(CDR X, CDR Y);
% LENGTH COMPUTES THE LENGTH OF L
  INTEGER FUNCTION LENGTH (L); SYMBOL L;
    BEGIN INTEGER K ← 0; SYMBOL L1;
      FOR L1 IN L DO K ← K+1;
      RETURN K;
    END;
```

† Abrahams *et al.* [AH66a], pp. 670–671.

592 MULTIPURPOSE LANGUAGES

and is not likely to be since work is at a virtual standstill due to funding problems, it is impossible to determine anything about machine and/or compiler independence.

Although strongly based on ALGOL, the language is not a true extension of ALGOL. Those familiar with ALGOL should have no difficulty in learning the deviations, but they must learn the LISP concepts themselves. At one public meeting, I asked one of the key LISP 2 designers why the name LISP 2 was given to the language when it really looked much more like ALGOL than LISP 1.5. The answer was that if the language was called LISP 2, then people would intuitively expect certain things from it; whereas if it were called ALGOL-LISP or something like that, people would expect something quite different. This seems to be a very valid answer. An interesting discussion of the overall concept of LISP is given by Weizenbaum [WZ67] in his review of the article by Abrahams *et al.* [AH66a].

The language definition was written by various people among those listed earlier. To a large extent, the same people were involved in the implementation, although there have been numerous reports of disagreement on techniques and objectives within the project itself. Since part of the implementation is being done by use of a syntax translator, namely META 2 (see Schorre [QT64]), the syntax has been formally defined in the second, third, and fourth items in the reference list at the end of this chapter.

The character set in LISP 2 consists of the 26 capital letters, the 10 digits, and the following symbols:

```
+   -   *   /   )   (   ]   [   '   #   .   ,   :   ;
=   >   <   ↑   ←   \   %   $       blank
```

Identifiers consist of a letter followed by any number of letters, numerals, and periods. Both data names and program unit labels are formed this way. Data names can have subscripts, which are shown in parentheses and separated by commas. Any expression (including conditional statements) which evaluates to an integer can be used as a subscript. Any number of subscripts are permitted, and they can be subscripted.

Since there is no data hierarchy, there is no qualification of that kind. However, LISP 2 introduces the concept of a section (similar to an assembly language section); names can be qualified with respect to the section name, using the $ to indicate that JOHN$DOE refers to the variable JOHN in section DOE.

A significant difference between LISP 2 and ALGOL is that the single characters in ALGOL such as **goto**, **if**, and **real** are not single characters in LISP 2; they are strings of characters and are analyzed as such by the compiler. They are considered reserved words, and data names must not conflict with them.

The operators include the five basic arithmetic ones and the relational operators =, <, >, /=, <=, and >=. There are logical operators AND, OR, and NOT; also there are operators carried over from LISP 1.5, namely CAR, CDR, ATOM, NULL, and CONS. Rules on punctuation, blanks, and noise words are the same as in ALGOL, except that blanks cannot be used within identifiers. Literals are defined through the use of the number sign, #, at the beginning and the end. The character ' within the string causes the character following it to be entered in the string; thus writing #A'#256## creates a string of six characters, namely A#256#.

The input form is similar in principle to that of ALGOL.

The program structure is similar to that of ALGOL, except for two significant differences. First, any expression other than a variable can be used as a statement; when an expression appears in a place where a statement is expected, the expression is evaluated but the value is discarded. Secondly, LISP 2 has functions but no procedures.

A block heading is in the form $D1\ D2\ D3\ S1, S2, \ldots, Sn$, where the Di represent declarations described later. Each of the Si is either the name of a variable or an assignment statement which gives an initial value for the variable; if no initial value is given, a default value, depending on the type, is used. A block declaration causes all the specified variables to be internal parameters of the block and to have the properties specified by the Di.

The delimiting of statements by semicolons, and of groups of statements by BEGIN ... END, is the same as in ALGOL. Since recursion is even more crucial to the LISP concepts than to ALGOL, it is of course permitted for functions. Parameter passage is of the call by value and call by location type; there is no call by name, although the use of a functional argument which remembers the context of its definition point together with a call by location achieve most of the call by name capability.

The most general form of LISP 2 data is an *S-expression*. *S*-expressions are built up from atoms which may be numbers, strings, identifiers, Booleans, functionals, and arrays. As in LISP 1.5, the class of *S*-expressions is defined recursively as follows: Every atom is an *S*-expression and if $e1$ and $e2$ are *S*-expressions, then ($e1$. $e2$) is an *S*-expression. Constant arrays are written by enclosing the elements in square brackets, e.g.,

[INTEGER 2 5 −1 4]

[SYMBOL [A B C] [A1 B1 C1]]

There are arithmetic variables, which can be integer, real, or octal, and Boolean variables. The value FALSE, the atom NIL, and the empty list ( ) are synonymous. The FUNCTIONAL data type is a LISP 2 function, and *type ARRAY* denotes an array whose elements are of the specified type. There is a *SYMBOL* type which includes all other classes of data. Except for *SYMBOL*, all the data classes include atomic data only. A symbolic constant is denoted by preceding an *S*-expression with a prime, e.g., 'ALPHA or '(L1L2).

The operators **CAR** and **CDR** have the same meaning as in LISP 1.5; e.g., CDR (A B C D) yields the list (B C D).

Arithmetic is performed in the normal ALGOL manner; the reverse slash, \, is used for computing integer remainder, and integer division is designated by a minus sign followed by a colon, −:. In the creation of expressions, the prefix operators *ATOM* and *NULL* have lower precedence than the relational operators but higher precedence than the Boolean operators. The infix operator for *CONS*, which is ⌀ . ⌀ has the lowest precedence of all.

There are three levels of binding which control the scope of data relative to the program structure.

Assignment statements are designated by a left-pointing arrow and are similar to ALGOL, except that there are additional operators such as *CAR* and *CDR*. The addition of these operators requires the user to be quite careful about precedence. Thus, the assignment statement A ← CAR C ← D.E means the same as A ← (CAR(C ← (D.E))).

There are no alphanumeric data-handling statements.

The unconditional control transfer is indicated by the single word **GO** followed by the label.

A *RETURN* statement may be used inside a block and its effect is to determine the value of a block expression. However, execution of the block is terminated either by executing a *RETURN* statement or by executing the last statement of the block without a transfer of control.

A conditional statement is of the form IF *E1* THEN *S1* [ELSE *S2*] where *E1* is an expression and *S1* is any kind of statement except a conditional statement or a *FOR* statement; *S2* is any kind of statement. By enclosing a conditional or *FOR* statement with *BEGIN* and *END* or with parentheses, it can be used in place of the *S1*, e.g.,

IF A < B THEN GO M ELSE IF A > B GO N ELSE IF B < 0 GO L;

IF A THEN BEGIN IF B THEN X ← 1 ELSE X ← 2 END ELSE GO L;

As usual, the expressions are evaluated from left to right until one is found whose value is other than *FALSE*; then the corresponding *THEN* statement is executed.

The *FOR* statement has the same format as in ALGOL, but it has some additional facilities. The user can write FOR *V* IN *Y* DO *S*, which causes the statement *S* to be executed for each element of the list *Y*, with *V* assuming the successive elements as its value in each execution of *S*. If *ON* is used instead of *IN*, *V* first assumes as values the entire list *Y*, then its successive terminal segments, CDR Y, CDDR Y, etc., until the list Y is exhausted. There are also WHILE and UNLESS expression clauses.

There is a TRY statement which causes control to be returned to itself if an error condition is detected during the execution of a statement within the TRY statement.

The CASE statement has the form CASE(S, E_1, E_2, ..., E_n), where S is an integer-valued expression known as the *selector*; if the value of S is between 1 and n, then the expression E_S is evaluated and is the resulting value of the CASE expression; if $S < 1$ or $S > n$, the value is E_n.

Although there are no direct list-handling statements in the LISP 2 source language, there are a number of list-handling functions, including CAR and CDR. There has been some consideration about including string- and/or pattern-handling statements but this has not been done yet.

The LISP input/output facilities are fairly general; every input or output operation references implicitly or explicitly a specific file. Although the user must associate a file with a particular device, the actual commands and operations available to the user are fairly device independent. The largest restrictions are that only one file can be active for input and one for output simultaneously; furthermore, in order to reduce buffering requirements, only one record for a given file can be in main memory at a time. (Since the main objective of LISP 2 is to handle problems which need list processing mechanisms and since such problems almost always have a shortage of storage, this restriction is not unreasonable.) It is possible to select a single file for both input and output simultaneously.

A file can be either inactive or available; in the latter case, it can be either selected or deselected and only one input and one output file can be selected at a given time. The commands for making an inactive file available or an available file inactive are, respectively, OPEN (*name, descr*) and SHUT (*name, descr*), where *name* and *descr* obviously represent the name and description of the file. The OPEN command establishes all the necessary communication linkages between LISP 2 and the (time-sharing) monitor or operating system. A variable FILES is used to maintain a list of all available file names and their descriptions. To select a file, the user writes INPUT (*name*) or OUTPUT (*name*). By using one of these commands in an assignment statement, the name of the previously selected file can be saved for subsequent reselection, e.g., ALPHA ← OUTPUT(ANSWER), and thus the user can later write OUTPUT(ALPHA) to reselect the file named ANSWER.

The user has reading and printing facilities which always operate on the currently selected input or output file. (The facilities are actually functions rather than statements.) The user either can use default conditions or program things such as page control and record control through the following functions:

READCH
: Reads a character from the currently selected input record; increments counters and positions for the next line or record if necessary.

PRINCH(*name*)
: The *name* is any expression evaluating to a one-character identifier that is entered in the line at the current column; counts and controls are automatically updated.

PRINATOM (*expression*)
: The expression must evaluate to an atom and it is entered into the output line starting at the current column.

PRINSTRING (*expression*)
: The expression must evaluate to a string which is taken literally as its print name.

PRIN (*expression*)
: The expression is printed as an S-expression.

READ
: Its value is the next S-expression in the file.

POSITION (*expr1, expr2*)
: The expressions must evaluate to the name of an opened file and an appropriate positioning action, respectively.

There are other functions for handling S-expressions, for dealing with binary files, for printing unusually formed strings, etc. There are also terminating functions which are affected as a result of page control variables and relative positioning of input records.

One of the more interesting or unusual facilities available through the LISP 2 input/output system is that the control variables can be changed or examined. These are fixed variable names which are qualified to the input/output section. For example, it is possible to define a new function that would advance the current column of the selected output file to some desired legal column and return the original column as output. There is a rather long list of these variables and what can be done with them, but naturally caution must be used.

The major emphasis in the development of LISP 2 has been at the level of the internal operation and facilities. This is in contrast to most languages, in which the language is first defined and then implemented. For a long time it was facetiously said of LISP 2 that it was "an implementation in search of a language". For these reasons, it is not surprising that there is significant interaction with the operating system and/or environment, although it is not really that directly available to the normal user. In particular, the storage allocation which is crucial in any list-handling system is done by garbage collection which is invoked automatically and can also be invoked by the user.

There is a CODE command which allows the user to write LAP assembly language.

With the exception of the input/output functions, there are no direct facilities for communicating with an operating system or a time-sharing system. On the other hand, nothing in LISP 2 makes it unusable within a time-sharing environment; in fact, it was designed to be so used.

There are three types of data declarations in LISP 2: *Type, storage mode,* and *transmission mode*. The types are *INTEGER, REAL, BOOLEAN, SYMBOL, OCTAL, FUNCTIONAL,* and *type ARRAY*. The data type *SYMBOL* includes all other classes of data and, except for that type, all the others include atomic data only. There are three types of storage mode, with one of them a default condition. The two types of transmission mode are LOC(ation) and by value, where the latter is the default condition. There is also a section type which does not change until the programmer changes it by means of another declaration; initially the section type is SYMBOL and if no type declaration is given, then the default condition is the section type. The section type determines the default data type for the section. It is also possible to have free declarations that are not made within functions or blocks but rather are made on the top level of LISP; free variables essentially have universal scope.

The file description consists of a number of specific, free-formatted entries which are actually on a file description list. The first is the *UNIT*, which specifies the device type, e.g., (UNIT . DISK) or (UNIT TAPE REELS). A second entry is the *FORM*, which can be either ASCII, BCD, or BINARY. There is also the potential of setting a connection flag by writing NEW or OLD so that the time-sharing system knows about the file. The NEW is the default condition. File security can be obtained by using the dotted pair (PROTECT . Z) to convey keys; the nature of the parameter Z depends on the protection scheme provided by the monitor.

The last characteristic of a file is its format, specifically its blocked and printed structure for input and output files, respectively. The parameter RECORD specifies the number of lines to be blocked in each record and is considered permanent for the life of the file. Page format is controlled by declarations *HORIZONTAL, VERTICAL,* and *OVERFLOW*. *HORIZONTAL* has a left margin, a right margin, and a maximum column parameter. *VERTICAL* has an upper-line boundary, a lower-line boundary, and a page boundary. All these declarations have default conditions.

There is a set of reserved variable names whose values are file descriptions for various input/output devices; although these are fixed, they can be modified. A few examples of these are

DISC. ((UNIT . DISC) (FROM . BCD) (RECORD . 50)
(HORIZONTAL . (1 73 80)) (VERTICAL . (1 51 50)))

CRT. ((UNIT . CRT) (FORM . BINARY) (RECORD . 680))

A key factor in the LISP 2 development is the creation and clear usage of two distinct language levels—source language (*SL*) and intermediate language (*IL*). The preceding discussion has been entirely about the source language, which obviously strongly resembles ALGOL 60. On the other hand, the syntax of *IL* is almost identical to LISP 1.5. *IL* is designed to retain the characteristic of allowing the program to have the same structure as data, so it can be manipulated by both the user and system programs. Thus LISP 2 loses at the user level the facility of having programs look like data, but it retains it at the intermediate language level. There is a macro expansion capability at the *IL* level but not at the source language level.

There are some machine dependent operations which are useful compiler directives.

LISP 2 can be used to bootstrap itself onto a new machine, and in fact the first LISP 2 system was written in *IL*. In any case, since earlier versions of LISP have been written in themselves, this facility can certainly be continued at least at the intermediate language level and, presumably, also at the source language level.

The most significant effect of the language on the compiler is the mapping into the intermediate language, because a person who merely looked at the source language and had no background whatsoever of earlier work on LISP would not necessarily tend to develop an internal system the way the designers did. One thing that has been done to improve efficiency is to store numbers directly as single words rather than to pack them. Because of the tradition of garbage collection in LISP, it was continued in LISP 2. (See the discussion of this issue in Weizenbaum [WZ67].)

Since LISP 2 has not received any real usage, no comment along those lines can be made. It appears that little—if any—further work will be done on it, but of course this is subject to immediate change.

REFERENCES

VIII.1. SCOPE OF CHAPTER

[IC62a] "General Panel Discussion: Is a Unification ALGOL-COBOL, ALGOL-FORTRAN Possible? The Question of One or Several Languages", *Symbolic Languages in Data Processing*, Gordon and Breach, New York, 1962, pp. 833–49.

[SM61] Sammet, J. E., "A Method of Combining ALGOL and COBOL", *Proc. WJCC*, Vol. 19 (1961), pp. 379–87.

VIII.3. JOVIAL

[AA67] "JOVIAL Usage and Compilers (as of 8/67)", *Appendix 1, Minutes of USASI X3.4 meeting* (Sept. 29, 1967).

REFERENCES

[AF67] *Standard Computer Programming Language for Air Force Command and Control System* (CED 2400), Air Force Manual AFM 100-24 (June, 1967).

[CE67] Clark, E. R., "On the Automatic Simplification of Source-Language Programs", *Comm. ACM*, Vol. 10, No. 3 (Mar., 1967), pp. 160–65.

[CO61] Coffman, E. G., Jr., *A Brief Description and Comparison of ALGOL and JOVIAL*, System Development Corp., FN-5618, Santa Monica, Calif. (June, 1961).

[KE62] Kennedy, P. R., *A Simplified Approach to JOVIAL* (A Training Document), System Development Corp., TM-780/000/00, Santa Monica, Calif. (Sept., 1962).

[KE65] Kennedy, P. R., *The TINT Users' Guide*, System Development Corp., TM-1933/000/02, Santa Monica, Calif. (Mar., 1965).

[KK65] Klein, S., "Automatic Paraphrasing in Essay Format", *Mechanical Translation*, Vol. 8, Nos. 3 and 4 (June, Oct., 1965), pp. 68–83.

[MD64] Marsh, D. G., "JOVIAL in Class", *Annual Review in Automatic Programming*, Vol. 4 (R. Goodman, ed.). Macmillan, New York, 1964, pp. 167–81.

[MI62] *FAST—FORTRAN Automatic Symbol Translator* (reference manual), MITRE Corp., SR-24, Bedford, Mass. (Jan., 1962).

[PE66] Perstein, M. H., *Grammar and Lexicon for Basic JOVIAL*, System Development Corp., TM-555/005/00, Santa Monica, Calif. (May, 1966).

[PE66a] Perstein, M. H., *The JOVIAL (J3) Grammar and Lexicon*, System Development Corp., TM-555/002/04, Santa Monica, Calif. (May, 1966).

[SC60] Schwartz, J. I., Petersen, K. E., and Olson, W. J., *JOVIAL and its Interpreter, A Higher Level Programming Language and an Interpretive Technique for Checkout*, System Development Corp., SP-165, Santa Monica, Calif. (Apr., 1960).

[SH00] Shaw, C. J., *Programming Languages and JOVIAL*, System Development Corp., BR-3/11-60, Santa Monica, Calif.

[SH61] Shaw, C. J., "System Development Corporation's Procedure-Oriented JOVIAL", *Datamation*, Vol. 7, No. 6 (June, 1961), pp. 28–32.

[SH61a] Shaw, C. J., *The JOVIAL Manual*, pt. 3, *The JOVIAL Primer*, System Development Corp., TM-555/003/00, Santa Monica, Calif. (Dec., 1961).

[SH63] Shaw, C. J., "A Specification of JOVIAL", *Comm. ACM*, Vol. 6, No. 12 (Dec., 1963), pp. 721–36.

[SH63a] Shaw, C. J., "JOVIAL—A Programming Language for Real-time Command Systems", *Annual Review in Automatic Programming*, Vol. 3 (R. Goodman, ed.). Pergamon Press, New York, 1963, pp. 53–119.

[SH63b] Shaw, C. J., "JOVIAL and Its Documentation", *Comm. ACM*, Vol. 6, No. 3 (Mar., 1963), pp. 89–91.

[SH64] Shaw, C. J., *A Comparative Evaluation of JOVIAL and FORTRAN IV*, System Development Corp., N-21169, Santa Monica, Calif. (Jan., 1964).

[SN65] Sandin, N. A. and Foote, E. B., *JTS User's Manual*, System Development Corp., TM-1577/000/01, Santa Monica, Calif. (Apr., 1965).

[ST66] Steel, T. B., Jr., *Some Observations on the Relationship Between JOVIAL and PL/I*, System Development Corp., TM-2930/000/01, Santa Monica, Calif. (May, 1966).

VIII.4. PL/I

[AL67] Alber, K., *Syntactical Description of PL/I Text and its Translation into Abstract Normal Form*, IBM Corp., TR 25.074, Vienna Lab., Vienna, Austria (Apr., 1967).

[AN66] Allen, C. D. et al., *An Abstract Interpreter of PL/I*, IBM Corp., TN 3004, Hursley, England (Nov., 1966).

[BA67] Bandat, K., *On The Formal Definition of PL/I*, IBM Corp., TR 25.073, Vienna Lab., Vienna, Austria (Mar., 1967).

[BC66] Beech, D. et al., *Concrete Syntax of PL/I*, IBM Corp., TN 3001, Hursley, England (Nov., 1966).

[BC66a] Beech, D., Nicholls, J. E., and Rowe, R., *A PL/I Translator*, IBM Corp., TN 3003, Hursley, England (Oct., 1966).

[BC67] Beech, D. et al., *Abstract Syntax of PL/I*, IBM Corp., TN 3002 (Version 2), Hursley, England (May, 1967).

[BU66] Burkhardt, W. H., "PL/I: An Evaluation", *Datamation*, Vol. 12, No. 11 (Nov., 1966), pp. 31–39.

[BZ67] Bond, E. R. and Cundall, P. A., "A Possible PL/I Extension for Mathematical Symbol Manipulation", *Symbol Manipulation Languages and Techniques, Proceedings of the IFIP Working Conference on Symbol Manipulation Languages* (D. G. Bobrow, ed.). North-Holland Publishing Company, Amsterdam, 1968, pp. 116–132.

[CQ65a] Christensen, C. and Mitchell, R., *Reference Manual for the NICOL 1 Programming Language*, 3rd ed., Computer Associates, CA-6511-3011, Wakefield, Mass. (Nov., 1965).

[CQ67] Christensen, C. and Mitchell, R., *Reference Manual for the NICOL 2 Programming Language*, Computer Associates, CA-6701-2611, Wakefield, Mass. (Jan., 1967).

[DL67] Donovan, J. J. and Ledgard, H. F., "A Formal System for the Specification of the Syntax and Translation of Computer Languages", *Proc. FJCC*, Vol. 31 (1967), pp. 553–69.

[DQ66] Dodd, G. G., "APL—A Language for Associative Data Handling in PL/I", *Proc. FJCC*, Vol. 29 (1966), pp. 677–84.

[EC66] *A Minimum PL/I Subset* (working paper), European Computer Manufacturers Association, ECMA/TC10/67/2, Geneva, Switzerland (Dec., 1966).

[GX67] Glass, R. L., *SPLINTER, A PL/I Interpreter Emphasizing Debugging Capability* (presented at ACM-sponsored PL/I Forum, Aug., 1967), Washington, D.C. (unpublished).

[IB00a] *Introduction to PL/I* (student text), IBM Corp., C20-1632, Data Processing Division, White Plains, N.Y.

[IB64] *NPL Technical Report*, IBM Corp., 320-0908, Data Systems Division, Poughkeepsie, N.Y. (1964).

[IB65] *PL/I Subset Language Specifications*, IBM Corp., C28-6809-1, Data Processing Division, White Plains, N.Y. (1965).

[IB65a] *A Guide to PL/I for FORTRAN Users* (student text), IBM Corp., C20-1637-1, Data Processing Division, White Plains, N.Y. (1965).

REFERENCES

[IB65b] *A PL/I Primer* (student text), IBM Corp., C28-6808-0, Data Processing Division, White Plains, N.Y. (1965).

[IB65e] *IBM System/360 Operating System: PL/I Language Specifications*, IBM Corp., C28-6571-0, Data Processing Division, White Plains, N.Y. (1965).

[IB66] *Formal Definition of PL/I*, IBM Corp., TR 25.071, Vienna Lab., Vienna, Austria (Dec., 1966).

[IB66a] *IBM System/360 Operating System PL/I(F): Programmer's Guide*, IBM Corp., C28-6594-0, Data Processing Division, White Plains, N.Y. (1966).

[IB66b] *IBM System/360 Operating System: PL/I Language Specifications*, IBM Corp., C28-6571-4, Data Processing Division, White Plains, N.Y. (Dec., 1966).

[IB66e] *IBM System/360 Operating System: PL/I Subroutine Library, Computational Subroutines*, IBM Corp., C28-6590-0, Data Processing Division, White Plains, N.Y. (1966).

[IB66f] *A Guide to PL/I for Commercial Programmers* (student text), IBM Corp., C20-1651-0, Data Processing Division, White Plains, N.Y. (1966).

[IB66g] *IBM System/360 Disk and Tape Operating Systems: PL/I Programmer's Guide*, IBM Corp., C24-9005-0, Data Processing Division, White Plains, N.Y. (1966).

[IB67] *PL/I Reference Data: Keywords and Character Sets*, IBM Corp., X20-1744-1, Data Processing Division, White Plains, N.Y. (1967).

[IB67d] *IBM System/360 PL/I Reference Manual*, IBM Corp., C28-8201-0, Data Processing Division, White Plains, N.Y. (1967).

[IB67f] *IBM System/360 PL/I Subset Reference Manual*, IBM Corp., C28-8202-0, Data Processing Division, White Plains, N.Y. (1967).

[LH67] Lawson, H. W., Jr., "PL/I List Processing", *Comm. ACM*, Vol. 10, No. 6 (June, 1967), pp. 358–67.

[MR64] McCracken, D. D., "The New Programming Language", *Datamation*, Vol. 10, No. 7 (July, 1964), pp. 31–36.

[PU67] Pursey, G., *Concrete Syntax of Subset PL/I*, IBM Corp., TN 3005, Hursley, England (Feb., 1967).

[QB67] Bates, F. and Douglas, M. L., *Programming Language/One*. Prentice-Hall, Inc., Englewood Cliffs, N.J., 1967.

[RG65] Radin, G. and Rogoway, H. P., "NPL: Highlights of a New Programming Language", *Comm. ACM*, Vol. 8, No. 1 (Jan., 1965), pp. 9–17.

[VK67] Walk, K. et al., *Abstract Syntax and Interpretation of PL/I* (draft for version 2), IBM Corp., TR 25.082, Vienna Lab., Vienna, Austria, (Dec., 1967).

[WC66] Weinberg, G. M., *PL/I Programming Primer*. McGraw-Hill, New York, 1966.

[XY64] *Report of the SHARE Advanced Language Development Committee* (Mar., 1964) (unpublished).

[XY64a] *Specifications for the New Programming Language* (Apr., 1964) (unpublished).

[XY64b]   *Report II of the SHARE Advanced Language Development Committee* (June, 1964) (unpublished).

[YH67]   Balzer, R. M., "Dateless Programming", *Proc. FJCC*, Vol. 31 (1967), pp. 535–44.

[ZG67]   McKeeman, W. M. et al., *Interim Report on the Stanford PL/I Compiler Generation Project* (presented at ACM-sponsored PL/I Forum, Aug., 1967), Washington, D.C. (unpublished).

VIII.5. FORMULA ALGOL

[FJ66]   Feldman, J. A., "A Formal Semantics for Computer Languages and its Application in a Compiler-Compiler", *Comm. ACM*, Vol. 9, No. 1 (Jan., 1966), pp. 3–9.

[IT66]   Iturriaga, R. et al., "Techniques and Advantages of Using the Formal Compiler Writing System FSL to Implement a Formula ALGOL Compiler", *Proc. FJCC*, Vol. 28 (1966), pp. 241–52.

[IT66a]   Iturriaga, R. et al., *The Implementation of Formula ALGOL in FSL*, Carnegie Inst. of Tech., Pittsburgh, Pa. (Oct., 1966).

[IT67]   Iturriaga, R., *Contributions to Mechanical Mathematics*, Carnegie Inst. of Tech., Pittsburgh, Pa. (Ph. D. thesis) (1967).

[PR64]   Perlis, A. J. and Iturriaga, R., "An Extension to ALGOL for Manipulating Formulae", *Comm. ACM*, Vol. 7, No. 2 (Feb., 1964), pp. 127–30.

[PR66]   Perlis, A. J., Iturriaga, R., and Standish, T. A., *A Definition of Formula ALGOL*, Carnegie Inst. of Tech., Pittsburgh, Pa. (Aug., 1966).

VIII.6. LISP 2

[AH66a]   Abrahams, P. W. et al., "The LISP 2 Programming Language and System", *Proc. FJCC*, Vol. 29 (1966), pp. 661–76.

[BO66]   Book, E., *The LISP 2 Syntax Translator*, System Development Corp., TM-2710/331/00, Santa Monica, Calif. (Apr., 1966).

[FI66]   Firth, D. and Kameny, S. L., *Syntax of LISP Tokens*, System Development Corp., TM-2710/210/00, Santa Monica, Calif. (Aug., 1966).

[KA66]   Kameny, S.L., *LISP 2 Source Language Syntax Specifications for Syntax Translator*, System Development Corp., TM-2710/230/00, Santa Monica, Calif. (Dec., 1966).

[LE64]   Levin, M., *Syntax of the New Language*, M.I.T., MAC-M-158, Project MAC, Cambridge, Mass. (May, 1964) (unpublished).

[LE66]   Levin, M. and Berkeley, E., *LISP 2 Primer*, System Development Corp., TM-2710/101/00 (draft), Santa Monica, Calif. (July, 1966).

[QT64]   Schorre, D. V., "META-II A Syntax-Oriented Compiler Writing Language", *Proc. ACM 19th Nat'l Conf.*, 1964, pp. D1.3-1–D1.3-11.

[WZ67]   Weizenbaum, J., "Review R67-22" (of *The LISP 2 Programming Language and System*), *IEEE Trans. Elec. Comp.*, Vol. EC-16, No. 2 (Apr., 1967), pp. 236–38.

IX SPECIALIZED LANGUAGES

IX.1. SCOPE OF CHAPTER

This chapter covers a whole category of languages which are classified as being used in specialized areas. In this context the word *specialized* is interpreted as *fairly narrow*, where this obviously implies a value judgment of the breadth by the author. It is possible, however, to be slightly less subjective than such a statement would imply. The majority of the categories in this chapter are those which either require very specialized knowledge and training (e.g., civil engineering) or are of interest to only a limited number of programmers (e.g., compiler writers).

This chapter has been divided into two major sections. The first deals with languages for special application areas, and the second describes languages which can be utilized in several application areas. The major distinction is that the first section requires (or assumes) knowledge of a particular technical discipline, whereas the languages in the second part could conceivably be used in a variety of applications. More specifically, the special application areas which are discussed are machine tool control, civil engineering, logical design, digital simulation of block diagrams, compiler writing, and some small miscellaneous applications. The second major section discusses languages for simulation, query (i.e., retrieval), graphics, computer-aided design, text editing, and on-line and operating systems. There is some overlap among various categories since no major attempt has been made to sharpen the distinctions. Finally, the list of languages discussed in each area is not necessarily exhaustive; the reference list at the end of this chapter is believed to be moderately complete.

Many of the languages in this chapter are those which have colloquially been called *problem-oriented*. However, as indicated in Chapter I, I prefer to reserve that phrase for a very wide area which essentially encompasses all programming languages. Because many of the languages (although certainly not all of them) contain terms and concepts which are meaningful only to those people with specialized knowledge, relatively little detail about each language is provided. In many cases, what has been done is merely to indicate the intended scope of the language, show an example or two, and, wherever appropriate, provide the list of basic statements or facilities in the language. Those interested in obtaining more detail about any particular area can do so by consulting the cited references, which are deemed to be the most significant.

One of the characteristics of these special application languages is that they tend to require more software support in the way of special application programs than the more general languages do. Thus, many of these special languages not only require a large majority of the facilities which are in the more general languages discussed in earlier chapters but need their own specialized routines as well. For example, most languages—even though they may be very specialized—provide some type of arithmetic, control transfers and some type of testing facilities. However, in addition to these, specialized routines may be required, such as computing a set of coordinates or representing an analogue computer integrator. In some cases, what has been done is first to provide a large set of subroutines; these might even be in a form which could be used with a *CALL* from FORTRAN or some other similar language. A language syntax is then placed on top to avoid the necessity of subroutine calls and notation. I do not mean to imply that the languages in this chapter are merely addenda to existing languages; with only a few exceptions, this is *not* the case. However, in a number of languages, they have been designed so that it is very easy to make additions from a language viewpoint to correspond to new facilities that might be desired or developed.

Most of the languages which are *not* additions to existing ones tend to have a fairly rigid format which resembles the macros in assembly programs more than the higher level languages in this book. Thus, the defining characteristics of a programming language which were given in Section I.4.2 have been interpreted much more loosely in this chapter than in any of the others, systems which would not have been included in the general categories discussed in Chapters IV–VIII are included here. I feel they are worth mentioning because they reflect a very significant trend in showing the need for specialized languages. For some of the application areas the syntactic flexibility of the more general languages is neither necessary nor desirable.

This entire area has received what I consider insufficient attention in developing generalized systems, but the proliferation of specialized languages attests to its importance. A discussion of the need for the latter is given by Licklider [LI65]. Early work in the development of tools for generating specialized languages was done in the SHADOW systems (see Barnett and Futrelle [BI62]). Although the term *syntax-directed compiler* was not used when the SHADOW work was originally started in the late 1950's, the designers were actually providing tools whereby a single system could handle the *syntax* for languages of widely differing kinds. (The SHADOW system needed specialized routines created to carry out the specialized tasks.) The syntax-directed compiling efforts discussed in Section IX.2.5 are more oriented toward generalized languages. Other attempts to provide systems whereby people can define their own specialized languages in narrow areas (and have them translated and executed) are being developed within the ICES system (see Section IX.2.2.3) and the AED work (see Section IX.3.4.2). As noted in Chapter XI, I personally feel this is an extremely important direction for future developments, and I actually started a project in 1962 with this objective. For various reasons, it was never completed.

IX.2. LANGUAGES FOR SPECIAL APPLICATION AREAS

IX.2.1. MACHINE TOOL CONTROL

The use of tools for cutting pieces of metal is well-known even for those not involved in the field known as *numerical control* (N/C). What is perhaps not so clear is that these tools can be controlled either manually or by means of a paper tape. The latter is presumably preferable from most viewpoints. The punched paper tape contains a steady stream of signals which provide appropriate direction to the tools involved. However, it is, of course, necessary for instructions to be prepared on the paper tape; the person who does this is commonly called a *part programmer* and the latter word has nothing to do with the programming associated with digital computers. The part programmer is the individual who takes a blueprint or other specification and prepares the punched paper tape. Since a tremendous amount of computation is involved in determining the coordinates of specified shapes (e.g., an ellipse), it seems natural to consider the use of a computer for doing this.

1. *APT*

In 1952, the M.I.T. Servomechanisms Laboratory (subsequently renamed Electronic Systems Laboratory) worked on a project sponsored by the Air Force Air Materiel Command to develop an automatic programming system for numerically controlled machine tools. In 1955 a prototype system was coded for the Whirlwind to demonstrate feasibility. This early version was restricted to two dimensions but it allowed parts to be programmed in terms of straight lines and circles in a variety of ways. In 1956 the APT (*A*utomatically *P*rogrammed *T*ools) Project was formed and in 1957 a group of Aerospace Industries Association member companies formed a joint effort under M.I.T. coordination to develop a more capable system, and better versions were produced.

A more advanced system (APT II) was prepared for the IBM 704 in 1958. This relieved the part programmer of the responsibility for computing successive cutter locations, and it enabled him to describe the curve in an artificial language with English words. A still more advanced system known as APT III was produced for the 7090 and released in December, 1961.

In September, 1961 the Aerospace Industries Association selected the Armour Research Foundation of the Illinois Institute of Technology to assume maintenance and validation responsibility and to direct the APT Long Range Program. The latter was to be a research and development effort involving the continued application and extension of the APT system. Member companies were to receive complete current systems, documentation, and information as they were developed. Work continued under that plan, and APT has been implemented on many different computers.

The APT system has three major sections. The first reads the program and does the necessary translation, compiles a sequence of numeric instructions, and does some geometric calculations to prepare the data in standard form. About 100 alternate geometric definitions are reduced to about 10 canonical forms. The second section interprets and controls the sequences of cutting instructions, and it computes the coordinate points through which the tools must pass to produce the specified part. The last section, known as the *postprocessor*, converts the control data into the proper format for a specific machine tool, compensates for machine tool dynamics, and adapts any previous generalized information to the peculiarities of individual cutting hardware. Figure IX-1 shows a part to be produced, the APT program to do this, and an explanation of that program.

The standardization of APT under USASI is well underway at the time of this writing; the work is being done by subcommittee X3.4.7, and consideration is also being given to international standardization. A definition of APT using Backus notation was given by Brown, Drayton, and Mittman [BP63]. The list of vocabulary words taken from that article is shown in Figure IX-2.

2. Others

A number of other computer systems have been developed for numerical control programming. Systems like AUTOSPOT (IBM) and PRONTO (GE) simplify tape preparation for point-to-point equipment. A subset of APT called ADAPT was defined. In 1956 an approach for handling three-dimensional milling was developed at the M.I.T. Servomechanisms Laboratory. Based on this approach, IBM developed a system called AUTO-PROMPT on the 709/90 to permit three-dimensional shapes to be programmed in terms of regions instead of space curves.

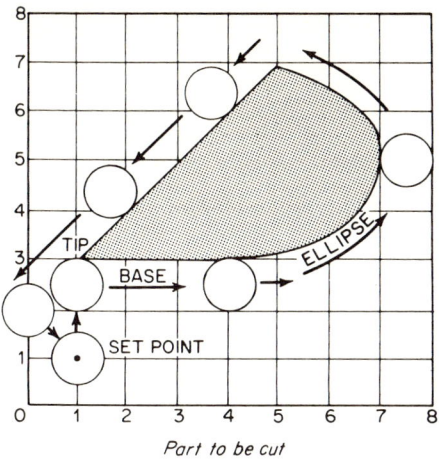

Part to be cut

Part Program	Explanation
CUTTER/1	Use a one inch diameter cutter.
TOLER/.005	Tolerance of cut is .005 inch.
FEDRAT/80	Use feedrate of 80 inches per minute.
HEAD/1	Use head number 1.
MODE/1	Operate tool in mode number 1.
SPINDL/2400	Turn on spindle. Set at 2400 rpm.
COOLNT/FLOOD	Turn on coolant. Use flood setting.
PT1=POINT/4,5	Define a reference point, PT1, as the point with coordinates (4,5).
FROM/(SETPT=POINT/1,1)	Start the tool from the point called SETPT, which is defined as the point with coordinates (1,1).
INDIRP/(TIP=POINT/1,3)	Aim the tool in the direction of the point called TIP, which is defined as the point with coordinates (1,3).
BASE=LINE/TIP, AT ANGL, 0	Define the line called BASE as the line through the point TIP which makes an angle of 0 degrees with the horizontal.
GOTO/BASE	Go to the line BASE.
TL RGT, GO RGT/BASE	With the tool on the right, go right along the line BASE.
GO FWD/(ELLIPS/CENTER, PT1, 3,2,0)	Go forward along the ellipse with center at PT1, semi-major axis = 3, semi-minor axis = 2, and major axis making an angle of 0 degrees with the horizontal.
GO LFT/(LINE/2,4,1,3,), PAST, BASE	Go left along the line joining the points (2,4) and (1,3) past the line BASE.
GOTO/SETPT	Go to the point SETPT in a straight line.
COOLNT/OFF	Turn off coolant flow.
SPINDL/OFF	Turn off spindle.
END	This is the end of the machine control unit operation,
FINI	and the finish of the part program.

Figure IX-1. Example of APT program for specific part to be cut.
Source: Hori, S., *Automatically Programmed Tools*, Armour Research Foundation of Illinois Institute of Technology, AZ-240, November, 1962. (Brochure.)

Figure IX-2.

Arithmetic Transfer Statements
 IF
 JUMPTO

Geometric Transfer Statements
 TRANTO

Termination Statements
 FINI

Statements
 LOOPST
 LOOPND

Procedure Statements
 CALL

Input-Output Control Statements
 PRINT
 READ
 TITLES
 PUNCH

Remarks
 REMARKS

Explicit Positioning Statements
 FROM
 GODLTA
 GOTO

Initial Continuous Motion Statements
 GO
 OFFSET

Intermediate Continuous Motion Statements
 GOLFT
 GORGT
 GOFWD
 GOBACK
 GOUP
 GODOWN

Function Designators
 DOTF
 LNTHF
 SQRTF
 SINF
 COSF
 EXPF
 LOGF
 ATANF
 ABSF

Geometric Expressions
 POINT
 LINE
 PLANE
 CIRCLE
 CYLNDR
 ELLIPS
 HYPERB
 CONE
 GCONIC
 LCONIC
 VECTOR
 MATRIX
 SPHERE
 QADRIC
 POLCON
 TABCYL

Post Processor Control Statement

END	PLABEL	MACHIN	COOLNT	SADDLE
STOP	PLUNGE	MCHTOL	SPINDL	LOADTL
OPSTOP	HEAD	PIVOTZ	TURRET	SELCTL
ISTOP	MODE	MCHFIN	ROTHED	CLEARC
RAPID	CLEARP	SEQNO	THREAD	CYCLE
SWITCH	TMARK	INTCOD	TRANS	DRAFT
RETRCT	REWIND	DISPLY	TRACUT	RITMIDI
DRESS	CUTCOM	AUXFUN	INDEX	PLOT
PICKUP	REVERS	CHECK	COPY	OVPLOT
UNLOAD	FEDRAT	POSTN	PREFUN	LETTER
PENUP	DELAY	TOOLNO	COUPLE	PPRINT
PENDWN	AIR	ROTABL	PITCH	PARTNO
ZERO	OPSKIP	ORIGIN	CLAMP	INSERT
CODEL	LEADER	SAFETY	ENDMDI	CAMERA
RESET	PPLOT	ARCSLP	ASLOPE	

Array Declarations
   RESERV

Coordinate Transformation Declarations
   REFSYS

Z-Surface Declarations
   ZSURF

Procedure Declarations
   MACRO
   TERMAC

Vocabulary Equivalence Declarations
   SYN

Direction Declarations
   INDIRP
   INDIRV

Tolerance Specifications
   TOLER
   INTOL
   OUTTOL

Cutter Specifications
   CUTTER

Calcuation Constant Controls
   CUT
   DNTCUT
   2DCALC
   3DCALC
   NDTEST
   TLAXIS
   MULTAX
   MAXDP
   NUMPTS
   THICK
   NOPS
   PSIS

Tool Position Declarations
   TLLFT
   TLRGT
   TLNDON
   TLON
   TANCRV
   TLONPS
   TLOFPS

Modifier Words

ATANGL	NOMORE	IPR	POSY	TYPE
CENTER	SAME	CIRCUL	POSZ	NIXIE
CROSS	MODIFY	LINEAR	RADIUS	LIGHT
FUNOFY	MIRROR	PARAB	RIGHT	FOURPT
INTOF	START	RPM	SCALE	TWOPT
INVERS	ENDARC	MAXRPM	SMALL	PTSLOP
LARGE	CCLW	TURN	TANTO	PTNORM
FEFT	CLW	FACE	TIMES	SPLINE
LENGTH	MEDIUM	BORE	TRANSL	RTHETA
MINUS	HIGH	BOTH	UNIT	THETAR
NEGX	LOW	XAXIS	XLARGE	XYZ
NEGY	CONST	YAXIS	XSMALL	TRFORM
NEGZ	DECR	ZAXIS	MIST	NORMAL
NOX	INCR	TOOL	TAPKUL	UP
NOY	ROTREF	AUTO	STEP	DOWN
NOZ	TO	FLOOD	MAIN	LOCK
IN	PAST	PARLEL	SIDE	SFM
OUT	ON	PERPTO	LINCIR	XCOORD
ALL	OFF	PLUS	MAXIPM	YCOORD
LAST	IPM	POSX	REV	ZCOORD

Figure IX-2. (cont. next page)

Figure IX-2. (cont.)

MULTRD	YLARGE	RED	RANGE	MILWAK
XYVIEW	YSMALL	GREEN	PSTAN	BENDIX
YZVIEW	YZPLAN	BLUE	CSTAN	DYNPAT
ZXVIEW	YZROT	INTENS	FRONT	TRW
SOLID	ZLARGE	LITE	REAR	ECS
DASH	ZSMALL	MED	SADTUR	CINCY
DOTTED	ZXPLAN	DARK	MILL	TRUTRA
CL	ZXROT	CHUCK	THRU	PRATTW
DITTO	3PT2SL	COLLET	DEEP	FOSDIK
PEN	4PT1SL	AAXIS	TRAV	BURG
SCRIBE	5PT	BAXIS	NORMPS	PROBOG
BLACK	INTERC	CAXIS	CONCRD	DVLIEG
XYPLAN	SLOPE	TPI	GECENT	
XYROT	SUNTRN	OPTION	SC4020	

Figure IX-2. APT vocabulary list. Each of these words has a meaning and is understood by people working in this field.
Source: Brown, Drayton, and Mittman [BP63], pp. 657–658. By permission of the Association for Computing Machinery, Inc.

There have been more than 100 *part programming* languages developed over the years, but relatively few have received widespread use. The trend appears to be to generalize the APT system to include all types of features from point-to-point to sophisticated geometric programming and then to extract subsets for various purposes.

IX.2.2. CIVIL ENGINEERING

The languages in this section are those specifically designed for use by people concerned with civil engineering. They certainly use languages such as FORTRAN, but the languages mentioned here are oriented specifically toward problems in the cited area. Among the most important characteristics of these languages is the minimum amount of computer expertise required. In other words, the objective is to allow the engineers to state the operations they wish to perform in a manner which is reasonably natural or logical for them. There is no question but that the same logical effect could be obtained by adding the necessary *subroutine packages* to a language such as FORTRAN or PL/I, but this would allow neither the syntactic flexibility that the engineer can obtain from a special language nor the scale of capabilities required to integrate the computer into the engineering process. Considered from the opposite point of view, it might be possible to develop systems whereby the engineer needs only to fill out a form to obtain his desired results. However, there is obviously less flexibility and freedom of development that way,

1. *COGO*

One of the most widely implemented and used languages for a special area is COGO. It has been implemented on at least the following computers: IBM 1620, 1130, 7040/44, 7090/94, 1410, 7070, and System/360; Burroughs 5000/5500; CDC 3600 and 160 A/1604; RCA Spectra 70; UNIVAC 1107; and under several time-sharing systems, e.g., CTSS (see Crisman [ZR65]) and QUIKTRAN. It was initially developed on the 1620 around 1960 under the direction of Professor C. Miller at M.I.T.'s Civil Engineering Department.

COGO (*CO*ordinate *GeO*metry) is a language to assist civil engineers

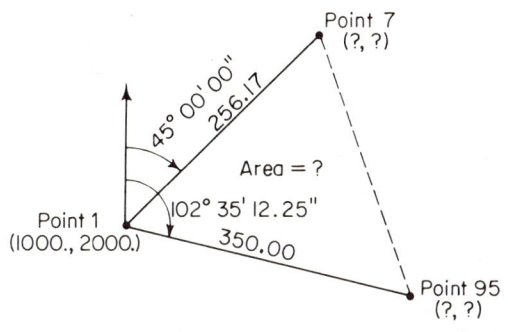

```
STORE            1           1000.    2000.
LOCATE/AZIMUTH   7    1      256.17   45   00      00
LOCATE/AZIMUTH   95   1      350.0    102  35      12.35
AREA             1    7      95
PAUSE
```

Figure IX-3. Small COGO program for figure shown. In the figure above, given the coordinates of point 1, the length and azimuth (clockwise angle from north) of lines 1–7 and 1–95, the COGO program shown computes the coordinates of points 7 and 95 and the area of the triangle. In the program, the second line reads: Locate point 7 by going from point 1 a distance of 256.17 at an azimuth of 45 degrees 00 minutes 00 seconds.
Source: Fenves [FE66], p. 48. Reprinted by permission from *Proceedings of the IBM Scientific Computing Symposium on Man-Machine Communication.* © 1966 by International Business Machines Corporation.

in doing plane geometry computations that are needed in surveying. Each command consists of a name and relevant data, e.g., DIVIDE/LINE J N M, which divides the line between defined points J and N into M equal parts and assigns intermediate point numbers J+1, J+2, ..., J+ (M−1). There is no interconnection between the commands except through the data. The engineer determines which commands to use, in which order, and their associated data. As an illustration of a COGO program, see Figure IX-3. A list of command names for the 7090 version is shown in Figure IX-4, and one of those commands is shown in more detail in Figure IX-5.

Long command name

ADJUST/ANG/LS	DISTANCE/SAVE	POINTS/INTERSECT
ALJUST/AZ/LS	DIVIDE/LINE	REDEFINE
ALIGNMENT	DIVIDE/STATION/LINE	RT/TRI/HYP
ANGLE	DUMP	RT/TRI/LEG
ANGLE/SAVE	EVEN/STATIONS	SEGMENT
ARC/ARC/INTERSECT	EXTERNAL/TANGENT	SEGMENT/MINUS
ARC/LINE/AZ	FINISH	SEGMENT/PLUS
ARC/LINE/POINTS	GIRDER/LENGTHS	SPIRAL/CURVE/INTER
AREA	INVERSE/AZIMUTH	SPIRAL/LINE/INTER
AREA/AZIMUTHS	INVERSE/BEARING	SPIRAL/SPIRAL/INTER
AREA/BEARINGS	LINES	START
AZ/INTERSECT	LOCATE/ANGLE	STATION/EL/POA
BR/INTERSECT	LOCATE/AZIMUTH	STATION/POA
CLEAR	LOCATE/BEARING	STORE
COORD/EL/OFFSET	LOCATE/DEFLECTION	STORE/SUPER
COORD/EL/POA	LOCATE/LINE	SUBGRADE
COORD/OFFSET	OFFSET/ALIGN	SUPER/EVEN
COORD/POA	OFFSET/ELEV	SUPER/SPECIAL
CROSS/TANGENT	OFFSET/EL/ALIGN	SURVEX/STATION
CURVE/SPIRAL	ORIGIN	TERMINUS
DEFINE/CURVE	PARALLEL/LINE	VERTICAL/SEGMENT
DISTANCE	PI	

Figure IX-4. List of 7090 COGO command names.
Source: Roos and Miller [RS64], extracts from pp. 27–29.

2. STRESS

STRESS (*ST*Ructural *E*ngineering *S*ystems *S*olver) is designed for use by engineers in analyzing framed structures. Work was started in the fall of 1962 at M.I.T. under the direction of Visiting Professor S. J. Fenves of the University of Illinois. STRESS has been implemented on at least the following computers: IBM 7090/7094; UNIVAC 1107; CDC 1604, 3400, and 3600; Burroughs 5500; under CTSS (see Crisman [ZR65]); and a subset on the IBM 1620 and 1130.

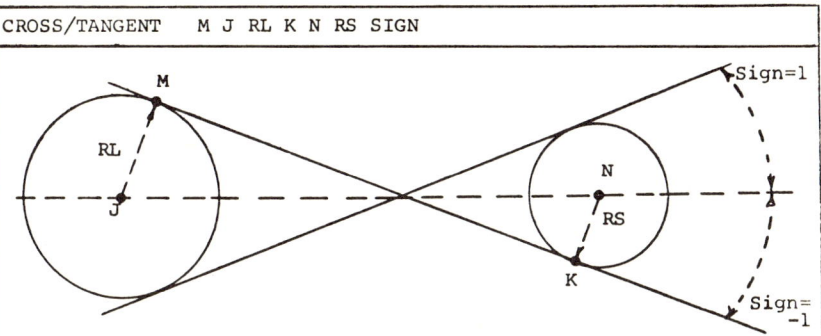

Figure IX-5. Specifications for a particular COGO command.
The figure is a copy of an actual page in the manual.
Source: Roos and Miller [RS64], p. 46.

The data in STRESS is more complicated than in COGO since COGO requires only geometric information, whereas stress also requires topological and mechanical properties and a great deal more other information. A sample STRESS problem, illustrating both the diagrammatic representation and the program for analysis of the space truss, is shown in Figure IX-6. A listing of many of the commands is given in Figure IX-7.

STRESS has been replaced and expanded by a system called STRUDL, an application subsystem of ICES.

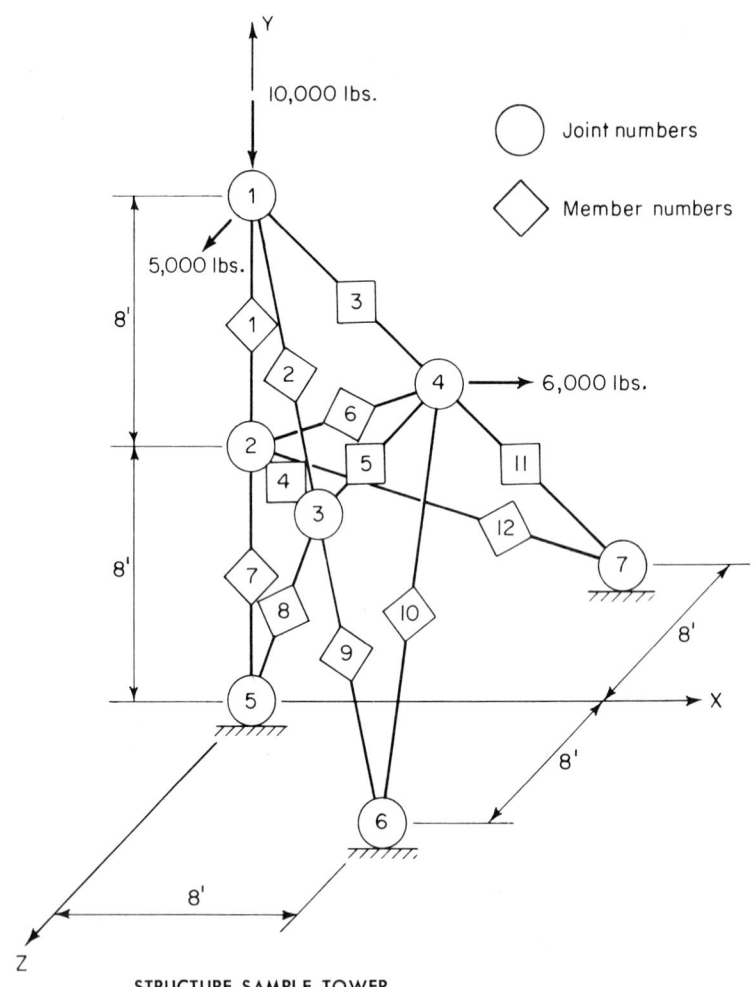

STRUCTURE SAMPLE TOWER
TYPE SPACE TRUSS
NUMBER OF JOINTS 7
NUMBER OF SUPPORTS 3
NUMBER OF MEMBERS 12
METHOD STIFFNESS
JOINT COORDINATES
1 X 0.0 Y 16.0 Z   0.0
2 X 0.0 Y 8.0 Z   0.0
3 X 4.0 Y 8.0 Z   4.0
4 X 4.0 Y 8.0 Z −4.0
5 X 0.0 Y 0.0 Z   0.0 SUPPORT
6 X 8.0 Y 0.0 Z   8.0 SUPPORT
7 X 8.0 Y 0.0 Z −8.0 SUPPORT

614

Figure IX-6. (cont.)

```
              MEMBER INCIDENCES
              1  2  1
              2  3  1
              3  4  1
              4  2  3
              5  3  4
              6  4  2
              7  5  2
              8  5  3
              9  6  3
             10  6  4
             11  7  4
             12  7  2
              MEMBER PROPERTIES, PRISMATIC
              1  AX  1.0
              2  AX  1.0
              3  AX  1.0
              4  AX  1.0
              5  AX  1.0
              6  AX  1.0
              7  AX  1.0
              8  AX  1.0
              9  AX  1.0
             10  AX  1.0
             11  AX  1.0
             12  AX  1.0
              NUMBER OF LOADINGS 1
              LOADING ARBITRARY
              TABULATE FORCES, JOINT DISPLACEMENTS, REACTIONS
              JOINT LOADS
              1 FORCE Y —10.0, FORCE Z 5.0
              4 FORCE X 6.0
              SOLVE THIS PART
```

Figure IX-6. STRESS program for analysis of space truss shown in the diagram.
Source: Fenves [FE66], p. 50. Reprinted by permission from *Proceedings of the IBM Scientific Computing Symposium on Man-Machine Communication.*
© 1966 by International Business Machines Corporation.

3. *ICES*

Following the successful systems described above, work to create a much more generalized system was started at the M.I.T. Civil Engineering Department under the supervision of Professor C. L. Miller. ICES (*I*ntegrated *C*ivil *E*ngineering *S*ystem) was first implemented on the IBM System/360

Structure Statement
    STRUCTURE Title

Loading Statement
    LOADING Title
    or
        Loading N

Size Descriptors
    NUMBER OF JOINTS N
    NUMBER OF SUPPORTS N
    NUMBER OF MEMBERS N
    NUMBER OF LOADINGS N

Tabulate Statement
    TABULATE
    followed by any of the following
    descriptors, singly or in a string:
    FORCES
    REACTIONS
    DISTORTIONS
    DISPLACEMENTS
    ALL

Members Properties Statement
    MEMBER PROPERTIES, List-of-common-properties
    M Properties-not-included-in-the-list-of-common-properties

Member Distortion Statement
    MEMBER DISTORTIONS
    M DISTORTION X α_1 Y α_2 Z α_3 ROTATION X α_4 Y α_5 Z α_6

Member End Load Statement
    MEMBER END LOADS
    M END FORCE X α_1 Y α_2 Z α_3 MOMENT X α_4 Y α_5 Z α_6
    START FORCE X α_7 Y α_8 Z α_9 MOMENT X α_{10} Y α_{11} Z α_{12}

Termination Statements
    SOLVE
    SOLVE THIS PART
    FINISH or FINISHED
    STOP

Selective Output Statements
    SELECTIVE OUTPUT
    PRINT DATA

Modification Descriptors
    ADDITIONS
    CHANGES
    DELETIONS

Joint Release Statement
    JOINT RELEASES
    J FORCE X Y X MOMENT X Y X, θ_1 θ_2 θ_3

Figure IX-7. Partial listing of STRESS commands. Source: Fenves [FE64], extracts from pp. 24–34.

around 1967. It was released in November 1967 and ordered by over 300 organizations. ICES consists of four major elements: (1) A set of engineering subsystems, each of which is designed to solve problems in a particular area of civil engineering, such as structural analysis and design, soil mechanics, and highway engineering; some specific systems are COGO, STRUDL, ROADS, BRIDGE, TRANSIT, PROJECT, (2) a set of special application area languages which are used to operate these subsystems, (3) facilities whereby people can relatively easily design new languages and embed them in the system and modify existing subsystems, and (4) an executive system which combines all these modules in the correct fashion. It is hoped that one of the most useful characteristics of ICES will be its facility to allow the user access to subroutines from more than one particular branch of engineering.

One of the facilities of ICES is a language called ICETRAN which has been used to write the subsystem programs. ICETRAN is an extension and expansion of FORTRAN IV. It contains a number of additions to FORTRAN which are similar in style and spirit and then some other additions which are quite dissimilar. Some of the statements in the former category are as follows:[1]

DYNAMIC ARRAY $a_1(k_1), a_2(k_2), a_3(k_3), \ldots, a_n(k_n)$
: The a_i are nonsubscripted array names which are to be treated dynamically. The k_i is optional and may be one of the letters I, R, or D to indicate integer, real, or double precision, respectively.

DEFINE $a(i_1, i_2, \ldots, i_n), s_1, s_2, \ldots, s_n, t, p, g$
: This essentially defines the structure of part or all of a dynamic array. The i_i may be integer expressions, the s_i are expressions, the t is optional and indicates data type, the p indicates priority for retention in core, and the g indicates automatic growth beyond the defined size.

RELEASE $a(i_1, i_2, \ldots, i_n), p$
: This releases arrays when they are not immediately needed but are to be saved for later use.

DESTROY $a(i_1, i_2, \ldots, i_n), \left\{ \begin{array}{l} \text{SIZE} \\ \text{REDUCE} \\ \text{blank} \end{array} \right\}$
: This destroys the arrays when they are no longer needed.

SWITCH (a, b)
: The a and b are dynamic array pointers which are to be interchanged by the command.

These statements are executable at object time to structure and control data according to the problem parameters. Memory is automatically managed by an executive routine.

There are statements that involve linking of modules within subsystems, namely LINK, TRANSFER, and BRANCH. A global COMMON makes data accessible to all modules.

There are statements to operate on a pushdown stack, namely

>ADD TO STACK (count, name)
>DELETE FROM STACK (count)
>COPY FROM STACK (count, name)
>TRANSFER TO STACK
>LINK TO STACK

---

[1] Extracted from [MT00], pp. 2-4–2-27, and 2-44.

The terms *count* and *name* refer to the number of items to be handled and the address of the stack.

There are a number of error processing statements such as *INHIBIT*, *ENABLE*, and *ERROR RETURN*. There are various statements to permit control of secondary storage. There is a matrix declaration which enables the user to write statements which are similar in structure to the regular ones but which operate on matrices. Thus, for example, the user could write

$$V(2)=V(2)-TRN(X(1))*B*V(2)*INV(TRN(X(1))*B*X(1))*X(1)$$

where *TRN* means matrix transposition and *INV* means inversion.

It should be stressed again that the purpose of ICETRAN is to permit programmers (who might themselves be the engineers) to write the necessary subsystem programs for the differing application areas.

One of the most interesting facets of ICES is its command definition language (CDL) which is used by the subsystem programmers to define and specify the processing of the language command for each subsystem. It is assumed that each of the application-oriented commands will consist of a word or a phrase defining some unit of engineering computation (e.g., *LOCATE AZIMUTH* and *STIFFNESS ANALYSIS*). The data for the command may be on the same or a following card. Phrases themselves may be data, and complex commands consisting of phrases mixed with data are permitted.

One common thing to do, of course, is to add to the vocabulary. This is done by means of the following type of statement:

REQUEST '*label*' MODE '*variable*' *treatment*

The '*label*' is the vocabulary word, the MODE represents the internal form of the data, the '*variable*' specifies the internal name of the variable or array, and *treatment* specifies whether the data is required or not. Commands are handled by means of the following:

$$\begin{Bmatrix} ADD \\ REPLACE \end{Bmatrix} \text{'command name' [MODE 'variable' treatment]}$$

where MODE specifies the internal form for the data. Other commands are shown in Figure IX-8. As an illustration of this, if the *STORE* command, used to define the X and Y coordinates of a known point, is to be added to the COGO subsystem, then the engineer might write

STORE POINT 10 X 1000.53 Y 960

Subsystem Specification
    SYSTEM 'subsystem-name' ['password']

Specification of Communication Variables
    COMMON
$$\left[\left\{\begin{matrix}\underline{\text{ADD}}\\ \underline{\text{REPLACE}}\\ \underline{\text{DELETE}}\end{matrix}\right\}\right]\text{'variable-1' location-1 ... 'variable-n' location-n}$$
    END COMMON

Data Descriptors
$$\left\{\begin{matrix}\underline{\text{ID}}\text{ 'label'}\\ \underline{\text{NO ID}}\end{matrix}\right\}\text{MODE 'variable' TREATMENT}$$

Existence Data
    EXISTENCE 'list' SET 'variable' TREATMENT

Presetting Storage Locations
    PRESET MODE 'variable' EQUAL value

Counting
    INCREMENT [MODE] 'variable' [BY constant]

Equating Variables
    MOVE [MODE] 'variable-1' to 'variable-2'

Ignorable Words
    IGNORE 'word-1' 'word-2' ...

Command Modifier
    MODIFIER 'label' [MODE 'variable' TREATMENT]
        ... requests ...
    OR MODIFIER 'label' [MODE 'variable' TREATMENT]
        ... requests ...
$$\underline{\text{END}}\left\{\begin{matrix}\text{MODIFIER}\\ \text{CONDITION}\end{matrix}\right\}\left\{\begin{matrix}\text{OPTIONAL}\\ \text{REQUIRED}\end{matrix}\right\}$$

Tests on COMMON Variables
$$\left\{\begin{matrix}\underline{\text{CONDITION}}\\ \text{OR }\underline{\text{CONDITION}}\end{matrix}\right\}\text{[MODE] 'variable' RELATION}\left\{\begin{matrix}\text{VARIABLE 'variable'}\\ \text{CONSTANT}\end{matrix}\right\}$$

Incremental Processing
$$\left\{\begin{matrix}\underline{\text{REPEAT}}\\ \underline{\text{REPEAT TABULAR}}\end{matrix}\right\}\text{['symbol']}$$
        ... requests ...
    END REPEAT [TABULAR]

Figure IX-8. Some of the CDL commands in ICES. The general notation is the COBOL metalanguage described in Section II.6.2. In addition, the use of ... requests ... represents any request that adds to a subsystem a vocabulary word which describes a data item.
Source: [MT00], extracts from pp. 3–5—3–19.

and this would be added to the COGO vocabulary by the following small program:[2]

```
SYSTEM 'COGO'
ADD 'STORE'
ID 'P' INTEGER 'NPOINT' REQUIRED
ID 'X' REAL 'XCOORD' STANDARD 0
ID 'Y' REAL 'YCOORD' STANDARD 0
EXECUTE 'STORE'
FILE
```

An example of how to define a language called STRUDL is worked out in Walter [WQ66].

There are many other reports describing ICES besides those listed in the references at the end of the chapter.

IX.2.3. Logical Design

The problem of debugging hardware is well-known to engineers just as the problem of debugging programs is well-known to programmers. Because of difficulties in eliminating the errors from a digital computer when it is built and the desirability of having a systematic way to describe its logical design, there have been a number of developments which provide languages for describing logical design of a digital computer. One of the hopes is that through the use of these languages the computer could be used to simulate the design of another machine and thus reduce the difficulties after the system is built. Not all the languages mentioned here are for the sole purpose of aiding the debugging; in some cases, they are merely to present an accurate picture of the actual logical design in a form different from that originally prepared. Not all these languages have been implemented.

1. *APL* (*Iverson*)

One of the early attempts along these lines is discussed in Chapter 2 of Iverson [IV62] (see Section X.4 in this book) where he describes instructions on an IBM 7090. This approach was carried further in providing a formal description of IBM System/360 (see Falkoff, Iverson, and Sussenguth [FA64]).

Unfortunately, since this language in that form was not implemented, it could not be used in a practical way for simulation.

---

[2] Roos [RS65], p. 154.

2. LOTIS

LOTIS is a formal language for describing machine *LO*gic, *TI*ming, and *S*equencing. It adapts features from both ALGOL 60 and APL. LOTIS *describes* parts of an object machine but *does not simulate* it; in order to do that, it would be necessary to provide commands to specify what the simulator is to do. A machine is defined completely by describing the structure and behavior of its data flow. The assignments provide this information. Among the facilities provided in this language are the following: The naming of memory locations or registers and the identification of individual positions within them (e.g., [31] represents the least significant position in a 32-bit register whose address is in *ar*); strings of consecutive positions can be defined and labeled; logical operators apply to single values and also to vectors; expressions are used to describe networks and conditional forms can be used; assignment statements specify register-to-register transfers; time declarations allow for differences in response time of a circuit; duration of transfers; and various mechanisms from languages such as ALGOL and FORTRAN (e.g., *GO TO, CALL*). An example of a portion of a program in this language is shown in Figure IX-9.

```
seq add, arit/
  1, core: call read/
  2, ready: md := sr/
  3, 8: ov := (ac[0] ≠ md[0]); ac := ac + md/
  4: if not ov then ov := (ac[0] ≠ md[0])else ov := 0;
     call instrfetch, 1/fin.
```

Figure IX-9. Portion of LOTIS program defining a hypothetical computer. In this program, *arit* contains sequences using the arithmetic registers *ac* and *md*; a fixed point add is shown, assuming negative numbers represented in 2's complement form. Step 3 is fixed-timed to account for carry propagation through the long adder by overriding the + operator-time declared for address-length adders. Step 4 completes forming the overflow indication in *ov* and activates *instrfetch* for the next instruction cycle.
Source: Schlaeppi [QY64], p. 448.

3. LDT

The *Logic Design Translator* (LDT) was designed to develop logic equations for a computer from the information contained in a systems diagram and the instruction repertoire of the machine. The system is composed of three programs (written in Burroughs ALGOL 58): The translator, a timing analysis, and a term development and logic equation generator. The language used to provide the system description is composed of three main sections: The *declarative*, which is a linguistic description of the block diagram of a machine; an *operational* section, which represents most of the processes used in doing the programming (e.g., arithmetic statements, conditional statements, etc.); and finally a *design* table, which is an inter-

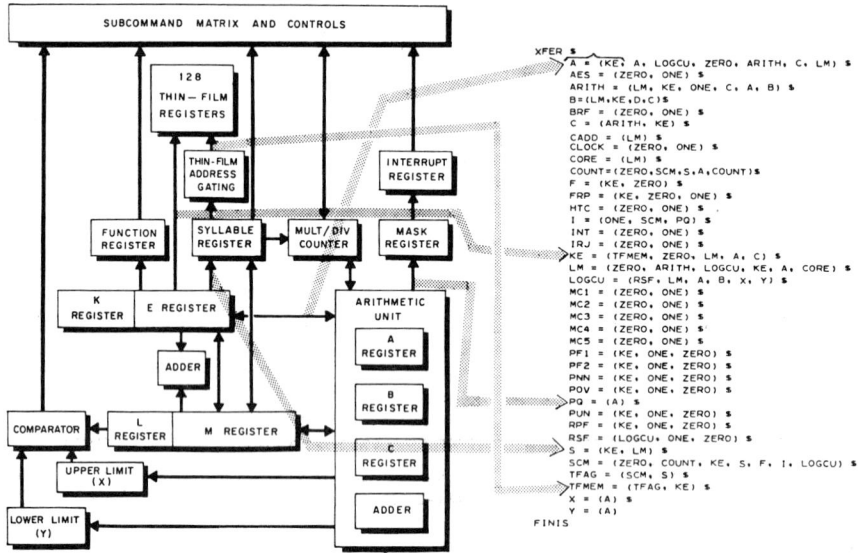

Figure IX-10. Relationship between information flow on a block diagram and LDT input.
Source: Proctor [PC64], p. 425.

mediate language used in translating a systems design block diagram into logic equations. An experiment was run where the object machine was the Burroughs D825. An example of a portion of the program is given in Figure IX-10.

4. Language for Simulating Digital Systems

Another running system is that described by McClure [MZ65] and implemented on the CDC 1604. His language contains sections for declarations, equations, and the executive. The language has a flavor very much like that of FORTRAN. Specific declarations which exist include *FLIPFLOPS*, *BOOLEAN*, *INTEGER*, and *REGISTER*. The equation section contains the input equations for the system flipflops and the equations for the intermediate signals. Control and testing information looks very straightforward; a short program for elementary testing might be the following:[3]

---

[3] McClure [MZ65], p. 18.

```
                READ X,ABC,N
LOOP ..         STEP
                PRINT X,ABC,N
                N = N - 1
                IF (N.GT.0) THEN GO TO LOOP
                STOP
```

5. Computer Compiler

An unimplemented system is that described by Metze and Seshu [MX66]. This proposal for a computer compiler allows the user to provide information about the design of the system at a higher level than the languages just described. The output of this computer compiler would be a description of the hardware with the logical conditions, flipflops, and their interconnections. The input language involves things such as register declarations, information about the decoding of instructions, transfer of information between registers, and branching statements.

The output of the system compiler is actually some preliminary information and then a string of microinstructions; the microlanguage is actually permissible in the source program as well.

An example of a little bit of the system design of a small digital computer is shown in Figure IX-11.

6. Computer Design Language

The work of Chu [CB65] is one of the two very ALGOL-like systems in this area, but it was also not claimed to be implemented. A subset was implemented on the 7094 for simulation purposes; it is described in Mesztenyi [YZ67]. Not only the flavor but much of the terminology and syntax is that of ALGOL. An example is shown in Figure IX-12.

7. SFD-ALGOL

The work of Parnas ([PN66] and [PN66a]) is a true extension of ALGOL for the purpose of allowing unambiguous descriptions of synchronous systems. It must be emphasized that the language provides a description of, rather than a program which would simulate, the system; Parnas indicates in [PN66] how the SFD-ALGOL description could be converted to an executable program relatively easily.

The major additions to ALGOL are as follows: A new delimiter, **time begin** (with a corresponding **end**), has been added. It is assumed that all statements within such delimiters are actions taken by the system during one clock pulse. These groups of statements are referred to as *time blocks*.

Two new declarations have been added, namely **input** and **output**. The procedure INPUT with one parameter must be declared in the outermost block.

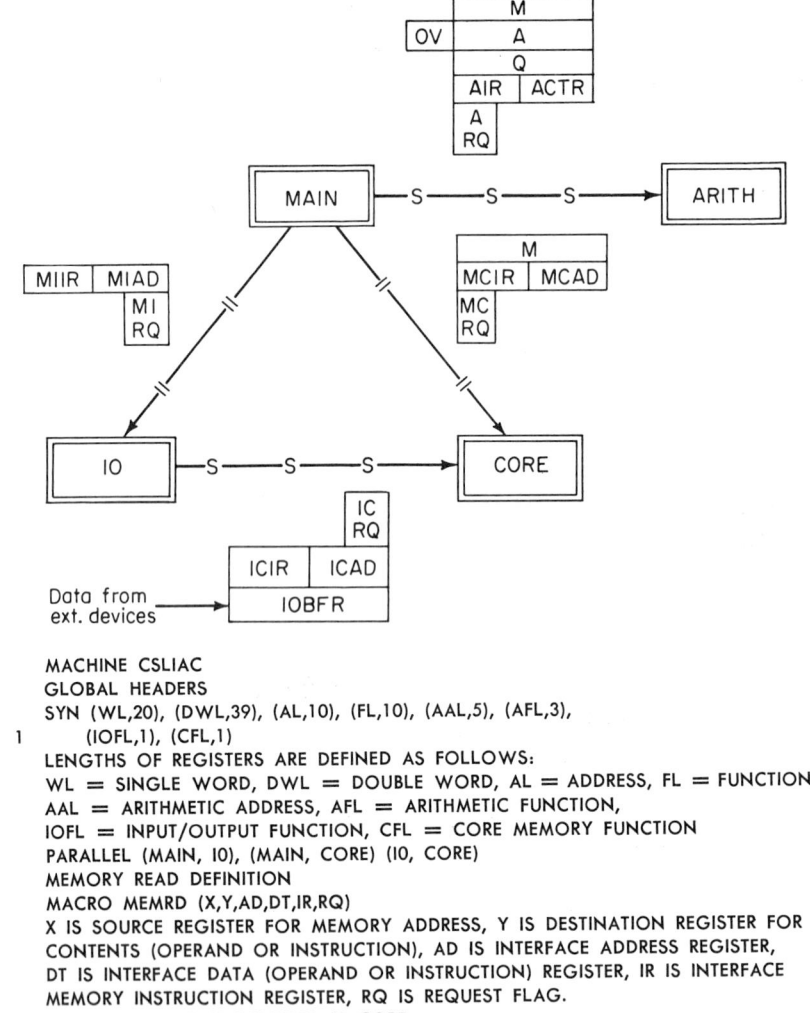

```
        MACHINE CSLIAC
*       GLOBAL HEADERS
        SYN (WL,20), (DWL,39), (AL,10), (FL,10), (AAL,5), (AFL,3),
    1   (IOFL,1), (CFL,1)
*       LENGTHS OF REGISTERS ARE DEFINED AS FOLLOWS:
*       WL = SINGLE WORD, DWL = DOUBLE WORD, AL = ADDRESS, FL = FUNCTION,
*       AAL = ARITHMETIC ADDRESS, AFL = ARITHMETIC FUNCTION,
*       IOFL = INPUT/OUTPUT FUNCTION, CFL = CORE MEMORY FUNCTION
        PARALLEL (MAIN, IO), (MAIN, CORE) (IO, CORE)
*       MEMORY READ DEFINITION
        MACRO MEMRD (X,Y,AD,DT,IR,RQ)
*       X IS SOURCE REGISTER FOR MEMORY ADDRESS, Y IS DESTINATION REGISTER FOR
*       CONTENTS (OPERAND OR INSTRUCTION), AD IS INTERFACE ADDRESS REGISTER,
*       DT IS INTERFACE DATA (OPERAND OR INSTRUCTION) REGISTER, IR IS INTERFACE
*       MEMORY INSTRUCTION REGISTER, RQ IS REQUEST FLAG.
*       NAME OF MEMORY CONTROL IS CORE.
        CALL CORE (RQ)
        AD = X
        IR = RCM
        ENDC
        IFF (Y = DT)
        Y = DT
        ENDM
        MEMORY WRITE DEFINITION. SAME ARGUMENT LIST
        MACRO MEMWR(X,Y,AD,DT,IR,RQ)
        CALL CORE (RQ)
```

Figure IX-11. System layout and data paths in a sample computer, and part of computer compiler program.
Source: Metze and Seshu [MX66], p. 259

The initial state of the system is defined to be the result of executing the first time block.

A new declaration, **nonstate**, modifies ALGOL in the sense that it must precede declarations for data which will remain fixed during the program. This is primarily to improve efficiency.

As an illustration, Figure IX-13 describes a pushdown stack which accepts one bit of information, pushing down any information which may already be in the stack to a maximum depth of five.

```
                    sequence MULTIPLICATION
                    comment begin This sequence describes
                    a multiplication which employs Booth's
                    algorithm for binary numbers in signed
                    two's complement representation. The
                    multiplier and multiplicand are initially
                    in register B and R respectively, and the
                    product is finally in the cascaded register
                    A&B[0-26]. end
                    register B[0-27], S[0-5], MULTIPLYALARM
                    operation MULTIPLY, RIGHTSHIFT
W*F05*T0:           R ← MA[C]; B ← A; A ← 0
W*F05*T1:           do PARITY; S ← 26
W*F05*T2:           G ← H[G]; do MULTIPLY
W*F05*T3:           H ← MB[G]; do RIGHTSHIFT; C ← D
W*F06*T0:           do MULTIPLY
W*F06*T1:           do RIGHTSHIFT
W*F06*T2:           do MULTIPLY; if S = 0 then G ← H[G]
W*F06*T3:           do RIGHTSHIFT; if S = 0 then H ← MB[G]
W*F07*T0:           do MULTIPLY; Y ← R[0]
W*F07*T1:           STOP ← MULTIPLYALARM ← A[0]*B[27]*Y
W*F07*T2:           G ← H[G];
W*F07*T3:           H ← MG[G]; W ← 0
MULTIPLY:           if B[27]*Y' = 1 then A ← A sub R
                    if B[27]'*Y = 1 then A ← A add R
RIGHTSHIFT:         A&B&Y ← shr A&B&Y;
                    if S≠0 then S ← S sub 1
```

Figure IX-12. Computer Design Language program to define multiplication. Source: Chu [CB65], p. 613. By permission of Association for Computing Machinery, Inc.

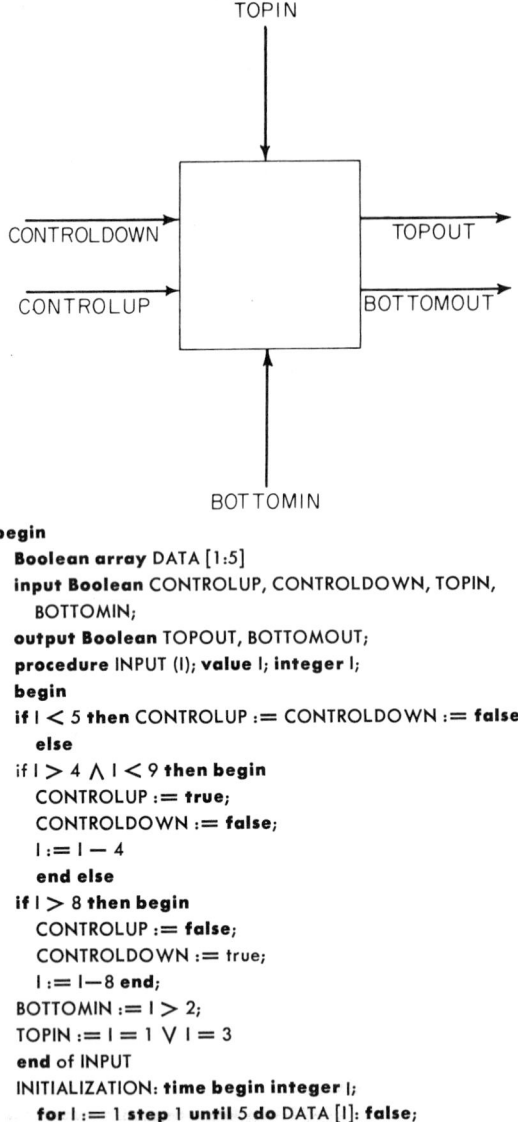

```
begin
  Boolean array DATA [1:5]
  input Boolean CONTROLUP, CONTROLDOWN, TOPIN,
    BOTTOMIN;
  output Boolean TOPOUT, BOTTOMOUT;
  procedure INPUT (I); value I; integer I;
  begin
  if I < 5 then CONTROLUP := CONTROLDOWN := false
    else
  if I > 4 ∧ I < 9 then begin
    CONTROLUP := true;
    CONTROLDOWN := false;
    I := I − 4
    end else
  if I > 8 then begin
    CONTROLUP := false;
    CONTROLDOWN := true;
    I := I−8 end;
  BOTTOMIN := I > 2;
  TOPIN := I = 1 ∨ I = 3
  end of INPUT
  INITIALIZATION: time begin integer I;
    for I := 1 step 1 until 5 do DATA [I]: false;
    TOPOUT := BOTTOMOUT := false
    end of initialization;
  WORK: time begin integer I;
  if CONTROLUP then begin
  for I := 4 step −1 until 1 do DATA [I+1] := DATA [I];
    DATA [1] := BOTTOMIN
    end else
    if CONTROLDOWN then begin
    for I := 4 step 1 until 4 do DATA [I] := DATA [I+1];
    DATA [5] := TOPIN;
    end;
```

Figure IX-13. (cont.)

```
        TOPOUT := DATA [5];
        BOTTOMOUT := DATA [1];
        go to WORK
        end of WORK time block
    end of description
```

Figure IX-13. SFD-ALGOL program to describe action of pushdown stack shown in the diagram. Any information which is already in the stack is pushed down to a maximum depth of 5.
Source: Parnas [PN66], pp. 74, 79. By permission of Association for Computing Machinery, Inc.

IX.2.4. Digital Simulation of Block Diagrams

1. *Introduction*

The phrase *digital simulation* as applied to languages actually has two meanings: One involves the simulation of block diagrams (primarily but not exclusively analog computers), whereas the other refers to discrete processes. The second category is discussed in Section IX.3.1 and the first is discussed here. The reason for making this distinction is that Section IX.2 is for those languages which require very specialized knowledge on the part of the user, which applies in this case because of the analogue computer or other scientific problem areas. The discrete simulation languages can be applied to very many technical disciplines, and the simulation techniques are relatively independent of the problem source.

> Simulation languages have been written for two more or less diverse reasons: to provide analog check cases, and to solve differential equations (in other words, replace an analog computer). ... The overwhelming majority of authors state that their language (or program) was designed to simulate systems that can be represented by block diagrams.[4]

A few of the systems which are apparently of great use in this area are not specifically mentioned because their *languages* are even less *problem-oriented* than the ones cited below. In this category are PACTOLUS (see Brennan and Sano [YR64]), MIDAS (see Petersen *et al.* [PS64]), and TELSIM (see Busch [QD66]). One of the earliest papers in this area is Selfridge [YD55]. An on-line system which is designed for use in a hybrid configuration is described by Cramer and Strauss [CR66]. Excellent summaries and reviews of the programs and/or languages in this field are given in Brennan and Linebarger [YR64a] and Clancy and Fineberg [ZZ65].

---

[4] Clancy and Fineberg [ZZ65], pp. 27–28.

2. DYANA

DYANA (*DY*namics *ANA*lyzer) is a language used for describing vibrational and other dynamics systems. It is one of the older *specialized languages*, and it was implemented on the IBM 704 around 1958 at the General Motors Research Laboratories. It is essentially an extension to FORTRAN, although a few minor restrictions are placed on naming variables and on other points.

Specifically defined variables are used to name the elements, excitations, and dependent and independent variables of dynamic systems described using DYANA. These variables have meaning in both FORTRAN and non-FORTRAN statements. As an illustration, the general form EμKγ (e.g., E15K07, E00K05) represent the spring element whose terminals are to points μ and γ when used in a non-FORTRAN statement, but they represent the value of the coefficient of damping when used in FORTRAN arithmetic or input/output statements. An example of a simple mechanical system and its DYANA program is shown in Figure IX-14.

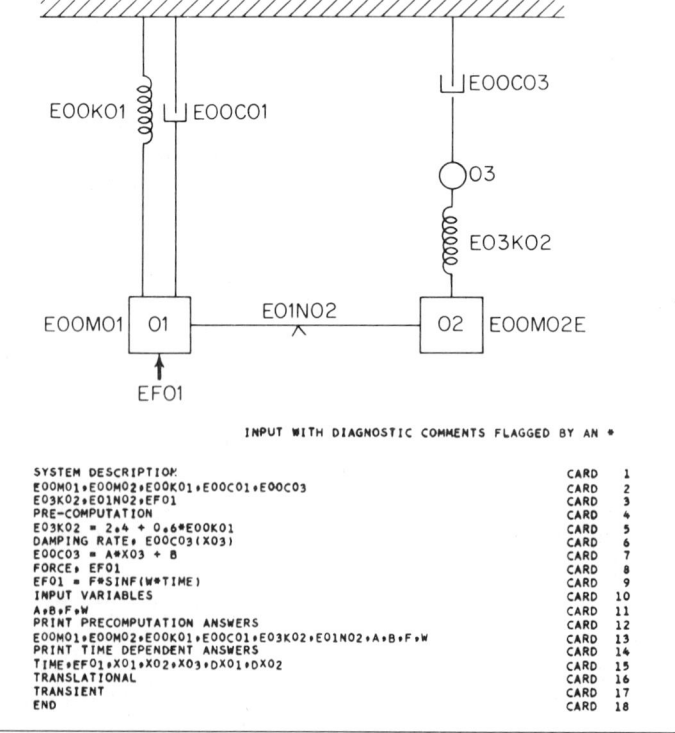

Figure IX-14. Mechanical system and corresponding DYANA program. Source: Theodoroff [TD58], p. 145.

3. DYSAC

DYSAC (*D*igital*Y* *S*imulated *A*nalog *C*omputer) is a language (implemented on the CDC 1604) where the user has available to him a large number of analog computer components which he can interconnect according to the needs of his problem. The list of components available to him is shown in Figure IX-15 and the various input data required to describe the

Components	Serial numbers
Integrators	N01 to N99
Adders	A01 to A99
Limiters	L01 to L50
Sq. Root Generators	Q01 to Q50
Sine Generators	S01 to S50
Cosine Generators	C01 to C50
Log. Generators	G01 to G50
Exp. Generators	X01 to X50
Function Generators	F01 to F50
Time Delay Units	T01 to T12
Dividers	D01 to D99
Relays	R01 to R99
Potentiometers	P01 to P99
,,	H01 to H99
,,	J01 to J99
,,	K01 to K99

Figure IX-15. DYSAC components.
Source: Hurley and Skiles [HJ63], p. 70.

Dysac Input Data Sections	*Control Options*
1. Problem title.	TITLE
	RETAIN TITLE
2. Patching. Description of connections between components.	PATCHING
	RETAIN PATCHING
3. Initial values for integrators.	INITIAL VALUES
	CLEAR INITIAL VALUES
	RETAIN INTEGRATOR VALUES
4. Potentiometer settings.	POT SETTING
	RETAIN POT SETTINGS
5. Function tables.	FUNCTION TABLES
	RETAIN FUNCTION TABLES
6. Headings.	HEADINGS
	RETAIN HEADINGS
7. Supplementary machine language instructions.	MACHINE LANGUAGE INSTRUCTIONS
	NO MACHINE LANGUAGE
	RETAIN MACHINE LANGUAGE

Figure IX-16. DYSAC statements for input data sections.
Source: Hurley and Skiles [HJ63], p. 71.

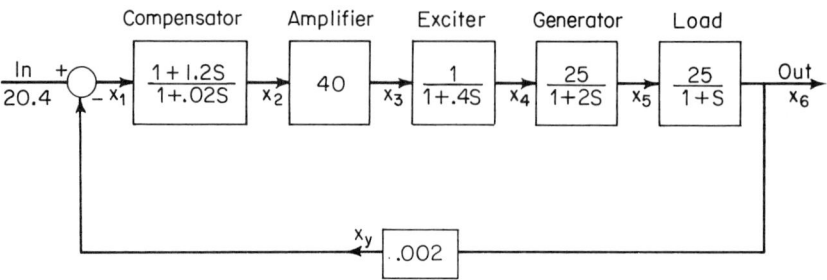

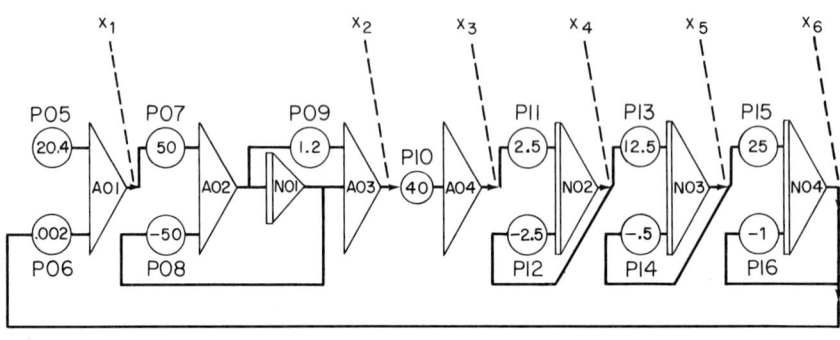

```
BCD  8  TITLE
BCD     CURRENT REGULATION SYSTEM
BCD  8  20.4 VOLT STEP INPUT
BCD  8  PATCHING
BCD     N01=A02.    N02=A04P11+N02P12.    N03=N02P13+ N03P14.
BCD     N04=P15N03+P16N04.
BCD     A01=P05+P06N04.    A02=P07A01+P08N01.    A03=P09A02+N01.
BCD     A04=P10A03.    OUT=P01.    OUT=A01.    OUT=A03.
BCD  8  OUT=A04.    OUT= N02.    OUT=N03.    OUT=N04.
RCD  8  CLEAR INITIAL VALUES
RCD  8  POT SETTINGS
DEC     0,.005,7,4
DEC     20.4,-.002,50,-50
DEC     1.20,40,2.5,-2.5
DEC  8  12.5,-.5,25,-1
BCD  B  RETAIN FUNCTION TABLES
BCD  8  HEADINGS
BCD         TIME, SEC        ERROR VOLTS        COMP. OUT,, V        AMP. VOLTS
RCD  B       EXCIT. VOLTS     GEN. VOLTS         LOAD AMPS
RCD  B  NO MACHINE LANGUAGE INSTRUCTIONS
END
```

Figure IX-17. Problem, diagram, and corresponding DYSAC program. Source: Hurley and Skiles [HJ63], p. 76.

actual analog simulation are shown in Figure IX-16. A block diagram of an illustrative problem and the corresponding DYSAC program is shown in Figure IX-17.

4. DAS

The DAS (*D*igital *A*nalog *S*imulator) system on the IBM 7090 also provides a number of components representing analog computer elements. A sample block diagram and equivalent program are shown in Figure IX-18.

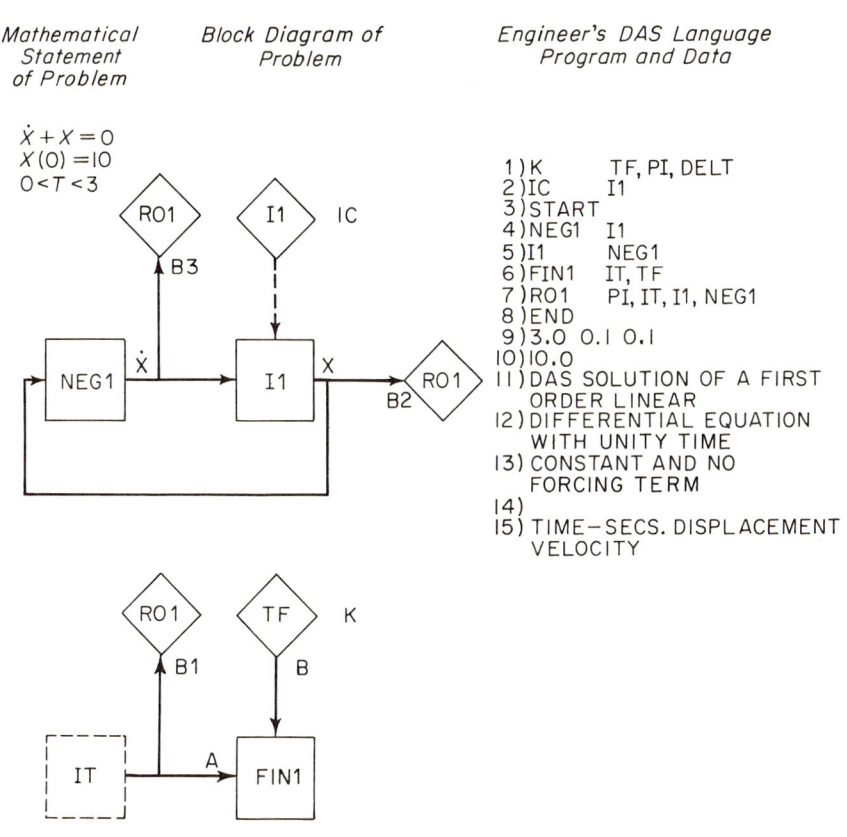

Figure IX-18. Mathematical problem, block diagram, and corresponding DAS program.
Source: Gaskill [GL64], p. 417.

5. DSL/90

A more recent system called DSL/90 is an extension of FORTRAN implemented on the IBM 7090/94. A few of the blocks, switching functions, and function generators available to the user are shown in Figure IX-19. A block diagram and the program to represent it are shown in Figure IX-20. The variable TIME is a system name representing the independent variable of integration and it can be renamed by the user. The right-hand sides of the assignment statements are merely FORTRAN expressions, since things such as INTGRL are merely new functions as far as FORTRAN is concerned.

FUNCTIONAL DESCRIPTION OF STANDARD DSL/90 BLOCKS

GENERAL FORM	FUNCTION
Y = INTGRL (IC, X) Y(0) = IC INTEGRATOR	$Y = \int_0^t X\,dt + IC$ EQUIVALENT LAPLACE TRANSFORM : $\frac{1}{s}$
Y = MODINT (IC, P_1, P_2, X) MODE-CONTROLLED INTEGRATOR	$Y = \int_0^t X\,dt + IC$ $P_1 = 1, P_2 = 0$ Y = IC $P_1 = 0, P_2 = 1$ Y = LAST OUTPUT $P_1 = 0, P_2 = 0$
Y = DELAY (N, P, X) P = TOTAL DELAY IN TERMS OF INDEPENDENT VAR. N = MAX NO. OF POINTS DELAY DEAD TIME (DELAY)	$Y(t) = X(t-P)$ $t \geq P$ $Y = 0$ $t < P$ EQUIVALENT LAPLACE TRANSFORM : e^{-Ps}

SWITCHING FUNCTIONS

Y = FCNSW (P, X_1, X_2, X_3) FUNCTION SWITCH	$Y = X_1$ $P < 0$ $Y = X_2$ $P = 0$ $Y = X_3$ $P > 0$
Y = INSW (P, X_1, X_2) INPUT SWITCH (RELAY)	$Y = X_1$ $P < 0$ $Y = X_2$ $P \geq 0$

FUNCTION GENERATORS

GENERAL FORM	FUNCTION
Y = AFGEN (FUNCT, X) ARBITRARY LINEAR FUNCTION GENERATOR	Y = FUNCT (X) $X_0 \leq X \leq X_n$ LINEAR INTERPOLATION Y = FUNCT (X_0) $X < X_0$ Y = FUNCT (X_n) $X > X_n$
Y = PULSE (P, X) PULSE GENERATOR WITH P AS TRIGGER	Y = 0 INITIAL Y = 1 $T_k \leq t < (T_k + X)$ Y = 0 OTHERWISE $k = 1, 2, 3, \ldots$ $T_k = t$ OF PULSE k, P_k
Y = NORMAL (P_1, P_2, P_3) NOISE GENERATOR (NORMAL DISTRIBUTION)	Y = GAUSSIAN DISTRIBUTION WITH MEAN, P_2, AND STANDARD DEVIATION, P_3 (P_1 = ANY ODD INTEGER)

Figure IX-19. Examples of some DSL/90 functional blocks, switching functions, and function generators.
Source: Syn and Linebarger [QP66], extracts from pp. 167–169.

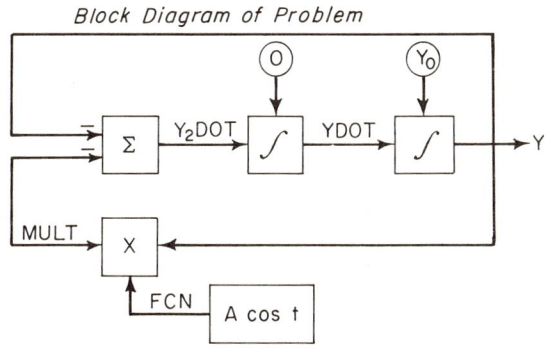

Mathematical Statement of Problem

$\ddot{y} + (1 + A \cos t) y = 0 \quad \dot{y}(0) = 0, \quad y(0) = Y_0$

Basic DSL/90 Program
```
Y2DOT =  -Y*(1.+ A*COS(TIME))
YDOT  = INTGRL (0.,Y2DOT)
Y     = INTGRL (YO,YDOT)
```

Figure IX-20. Differential equation, diagram, and corresponding DSL/90 program.
Source: Syn and Linebarger [QP66], p. 169.

There are also control statements, e.g., *PRINT*, *PREPAR*, *GRAPH*, *LABEL*, and *ARRANGE*. Control statements can be used to set error bounds or cutoff conditions, integration methods, etc. (e.g., **CONTRL** and **CONTIN**).

One of the interesting characteristics of DSL is that it has a certain amount of nonprocedural aspect to it since it will alter the sequence of input statements if desirable, and any operational statement is considered properly sequenced if all its inputs are available.

IX.2.5. Compiler Writing

1. Introduction

It may come as somewhat of a shock to some readers when they realize that compiler writing is being considered a special application, in the same category as the other sections in this chapter. However, this is a perfectly justifiable viewpoint, since the preparation of a compiler is a relatively narrow task from a technical point of view, although it may actually be of more interest to many readers than several of the other specialized application areas discussed here. It is essential to realize that this is definitely *not* a section on writing compilers, nor does the reference list at the end of this chapter provide a complete bibliography of compilation techniques. An overall view of compilation techniques from several (but not all) viewpoints is given by Feldman and Gries [FJ68]; that report also contains a good bibliography.

To reemphasize the difference between compiler-writing techniques and languages, we can consider the analogies between techniques used for solving differential equations and FORTRAN, or the systems design and analysis for an accounts receivable application and the language (e.g., COBOL), in which this application is actually written. Unfortunately, one major class of techniques for writing compilers, namely the syntax-directed (sometimes called *syntax-oriented* or *table-driven*), tends to blur the distinction between the technique and the language.

It is ironic that many compiler writers have tended to be very resistant to the use of higher-level languages for their own application. There has been a common attitude of "it is all right for people writing application programs to use my compilers, but I cannot afford any potential inefficiency in the compiler itself". This is probably one of the reasons that there has been less development along the lines of languages for writing compilers than one might expect.

There are really two major directions in which compiler-writing languages have progressed. The second has just been mentioned, and the first one is not very prevalent currently. The first, and actually classical, approach to a compiler-writing language is simply to design a language which provides in a convenient way those facilities deemed necessary by a compiler writer, independently of the types of algorithms that he might wish to use for various phases of the compiler-writing activity such as scanning, storage allocation, and code generation. One of the earliest of these languages is the CLIP work (see Section IX.2.5.2) done at the System Development Corporation; this was mentioned in Section VIII.3.1 as part of the history of JOVIAL.

There does not seem to be any other significant and successful development in the area of providing or defining a language specifically useful for writing compilers, although various attempts have been made. The relatively recent emergence of PL/I may cause it to become a strong contender for use in this area since it contains essentially all the provisions and features that are needed. It has in fact been used to write much of a large time-sharing system (i.e., MULTICS) which is an application with some similarity to compiler writing.

The modern techniques for compiler writing are primarily based on the concept of syntax-directed compilation, sometimes called table-driven compilers. The foundation work in this technique was done by Irons [IR61], Glennie [GC60], and Brooker and Morris [BX62]. Even earlier—though unpublished—work by Barnett at M.I.T. on the SHADOW systems had many of the rudiments of this concept (see Barnett and Futrelle [BI62]). It is completely beyond the scope of this book to provide a description of what is meant by syntax-directed compiling. A good reference for obtaining fundamental information about this subject is the article by Cheatham and Sattley [CH64]; the article by Floyd [FL64] also provides useful background. Unfortunately, the distinction between the syntax-directed compiling technique and the languages that have been developed along these lines is an extremely subtle one which cannot be described here. A further and even more difficult subtlety is the distinction between the metalanguage used to define the language syntax (as discussed in Section II.6.2.2) and the language used to write programs to handle the definitions. In some cases mentioned here, this difference is blurred or ignored for the sake of simplicity. As in the other sections in this chapter, only a few examples can be given to show the flavor of what is involved. The essential point is that given a set of syntactic definitions for a language, there needs to be a program written to handle them, and a language might be designed to be useful in writing programs of that type. Furthermore, there is increasing recognition of the need for defining the semantics of the language, and again a possible need for a language to write either the semantic definitions or the programs to handle such definitions. As the simplest illustration of this concept, we must recognize that a syntactic definition can specify that A+B is legitimate, but we need further information to know that the machine code for such an expression might be something of the form CLA A, ADD B. One of the difficulties in discussing these languages is that they presuppose a particular form for the syntactic definition of the language. These are not described.

Various attempts at compiler-writing language systems, not discussed in later sections, are listed in the references for this section at the end of the chapter.

2. CLIP

The CLIP work was literally an early attempt to define a language which would be useful for writing compilers; however, the designers rapidly reached the conclusion that this was not an application significantly different from a more general information processing problem, hence the acronym CLIP is for *C*ompiler *L*anguage for *I*nformation *P*rocessing. The designers used ALGOL 58 (née IAL), but they made the essential and obvious additions to it in the area of data manipulation and declarations and input/output. Specifically, they added a type declaration to specify the type and size of unsubscripted variables. For example,

Type(10,A,B: 6H,C: D,E,F)

declares A and B as integers less than the value 10; C is an alphanumeric symbol of 6 characters; and D, E, and F are Boolean variables. In addition to this, a table declaration is used to specify subscripted variables. A string declaration specifies up to 120 character positions. Finally, there is an origin declaration to specify the arrangement of tables, string and/or unsubscripted variables, in machine storage. READ and WRITE statements are used for input and output.

CLIP was one of the first illustrations of a compiler used to write itself since portions of it were written in CLIP and hand-translated to 709 machine code. As discussed earlier, in the description of JOVIAL, the latter was an outgrowth of CLIP, and JOVIAL itself has been used to write many of its own compilers. CLIP has thus served its purpose and faded away.

3. *TMG*

One of the apparently successful attempts at a compiler-writing system is TMG. According to its developer, McClure, "The original objective of this system was to make it as easy as possible to construct a simple one pass translator for some specialized language."[5] An example of a statement in TMG is the following:

INTEGER .. ZERO* MARKS DIGIT DIGIT* INSTALL

This statement says that the scanning mechanism should skip over an arbitrary number of leading zeros, mark the start of the string, find at least one digit and then space over all additional consecutive digits, and put the symbol in the symbol table and the output tree. Arithmetic can be done during compilation using a function COMPUTE which has an assignment statement as its argument, e.g., COMPUTE (NEXT−VALUE = LAST−ONE+2). A conditional statement can also be written, e.g., IF (INTVAL .LE. LAST−LABEL). There are a number of built-in functions in TMG; e.g., ARBNO ( ) looks for an arbitrary number of occurrences of the syntactic unit which is its argument, GLOT spaces over the input string to the next card boundary, etc. The reader interested in pursuing this system in detail will find a completely worked out example in [MZ65a]. A small section of this is shown in Figure IX-21.

TMG was used to write ALTRAN (see Section VII.5). It has been implemented on a few machines.

---

[5] McClure [MZ65a], p. 262.

```
STOP-STATEMENT..  $STOP$                    =$(    TRA   S.JXIT    //$)
GO-TO-STATEMENT..  $GO$ $TO$ LABEL = $(     TRA    $P1//$)
IF-STATEMENT..  $IF$ $($ BOOLEAN-EXPRESSION $)$ $,$/*+1 UNLABELED-STATEMENT =
    (1)$($P2(BV1)$P1$Q1 BSS     0     //$)
ASSIGNMENT..  VARIABLE $=$ SUM = $($P1$P2(STO)$)
BOOLEAN-EXPRESSION..  SUM RELATIONAL-OPERATOR SUM = (1)$($P1      STO       R.TMP
$P3       FSB     R.TMP     //$P2(BV1)$)
RELATIONAL-OPERATOR..  $.NE.$/RO1 = (1)$(    TZE     $Q1    //$)
RO1..          $.EQ.$/RO2    = (1)$(    TNZ     $Q1    //$)
RO2..          $.GT.$/RO3    = (1)$(/TZE/$Q1///TPL/$Q1//$)
RO3..          $.LT.$         = (1)$(/TZE/$Q1///TMI/$Q1//$)
*
*
SUM..      TERM = $($P1$)/SUMA
SUMA..AOP     TERM = $($P1     STO    A.TMP+$F3(T)$F1(T)
$P3     $P2    A.TMP+$F2(T)$F3(T)//$1/SUMA
AOP..  $+$/($-$ = $(FSB$)) = $(FAD$)
*
TERM..     PRIME  = $($P1$)/TERMA
TERMA..   $*$/TMA1 PRIME = $($P1      STO    A.TMP+$F3(T)$F1(T)
$P2     XCA       //      FMP    A.TMP+$F2(T)$F3(T)//$)/TERMA
TMA1..  $/$    PRIME = $($P1     STO    A.TMP+$F3(T)$F1(T)
$P2     FDP    A.TMP+$F2(T)$F3(T)    //     XCA    //$)/TERMA
*
PRIME..  VARIABLE/PR1 = $($P1(CLA)$)
PR1..  CONSTANT/PR2 = $(    CLA   =$P1//$)
PR2..($/PR3     SUM     $)$ = $($P1$)
PR3..SIMPLE-VARIABLE $($ SUM $)$ = $($P1     STO    A.TMP+$F3(T)
    TSX     $P2,4    //     PZE A.TMP+$F3(T) //$)
VARIABLE..  SIMPLE-VARIABLE $($/V1 CHECKFLAG(ARRAY)/V2 SUM $)$ =
    (1)$($P1    UFA    =0232000000001    //    ARS    1//    PAC    ,4///$Q1/$P2-1,4//$)
V2..  SETFLAG(FUNC) **
V1..  NOFLAGS = (1)$(     $Q1     $P1//$)
*
SIMPLE-VARIABLE..  MARKS LETTER   LETTER-OR-DIGIT* INSTALL = $($P1$)
INTEGER..  ZERO* MARKS DIGIT DIGIT* INSTALL = $($P1$)
*
IO-LIST..  IO-VAR      ($,$ IO-VAR )* = (1)$(    TSX     $Q1,4
$P2$P1    TSX    X.IEND,4    //$)
IO-VAR..  VARIABLE = $($PI(STR)$)
CONSTANT..  MARKS DIGIT DIGIT* PERIOD/CN1 DIGIT* INSTALL = R1
CN1..  INSTALL = $($P1..$)
               .DEFINITIONS.
READ   =   $(X.READ$)
PRNT   =   $(X.PRNT$)
FAD    =   $(FAD$)
FSB    =   $(FSB$)
CLA    =   $(CLA$)
STO    =   $(STO$)
BV1    =   $($Q1$)
STR    =   $(STR$)
EMPTY  =   $($)
R1     =   $($P1$)
           END
```

Figure IX-21. Example of part of TMG program.
Source: McClure [MZ65a], p. 273. By permission of Association for Computing Machinery, Inc.

4. COGENT

The COGENT (*CO*mpiler and *GEN*eralized *T*ranslator) programming system on the CDC 3600 and 3800 is an attempt to combine elements from syntax-directed compiling and list structures. Quoting from its developer, Reynolds, "Thus a program written in the COGENT language is a list processing program in which the list structures normally represent phrases of one or more *object* languages (i.e., the input and output languages to be processed by the program), in a representation determined by the syntax of these languages. Correspondingly, the COGENT language itself contains two major constructions: *productions*, which define the object language syntax, and *generator definitions*, which define list processing subroutines called generators."[6] The productions are similar to those of the Backus Normal Form discussed in Section II.6.2. A generator is actually a subroutine for manipulating list structures and has a language form similar to that of an ALGOL procedure. However, it differs conceptually in that the values of the variables for the generators are list structures, usually representing phrases of the object language. The meaning of variables, expressions, and simple assignment statements is the same in COGENT as in most programming languages. However, there are synthetic and analytic assignment statements to deal with list structures. As an example of an analytic assignment statement, if Z has the value (TERM/ABE*BED) then the statement

$$Z \;/= \;(TERM/(FACTOR)*(FACTOR)),X,Y.$$

will give X the value (FACTOR/ABE) and Y the value (FACTOR/BED). As an example of a conditional statement, preceding the statement above by

$$+10 \;\; IF$$

will cause X and Y to be assigned the factors of Z if it has any, and it will transfer control to statement 10; otherwise control goes to the next statement without changing X and Y.[7]

A complete example for translating algebraic expressions from Polish prefix notation into conventional infix notation is given in Reynolds [RE65a]. A small section of this is shown in Figure IX-22.

5. META 5

On the West Coast, an active working group of the ACM Los Angeles Chapter and a number of specific individuals have been involved in producing

---

[6] Reynolds [RE65a], p. 422.
[7] Reynolds [RE65a], extracted from pp. 426–27. By permission of Association for Computing Machinery, Inc.

```
$SECSYN
    (FACTOR) = (VAR),(()(EXP)()).
    (TERM)   = (FACTOR),(TERM)(MOP)(FACTOR).
    (EXP)    = (TERM),-(TERM),(EXP)(AOP)(TERM).
$PROGRAM
$GENERATOR GVAR ((X)
    X /= (EXP/VAR)),X. $RETURN(X). ).
$GENERATOR GMULT((OP,X,Y)
    +1 IF X =/ (EXP/TERM)),X.
    X /= (TERM/(()(EXP)())),X.
1/  +2 IF Y =/ (EXP/(FACTOR)),Y.
    Y /= (FACTOR/(()EXP)())),Y.
2/  X /= (EXP/(TERM)(MOP)(FACTOR)),X,OP,Y.
    $RETURN(X), ).
```

Figure IX-22. Example of part of COGENT program.
Source: Reynolds [RE65a], p. 434. By permission of Association for Computing Machinery, Inc.

a series of syntax-directed compiling systems known as the *META* compilers. Three references to differing phases in this development are listed at the end of this chapter. The META 5 version has proved to be fairly powerful, and it has been used for the translation of Basic FORTRAN to JTS (one of the versions of JOVIAL), and JS (another version of JOVIAL) to PL/I. The META 5 language has a number of predefined functions in the areas of data manipulation, arithmetic, relations, and input/output. The language definition itself uses essentially Backus' notation. The example shown in Figure IX-23 converts constants written in JOVIAL format to those of

```
            'ABC','(('"))'
            **THIS IS A DEMONSTRATION PROGRAM
                IN META5 **                              (1)
            .META5 DEMO SEQCON                           (2)
            .ITEM CNT *V;                                (3)
            SEQCON = 1$(.BK HOLCON .OUT(-*/));           (4)
            HOLCON = .NUM 'H('.SET(*V).PUT(*,CNT)
                .PUT('"',*)                              (5)
            $1(.GR(CNT,0)(-'".CHAR/.CHAR.PUT('"',*))
                .SOR(CNT))                               (6)
            ')'.PUT('"',*);                              (7)
            .END                                         (8)
```

Figure IX-23. META program to convert JOVIAL constants to PL/I constants.
Source: Oppenheim and Haggerty [OP66], p. 467. By permission of the Association for Computing Machinery, Inc.

PL/I format. In the former, the actual format is nH (n characters), whereas in PL/I the characters are simply enclosed within apostrophes. Thus, 3H(ABC) is converted to 'ABC'; the conventions on apostrophes in PL/I cause apostrophes to be duplicated within a constant if they occur once, so 5H((('))) converts to '((''))' in PL/I.

6. TRANDIR

One of the most prolific groups in the field of compiler writing is the small firm, Massachusetts Computer Associates, Inc. (COMPASS). They have developed a series of compilers and compiling techniques, most of which are beyond the scope of this section. However, the TGS-II (*T*ranslator *G*enerator *S*ystem) involves a significant language embedded within an overall system. The language is called TRANDIR, and the system is called TRANGEN; the latter is essentially an interpreter which executes programs written in TRANDIR to describe a specific translation process. TRANDIR contains primarily an algebraic section, a pattern-matching section, and a number of built-in functions. The TRANDIR language is used for all phases of the compiler, as distinguished from the FSL language to be discussed later. TRANDIR contains a number of action operators, such as CYCLE, EMIT, and SCAN, which involve system-defined tables. There are

```
U1...$IF REL $THEN USTAT //   $ELSE    //        Z = FAIL. FAIL = NEWLBL.
                                                 EMIT(TRANS,FAIL).
                                                 EMIT(LABEL,Z).
                                           Z1.. EXCISE. SCAN(1).TRY(10).


U2...$IF REL $THEN USTAT //   $.         //   Z2.. EMIT(LABEL,FAIL).
                                              Z3.. CYCLE. TO(Z1).


       ...$ELSE USTAT //     $.          //        TO(Z2).


          ...USTAT //        $.          //        TO(Z3).


          ...USTAT //                              PRINT(‡ERROR 3‡).
                                                   RELEASE.
                                                   CURSYM = $. . TRY(U2).
```

Figure IX-24. Example of part of TRANDIR program.
Source: Dean [DE64], p. 73.

symbol descriptor variables as well as integer and statement label variables. The symbol descriptor variables refer to locations in a communications area. As an illustration, one of the built-in functions is COPER, which has as an argument a symbol descriptor value which represents a terminal symbol of the language being translated. The action statements listed earlier are really a set of functions that specify the operations to be performed to execute a given translation. For example the EMIT transfers a sequence of symbol descriptors from one table to another; PHRASE replaces the syntactic structure just analyzed with a symbol descriptor value specified by the argument of the phrase. A pattern test can be used to recognize specific syntactic patterns for macro sequences. A small portion of a TRANDIR program to translate a simple language is given in Figure IX-24.

7. *FSL*

The FSL (*Formal Semantic Language*) work on the G-20 by Feldman [FJ66] and its subsequent use in the VITAL system (see Mondshein [LQ67]) is a specific example of a well-defined language for expressing semantics. Its actual usage for the development of Formula ALGOL (see Section VIII.5) is described by Iturriaga *et al.* [IT66]. In Feldman's system, the overall compiler-writing system uses two formal languages to describe a compiler. The first is the *production language* which is used to define the syntax analyzer, and the second is the Formal Semantic Language which is used to write a set of routines to define the semantics of the source language. The basic unit in an FSL program is a labeled statement. Whenever a production contains something of the form EXEC *n* then the semantic routine labeled *n* is executed. For example, whenever a double-precision real variable is declared in a source program, the following semantic routine is executed:

TO ← STORLOC; SET[TO, DOUBLE]; ENTER[SYMB; LEFT2, TO, REAL, LEV]; STORLOC ← STORLOC +2 ↓

This causes the current value of STORLOC to be placed in a temporary location with bits marking it as a double-precision operand. A description of the variable, containing its name, the tagged address, the word REAL, and the current level, is placed in the SYMBol table. The storage pointer is then increased by 2. An illustration of part of a semantic routine to check the types of A and B in a sum A+B is shown in Figure IX-25.

8. *AED (cross-reference only)*

The AED system is discussed in Section IX.3.4.2 because the key developers tend to continually describe it as being related to computer-aided design. However, many of the concepts and techniques are directly related to compiler writing, and it is worth noting that fact here.

```
100 ↓ TEST[LEFT4,BOOLEAN] V TEST[LEFT2,BOOLEAN] →
        FAULT 100 : RIGHT2 ← RIGHT2 ∧ X7 ;
      TEST[LEFT4,FORMULA] V TEST[LEFT2,FORMULA] →
        MACHINE ← 2 ;
      ∼ TEST[LEFT4,FORMULA] →
          CODE(CONSTRUCTFORMULA[LEFT4]) $;
      ∼ TEST[LEFT2,FORMULA] →
          CODE(CONSTRUCTFORMULA[LEFT2]) $:
      MACHINE ← 1 ;
      TEST[LEFT2,REAL] V TEST[LEFT4,REAL] →
        SET[RIGHT2,REAL] : SET[RIGHT2,INTEGER] $$$
```

Figure IX-25. Example of part of FSL program.
Source: Iturriaga *et al.* [IT66], p. 248.

IX.2.6. MISCELLANEOUS

The previous sections in this chapter have discussed languages for the indicated application areas. The purpose of this last section for special application areas is to mention briefly some languages which do not conveniently (or by any stretch of the imagination) fall into one of the other sections. Except for the Matrix Compiler, the languages here are relatively recent and pretty much "one of a kind". While there are other languages which could be considered to fall into special categories, the ones mentioned here appear to be the most unusual or interesting.

1. *Matrix Computations: Matrix Compiler*

One of the earliest of these *specialized* systems was the UNIVAC Matrix Compiler. It was highly oriented to the 12-character UNIVAC word and thus violates some of the criteria in Section I.4.2. Its interest is historical since it provided the user with language and facilities for performing a number of operations on matrices, including addition, multiplication, inversion, computing the norm, and transposition. For example, if the user wrote I+MULTY00IJ0 this meant that the matrices on tapes I and J were to be multiplied together and have the identity matrix added to them to produce the result. This was followed by another 12-character operation to specify information about the output.

2. *Cryptanalysis: OCAL*

The OCAS(*O*n-line *C*ryptanalytic *A*id *S*ystem) of Edwards [ED66] is an on-line system designed to ease the work of a cryptanalyst. It contains a display generator, control program, and also a language OCAL (*O*n-Line *C*ryptanalytic *A*id *L*anguage). The language requirements include the need for manipulating strings of alphabetic characters and also for doing algebraic computations, including matrix operations. OCAL is primarily a

```
PERIOD:  PROCEDURE (CRYPT,PER,N,M)
*
*   PARAMETERS ARE:
*     CRYPT - A STRING GIVING THE CRYPTOGRAM
*     PER - AN INTEGER VECTOR WITH SUBSCRIPT RANGE N TO M
*           THE WEIGHTED TALLIES ARE RETURNED IN THIS VECTOR
*     N - AN INTEGER GIVING THE LOWEST PERIOD TO BE TESTED
*     M - INTEGER GIVING HIGHEST PERIOD TO BE TESTED
*
*   DECLARATIONS NEXT
*
        STRING CRYPT
        CHARACTER C
        INTEGER PER,N,M
        INTEGER DIST,INDEX,ALPS,K,L1,L2
        INTEGER SEP,TR
        READER R
        DECLARE PER(*)
*
*   THE VECTOR PER IS DIMENSIONED IN THE CALLING PROCEDURE
*
        DECLARE DIST(LENGTH.(CRYPT)/5)
*
*   DECLARING THE LOCAL VECTOR DIST
*
        READER R
*
*   THE ACTUAL PROCEDURE BEGINS HERE
*
        ATTACH.(R,CRYPT)
        ALPS = SIZE.(ALPHABET.(CRYPT))
        INDEX = 1
        DO WHILE INDEX .LE. ALPS, LOOP1: BEGIN
*
*   ITERATE OVER THE SIZE OF THE ALPHABET.
*
            C =$NC.(R)
*
*   RETURN READER TO HEAD OF STRING AND READ FIRST CHARACTER
*
            K = 1
            DO UNTIL ENDSTRING, LOOP2: BEGIN
*
*   READ THE STRING CRYPT CHARACTER BY CHARACTER
*
                IF C .E. INDEX, COND1: BEGIN
                    DIST(K) = $V.(R)
*
*   RECORD DISTANCE FROM HEAD OF STRING
*
                    K = K + 1
                END COND1
                C = $IC.(R)
*
*   INCREMENT THE READER AND READ NEXT CHARACTER
*
            END LOOP2
            L1 = 1
            DO UNTIL L1 .E. K, LOOP3: BEGIN
*
*   COMPUTE THE CHARACTER DISTANCE BETWEEN EACH OCCURENCE
*
                L2 = L1 + 1
                DO UNTIL L2 .G. K, LOOP4: BEGIN
                    SEP = DIST(L2) - DIST(L1)
                    TR = N
                    DO UNTIL TR .G. M, LOOP5: BEGIN
*
*   TEST EACH PERIOD FROM N TO M FOR REMAINDER 0
*
                        IF (SEP .R. TR) .E. 0,
                            PER(TR) = PER(TR) + 1
                        TR = TR + 1
                    END LOOP5
                    L2 = L2 + 1
                END LOOP4
                L1 = L1 + 1
            END LOOP3
            INDEX = INDEX + 1
        END LOOP1
*
*   NOW WEIGHT EACH ITEM IN PER BY THE RESPECTIVE PERIOD
*
        K = 1
        DO UNTIL K .G. M, LOOP6: BEGIN
            PER(K) = PER(K) + 1
            K = K + 1
        END LOOP6
*
*   ALL DONE
*
        END PERIOD
```

Figure IX-26. OCAL program for solving a cryptanalysis problem.
Source: Edwards [ED66], pp. 51–52.

644 SPECIALIZED LANGUAGES

synthesis of MAD and SNOBOL, together with some ideas from SLIP and PL/I. OCAL includes declarations and statements for sequence control, string pattern-matching, assignment, and error control. An example of an OCAL program is shown in Figure IX-26.

3. *Movie Creation: Animated Movie Language and BUGSYS*

The very interesting work of Knowlton [KO64] in using a computer to produce movies involved the use of two languages—*scanner* and *movie*—on the 7090. Scanner is based on a conceptual framework which includes a number of scanners which the programmer imagines sitting on various squares of the surfaces. An example[8] of an instruction in the scanner language is

IFANY (B,R,10) (B,A,C) (A,E,7)T(A,T,B) (A,U,2) (A,W,3) LOC5

This says that if scanner B is to the Right of X=10, or scanner B is Above

```
         SURF   MACRO   X1,Y1,WIDTH,HEIGHT,NAME,SIZELN,SHIFT (SHIFT OPTIONAL)
                PLACE   B,BB,X1,Y1
                THEN    (A,T,B) (A,R,WIDTH) (A,L,1) (A,U,HEIGHT) (A,D,1)
                PAINT   A,B,WRITE,2              PAINT RECTANGULAR AREA
                                                 WITH 2'S
                THEN    (C,T,B,) (C,U,1) (C,R,1) (D,T,B) (D,D,8)
                TYPE    NAME,C,8X11,1,1,WRITE,0  TYPE NAME WITH 0's
                CENTER  A,B                      CENTER IT
                IFF     0,/CRS/SHIFT
                THEN    (D,R,SHIFT)              SHIFT SIZE LABEL RIGHT
                TYPE    SIZELN,D,5X7,1,1,WRITE,7 TYPE SIZE LABEL
                ENDM                             END OF SURF DEFINITION
   *                                             END OF DEFINITIONS OF NEW
                                                 MACROS
                FILM                             BEGINNING OF PROGRAM
                FRAMES  0                        (NO OUTPUT DESIRED DURING
                                                 TYPING)
                PAINT   BB,O,WRITE,O             CLEAR SURFACE BB TO 0'S
                BENTLN  46,173( (L,3,D,2) (D,18) (L,3,D,2) (R,3,D,2) (D,17) (R,3,D,2) )
                BENTLN  95,173( (R,3,D,2) (D,46) (R,3,D,2,) (L,3,D,2,) (D,40) (L,3,D,2) )
                PLACE   C,BB,8,167               (TWO BRACES NOW DONE)
                TYPE    ORSUR,C,5X7,1,2,WRITE,7  TYPE "OR SURFACE ..." NOTE
                PLACE   C,BB,10,60
                TYPE    FIG2A,C,5X7,1,3,WRITE,7  TYPE "FIG. 2A..." CAPTION
                PLACE   C,BB,103,125
```

Figure IX-27. Part of movie animation program.
Source: Knowlton [KO64], p. 86.

---

[8] Knowlton [KO64], p. 69.

(i.e., in a higher line than) scanner C, or scanner A is sitting on a number Equal to 7, then the following things are to be done: Scanner A moves To the same surface and the same square as scanner B, scanner A moves Up 2 squares and Writes the number 3; control then goes to the line of coding labeled LOC5. Operations include tests on the position of the scanner relative to particular coordinates, tests on size, etc. There are operations for moving horizontally, vertically, up and down, and left and right for a specified number of squares, and provisions for doing simple arithmetic. The flavor of this language is very similar to that of L^6 (discussed in Section VI.4); this is hardly surprising since both languages were developed by the same person.

Instructions of the movie language fall into three categories: (1) Controlling the output or temporary storage of pictures, e.g., CAMERA, FRAMES, and UNTIL; (2) performing drafting and typing operations, e.g., entire scanner language, LINE, and ARC; and (3) modifying rectangular areas, e.g., BORDER, GROW, DISOLV, ZOOMIN. A portion of a complete program is shown in Figure IX-27.

The BUGSYS system on the 7094 is based on the work just described;

```
DOWN3   MOVE    AA, DOWN, 1
        GO.TO   ERROR
        TEST    AA, EQUAL, 6
        GO.TO   DOWN3
        CENTER  AA, COLUM, 12, 6
        GO.TO   ERROR
        GO.TO   ERROR
        PLACE   ORIGIN, A, AA—1
RIGHT   MOVE    AA, RIGHT, 1
        GO.TO   ERROR
        TEST    AA, EQUAL, 6
        GO.TO   ERROR
        CENTER  AA, COLUMN, 12, 6
        GO.TO   ERROR
        GO.TO   BACK
        GO.TO   RIGHT
BACK    MOVE    AA, LEFT
        PLACE   BB, AA, AA—1
UP1     MOVE    BB, UP, 1
        GO.TO   ERROR
        TEST    BB, EQUAL, 0
        GO.TO   UP1
```

Figure IX-28. Example of part of BUGSYS program.
Source: Ledley, Jacobsen, and Belson [LL66], p. 82. By permission of Association for Computing Machinery, Inc.

it can be used for varying types of applications, including the analysis of photomicrographs of neuron dendrites. The concept of the system is the use of a set of figures which are visualized as a family of *bugs*. A bug can be placed and then moved. It can also change the gray-level value of the spot on which it is located and can lay down a stick across a thick line in a picture as an aid to locating the middle of the line. The *PLACE* statement assigns a name and initial coordinates to a bug. The *MOVE* moves it a specified distance either horizontally or vertically. There is a series of test statements which examine the gray-level value of the picture at the location of the bug. The gray-level value may be changed by the *change* statement. A small section of a program is shown in Figure IX-28.

4. *Social Science Research: DATA-TEXT*[9]

The DATA-TEXT System, developed in the Department of Social Relations at Harvard University, is a system to aid people who are doing social science research. It was implemented on the IBM 7090/94. The user is allowed to specify information about the data and invoke a number of specific routines to do calculations for him. In some cases, the raw data (assumed to be on punch cards) can be used directly; whereas in other cases it must be modified somewhat before being used as input to do statistical analyses. For example, if the subject concerns the attitude of college students toward education, a partial indication of what the variables look like is

```
*VAR(1)  =  X(1)         = SEX OF SUBJECT (MALE/FEMALE)
*VAR(2)  =  X(2)         = YEAR AT COLLEGE (FIRST/SECOND/THIRD/FOURTH)
*VAR(3)  =  X(17)        = HOME REGION (NORTH/MIDWEST/WEST/SOUTH)
*VAR(4)  =  X(6)+1       = PARENTS EDUCATION (GRADE/HIGH SCHOOL/COLLEGE)
*VAR(5)  =  ORDER X(4)   = VERBAL IQ  (1-4=LOW/5-7=AVERAGE/OVER 7=HIGH)
*VAR(6)  =  ORDER X(5)   = MATH IQ    (1-4=LOW/5-7=AVERAGE/OVER 7=HIGH)
*VAR(7)  =  X(4)/X(5)    = RELATIVE VERBAL/MATH ABILITY
*VAR(8)  =  X(4)+X(5)    = TOTAL IQ SCORE
*VAR(9)  =  X(7)**2      = SQUARED ACTIVITIES INDEX
*VAR(10) =  X(8)*4 + X(4)*2 + X(9) = SOCIAL-ECONOMIC STATUS
*VAR(11) =  (X(3)+X(5)) / (X(11)+X(12)) = RATIO OF STUDY TIME
*VAR(12) =  SQRT(X(27)**2+X(28)**2) = MEAN SQUARE HOURS
*VAR(13) =  LOG(ABS X(35)/ABS X(36)) = DEVIATION INDEX
*VAR(14) =  INT X(43) = GRADE INDEX (A/B/C/D/E)
*VAR(15) =  ARCSIN X(29) = PERCENTAGE INDEX
```

Figure IX-29. Example of DATA-TEXT input data.
Source: Couch, A. S. [DF67], p. 2.

[9] This has no connection with the IBM on-line system for text editing called DATATEXT which is mentioned in Section IX.3.5.

given in Figure IX-29. In some cases it is necessary to transform the data, as illustrated in Figure IX-30. The user can also cause the reading of text material and request various kinds of content analyses as shown in Figure IX-31.

```
*VAR(51)     =  1  IF X(81) = 5        = FIELD OF CONCENTRATION
*OR          =  2  IF X(81) = 6 AND IF X(82) LESS THAN 3
*OR          =  3  IF X(83) GREATER THAN 9 OR IF X(84) = BLANK
*VAR(52)     =  X(25)+X(26)+X(27)  IF X(30) = 0 = CONDITIONAL INDEX
*OR          =  X(28)*50-X(29)     IF X(30) = X(31)
*OR          =  SUM X(16—20)       IF X(30) = BLANK
*VAR(53—60)  =  X(53—60)           IF VAR(1)=1-ARRAY OF SCORES
*OR          =  X(63-70)           IF (VAR(10)+VAR(11)) LESS THAN VAR(12)
*OR          =  SQRT X(83—90)      IF X(9)=2 AND IF X(8)=5 AND IF X(7)=1
```

Figure IX-30. Example of defining new variables based on conditions. Source: Couch, A. S. [DF67], p. 3.

```
*DECK     COMPARATIVE ANALYSIS OF INTERVIEW MATERIAL
*FORMAT (A6,74A1)/ UNIT,TEXT(74)
*CONCEPTS
 FAMILY  = MOTHER, FATHER, SON, DAUGHTER, CHILDREN
 SELF    = I, MYSELF, ME, MINE, MY
 MALE    = MAN, MASCULINE, HE, HIM, HIS, MEN, BOY
 FEMALE  = WOMAN, FEMININE, SHE, HER, WOMEN, GIRL
*CONCEPTS END
*COMPUTE WORD FREQUENCIES
*COMPUTE CONCEPT FREQUENCIES, COMPARE(UNIT)
```

Figure IX-31. DATA-TEXT program. Source: Couch, A. S. [DF67], p. 4.

5. *Equipment Check-out: STROBES and DIMATE*

To assist in the repair of electronic systems, STROBES (*S*hared *T*ime *R*epair *O*f *B*ig *E*lectronic *S*ystems) involves hardware and also a means for the user to communicate with the hardware. The language statements are generally of the form

label, opcode, parameter, parameter; comments

Other slightly more powerful forms are available. A few of the operations available include *TON* (turn the oscilloscope trace on), *TPI* (type in a set of integers starting at a specified location), *INI* (clear the temporary library), and *ATO* (type the alphanumeric contents of specified number of words

```
T TSK, MEM, 2, 1;
0000  ENT; SPACE FOR THE MARK
0001  PAR, 1; BRING THE FIRST PARAMETER
0004  STL,, L0; STORE IT IN LOCATION L0
0005  PAR, 2; BRING THE SECOND PARAMETER
0010  STL,, L1; STORE IT IN LOCATION L1
0013  STZ, 1, L0; STORE ZEROS IN THE LOCATION SPECIFIED BY L0
0014  CAL,, L1; CLEAR ADD THE PATTERN IN LOCATION L1
0015  STL, 1, L0; STORE THE PATTERN IN THE LOCATION SPECIFIED BY L0
0016  TOF; TURN OFF THE OSCILLOSCOPE TRACE
0017  RTM; RETURN TO MARK
0020  L0, 0; DECLARE L0 AND ITS CONTENTS
0021  L1, 0; DECLARE L1 AND ITS CONTENTS
0022  END; TERMINATE ASSEMBLY
```

Figure IX-32. Example of STROBES program. The sequence numbers are actually typed by the STROBES system.
Source: Quatse [QU65], p. 1071.

beginning at the specified location). An illustration of what the maintenance engineer might do is shown in Figure IX-32.

Another language to assist in check-out (see Scheff [SD66] and [SD66a]) is incorporated as part of DIMATE (*D*epot *I*nstalled *M*aintenance *A*utomatic *T*est *E*quipment) to assist in conducting equipment tests. The complete language can be divided into three major categories: Basic test processes (e.g., establishing proper time delays and evaluating measured values against specified limits), internal test program control and arithmetic computation (e.g., conditional jumps), and compiler processing. For example,[10] the artificial language statement

$$\text{CONNECT } 100 \text{ VDC } J80-19 \text{ } J80-12$$

would be written to mean *connect a* 100-*volt DC stimulus source to test points J*80-19 *and J*80-12. Similarly, the sentence *measure* 30.3 *ohms with a tolerance of* 12 *percent between joints J*1-20 *and P*1-20 would be written as

$$\text{MEASURE } 30.3 \text{ OHMS } \pm 12\% \text{ } J1-20 \text{ } P1-20$$

Stimuli available include voltage (both AC and DC), sine waves (both AM and FM), and pulse train. Measurement capabilities include voltage, resistance, frequency, and pulse width. Stimulus statements include relay supply, signal ground, and chassis ground. Measurement statements include AC and DC voltage, frequency, and time. A sample of a portion of a test program is given in Figure IX-33.

---

[10] Scheff [SD66], p. 261. By permission of Association for Computing Machinery, Inc.

```
        COMPILE VALIDATE9075
        PRINT      C. HUGHES TEST PROGRAM. UUT NO. 9075.
  T001  CONNECT 6,3    VF1L     60CPS    J10-1    J10-2
        CONNECT 120    VP3      400CPS   J10-3    J10-4   J10-5   J10-6
        MEASURE 250    VDC      25       J10-7    J10-8
        GOTO           AP001    A002
        PRINT      VOLTAGE TOO HIGH.
        HALT
  P001  PRINT      VOLTAGE TOO LOW._
  T002  CONNECT 45     VRMS     100CPS   ------- J10-9
        PRINT      CONNECT DC/AC PROBE TO  J10-10. PRESS PROCEED.
        PAUSE
        *MEASURE1,5    VRF      -,5      A
        CONNECT C      SWITCH   J10-11   J10-12   -------
        MEASURE
        GOTO                    B003
        PRINT      (MSG1)*
  T003  DISCONN 45     VRMS     100CPS   ------- J10-9
        DELAY   2      SEC      J10-13
        CONNECT 5,0    VDC      J10-14   J10-6
        TRIGGER
        MEASURE 0      KOHMS    .01      J10-15   J10-16
******  LOWER LIMIT IS OUT OF RANGE
        GOTO    -      A-       A004
        PRINT      (MSG1)*
  T004  DISCONN C      SWITCH
        CONNECT 2,0    VRMS     5KC      -10DB    J10/
******  ATTEN OUTPUT AT A13J14, CHECK PRINT MESSAGE
        DELAY   20     MSEC
        MEASURE 1      KOHMS    .01      J10-15   J10-16
        GOTO           P005     P004
        HALT
  P004  PRINT      END OF PROGRAM.
        HALT
  MSG1  MESSAGE    TEST OUT OF TOLERANCE.
        END     9075                     1        1
```

Figure IX-33. Program for testing equipment.
Source: Scheff [SD66], p. 266. By permission of Association for Computing Machinery, Inc.

IX.3. SPECIALIZED LANGUAGES ACROSS APPLICATION AREAS

IX.3.1. DISCRETE SIMULATION

1. Introduction

The subject of simulation languages is probably the one field that is receiving insufficient space in this book. That complaint might be rendered on the basis of the significant number of simulation languages that have been developed, the numbers of computers on which they have been implemented, and the amount of their usage, which might be as widespread as the string and list processing languages or the formal algebraic manipulation languages. The justification for including the simulation languages in this more specialized area with limited descriptions is that their usage is unique and presently does not appear to represent or supply much carry-over into other fields. In the areas of list and string processing, the techniques (if not the individual languages) have received significant usage in a wide variety of places and languages. The formal algebraic manipulation languages are either newer or less generally available, but they represent a potentially major usage for the entire field of scientific problems; most people with mathematical training should be able to determine for themselves the meaning of these languages and perhaps form an opinion on their potential future value. In the case of simulation languages, unless an individual has actually tried to simulate a system or a process, he is not apt to appreciate or completely understand the importance or subtleties of the facilities being provided.

There are two major subcategories within the simulation languages: Continuous and discrete. The continuous, and particularly the simulation of analog computers, was discussed in Section IX.2.4 rather than here because the languages require specialized knowledge of analog computers, whereas those in this section require only detailed knowledge of any problem area to be simulated. The discrete simulation languages themselves are again subdivided into flowchart-oriented and statement-oriented languages. In the first case, the user is defining blocks (and then flowcharts or equivalent diagrams) and assembling these blocks into a program structure that represents the system to be simulated. The statement description languages use programming statements to define conditions that must apply before certain actions can take place and to describe the results of these actions (including use of the time relationships between the system elements and activities). The most notable example of the *block diagram* approach is GPSS, while SIMSCRIPT exemplifies the *event* school. Stated another way, GPSS is fundamentally different from most of the other languages because it describes the structure and action of a system by using block diagrams in which each block represents a step in the action of the system. This is higher level but less flexible than the other languages which consist of statements combined to form subprograms. Except for GPSS and DYNAMO, each language requires some knowledge or understanding of a particular general purpose language.

A still further distinction that can be made is whether the languages are *stand-alone* or additions to existing languages. SIMULA is a direct extension of ALGOL; the others are independent, except that SIMSCRIPT is an extended FORTRAN-like language which in its first version used the implementation technique of translating to FORTRAN statements.

One other characteristic of the simulation languages is important to note. The latter represent a field in which the existence of a *language* spurred a great deal of activity, whereas the previous sets of subroutine packages had been used but not as extensively. The most widely used languages are GPSS and SIMSCRIPT; most of the others are research-oriented.

For the reader who is interested in pursuing the differences and similarities of these systems in a general way, there are several excellent articles, in particular those by Teichroew and Lubin [TE66], Krasnow [KQ67], and Kiviat [KW66]. I would like to specifically acknowledge the invaluable background given to me by those articles in preparing much of this section.

2. DYNAMO

One of the earliest of the simulation languages, and the only one discussed here which is continuous, is DYNAMO; this was developed at M.I.T. and completed as early as 1959 on the IBM 704. Versions were developed on the 709 and 7090 computers. A significantly improved version, DYNAMO II, was written using AED-0 (see Section IX.3.4.2) and was running on the 7094. Plans exist for making DYNAMO II available on the IBM System/360 and GE 645. Only DYNAMO (I) is discussed here.

DYNAMO is a continuous system, meaning that every basic variable is continuous and has a first derivative with respect to time. DYNAMO actually approximates the continuous process by a set of first-order difference equations. In this sense it is similar to the languages discussed in Section IX.2.4, but the area of usage is the same as the others in this section. Level equations are used to describe the basic variables. A level at TIME K depends only upon the preceding time instant (TIME J) and the rates of change over the time interval JK (K = J + DT), where DT is a standard time increment. For example, an inventory at TIME K (=IAR.K) is equal to the inventory at TIME J (=IAR.J), plus what was received during the interval JK (=SRR.JK), minus what was shipped during the interval JK (=SSR.JK). This is represented by the level equation

$$IAR.K = IAR.J + (DT)(SRR.JK - SSR.JK)$$

While that happens to be equivalent to the exact language format, most equations are represented differently than normally written. For example,

$$V = V + (DT)((P+Q)/Y)$$

is actually punched exactly as

$$V.K = V.J+(DT)(1/\pm Y)(\pm P+Q)$$

```
*       M478-248,DYN,TEST,1,1,0,0
RUN     2698JP
NOTE    MODEL OF RETAIL STORE
NOTE
1L      IAR.K=IAR.J+(DT) (SRR.JK-SSR.JK)           INVENTORY ACTUAL
1L      UOR.K=UOR.J+(DT) (RRR.JK+SSR.JK)           UNFILLED ORDERS
20A     NIR.K=IAR.K/DT                             NEGATIVE INVENTORY
20A     STR.K=UOR.K/DFR                            SHIPMENTS TRIED
54R     SSR.KL=MIN(STR.K,NIR.K)                    SHIPMENTS SENT
40R     PSR.KL=RRR.JK+(1/DIR) (IDR.K-IAR.K)        PURCHASE ORDERS SENT
12A     IDR.K=(AIR) (RSR.K)                        INVENTORY DESIRED
3L      RSR.K=RSR.J+(DT) (1/DRR) (RRR.JK-RSR.J)    REQUISITIONS SMOOTHED
39R     SRR.KL=DELAY3(PSR.JK,DTR)                  SHIPMENTS RECEIVED
NOTE
NOTE    INITIAL CONDITIONS
NOTE
12N     UOR=(DFR) (RRR)
6N      RSR=RRR
6N      IAR=IDR
NOTE
NOTE    INPUT
NOTE
7R      RRR.KL=RRI+RCR.K                           REQUISITIONS RECEIVED
45A     RCR.K=STEP(STH,5)                          REQUISITION CHANGE
NOTE    .
NOTE    CONSTANTS
NOTE
C       AIR=8 WKS                                  CONSTANT FOR INVENTORY
C       DFR=1 WK                                   DELAY IN FILLING ORDERS
C       DIR=4 WKS                                  DLY REFILLING INVENTORY
C       DRR=8 WKS                                  REQUISITION SMTHNG T C
C       DTR=2 WKS                                  DELAY IN TRANSIT
C       RRI=1000 ITEMS/WK                          REQ. RECEIVED INITIALLY
C       STH=100 ITEMS/WK                           STEP HEIGHT
NOTE
PRINT   1)IAR,IDR/2)UOR/3)RRR,SSR/4)PSR,SRR
PLOT    IAR=I,UOR=U/RRR=R,SSR=S,PSR=P,SRR=Q
SPEC    DT=0.1/LENGTH=10/PRTPER=5/PLTPER=0
```

Figure IX-34. DYNAMO example of model of retail store.
Source: Pugh [PG63], p. 17.

Multiplication is generally noted implicitly, i.e., without any intervening operator.

DYNAMO assumes that it is dealing with a continuous process with each state variable having its level and all derivatives with respect to time existing and known at each instant of time. DYNAMO also provides auxiliary equations to allow more complicated variables, initial value equations, constants, and tables for defining arbitrary functions. DYNAMO obtains results by performing a sequential solution of all the equations describing the system to be simulated. An example of a DYNAMO program is given in Figure IX-34.

3. *GPSS*

One of the earliest simulation languages is GPSS, first described publicly in 1961 (see Gordon [GG61]). It was originally implemented on the IBM

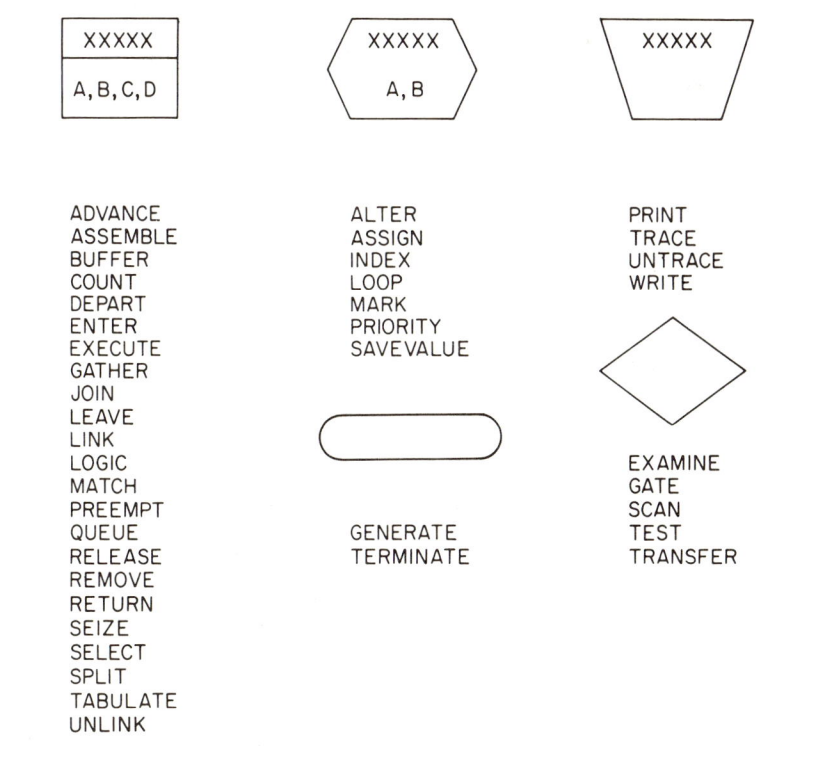

Figure IX-35. Some of the GPSS block types and corresponding operations. Source: [IB67j], p. 73. Reprinted by permission from *General Purpose Simulation System/360 Introductory User's Manual.* © 1967 by International Business Machines Corporation.

7090 and since then newer versions have been implemented on other machines, e.g., IBM 7040/44, 7090/94, System/360, RCA Spectra 70, and the UNIVAC 1107.

Unlike most of the other simulation languages, GPSS programs are based on a block diagram drawn by the user to represent the system he wishes to simulate. The principal data elements in GPSS are *transaction* (representing units of traffic), *equipment* acted upon by transactions (including *facilities, storages,* and *logic switches*), and *blocks* (specifying the logic of the system). Other elements are provided for statistical measurements,

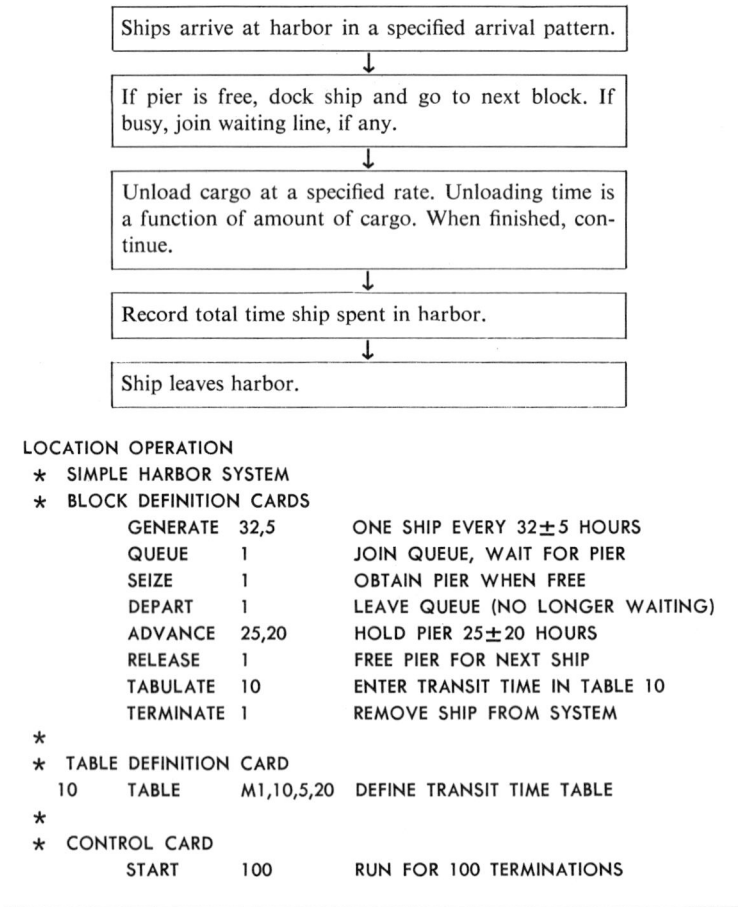

Figure IX-36. Example of harbor arrival problem and GPSS solution. Source: [IB67j], pp. 1 and 6. Reprinted by permission from *General Purpose System/360 Introductory User's Manual.* © 1967 by International Business Machines Corporation.

computation, data referencing, and chaining. A set of subroutines is associated with each type of standard block. The system is represented in terms of these blocks, and the program then creates transactions, moves them to the specified blocks, and executes the actions associated with each block. Some of the block formats and the symbols associated with them are shown in Figure IX-35. An example of a simple harbor system is shown in Figure IX-36.

GPSS appears to be easier to learn than the statement-oriented simulation languages, although it is somewhat less flexible.

4. *SIMSCRIPT*

SIMSCRIPT was developed at the RAND Corporation in the early 1960's, based on other work in developing programming packages. It has undergone improvement, including the development of SIMSCRIPT I.5 and II. SIMSCRIPT I.5 has been widely implemented and runs on at least the following machines: IBM 7090/94, System/360, CDC 3600, 3800, 6400, 6600, and 6800; Philco 210, 211 and 212; UNIVAC 490, 1107, and 1108.

The basic idea in SIMSCRIPT is that a system can be described in terms of *entities, attributes* (i.e., the properties associated with entities), and *sets*

Entity Operations	*Decision Commands*
CREATE	IF
DESTROY	GO TO
CAUSE	FIND
CANCEL	*Input/Output Commands*
FILE	SAVE
REMOVE	READ
	WRITE
Arithmetic and Control Commands	ADVANCE
LET	BACKSPACE
STORE	REWIND
DO TO	ENDFILE
LOOP	LOAD
REPEAT	RECORD MEMORY
	RESTORE STATUS
Control and Decision Phrases	*Miscellaneous Commands*
FOR	ACCUMULATE
WITH	COMPUTE
OR	STOP
AND	DIMENSION
WHERE	FORTRAN Inserts

Figure IX-37. Names of SIMSCRIPT commands and phrases. Source: Markowitz, Hausner, and Karr [MA63].

(groups of entities). For example, an entity might be **PERSON**; an attribute might be **ADDRESS**, **AGE**, and **SEX**; and sets might be **ACM**, **IEEE**, and **SIAM**. Unlike the block diagram approach in GPSS, the user writes a set of statements to indicate the action of the system. For example, if an order is shipped to a base, its arrival following a transit-time delay might be caused by the following statements:[11]

```
        CREATE ARRVL
        STORE ORDER IN ITEM(ARRVL)
        STORE BASE IN DESTN(ARRVL)
        CAUSE ARRVL AT TIME + TRANT
```

The language is very FORTRAN-like in appearance; in fact, its initial implementation was by means of a translation to FORTRAN programs. A list of the names of available commands and phrases is shown in Figure IX-37. An example is given in Figure IX-38.

```
            ENDOGENOUS EVENT EPROC
            STORE ORDRP(EPROC) IN ORDER
            STORE MGPRC(EPROC) IN MG
            DESTROY EPROC
  C         — DISPOSITION OF THE ORDER —
            IF ROUT(ORDER) IS EMPTY, GO TO 10
            CALL ARRVL(ORDER)
            GO TO 20
     10     LET CUMCT = CUMCT + TIME — DATE(ORDER)
            LET NORDR = NORDR + 1.0
            DESTROY ORDER
  C         — DISPOSITION OF THE MACHINE —
     20     IF QUE(MG) IS EMPTY, GO TO 30
            REMOVE FIRST ORDER FROM QUE(MG)
            CALL ALLOC(MG, ORDER)
            ACCUMULATE NINQ(MG) INTO CUMQ(MG) SINCE TMQ(MG),
      X         POST NINQ(MG) — 1.0
            RETURN
     30     LET NOAVL(MG) = NOAVL(MG) + 1
            RETURN
            END
```

Figure IX-38. SIMSCRIPT program for order in machine shop.
Source: Markowitz, Hausner, and Karr [MA63], p. 27.

5. *SOL*

SOL (*S*imulation *O*riented *L*anguage) is an attempt to combine the best features of GPSS and SIMSCRIPT but to cast them in an ALGOL-like notation. It has apparently been implemented on the Burroughs B5000/5500 and the UNIVAC 1107.

---

[11] Markowitz, Hausner, and Karr [MA63], p. 7.

The fundamental elements in SOL are local and global variables, transactions, facilities, and stores. Each of these may have a value. A formal definition of the syntax of SOL is given in Knuth and McNeley [KN64a]. An example involving the simulation of a communication system with multiple on-line consoles is given in Figure IX-39. Note that although SOL is very ALGOL-like, it is not an extension of ALGOL as is SIMULA, discussed in Section IX.3.1.7.

6. *MILITRAN*

MILITRAN was developed on the 7090/94 to provide a language specifically useful in simulating the analysis of military systems, although it is certainly not restricted to that. It is FORTRAN-like in its notational approach and in some of its facilities. In addition to providing simulation capability, it also provides well-defined list processing operations. A list of the statements is given in Figure IX-40. An example is shown in Figure IX-41.

7. *SIMULA*

SIMULA (*SIMU*lation *LA*nguage) is a true extension of ALGOL 60. It was originally implemented on the UNIVAC 1107 early in 1965. The basic idea is to add to ALGOL the concept of a collection of programs called *processes* conceptually operating in parallel. The processes perform

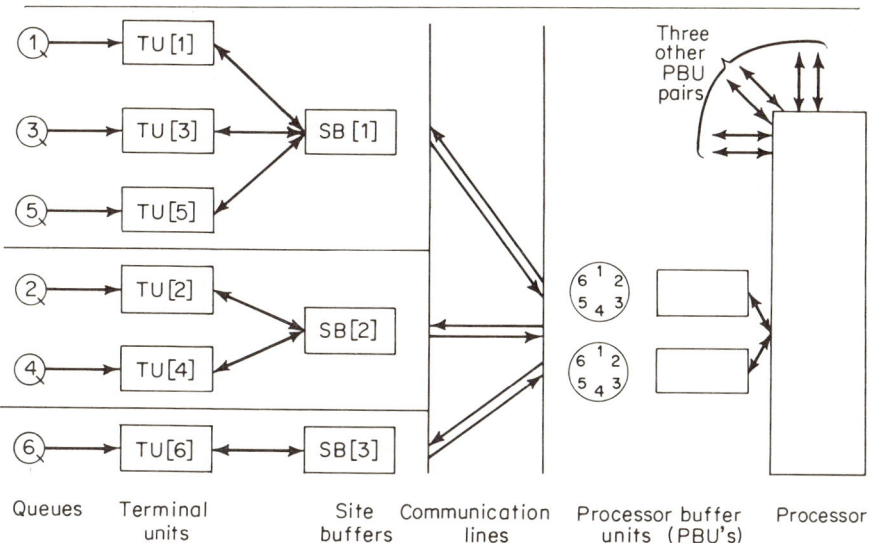

Figure IX-39. (*cont. next page*)

Figure IX-39. (cont.)

```
begin
facility TU[6], SB[3], LINE, COMPUTER;
store 10 QUEUE[6];
integer TUSTATE[6], SBNUMBER[6],
    TUMESSAGE[6];
table (2000 step 500 until 15000) TABLE[6];
process MASTER CONTROL;
begin SBNUMBER[1]←1; SBNUMBER[2]←2;
    SBNUMBER[3]←1; SBNUMBER[4]←2;
    SBNUMBER[5]←1; SBNUMBER[6]←3;
wait 60×60×1000; stop end;
process USERS;
begin integer Q, START TIME, MESSAGE TYPE;
new transaction to START; new transaction
    to START;
ORIGIN: new transaction to START; wait
    0:5000; go to ORIGIN;
START: Q←1:6; enter QUEUE[Q];
MESSAGE TYPE←(1,1,2,2,2,2,2,3,3,3);
seize TU[Q];
TUMESSAGE[Q]←MESSAGE TYPE;
wait 6000:8000;
START TIME←time;
output #TU#, Q, #SENDS MESSAGE#,
    MESSAGE TYPE, #AT TIME#, time;
TUSTATE[Q]←1;
wait until TUSTATE[Q]=0;
release TU[Q]; leave QUEUE[Q];
tabulate (time−START TIME) in TABLE[Q];
output #TU#, Q, #RECEIVES REPLY AT
    TIME#, time;
cancel end;
process PBU; begin integer S, T, WORDS;
new transaction to SCAN; T←3;
SCAN: T←T+1; if T>6 then T←1; wait 1;
S←SBNUMBER[T];
seize LINE;
wait 5; if SB[S] busy then (wait 80; release
    LINE; go to SCAN);
seize SB[S]; wait 15; if TUSTATE[T]≠1 then
    (wait 65; release LINE; release SB[S]; go to
    SCAN);
wait 225; SEND: wait 170; if pr(0.02) then
    (wait 20; go to SEND);
new transaction to COMPUTATION; wait 20;
    release SB[S];
release LINE; TUSTATE[T] ←2; cancel;
COMPUTATION: seize COMPUTER;
    WORDS←TUMESSAGE[T]+2;
wait (if WORDS=3 then 250 else if WORDS
    =4 then 300 else 400);
release COMPUTER;
OUTPUT: wait 1; seize LINE; wait 5;
if SB[S] busy then (wait 80; release LINE;
    go to OUTPUT);
seize SB[S]; wait 75;
RECEIVE: wait 80; if pr(0.01) then (wait 20;
    go to RECEIVE);
release LINE;
WORDS←WORDS−1;
if WORDS=0 then new transaction to
    SCAN;
wait 325; release SB[S]; wait 170;
if WORDS>0 then go to OUTPUT;
TUSTATE[T]←0; cancel end;
process OTHER PBUS;
begin integer I; I←6;
CREATE: new transaction to COMPUTE;
I←I−1; if I>0 then go to CREATE; cancel;
COMPUTE: wait 3200: 5000; seize COMPUTER;
wait (250, 250, 300, 300, 300, 300, 300, 400,
    400, 400);
release COMPUTER; go to COMPUTE end;
end.
```

Figure IX-39. Complete SOL program for the multiple console on-line communication system shown in the diagram on the preceding page.
Source: Knuth and McNeley [KN64], p. 403.

Substitution Statement
  $a = b$

Control Statements
  GO TO s
  PAUSE j
  STOP
  IF $(b), st, sf$
  UNLESS $(b), sf, st$
  DO (s) UNTIL b, $n = e_1, e_2$
  DO (s) FOR $a.IN.b$
  CONTINUE

Procedure Statements
  PROCEDURE n
  PROCEDURE $n(a_1, a_2, \ldots, a_n)$
  EXECUTE n
  EXECUTE n $(a_1, a_2, \ldots, a_n)$
  RETURN
  RETURN a

Input/Output Statements
  FORMAT (format specification)
  READ (t, s) list
  WRITE (t, s) list
  READWRITE (t_1, s_1, t_2, s_2) list
  BINARY READ (t) list
  BINARY WRITE (t) list
  END FILE RETURN (s)
  ENDR ECORD RETURN (s)
  BACKSPACE (t)
  BACKSPACE FILE (t)
  END FILE (t)
  REWIND (t)
  UNLOAD (t)

Environment Declarations
  REAL $n(i_1, i_2, \ldots, i_k), \ldots, n_m(i_1, i_2, \ldots, i_j)$
  INTEGER $n_1(i_1, i_2, \ldots, i_k), \ldots, n_m(i_1, i_2, \ldots, i_j)$
  LOGICAL $n_1(i_1, i_2, \ldots i_k), \ldots, n_m(i_1, i_2, \ldots, i_j)$
  OBJECT $n_1(i_1), n_2(i_2), \ldots, n_m(i_m)$
  PROGRAM OBJECT $n_1(i_1, i_2, \ldots, i_k), \ldots, n_m(i_1, i_2, \ldots, i_j)$
  CLASS (c) CONTAINS $a_1, a_2, \ldots, a_m$
  NORMAL MODE $m_1(a_1, a_2, \ldots, a_k), m_2(b_1, b_2, \ldots, b_r)$
  VECTOR $N((a_1, a_2, \ldots, a_i), d_1, d_2, \ldots, d_i)$
  COMMON $n_1, n_2, \ldots, n_i$

List Processing Statements
  LIST n $((c_1, c_2, \ldots, c_i), d)$
  LENGTH (n)
  RESET LENGTH (n) to p
  PLACE $(e_1, e_2, \ldots, e_i)$ IN n
  REMOVE ENTRY $n(k)$
  PLACE ENTRY $m(j)$ IN n
  REPLACE ENTRY $n(k)$ BY $(e_1, e_2, \ldots, e_i)$
  REPLACE ENTRY $n(k)$ BY ENTRY $m(j)$
  REMOVE $(b_1, b_2, \ldots, b_i)$ FROM n
  REPLACE $(b_1, b_2, \ldots, b_i)$ BY $(e_1, e_2, \ldots, e_j)$ IN n
  REPLACE $(b_1, b_2, \ldots, b_i)$ BY ENTRY $m(j)$ IN n
  MINIMUM INDEX $(n(b_1, b_2, \ldots, b_i, b_x), s)$
  RANDOM INDEX $(n(b_1, b_2, \ldots, b_i, b_x), s)$

Event Statements
  PERMANENT EVENT $n((a_1, a_2, \ldots, a_i), d)$
  CONTINGENT EVENT $n((a_1, a_2, \ldots, a_i), d)$
  NEXT EVENT
  NEXT EVENT $(n_1, n_2, \ldots, n_i)$
  NEXT EVENT EXCEPT $(n_1, n_2, \ldots, n_i)$
  END
  END CONTINGENT EVENTS (s)

Figure IX-40. List of MILITRAN statements.
Source: *MILITRAN . . . A Technical Summary*, Gulton Systems Research Group, Inc., Mineola, N.Y., p. 6. (Booklet).

their operations in groups called *active phases* or *events*. A process carries data and executes actions. Some of the extensions are purely syntactic, while others simply provide a set of relevant and necessary procedures.

It is particularly worth noting that significant elements of list processing are provided.

INPUT	READ (5,INPUT) ARRIVAL RATE, SERVICE RATE, DAYS PER RUN,
	1 PROFIT PER TRUCK, COST PER SPACE PER DAY
	FORMAT (5F10.3)
	PLACE (DAYS PER RUN) IN END OF RUN
	PLACE (−LOG(RANDOM)/ARRIVAL RATE) IN ARRIVAL
	NEXT EVENT ... SIMULATION BEGINS
	CONTINGENT EVENT ARRIVAL ((ARRIVAL TIME), 1)
	IF(LENGTH(SERVICE QUEUE).GE.SPACES), NEXT TRUCK
	SERVE TIME = MAX(TIME,SERVE TIME) − LOG(RANDOM)/SERVICE RATE
	PLACE (SERVE TIME) IN SERVICE QUEUE
NEXT TRUCK	REPLACE ENTRY ARRIVAL (1) BY (TIME−LOG(RANDOM)/ARRIVAL RATE)
	END ... THIS STATEMENT IS ALSO INTERPRETED AS 'NEXT EVENT'
	CONTINGENT EVENT SERVICE QUEUE ((SERVICE TIME), SPACES)
	PROFIT = PROFIT + PROFIT PER TRUCK
	REMOVE ENTRY SERVICE QUEUE (INDEX)
	END
	CONTINGENT EVENT END OF RUN ((END TIME), 1)
	WRITE (6, OUTPUT) PROFIT − SPACES∗TIME∗COST PER SPACE PER DAY
OUTPUT	FORMAT (9H PROFIT = F7.2)
	STOP
	END

Figure IX-41. MILITRAN program for finite length queue. Source: *MILITRAN ... a Technical Summary*, Gulton Systems Research Group, Inc. Mineola, N.Y., p. 7. (Booklet.)

An example of a skeletal description of a classical *job shop* system is shown in Figure IX-42.

8. OPS

OPS (which was originally an acronym for *O*n-*L*ine *P*rocess *S*ynthesizer) is the one simulation language designed for use with (and embedded in) an on-line system (namely M.I.T.'s CTSS, see Crismann [ZR65]). It went through several versions, under the direction of Professor M. Greenberger of M.I.T., starting in 1964; the one described here is OPS-3. Ideas for improvements in OPS-4 are mentioned briefly later. OPS is considered to be experimental in the sense that it is trying to explain how to do on-line simulations effectively. Because of the on-line environment, the user has many advantages in changing and working on parts of his model.

The two main concepts in OPS are the *KOP* (an ordered set of new operators written in terms of old ones) and the *AGENDA*, which is a schedule of activities and is available for modification by the user. Activities are scheduled on the *AGENDA* for execution at a specified time or as the result of meeting a particular condition. Delays and waits are used to combine events to define activities.

```
begin integer nmg; read(nmg);
SIMULA begin integer array available [1 : nmg];
  set arrary que [1 : nmg];
activity order(n); integer n;
begin integer i, mg; integer array mgroup[1 : n];
  array ptime[1 : n];
  read(mgroup, ptime);
  for i:=1 step 1 until n do
  begin mg := mgroup [i];
  if available[mg] = 0 then
  begin wait (que[mg]);  remove(current) end
  else available[mg] := available[mg] -1;
  hold(ptime[i]);
  if empty(que[mg]) then available[mg] := available[mg] +1
  else activate first (que[mg])
  end path through shop
  end order;
integer n;  real T;
read(available);
next: read(n, T);  reactive current at T;
if n > 0 then begin activate new order(n); go to next end
end SIMULA end program
```

Figure IX-42. Skeleton SIMULA description of job shop system.
Source: Dahl and Nygaard [DH66], p. 676. By permission of Association for Computing Machinery, Inc.

A *SCHED* operator has three options (to denote time *T* or the current time or to meet a condition), and parameters may be passed to an activity when it is called. Conditions are (unfortunately) specified in prefix notation, but they may be quite complex, e.g., SCHED JOE WHEN OR LESS A B GREATER C 25, which places a conditional call to the activity named JOE on the AGENDA with stipulations on the relation to the variable TIME and with the condition that execution should take place when either A is less than B or C is greater than 25.

Other operators include *DELAY, WAIT, RETRNA, PRINTK, CALLAK, LOCAL, DRAW, CANCEL, RSCHED, PRINT,* and *READ*. An example is shown in Figure IX-43.

Plans for OPS-4 are described in Jones [JM67]. The key points are as follows: It will use the MULTICS system being implemented on the GE 645 at Project MAC; PL/I will be used as the basic language, and special data types (e.g., sets, queues, and tables) will be added; the activity representation of SOL and SIMULA will be used rather than the event orientation of SIMSCRIPT.

```
              KOP ARRIVE
        10    SET Q = Q + 1
        20    DRAW DT EXPONE 5
        30    SCHEDK ARRIVE AT SUM TIME DT
        40    SCHEDK STAT IMMED
        50    RETRNA
              KOP STAT
        10    IF Q .LE. QMAX
        20    GOTO 50
        30    SET QMAX = Q
        40    PRINT QMAX
        50    IF S .GE. 5
        60    WAIT LESS S 5
        70    SET S = S + 1
        80    SCHEDK SERVIC IMMED
        90    RETRNA
              KOP SERVIC
        10    SET Q = Q - 1
        20    DRAW STIME RANDOM 5 50
        30    DELAY STIME
        40    SET S = S - 1
        50    RETRNA
```

Figure IX-43. OPS-3 program for multi-server queuing model.
Source: Greenberger and Jones [YP66], pp. 135-136. By permission of Association for Computing Machinery, Inc.

IX.3.2. QUERY

1. *Introduction*

The subject area encompassed by the phrase *query language* is an extremely wide one, with a number of different concepts shading from one to another in a hazy fashion. Some of the terms which tend to become intermingled in this way are *query languages, information retrieval languages, data base* and *file management systems, natural languages,* and *question-answering systems.* Conceptually, these range from one extreme involving a fixed format file and a few very rigid ways in which to extract information from it, to the other extreme which includes the general question-answering systems from English text. Two useful surveys are given by Simmons, [SE65] and [SE66a]. Part of the difficulty in sorting out these concepts is that it depends considerably on the viewpoint of the user. In one case, the prime interest (perhaps technically if not administratively) is in the structure of the file. In such a case, the main interest lies in determining how complex a data structure can be incorporated, how easily it can be updated and changed, and how complex are the queries which can be addressed to it. In this instance, the complexity of the query applies to the substance of the question itself and not to the way in which it is phrased. For example, a statement saying FIND SHIPS/BH RH : NOT NY/6M, might have the same logical meaning as FIND ALL THE SHIPS LOCATED IN BOSTON HARBOR WHO HAVE RED-HEADED MEN ABOARD AND HAVE NOT BEEN IN NEW YORK WITHIN THE PAST SIX MONTHS.

Although the format is considerably different in a case like this, the retrieval is based on a very simple set of Boolean conditions, while the language in which it is phrased may be simple or complex. On the other hand, if we ask WHAT NEW YORK CITIES HAVE A LARGER POPULATION THAN PODUNK?, this is actually more complex from a data retrieval point of view because New York can be both a city and state and the question implies that the system is able to ascertain which is being referred to from the context of the question.

Most of these systems really have two languages associated with them; one for file (or data) definition and updating, and the other for use in retrieving information. The latter is usually a stable and well-defined language, i.e., one that makes an attempt to look *natural* but is in reality very restricted and formalized like the languages in earlier chapters. (No attempt is made here to discuss the file management language unless it coincides with the retrieval language.) Some of these query languages may or may not have computational facilities, i.e., permit counting, tabulating, or simple formula evaluation. It is no secret (or surprise) that a number of the efforts in this area have been motivated by the needs of the military command and control environment. In a military situation, it is considered desirable by many (although not all) people to allow the military commander direct access to the machine to obtain the information he wants. Unfortunately, there may be inherent ambiguities in the questions posed that are understood only by the people preparing the system. For example, if somebody were to query a system and ask HOW MANY PLANES WERE LOST DURING WORLD WAR II?, the answer would really depend on the definition of the word *lost* and this might only be known to the people who created the files. Even more undesirable is the possibility that the person making the request might not realize the potential ambiguity and misapply the resulting answers.

Most systems concerned with data management require some capability to do logical processing, computations, and input/output. This is not unlike any programming system, but there is an added need here for good file manipulation capabilities. For the sake of those readers who may be interested in the more general file management problem, a number of items are listed in the references at the end of the chapter, even though they have little or no interest from a language point of view. Those systems that I deem to be the most interesting or typical, from a language viewpoint, are discussed below.

The phrase *question-answering system* is applied primarily to those systems which accept a natural-looking English statement or question as input, and provide answers based on files whose format might be rigidly fixed as in the data base systems, or general text from an encyclopedia, or somewhere in between. Almost all these systems, of necessity, concern themselves with the three aspects of syntax, semantics, and fact retrieval. In the syntax, the sentence is analyzed to obtain its grammatical structure; the semantics determines which of several possible meanings is most relevant in the given case; and the fact retrieval portion actually does the work that is required. This last aspect becomes related to the file structure problem. If we assume that the information stored is well structured but sufficiently general to be capable of being used to answer a number of questions, then its method of storage and retrieval becomes a significant problem. On the other hand, some of the systems use English text as a data base from which to find answers. Naturally, the flexibility of the English language input is quite significant. Some of the systems, e.g., STUDENT (see Bobrow [BB64], [BB64c]), operate primarily on recognizing key words. Those mentioned later perform grammatical analysis to varying levels of complexity. Those readers who are particularly interested in this aspect of the query language problem will find additional useful references in Section X.5.

2. *COLINGO and C-10*

One of the systems in the general category of higher artificial languages with a wide data base is COLINGO, developed at the MITRE Corporation and implemented on the IBM 1401. More accurately, COLINGO is a system of programs called by an interpreter for the COLINGO control language. The files are described using a language based heavily on the COBOL DATA DIVISION. Typical statements in COLINGO are the following:

GET FORCE−STATUS−FILE IF UNIT EQ 82−ABN COMPUTE FORCE−RATIO TOTAL−OFFICERS/ TOTAL−EM

GET A−FILE IF STRENGTH/AUTH GR 500 EXECUTE 01 02 03 IF/NOT PRINT ALL

The basic command list in COLINGO is shown in Figure IX-44.

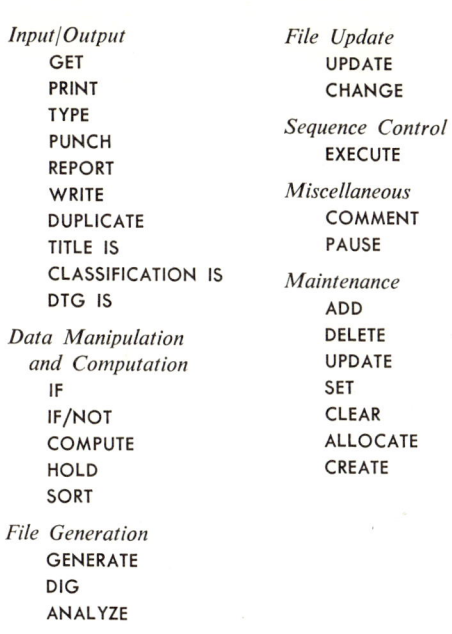

Figure IX-44. List of basic commands in COLINGO.
Source: Spitzer, Robertson, and Neuse [SZ65], p. 43. Reprinted from INFORMATION SYSTEM SCIENCES: Proceedings of the Second Congress, Ed. by Joseph Spiegel and Donald E. Walker, Spartan Books, 1964.

Attempts to improve COLINGO by putting it on a larger machine and obtaining greater flexibility lead to the C-10 system developed on the 1410. A small sample of a C-10 program is shown in Figure IX-45. Note that it really looks very much like the style of programming languages discussed in earlier chapters.

Both COLINGO and C-10 are more *systems* than *languages*, and the file structure in C-10 is more developed than in COLINGO.

3. *473L Query*

Another system with similar principles but which allows more complicated (although still formally defined) sentences is the query language developed for the Air Force 473L system. A typical statement in this system

```
BEGIN;                                  ''REPORT IS TO COVER DIVISIONS,DEPARTMENTS
   VARIABLE T1, T2, T.TITLE, T.QUANT'', GROUPS, OR SECTIONS; AND N, INDICATING
            , T.SAL, T.CODE, T.UNIT, ''HOW MANY UNITS TO REPORT ON.
            IND, TN, T3, INC, INC1,  ''
            LIMIT, LIMIT1, U.TITLE;  ''DECLARE VARIABLES.
    IND = Ø;                             ''
    TN  = Ø;                             ''
    IF LEVEL = 'SECTION'; INC = Ø;       ''
    IF LEVEL = 'GROUP';                  ''INITIALIZE.
       BEGIN; INC = 9; INC1 = 1; END;    ''
    IF LEVEL = 'DEPARTMENT';             ''INCREMENTS HAVE TO DO WITH THE
       BEGIN; INC = 99; INC1 = 1Ø;END;  ''ASSIGNMENT OF NUMBERS TO UNITS.
    IF LEVEL = 'DIVISION';               ''(PERSONNEL FILE HAS BEEN SORTED BY
       BEGIN; INC = 999; INC1=1ØØ;END;  ''               UNIT NUMBER)
    OUTPUT TO PRINTER;                   ''DECLARE PRINTER AS OUTPUT DEVICE.
    DEFINE PAGE FOR OUTPUT AS 33*132;   ''DEFINE SIZE OF PRINTER PAGE.
    DO WRITE.TITLE(LEVEL);               ''WRITE TITLE
    READ PERSONNEL; ELSE RETURN;         ''
ORG:READ ORGANIZATION; ELSE RETURN;     ''FIND ORGANIZATION UNIT OF APPROPRIATE
    IF ORG.LEVEL NE LEVEL; GO TO ORG;   ''LEVEL IN ORGANIZATION FILE.
    TN = TN+1;                           ''
    IF TN > N; RETURN;                   ''RETURN IF ENOUGH UNITS PROCESSED.
    LIMIT = ORG.UNIT+INC;                ''
    DO WRITE.HEADING(ORG.UNIT,          ''WRITE HEADINGS. AUTHORIZED COMPLEMENT
       REPORTS.TO, ORG.NAME);           ''IS PRINTED FIRST.
    T1 = Ø;                              ''
    T2 = Ø;                              ''INITIALIZE TOTALS.
    T3 = Ø;                              ''
O.JOB:READ JOB; ELSE GO TO O.UNIT;      ''
    DO WRITE.LINE (Ø, JOB(CODE),        ''WRITE LINE OF REPORT FOR EACH JOB.
       JOB(TITLE), JOB(QUANT),          ''
       JOB(SALARY));                    ''
    T1 = T1 + JOB(SALARY);              ''INCREMENT TOTALS.
    T3 = T3 + JOB(QUANT);               ''
    GO TO O.JOB;                        ''
```

Figure IX-45. Portion of a C-10 program.
Source: Steil [QZ67], p. 208.

is the following:

RETRIEVE FORCE STATUS WITH COMMAND EQUAL SAC, UNIT EQUALS 413ATW THEN RETAIN ACFT POS: RETRIEVE FORCE STATUS WITH COMMAND = SAC, ACFT POS > [R1, ACFT POS, OR] THEN LIST UNIT, ACFT POS

The general form of a statement is as follows:

program-indicator file-indicator qualifier-conjunction qualifier selector-conjunction output-director output-selector

The *program-indicator* provides a beginning for the sentence, e.g., RETRIEVE, FIND. The *file-indicator* is a *file name*. The *qualifier-conjunction* is usually WITH. The *qualifier* describes the data to be retrieved and usually consists of an attribute, a comparator, and a value, e.g., RUNWAY LENGTH > 5000. The *selector-conjunction* is usually THEN. The *output-director* is usually a single word defining the output device and format. The *output-selector* contains the names of the output variables and details on their format.

Punctuation

' * [ ] : ,
= ≠ > <
¬ □ ( ) ?

Key Words

ALL	PUNCH
AND	READ
ANY	RETAIN
BY	RETRIEVE
CARDS	RN (Retained File Number)
DECR (Decreasing)	SAME
GCD (Great Circle Distance)	SAVE
GET	SUM
GREATEST	TAPE
H (Horizontal)	THEN
INCR (Increasing)	TITLE
LEAST	TOTAL
LIST	TRANSFER
MAX	UPDATE
MIN	V (Vertical)
OR	WITH
PRINT	

Figure IX-46. Punctuation characters and key words in 473L Query language. Source: Barlow and Cease [BL65], extracts from pp. 81–86. Reprinted from INFORMATION SYSTEM SCIENCES: Proceedings of the Second Congress, Ed. by Joseph Spiegel and Donald E. Walker, Spartan Books, 1964.

The punctuation (which is definitely significant) and key vocabulary words are shown in Figure IX-46.

4. *ADAM*

Still another system developed at MITRE is ADAM (*A* generalized *DA*ta *M*anagement system), which runs on the IBM 7030 (STRETCH) and has been operational since early 1965. As with other systems in this category, a significant part of the attention is devoted to the format of the file and the methods for updating it. From a language point of view, the most interesting aspect about ADAM is that it has a syntax-directed translator which enables the user to define other languages. The user also has available a substitution macro facility whereby he can abbreviate complex expressions. For example,[12]

    LET NONSTOP MEAN (IF NR OF STOPS EQ 0)

---

[12] Connors [CY66], p. 198.

and

> LET SKED MEAN (FOR DESTINATION/2/./3/ TYPE ORIGIN/1/FLIGHTS) USING REINSERT

define substitutions, and the message

> SKED BOSTON CHICAGO NONSTOP

would be transformed to

> FOR DESTINATION CHICAGO.IF NR OF STOPS EQ 0, TYPE ORIGIN BOSTON FLIGHTS

ADAM is an on-line system. Its language contains common arithmetic, assignment, conditional, and control statements. A specific language for defining files exists.

5. *BASEBALL*

The development of *questioning-answering systems* has been underway since 1959 if not earlier. One of the first was the BASEBALL system in which the user was able to write such things as

WHERE DID THE RED SOX PLAY ON JULY 7?

WHO DID THE YANKEES LOSE TO ON AUGUST 8?

The input sentences are restricted to single clauses and do *not* permit logical connectives such as *and* and *or*. Relation words such as *most* or *highest* are also not permitted.

BASEBALL is programmed in IPL-V (see Section VI.3) and organizes the data into list structures. The first part of the program uses a dictionary, parsing routines and semantic analysis routines to translate the input question into a specification list similar in format to that of the data. This permits retrieval of the answers.

6. *DEACON*

DEACON (*D*irect *E*nglish *A*ccess and *CON*trol) has been under development at GE TEMPO since at least 1963. Some of the allowable statements in DEACON strongly resemble those in the systems discussed earlier, but that is because they are related to military subjects rather than because of their structure, which is far more general. In systems such as COLINGO and the 473L Query Language, the languages are definitely artificial with limited or no syntactic flexibility, whereas DEACON actually uses linguistic techniques to analyze sentences which are far more complex, e.g.,

HAS THE 25TH BATALLION ARRIVED IN TEXAS SINCE 3 P.M.? IS THE 100TH SCHEDULED TO ARRIVE AT FT. LEWIS BEFORE THE 200TH LEAVES FT. LEWIS?

The data is stored in ring-type list structures. It includes time-dependent data. "Thompson hypothesizes that English essentially becomes a formal language as defined if its subject matter is limited to 'material whose interrelationships are specifiable in a limited number of precisely structured categories [memory structures].' ... Because these programs are written in terms of structural categories (independent of content), the interpretation rules apply to any subject matter that is stored in these categories."[13] The authors state in a footnote that "This is the major advance of DEACON over Green's BASEBALL ... and Lindsay's SAD SAM."[14]

Work on this project seems either dormant or terminated.

7. *Protosynthex*

The Protosynthex system is based on natural English text since it is an attempt to answer questions (phrased in natural English) from an encyclopedia. As such, it is the only system described in this section which does not have a structured data base. The basic principle is to use synonyms, intersection logic, and a simple scoring function to find the sentences and paragraphs which most closely resemble the question. Statements thus retrieved which do not match the structure of content words of the question are rejected. For example, the question **What animals live longer than men?** might cause the following list of content words to be used in searching the index:[15]

Word	Words of Related Meaning
animals	mammals, reptiles, fish
live	age
longer	older, ancient
men	person, people, women

The smallest unit with the greatest intersection with these words is then found, where some weighting is applied to certain words in the search.

---

[13] Craig *et al.* [CJ66], p. 365.
[14] *Op. cit.*,
[15] Simmons [SE65], p. 60.

The highest scoring units are selected and then both the question and the proposed answer are parsed to make sure that the structures are the same.

An example of input data, questions, and answers is given in Figure IX-47.

```
ENTER(((HORSE . NOUN)(IS . ISVB)(ANIMAL . NOUN)(LIVE . VERB)
(ON . PREP)(FARM . NOUN)(STRONG . ADJ)(PULL . VERB)
(WAGON . NOUN)(MOST . ADJ)(ARE . ISVB)(EAT . VERB)(HAY . NOUN)
(ALL . ADJ)(RED . ADJ)(MANY . ADJ)(HAVE . HAVB)(LONG . ADJ)
(MANE . NOUN)(SOME . ADJ)(PERSON . NOUN)(RACE . VERB)))
DICTIONARY
PLURALS(((HORSE . HORSES)(ANIMAL . ANIMALS)(FARM . FARMS)
(WAGON . WAGONS)(MANE . MANES)(PERSON . PEOPLE)))
OK
START()
READY
A HORSE IS AN ANIMAL / HORSES LIVE ON A FARM / STRONG HORSES
PULL WAGONS / MOST HORSES ARE STRONG .
  LEARNED
  READY
HORSES EAT HAY / ALL HORSES ARE RED / MANY HORSES HAVE LONG
MANES / SOME PEOPLE RACE HORSES .
  LEARNED
  READY
FORGET(((ALL HORSES ARE RED))
  FORGOTTEN
  READY
WHAT IS A HORSE ?
  HORSE IS AN ANIMAL
  READY
WHAT DO HORSES DO ?
  HORSES LIVE ON A FARM AND EAT HAY / STRONG HORSES
  PULL WAGONS
  READY
WHAT DO STRONG HORSES DO ?
  STRONG HORSES PULL WAGONS
  READY
DO MANY HORSES HAVE LONG MANES ?
  YES
  READY
DO ALL HORSES HAVE LONG MANES ?
  TO MY KNOWLEDGE- NO
  READY
DESCRIBE HORSES ?
  HORSES ARE AN ANIMAL / MOST HORSES ARE STRONG / MANY
  HORSES HAVE LONG MANES / HORSES LIVE ON A FARM AND EAT
  HAY / STRONG HORSES PULL WAGONS / SOME PEOPLE RACE HORSES
  READY
WHAT DO HORSES EAT ?
  HORSES EAT HAY
  READY
```

Figure IX-47. Examples of control statements, questions, and answers from Protosynthex II.
Source: Simmons [SE66a], p. 69.

8. *AESOP*

The advent of on-line techniques naturally is playing an important role in the development of query systems. One of the interesting examples of such a system is AESOP (*A*n *E*volutionary *S*ystem for *O*n-Line *P*rocessing), which was developed at the MITRE Corporation with a prototype running on the IBM 7030 (STRETCH). Heavy use is made of the display screen and a light pen. The latter is used both for causing immediate execution of system functions and for engaging in a preparatory interactive dialogue.

AESOP is primarily oriented toward retrieving data from a data base, acting upon it, and restoring it. Input is from typewriter or light pen. The light pen is used to define the system mode, to select displayed abbreviations for frequently used simple commands, and to compose complex messages for data retrieval and file modifications. For example, a portion of the screen might display the mode names **TABULAR, TREE, FILE MANIP, ERASE,** and **COPY**. Another part of the screen shows **SET FILES, CLEAR FILES, ADD FILES,** and **FINISH**. If the user points to **SET FILES**, for example, the screen will display the names of all files in the system. Pointing to the desired names causes the system to display on the screen only those pointed to. Pointing to the name in its new position causes the first part of the file to be displayed.

The *TREE* mode causes selected information to be displayed in the form of a tree. The user language tree is used to compose messages at the top of the screen. This technique greatly reduces the likelihood of composing illegal input messages. In particular, one such tree is part of the user language and is shown in Figure IX-48. The selection of the *ERASE, COPY,* or *FILE MANIP* modes results in the display of skeleton user messages at the

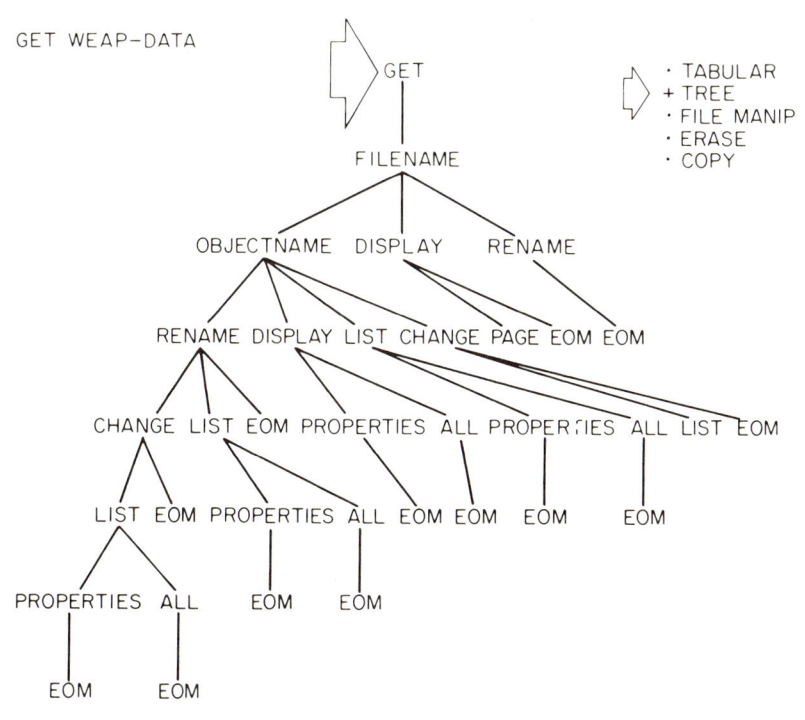

Figure IX-48. Communication tree in AESOP.
Source: Summers and Bennett [UT67], p. 74.

DATA RETRIEVAL MESSAGES

Display Request Messages
GET filename DISPLAY [PAGE number [SECTION number]]
DISPLAY PAGE number
DISPLAY SECTION number
RESTORE
GET filename object DISPLAY $\left[\begin{Bmatrix} \text{ALL} \\ \text{propertylist} \end{Bmatrix} \right]$
DISPLAY LINE number $\left[\begin{Bmatrix} \text{ALL} \\ \text{propertylist} \end{Bmatrix} \right]$

Hardcopy Output Request Messages
GET filename object LIST $\left[\begin{Bmatrix} \text{ALL} \\ \text{propertylist} \end{Bmatrix} \right]$

Booster System Typewriter Input Messages
PRINT DISPLAY
GET filename PRINT
FIND DISTANCE FROM $\begin{Bmatrix} \text{latitude longitude TO } \begin{Bmatrix} \text{latitude longitude} \\ \text{filename object} \end{Bmatrix} \\ \text{filename object TO } \begin{Bmatrix} \text{latitude longitude} \\ \text{[filename] object} \end{Bmatrix} \end{Bmatrix}$

FILE MODIFICATION MESSAGES

GET filename object CHANGE propertyname$_1$ propertyvalue$_1$ ... propertyname$_n$ propertyvalue$_n$ [RENAME newobjectname] [LIST...]
GET filename $\begin{Bmatrix} \text{object RENAME} \\ \text{RENAME object} \end{Bmatrix}$ newobjectname [CHANGE...] [LIST...]
GET filename RENAME object$_1$ newobjectname$_1$ ... object$_n$ newobjectname$_n$
GET filename objectspec CHANGE propertyname value$_1$ ... value$_n$
GET filename object $\begin{Bmatrix} \text{ADD} \\ \text{SUB} \end{Bmatrix}$ integer$_1$ propertyname$_1$ ... integer$_n$ propertyname$_n$ [LIST...]

ERASE MESSAGE

GET filename objectspec ERASE $\begin{Bmatrix} \text{ALL} \\ \text{ALLPROPS} \\ \text{OBJECTNAME} \\ \text{propertylist} \end{Bmatrix}$

LINE MOVE OPERATIONS

GET filename SWAP $\begin{Bmatrix} \text{LINE} \\ \text{LINES} \end{Bmatrix}$ $\begin{Bmatrix} \text{\# AND \#} \\ \text{\#} \rightarrow \text{\# AND \#} \rightarrow \text{\#} \end{Bmatrix}$
GET filename REORDER $\begin{Bmatrix} \text{LINE} \\ \text{LINES} \end{Bmatrix}$ \# $\rightarrow$ \# AS \#\#\#...
GET filename INSERT $\begin{Bmatrix} \text{LINE} \\ \text{LINES} \end{Bmatrix}$ $\begin{matrix} \text{\#} \\ \text{\#} \rightarrow \text{\#} \\ \text{\#\#\#...} \end{matrix}$ $\begin{Bmatrix} \text{BEFORE} \\ \text{AFTER} \end{Bmatrix}$ \#

Figure IX-49. (cont.)

COLUMN MOVE OPERATIONS

GET filename objectspec SWAP $\begin{Bmatrix} \text{COL} \\ \text{COLS} \end{Bmatrix}$ propertyname AND propertyname

GET filename objectspec INSERT $\begin{Bmatrix} \text{COL} \\ \text{COLS} \end{Bmatrix}$ propertylist $\begin{Bmatrix} \text{BEFORE} \\ \text{AFTER} \end{Bmatrix}$ propertyname

COPY MESSAGE

COPY FROM filename (objectspec) $\begin{Bmatrix} \text{ALLPROPS} \\ propertylist \end{Bmatrix}$

INTO filename (objectspec) $\begin{Bmatrix} \text{ALLPROPS} \\ propertylist \end{Bmatrix}$

COPY WITH OBJECTNAMES FROM filename (objectspec)
INTO filename (objectspec)

COPY WITH OBJECTNAMES FROM filename (objectspec) $\begin{Bmatrix} \text{ALLPROPS} \\ propertylist \end{Bmatrix}$

INTO filename (objectspec) $\begin{Bmatrix} \text{ALLPROPS} \\ propertylist \end{Bmatrix}$

COPY WITH PROPNAMES FROM filename (objectspec) $\begin{Bmatrix} \text{ALLPROPS} \\ propertylist \end{Bmatrix}$

INTO filename (objectspec) $\begin{Bmatrix} \text{ALLPROPS} \\ propertylist \end{Bmatrix}$

COPY WITH $\begin{Bmatrix} \text{OBJECTNAMES AND PROPNAMES} \\ \text{PROPNAMES AND OBJECTNAMES} \end{Bmatrix}$ FROM filename (objectspec) $\begin{Bmatrix} \text{ALLPROPS} \\ propertylist \end{Bmatrix}$ INTO filename (objectspec) $\begin{Bmatrix} \text{ALLPROPS} \\ propertylist \end{Bmatrix}$

SELECTIVE RETRIEVAL MESSAGE

GET filename [objectspec] IF propertyname $\begin{Bmatrix} \text{LT} \\ \text{LEQ} \\ \text{EQ} \\ \text{GT} \\ \text{GREQ} \end{Bmatrix}$ value $\left[\begin{Bmatrix} \text{AND} \\ \text{OR} \end{Bmatrix} \text{propertyname} \begin{Bmatrix} \text{LT} \\ \text{LEQ} \\ \text{EQ} \\ \text{GT} \\ \text{GREQ} \end{Bmatrix} \text{value} \dots \right]$

STATEMENT MESSAGE

DO filename object [$parameter_1$ ... $parameter_n$]

MISCELLANEOUS MESSAGES

DATE dd–dd
SAVE

Figure IX-49. AESOP commands used on typewriter. The general notation is the COBOL metalanguage described in Section II.6.2. Certain unique items for this figure include the following: # represents a line number; #→# represents a range of line numbers; an *objectspec* specifies what objects are to be operated on (e.g. *name* and/or *linenumber*); *propertylist* is a variable length list of property names; LIST ... and CHANGE ... represent the longer expressions shown in earlier messages within the figure. Source: Summers and Bennett [UT67], pp. 85–86.

top of the screen. The variable positions of these messages are filled in by using the light pen to select data from other parts of the screen. Any reader interested in more details on this system will find an excellent set of figures of many screen displays in the paper by Summers and Bennett [UT67].

The language used for input from a typewriter is shown in Figure IX-49.

IX.3.3. GRAPHIC AND ON-LINE DISPLAY LANGUAGES

The dividing line between languages used to provide graphical or other types of display output on a scope attached to a computer and systems (and languages) used for computer-aided design is a hazy one. I have attempted to draw the distinction by including in this section those languages which seem to be potentially of somewhat more generality than those primarily aimed at computer-aided design, although some in that category are also very general. Hence the reader is requested not to worry too much about the distinction between languages in this section and in Section IX.3.4.

The DIALOG system described in Section IV.6.10 is specialized to the solution of numerical scientific problems, but it has some interesting features relative to on-line display systems. CORAL, discussed in Section VI.9.3.2, can be used for creating graphic displays.

Discussions of some of the broader issues in graphics are given in some of the references listed at the end of this chapter.

1. *GRAF*

From a language point of view, the GRAF (*GR*aphic *A*dditions to *FORTRAN*) system is a particularly clean-cut one. A new data type, namely a *display variable*, is added to FORTRAN. Its value is really a string of orders which are capable of generating a display when transmitted to an appropriate device. Display variables are named the same way as other FORTRAN variables, must be declared by writing **DISPLAY**, can be dimensioned, can appear in **EQUIVALENCE** and **COMMON** statements, and can be passed to subroutines as arguments. A set of built-in display functions

```
DISPLAY A, B, SQU, POLE(11), K99 (3, 2, 4)
A = B
A = A + B
A = B + A
SQU = POLE(5) + POLE(3)
POLE(1) = PLACE(RX,0)
POLE(1) = PLACE(RX, 0) + LINE(X3, Y7)
K99 = PLACE(0, 0) + PRINT 14, (ZK(I), I = 1, 7) + PLACE(2000,2000)
```

Figure IX-50. Small GRAF program.
Source: Hurwitz, Citron, and Yeaton [HW67], p. 554.

include *POINT, LINE, PLACE, CHAR,* and *PRINT,* each of which has a value which is a string of graphic orders, e.g., LINE(X, Y) generates orders for plotting a line. A *display expression* is a sequence of display variables and display functions separated by plus signs; its value is the string of graphic orders obtained by concatenating the values from left to right, e.g., BOX = PLACE (X, Y)+LINE(U, Y)+LINE(U, V)+LINE(X, Y). An assignment statement can be used to give a value to a display variable. The functions *PLOT* and *UNPLOT,* respectively, transmit or remove the values of the statement arguments to the device buffer. A cursor can be set in a display variable by the *SETCUR* subroutine and removed by *RMVCUR*. Other functions are available. A sample program is shown in Figure IX-50. This system was implemented on the IBM System/360 using FORTRAN (E level).

Definition

† POINT/NAME/X,Y/OP where OP = N for NAME, and
† LINE/NAME/PT1/PT2 T for TEXT
  ARC/NAME/PT1/PT2/PT3
  PARC/NAME/F/V/PT1/PT2
  EARC/NAME/F1/F2/PT1/PT2

Manipulation (level 1)

† COIN/PT1/PT2
† MERGE/PT1/PT2
† COPY/NAME/L/PT
† INTPT/NAME/L/X,Y
  SPIN/L1/L2/PT/DEGREE

Transformation

† MOVE/NAME/OP/PT/QUAL where OP = P for point
† ROTATE/NAME/OP/PT/DEGREE/QUAL N for node
† SCALE/NAME/OP/PT/DEGREE,XX,YY/QUAL S for subpicture
  D for display, i.e.,
  the screen level
  picture

Control

† CLEAR
† DELETE/NAME/OP/QUAL where OP = P for point
† ASSIGN/NAME/NODE1/.../NODEn N for node
† USE/NAME/QUAL L for line
† SHOW/NAME S for subpicture
† LOCATE/X,Y;   ENDLOC U for uniformly
† TEXT/PT/OP1/H/W/OP2/---...$------$...$$ where OP1 = TL for top left
  TR for top right
  BL for bottom left
  BR for bottom right
  OP2 = L for left
  R for right

†Those primitives included in the present version [see source] of PENCIL are identified by†. This set provides all basic capabilities, including hierarchical assembly of pictures.

Figure IX-51. List of PENCIL primitives.
Source: Van Dam and Evans [VD67], p. 602.

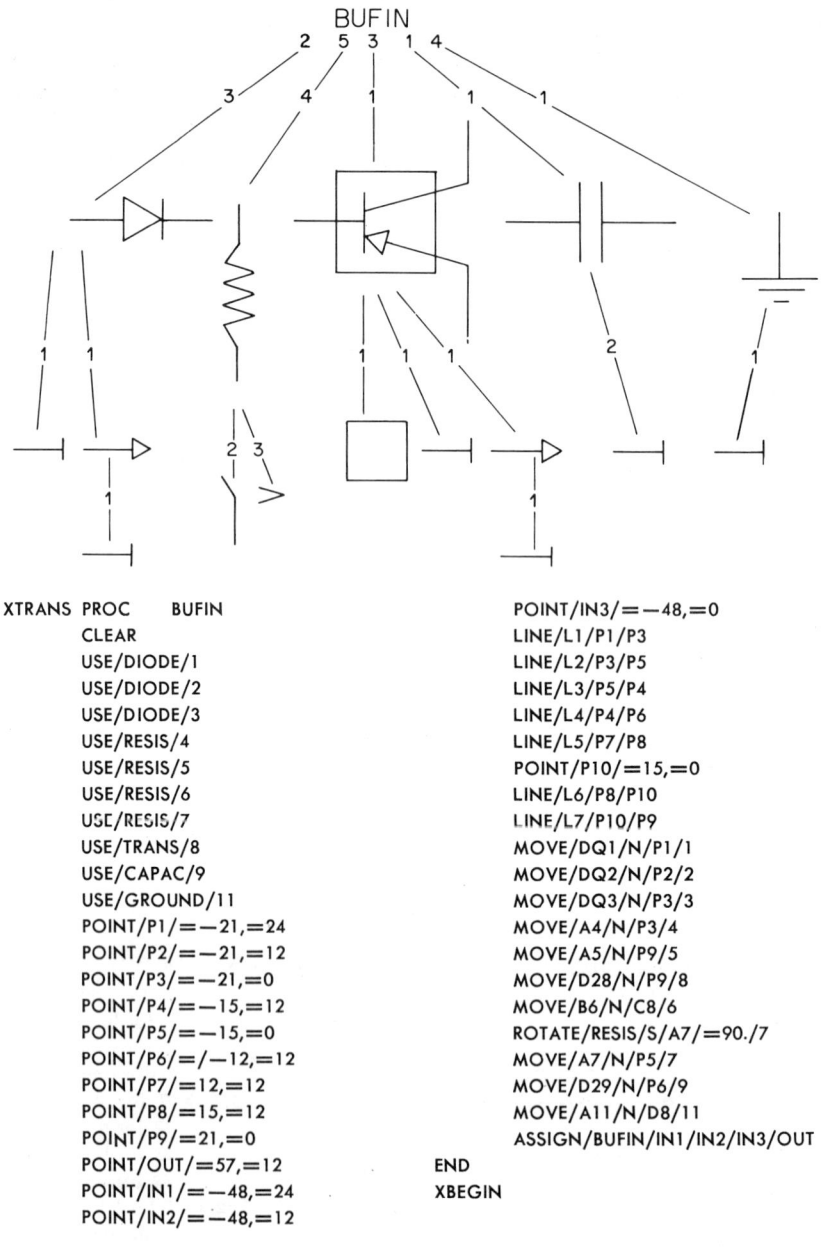

```
XTRANS PROC    BUFIN                POINT/IN3/=-48,=0
       CLEAR                        LINE/L1/P1/P3
       USE/DIODE/1                  LINE/L2/P3/P5
       USE/DIODE/2                  LINE/L3/P5/P4
       USE/DIODE/3                  LINE/L4/P4/P6
       USE/RESIS/4                  LINE/L5/P7/P8
       USE/RESIS/5                  POINT/P10/=15,=0
       USE/RESIS/6                  LINE/L6/P8/P10
       USE/RESIS/7                  LINE/L7/P10/P9
       USE/TRANS/8                  MOVE/DQ1/N/P1/1
       USE/CAPAC/9                  MOVE/DQ2/N/P2/2
       USE/GROUND/11                MOVE/DQ3/N/P3/3
       POINT/P1/=-21,=24            MOVE/A4/N/P3/4
       POINT/P2/=-21,=12            MOVE/A5/N/P9/5
       POINT/P3/=-21,=0             MOVE/D28/N/P9/8
       POINT/P4/=-15,=12            MOVE/B6/N/C8/6
       POINT/P5/=-15,=0             ROTATE/RESIS/S/A7/=90./7
       POINT/P6/=/-12,=12           MOVE/A7/N/P5/7
       POINT/P7/=12,=12             MOVE/D29/N/P6/9
       POINT/P8/=15,=12             MOVE/A11/N/D8/11
       POINT/P9/=21,=0              ASSIGN/BUFIN/IN1/IN2/IN3/OUT
       POINT/OUT/=57,=12     END
       POINT/IN1/=-48,=24    XBEGIN
       POINT/IN2/=-48,=12
```

Figure IX-52. PENCIL program for diagram shown.
Source: Van Dam and Evans [VD67], p. 603.

2. PENCIL

PENCIL (*P*ictorial *ENC*od*I*ng *L*anguage) is a language which applies to a relatively simple data structure, as distinguished from the more complicated ones existing in some of the systems described or referenced in Section IX.3.4. PENCIL was implemented on the IBM 7040 under the MULTILANG system (see Wexelblat and Freedman [WP67]). This list of elements in the system is shown in Figure IX-51. A specific program to match a complex diagram is shown in Figure IX-52.

3. Graphic Language

Another completely independent language is that developed by Schwinn [UZ67]. The language consists of statements in the following categories: *User request, label request, data dimension, scan request, plot request, draw*

```
USER HEADING: SCHWINN;
VARIABLE LABELS: X(1)=LABEL1, LABEL2,
   LABEL3, LABEL4, LABEL5;
DATA: 4 VARIABLES, 5 OBSERVATIONS:
PLOT(X(1),X(2)) SAVE RAW DATA IN FILE1,
               HORIZONTAL INCHES = 3.0,
               VERTICAL INCHES = 4.0,
               GRAPH LABEL = ONE, SQUARE,
               SYMBOL SIZE = 0.15;
FORMAT(F2.1,F2.1,2F1.1);
004004;
103013;
302933;
701071;
951792;
PLOT(X(1),X(2)) READ RAW DATA FROM FILE1,
               HORIZONTAL INCHES = 5.0,
               VERTICAL INCHES = 7.0;
               GRAPH LABEL = TWO,
               DRAW LEAST SQUARES LINE;
PLOT(X(3),X(2)) READ RAW DATA FROM FILE 1,
               HORIZONTAL INCHES = 8.0.
               VERTICAL INCHES = 2.0,
               GRAPH LABEL = THREE.X,
               NUMBER OF HORIZONTAL INTERVALS = 9.0;
END OF RUN
```

Figure IX-53. Example in Graphic language.
Source: Schwinn [UZ67], p. 476. By permission of Association for Computing Machinery, Inc.

request, and a few miscellaneous items. Some of the facilities available under the *plot request* include DRAW LEAST SQUARES LINE, DRAW LINE BETWEEN THE POINTS, HORIZONTAL MAXIMUM FROM DATA, VERTICAL MINIMUM FROM DATA, and SAME HORIZONTAL MAXIMUM. Some of the optional specifications available to the user include the following: NUMBER OF HORIZONTAL INTERVALS, DRAW ALL AXES, DRAW HORIZONTAL AXIS, NUMBER OF PERPENDICULAR INTERVALS, SUPPRESS PLOT OUTPUT, and SAME PERPENDICULAR SCALE. In order to combine these facilities with calculations which might be required, the user is permitted to include a breakout character which allows him to use whatever other language he wants. Figure IX-53 shows an example of a program written in this language, assuming FORTRAN as the "other" language.

4. *DOCUS*

DOCUS (*D*isplay *O*riented *C*omputer *U*sage *S*ystem) appears to be one of the most general in this area. It is based entirely upon a push-button scheme with responses shown on the scope. The overall system consists of a number of subsystems languages, specifically the *G*eneral *O*perating *L*anguage (GOL), a *P*rocedure *I*mplementation *L*angugae (PIL), *D*isplay-*O*riented *L*anguage (DOL), FORTRAN, CODAP, a debugging language, and some applications for file management, text manipulation, computation, and graphics. The system is implemented on a CDC 1604B. A list of the control functions in these various categories is given in Figure IX-54.

5. *AESOP* (*cross-reference only*)

AESOP is an on-line data management system with powerful graphical facilities for its own operations. See Section IX.3.2.8.

IX.3.4. Computer-Aided Design

1. *General*

The area of computer-aided design is a complex one, of which the language capability is only one portion, and a relatively small portion at that. The *design* being referred to usually involves line (or three-dimensional) drawings which might be from areas such as circuit design, mechanical engineering, and automotive design. Much of the conceptual work that has been done in this area has pertained to on-line systems in general and to the use of graphics in particular. The early work of Sutherland on SKETCHPAD [QW63] was an illustration of the use of a computer to do design work. The tools developed for doing this, in particular the ring (list)

General Operating Language

General communication
   SEND MESSAGE
   SHOW MESSAGE

DOCUS library control
   LIST DISPLAYS
   LIST KEY PROGRAMS
   END DISPLAY
   END PAGE
   CHANGE PAGE
   DELETE PAGE
   INSERT PAGE
   ADD DISPLAY TO LIBRARY
   DELETE DISPLAY FROM LIBRARY
   COMBINE
   LIST COMBINE

Output control
   LIST OUTPUT

General control
   START/STOP
   HALT/PROCEED
   SELECT
   REJECT
   ENTER

Display manipulation
   SEEK AND SHOW
   SAVE PAGE
   RESTORE PAGE
   NEXT PAGE
   PREVIOUS PAGE
   SCAN FORWARD
   SCAN BACKWARDS

Sample of DOL Statements

Controlling devices on console
   LIGHTS ON
   LIGHTS OFF
   ALARM ON
   ALARM OFF
   SAVE LIGHTS
   RESTORE LIGHTS

   NAME LIGHT
   DEFINE LIGHT GROUP
   TEST IF LIGHTS ON
   READ CURSOR
   PROCESS LIGHT
   PEN MESSAGE

Composition of textual displays
   DEFINE AREA
   LITERAL AREA
   DEFINE COMPOSITE
   DEFINE STRING
   LITERAL STRING

   MOVE AREA
   ROTATE AREA
   FILL AREA
   MOVE STRING

Composition of graphical displays
   DRAW LINE
   MOVE POINT
   ROTATE FIGURE
   EASE LINE

   JOIN LINE
   EXPAND FIGURE
   CONTRACT FIGURE
   COMPOSE PICTURE

Unique PIL Functions

   BEGIN PROGRAM
   INSERT CODE
   DELETE CODE
   CHANGE CODE
   END PROGRAM

   KEEP SOURCE PROGRAM
   COMPILE/ASSEMBLE
   ADD PROGRAM TO LIBRARY
   DELETE PROGRAM FROM LIBRARY
   LIST PROGRAM DESCRIPTION

Figure IX-54. GOL, DOL, and PIL operations in DOCUS.
Source: Corbin and Frank [CF66], extracts from pp. 520–522. By permission of Association for Computing Machinery, Inc.

structure and CORAL (see Section VI.9.3.2), have somewhat wider implications than that of computer-aided design. On the other hand, much of the work done at the General Motors Laboratories to provide computer aid to designers has little or nothing to do with languages as such; the primary concern is with the types of items that the user can display on a screen and change by operating with a light pen.[16]

From a purely language point of view, the only major system in this field (aside from possible consideration of the specialized list systems just mentioned) is the AED work at M.I.T.

2. *AED*

The work on AED (*A*utomated *E*ngineering *D*esign or *A*LGOL *E*xtended for *D*esign) has been carried on at M.I.T. since 1959 under the direction of D. T. Ross. In spite of the fact that representatives from 20 industrial organizations which would have a use for this system have each made contributions of one person for a year, AED has not received widespread usage outside M.I.T. This is partly due to poor timing relative to computer conversions within the companies, massive and difficult documentation, and perhaps a lack of appreciation of its good qualities; it does not necessarily represent a repudiation of the system concepts or implementations. On the other hand, in my personal view, a discussion of AED more properly belongs in a compiler-writing section or perhaps even in the multipurpose language chapter rather than in a section on computer-aided design. It was placed here in order to comply with the general tenor of the remarks and literature of the developing group, who continually refer to it as a foundation for computer-aided design and man-machine problem-solving.

AED has a strong philosophical foundation relating to problem-solving and associated models, and this is considered fundamental to the work. However, this aspect of it is beyond the scope of this book. Fortunately, it is not necessary either to understand or to know this facet of the work in order to understand those parts which are most significant from a language viewpoint.

The AED-0 language is an ALGOL-like extended subset. Some of the omissions or changes to ALGOL are as follows: Only one subscript is allowed, the parameter transfer mechanism employs *references* rather than *name* and *value*, block structure controls scope but is not used for storage allocation and control, recursive procedures must be declared recursive, and various syntactic forms are modified slightly.

There are several new features of interest. The most significant innovation is the incorporation of *n-component elements* to permit *plex* programming

---

[16] See *Proc. FJCC* (1964) for a number of articles describing this system.

(a generalization of the early concepts of list processing introduced in a paper by Ross [RD61]). The data structure of a plex is a combination of elements interconnected in a network by multiple pointer components (i.e., each node of the network can have pointers to many other nodes). In AED-0, integer, real, Boolean, or *pointer components* occupying full or partial computer words may be declared and can be included as items in the *n*-component element. The system supplies routines which allow multiword elements to be created and destroyed in various ways. (A detailed discussion of the implementation techniques is given in Ross [RD67].) Component reference is identical to functional reference, and arbitrarily nested reference expressions, e.g., A(B(C(D))), may be made. In addition to the compiler, AED includes an elaborate macro preprocessor, and the philosophy of using the system and language puts heavy emphasis on the ability to improve efficiency by automatically changing the method of implementation of certain notational forms merely by altering declarations. For example, the notation A(B) may be mapped into various implementations, depending on the declarations of A and B. Thus A(B) may represent a function of an argument, a component of an element, an array with index, or a macro call with argument, but it produces the same results in all cases.

Another feature of AED-0 is called *phrase substitution* (which is equivalent to or more general than what I have called embedding, depending on usage). The language and its compiler are designed such that any phrase or syntactic unit of an appropriate *type* may be used wherever a single identifier of that type is allowed. Assignment statements take on the value assigned. This enables such features as embedded assignment statements, e.g.,

$$Y = Z + (Z = A + B)$$

(causing the value of Y to become 2(A + B)), and *valued blocks* of the general form

INTEGER BEGIN . . . END

which permit an entire program to replace a single identifier of the corresponding type.

AED-0 also includes automatic stack declaration and manipulation, bit manipulation, character string manipulation, an operator LOC which provides pointers to values of expressions, a *synonym* feature which permits all spellings including those of reserved words to be altered, as well as a PRESET facility for initializing variables and arrays at compile time.

Modification of AED-0 as a language ceased in 1964 and further development of language features has been made in the form of *integrated packages* of procedures; these represent the raw semantics of language features which will be given appropriate syntax in future AED languages. The "culture" of AED usage depends heavily upon these packages, several of which interlock directly with the compiled features of the language itself. For example, an *ISARG* package permits the use of optional arguments in procedure calls; the *DOIT* package permits procedure names to be stored in and executed from data structures, including dynamic loading so that program control and data structures are interlocked (enabling generalizations of such techniques as *coroutines* to be easily performed); the *GENCAL* package permits dynamic compilation of molecular procedure calls at run time. Various techniques for using these features, in combination with the ability to nest procedure definitions and declare procedures separately from their definitions, permit many sophisticated control structures such as multientry procedures to be developed. Other packages provide facilities for system building. The *RWORD* package gives sophisticated free-format input and the *ASMBL* package gives free-format output of character streams; the *generalized string* package creates and manipulates arbitrary *string* structures of arbitrary elements in any combination, including ordered and unordered uni- or multidirectional lists and rings, stacks, queues, hash-coded tables, or other more elaborate specialized data structures; the *IOBCP* (input/output buffer control package) provides a uniform operating system interface for word- and character-oriented files for arbitrary storage or input/output devices, including control of logical and physical records, buffering, and timing; the *delayed merge* package and the *generalized alarm* package permit segmentation of large program actions, including alteration of control and a wide spectrum of error-handling facilities. Although many of these packages are more closely related to software design than language, their use is so integral with the direct linguistic features of AED-0 that they form a significant part of the pragmatics of AED-0 as a language.

The pragmatics of AED are completed by referring to Figure IX-55, which indicates the lexical, parsing, modeling, and analysis phases of a system constructed according to the AED philosophy (see Ross [RD67a]), using the machine and language independent AED-1 compiler as an example. (Initially that compiler processed the AED-0 language and there was no specific AED-1 language.) The *RWORD* system constructs a finite-state machine lexical processor in accordance with *item* descriptions expressed in the language of regular expressions; the AEDJR system (a syntax-driven parser based on Ross's *algorithmic theory of language* [RD62]) constructs tables for parsing a specified language; and the modeling and analysis phases (shown in Figure IX-55 for the special case of program compilation), may also be set up to handle specialized problem-oriented languages, though not yet in as systematic a fashion as the other phases. Because of the interrelationships of the elements in the compiling process, it is easy to intermingle graphical elements (e.g., push buttons and light pen responses at a

IX.3.4.2. AED

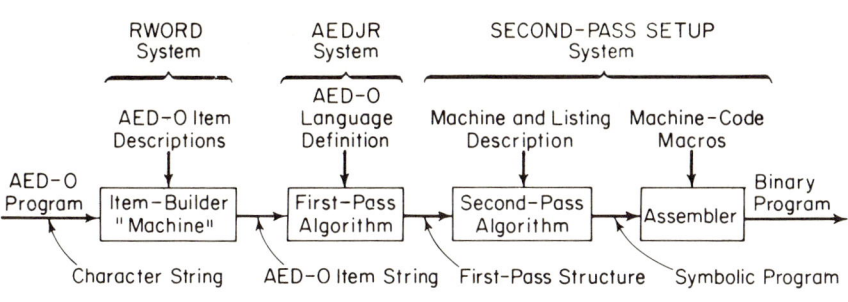

Figure IX-55. General structure of AED-1 compiler.
Source: Ross [RD67a], p. 372. By permission of Association for Computing Machinery, Inc.

display console) with verbal elements (e.g., **+**, **−**, **for**) in a single language. Extensions of the systematic treatment of system building to include these mixed language forms as well as generalized modeling were being incorporated during 1967 into the CADET (*C*omputer-*A*ided *D*esign *E*xperimental *T*ranslator) system.

Various combinations of the "system of systems" which constitute AED have been successfully used in a variety of applications, including a system for formal algebraic manipulation (Wolman [YW66]), a translator from MAD to AED-0 (Lapin [LB67]), a new processor for the DYNAMO language, as well as systems for information retrieval, econometric modeling, nonlinear circuit simulation, stress analysis, differential equations, ship design, geometric modeling, chemical process design, and other applications.

In 1967 the AED system was being bootstrapped from the Compatible Time-Sharing System at M.I.T. (see Crisman [ZR65]) onto the IBM System/360, IBM 7094, and Univac 1108 computers. (An earlier batch-processing version had also been distributed for the IBM 7094 in previous years.)

At the time of this writing, AED and its concepts do not play a major role in the computing community, either from a theoretical viewpoint or in terms of practical utility. I personally think this is due more to the difficulties and/or lack of documentation and training rather than to the system itself. Although not widely publicized, an AED technical meeting held at M.I.T. in January, 1967 was attended by over 400 people representing over 100 organizations. In view of this level of interest and the acknowledged impact of the APT system (developed earlier by the same group—see Section IX.2.1.1) in the field of numerical control, AED may yet become a significant force in the computing world.

IX.3.5. Text Editing and Processing

The increasing use of on-line systems has made it possible to develop systems which permit a user to modify text (e.g., material such as this book, letters, or reports) which is stored in some type of device accessible to a central computer. As the Acknowledgements indicate, the IBM ATS (*A*dministrative *T*erminal *S*ystem) (originally designed by M. Nekora), was used in processing this book in its manuscript form. More specifically, nine chapters went through several revisions on ATS. A version of ATS called DATATEXT may be leased from IBM's Information Marketing Department. The functions that such systems need to perform apply to small units of information (e.g., a few words) or larger units (e.g., lines, sentences, paragraphs). The operations include items such as deletion, insertion, replacement, movement of some units to a different place in the text, printouts at a terminal, provision for saving and obtaining files, and the preparation of the text for use with high-speed printing or photocomposing devices. One of the key technical issues in this type of system is how much information will be defined by a physical line number and how much is to be specified by context. In the latter case, we might have something as

---

Delete and close from the fifth paragraph through the second occurrence of a paragraph beginning with the name "Alice".
Delete and close the expression "and then unrolled the parchment scroll".
Delete "she made some tarts. All on a".
Insert "and precisely" before the fourth line in the eighth paragraph.
Run on the first paragraph.
Start a new line with the last sentence in the seventeenth paragraph
Start a new paragraph with the last two words in the fourth line of the seventh paragraph.
Start a new page with the paragraph ending with the expression "the king".
Indent the second sentence in the eighth paragraph.
Start a new line with the fourth word in the twentieth paragraph.
Exchange the last four words in the second line of the sixth paragraph with the first two words in the paragraph ending in the word "talking".
Exchange the eighth occurrence of the word "King" with the word "Hatter" in the twenty-first paragraph.
Exchange the word "Hatter" with the word "king" in the twenty-third paragragh.
Exchange the tenth paragraph through the paragraph ending in "away" with the thirteenth paragraph.
Exchange "'Not yet, not yet!'" with "the Rabbit hastily interrupted.".
Replace the word "the" in the thirteenth paragraph with "The".
Replace "!" in the thirteenth paragraph with ".".

Figure IX-56. Examples of editing statements in ES-1.
Source: Barnett and Kelley [BI63], p. 101.

elaborate as, DELETE THE THIRD OCCURRENCE OF 'ANIMAL' IN THE SENTENCE BEGINNING 'THE QUICK BROWN FOX'. Only the ES-1 system (see Barnett and Kelley [BI63]) permits anything as general as this, although it was used in a batch environment rather than on line.

Since most of these systems have a primitive language at best, no attempt has been made to survey many of them; just a few are illustrated. Figure IX-56 shows an example of ES-1 commands. Figure IX-57 contains a summary list of commands for the IBM DATATEXT system, while Figure IX-58 shows an example of their use. Commands from the editing system available under M.I.T.'s CTSS (see Crisman [ZR65]) are shown in Figure IX-59.

A system called SAFARI has been developed at the MITRE Corporation both to edit text and to perform a sophisticated linguistic and logical processing that structures the information content. The texts are presented on a graphic display: A light gun is used in conjunction with light button commands shown in the margin and with a typewriter; the latter is used to select, rearrange, paraphrase displayed material, insert new material, and to initiate various processing steps.

Instruction[†]	Meaning	Instruction[†]	Meaning
a	Automatic mode	t	Transmit message
as	Automatic mode—suppress space between paragraphs	u	Uncontrolled mode
		uc	Line centering mode
c	Clear working storage	+	Additional line request
d	Delete permanent storage	!f	Form letter mode
e	Erase lines and/or units	!g	Storage report
f	Footing mode	!p	Page numbering begins with number following p
g	Get document from permanent storage		
		!r	Reprint action
h	Heading mode	!-t---t	Tab settings
i	Insert	!number	Number of lines to be printed per page
j	Justify (printout)		
jn	Justify with unit numbers	t0	Transmit to customer assistance terminal
m	Message retrieval		
n	Request for next unit number	t96	Transmit to upper- and lower-case printer
p	Print		
p0	Print as entered	t97	Transmit to tape from working storage
p0n	Print as entered with unit numbers		
		t98	Transmit to printer
pn	Print with unit numbers	t99	Transmit to card punch
s	Store in permanent storage		

[†]Each instruction is preceded by hitting the Attention key on the terminal.

Figure IX-57. List of DATATEXT commands.
Source: [IB67b], p. 37. Reprinted by permission from *DATATEXT Operator's Instruction Guide*. © 1967 by International Business Machines Corporation.

a
AUTOMATIC MODE

        This book was typed using ATS and I am
now merely demonstrating some of the facilities
that it has.

Soppose I mkae an error leave out a word.

I will make the corrections and also insert a sentence.

!10

jn65
DBL ATTN EA PAGE
        This book was typed using ATS and I am now merely   3
demonstrating some of the facilities that it has. Soppose I mkae   4
an error leave out a word. I will make the corrections and also   5
insert a sentence.

4   Soppose   Suppose
Suppose I mkae an error leave out a word.

4   mkae an error   make an error and
Suppose I make an error and leave out a word.

i4
Suppose I make an error

This is the inserted sentence, after the one beginning Suppose.

jn65
DBL ATTN EA PAGE
        This book was typed using ATS and I am now merely   3
demonstrating some of the facilities that it has. Suppose I make   4
an error and leave out a word. This is the inserted sentence,   5
after the one beginning Suppose. I will make the corrections and   6
also insert a sentence.

Figure IX-58. Sample of DATATEXT usage.

FIND	LOCATE	NEXT	DELETE
PRINT	RETYPE	TOP	BOTTOM
INSERT	CHANGE	BLANK	OVERLAY
VERIFY	BRIEF	CLIP	SERIAL
COLON	TABSET	FILE	

Figure IX-59. Names of commands in CTSS context editor.
Source: Crisman [ZR65], extracts from Section AH.3.02, pp. 4–9.

IX.3.6. CONTROL LANGUAGES FOR ON-LINE AND OPERATING SYSTEMS

The growth of the complicated and powerful batch operating systems and the increased interest and use of on-line systems have had many profound effects on the computing industry. The one which is being addressed here is naturally that which pertains to languages. There are two major philosophical problems which make this issue (and indeed this section) anomalous. First of all, it is not clear that the items discussed here are really *languages* in the sense of Chapter I since they violate some of the defining characteristics given there. Secondly, there is a legitimate question in the case of some of the on-line systems as to whether there is really a separate *terminal language* that differs from the problem language. I now wish to consider both of these issues briefly.

It is no longer sufficient (or even possible) for a user who wishes to compile and execute a FORTRAN program to take his deck to a machine room and run it. Similarly, he is not able merely to type in a FORTRAN program at an on-line system and have it compiled and executed. Between the user's problem statement and his desired results there is an additional barrier beyond writing the solution to his problem, namely the method by which he communicates his desires to the system. Lest any casual reader wonder why this is so, it should suffice to point out first that in an on-line system there are parallel wishes of other users for differing languages and actions to take place. In the batch operating systems, the preceding user may be asking for an entirely different language compilation, with data coming from many different directions. Thus there has grown up a rather complex sequence of control characters (i.e., a control language) which is needed to bridge the gap between the user who has finally stated his problem in a higher level language, and the computer on which it is to be run. I use the phrase *sequence of control characters* rather than *control language* because the real issue is whether this is a higher level language in the sense of this book. It is true that machine code knowledge is unnecessary, but the potential for conversion to other computers is relatively small unless the other computer has duplicated the operating or time-sharing system. (Thus it is theoretically easy to convert, but it is difficult to do on a practical basis.) There is definitely an instruction explosion, but it is somewhat doubtful that the notation is particularly problem-oriented. Thus this *sequence of control characters* satisfies some—but not all—of the defining characteristics for a programming language. *If* we consider the sequence a *superstructure* on the basic language, then the second philosophical question involved is whether this superstructure (i.e., control language) is *really* part of the problem language or not. I personally feel that the language used to communicate with an operating or on-line system is *not* now part of the language which is used to state solutions to problems; however, there are good arguments that can be given on both sides.

Another way to consider this problem is to say that this control language *should* be, to the largest extent possible, the same as the actual language used to write the solutions to problems. This means that whenever the same functional capabilities are needed (e.g., input/output, tests, and obtaining data), the syntax should be the same as the problem language.

Space does not permit a detailed description of the various types of commands that exist in operating and on-line systems. It will be more useful to provide general principles with specific illustrations and let the user draw on his own experience.

In considering the on-line systems, we must distinguish carefully between those which are general purpose and those which are dedicated to one, or at most, a few specific activities. In the first category exist the Project MAC system (CTSS) at M.I.T. (Crisman [ZR65]) and the time-sharing system at the System Development Corporation (Coffman, Schwartz, and Weissman [CO64]). In the second category are those systems described in Section IV.6, plus a few scattered others. In all these systems there is a provision for some type of sign in (usually a *LOGIN* command) and some type of sign off (e.g., *LOGOUT*). These serve the purpose of identifying the user and permitting the necessary accounting to take place.

In the general purpose system with several higher level language compilers available, it becomes necessary to specify in some way what the specific task is that the user wishes done. For example, he may want to type in a FORTRAN program and have it compiled, or he may wish to retrieve from some storage area a program (in another language) he had been working on, change it, and then have it compiled and executed. This type of facility is usually provided by specifying the exact name of the language involved (e.g., in CTSS the user wrote mad in order to cause compilation of a MAD program). Editing facilities are strongly required, both for changing individual lines and for modifying large units of information. For example, when the user is typing in, he must have some immediate way to correct a typing error; this is usually handled by striking a special key which indicates where the error is, or how far back to go, and then permits retyping. On a larger level, it is often necessary (and, in fact, one of the valuable features of an on-line system) to insert, delete, or change individual lines or portions of a program. Thus there is a need for specific commands which can perform editing either by accessing numbered lines or by context. In the former case, the system is usually supplying specific line numbers as the user types in his program. In the latter, the user wants to say something like *change the $+$ in $A + B$ to $-$*. A more likely notation might be

$$\text{ch } A + B \; A - B$$

The final category of commands that are essential in a general purpose on-line system are those which save files or programs and permit them to

IX.3.6. CONTROL LANGUAGES FOR ON-LINE AND OPERATING SYSTEMS

be obtained subsequently. This also requires provisions for deleting elements from files, for printing out their contents, etc. One of the major issues, which is only tangentially a language one, is how an individual's files are protected from unauthorized access by another person, where the access can be either in terms of deletion or examination of the information.

An example of a portion of a console session on M.I.T.'s CTSS is shown in Figure IX-60.

The commands that are provided for the on-line systems of the types discussed in Section IV.6 are simpler in the sense that there is little or no need for identification of the higher level language nor for as many operations. Thus people who are using QUIKTRAN, JOSS, or CPS know exactly what they are running, and it is not necessary to specifically request a subsystem. On the other hand, they have similar requirements for signing in and out, editing, or saving and retrieving information from files. The condensed manual for CPS shown in Section IV.6.5 as Figure IV-16 summarizes the commands available for that particular system (which also provides for a Remote Job Entry). For example, *SAVE* (with and without password) and *LOAD* are used, respectively, to store and retrieve programs from a file. Stored programs can be deleted by using *LIB SCRATCH*, and a listing of the names can be obtained by using *LIB LIST*. An interesting safety feature is the *AUTOSAVE PARAMETERS* option which is used with *SAVE* to request that the state of the user's work be saved periodically (and automatically) for later retrieval if there is an unexpected system shutdown. The user starts and terminates by using *LOGIN* and *LOGOUT*, respectively (with appropriate account numbers).

In QUIKTRAN there are seven commands involving communication with the system. Each is preceded by a semicolon for identification. The *USER* identifies the user to the system, and the *CONSOLE* command sets up the terminal for conversational operations. The *EXIT* deactivates the user's identification code, and *FINISH* terminates the operation. *EJECT* causes the terminal printer to skip to a new page, and *SEND* transmits a message to the operator. An interesting command is the *ECHO (test pattern)*. In case of suspected system malfunction, the system reads any *test pattern* typed by the user and *echoes* it back on the terminal. If they do not match, the fault is probably a system difficulty; if they do, then the user is probably in error.

DIAMAG, which is an extension to ALGOL for on-line usage on the IBM 7044 with a satellite computer for the teletypes, has a fairly powerful set of control commands, considering that this is a *dedicated* system. It can be used in a desk calculator mode, a conversational mode, and a batch processing mode. The control language acts on a file which is a sequence of elements of the following four types: Bits, characters, machine words, and lines. The basic operations on the files are insertion, extraction, and

```
login m1416 1591
WAIT,
 M1416  1591 LOGGED IN    5/27 1112.9
READY.
listf 5 20 63
WAIT,
    10 FILES     20 TRACKS USED
 DATE           NAME          MODE    NO. TRACKS
  5/20/63       MAIN    MAD    P         15
  5/17/63       DPFA    SYMTB  P          1
  5/17/63       DPFA    BSS    P          1
  5/17/63       DPFA    FAP    P          2
READY.
input
WAIT,
 00010              entry    recoup
 00020 recoup      tra      *+1
 00030             cal      1,4
 00040             sto      recoup
 00050             trs      2,4
 00060             end
 00070 π
 MAN. 40           sta      recoup
 MAN. file subr fap
WAIT,
READY.
fap subr
WAIT,
 0    00005  .000 00 4 00002       TRS     2,4             00000050
 00006  FIRST LOCATION NOT USED
 FAILED
READY.
.
.
.
start
WAIT,
 FILE   TEST  DATA NOT FOUND.
 NO ERROR RETURN SPECIFIED
READY.
pm lights
WAIT,
 PROG SEEK   STOP=   112 REL., 14273 ABS.  TSX  007400414161
 AC = 000014000000, S =0, Q =0 MQ = 000010000000 SI = 400004000000
 IX1 =    2 IX2 =    14 IX4 = 63505 SENSE LIGHTS ON          4
 FPT ON ,DCT OFF, ACOF OFF
READY.
save may27
WAIT,
READY.
listf
WAIT,
    16 FILES     66 TRACKS USED
 DATE          NAME         MODE    NO. TRACKS
  5/27/63      MAY27   SAVED   P       31
  5/27/63      MAIN    BSS     P        5
  5/27/63      MAIN    MADTAB # QUIT,
READY.
logout
WAIT,
 M1416  1591 LOGGED OUT  5/27 1140.3
  TOTAL TIME USED=   01.6 MIN.
READY.
```

Figure IX-60. Sample of actual CTSS session. The user types in lower case and the computer responds in upper case.
Source: Corbato [ZV63], pp. 89–91.

IX.3.6. CONTROL LANGUAGES FOR ON-LINE AND OPERATING SYSTEMS 691

concatenation, with commands for these and other operations, e.g., *INSERT, EXTRACT, CONCAT, LIST, ERASE, COPY, STORE, DELETE, COMPILE, SAVECOMP, TRANSFORM,* and *INCLUDE.*

The concept whereby people enter jobs at a terminal but expect them to be executed at some future time, presumably in a batch environment or as background, is becoming increasingly important. The terms *remote job entry* (RJE) or *foreground initiated background* (FIB) are often used; regardless of the term, the user has entered information from a terminal but is *not* waiting for instantaneous response.

At least two different systems (see Pyle [PY65] and Bequaert [BV67]) have a hazy border line between the actual problem language and the means of communication with the specific system. In both cases, the system asks the user for information rather than having the user supply it directly. For example, in Pyle's QUIN system, the user replies in lower case to the upper-case system query:[17]

> WHICH CALCULATION DO YOU WANT? what do you have
> THE FOLLOWING CALCULATIONS ARE NOW AVAILABLE
>   KINET
> WHICH CALCULATION DO YOU WANT? kinet
>   . . .
> PLEASE GIVE INPUT VALUES NOW
>   . . .
> YOU MAY EDIT NOW
> EDIT? change gen
> THIS CHANGE MAY AFFECT INICON REACT

When the on-line system involves the use of graphic display, then some differing types of control functions are either needed or useful. The ability to point with a light pen rather than type in information provides a different framework within which to work. Relevant examples are given in Section IX.3.3.

Turning now from the subject of on-line systems, let us consider a complex operating system such as that for the IBM System/360 (OS/360). The *Job Control Language* (JCL) is the method by which the user communicates with the operating system. It provides him with facilities for retrieving data sets of varying kinds and optimizing the use of input/output equipment, as well as the basic requirement for specifying just what is to be done with the program. There are six types of statements involved in JCL: *job, execute, data definition, command, delimiter,* and *null.* The general format for the first four of these statements is

---

[17] Pyle [PY65], extracts from pp. 223–24.

692 SPECIALIZED LANGUAGES

// name opcode parameters comments

where the *name* is optional in some cases. The operation codes for the first three statement types are JOB, EXEC, and DD. Some of the parameters can appear in an arbitrary order. The purpose of *JOB* is to identify the job and the user. A *COND* parameter permits the user to test the completion of previous jobs and use this to control what is to happen next; this is done by means of a code number returned by the operating system and tests

```
//LOOKUP        EXEC    PGM=SEARCH
//IN1           DD      DSNAME=A.B.C,DISP=OLD
//OUT1          DD      UNIT=2311,SPACE=(TRK,(10,2)),=DISP(,PASS)
//REDUCE        EXEC    PGM=TRUNCATE
//IN2           DD      DSNAME=*.LOOKUP.OUT1,DISP=(OLD,DELETE)
//WORK          DD      UNIT=TAPE
//OUT2          DD      UNIT=2311,SPACE=(TRK,(5,1)),DISP=(,PASS)
//DISPLAY       EXEC    PGM=PRINT
//IN3           DD      DSNAME=*.REDUCE.OUT2,DISP=(OLD,DELETE)
//OUT3          DD      SYSOUT=A
```
(a)

```
//STEP1         EXEC    ANALYSIS
//LOOKUP.OUT1   DD      UNIT=2400,DISP=
//REDUCE.WORK   DD      UNIT=180
//REDUCE.XTRA   DD      UNIT=181
//DISPLAY.IN3   DD      DISP=(OLD,KEEP)
```
(b)

```
//LOOKUP        EXEC    PGM=SEARCH
//IN1           DD      DSNAME=A.B.C,DISP=OLD
//OUT1          DD      UNIT=2400,SPACE=(TRK,(10,2))
//REDUCE        EXEC    PGM=TRUNCATE
//IN2           DD      DSNAME=*.LOOKUP.OUT1,DISP=(OLD,PASS)
//WORK          DD      UNIT=180
//XTRA          DD      UNIT=181
//OUT2          DD      UNIT=2311,SPACE=(TRK,(5,1)),DISP=(,PASS)
//DISPLAY       EXEC    PGM=PRINT
//IN3           DD      DSNAME=*.REDUCE.OUT2,DISP=(OLD,KEEP)
//OUT3          DD      SYSOUT=A
```
(c)

Figure IX-61. Uses of Job Control Language: (a) catalogued procedure, (b) changes to catalogued procedure, and (c) result of changing catalogued procedure.
Source: [IB67g], pp. 63–64. Reprinted by permission from *IBM System/360 Operating System Job Control Language*. © 1967 by International Business Machines Corporation.

involving relational operators, e.g.,

// PAYROLL JOB 5048321,A.USER,COND((12,LE),(8,EQ), . . .

EXEC primarily identifies the program to be executed. In some cases the user can specify the maximum amount of time for completion of the job since this is useful information in making assignments in a multiprogramming system. The data (*DD*) statements specify the required information about retrieving and storing data; this is essential since the data can occur in a number of different places. This is probably the most complex part of JCL. The command statements are inserted by the operator and provide for things such as *DISPLAY, MOUNT, START, STOP,* and *UNLOAD.* The delimiter statement is simply used to mark the end of a data set in the overall input stream to the operating system. The null statement is simply used to mark the end of a certain job.

Applications requiring many control statements can be stored as *catalogued procedures* and then retrieved and modified by other control cards. Figure IX-61a,b, and c shows (1) an example of a catalogued procedure, (2) the cards to change it, and (3) the result.

REFERENCES

IX.1. SCOPE OF CHAPTER

[LI65] Licklider, J. C. R., "Languages for Specialization and Application of Prepared Procedures", *Information System Sciences: Proceedings of the Second Congress.* Spartan Books, Washington, D.C., 1965, pp. 177–87.

IX.2.1. MACHINE TOOL CONTROL

[AI61] *APT Concept and Application*, Aerospace Industries Association of America, Inc. (1961).

[AI61a] *APT Introduction to Part Programming*, Aerospace Industries Association of America, Inc. (1961).

[BP63] Brown, S. A., Drayton, C. E., and Mittman, B., "A Description of the APT Language", *Comm. ACM*, Vol. 6, No. 11 (Nov., 1963), pp. 649–58.

[IB67k] *System/360 APT Numerical Control Processor (360A-CN-10X)—Part Programming Manual*, IBM Corp., H20-0309-0, Data Processing Division, White Plains, N.Y. (1967).

[II67] *APT Part Programming.* McGraw-Hill, New York, 1967.

[ZC67] Mittman, B., "Development of Numerical Control Programming Languages in Europe", *Proc. ACM 22nd Nat'l Conf.*, 1967, pp. 479–82.

IX.2.2. CIVIL ENGINEERING

[FE66] Fenves, S. J., "Problem-Oriented Languages for Man-Machine Communication in Engineering", *Proceedings of the IBM Scientific Computing Symposium on Man-Machine Communication*, IBM Corp., 320-1941-0, Data Processing Division, White Plains, N.Y. (1966), pp. 43–56.

[WQ66] Walter, R. A., "A System for the Generation of Problem-Oriented Languages", *Proc. 5th Nat'l. Conf., The Computer Society of Canada* (May–June, 1966), pp. 351–55.

IX.2.2.1. *COGO*

[EI67] *Engineer's Guide to ICES COGO I* (first edition), M.I.T., R67-46, Dept. of Civil Engineering, Cambridge, Mass. (Aug. 1967).

[RS64] Roos, D. and Miller, C. L., *COGO-90: An Engineering User's Manual*, Dept. of Civil Engineering, Research Report R64-12, M.I.T., Cambridge, Mass. (Apr., 1964).

IX.2.2.2. *STRESS*

[FE64] Fenves, S. J. et al., *STRESS: A User's Manual*. M.I.T. Press, Cambridge, Mass., 1964.

[FE65] Fenves, S. J. et al., *STRESS: A Reference Manual*. M.I.T. Press, Cambridge, Mass., 1965.

[LG67] Logcher, R. D., et al., *ICES STRUDL-I, The Structural Design Language Engineering User's Manual* (first edition), M.I.T., R67-56, Dept. of Civil Engineering, Cambridge, Mass. (Sept., 1967).

IX.2.2.3. *ICES*

[MT00] *ICES Programmers' Guide* (preliminary edition), M.I.T., Dept. of Civil Engineering, Cambridge, Mass.

[RS65] Roos, D., "An Integrated Computer System for Engineering Problem Solving", *Proc. FJCC*, Vol. 27, pt. 2 (1965), pp. 151–59.

[RS67] Roos, D., *ICES Systems Design* (2nd ed., revised). M.I.T. Press, Cambridge, Mass., 1967.

[RS67a] Roos, D. (ed.), *ICES System: General Description*, M.I.T., R67-49, Dept. of Civil Engineering, Cambridge, Mass. (Sept., 1967).

[WQ66] Walter, R. A., "A System for the Generation of Problem-Oriented Languages", *Proc. 5th Nat'l Conf., The Computer Society of Canada* (May–June, 1966), pp. 351–55.

IX.2.3. LOGICAL DESIGN

IX.2.3.1. *APL (Iverson)*

[FA64] Falkoff, A. D., Iverson, K. E., and Sussenguth, E. H., "A Formal Description of System/360", *IBM Systems Jour.*, Vol. 3, Nos. 2 and 3 (June, 1964), pp. 198–262.

[IV63] Iverson, K. E., "Programming Notation in Systems Design", *IBM Systems Jour.*, Vol. 2 (June, 1963), pp. 117–28.

IX.2.3.2. *LOTIS*

[QY64] Schlaeppi, H. P., "A Formal Language for Describing Machine Logic, Timing, and Sequencing (LOTIS)", *IEEE Trans. Elec. Comp.*, Vol. EC-13, No. 4 (Aug., 1964), pp. 439-48.

IX.2.3.3. *LDT*

[PC64] Proctor, R. M., "A Logic Design Translator Experiment Demonstrating Relationships of Language to Systems and Logic Design", *IEEE Trans. Elec. Comp.*, Vol. EC-13, No. 4 (Aug., 1964), pp. 422-30.

IX.2.3.4. *Language for Simulating Digital Systems*

[MZ65] McClure, R. M., "A Programming Language for Simulating Digital Systems", *J. ACM*, Vol. 12, No. 1 (Jan., 1965), pp. 14-22.

IX.2.3.5. *Computer Compiler*

[MX66] Metze, G., and Seshu, S., "A Proposal for a Computer Compiler", *Proc. SJCC*, Vol. 28 (1966), pp. 253-63.

IX.2.3.6. *Computer Design Language*

[CB65] Chu, Y., "An ALGOL-Like Computer Design Language", *Comm. ACM*, Vol. 8, No. 10 (Oct., 1965), pp. 607-15.

[YZ67] Mesztenyi, C. K., *Translator and Simulator for the Computer Design and Simulation Program (CDSP), Version 1.* University of Maryland, TR-67-48, Computer Science Center, College Park, Md.

IX.2.3.7. *SFD-ALGOL*

[PN66] Parnas, D. L., "A Language for Describing the Functions of Synchronous Systems", *Comm. ACM*, Vol. 9, No. 2 (Feb., 1966), pp. 72-76.

[PN66a] Parnas, D. L., "State Table Analysis of Programs in an ALGOL-Like Language", *Proc. ACM 21st Nat'l Conf.*, 1966, pp. 391-400.

IX.2.4. DIGITAL SIMULATION OF BLOCK DIAGRAMS

IX.2.4.1. *Introduction*

[CR66] Cramer, M. L. and Strauss, J. C., "A Hybrid-Oriented Interactive Language," *Proc. ACM 21st Nat'l Conf.*, 1966, pp. 479-88.

[PS64] Petersen, H. E. *et al.*, "MIDAS—How It Works and How It's Worked", *Proc. FJCC*, Vol. 26 (1964), pp. 313-24.

[QD66] Busch, K. J., "TELSIM, A User-Oriented Language for Simulating Continuous Systems at a Remote Terminal", *Proc. FJCC*, Vol. 29 (1966), pp. 445-63.

[YD55] Selfridge, R. G., "Coding a General-Purpose Digital Computer to Operate as a Differential Analyzer", *Proc. WJCC* (1955), pp. 82–84.

[YR64] Brennan, R. D. and Sano, H., "PACTOLUS—A Digital Analog Simulator Program for the IBM 1620", *Proc. FJCC*, Vol. 26 (1964), pp. 299–312.

[YR64a] Brennan, R. D. and Linebarger, R. N., "A Survey of Digital Simulation: Digital Analog Simulator Programs", *Simulation*, Vol. 3, No. 6 (Dec., 1964), pp. 22–36.

[ZZ65] Clancy, J. J. and Fineberg, M. S., "Digital Simulation Languages: A Critique and a Guide", *Proc. FJCC*, Vol. 27, pt. 1 (1965), pp. 23–36.

IX.2.4.2. *DYANA*

[OZ58] Olsztyn, J. T., "DYANA: Dynamics Analyzer-Programmer, Part II, Structure and Function", *Proc. EJCC* (1958), pp. 148–52.

[TD58] Theodoroff, T. J., "DYANA: Dynamics Analyzer-Programmer, Part I, Description and Application", *Proc. EJCC* (1958), pp. 144–47.

IX.2.4.3. *DYSAC*

[HJ63] Hurley, J. R. and Skiles, J. J., "DYSAC: A Digitally Simulated Analog Computer", *Proc. SJCC*, Vol. 23 (1963), pp. 69–82.

IX.2.4.4. *DAS*

[GL63] Gaskill, R. A., Harris, J. W., and McKnight, A. L., "DAS—A Digital Analog Simulator", *Proc. SJCC*, Vol. 23 (1963), pp. 83–90.

[GL64] Gaskill, R. A., "A Versatile Problem-Oriented Language for Engineers", *IEEE Trans. Elec. Comp.*, Vol. EC-13, No. 4 (Aug., 1964), pp. 415–21.

IX.2.4.5. *DSL/90*

[QP66] Syn, W. M. and Linebarger, R. N., "DSL/90—A Digital Simulation Program for Continuous System Modeling", *Proc. SJCC*, Vol. 28 (1966), pp. 165–87.

IX.2.5. COMPILER WRITING

IX.2.5.1. *Introduction*

[BX62] Brooker, R. A. and Morris, D., "A General Translation Program for Phrase Structure Languages", *J. ACM*, Vol. 9, No. 1 (Jan., 1962), pp. 1–10.

[BX63] Brooker, R. A. *et al.*, "The Compiler Compiler", *Annual Review in Automatic Programming*, Vol. 3 (R. Goodman, ed.). Pergamon Books, New York, 1963, pp. 229–76.

[CH64]   Cheatham, T. E., Jr. and Sattley, K., "Syntax Directed Compiling", *Proc. SJCC*, Vol. 25 (1964), pp. 31–57. (Also in [RO67].)

[FJ68]   Feldman, J. A. and Gries, D., "Translator Writing Systems", *Comm. ACM*, Vol. 11, No. 2, (Feb., 1968), pp. 77–113.

[FL64]   Floyd, R. W., "The Syntax of Programming Languages—A Survey", *IEEE Trans. Elec. Comp.*, Vol. EC-13, No. 4 (Aug., 1964), pp. 346–53. (Also in [RO67].)

[GC60]   Glennie, A. E., *On the Syntax Machine and the Construction of a Universal Compiler*, Tech. Report No. 2, Carnegie Inst. of Tech. Computation Center (AD-240512) (July, 1960).

[GW64a]   Garwick, J. V., "GARGOYLE, A Language for Compiler Writing", *Comm. ACM*, Vol. 7, No. 1 (Jan., 1964), pp. 16–20.

[IC62b]   "Panel Discussion: Languages for Aiding Compiler Writing", *Symbolic Languages in Data Processing*. Gordon and Breach, New York, 1962, pp. 187–203.

[IR61]   Irons, E. T., "A Syntax Directed Compiler for ALGOL 60", *Comm. ACM*, Vol. 4, No. 1 (Jan., 1961), pp. 51–55. (Also in [RO67].)

[IR63]   Irons, E. T., "The Structure and Use of the Syntax Directed Compiler", *Annual Review in Automatic Programming*, Vol. 3 (R. Goodman, ed.). Pergamon Press, New York, 1963, pp. 207–28.

[PT66]   Pratt, T. W. and Lindsay, R. K., "A Processor-Building System for Experimental Programming Languages", *Proc. FJCC*, Vol. 29 (1966), pp. 613–21.

[QS61]   Sibley, R. A., "The SLANG System", *Comm. ACM*, Vol. 4, No. 1 (Jan., 1961), pp. 75–84.

[RO64]   Rosen, S., "A Compiler-Building System Developed by Brooker and Morris", *Comm. ACM*, Vol. 7, No. 7 (July, 1964), pp. 403–14. (Also in [RO67].)

IX.2.5.2. *CLIP*

[BO60]   Book, E. and Bratman, H., *Using Compilers to Build Compilers*, System Development Corp., SP-176, Santa Monica, Calif. (Aug., 1960).

[BR59]   Bratman, H., *Project CLIP (The Design of a Compiler and Language for Information Processing)*, System Development Corp., SP-106, Santa Monica, Calif. (Sept., 1959).

[EG61]   Englund, D. and Clark, E., "The CLIP Translator", *Comm. ACM*, Vol. 4, No. 1 (Jan., 1961), pp. 19–22.

[IS59]   Isbitz, H. M., *CLIP: A Compiler Language for Information Processing*, System Development Corp., SP-117, Santa Monica, Calif. (Oct., 1959).

IX.2.5.3. *TMG*

[MZ65a]   McClure, R. M., "TMG—A Syntax Directed Compiler", *Proc. ACM 20th Nat'l Conf.*, 1965, pp. 262–74.

IX.2.5.4. *COGENT*

[RE65]   Reynolds, J. C., *COGENT Programming Manual*, Argonne Nat'l Lab., ANL-7022, Argonne, Ill. (Mar., 1965).

[RE65a] Reynolds, J. C., "An Introduction to the COGENT Programming System", *Proc. ACM 20th Nat'l Conf.*, 1965, pp. 422–36.

IX.2.5.5. *META 5*

[OP66] Oppenheim, D. K. and Haggerty, D. P., "META 5: A Tool to Manipulate Strings of Data", *Proc. ACM 21st Nat'l Conf.*, 1966, pp. 465–68.

[QT64] Schorre, D. V., "META-II—A Syntax-Oriented Compiler Writing Language", *Proc. ACM 19th Nat'l Conf.*, 1964, pp. D1.3-1–D1.3-11.

[QV64] Schneider, F. W. and Johnson, G. D., "META-3—A Syntax-Directed Compiler-Writing Compiler to Generate Efficient Code", *Proc. ACM 19th Nat'l Conf.*, 1964, pp. D1.5-1–D1.5-8.

IX.2.5.6. *TRANDIR*

[CH66a] Cheatham, T. E., Jr., "The TGS-II Translator Generator System", *Proceedings of the IFIP CONGRESS 65*, Vol. 2. Spartan Books, Washington, D.C., 1966, pp. 592–93.

[DE64] Dean, A. L., Jr., *Some Results in the Area of Syntax Directed Compilers*, Massachusetts Computer Associates, CA-6412-011, Wakefield, Mass. (Dec., 1964).

IX.2.5.7. *FSL*

[FJ64] Feldman, J. A., *A Formal Semantics for Computer-Oriented Languages*, Carnegie Inst. of Tech., Pittsburgh, Pa. (Ph. D. thesis) (May, 1964).

[FJ66] Feldman, J. A., "A Formal Semantics for Computer Languages and its Application in a Compiler-Compiler", *Comm. ACM*, Vol. 9, No. 1 (Jan., 1966), pp. 3–9.

[IT66] Iturriaga, R. *et al.*, "Techniques and Advantages of Using the Formal Compiler Writing System FSL to Implement a Formula ALGOL Compiler", *Proc. SJCC*, Vol. 28 (1966), pp. 241–52..

[LQ67] Mondshein, L. F., *VITAL: Compiler-Compiler System Reference Manual*, M.I.T. Lincoln Lab., Tech. Note 1967-12, Lexington, Mass. (Feb., 1967).

IX.2.6. MISCELLANEOUS

IX.2.6.1. *Matrix Computations: Matrix Compiler*

[MF57] McGinn, L. C., "A Matrix Compiler for UNIVAC", *Automatic Coding, Jour. Franklin Inst., Monograph No. 3*, Philadelphia, Pa. (Apr., 1957), pp. 71–83.

IX.2.6.2. *Cryptanalysis: OCAL*

[ED66] Edwards, D. J., *OCAS—On-Line Cryptanalytic Aid System*, M.I.T., MAC-TR-27, Project MAC, Cambridge, Mass. (May, 1966).

IX.2.6.3. *Movie Creation: Animated Movie Language and BUGSYS*

[KO64]   Knowlton, K. C., "A Computer Technique for Producing Animated Movies", *Proc. SJCC*, Vol. 25 (1964), pp. 67–87.

[LL66]   Ledley, R. S., Jacobsen, J., and Belson, M., "BUGSYS: A Programming System for Picture Processing—Not for Debugging", *Comm. ACM*, Vol. 9, No. 2 (Feb., 1966), pp. 79–84.

IX.2.6.4. *Social Science Research: DATA-TEXT*

[DF67]   Couch, A. S., *The DATA-TEXT System* (Presented at SHARE meeting, Aug., 1967, unpublished).

[HD67]   *DATA-TEXT Manual* (preliminary manual), Dept. of Social Relations, Harvard U., Cambridge, Mass. (Mar., 1967).

IX.2.6.5. *Equipment Check-out: STROBES, DIMATE*

[QU65]   Quatse, J. T., "Strobes—Shared Time Repair of Big Electronic Systems", *Proc. FJCC*, Vol. 27, pt. 1 (1965), pp. 1065–71.

[SD66]   Scheff, B. H., "A Simple User-Oriented Compiler Source Language for Programming Automatic Test Equipment", *Comm. ACM*, Vol. 9, No. 4 (Apr., 1966), pp. 258–66.

[SD66a] Scheff, B. H., "Bypassing Professional Programmers", *Datamation*, Vol. 12, No. 10 (Oct., 1966), pp. 65–81.

IX.3.1 DISCRETE STIMULATION

IX.3.1.1. *Introduction*

[JM67]   Jones, M. M., "On-Line Simulation", *Proc. ACM 22nd Nat'l Conf.*, 1967, pp. 591–99.

[KQ67]   Krasnow, H. S., "Computer Languages for System Simulation", *Digital Computer User's Handbook* (M. Klerer and G. A. Korn, eds.). McGraw-Hill, New York, 1967, pp. 1-258–1-277.

[KW66]   Kiviat, P. J., "Development of New Digital Simulation Languages", *Jour. Ind. Eng.*, Vol. XVII, No. 11 (Nov., 1966), pp. 604–609.

[RP67]   Reitman, J., "The User of Simulation Languages—The Forgotten Man", *Proc. ACM 22nd Nat'l Conf.*, 1967, pp. 573–79.

[TE66]   Teichroew, D. and Lubin, J. F., "Computer Simulation—Discussion of the Technique and Comparison of Languages", *Comm. ACM*, Vol. 9, No. 10 (Oct., 1966), pp. 723–41.

IX.3.1.2. *DYNAMO*

[PG63]   Pugh, A. L., *DYNAMO User's Manual* (2nd ed.). M.I.T. Press, Cambridge, Mass., May, 1963.

IX.3.1.3. *GPSS*

[EF64] Efron, R. *et al.*, "A General Purpose Digital Simulator and Examples of Its Application, pts. I, II, III, and IV", *IBM Systems Jour.*, Vol. 3, No. 1 (1964), pp. 21–56.

[GG61] Gordon, G., "A General Purpose Systems Simulation Program", *Proc. EJCC*, Vol. 20 (1961), pp. 87–104.

[GG62] Gordon, G., "A General Purpose Systems Simulator", *IBM Systems Jour.*, Vol. 1 (Sept., 1962), pp. 18–32.

[HK65] Herscovitch, H. and Schneider, T. H., "GPSS III—An Expanded General Purpose Simulator", *IBM Systems Jour.*, Vol. 4, No. 3 (1965), pp. 174–83.

[IB67i] *General Purpose Simulation System/360 User's Manual*, IBM Corp., H20-0326, Data Processing Division, White Plains, N.Y. (1967).

[IB67j] *General Purpose Simulation System/360 Introductory User's Manual*, IBM Corp., H20-0304, Data Processing Division, White Plains, N.Y. (1967).

IX.3.1.4. *SIMSCRIPT*

[DC64] Dimsdale, B. and Markowitz, H. M., "A Description of the SIMSCRIPT Language", *IBM Systems Jour.*, Vol. 3, No. 1 (1964), pp. 57–67.

[KW66a] Kiviat, P. J., *Introduction to the SIMSCRIPT II Programming Language*, RAND Corp., P-3314, Santa Monica, Calif. (Feb., 1966).

[MA63] Markowitz, H. M., Hausner, B., and Karr, H. W., *SIMSCRIPT—A Simulation Programming Language*. Prentice-Hall, Inc., Englewood Cliffs, N.J., 1963.

[TN65] Tonge, F. M., Keller, P., and Newell, A., "QUICKSCRIPT—A SIMSCRIPT-Like Language for the G 20", *Comm ACM*, Vol. 8, No. 6 (June, 1965), pp. 350–54.

IX.3.1.5. *SOL*

[KN64] Knuth, D. E. and McNeley, J. L., "SOL—A Symbolic Language for General-Purpose Systems Simulation", *IEEE Trans. Elec. Comp.*, Vol. EC-13, No. 4 (Aug., 1964), pp. 401–408.

[KN64a] Knuth, D. E. and McNeley, J. L., "A Formal Definition of SOL", *IEEE Trans. Elec. Comp.*, Vol. EC-13, No. 4 (Aug., 1964), pp. 409–14.

IX.3.1.6. *MILITRAN*

[YC64] *MILITRAN Reference Manual*, DDC #AD601-794, Gulton Systems Research Group, Inc., Mineola, N.Y. (June, 1964).

[YC64a] *MILITRAN Programming Manual*, DDC #AD601-796, Gulton Systems Research Group, Inc., Mineola, N.Y. (June, 1964).

IX.3.1.7. *SIMULA*

[DH66] Dahl, O. and Nygaard, K., "SIMULA—An ALGOL-Based Simulation Language", *Comm. ACM*, Vol. 9, No. 9 (Sept., 1966), pp. 671–82.

IX.3.1.8. *OPS*

[JM67] Jones, M. M., "On-Line Simulation", *Proc. ACM 22nd Nat'l Conf.*, 1967, pp. 591–99.
[YP65] Greenberger, M. et al., *On-line Computation and Simulation: The OPS-3 System*. M.I.T. Press, Cambridge, Mass., 1965.
[YP66] Greenberger, M. and Jones, M. M., "On-Line Simulation in the OPS System", *Proc. ACM 21st Nat'l Conf.*, 1966, pp. 131–38.

IX.3.2. QUERY

IX.3.2.1. *Introduction*

[BB64]   Bobrow, D. G., *Natural Language Input for a Computer Problem Solving System*, M.I.T., MAC-TR-1, Project MAC, Cambridge, Mass. (Sept., 1964) (Ph. D. thesis).
[BB64c] Bobrow, D. G., "A Question-Answering System for High School Algebra Word Problems", *Proc. FJCC*, Vol. 26, pt. 1 (1964), pp. 591–614.
[CC62a] "Discussion—The Pros and Cons of a Special IR Language", *Comm. ACM*, Vol. 5, No. 1 (Jan., 1962), pp. 8–10.
[CD63]   Climenson, W. D., "RECOL—A Retrieval Command Language", *Comm. ACM*, Vol. 6, No. 3 (Mar., 1963), pp. 117–22.
[CH62]   Cheatham, T. E., Jr. and Warshall, S., "Translation of Retrieval Requests Couched in a 'Semiformal' English-Like Language", *Comm. ACM*, Vol. 5, No. 1 (Jan., 1962), pp. 34–39.
[CV64]   Cooper, W. S., "Fact Retrieval and Deductive Question-Answering Information Retrieval Systems", *J. ACM*, Vol. 11, No. 2 (Apr., 1964), pp. 117–37.
[CV65]   Cooper, W. S., *Automatic Fact Retrieval*, I.B.M. Corp., RJ 326, Research Division, San Jose, Calif. (Jan., 1965).
[DI67]   Dixon, P. J. and Sable, J. D., "DM-1, a Generalized Data Management System", *Proc. SJCC*, Vol. 30 (1967), pp. 185–98.
[FO65]   Foster, D. C., "The Information Processing System for the AN/FYK1(V) Data Processing Set", *Information System Sciences: Proceedings of the Second Congress*. Spartan Books, Washington, D.C., 1965, pp. 49–55.
[FW66]   Franks, E. W., "A Data Management System for Time-Shared File Processing Using a Cross-Index File and Self-Defining Entries", *Proc. SJCC*, Vol. 28 (1966), pp. 79–86.
[GH65]   Grant, E. E., *The LUCID Users' Manual*, System Development Corp., TM-2354/001/00, Santa Monica, Calif. (June, 1965).

[GR62] Grems, M., "A Survey of Languages and Systems for Information Retrieval", *Comm. ACM*, Vol. 5, No. 1 (Jan., 1962), pp. 43–46.

[HV63] Haverty, J. P. and Patrick, R. L., *Programming Languages and Standardization in Command and Control*, RAND Corp., Memo No. RM-3447-PR, Santa Monica, Calif. (1963).

[KG66] Kellogg, C. H., *An Approach to the On-Line Interrogation of Structured Files of Facts Using Natural Language*, System Development Corp., SP-2431/000/00, Santa Monica, Calif. (Apr., 1966).

[KG67] Kellogg, C. H., *On-Line Translation of Natural Language Questions into Artificial Language Queries*, System Development Corp., SP-2827/000/00, Santa Monica, Calif. (Apr., 1967).

[MW66] Meadow, C. T., and Waugh, D. W., "Computer Assisted Interrogation", *Proc. FJCC*, Vol. 29 (1966), pp. 381–94.

[NB67] Nelson, D. B., Pick, R. A., and Andrews, K. B., "GIM-1, a Generalized Information Management Language and Computer System", *Proc. SJCC*, Vol. 30 (1967), pp. 169–73.

[QE66] Bryant, J. H. and Semple, P., Jr., "GIS and File Management", *Proc. ACM 21st Nat'l Conf.*, 1966, pp. 97–107.

[RA64] Raphael, B., "A Computer Program Which 'Understands'", *Proc. FJCC*, Vol. 26, pt. 1 (1964), pp. 577–89.

[SE65] Simmons, R. F., "Answering English Questions by Computer: A Survey", *Comm. ACM*, Vol. 8, No. 1 (Jan., 1965), pp. 53–69.

[SE66] Simmons, R. F., Burger, J. F., and Long, R. E., "An Approach Toward Answering English Questions from Text", *Proc. FJCC*, Vol. 29 (1966), pp. 357–63.

[SE66a] Simmons, R. F., "Natural-Language Processing", *Datamation*, Vol. 12, No. 6 (June, 1966), pp. 61–72.

[SK67] Savitt, D. A., Love, H. H., Jr., and Troop, R. E., "ASP: A New Concept in Language and Machine Organization", *Proc. SJCC*, Vol. 30 (1967), pp. 87–102.

[SL65] Slagle, J. R., "Experiments with a Deductive Question-Answering Program", *Comm. ACM*, Vol. 8, No. 12 (Dec., 1965), pp. 792–98.

[UW67] Summit, R. K., "DIALOG—An Operational, On-Line Reference Retrieval System", *Proc. ACM 22nd Nat'l Conf.*, 1967, pp. 51–56.

[UY63] Swets, J. A. *et al.*, *The Socratic System: A Computer System to Aid in Teaching Complex Concepts*, Bolt, Beranek, and Newman, Report No. 1007, Cambridge, Mass. (Apr., 1963).

[ZT67] Chapin, P. G. *et al.*, *SAFARI, An On-Line Text-Processing System User's Manual*, MITRE Corp., MTP-60, Bedford, Mass. (Mar., 1967).

IX.3.2.2. *COLINGO and C-10*

[QZ67] Steil, G. P., Jr., "File Management on a Small Computer: The C-10 System", *Proc. SJCC*, Vol. 30 (1967), pp. 199–212.

[SZ65] Spitzer, J. F., Robertson, J. G., and Neuse, D. H., "The COLINGO System Design Philosophy", *Information System Sciences: Proceedings of the Second Congress*. Spartan Books, Washington, D.C., 1965, pp. 33–47.

IX.3.2.3. *473L Query*

[BL65] Barlow, A. E. and Cease, D. R., "Headquarters, U.S. Air Force Command and Control System Query Language", *Information System Sciences: Proceedings of the Second Congress*. Spartan Books, Washington, D.C., 1965, pp. 57–76.

IX.3.2.4. *ADAM*

[CY66] Connors, T. L., "ADAM—A Generalized Data Management System", *Proc. SJCC*, Vol. 28 (1966), pp. 193–203.

IX.3.2.5. *BASEBALL*

[GB61] Green, B. F., *et al.*, "BASEBALL: An Automatic Question-Answer", *Proc. WJCC*, Vol. 19 (1961), pp. 219–24. (Also in [FG63].)

IX.3.2.6. *DEACON*

[CJ64] Craig, J. A., Pruett, J., and Thompson, F., *DEACON Breadboard Grammar*, General Electric Co., RM64TMP-14, TEMPO, Santa Barbara, Calif. (1964).

[CJ66] Craig, J. A., *et al.*, "DEACON: Direct English Access and CONtrol", *Proc. FJCC*, Vol. 29 (1966), pp. 365–80.

[TH63] Thompson, F. B., *The Semantic Interface in Man-Machine Communications*, General Electric Co., RM63TMP-35, TEMPO, Santa Barbara, Calif. (1963).

[TH64] Thompson, F. B. *et al.*, *DEACON Breadboard Summary*, General Electric Co., RM64TMP-9, TEMPO, Santa Barbara, Calif. (1964).

[TH64a] Thompson, F. B., *The Application and Implementation of DEACON Type Systems*, General Electric Co., RM64TMP-11, TEMPO, Santa Barbara, Calif. (1964).

[TH66] Thompson, F. B., "English for the Computer", *Proc. FJCC*, Vol. 29 (1966), pp. 349–56.

IX.3.2.7. *Protosynthex*

[SE63] Simmons, R. F. and McConlogue, K. L., "Maximum-Depth Indexing for Computer Retrieval of English Language Data", *Amer. Documentation*, Vol. 14, No. 1 (1963), pp. 68–73.

[SE64] Simmons, R. F., Klein, S., and McConlogue, K. L., "Indexing and Dependency Logic for Answering English Questions", *Amer. Documentation*, Vol. 15, No. 3 (1964), pp. 196–204.

[SE65] Simmons, R. F., "Answering English Questions by Computer: A Survey", *Comm. ACM*, Vol. 8, No. 1 (Jan., 1965), pp. 53–69.

IX.3.2.8. *AESOP*

[QA65] Bennett, E., Haines, E. C., Jr., and Summers, J. K., "AESOP: A Prototype for On-Line User Control of Organizational Data Storage, Retrieval and Processing", *Proc. FJCC*, Vol. 27, pt. 1 (1965), pp. 435–55.

SPECIALIZED LANGUAGES

[UT67] Summers, J. K. and Bennett, E., "AESOP—A Final Report: A Prototype On-Line Interactive Information Control System", *Information System Science and Technology* (D. Walker, ed.). Thompson Book Co., Washington, D.C., 1967, pp. 69–86.

IX.3.3. GRAPHICS

[QW63] Sutherland, I. E., *Sketchpad: A Man-Machine Graphical Communication System*, M.I.T. Lincoln Lab., Tech. Report No. 296, Lexington, Mass. (Jan., 1963).

[QW66] Sutherland, I. E., "Computer Graphics; Ten Unsolved Problems", *Datamation*, Vol. 12, No. 5 (May, 1966), pp. 22–27.

[QX66] Skinner, F. D., "Computer Graphics—Where Are We?", *Datamation*, Vol. 12, No. 5 (May, 1966), pp. 28–31.

[RB65] Roberts, L. G., "Graphical Communication and Control Languages", *Information System Sciences: Proceedings of the Second Congress*. Spartan Books, Washington, D.C., 1965, pp. 211–17.

[SU66] Sutherland, W. R., *On-Line Graphical Specification of Computer Procedures*, M.I.T. Lincoln Lab., Tech. Report No. 405, Lexington, Mass. (May, 1966).

[SU67] Sutherland, W. R., "Language Structure and Graphical Man-Machine Communication", *Information System Science and Technology* (D. Walker, ed.). Thompson Book Co., Washington, D.C., 1967, pp. 29–31.

[YF67] Morrison, R. A., "Graphic Language Translation with a Language Independent Processor", *Proc. FJCC*, Vol. 31 (1967), pp. 723–31.

IX.3.3.1. *GRAF*

[HW67] Hurwitz, A., Citron, J. P., and Yeaton, J. B., "GRAF: Graphic Additions to FORTRAN", *Proc. SJCC*, Vol. 30 (1967), pp. 553–57.

IX.3.3.2. *PENCIL*

[VD67] Van Dam, A. and Evans, D., "A Compact Data Structure for Storing, Retrieving and Manipulating Line Drawings", *Proc. SJCC*, Vol. 30 (1967), pp. 601–610.

[WP67] Wexelblat, R. L. and Freedman, H. A., "The MULTILANG On-line Programming System", *Proc. SJCC*, Vol. 30 (1967), pp. 559–69.

IX.3.3.3. *Graphic Language*

[UZ67] Schwinn, P. M., "A Problem Oriented Graphic Language", *Proc. ACM 22nd Nat'l Conf.*, 1967, pp. 471–77.

IX.3.3.4. *DOCUS*

[CF66] Corbin, H. S. and Frank, W. L., "Display Oriented Computer Usage System", *Proc. ACM 21st Nat'l Conf.*, 1966, pp. 515–26.

IX.3.4. COMPUTER-AIDED DESIGN

IX.3.4.1. General

[JG67] Gray, J. C., "Compound Data Structure for Computer Aided Design; A Survey", *Proc. ACM 22nd Nat'l Conf.*, 1967, pp. 355–65.

[QW63] Sutherland, I. E., *Sketchpad: A Man-Machine Graphical Communication System*, M.I.T. Lincoln Lab., Tech. Report No. 296, Lexington, Mass. (Jan., 1963).

[UU63] Stotz, R. H., "Man-Machine Console Facilities for Computer-Aided Design", *Proc. SJCC*, Vol. 23 (1963), pp. 323–28.

IX.3.4.2. AED

[LB65] Lapin, R. B., Ross, D. T. and Wise, R. B., *Some Experiments with an Algorithmic Graphical Language*, M.I.T., ESL-TM-220, Electronic Systems Lab., Cambridge, Mass. (Aug., 1965).

[LB67] Lapin, R. B., *Translation Between Artificial Programming Languages*, M.I.T., ESL-R-306, Electronic Systems Lab., Cambridge, Mass. (Apr., 1967).

[MT00a] *AED Kit*, (unpublished).

[RD61] Ross, D. T., "A Generalized Technique for Symbol Manipulation and Numerical Calculation", *Comm. ACM*, Vol. 4, No. 3 (Mar., 1961), pp. 147–50.

[RD62] Ross, D. T., *An Algorithmic Theory of Language*, M.I.T., ESL-TM-156, Electronic Systems Lab., Cambridge, Mass. (Nov., 1962).

[RD63] Ross, D. T. and Rodriguez, J. E., "Theoretical Foundations for the Computer-Aided Design System", *Proc. SJCC*, Vol. 23 (1963), pp. 305–322.

[RD67] Ross, D. T., "The AED Free Storage Package", *Comm. ACM*, Vol. 10, No. 8 (Aug., 1967), pp. 481–92.

[RD67a] Ross, D. T., "The Automated Engineering Design (AED) Approach to Generalized Computer-Aided Design", *Proc. ACM 22nd Nat'l Conf.*, 1967, pp. 367–85.

[YW66] Wolman, B. L., "Operators for Manipulating Language Structures", (summary only), *Comm. ACM*, Vol. 8, No. 9 (Aug., 1966), pp. 553–54.

IX.3.5. TEXT EDITING

[BI63] Barnett, M. P. and Kelley, K. L., "Computer Editing of Verbal Texts, pt. 1. The ES1 System", *Amer. Documentation*, Vol. 14, No. 2 (Apr., 1963), pp. 99–108.

[IB67b] *DATATEXT Operator's Instruction Guide*, IBM Corp., J20-0010, Data Processing Division, White Plains, N.Y. (1967).

[WV67] Walker, D. E., "SAFARI, an On-line Text-Processing System", *Proceedings of the American Documentation Institute Annual Meeting*, Vol. 4. Thompson Book Co., Washington, D.C., 1967, pp. 144–47.

IX.3.6. CONTROL LANGUAGES FOR ON-LINE AND OPERATING SYSTEMS

[AU67]   Auroux, A., Bellino, J., and Bolliet, L., "DIAMAG: A Multi-Access System for On-Line ALGOL Programming", *Proc. SJCC*, Vol. 30 (1967), pp. 547–52.

[BV67]   Bequaert, F. C., "RPL: A Data Reduction Language", *Proc. SJCC*, Vol. 30 (1967), pp. 571–75.

[CO64]   Coffman, E. G., Jr., Schwartz, J. I., and Weissman, C., "A General-Purpose Time-Sharing System", *Proc. SJCC*, Vol. 25 (1964), pp. 397–411.

[FZ67]   Feingold, S. L., "PLANIT: A Flexible Language Designed for Computer-Human Interaction", *Proc. FJCC*, Vol. 31 (1967), pp. 545–52.

[GQ67]   Gross, L. N., *On-Line Programming System User's Manual*, MITRE Corp., MTP-59, Bedford, Mass. (Mar., 1967).

[IB67g]   *IBM System/360 Operating System Job Control Language*, IBM Corp., C28-6539-4, Data Processing Division, White Plains, N.Y. (Mar. 1967).

[PY65]   Pyle, I. C., "Data Input by Question and Answer", *Comm. ACM*, Vol. 8, No. 4 (Apr., 1965), pp. 223–26.

[ZV62]   Corbato, F. J., Merwin-Daggett, M., and Daley, R. C., "An Experimental Time-Sharing System", *Proc. SJCC*, Vol. 21 (1962), pp. 335–44.

[ZV63]   Corbato, F. J. *et al.*, *The Compatible Time-Sharing System, A Programmer's Guide*. M.I.T. Press, Cambridge, Mass., 1963.

X SIGNIFICANT UNIMPLEMENTED CONCEPTS

X.1. SCOPE OF CHAPTER

This very short chapter serves the purpose of providing a space for a few ideas that either have not been mentioned in any earlier chapter or, alternatively, deserve somewhat more attention in a more general content.

It was stated in the preface to this book that only languages which had been implemented were going to be described. The reason for this restriction is that the number of proposed languages is probably at least as great as those already in existence, if one considers significant additions to existing languages as new ones. In spite of this previous stipulation, there are several ideas, concepts, or unimplemented languages (the term depends on your viewpoint) which have been around for differing lengths of time and which seem to be of greater than average importance or interest. Whether any of these can or ever will come into practical existence is highly questionable; they are currently of theoretical interest only but it seems essential that the concepts at least be recorded. Clearly, my personal biases and interests show through strongly here.

Brief sections are devoted to (1) one of the earliest—but still unsolved—problems, namely that of a Universal Computer-Oriented Language to bridge the gap between higher level languages and machine code; (2) a theoretical approach to data processing, the Information Algebra; (3) the programming language of Iverson, beyond the implemented subset in Section IV.6.8; (4) the use of English as a programming language; and (5) attempts to consider building hardware to accept higher level languages directly.

X.2. UNCOL

The first of the languages to be discussed which is of theoretical interest only is in fact one of the oldest, namely that known as UNCOL (*UN*iversal *C*omputer *O*riented *L*anguage). The motivation for this language unfortunately still exists and here we have a dual problem. In the other parts of this chapter there are *languages* which have *not* been *implemented*; here we are faced with a situation in which the *language* itself does *not exist*, except as a concept. The basic problem which motivated this work is one that still remains with us, namely that we may have M machines and N languages which we would like to have translated into these M machines. Simple arithmetic shows that this requires $M \times N$ compilers, which is a prohibitive cost. The proposed solution to this is illustrated in Figure X-1, which shows

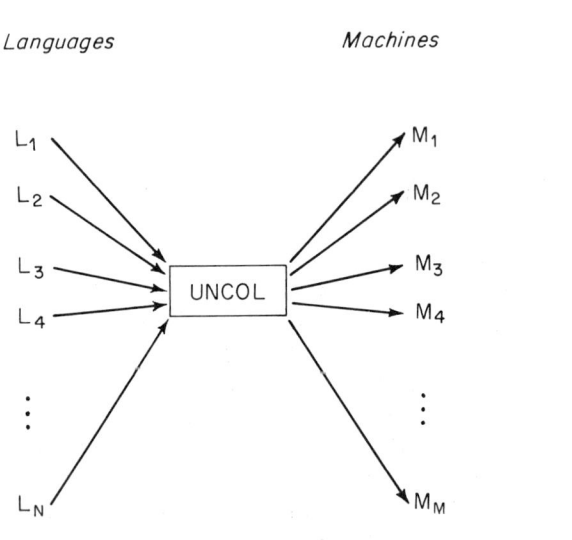

Figure X-1. Use of UNCOL to reduce the number of compilers in going from N languages to M machines.

a language that would be neither a programming language as we understand the term nor a machine language but something which is intuitively between them. The idea is to translate each higher level language to UNCOL and write a translator of UNCOL to each machine language. Simple arithmetic shows that in this case far less than $M \times N$ compilers are required—it is essentially $M + N$ "half compilers." If such a language existed and a new higher level language were defined, it would be necessary only to write the first half of a compiler, namely translating the new language to UNCOL;

this language would then be available on all computers because the UNCOL machine program existed. Similarly, if a new computer were developed, it would only be necessary to write the second half of the compiler, namely the UNCOL-to-machine-language translator. These first and second halves could be joined together in any appropriate fashion. This concept is described in great detail in the referenced papers by Strong *et al.* [QR58] and [QR58a]. Unfortunately, in my opinion, the people proposing this concept (which I think is a valid one) did a disservice in proposing the word *generator* to mean the first half of the compiler and *translator* to mean the second half. This is contrary to the intuitive notions of these terms held by many people. To a limited extent, people involved in writing JOVIAL compilers have provided an intermediate language for their own use. However, since they have a fixed front end, this is not really a useful UNCOL for a variety of languages. A few attempts at specifying UNCOL are given in the reference list at the end of this chapter.

X.3. INFORMATION ALGEBRA

The *L*anguage *S*tructure *G*roup (LSG) of the Development Committee of CODASYL (see Section V.3.1 for discussion of CODASYL) was formed in July, 1959 to study the structure of programming languages for data processing and to make recommendations for future developments. Recognizing the limitations of programming languages in the area of machine dependence, the need for sequencing procedural statements, and problems of dividing a large data processing application into specific runs, the LSG decided to try to devise some language to serve as a theoretical basis for truly automatic programming. The LSG did not produce a user-oriented language for defining problems, nor was it able to specify even in general terms an algorithm for translating statements in the Information Algebra that was developed into machine language programs. On the other hand, the Information Algebra which the LSG defined in [CC62], based on work of R. Bosak, is an attempt to provide a theoretical framework for concepts that have been understood for years by business systems analysts. The concepts are mathematical; particularly, they include notions of set theory.

The algebra has been built on three undefined concepts and three postulates. The concepts are *property*, *value*, and *entity*. The first postulate says that each property has one and only one set of values (called the *property value set*) assigned to it. The second postulate states that every entity has one and only one value assigned to it from each property value set. The third postulate states that each property has at least the values *undefined* and *missing* assigned to it if necessary. As an illustration, in a payroll application, one class of entities is the employees. Some of the properties which

Properties	Value set	Areas			
		Old pay file OP	Daily work file DW	New employee file NE	New pay file NP
q_1 = File ID	PF, DW, NE	X (always PF)	X (always DW)	X (always NE)	X (always PF)
q_2 = Man ID	00000 … 99999	X	X	X	X
q_3 = Name	20 alphabetic characters	X	Ω	X	X
q_4 = Rate	00.00 … 99.99	X	Ω	X	X
q_5 = Hours	00 … 24	Ω	X	Ω	Ω
q_6 = Day #	0 … 7	Ω	X	Ω	Ω
q_7 = Total salary	00000.00 … 99999.99	X	Ω	Ω	X
q_8 = Pay period #	00 … 52	X	Ω	X	X
q_9 = Salary	000.00 … 999.00	X	Ω	Ω	X

Figure X-2. System information for payroll problem written in Information Algebra. Items designated by Ω are undefined.
Source: [CC62], p. 202. By permission of Association for Computing Machinery, Inc.

might be selected are identification number, name, sex, and pay rate. A payroll file would contain a set of values of these properties for each entity. Various ways of grouping data must be provided and these are done through specifically defined concepts called *line*, *area*, and *bundle* and the definitions of the concept of *functions of lines*, *areas*, and *bundles*. In a very rough intuitive

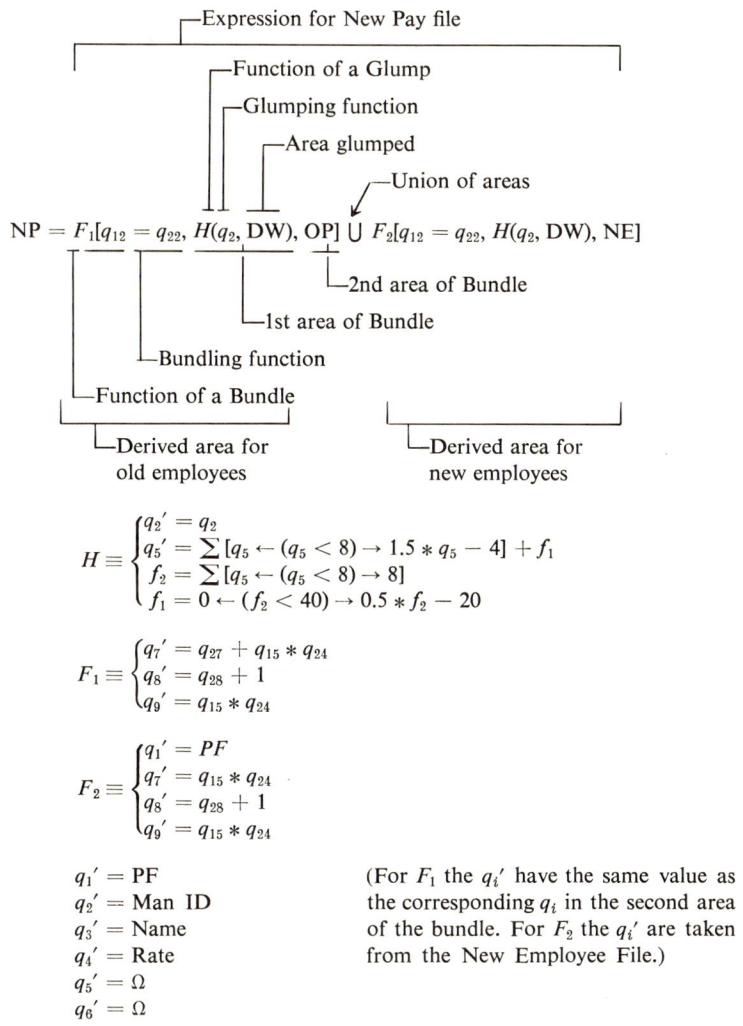

Figure X-3. Payroll program written in Information Algebra.
Source: [CC62], p. 203. By permission of the Association for Computing Machinery, Inc.

sense, an area is analogous to a file, and a bundle provides ways of matching. A third rigorously defined concept called a *glump* is introduced to provide ways of summarizing. As an illustration of what statements in this language actually look like, Figure X-2 defines the information needed in a payroll problem and Figure X-3 shows what the actual program would look like in the Information Algebra. An even more interesting application or illustration, particularly because it was not done by any of the people on the committee, is that described by Katz and McGee [KZ63] in which they show how to express an assembly program using the Information Algebra.

The committee which prepared the report consisted of R. Bosak (SDC), R. Clippinger (Honeywell), C. Dobbs (UNIVAC), R. Goldfinger (IBM) (chairman), R. B. Jasper (Navy Management Office), W. Keating (NCR), G. Kendrick (GE), and J. E. Sammet (IBM). They, plus a few others, voted in the summer of 1964 to disband (probably one of the few committees in history to do so) because the members did not really have the time to devote to the activity and, frankly, did not have any significant approaches about how to try to implement such a system, even in very broad terms.

The greatest significance of this work hopefully will show up when more is understood about the theory and science of programming in general, and data processing in particular. An even earlier attempt to formalize some concepts in the data processing area is that of Young and Kent [YJ58]. Other individuals (e.g., Lombardi [LM62]) and groups (e.g., SHARE Committee on Theory of Information Handling) have also done work in this area.

X.4. APL (IVERSON)

The language developed by Iverson (see the various references at the end of this chapter, with his book [IV62] as the definitive work) has been known in public form since at least 1962 and had then been under development for several years. It has more recently been referred to as APL (*A Programming Language*), which means the language, as distinct from APL/360, which is an implemented subset (described in Section IV.6.8.) The computer world seems to be divided into an inside and an outside on this particular subject; namely those who think it is the solution to all problems and those who see no use whatsoever for it. I am not trying to evade the issue in taking a middle ground because I can see many of the advantages and the power of this language, while still recognizing its practical limitations. To quote Iverson, "The language is based on a consistent unification and extension of existing mathematical notations, and upon a systematic extension of a small set of basic arithmetic and logical operations to vectors, matrices, and trees."[1] I think this is a very fair statement, but it overlooks what many people find

---

[1] [IV62a], p. 345.

	Operation	Notation ①	Definition ②	
OPERANDS	Scalar	x		
	Vector	$\mathbf{x}$	$\mathbf{x} \equiv x_0, x_1, \cdots, x_{(\nu \mathbf{x})-1}$	$\nu \mathbf{x}$ = number of components
	Matrix	X	$X \equiv \begin{bmatrix} X_0^0 & \cdots & X_{(\nu X)-1}^0 \\ \vdots & & \vdots \\ X_0^{(\mu X)-1} & \cdots & X_{(\nu X)-1}^{(\mu X)-1} \end{bmatrix}$	$X^i \equiv i$th row vector $X_j \equiv j$th column vector μX = number of rows νX = number of columns
LOGICAL and NUMERICAL	Arithmetic	$+ - \times \div$	Usual definitions	All operations are extended component-by-component to dimensionally compatible vectors and matrices. If one of the operands is a scalar, it is treated as a vector or matrix of appropriate dimension whose components are all equal. Examples: $z \leftarrow x + y$ $z \leftarrow x \times y$ $W \leftarrow U \wedge V$ $w \leftarrow x \neq y$ $w \leftarrow x < y$
	Absolute value	$z \leftarrow \mid x$	$z \equiv$ maximum of x and $-x$	
	Floor	$k \leftarrow \lfloor x$	$k \leq x < k+1$	
	Ceiling	$k \leftarrow \lceil x$	$k-1 < x \leq k$	k, m, n, q integers
	Residue modulo m	$k \leftarrow m \mid n$	$n \equiv (m \times q) + k,$ $0 \leq k < m$	
	And	$w \leftarrow u \wedge v$	$w \equiv 1$ if and only if	$u \equiv 1$ and $v \equiv 1$
	Or	$w \leftarrow u \vee v$		$u \equiv 1$ or $v \equiv 1$
	Negation	$w \leftarrow \bar{u}$		$u \equiv 0$
		$w \leftarrow \sim u$		$u \equiv 0$
	Relation	$w \leftarrow x \mathcal{R} y$		$x \mathcal{R} y$ is true
REDUCTION	Reduction	$z \leftarrow \odot/\mathbf{x}$	$z \equiv x_0 \odot x_1 \odot \cdots \odot x_{(\nu \mathbf{x})-1}$	$\odot$ is any binary operator or relation. The case $X \overset{+}{\underset{\times}{}} Y$ is the ordinary matrix product. The expressions $X \overset{\odot_1}{\underset{\odot_2}{}} \mathbf{y}$, $\mathbf{x} \overset{\odot_1}{\underset{\odot_2}{}} Y$, and $\mathbf{x} \overset{\odot_1}{\underset{\odot_2}{}} \mathbf{y}$ are treated as in matrix algebra. Thus $\mathbf{x} \overset{+}{\underset{\times}{}} \mathbf{y}$ is the scalar product.
	Row reduction	$\mathbf{z} \leftarrow \odot/X$	$z_i \equiv \odot/X^i$	
	Column reduction	$\mathbf{z} \leftarrow \odot//X$	$z_j \equiv \odot/X_j$	
	Matrix product	$Z \leftarrow X \overset{\odot_1}{\underset{\odot_2}{}} Y$	$Z_j^i \equiv \odot_1/X^i \odot_2 Y_j$	
BASE VALUE	Base 10 value	$z \leftarrow 10 \perp \mathbf{x}$	z is the base-10 value of the vector $\mathbf{x}$	
	Base 2 value	$z \leftarrow \perp \mathbf{u}$	z is the base-2 value of the vector $\mathbf{u}$	
		$\mathbf{z} \leftarrow \perp U$	$z_i \equiv \perp U^i$	
	Representation base 10	$\mathbf{z} \leftarrow 10(n) \top j$	$\nu \mathbf{z} = n$ and $10 \perp \mathbf{z} = 10^n \mid j$	
	base 2	$\mathbf{u} \leftarrow (n) \top j$	$\nu \mathbf{u} = n$ and $\perp \mathbf{u} = 2^n \mid j$	
SELECTION	Catenation	$\mathbf{z} \leftarrow \mathbf{x}, \mathbf{y}$	$\mathbf{z} \equiv x_0, x_1, \cdots, x_{(\nu \mathbf{x})-1}, y_0, y_1, \cdots, y_{(\nu \mathbf{y})-1}$	
	Row catenation	$Z \leftarrow X \oplus Y$	$Z_0 \equiv \mathbf{x}; Z_1 \equiv \mathbf{y}$	
	Compression vector	$\mathbf{z} \leftarrow \mathbf{u}/\mathbf{x}$	$\mathbf{z}$ obtained by suppressing from $\mathbf{x}$ each x_i for which $u_i \equiv 0$	
	row	$Z \leftarrow \mathbf{u}/X$	$Z^i \equiv \mathbf{u}/X^i$	
	column	$Z \leftarrow \mathbf{u}//X$	$Z_j \equiv \mathbf{u}/X_j$	
	row list	$\mathbf{z} \leftarrow \mathrm{E}/X$	$\mathbf{z} \equiv X^0, X^1, \cdots, X^{(\mu X)-1}$	
	Row list expansion	$X \leftarrow \mathrm{E}(m, n) \backslash \mathbf{z}$	$\mu X \equiv m, \nu X \equiv n$, and $\mathrm{E}/X \equiv \mathbf{z}$. Thus $X \equiv \begin{bmatrix} z_0 & \cdots & z_{n-1} \\ z_n & \cdots & \cdots \\ \cdots & z_{(m \times n)-1} \end{bmatrix}$	
	Mask	$\mathbf{z} \leftarrow /\mathbf{x}; \mathbf{u}; \mathbf{y}/$	$\bar{\mathbf{u}}/\mathbf{z} \equiv \bar{\mathbf{u}}/\mathbf{x}; \quad \mathbf{u}/\mathbf{z} \equiv \mathbf{u}/\mathbf{y}$.	
	Indexing	$z \leftarrow x_m$	$z_i \equiv z_{m_i}$	
		$Z \leftarrow X^\mathbf{m}$	$Z^i \equiv X^{m_i}$	
		$Z \leftarrow X_\mathbf{m}$	$Z_i \equiv X_{m_i}$	
	Maximum prefix	$\mathbf{w} \leftarrow \alpha/\mathbf{u}$	$\mathbf{w} \equiv \alpha^j(\nu \mathbf{u})$ and j is maximum for which $\wedge/\mathbf{w}/\mathbf{u} \equiv 1$	
SHIFTING	Left rotation	$\mathbf{z} \leftarrow k \uparrow \mathbf{x}$	$z_i \equiv x_j; j \equiv (\nu \mathbf{x}) \mid i + k$	cyclic left (right) rotation of $\mathbf{x}$ by k places.
	Right rotation	$\mathbf{z} \leftarrow k \downarrow \mathbf{x}$	$z_i \equiv x_j; j \equiv (\nu \mathbf{x}) \mid i - k$	
	Left shift	$\mathbf{z} \leftarrow k \overset{\circ}{\uparrow} \mathbf{x}$	$\mathbf{z} \equiv \bar{\alpha}^k \wedge k \uparrow \mathbf{x}$	left (right) shift bringing zeros into evacuated positions
	Right shift	$\mathbf{z} \leftarrow k \overset{\circ}{\downarrow} \mathbf{x}$	$\mathbf{z} \equiv \bar{\alpha}^k \wedge k \downarrow \mathbf{x}$	
SPECIAL VECTORS	Full	$\mathbf{w} \leftarrow \epsilon(n)$	$w_i \equiv 1$	
	Characteristic	$\mathbf{w} \leftarrow \epsilon^j(n)$	$w_i \equiv (\vee/i = j)$	
	Prefix	$\mathbf{w} \leftarrow \alpha^j(n)$	$w_i \equiv (i < j)$	Dimension of $\mathbf{w}$ is n. The n may be omitted if it is clear from context.
	Suffix	$\mathbf{w} \leftarrow \omega^j(n)$	$w_i \equiv ((n - i) \leq j)$	
	Random	$w \leftarrow ?$	$w \equiv 0$ or 1 (arbitrary)	
		$\mathbf{w} \leftarrow ?(n)$	$w_i \equiv 0$ or 1	
		$\mathbf{w} \leftarrow ?^j(n)$	$w_i \equiv 0$ or 1 but $+/\mathbf{w} \equiv j$	
	Interval	$\mathbf{z} \leftarrow \iota^j(n)$	$\mathbf{z} \equiv j, j+1, \cdots, j+n-1$	

Figure X-4. Partial list of APL notation.
Source: Falkoff, Iverson, and Sussenguth [FA64], p. 200. Reprinted by permission from *IBM Systems Journal*. © June, 1964 by International Business Machines Corporation.

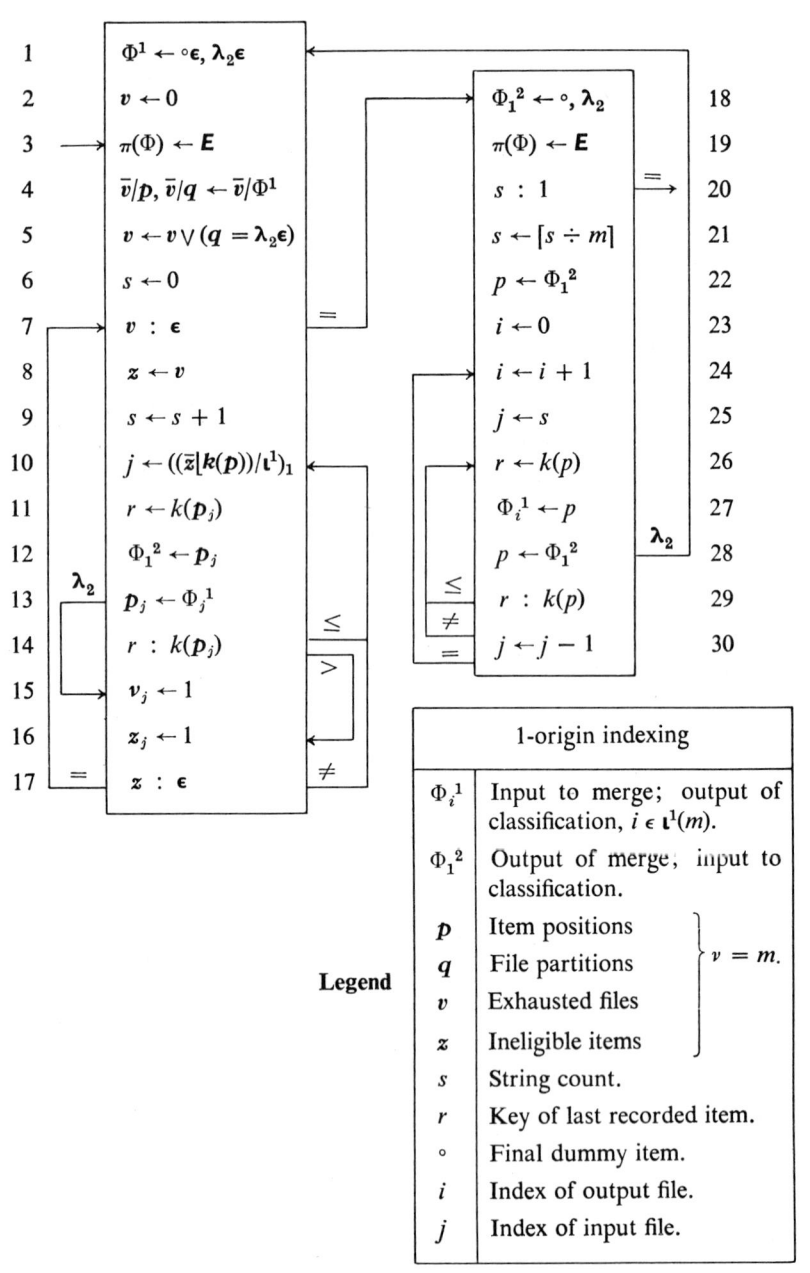

Figure X-5. m-way merge sort written in APL.
Source: Iverson [IV62a], p. 182.

to be its greatest difficulty, namely the notation involved. A subset of the language notation is shown in Figure X-4, and a program to do sorting is shown in Figure X-5. Programs have been very successfully written to deal with scientific applications, systems programming, data processing, and descriptions of hardware. Unfortunately, current hardware does not permit the practical usage of such a language. The subset which has been implemented (see Section IV.6.8) is proving to be of extreme interest to a number of people.

The major characteristics are the heavy use of operators on vectors and matrices; thus, for example, the need for a loop-control statement is eliminated in many cases. Other operators also contribute to the conciseness of the language. On the other hand, the question has been raised as to whether this is a *language* or a *notation*.

Since I stated the view in Chapter III that the character set is of prime importance in the design of a programming language, I cannot become enthusiastic about a language which has this notational complexity. Perhaps a good use of Iverson's language would be to serve as a canonical form for use in defining the meaning of many (although certainly not all) statements in other languages.

X.5. ENGLISH

It may surprise some readers to find *English* being listed as a language in this section. The reasons for it are discussed to a large extent in Chapter XI, where an argument is given in favor of allowing the user to converse with a computer in his own natural language. For the sake of convenience, I choose to call this English, but it should definitely be understood that it is a shorthand way of saying *any natural language*. It must be emphasized that a natural language includes scientific notation wherever appropriate, in one or two dimensions. For example, the input of two-dimensional diagrams for chemical structures is a perfectly appropriate language and is legitimately included in my meaning of *English*.

A number of arguments in favor of the use of English as a programming language are given in Fraser [FV67], Sammet [SM66b], Halpern [HL66], and Thompson [TH66]. In addition, Section IX.3.2 described a number of query and retrieval systems, some of which are very *English-like*. It is not the purpose of this section to repeat those arguments, some of which are summarized in Chapter XI. The main reason for reintroducing the concept here is to point out and emphasize the difference between the use of English for retrieval and/or query, and its use as a specific programming language. In the former case, as illustrated earlier, the user tends to specify one or two simple sentences or questions, which might even have conditional statements within them. He then expects to obtain results. For example, he might simply be requesting specific information from a system, such as **FIND ALL OF THE TRAINS FROM PODUNK TO OSHKOSH WHICH ARRIVE AFTER 6:00 P.M., BUT BEFORE 8:00 P.M. AND WHICH HAVE DINING CARS**. This is something far different from a specific program in which an individual may wish to actually solve a detailed problem stated in English. The main point to be made here is that there is a matter of degree involved from several different points of view. One involves the actual syntactic and semantic analysis of English. This may not really be any worse for full English programs than it is for the query systems, assuming that the programs themselves are known to be within some reasonable area of application. However, we would of course like to be able to have one program for handling all programs written in English, rather than have to have a separate one for each data base or each application. The second place in which a matter of degree occurs is the actual size of the program. While this is not necessarily of great theoretical interest, it tends to have many practical implications as far as implementation is concerned. Thus, even if we have the mechanism for parsing English sentences and understanding what to do with them, the fact that there might be several hundred of these in a single program, together with a large number of data variables, control transfer statements, complex calculations, etc., presents a problem which transcends the mere syntactic and semantic analysis that we already know is a major difficulty. For example, consider the following two problems:

(1) A deck of cards is given to be read in format ($I3, F10.2$), one card for each person in a certain village. The number in the first field is the person's age; the number in the second is his income for 1960. Following this deck is a card with -1 in the first field; this information will be used to test for the end of deck.

Find the average salary of the people in each 5-yr. age group, i.e., 0–4, 5–9, 10–14, ..., 95–99. Print out the lower age limit for each group, i.e., $0, 5, 10, ..., 95$, the average salary for that group, and the number of people in each group. Precautions should be taken to avoid division by zero in case there are no people in an age group.[2]

(2) Determine the current in an AC circuit consisting of resistance, inductance, and capacitance in series, given a number of sets of values of resistance, inductance, and frequency. The current is to be determined for a number of equally spaced values of the capacitance (which lie between specified limits) for voltages of $1.0, 1.5, 2.0, 2.5,$ and 3.0 volts. The equation for determining the current flowing through such a circuit is

$$i = \frac{E}{\sqrt{R^2 + (2\pi f L - 1/2\pi f C)^2}}$$

---

[2] [IB61], p. 79. Reprinted by permission from *FORTRAN II General Information Manual*. © 1961 by International Business Machines Corporation.

where i = current, amperes $\quad C$ = capacitance, farads
E = voltage, volts $\qquad\quad f$ = frequency, cycles per second
R = resistance, ohms $\quad\:\pi$ = 3.1416[3]
L = inductance, henrys

These examples involve difficulties which are orders of magnitude beyond those involved in executing just a few commands or merely retrieving information. Most of the work shown in the references appears *superficially* to contribute toward this goal, but very little involves actual programs rather than query answering.

X.6. HARDWARE IMPLEMENTATION OF PROGRAMMING LANGUAGES

Considering the large number of programming languages which are discussed in this book, to say nothing of their variations and dialects which are not covered, it is almost surprising that nobody (as far as I know) has actually built a machine to directly handle one or more of these languages. It seems worth mentioning a few of the suggestions that have been made over a period of time for actually doing so. No discussion of the economic or technical feasibility of this approach is intended.

The first attempt to build a computer with a *programming language order code* that I am aware of, and actually the only one I know that was ever really implemented, was the NCR 304. A justification for that machine in terms of automatic coding as it was understood in *1957* is given by Yowell [YO57]. He stated that the objective of the 304 was to reduce the cost of many business applications by incorporating into the hardware some of the features found in automatic coding systems. However, the "highest level" that was really receiving *major* public consideration at the time were languages of the Autocoder level; hence the 304 became a three-address machine with powerful commands, including *Merge, Edit* (with check protection), and *Summarize*. When COBOL was developed, it actually did not contain (at least in its earliest versions) an instruction as powerful as that on the 304, namely the *Merge*. Furthermore, because some of the instructions on the 304 were relatively high level, it was difficult to break them down to perform equivalently high level operations which were specified in COBOL but with very different rules. Having heard of this particular situation, I personally have been a nonadvocate of machine hardware implementation of higher level languages since then. However, some of the more recent work might turn out to be more useful, particularly if applied to *standardized* languages (e.g., FORTRAN).

---

[3] [IB61], p. 86. Reprinted by permission from *FORTRAN II General Information Manual*. © 1961 by International Business Machines Corporation.

The Burroughs B-5000 included hardware to directly execute Polish notation strings, which are often used as the output from the first pass of a compiler.

The *proposed* computer described by Anderson [AJ61] was actually meant to implement much of ALGOL 60. His design included a large part of ALGOL 60, but with the major omission of **comment**.

Far more recently, a *proposal* has been made by Bashkow, Sasson, and Kronfeld [UF67] to actually implement a FORTRAN machine. Their proposal shows a subset that consisted essentially of the assignment statement, the GO TO and the computed or assigned GO TO, IF, PAUSE, DO, CONTINUE, END, READ, PRINT, and DIMENSION. This is roughly equivalent to the FORTRAN for the IBM 1620. The major concept missing from their proposed design is that of subroutines, either programmer-invoked or built-in.

Another *proposed* FORTRAN hardware computer is described by Melbourne and Pugmire [ZM65]. They discuss using micro-programs and say that the results of various simulations showed execution times greater than hand-coded programs but less than compiled programs.

The CDC 3600 was modified to implement IPL-V and used for checker playing. (See Hodges [HX64] and Cowell and Reed [ZN65], respectively.)

The approaches just mentioned attempt to consider the design of hardware to implement an existing language. Still another *proposed* approach is to create a new higher level language and then to try to build hardware to implement it. This was the approach taken in considering ADAM as described by Mullery, Schauer, and Rice [MH63] and also in Mullery [MH64]. They postulate the existence of data of completely variable length, with up to eight identifiers to indicate the data structure. These correspond to *character, symbol, phrase, sentence, paragraph, chapter, book,* and *library.* The language itself is meant to consist of verbs, nouns, and modifiers. They say, for example, that an instruction might be written as follows:

$$v \text{ Start } v \text{ ③} A + B \rightarrow C \text{ ③}.$$

where ③ is meant to show one of the data levels. The verbs that they describe include *define* (shown as a right arrow), *insert, join,* and *delete.* Verbs which operate on data strings as sets include *intersect, union,* and *difference.* Normal arithmetic verbs are allowed and *set* and *truncate* are provided; they may be modified by the adverb *round.* Verbs to change structure, to do editing, and for control and conditional statements are also described. This machine was not implemented.

In machines which permit micro-programming, there is much more potential for this type of thing to be done. CPS (see Section IV.6.5) has a version in which the micro-programming facility of the IBM System/360 was used to provide more efficiency in some of the frequently used operations. A micro-programmed implementation of EULER (a generalization of ALGOL) on the IBM System/360 Model 30 has been done (see Weber [YB67]). Whether micro-programming of the future will permit significant building of new languages in a hardware-software mixture is unforeseeable at this time. Many of the elements discussed in Chapter II along with economic considerations, rather than technical feasibility, may be the determining factors in the decisions to build such systems.

REFERENCES

X.2. UNCOL

[BR61]   Bratman, H., "An Alternate Form of the 'UNCOL Diagram'", *Comm. ACM*, Vol. 4, No. 3 (Mar., 1961), p. 142.

[CI58]   Conway, M. E., "Proposal for an UNCOL", *Comm. ACM*, Vol. 1, No. 10 (Oct., 1958), pp. 5-8.

[ST61]   Steel, T. B., Jr., "UNCOL: The Myth and the Fact", *Annual Review in Automatic Programming*, Vol. 2 (R. Goodman, ed.). Pergamon Press, New York, 1961, pp. 325-44.

[ST61a]   Steel, T. B., Jr., "A First Version of UNCOL", *Proc. WJCC*, Vol. 19 (1961), pp. 371-78.

[QR58]   Strong, J. *et al.*, "The Problem of Programming Communication with Changing Machines: A Proposed Solution", *Comm. ACM*, Vol. 1, No. 8 (Aug., 1958), pp. 12-18.

[QR58a]   Strong, J. *et al.*, "The Problem of Programming Communication with Changing Machines: A Proposed Solution, Part 2", *Comm. ACM*, Vol. 1, No. 9 (Sept., 1958), pp. 9-16.

X.3. INFORMATION ALGEBRA

[CC62]   "An Information Algebra" (Phase 1 Report, Language Structure Group of the CODASYL Development Committee)", *Comm. ACM*, Vol. 5, No. 4 (Apr., 1962), pp. 190-204.

[KZ63]   Katz, J. H., and McGee, W. C., "An Experiment in Non-Procedural Programming", *Proc. FJCC*, Vol. 24 (1963), pp. 1-13.

[LM62]   Lombardi, L. A., "Mathematical Structure of Nonarithmetic Data Processing Procedures", *J. ACM*, Vol. 9, No. 1 (Jan., 1962), pp. 136-59.

[YJ58]   Young, J. W., Jr. and Kent, H. K., "Abstract Formulation of Data Processing Problems", *Jour. Ind. Eng.*, Vol. IX, No. 6 (Nov.-Dec., 1958), pp. 471-79.

X.4. APL (IVERSON)

[FA64]   Falkoff, A. D., Iverson, K. E., and Sussenguth, E. H., "A Formal Description of System/360", *IBM System Jour.*, Vol. 3, Nos. 2 and 3 (June, 1964), pp. 198-262.

[IV62] Iverson, K. E., *A Programming Language*. John Wiley & Sons, New York, 1962.

[IV62a] Iverson, K. E., "A Programming Language", *Proc. SJCC*, Vol. 21 (1962), pp. 345–51.

[IV62b] Iverson, K. E., "A Common Language for Hardware, Software, and Applications", *Proc. FJCC*, Vol. 22 (1962), pp. 121–29.

[IV64a] Iverson, K. E., "Formalism in Programming Languages", *Comm. ACM*, Vol. 7, No. 2 (Feb., 1964), pp. 80–88.

See also entries under IV.6.8 and IX.2.3.1.

X.5. ENGLISH

[BB63] Bobrow, D. G., "Syntactic Analysis of English by Computer—A Survey", *Proc. FJCC*, Vol. 24 (1963), pp. 365–87.

[FV67] Fraser, J. B.,"The Role of Natural Language in Man-Machine Communication", *Information System Science and Technology* (D. Walker, ed.). Thompson Book Co., Washington, D.C., 1967, pp. 21–28.

[HL66] Halpern, M. I., "Foundations of the Case for Natural-Language Programming", *Proc. FJCC*, Vol. 29 (1966), pp. 639–49.

[IB61] *FORTRAN II General Information Manual*, IBM Corportion, F28-8074-3, Data Processing Division, White Plains, N.Y. (1961).

[KG67] Kellogg, C. H., *On-Line Translation of Natural Language Questions into Artificial Language Queries*, System Development Corp., SP-2827/000/00, Santa Monica, Calif. (Apr., 1967).

[KI64] Kirsch, R. A., "Computer Interpretation of English Text and Picture Patterns", *IEEE Trans. Elec. Comp.*, Vol. EC-13, No. 4 (Aug., 1964), pp. 363–76.

[MM00] McMahon, L. E., *FASE: A Fundamentally Analyzable Simplified English*, Bell Telephone Lab., Murray Hill, N.J. (unpublished).

[SE65] Simmons, R. F., "Answering English Questions by Computer: A Survey", *Comm. ACM*, Vol. 8, No. 1 (Jan., 1965), pp. 53–69.

[SE66] Simmons, R. F., Burger, J. F., and Long, R. E., "An Approach Toward Answering English Questions from Text", *Proc. FJCC*, Vol. 29 (1966), pp. 357–63.

[SE66a] Simmons, R. F., "Natural-Language Processing", *Datamation*, Vol. 12, No. 6 (June, 1966), pp. 61–72.

[SM66b] Sammet, J. E., "The Use of English as a Programming Language", *Comm. ACM*, Vol. 9, No. 3 (Mar., 1966), pp. 228–30.

[TH66] Thompson, F. B., "English for the Computer", *Proc. FJCC*, Vol. 29 (1966), pp. 349–56.

[TR67] Tabory, R. and Peters, P. S., Jr., *Can One Instruct Computers in English? A Feasibility Study Concerning the Use of English in User/Computer Communication*, IBM Corp., TM 48.67.003, Federal Systems Division, Cambridge, Mass. (Oct., 1967).

[WZ67a] Weizenbaum, J., "Contextual Understanding by Computers", *Comm. ACM*, Vol. 10, No. 8 (Aug., 1967), pp. 474–80.

See also entries under IX.3.2.

X.6. HARDWARE IMPLEMENTATION OF PROGRAMMING LANGUAGES

[AJ61] Anderson, J. P., "A Computer for Direct Execution of Algorithmic Languages", *Proc. EJCC*, Vol. 20 (1961), pp. 184–93.

[HX64] Hodges, D., *IPL-VC, A Computer System Having the IPL-V Instruction Set*, Argonne Nat'l Lab., ANL-6888, Applied Mathematics Division, Argonne, Ill. (May, 1964).

[MH63] Mullery, A. P., Schauer, R. F., and Rice, R., "ADAM—A Problem-Oriented Symbol Processor", *Proc. SJCC*, Vol. 23 (1963), pp. 367–80.

[MH64] Mullery, A. P., "A Procedure-Oriented Machine Language", *IEEE Trans. Elec. Comp.*, Vol. EC-13, No. 4 (Aug., 1964), pp. 449–55.

[UF67] Bashkow, T. R., Sasson, A., and Kronfeld, A., "System Design of a FORTRAN Machine", *IEEE Trans. Elec. Comp.*, Vol. EC-16, No. 4 (Aug., 1967), pp. 485–99.

[YB67] Weber, H., "A Microprogrammed Implementation of EULER on the IBM System/360 Model 30", *Comm. ACM*, Vol. 10, No. 9 (Sept., 1967), pp. 549–58.

[YO57] Yowell, E. C., "A Mechanized Approach to Automatic Coding", *Automatic Coding, Jour. Franklin Inst.*, Monograph No. 3, Philadelphia, Pa. (Apr., 1957), pp. 103–111.

[ZG67a] McKeeman, W. M., "Language Directed Computer Design", *Proc. FJCC*, Vol. 31 (1967), pp. 413–17.

[ZM65] Melbourne, A. J. and Pugmire, J. M., "A Small Computer for the Direct Processing of FORTRAN Statements", *Computer Jour.*, Vol. 8, No. 1 (Apr., 1965), pp. 24–27.

[ZN65] Cowell, W. R. and Reed, M. C., *A Checker-Playing Program for the IPL-VC Computer*, Argonne Nat'l Lab., ANL-7109, Applied Mathematics Division, Argonne, Ill. (Oct., 1965).

XI FUTURE LONG-RANGE DEVELOPMENTS

XI.1. INTRODUCTION

Any reader who expects this chapter to provide a crystal ball which will specifically foretell the future is doomed to disappointment. Without the facilities of Nostradamus or more modern people with similar abilities, it is only possible for me to extrapolate from present knowledge. Considering the normal delay between writing and physical publication, it is highly likely that some (although hopefully not all) of the material in this chapter will be either irrelevant or obsolescent by the time it is actually being read. In spite of these disclaimers, it does seem worthwhile to attempt to forecast what is likely to be done and to indicate some *purely personal opinions* on what should be done. No particular attempt is being made to distinguish between what is likely to, and what should, be done although in some places an indication will be given. In both cases, the prognostication is basically *long-range* rather than something which involves the next few years.

Pertinent references are listed at the end of this chapter, but without any specific citations of them in the text since the titles are generally self-explanatory relative to the topics involved.

It seems likely that the future development of programming languages will fall primarily into two major categories, one of which can be labeled as *theory-oriented*, while the other can be called *user-oriented*. The first area is but a subset of what hopefully will be forthcoming within the next 10 years, namely a real theory of programming. Right now, the analysis of a problem, its statement in some type of machine-understandable form, and the production of the desired results in a reasonable amount of time and at a nonprohibitive cost is very much of an art with very little associated science. Among the two most important aspects of this general theory of programming which will *not* be discussed in this chapter are the general problem of the tradeoff between time and space in a program and a scientific way of measuring tradeoffs between hardware and software. These are problems which need general solutions; when these solutions are available, they can be applied to programming languages and their implementation.

The *user-oriented* category naturally contains those facets which are directly related to making it easier for the user to communicate with a computer to solve his problem.

It is no accident that the development of a specially created *universal programming language* is omitted from this chapter. It is my firm opinion that not only is such a goal unachievable in the foreseeable future, but it is not even a desirable objective. It would force regimentation of an undesirable and impractical kind and either would prevent progress or, alternatively, would surely lead to deviations. However, the possibility of a single programming language with powerful enough features for self-extension to transform it into *any desired form* is interesting to consider. (By *any desired form*, I mean all the languages in this book, plus any others which are developed subsequently.) The techniques for this development are clearly unknown currently, but they could conceivably be found in the future.

XI.2. THEORY-ORIENTED CATEGORY

The theory-oriented aspect of future programming language work can be divided into five major categories: (1) language definition, translation, and creation; (2) next major conceptual step; (3) nonproceduralness; (4) problem-describing; and (5) use of mathematical concepts.

XI.2.1. LANGUAGE DEFINITION, TRANSLATION, AND CREATION

As of the end of 1967, we do not have an adequate tool for defining the syntax of several programming languages in a form which can be automatically—but effectively—handled by some type of translating program. This comment seems valid in spite of the vast work and progress that has been made in the whole field of developing syntax-directed compilers. Methods that work for defining one (or perhaps two) language(s) exist, but the moment dissimilar languages (e.g., ALGOL and COBOL) are considered, then the techniques either collapse completely or are so inefficient as to render them useless, Part of the difficulty is that the border line between syntax and semantics of a programming language is not clearly understood or defined. Furthermore, the formalism for defining the semantics is only in a very preliminary stage at the moment, although certainly some of the developments are encouraging; there is work underway both in formalizing PL/I semantics (see references for Section VIII.4) and in providing better formal tools for implementing the semantics. Finally, there is no way to record in any formal fashion the view of the actual user relative to what he means by a particular language (i.e., the pragmatics).

One of the many currently unsolved problems is how to determine when a compiler actually translates correctly a source program in a given programming language. The reason this is a major difficulty is because we have two—not just one—unsolved problems here. The first relates to the lack of complete rigor and clarity of the actual language definition itself. The second, however, is a more fundamental problem that transcends just the area of programming languages, namely the fact that we do not in general know how to determine when a program does what we want it to do. (In this situation it is the compiler, not the source program, whose validity is being questioned.) Any measures of this latter characteristic are currently default or intuitive judgments; thus if we obtain answers that satisfy the right people or seem to resolve some particular problem or provide information to somebody that he accepts as being reasonable, then we say that a program is working. However, we have no formal way of defining the intent of a program; thus there is certainly no rigorous measure of success, even for a substantial—although incomplete—set of cases. We do not have any techniques for knowing when a program is completely debugged; in fact, the complexity of current programs seems to make it impossible to ascertain this on a practical basis at all. Although on a digital computer the number of paths in a program may be finite and the variations of input data may also be finite, the practical situation indicates that the number of possibilities in each of these cases by itself is so large that no practical adequate testing procedure can be devised. Furthermore, the necessity for combining program paths and data just increases the difficulty of the problem at an exponential rate. Thus, in summary of this aspect, we do not know how to rigorously define languages and, even if we did, we have no techniques for determining when a compiler provides a complete and correct translation of all possible legal programs written in that language.

Related to this facet of the problem is the one of compiler efficiency, which is (only) partially a language problem. The user should have available to him the ability to specify whether he wants rapid compilation or maximally efficient object programs or whether he wants to specify a particular compromise between these. While this is primarily an implementation rather than a language problem, nevertheless there are features and clues in the language design which could help.

Another facet of the general theory-oriented problem is the actual language design methodology itself. Up until now, virtually all languages have been designed on a completely *ad hoc* basis, even though the designers may in fact have had (or least thought they had) rational reasons for their decisions. Thus, if a specific objective is defined, there is no algorithm for designing a programming language to meet that objective; there is not even any clear delineation of various tradeoffs involved, let alone any quantitative measures of them. As a result, language designers continue to argue over the merits (or lack thereof) of things such as requiring data names to begin with a letter, allowing them to coincide with fixed words in the language, or having fewer than N characters. Each of these decisions is made on an arbitrary—although often well-founded—basis, usually with some concern for tradeoffs involved. However, there is no general overall approach to language design which would say that to achieve certain objectives the following well-defined tradeoffs exist, and the language designers need only go down the list and check what they are least willing to sacrifice and what facets or facilities are most important to them. The ultimate in this direction would be a true *language generator* whose input parameters would be the complete characterization of all objectives to be achieved by the proposed language. This is a utopia that does not seem likely to be achieved for many decades—if ever.

XI.2.2. Next Major Conceptual Step

Because there is no theory, there is a need for both a definition of and a concern for the next major conceptual step forward in programming languages. There have been only a few schools of language development to date, although the absolute improvements and advantages from beginning to end within each line are considerable. One of these schools is exemplified by FORTRAN, COBOL, and ALGOL at its base, and PL/I and its extensions at its summit. Except from a very pragmatic point of view, we probably cannot push this line of development much further.

Another line of development is that exemplified by the LISP school. However, the advocates of that type of formalism seem to have learned the hard way that while the logical concepts and notation which were introduced in LISP 1 and LISP 1.5 contribute to clarity of logic and easy solution of certain kinds of problems, nevertheless the notation cannot really exist very well in the practical world. This is clearly demonstrated by the development of LISP 2, which is much more ALGOL-like in its approach to the user than it is LISP-like.

It is possible that the work done by Iverson (see Sections IV.6.8 and X.4) represents an independent school, but this is not my current view. It is really a side branch of the procedural languages exemplified by FORTRAN, COBOL, ..., PL/n.

Hence what is missing—and unpredictable—is what the next major conceptual step in programming languages will be. The closest I care to come in predicting this is in the area of fewer requirements for stating detailed sequential procedures.

XI.2.3. Nonprocedural Languages

It seems clear that one of the necessary directions for theoretical development is to pinpoint more specifically the concept of proceduralness. It was pointed out in Chapter I that *nonprocedural* is really a relative term that changes with the state of the art. It would be very desirable to have some way of making these concepts fixed rather than relative by assigning a *proceduralness factor*. Thus we might establish machine code as a base at one end and a true problem-describing language (which is discussed in the next section) at the other end; however, while it is relatively easy to establish the end points, there does not appear to be any way at the moment to establish fixed points along the scale. Certainly the number of primitives or executable commands in the language is a factor, but it is far from clear how one compares the statement *integrate $F(x)$ dx from A to B* with *calculate the square root of the prime numbers from 3 to 97 and print in three columns*. In an intuitive sense, the first statement corresponds to what we think of as a subroutine, whereas the second is essentially a concatenation of a series of individual statements, of which some may be real subroutines and others may be basic components of any language (e.g., looping facilities). However, it is not clear which should be considered more procedural than the other. The theoretical problem involved is really to establish some type of measuring scale by which the amount of proceduralness to be specified by the user can be defined for any given language.

It seems to me that one of the most likely candidates for the next major conceptual step forward lies along the lines of defining what is meant by more *nonprocedural* languages and then designing and implementing them. The Information Algebra (see Section X.3) was an attempt at this which collapsed at the first attempt to develop an approach to implementation.

XI.2.4. Problem-Describing Languages

Another aspect of the theory-oriented future of programming languages is in the direction of what I choose to call true *problem-describing languages*. In the ultimate and in the simplest illustration, the user would write *calculate the payroll for the XYZ Company* and the computer would do the rest. We may subdivide this situation into two different cases. The first occurs the initial time that this command is given; in that situation, we must assume that the computer will enter into a dialogue (in which the physical means of doing this is irrelevant theoretically, although significant from a practical viewpoint) in order to obtain the necessary information. A key point, however, is that the computer would have stored within it enough background and enough knowledge of what a payroll and a company are to ask the appropriate questions.

A significant aspect of this mythical problem-describing language involves the data structure. It is much easier to envision a system which could understand or generate the necessary commands, or ask the necessary questions to establish them for a given task, than it is to envision the system understanding the general problem of data structure.

XI.2.5. Use of Mathematical Concepts

The final aspect of the theory-oriented approach is the potential application of mathematical notations and concepts. While this might be considered implicit within any definition of a theoretical approach to programming languages, it seems important enough to make the category explicit. An early attempt at this was made with the Information Algebra (see Section X.3) which several people have said was 10 years ahead of its time (when published in 1962). The fact that it was extremely difficult to work with is perhaps some indication of the validity of that comment. (Simpler uses of fundamental mathematical concepts or notations appeared in LISP and in Iverson's work.) Regardless of whether that particular development (i.e., the Information Algebra) was or was not significant, there is clearly a wide open field for the application of mathematical structures to data processing in general and programming languages in particular. It may be that programming languages of the future will consist of mathematical descriptions of operations to be performed, that are "sugarcoated" to create an actual language which is suitable for use by a reasonably large number of people.

XI.3. USER-ORIENTED CATEGORY

There are five main aspects to the user-oriented developments: (1) User-defined languages, (2) natural language usage, (3) utilization of, and communication with, the hardware and executive software systems, (4) languages for new application areas, and (5) languages for creating software.

XI.3.1. User-Defined Languages

It seems to me that from several points of view the area of user-defined languages is the most likely and the most promising future development. Before discussing these, it is of course necessary to define what is meant by this phrase. The one thing that has become clear over the relatively short life of the computer industry is that there are almost as many ways to say *add a to b* as there are people capable of saying it. Furthermore, while each of these different ways achieves the same result, almost every person has a specific reason for preferring his notation to somebody else's. In other words, we have a very heavy built-in personal prejudice to either satisfy or counteract. Trying to satisfy people has not worked too well; for every group that is pacified with a particular language, another group that is trying to do similar kinds of problems finds itself unhappy and either develops a new language or makes significant changes in the existing one. Some of the ways in which these people could be satisfied are indeed trivial; for example, these changes might range from something as elementary as allowing the user to specify whether he wants to say **GO TO**, **JUMP**, or **TRANSFER TO** to something which provides him with some type of simple language extension and, therefore, would permit him to add *integrate* to his list of commands (at the language level, not just as a subroutine) if it were not available to him initially. However, in this case there is a great efficiency problem, particularly in this conceptually trivial area of providing alternative ways of saying the same thing. Of more fundamental significance, and correspondingly harder to satisfy, is the need for differing executable packages. Thus, the mathematically oriented people are concerned with integration, matrix inversion, etc., and they need not only the language facilities for requesting these operations but the routines which will carry them out. On the other hand, the insurance or banking people will be very concerned with report writing, tabulations, etc. Although we have attempted to develop languages which are fairly general purpose for certain broad categories (as defined in Chapters IV to VIII), the very number of them indicates that we have not succeeded. It therefore seems appropriate to look very hard at techniques and systems for permitting users to define their own artificial languages. In order for this to be meaningful and practical, there must be some type of simple language-translating-mechanism creator; i.e., it is not enough for the user to be able to easily define his own language; he must also have some type of system which will have it implemented for him very easily. To accomplish this, there needs to be great assistance from some of the aspects under the theory-oriented future developments, namely improved means of defining languages, having them translated, and knowing that we have actually created a translator for them. One very broad approach to this problem of user-defining languages is of course to provide large sets of subroutines which will actually do the work and then provide a language on top which allows the user to specify the way in which he wants to invoke these facilities.

The advantage of allowing the user to define his own language is that he can insist that the computer (and compiler) adapt itself to his terminology and method of expression and not vice versa as the situation has been up until now. Another reason for this particular approach being extremely important is that there are now, and will surely continue to be, very diverse opinions on the overall types of languages that should exist. As noted in the next section, one school of thought holds that future language development should be toward the direction of natural language input; on the other hand, there is a vociferous school that says this is very bad because users need to be forced to use the rigor and clarity that come with formal notations and languages. There does not appear to be any potential resolution of these diverse viewpoints on the horizon; by permitting users to choose the course they prefer, it may be that one or the other of these approaches will show itself to be more valid.

A short-range goal and approach is the inclusion of effective macro facilities in programming languages. Just as assembly programs with powerful macro capabilities begin to approach the higher level languages, so the latter might really begin to approach *user-defined* languages. The original PL/I macro facility was an excellent step in the right direction, but naturally it could not be expected to provide the full capability of this type which is needed. Thus we can expect to see an increase in the development of languages which are relatively simple but provide very good facilities for extensions at the same level as the base language. However, just as macros added to assembly languages do not cause them to become full-fledged programming languages, we cannot necessarily expect programming languages with macros to become full-fledged user-defined languages.

XI.3.2. Use of Natural Language

I have long favored the use of natural language (and for myself that means English) as a programming language, where this concept *definitely includes* mathematical or any other relevant scientific notation. An increasingly large number of programmers will be needed to keep the computers of future generations busy, where the term *future generation* applies to hardware and not to people. Since we have had new computer generations roughly every 5 years and since the engineers seem to be able to design bigger and faster machines more rapidly than the programmers can keep them occupied, it is clear that the number of programmers, or at least programmer productivity, must be significantly improved. There are generally conceded to be two approaches to improving this productivity. One is to give the professional programmer better tools with which he can perform his work; whether these tools are better languages, better subroutine packages, or on-line systems, is almost irrelevant. The significant thing is that improved facilities are given to those people whose fundamental training and mission is to take some kind of problem statement and put it into a form which can be understood by a computer and hence solved. In order to indicate—if not prove—the fallacy of this approach, I must again use the oft-quoted analogy with the telephone company. It has been said that if the telephone company had not gone to dial telephoning, then every woman between the ages of 20 and 50 would have been forced to become a telephone operator in order to keep up with the demands. Similarly, at the rate the hardware and applications are both developing, it may be necessary for vast numbers of people to become programmers. In fact, the programmers may well have to outnumber significantly the number of people who have problems to solve, although the problems themselves become more and more complex.

The other method of improving programmer productivity is to essentially make everybody his own programmer, just as almost everybody drives his own car rather than obtaining the services of a professional driver. In order to make this practical, it is obviously necessary to make it very easy for people to communicate with the computer. The physicist or insurance broker who has a problem he wishes to have solved on the computer would like to do this with maximum attention to his problem and minimum attention to the annoying intricacies of the computer itself. Obviously he cannot do this without learning something, just as to use an adding machine, a slide rule, or any other tool, a person must learn how to operate it. One can consider three major levels of ease (or lack thereof) of usage. The first is direct usage of machine language, even through a symbolic assembly program. The second is through the use of artificial languages of the type discussed in this book. The third is through a language which is natural to the user. It seems to me that the third choice is by far the most desirable course. It is essential to note that there is a *very* significant difference between this approach and the one which was discussed somewhat earlier, namely giving the user the ability to define his own artificial language. In that case, the user has a very definite responsibility for creating a language which is specifically tailored to his needs, and of course it is assumed that he would be given some mechanism for easily implementing this. In the case of the natural language approach, there is no artificial language per se defined; the user communicates with the computer in the same way that he communicates with people. When he is ambiguous, then the computer will query him to determine more specifically the meaning of what he says. If he uses unfamiliar terms, then again the computer must request further enlightenment. If he asks for something that seems unreasonable, either because it is in conflict with other directives or because it might take too much of the computer facilities, then the computer might object, just as a single engineer in a large corporation who asks for 25 typists, 17 technicians, 15 engineering aids, and several computers is likely to be told that his demands upon the resources of the organization are quite unreasonable. Among the objections to this natural language approach are that it encourages lack of rigor and that the user must understand the computer in order to use it. To dispose of the latter point first, most people are today using equipment whose inner workings they do not understand at all; it is a rare and unusual person who knows very much about what goes on under the hood of a car, inside a television set, or even inside something as simple as a toaster. With regard to the lack of rigor, I have never noticed that a programming language really forced rigor onto anybody. An individual is just as likely to omit logically necessary steps from a FORTRAN program as he is from stating the problem in English; in fact, he is more likely to omit them from the FORTRAN than from the English because in the former case he must remember a whole host of notational devices which are irrelevant to his problem and detract from his thoughts about the application. In other words, when we made the step from assembly languages to the higher level languages discussed in this book, one of the great advantages of the programming languages was that the user could concentrate upon the logic of his program and not have to worry about intricate details of modifying index registers appropriately. Similarly, we are ready to take the step forward to again allow an order-of-magnitude improvement in the ability to think constructively about the problem rather than about the means of stating the way of solving it.

One of the objections to this approach has been its potential cost and difficulty. It is clearly not within the current state of the art; it is not even within any reasonable cost factor relevant to current technology. However, that has never deterred us before and it does not seem a sufficient reason to stop us now.

There are two main ways in which this approach can be brought to fruition, and both seem worthy of support. One is the approach which can be called *bottom up*, in which artificial languages are developed to become increasingly close to natural language; the opposing, or *top down* view is that exemplified by the work of the linguists who are concentrating on understanding English grammar and developing generative and recognition procedures for increasingly large subsets of it. Neither of these approaches seems to have any major advantage over the other and it is probable that both will continue to proceed and perhaps meet near the middle to solve this problem.

In summary, it seems essential that a major area of future language work is to allow people to communicate with the computer in the same way that they communicate with each other, at least from a language point of view. Naturally it would be desirable to have hardware developments go along with this so that a person could indeed *talk* to the computer rather than *writing* what he wants to say, but this is a factor not within the scope of this discussion.

XI.3.3. Communication with Hardware and Software

The third area of the user-oriented developments of the future concerns itself with the language interaction between the person and the system. *System* is used to encompass both the hardware and the software systems. Let us consider each of these separately.

In the first place, the user needs better facilities for making use of specific hardware devices that he has or for overcoming the difficulties of the ones that exist. For example, relatively new graphic equipment needs languages which can suitably manipulate this hardware without the user worrying about the details of the equipment. In a similar vein, anybody who has ever written an equation on a typewriter, let alone a keypunch machine, has been plagued by the necessity of linearizing what is essentially two-dimensional mathematical notation. In Section IV.7 there was a discussion of several systems which were designed to overcome this difficulty; in each of those cases, special hardware was needed. Since that hardware has been known to exist and since its availability makes it easier for users to communicate certain types of problems, further mutual development for language and hardware is certainly needed.

Another future development will be the creation of languages (or features in them) to permit effective utilization of multiprocessors, parallel computation, and special machine configurations. Until there are some languages which truly eliminate the need for specifying sequential steps, the language will have to provide the compiler with the information needed to utilize some of the hardware configurations. Some ideas along these lines have been described.

In spite of the comments above, the largest need in this area is for better language developments for the user to communicate with the software system. Let us consider the batch environment first. With today's complexity of operating systems, the user must not only write his program in some programming language, but then he must worry about a whole separate language which is going to tell the operating system about tapes and other peripheral units and a host of information which has even less to do with his problem than some of the requirements of the programming language. Some way must be developed for the user to be able to take his problem, written in a programming language, and place it directly into a complex operating system without having to learn an additional (new) language.

The advent of time-sharing has of course opened an area of interesting problems and opportunities in languages. There are three distinctions which must be made here. The first is the availability of languages under a time-sharing system which are also available under a batch system. For example, FORTRAN, which is naturally a language developed for a batch environment, is no different as a language under time-sharing, although the on-line facilities indeed make it easier to debug the program and/or obtain fast response. However, there is nothing at all significant about the language elements here. The second distinction involves a separate language development which provides the user with specific facilities which are only meaningful in an on-line environment. Some of the facilities in QUICKTRAN, for example, are not meaningful unless the user is at a computer terminal. Similarly, in the field of algebraic manipulation, certain facilities which are meaningful and essential in an on-line environment are out of place or irrelevant in a batch situation. Thus we need the further development of languages which are specifically oriented to the user who is operating in an on-line situation and whose characteristics relate to the problem solution rather than the control functions. Finally, the last facet actually does relate to the control function. In other words, the fact that the user is on line gives him both the duty and the right to instruct the system to carry out certain tasks for him. These are the kinds of facilities that were mentioned in Section IX.3.6. These might be along the lines of editing or correcting, etc., and have relatively little to do with the problem solution itself.

It is possible to develop control functions for use with a time-sharing system that is combined with a standard language. This has been done and described in some of the existing developments, e.g., QUIKTRAN.

One interesting combination and/or usage of time-sharing and programming languages is in the area of a teaching machine. It is not the purpose of this book to go into a discussion of teaching machines per se. However, to the extent that a programming language is not a natural language and therefore must be learned, the on-line situation may be ideally suited to help in this. Some systems have already developed simple facilities in this direction (e.g., MAP, described in Section IV.6.6). For the inexperienced user, long and elaborate directions are given on what to type in and how, as well as for the interpretation of the response. These same systems provide the knowledgeable or sophisticated user with shortcuts, both in terms of input and shorter responses, thus making the entire time span much shorter. This type of facility should be increased and improved in future developments.

XI.3.4. Languages for New Application Areas

Another facet of future developments which is user-oriented is the opening of new application areas. While in theory there is certainly nothing to prevent any user or group thereof from utilizing the computer for any application that is meaningful, nevertheless the presence or absence of a suitable language may be the determining factor. A clear illustration occurs in considering the general area of formal algebraic manipulation. Packages to do certain specific tasks, e.g., differentiation, have existed since 1953, but people did not rush to use a computer in this area until suitable languages became available. In other words, it is not necessarily sufficient to have the subroutines which will carry out the desired operations; the user may either want or need, in order to work effectively, a better means of communication with these packages than simply invoking them as subroutines. Thus, one of the functions of future language development must be to look for areas in which the nonexistence of a suitable language is a barrier or at least a hindrance in using the computer in this particular direction.

XI.3.5. LANGUAGES FOR WRITING SOFTWARE

One class of computer users which usually is overlooked in discussions like this is the writer of the computer software. By software I mean the programs which assist the user in solving his problem but do not in and of themselves produce useful results. The obvious examples of this software are the compilers themselves, the operating and time-sharing systems, the sorts and report writers, etc. This is becoming an increasingly important part of the overall programming effort; some people have said facetiously that if all the people who are doing software development would discontinue that and instead devote their time to solving the practical problems, the industry would be in a much better position. It should be clear that I do not share that view. Software development is simply another type of computer usage which should be made more efficient. Again, although there was a discussion in Section IX.2.5 of languages which were designed to help make it easier to write compilers, there is no really practical or widely used system for that purpose today, and one is desperately needed. This is obviously connected with not only the problem of language definition and compiler interaction but with techniques for writing compilers and also with the on-line and operating systems functions. If we consider the software developer as one of the computer users, then he also needs languages which are suitable for his purpose. The biggest problem here, however, lies in somewhat contradictory objectives. In most applications, the objective of the programming language is to remove any worry about machine operation and machine facilities from the programmer. However, in the case of the programmer who is writing operating and time-sharing systems, obviously there must be access to the physical hardware. The trick is to do this in such a way that the programmer has all the advantages of the higher level language without losing any control over the actual hardware that he must manipulate.

XI.4. INTERRELATIONSHIPS AMONG SOME OF THESE CONCEPTS

There tends to be considerable confusion about the connections and differences among *user-defined* languages, *natural language* systems, and *non-*

procedural languages. This section attempts to point out the relationships among these three ideas.

For any given artificial language, it is possible to introduce more English-like terminology to express the same ideas. For example, the simple assignment statement X = Y + Z could be written as **ASSIGN THE VALUE OF Y + Z TO X**; similarly, a **GO TO K** could be written as **TRANSFER CONTROL TO STATEMENT LABEL K**. There are very few people who would claim that this *Anglicized* version was better or more natural.

On the other hand, if one considers the short program (in a mythical language)

```
DO I = 3, 95, 2
CALL PRIME (I)
IF T, CALL SQURT(I); ELSE RETURN
PRINT I, SQURT(I)
RETURN
```

then many people would claim that its replacement by

COMPUTE THE SQUARE ROOT OF THE PRIME NUMBERS BETWEEN 3 AND 95 AND PRINT IN TWO COLUMNS

was an improvement in clarity for both reading and writing. In this case, however, two different concepts are being combined. The first is the use of a more natural (i.e., English-like) language. The second concept is the combination into a single sentence of all the actions to be performed, thus eliminating the requirement for the sequential information inherent in the program steps. Hence, one is dealing with the *natural* language aspect, while the other deals with the *nonprocedural* aspect; they are clearly not the same.

If we assume the desirability of allowing people to define their own artificial languages, with some type of general system for generating the necessary processors, then obviously some type of common core is necessary. The one common core among all applications which can be utilized is English. Thus, the approach to developing a system which permits people to define their own application languages goes hand in hand with certain developments in the use of English as a programming language. If each language is highly symbolic, then there is no common central core and the development of a general generating system requires much more work by the language builder. On the other hand, improved knowledge of how to handle English-like input for a particular application probably sheds very little (if any) light on the problems of developing a system in which people can define their own languages.

The issue of developing nonprocedural languages is separate from the two problems above. In addition to the examples given earlier, the key factor is that nonprocedural refers to the amount of specific detail (including sequencing information) that an individual must supply about the problem he wants to solve or its method of solution. The form in which he states this information is almost irrelevant. For example, writing DIFFERENTIATE THE FUNCTION SIN (X * COS(X)) WITH RESPECT TO X does not convey any more meaning than the symbols DIFF SIN (X*COS (X)), X. Within a given application area, the user might have a language which is quite nonprocedural but very unnatural (e.g., Information Algebra) or one which was nonprocedural and natural but shed no light on the problem of easily creating languages (and their processors) for a given application area.

XI.5. CONCLUSIONS AND SUMMARY

One of the philosophical considerations that must be kept in mind in viewing the future developments of programming languages is the role that the computer will play in society. If the computer is considered no more than a tool such as the slide rule or an adding machine, then there is no doubt but that programming languages must develop into the area of naturalness for the user. If, on the other hand, the computer is considered a very special device whose usage is to be limited to a relatively few specialists, then it is legitimate to require a higher level of actual computer skill for the use of this equipment.

The two main directions for most likely and most needed future developments are theoretical and user-oriented. While these are clearly not completely independent, their points of overlap are surprisingly small. This is actually quite encouraging because it means that some of the practical user-oriented problems are not really being completely held up by lack of appropriate theoretical development. On the other hand, some of the theoretical developments will need somewhat more input from the user-oriented side before they can come to fruition. Therefore, it seems essential that a two-pronged attack be made along the indicated lines.

REFERENCES

XI. FUTURE LONG-RANGE DEVELOPMENTS

[AJ65] Anderson, J. P., "Program Structures for Parallel Processing", *Comm. ACM*, Vol. 8, No. 12 (Dec., 1965), pp. 786–88.

[CH66] Cheatham, T. E., Jr., "The Introduction of Definitional Facilities into Higher Level Programming Languages", *Proc. FJCC*, Vol. 29 (1966), pp. 623–37.

[DJ63]   Dijkstra, E. W., "On the Design of Machine Independent Programming Languages", *Annual Review in Automatic Programming*, Vol. 3 (R. Goodman, ed.). Pergamon Press, New York, 1963, pp. 27–42.

[FV67]   Fraser, J. B., "The Role of Natural Language in Man-Machine Communication", *Information System Science and Technology* (D. Walker, ed.). Thompson Book Co., Washington, D.C., 1967, pp. 21–28.

[HL66]   Halpern, M. I., "Foundations of the Case for Natural-Language Programming", *Proc. FJCC*, Vol. 29 (1966), pp. 639–49.

[ML60]   McIlroy, M. D., "Macro Instruction Extensions of Compiler Languages", *Comm. ACM*, Vol. 3, No. 4 (Apr., 1960), pp. 214–20.

[OA65]   Opler, A., "Procedure-Oriented Language Statements to Facilitate Parallel Processing", *Comm. ACM*, Vol. 8, No. 5 (May, 1965), pp. 306–307.

[RC65]   Rice, R. *et al.*, "Promising Avenues for Computer Research", *Proc. FJCC*, Vol. 27, pt. 2 (1965), pp. 85–100.

[SM66b] Sammet, J. E., "The Use of English as a Programming Language", *Comm. ACM*, Vol. 9, No. 3 (Mar., 1966), pp. 228–30.

[WA67]   Waite, W. M., "A Language-Independent Macro Processor", *Comm. ACM*, Vol. 10, No. 7 (July, 1967), pp. 433–40.

[ZY62]   Cantrell, H. N., "Where Are Compiler Languages Going?", *Datamation*, Vol. 8, No. 8 (Aug., 1962), pp. 25–28.

A BIBLIOGRAPHY ARRANGEMENTS AND AUTHOR LIST

In deciding which works to include in the bibliography, I decided not to make a distinction between those items specifically referred to in the text and those items which might be of use to people interested in pursuing the subject further. With the exception of "GENERAL" and "RELATED INFORMATION" lists, which may be found in the BIBLIOGRAPHY (pp. 739-741), all citations have been placed at the end of the relevant chapter, grouped by subsection (which usually corresponds to a single language). Those "REFERENCES" are actually combined Reference-Bibliography lists. The material included ranges from *essential* to *useful* to items that are not particularly good but represent the *only material* on the subject known to me or present some *unusual* facet of the work. (I have personally seen most—but not all—of the items listed.) No particular attempt has been made to include *all* material relevant to each language, especially for the major ones. In many cases, the items describe problems for which the language was used rather than the language itself. A few items which were referenced from, or should be included in, many sections were placed in the GENERAL list; readers not finding a particular citation at the end of the chapter they are reading should consult the GENERAL list.

A four- or five-character code has been assigned to each *citation*. In most cases, this consists of the first two letters of the leading author's name, but where this would have involved duplication or where there was no author, arbitrary letters were chosen. The two digits generally represent the year of publication. Wherever there was more than one reference from the same source during one year, the letters a, b, etc., were used to distinguish them. The digits 00 were used when no date was clearly given on the document. The references for each section have been arranged alphabetically by *code number*, not by author, to expedite lookup from the text where there was no author. Much effort has been used to try to provide full citations of the source documents, in order to help readers with differing methods of access to the literature.

The GENERAL list of references consists of either items which are referred to in the text many times, or material which might be of interest to readers of this book even though not necessarily referred to in the text. The RELATED INFORMATION list contains items which do not logically fit anywhere else but might be of interest to some readers, or appeared too late to be put in a more appropriate place.

Following the RELATED INFORMATION references, there is an AUTHOR LIST containing an alphabetic listing of the author for each item, together with the code numbers assigned to each of his references and the specific page(s) in the book where the full citation occurs. Documents by a single author are shown separately from those by a leading author or editor. The documents without specific authors are not included in this list. The major issuer of such material is the IBM Corporation, whose manuals are identified by the code letters IB.

It is hoped that this arrangement will provide the reader with all the relevant material on a subject grouped together in the appropriate chapter, and also a listing of authors so that a particular document may be easily found if the author is known.

The following abbreviations for the major sources have been used:

Comm. ACM: Communications of the Association for Computing Machinery.
J. ACM: Journal of the Association for Computing Machinery.
Proc. SJCC: Proceedings of the Spring Joint Computer Conference.
Proc. FJCC: Proceedings of the Fall Joint Computer Conference.
Proc. EJCC: Proceedings of the Eastern Joint Computer Conference.
Proc. WJCC: Proceedings of the Western Joint Computer Conference.

For the 1962 *FJCC* through the 1966 *FJCC*, the publisher is Spartan Books, Washington, D.C.; this is *not* shown in the individual citations. The 1967 *SJCC* and *FJCC* were published by Thompson Books, Washington, D.C. (also not shown in the citations). Earlier publishers are not listed since the volume could be found in a library without knowing the publisher or the reader may have a copy or be able to find information through AFIPS.

BIBLIOGRAPHY

GENERAL

[AC00]   *ACM SIGPLAN Notices*, ACM, New York, N.Y. (appears monthly).

[BI62] Barnett, M. P. and Futrelle, R. P., "Syntactic Analysis by Digital Computer", *Comm. ACM*, Vol. 5, No. 10 (Oct., 1962), pp. 515–26.

[BU65a] Burkhardt, W. H., "Universal Programming Languages and Processors: A Brief Survey and New Concepts", *Proc. FJCC*, Vol. 27, pt. 1 (1965), pp. 1–21.

[CC63] "Survey of Programming Languages and Processors", *Comm. ACM*, Vol. 6, No. 3 (Mar., 1963), pp. 93–99.

[DT65] "The RAND Symposium, Part 2", *Datamation*, Vol. 11, No. 9 (Sept., 1965), pp. 66–73.

[DV66] Davis, R. M., "Programming Language Processors", *Advances in Computers* (F. L. Alt and M. Rubinoff, eds.). Academic Press, New York, 1966, pp. 117–80.

[FG63] Feigenbaum, E. A. and Feldman, J. (eds.), *Computers and Thought*. McGraw-Hill, New York, 1963.

[FX66] Fox, L. (ed.), *Advances in Programming and Non-Numerical Computation*. Pergamon Press, New York, 1966.

[GD61] Goodman, R. (ed.), *Annual Review in Automatic Programming*, Vol. 2. Pergamon Press, New York, 1961.

[GD63] Goodman, R. (ed.), *Annual Review in Automatic Programming*, Vol. 3. Pergamon Press, New York, 1963.

[HG67] Higman, B., *A Comparative Study of Programming Languages*. American Elsevier Publishing Co., Inc., New York, 1967.

[IC62] *Symbolic Languages in Data Processing* (Proceedings of the Symposium organized and edited by the International Computation Centre, Rome, March 26–31, 1962). Gordon and Breach, New York, 1962.

[IE64] *IEEE Trans. Elec. Comp.*, Vol. EC-13, No. 4 (Aug., 1964).

[PA65] Pantages, A., "Language in the Sixties" (report on SHARE-JUG Workshop on programming languages), *Datamation*, Vol. 11, No. 11 (Nov., 1965), pp. 141–42.

[PW68] Wegner, P., *Programming Languages, Information Structures, and Machine Organization*. McGraw-Hill, New York, 1968.

[RO64a] Rosen, S., "Programming Systems and Languages—A Historical Survey", *Proc. SJCC*, Vol. 25 (1964), pp. 1–15. (Also in [RO67].)

[RO67] Rosen, S. (ed.), *Programming Systems and Languages*. McGraw-Hill, New York, 1967.

[SM68] Sammet, J. E., "Roster of Programming Languages—1968", *Comp. and Automation*, Vol. 17, No. 6 (June, 1968), pp. 120–23.

[SP65] Spiegel, J. and Walker, D. E. (eds.), *Information System Sciences: Proceedings of the Second Congress*. Spartan Books, Washington, D.C., 1965.

[ZR65] Crisman, P. A. (ed.), *The Compatible Time-Sharing System, A Programmer's Guide*, Second Edition. M.I.T. Press, Cambridge, Mass., 1965.

RELATED INFORMATION

[AF65] *Advanced Programming Developments: A Survey*, Air Force Systems Command, ESD-TR-65-171, Electronic Systems Division, Directorate of Computers, Bedford, Mass. (Feb., 1965).

[FF63] Carracciolo di Forino, A., "Some Remarks on the Syntax of Symbolic Programming Languages", *Comm. ACM*, Vol. 6, No. 8 (Aug., 1963), pp. 456–60.

[GI67] Ginsberg, A. S., Markowitz, H. M., and Oldfather, P. M., "Programming by Questionnaire", *Proc. SJCC*, Vol. 30 (1967), pp. 441–46.

[GO62] Gorn, S., "Mechanical Pragmatics: A Time-Motion Study of a Miniature Mechanical Linguistic System", *Comm. ACM*, Vol. 5, No. 12 (Dec., 1962), pp. 576–89.

[GO67] Gorn, S., "Handling the Growth by Definition of Mechanical Languages", *Proc. SJCC*, Vol. 30 (1967), pp. 213–24.

[GU62] Grau, A. A., "A Translator-Oriented Symbolic Programming Language", *J. ACM*, Vol. 9, No. 4 (Oct., 1962), pp. 480–87.

[GY66] Gosden, J. A., "Explicit Parallel Processing Description and Control in Programs for Multi- and Uni-Processor Computers", *Proc. FJCC*, Vol. 29 (1966), pp. 651–60.

[LD66] Landin, P. J., "The Next 700 Programming Languages", *Comm. ACM*, Vol. 9, No. 3 (Mar., 1966), pp. 157–66.

[NR67] Narasimhan, R., "Programming Languages and Computers: A Unified Metatheory", *Advances in Computers*, Vol. 8 (F. L. Alt and M. Rubinoff, eds.). Academic Press, New York, 1967, pp. 188–245.

[OA66] Opler, A., "Requirements for Real-Time Languages", *Comm. ACM*, Vol. 9, No. 3 (Mar., 1966), pp. 196–99.

[PN67] Parnas, D. L. and Darringer, J. A., "SODAS and a Methodology for System Design", *Proc. FJCC*, Vol. 31 (1967), pp. 449–74.

[PR64a] Perlis, A. J., "A Format Language", *Comm. ACM*, Vol. 7, No. 2 (Feb., 1964), pp. 89–97.

[QC65] Bauer, F. L. and Samelson, K., "Sequential Formula Translation", *Comm. ACM*, Vol. 3, No. 2 (Feb., 1960), pp. 76–82. (Also in [RO67].)

[ST64] Steel, T. B., Jr., "Beginnings of a Theory of Information Handling", *Comm. ACM*, Vol. 7, No. 2 (Feb., 1964), pp. 97–103.

[VR58a] Voorhees, E. A., "Algebraic Formulation of Flow Diagrams", *Comm. ACM*, Vol. 1, No. 6 (June, 1958), pp. 4–8.

AUTHOR LIST

Page numbers set in **bold face** indicate the pages on which the full citation appears. Page numbers set in regular type indicate pages in the text which contain a reference to the document specified. Sets of page numbers are subdivided by chapters.

Abrahams, P. W. [AH63], **467**; [AH66], **126**, 109; [AH67], **464**.
Abrahams, P. W., *et al.* [AH66a], **602**, 590, 591, 592; 761.
Adams, C. W., *et al.* [AD54], **26**, 132, **300**.
Alber, K. [AL67], **62**, **600**.
Allen, C. D., *et al.* [AN66], **62**, **600**.
Allen, J. J., *et al.* [AX63], **62**, 42; **302**, 147; 153, 164.
Alt, F. L. [AT64], **62**.
Anderson, J. P. [AJ61], **721**, 718; [AJ65], **736**.
Arden, B. W., *et al.* [AR61], **304**; [AR61a], **308**, 206.
Ash, R., *et al.* [AS57], **301**; 135.

Auroux, A., et al. [AU67], **304**, 195; **706**; 757.
Ayers, J. A. [AY63], **302**, 159.
Bachman, C. W. [QJ65], **381**.
Bachman, C. W., et al. [QJ64], **381**; 760.
Backus, J. W. [BS54a], **300**; [BS60], **62**, 53; **304**, 175.
Backus, J. W., et al. [BS54], **300**; [BS57], **302**; 144; [BS64], **302**.
Baker, C. L. [BK56], **26**, 8; [BK64], **308**, 219; [BK66], **308**, 217.
Balke, K. G., et al. [BQ62], **312**, 265, 66, 268, 269, 270; 756.
Balzer, R. M. [YH67], **602**.
Bandat, K. [BA67], **62**; **600**, 548.
Barlow, A. E., et al. [BL65], **703**, 667; 758.
Barnett, M. P., et al. [BI62], **739**; 448, 605, 635; [BI63], **705**, 684, 685.
Bashkow, T. R., et al. [UF67], **721**, 718.
Bastian, A. L., et al. [UB62], **468**.
Bates, F., et al. [QB67], **601**, 548.
Bauer, F. L., et al. [QC65], **741**.
Baumann, R., et al. [BN64], **304**, 181.
Beech, D., et al. [BC66], **62**; **600**, 548; [BC66a], **62**; **600**, 548; [BC67], **62**; **600**, 548.
Bemer, R. W. [QF57], **301**.
Bennet, N. W. [BE65], **379**, 335.
Bennett, E., et al. [QA65], **470**, 461; **703**; 754.
Bequaert, F. C. [BV67], **706**, 691.
Bergin, G. P. [BG64], **126**.
Berkeley, E. C., et al. [BY66], **467**, 410; 761.
Berman, R., et al. [BF62], **379**, 344.
Bernick, M. D., et al. [BM61], **520**, 473; 754.
Blackwell, F. W. [QK67], **311**, 255; **470**, 464; 761.
Bleiweiss, L., et al. [JL66], **521**, 475.
Bobrow, D. G. [BB63], **720**; [BB64], **701**, 664; [BB64c], **467**; **701**, 664; [BB67a], **465**; **520**.
Bobrow, D. G., et al. [BB64a], **465**, 386; [BB64b], **465**, 388; [BB67], **467**.
Bond, E. R., et al. [BZ64], **520**, 476; 758; [BZ67], **521**, 476; **600**, 546.
Book, E. [BO66], **602**, 590.
Book, E., et al. [BO60], **697**.
Bottenbruch, H. [BH62], **304**, 181.
Bouricius, W. G., et al. [UD67], **310**.
Boyer, M. C. [QI65], **312**, 313, 264, 299.
Bradford, D. H., et al. [BD61], **312**.
Bratman, H. [BR59], **697**; [BR61], **719**.
Brennan, R. D., et al. [YR64], **696**, 627 ; [YR64a], **696**, 627.
Brody, T. A. [UQ65], **468**.
Bromberg, H. [BJ61], **379**, 332; [BJ67], **379**.

Brooker, R. A., *et al.* [BX62], **62**, 57; **696**, 635; [BX63], **696**.
Brown, P. J. [UH67], **127**.
Brown, S. A., *et al.* [BP63], **693**, 606, 610; 755.
Brown, W. S. [BW66], **522**; [BW66a], **522**, 502; 754.
Brown, W. S., *et al.* [BW63], **522**.
Bryan, G. E. [UJ67], **308**.
Bryan, G. E., *et al.* [UJ67a], **308**; 760; [UJ67b], **308**.
Bryant, J. H., *et al.* [QE66], **702**.
Budd, A. E. [QH66], **28**, 23; **303**; **380**.
Burkhardt, W. H. [BU65], **62**; **302**; 156; [BU65a], **740**; [BU66], **600**, 548.
Busch, K. J. [QD66], **695**, 627.
Callahan, M. D.,*et al.* [ZQ67], **381**, 335.
Cameron, S. H., *et al.* [CS66], **311**, 255; [CS67], **311**, 255, 257; 757.
Cantrell, H. N. [ZY62], **737**.
Carleton, J. T., *et al.* [CT64], **302**, 170, 171, 172; 763.
Carracciolo di Forino, A. [FF63], **740**.
Carter, G. L., *et al.* [CA63], **312**, 265.
Chanon, R. N. [ZS66], **469**, 435.
Chapin, N. [CZ64], **466**; [CZ65], **27**, 23.
Chapin, P. G., *et al.* [ZT67], **470**; **702**.
Cheatham, T. E., Jr. [CH66], **127**; **736**; [CH66a], **698**.
Cheatham, T. E., Jr., *et al.* [CH62], **701**; [CH64], **63**, 57; **697**, 635.
Christensen, C. [CQ64], **470**, 454, 457; 754; [CQ65], **470**, 455; 754; [CQ66], **470**, 457.
Christensen, C., *et al.* [CQ65a], **600**, 542; [CQ67], **600**.
Chu, Y. [CB65], **695**, 623, 625, 656.
Church, A. [ZP41], **468**, 413.
Clancy, J. J., *et al.* [ZZ65], **696**, 627.
Clapp, L. C., *et al.* [CL63], **522**, 510, 511, 512; [CL66], **522**, 512; **522**.
Clark, E. R. [CE67], **599**, 530.
Clem, P. L., Jr., *et al.* [CM66], **311**, 258.
Climenson, W. D. [CD63], **701**.
Clippinger, R. F. [CP61], **378**, 327.
Coffman, E. G., Jr. [CO61], **599**, 530.
Coffman, E. G., Jr., *et al.* [CO64], **706**, 688.
Cohen, L. J. [ZX67], **127**.
Connors, T. L. [CY66], **703**, 667, 703.
Conway, M. E. [CI58], **719**.
Conway, R. W., *et al.* [CN63], **313**, 756; [CN65], **470**; 755.
Cooper, W. S. [CV64], **701**; [CV65], **701**.
Corbato, F. J., *et al.* [ZV62], **706**; [ZV63], **706**, 690.
Corbin, H. S., *et al.* [CF66], **704**, 679; 757.

Couch, A. S. [DF67], **699**, 646, 647.
Cowan, R. A. [CW64], **379**, 345.
Cowell, W. R., *et al.* [ZN65], **467**, 394; **721**, 718.
Craig, J. A., *et al.* [CJ64], **703**; [CJ66], **703**, 669; 757.
Cramer, M. L., *et al.* [CR66], **695**, 627.
Crisman, P. A. (ed.) [ZR65], **740**; 216, 240, 406, 417, 438, 475, 491, 611, 612, 660, 683, 685, 686, 688.
Culler, G. J. [CU67], **311**; 757.
Culler, G. J., *et al.* [CU62], **311**, 254; [CU63], **311**, 253; [CU65], **311**, 253, 254.
Cunningham, J. F. [CG63], **379**, 345.
Cuthill, E., *et al.* [CX65], **521**.
d'Agapeyeff, A., *et al.* [DG63], **377**.
Dahl, O., *et al.* [DH66], **701**, 661; 763.
Darlington, J. L. [DA65], **468**, 435.
Davis, R. M. [DV66], **740**.
Dean, A. L., Jr. [DE64], **698**, 640; 764.
de Bakker, J. W. [UE67], **64**; 306.
Desautels, E. J., *et al.* [DS67], **469**.
Dijkstra, E. W. [DJ63], **61**, 35; **737**; [DJ64], **312**.
Dimsdale, B., *et al.* [DC64], **700**.
Dixon, P. J., *et al.* [DI67], **701**.
Dodd, G. G. [DQ66], **465**; **600**, 546.
Donally, W. L. [DO62], **379**.
Donovan, J. J., *et al.* [DL67], **600**.
Duby, J. J. [DB67], **521**.
Dunn, T. M., *et al.* [DM64], **309**, 226.
Dupchak, R. [DU63], **466**, 393; [DU65], **466**, 394.
Edwards, D. J. [ED66], **698**, 642, 643; 762.
Efron, R., *et al.* [EF64], **700**.
Emery, J. C. [EM62], **379**.
Engelman, C. [EN65], **467**; **521**; 761.
Englund, D., *et al.* [EG61], **697**; 755.
Evans, T. G. [EV64], **467**, 410.
Evans, T. G., *et al.* [EV66], **126**, 114.
Falkoff, A. D., *et al.* [FA64], **694**, 620; **719**, 713; 755; [FA67], **310**, 247, 249, 250; [FA67a], **310**, 247, 248; 755.
Farber, D. J., *et al.* [FB64], **469**, 436; [FB66], **469**, 436, 437; 763.
Feigenbaum, E. A. [FG61], **466**, 393.
Feigenbaum, E. A., *et al.* [FG63], 385; **740**.
Feingold, S. L. [FZ67], **706**.
Feldman, J. [FN62], **466**.
Feldman, J. A. [FJ64], **698**; 759; [FJ66], **602**, 584; **698**, 641; 759.
Feldman, J. A., *et al.* [FJ68], **697**, 634.

Fenves, S. J., *et al.* [FE64], **694**; 616; 764; [FE65], **694**; 764; [FE66], **694**, 611, 615.
Ferguson, H. E., *et al.* [FR63], **27**.
Fimple, M. D. [FP64], **302**, 157; **379**.
Firth, D., *et al.* [FI66], **602**, 590.
Floyd, R. W. [FL64], **63**, 52, 53, 57; **697**, 635.
Forest, B. [FS61], **305**, 174.
Forgie, J. W. [FY65], **310**, 245.
Forte, A. [FT67], **469**; [FT67a], **469**.
Foster, D. C. [FO65], **701**.
Fowler, M. E., *et al.* [FH64], **302**, 164.
Fox, L. (ed.) [FX66], **740**.
Franks, E. W. [FW66], **701**.
Fraser, J. B. [FV67], **720**, 715; **737**.
Fredericks, D. S., *et al.* [FD65], **379**, 338.
Freeman, D. N. [FM64], **313**.
Fried, B. D. [FQ66], **311**, 253.
Fried, B. D., *et al.* [FQ64], **311**, 253.
Galler, B. A., *et al.* [GA67], **127**; **305**, 192.
Garwick, J. V. [GW64], **305**; [GW64a], **697**.
Garwick, J. V., *et al.* [GW00], **305**, 192, 195; 759.
Gaskill, R. A. [GL64], **696**, 631.
Gaskill, R. A., *et al.* [GL63], **696**; 757.
Gawlik, H. J. [GK63], **312**, 281, 283; 762.
Gawlik, H. J., *et al.* [GK67], **312**, 281; 762.
Gelernter, H., *et al.* [GE60], **465**, **467**, 388, 406.
Ginsberg, A. S. [GI67], **741**.
Glass, R. L. [GX67], **600**, 542.
Glennie, A. E. [GC60], **63**, 57; **697**, 635.
Goodman, R. (ed.) [GD61], **740**; [GD63], **740**.
Goodstat, P. B. [GS67], **62**, 46.
Gordon, G. [GG61], **700**, 653; [GG62], **700**.
Gordon, R. M. [GN63], **380**.
Gorn, S. [GO61], **63**; [GO61a], **63**, 57; [GO62], **741**; [GO67], **741**.
Gosden, J. A. [GY66], **741**.
Grad, B. [UC61], **29**, 11.
Graham, R., *et al.* [GM00], **301**, 142; 759.
Grant, E. E. [GH65], **701**.
Grau, A. A. [GU62], **741**.
Gray, J. C. [JG67], **465**, **705**.
Green, B. F. [GB61a], **465**.
Green, B. F., *et al.* [GB61], **466**; **703**; 755.
Green, J. [GT59], **305**, 174.
Greenberger, M., *et al.* [YP65], **701**; 762; [YP66], **701**, 662; **762**.

Greene, I. [GV62], **380**.
Grems, M. [GR62], **702**.
Grems, M., et al. [GR55], **301**, 133; 755.
Griswold, R. E. [GF65a], **469**.
Griswold, R. E., et al. [GF65], **469**; [GF67], **469**, 447; [GF68], 763.
Gross, L. N. [GQ67], **470**; **706**.
Gullahorn, J. T., et al. [ZK63], **467**.
Guzman, A., et al. [ZH66], **466**, **468**, **469**, 388, 416, 435.
Haines, E. C., Jr. [HA65], **470**, 457, 458; [HA67], **470**; 764.
Halpern, M. I. [HL64], **27**, 8; [HL65], **62**; [HL66], **61**, 35; **720**, 715; **737**; 758.
Halstead, M. H. [HS62], **307**, 197, 198, 200; 762; [HS63], **307**, 197, 198; [HS67a], **307**.
Halstead, M. H., et al. [HS67], **307**, 205.
Haverty, J. P. [HV64], **27**, 23.
Haverty, J. P., et al. [HV63], **702**.
Haynam, G. E. [HN65], **305**, 181.
Heising, W. P. [HE63], **302**, 146, 156; [HE64], **302**, 154, 156; [HE64a], **302**, 152.
Hellerman, H. [HH64], **310**, 253; 762.
Herscovitch, H., et al. [HK65], **700**.
Hicks, W. [HI62], **380**.
Higman, B. [HG67], **740**.
Hilsenrath, J., et al. [HR66], **313**, 297; 762.
Hilton, W. R., et al. [HT66], **468**, 419.
Hodges, D. [HX64], **466**, 394; **721**, 718.
Holberton, F. E. [HF54], **27**, 4; **378**, 322.
Homer, E. D. [HM66], **27**; **313**, 300.
Hopper, G. M. [HP53], **27**, 12; [HP55], **27**, 4.
Hopper, G. M., et al. [HP53a], **27**, 12.
Howard, J. C. [HQ67], **521**.
Humby, E. [HY62], **61**, 35; **380**, 338; [HY63], **61**, 35; **380**, 338.
Hunt, E. B., et al. [HB63], **466**.
Hurley, J. R., et al. [HJ63], **696**, 629, 630; 758.
Hurwitz, A., et al. [HW67], **704**, 674; 759.
Huskey, H. D., et al. [HU60], **307**; [HU61], **305**, 192; [HU63], **307**, 197, 198.
Irons, E. T. [IR61], **63**, 57; **305**; **697**, 635; [IR63], **697**.
Irons, E. T., et al. [IR59], **305**, 174.
Isbitz, H. M. [IS59], **697**; 755.
Iturriaga, R. [IT67], **602**, 584.
Iturriaga, R., et al. [IT66], **602**, 584, **698**, 641, 642; [IT66a], **602**, 584.
Iverson, K. E. [IV62], **310**, 247; 620; **720**, 712; 755; [IV62a], **720**, 712, 714; [IV62b], **720**; [IV63], **694**, [IV64], **63**; **305**; [IV64a], **720**.
Johnsen, R. F., Jr. [JO60], **307**.
Jones, M. M. [JM67], **699**, **701**, 661.

Junker, J. P., *et al.* [JU65], **303**; 380.
Kahrimanian, H. G. [KD54], **520**, 472.
Kameny, S. L. [KA66], **602**, 590.
Kanner, H. [KF59], **305**, 174.
Kaplow, R., *et al.* [KP66], **309**, 240, 241, 242, 244; 761; [KP66a], **310**, 240; 761.
Kapps, C. A. [KC67], **470**, 462; 764.
Katz, C. [KX62], **379**, 328, 329.
Katz, J. H., *et al.* [KZ63], **719**, 712.
Kavanagh, T. F. [KV60], **28**, 11.
Keller, J. M., *et al.* [KR64], **28**, 12; **309**, 226.
Kellogg, C. H. [KG66], **702**; [KG67], **702**; **720**.
Kemeny, J. G., *et al.* [KM66], **309**, 229; [KM67], **309**; 755.
Kennedy, P. R. [KE62], **599**, 529; [KE65], **599**, 528.
Kesner, O. [KS62], **380**.
Kinzler, H. M., *et al.* [KB57], **378**, 316.
Kirsch, R. A. [KI64], **720**.
Kiviat, P. J. [KW66], **699**, 651; [KW66a], **700**.
Klein, S. [KK65], **599**, 530.
Klerer, M., *et al.* [KL64], **313**, 294; [KL65], **313**; [KL65a], **313**, 285, 292, 293; 760; [KL65b], **313**, 286, 287; 760; [KL66], **313**, 288; [KL67], **28**, 20; **313**, 291; [KL67a], **313**.
Knowlton, K. C. [KO64], **699**, 644; 755; [KO65], **467**; [KO66], **467**, 400, 404; 760.
Knuth, D. E. [KN67], **305**.
Knuth, D. E., *et al.* [KN61], **305**; [KN64], **700**, 658; 764; [KN64a], **700**, 657; 764.
Krasnow, H. S. [KQ67], **699**, 651.
Landin, P, J. [LD65], **305**; [LD65a], **305**; [LD66], **741**.
Landweber, P. S. [LW64], **63**.
Laning, J. H., *et al.* [LA54], **28**, 301, 131; 760.
Lapin, R. B. [LB67], **705**, 683.
Lapin, R. B., *et al.* [LB65], **705**.
Lawson, H. W., Jr. [LH67], **601**, 567.
Leavenworth, B. M. [LV66], **127**.
Ledley, R. S., *et al.* [LL66], **699**, 645; 755.
Leroy, H. [LR67], **127**; **305**.
Leslie, H. [LS67], **377**, 314.
Levin, M. [LE64], **602**, 590.
Levin, M., *et al.* [LE66], **602**.
Licklider, J. C. R. [LI65], **693**, 604.
Lippitt, A. [LP62], **380**.
Logcher, R. D., *et al.* [LG67], **694**.
Lombardi, L. A. [LM62], **719**, 712.
Longo, L. F. [LN62], **377**, **380**, 314.

Madnick, S. E. [VB67], **465**.
Makinson, T. N. [MN61], **380**.
Manove, M., et al. [MV67], **522**.
Markowitz, H. M., et al. [MA63], **700**, 655, 656; 763.
Marks, S. L., et al. [ZL67], **308**, 225.
Marsh, D. G. [MD64], **599**, 530.
Martin, W. A. [ZB67], **522**, 514, 515, 516, 517; 764.
Masterson, K. S., Jr. [MS60], **307**.
McCarthy, J. [MC60a], **467**, 405.
McCarthy, J., et al. [MC60], **467**, 405; [MC62], **468**, 406, 408, 409.
McClure, R. M. [MZ65], **695**, 622; 763; [MZ65a], **522**, 503; **697**, 636, 637; 764.
McCracken, D. D. [MR61] 305; [MR64], **601**, 548; [MR65], **303**, 152.
McGee, W. C. [MG59], **377**; [MG63], **377**.
McGee, W. C., et al. [MG60], **377**.
McGinn, L. C. [MF57], **698**; 762.
McIlroy, M. D. [ML60], **127**, 121; 737.
McIlroy, M. D., et al. [ML66], **522**, 505; 754.
McKeeman, W. M. [ZG67a], **721**.
McKeeman, W. M., et al. [ZG67], **602**, 542.
McMahon, J. T. [YS62], **304**, **307**.
McMahon, L. E. [MM00], **720**.
Meadow, C. T., et al. [MW66], **702**.
Mealy, G. H. [YM67], **126**.
Melbourne, A. J., et al. [ZM65], **721**, 718.
Mesztenyi, C. K. [YZ67], **695**, 623.
Metze, G., et al. [MX66], **695**, 623, 624; 756.
Millen, J. K. [ZJ67], **522**, 501.
Miller, E. R., et al. [YL59], **378**.
Mittman, B. [ZC67], **693**.
Mondshein, L. F. [LQ67], **698**, 641.
Mooers, C. N. [ME66], **469**, 449, 451; 764; [ME67], **470**, 449, 453.
Mooers, C. N., et al. [ME65], **469**, 449.
Morris, A. H., Jr. [MB67], **522**, 507; 758.
Morrison, R. A. [YF67], **704**.
Morrissey, J. H. [MJ65], **309**.
Moulton, P. G., et al. [MO67], **303**.
Mullery, A. P. [MH64], **721**, 718.
Mullery, A. P., et al. [MH63], **721**, 718.
Mullin, J. P. [MU62], **380**.
Myer, T. H. [MY66], **308**.
Naftaly, S. M. [NF64], **380**.
Narasimhan, R. [NR67], **741**.

Naur, P. (ed.) [NA60], 53; **305**, 176, 181; [NA63], 53; **126**, 103; **306**, 177, 181, 182, 191, 754; [NA63a], **306**, 181.
Neidleman, L. D. [ND67], **521**.
Nelson, D. B., *et al.* [NB67], **702**.
Nelson, E. A., *et al.* [NE65], **28**, 19.
Newell, A. [NW63], **466**, 393.
Newell, A., *et al.* [NW56], **465**, **466**, 382, 389, 391; [NW57], **466**, 389; [NW57a], **466**, 389; [NW60], **466**, 389; [NW61], **466**, 393; [NW65], **466**, 389, 391, 393, 395, 398; 765.
Nolan, J. [NO53], **520**, 472.
Olsen, T. M. [OL65], **62**, 42; **303**, 153.
Olsztyn, J. T. [OZ58], **696**; 757.
Opler, A. [OA65], **737**; [OA66], **741**.
Oppenheim, D. K., *et al.* [OP66], **698**, 639; 762.
O'Sullivan, T. C. [OU67], **308**, 216.
Oswald, H. [OS64], **303**, 147, 152.
Pantages, A. [PA65], **740**.
Parnas, D. L. [PN66], **695**, 623, 627; 763; [PN66a], **695**, 623.
Parnas, D. L., *et al.* [PN67], **741**.
Peck, J. E. [PK67], **306**, 194.
Perlis, A. J. [PR64a], 741; [PR65], **126**, 65, 104.
Perlis, A. J., *et al.* [PR57], **302**, 139; 760; [PR57a], **302**, 139; [PR58], **28**, 22; **306**, 173; [PR64], **602**, 583; [PR66], **602**, 584; 758.
Perstein, M. H. [PE66], **599**, 529; [PE66a], **599**, 529, 539; 760.
Petersen, H. E., *et al.* [PS64], **695**, 627.
Pratt, T. W., *et al.* [PT66], 697.
Proctor, R. M. [PC64], **695**, 622; 760.
Pugh, A. L. [PG63], **699**, 652; 758.
Pursey, G. [PU67], **63**; 601.
Pyle, I. C. [PY65], **706**, 691.
Quatse, J. T. [QU65], **699**, 648; 764.
Rabinowitz, I. N. [RN62], **303**, 156.
Radin, G., *et al.* [RG65], **601**, 548.
Raphael, B. [RA64], **702**; [RA66], **126**, 65; [RA66a], **465**, 385; [RA66b], **465**, 385.
Raphael, B., *et al.* [RA67], **126**, 65; **465**, 387; **520**.
Reinfelds, J., *et al.* [RF66], **311**, 258; 754.
Reitman, J. [RP67], **699**.
Reynolds, J. C. [RE65], **697**; 756; [RE65a], **698**, 638, 639; 756.
Rice, J. R., *et al.* [RI66], **28**, 20; **313**, 299.
Rice, R., *et al.* [RC65], **737**.
Robbins, D. K. [RM62], **28**, 14; **303**, 157, 164.
Roberts, L. G. [RB65], **470**, 462; **704**; 756.
Rochester, N. [RT53], **29**, 2; [RT66], **63**, 56.

Roos, D. [RS65], **694**, 620; 759; [RS67], **694**; [RS67a], **694**; 759.
Roos, D., *et al.* [RS64], **694**, 612, 613; 756.
Rose, A. J. [RJ66], **310**, 247.
Rosen, S. [RO61], **303**; 153; [RO64], **697**; [RO64a], **740**; [RO67], **740**.
Rosen, S., *et al.* [RO65], **303**, 164.
Ross, D. T. [RD61], **705**, 681; [RD62], **705**, 682; [RD67], **705**, 681; [RD67a], **705**, 682, 683.
Ross, D. T., *et al.* [RD63], **705**.
Roth, J. P., *et al.* [RQ67], **310**, 247.
Ruyle, A., *et al.* [RU67], **308**, 217, 240, 245, 254, 258.
Sakoda, J. M. [SA65], **303**, 164; **465**, 388.
Sammet, J. E. [SM61], **306**; **379**, **380**, 330; **598**; [SM61a], **63**, 56; **380**, 340, 344; [SM61b], **380**, 332; [SM61c], **381**, 332; [SM62], **381**, 345; [SM66], **465**, 385; **520**, 472; [SM66a], **520**, 471, 472; [SM66b], **61**, 35; **720**, 715; **737**; [SM67], **520**, **521**, 471, 472; [SM67a], **465**, 385; **520**, 472; [SM68], **740**, 753.
Sammet, J. E., *et al.* [SM64], **521**, 476, 486.
Sanders, N., *et al.* [SS63], **306**.
Sandin, N. A., *et al.* [SN65], **599**, 528.
Satterthwait, A. C. [SR66], **465**.
Savitt, D. A., *et al.* [SK67], **702**.
Saxon, J. A. [SX63], **381**, 345.
Saxon, J. A., *et al.* [SX66], **307**.
Scheff, B. H. [SD66], **699**, 648, 649; 757; [SD66a], **699**, 648.
Schlaeppi, H. P. [QY64], **695**, 621; 761.
Schlesinger, S., *et al.* [QL67], **28**, 20; **313**, 299.
Schneider, F. W., *et al.* [QV64], **698**.
Schorré, D. V. [QT64], **602**, 592; **698**.
Schwalb, J. [SB63], **379**, 329.
Schwartz, J. I. [SC65], **29**, 23.
Schwartz, J. I., *et al.* [SC60], **599**, 528.
Schwarz, H. R. [QN62], **306**, 181.
Schwinn, P. M. [UZ67], **704**, 677; 759.
Sconzo, P., *et al.* [SO65], **521**, 474.
Seitz, R. N., *et al.* [UV67], **311**, 258, 259, 263; 754.
Selfridge, R. G. [YD55], **696**, 627.
Shantz, P. W., *et al.* [SY67], **303**, 164.
Shaw, C. J. [SH00], **599**, 525, 529; [SH61], **599**, 529; [SH61a], **599**; [SH62], **29**, 23; **126**, 65; [SH63], **599**; 760; [SH63a], **599**, 529; [SH63b], **599**, 524, 525, 528, 529, 530; [SH64], **599**, 530; [SH64a], **29**; [SH65], **377**; [SH66], **29**, 19, 23.
Shaw, J. C. [JC64], **308**, 217, 218; 760; [JC65], **308**, 220.
Shaw, J. C., *et al.* [JC58], **466**, 389.
Sibley, R. A. [QS61], **697**.
Siegel, M., *et al.* [SG62], **380**.

Simmons, R. F. [SE65], **702**, **703**, 662, 669; **720**; [SE66a], **702**, 662, 670; **720**.
Simmons, R. F., *et al.* [SE63], **703**; [SE64], **703**; 763; [SE66], **702**, **720**.
Singman, D., *et al.* [SI65], **307**, 205.
Skinner, F. D. [QX66], **704**.
Slagle, J. R. [SL61], **468**, 410; [SL65], **702**.
Spiegel, J., *et al.* [SP65], **740**.
Spitzer, J. F., *et al.* [SZ65], **702**, 665; 756.
Standish, T. A. [QM67], **126**.
Steel, T. B., Jr. [ST57], **29**, 8; [ST61], **719**; [ST61a], **719**; [ST64], **741**; [ST66], **599**, 525, 528, 530 [ST66a], **63**, 52; [ST67], **62**, 46.
Stefferud, E. [SF63], **467**, 393.
Steil, G. P., Jr. [QZ67], **702**, 666; 755.
Stone, H. S. [TS67], **306**, 192.
Stone, P. J., *et al.* [SJ63], **468**, 435.
Stotz, R. H. [UU63], **310**, 240; **705**.
Stotz, R. H., *et al.* [UU67], **310**, 240.
Stowe, A. N., *et al.* [SW66], **310**, 245, 246; 761.
Strachey, C. [SQ65], **470**, 450.
Strachey, C., *et al.* [SQ61], **126**, 90; **306**, 181, 194.
Strong, J., *et al.* [QR58] and [QR58a], **709**; **719**; 764.
Summers, J. K., *et al.* [UT67], **704**, 671, 673; 674; 754.
Summit, R. K. [UW67], **702**.
Sutherland, I. E. [QW63], **470**, 462; **704**, **705**, 678; [QW66], **704**.
Sutherland, W. R. [SU66], **470**, 463; **704**; [SU67], **704**.
Swets, J. A., *et al.* [UY63], **702**.
Swigert, P. [SV66], **521**.
Syn, W. M., *et al.* [QP66], **696**, 632, 633; 757.
Tabory, R., *et al.* [TR67], **720**.
Taylor, A. [TB60], **301**; 378.
Taylor, W., *et al.* [TA61], **306**.
Teichroew, D., *et al.* [TE66], **699**, 651.
Theodoroff, T. J. [TD58], **696**, 628; 757.
Thompson, C. E. [TM56], **29**, 5.
Thompson, F. B. [TH63], **703**; [TH64a], **703**; [TH66], **703**; **720**, 715; 758.
Thompson, F. B., *et al.* [TH64], **703**.
Tobey, R. G. [TO66], **521**.
Tonge, F. M. [TN60], **467**, 393.
Tonge, F. M., *et al.* [TN65], **700**.
Van Dam, A., *et al.* [VD67], **704**, 675, 676; 762.
van Wijngaarden, A. [VW63], **306**; [VW68], **306**, 178.
von Sydow, L. [VS67], **306**, 193.
Voorhees, E. A. [VR58a], **741**.
Waite, W. M. [WA67], **127**; 737.
Waks, D. J. [WK67], **308**.

Walk, K., *et al.* [VK67], **601**, 548.
Walker, D. E. [WV67], **705**.
Walter, R. A. [WQ66], **694**, 620.
Walton, J. J. [WB67], **521**.
Watt, J. B., *et al.* [WJ62], **307**, 205.
Weber, H. [YB67], **721**, 719.
Wegner, P. [PW68], **740**.
Wegstein, J. H., *et al.* [WG62], **306**, 194.
Weil, R. L., Jr. [WL65], **126**, 91; **306**.
Weinberg, G. M. [WC66], **601**, 548.
Weissman, C. [WE67], **468**, 410; 761.
Weizenbaum, J. [WZ63], **304**, 164; **466**, 387, 388; [WZ67], **468**, 416; **602**, 592, 598; [WZ67a], **720**.
Wells, M. B. [WS61], **312**, 280; [WS63], **312**; 761; [WS64], **312**, 278; [WS64a], **313**.
Wexelblat, R. L., *et al.* [WP67], **704**, 677.
Whiteman, I. R. [WF66], **29**, 20.
Whitmore, A. J. [WH62], **381**, 345.
Wiesen, R. A., *et al.* [WU67], **310**.
Wilkes, M. V. [WI64], **29**, 20; [WI64a], **465**, **468**, 409.
Wilkes, M. V., *et al.* [WI51], **29**, 4.
Willey, E. L., *et al.* [WY61], **381**; [WY61a], **377**, 315.
Winiecki, K. (ed.) [WN66], **311**, 253.
Wirth, N. [WT63], **307**; [WT68], **29**, 9.
Wirth, N., *et al.* [WT66], **307**; [WT66a], **307**, 194; [WT66b], **307**, 194.
Wolman, B. L. [YW66], **705**, 683.
Wood, L. H., *et al.* [WD66], **312**, 258.
Woodger, M. [WO64], **307**.
Woodward, P. M. [WX66], **468**, 409.
Wright, D. L. [WR66], **304**, 152.
Yershov, A. P. [YE66], **307**.
Yngve, V. H. [YN57], **468**, 416; [YN58], **469**; [YN62], **469**, 435; [YN63b], **469**, 420, 421; [YN66], **469**, 423.
Yngve, V. H., *et al.* [YN63], **64**, 58; [YN63a], **64**, 58.
Young, J. W., Jr. [YJ65], **29**, 20.
Young, J. W., Jr., *et al.* [YJ58], **719**, 712.
Yowell, E. C. [YO57], **721**, 717.
Zemanek, H. [ZE66], **64**, 51.

B LANGUAGE SUMMARY

This appendix contains a list of every language specifically listed in the outline, together with the acronym and a very brief description. The chapter (and subsection) in which the language is described is shown and the code numbers of the one or two best references for the language, with the page on which the full citation can be found, are listed. In cases where the best reference is less accessible than a slightly inferior document, both have been given. (It should be obvious that old reports are probably unavailable; they are being listed only for the sake of completeness.)

Languages considered as specialized (and discussed in Chapter IX) have been marked with a †. Languages with a ‡ are of *primarily* historical interest, even though they might still be in use somewhere.

Readers who are interested in contacting specific individuals about a particular language should—with caution—see Sammet [SM68], which is similar (but not identical) to this Appendix. That article contains the names of individuals or organizations to contact for information. Those names have not been included here because they are valid as of the spring of 1968 but not necessarily much beyond that.

A-2 and A-3 ‡
   An early language on UNIVAC for doing mathematical problems. Basically a three-address code.
   IV.2.1.4   [RR55], p. 301.

ADAM (*A DA*ta *M*anagement System) †
   A data-base handling system with facilities for some user defining of source language.
   IX.3.2.4   [CY66], p. 703.

754 LANGUAGE SUMMARY

AED (*A*utomated *E*ngineering *D*esign or *A*LGOL *E*xtended for *D*esign) †
  A generalized language system and a set of concepts; includes AED JR., AED-0, and AED-1. Based on ALGOL and an algorithmic theory of language. Provides techniques for building processors for new languages.
  IX.3.4.2   [RD67a], p. 705; [MT00a], p. 705.

AESOP (*A*n *E*volutionary *S*ystem for *O*n-Line *P*rocessing) †
  An on-line query system, based primarily on the use of a display screen and light pen.
  IX.3.2.8   [QA65], p. 704; [UT67], p. 703.

AIMACO (*AI*r *MA*teriel *CO*mmand Compiler) ‡
  An improvement and modification of FLOW-MATIC. Supplanted by COBOL.
  V.2.2   [AM58], p. 378.

ALGOL (*ALGO*rithmic *L*anguage)
  A language developed jointly by people in the United States and Europe. Suitable for expressing solutions to problems requiring numeric computations and some logical processes. Second version, ALGOL 60 (which had a few revisions in 1962) is the current language. It has no officially defined input/output. Revised ALGOL 60 (with input/output specifications added) has been approved for all practical purposes as an ISO Standard.
  IV.4   [NA63], p. 306.

ALGY ‡
  One of the first attempts at an independent language for doing formal algebraic manipulation.
  VII.2.1   [BM61], p. 520.

ALTRAN
  An extension to FORTRAN to do formal algebraic manipulation of rational functions. Uses ALPAK subroutines.
  VII.5   [ML66], p. 522; [BW66], p. 522.

AMBIT (*A*lgebraic *M*anipulation *B*y *I*dentity *T*ranslation)
  A string manipulation language based on a replacement rule involving a pointer.
  VI.9.1   [CQ64], p. 470; [CQ65], p. 470.

AMTRAN (*A*utomatic *M*athematical *TRAN*slation)
  An on-line keyboard system allowing input and output of equations in a seminatural format, and output of graphical and numerical solutions on a scope or typewriter.
  IV.6.11   [RF66], p. 311; [UV67], p. 311.

Animated Movie †
: A language to assist in preparing animated movies.
: IX.2.6.3   [KO64], p. 699.

APL/360 (*A* *P*rogramming *L*anguage on *360*)
: An on-line version of a subset of APL.
: IV.6.8   [FA67a], p. 310.

APL (*A* *P*rogramming *L*anguage)
: A general but unimplemented language with complex notation and unusual but powerful operations. See also APL/360 and PAT.
: IX.2.3.1   [FA64], p. 694; X.4   [IV62], p. 720.

APT (*A*utomatically *P*rogrammed *T*ools) †
: A language for numerically controlled machine tools. A USASI standard is being developed.
: IX.2.1.1   [BP63], p. 693; [II67], p. 693.

B-∅ ‡
: See FLOW-MATIC.

BACAIC (*B*oeing *A*irplane *C*ompany *A*lgebraic *I*nterpreter *C*oding System) ‡
: One of the early languages for mathematical problems, i.e., a pre-FORTRAN system on the 701.
: IV.2.1.5   [GR55], p. 301.

BASEBALL †
: A question-answering system whose data base contains information about baseball.
: IX.3.2.5   [GB61], p. 703.

BASIC (*B*eginner's *A*ll-Purpose *S*ymbolic *I*nstruction *C*ode)
: A very simple language for use in solving numerical problems developed in an on-line system.
: IV.6.4   [KM67], p. 309.

BUGSYS †
: A language for use in preparing animated movies.
: IX.2.6.3   [LL66], p. 699.

C-10 †
: An improved version of COLINGO.
: IX.3.2.2   [QZ67], p. 702.

CLIP (*C*ompiler *L*anguage for *I*nformation *P*rocessing) † ‡
: A language based on ALGOL 58, useful for writing compilers. JOVIAL is an outgrowth of CLIP.
: IX.2.5.2   [IS59], p. 697; [EG61], p. 697.

CLP (*C*ornell *L*ist *P*rocessor)
: An extension of CORC to do list processing.
: VI.9.3.1   [CN65], p. 470.

756 LANGUAGE SUMMARY

COBOL (*CO*mmon *B*usiness *O*riented *L*anguage)
   An English-like language suitable for business data processing problems. Developed and maintained by a committee of representatives from manufacturers and users. It has been implemented on most computers. A USASI Standard has been approved.
   V.3   [US65], p. 381; [XB67], p. 381.

COGENT (*CO*mpiler and *GEN*eralized *T*ranslator) †
   A compiler-writing language with strong elements of list processing.
   IX.2.5.4   [RE65], p. 697; [RE65a], p. 698.

COGO (*CO*ordinate *GeO*metry) †
   A specialized language for solving coordinate geometry problems in civil engineering.
   IX.2.2.1   [RS64], p. 694; [EI67], p. 694.

COLASL
   A language for numerical mathematical problems, based on use of a special typewriter which permits two-dimensional (i.e., natural) input of mathematical expressions.
   IV.7.2   [BQ62], p. 312.

COLINGO (*C*ompile *O*n *LIN*e and *GO*) †
   A formalized English-like query system for command and control applications.
   IX.3.2.2   [SZ65], p. 702.

COMIT
   The first significant string-handling and pattern-matching language.
   VI.6   [MT61], p. 468; [MT61a], p. 468.

Commercial Translator ‡
   An English-like language for doing business data processing problems. Supplanted by COBOL.
   V.2.3   [IB60a], p. 378.

Computer Compiler †
   Proposed language for describing the design of a computer.
   IX.2.3.5   [MX66], p. 695.

Computer Design †
   An unimplemented ALGOL-like language for describing the design of a computer.
   IX.2.3.6   [CB65], p. 695.

CORAL (*C*lass *O*riented *R*ing *A*ssociated *L*anguage)
   A language on the TX-2 for handling certain ring types of lists.
   VI.9.3.2   [RB65], p. 470.

CORC (*CO*rnell *C*ompiler)
   A simple language for use by students in doing mathematical problems.
   IV.8.1   [CN63], p. 313.

CPS (*C*onversational *P*rogramming *S*ystem)
   An on-line PL/I-like extended subset.
   IV.6.5   [IB67a], p. 309.
Culler-Fried
   An on-line system for doing mathematics, based on the use of a special keyboard for ease in building up arbitrary combinations of operations. Normally known by the names of the developers.
   IV.6.9   [CU67], p. 311.
DAS (*D*igital *A*nalog *S*imulator) †
   A language to provide representations of the components in an analog computer.
   IX.2.4.4   [GL64], p. 696.
DATA-TEXT (Harvard) †
   A language for use by social scientists in doing their numerical computations and analyses.
   IX.2.6.4   [HD67], p. 699.
DEACON (*D*irect *E*nglish *A*ccess and *CON*trol) †
   A query system with fairly natural English input for command and control applications.
   IX.3.2.6   [CJ66], p. 703.
DIALOG
   An on-line system for doing numerical mathematical computations by using a light pen pointing at a screen to create the program.
   IV.6.10   [CS67], p. 311.
DIAMAG
   An on-line extension of ALGOL.
   IV.4.5.6   [AU67], p. 304.
DIMATE (*D*epot *I*nstalled *M*aintenance *A*utomatic *T*est *E*quipment) †
   Contains a language to assist in conducting automatic equipment tests.
   IX.2.6.5   [SD66], p. 699.
DOCUS (*D*isplay *O*riented *C*omputer *U*sage *S*ystem) †
   An on-line system based entirely upon push buttons, with responses shown on the scope.
   IX.3.3.4   [CF66], p. 704.
DSL/90 (*D*igital *S*imulation *L*anguage on 70*90*) †
   An addition to FORTRAN which provides representation of blocks, switching functions, and function generators similar to those available with an analog computer.
   IX.2.4.5   [QP66], p. 696.
DYANA (*DY*namics *ANA*lyzer) † ‡
   One of the early specialized languages. Used for describing vibrational and other dynamics systems.
   IX.2.4.2   [OZ58], p. 696; [TD58], p. 696.

758 LANGUAGE SUMMARY

DYNAMO †
: One of the first continuous simulation languages.
IX.3.1.2   [PG63], p. 699.

DYSAC (*Digitally Simulated Analog Computer*) †
: A language to provide representation of a number of analog computer components.
IX.2.4.3   [HJ63], p. 696.

English
: The concept of using a natural language (e.g., English) as a programming language.
X.5   [HL66], p. 720; [TH66], p. 720.

Extended ALGOL
: A specific set of additions to ALGOL.
IV.4.5.8   [QG66], p. 306.

473L Query †
: A formalized English-like query system for use in the Air Force 473L system.
IX.3.2.3   [BL65], p. 703.

FACT (*Fully Automatic Compiling Technique*) ‡
: An English-like language suitable for business data processing. Largely supplanted by COBOL.
V.2.4   [HO61], p. 379.

FLAP
: A program to do symbolic mathematics.
VII.6   [MB67], p. 522.

FLOW-MATIC ‡
: The first English-like language for doing business data processing; implemented on UNIVAC I. Supplanted by COBOL.
V.2.1   [RR59a], p. 378.

FORMAC (*FORmula MAnipulation Compiler*)
: Originally an extension of FORTRAN to do formal algebraic manipulation. PL/I-FORMAC uses similar concepts and makes many improvements and additions.
VII.3   [BZ64], p. 520; [IB65c], p. 521; [IB67c], p. 521.

Formula ALGOL
: An extension of ALGOL which provides basic operations for doing formal algebraic manipulation, list processing, and string manipulation.
VIII.5   [PR66], p. 602.

FORTRAN (*FOR*mula *TRAN*slator)
   The first language to be used widely for solving numerical problems. Originally developed by IBM on the 704, it has existed in many versions and has been implemented on almost all computers of most manufacturers. The first language to become a USASI Standard. There are actually two standards: Basic FORTRAN and FORTRAN, which correspond approximately to what are known as FORTRAN II and FORTRAN IV, respectively.
   IV.3   [AA66], p. 302; [AA66a], p. 302; [CC64], p. 302.

FORTRANSIT (*FO*rmula *TRANS*lator *I*nternal *T*ranslator)   ‡
   A subset of FORTRAN which was translated into IT on the IBM 650.
   IV.2.2.3   [IB57a], p. 301.

FSL (*F*ormal *S*emantics *L*anguage)   †
   A language for use in defining semantics needed for compiler writing.
   IX.2.5.7   [FJ64], p. 698; [FJ66], p. 698.

GAT (*G*eneralized *A*lgebraic *T*ranslator)   ‡
   An improved version of IT.
   IV.2.2.3   [GM00], p. 301.

GECOM (*GE*neralized *COM*piler)   ‡
   A language with a syntax similar in spirit to that of COBOL, but with some facilities from ALGOL added.
   V.2.5   [GZ61], p. 379.

GPL (*G*eneralized *P*rogramming *L*anguage)
   A relatively new attempt at defining a general language, also containing self-extending capabilities; similar in spirit to ALGOL.
   IV.4.5.7   [GW00], p. 305.

GPSS (*G*eneral *P*urpose *S*ystems *S*imulator)   †
   A language for discrete simulation problems based on a block diagram approach.
   IX.3.1.3   [IB67i], p. 700; [HK65], p. 700.

GRAF (*GR*aphic *A*dditions to *FORTRAN*)   †
   A language which adds a graphic data-type to FORTRAN to facilitate the use of graphics on the computer.
   IX.3.3.1   [HW67], p. 704.

Graphic Language   †
   A specific language for specifying graphic operations.
   IX.3.3.3   [UZ67], p. 704.

ICES (*I*ntegrated *C*ivil *E*ngineering *S*ystem)   †
   A generalized system for civil engineering, including specific languages (e.g., COGO) and some facilities for defining new languages.
   IX.2.2.3   [RS65], p. 694; [RS67a], p. 694.

IDS (*I*ntegrated *D*ata *S*tore)
    An extension to COBOL to permit data to be represented in ringtype-lists.
    V.4.1.1   [QJ64], p. 381.

Information Algebra
    An abstract and theoretical approach to defining data processing; it has not been implemented.
    X.3   [CC62], p. 719.

IPL-V (*I*nformation *P*rocessing *L*anguage *V*)
    The fifth version of a language to do list processing, in which the instructions are conceptually at an assembly language level.
    VI.3   [NW65], p. 466.

IT (*I*nternal *T*ranslator) ‡
    An early language used for mathematical computations on the 650.
    IV.2.2.3   [PR57], p. 302.

JOSS (*J*OHNNIAC *O*pen *S*hop *S*ystem)
    One of the first on-line systems for doing numerical computations.
    IV.6.2   [JC64], p. 308; [UJ67a], p. 308.

JOVIAL (*J*ules' *O*wn *V*ersion of *IAL*)
    A language containing facilities for numerical computations and some data processing. Most widely used for command and control applications.
    VIII.3   [PE66a], p. 599; [SH63], p. 599.

Klerer-May
    A language for numerical mathematical problems, based on the use of a special typewriter which permits two-dimensional (i.e., natural) input of mathematical expressions.
    IV.7.5   [KL65a], p. 313; [KL65b], p. 313.

L^6 (*B*ell *T*elephone *L*aboratories' *L*ow-*L*evel *L*inked *L*ist *L*anguage)
    A list processing language which allows the user to define the types and sizes of his lists.
    VI.4   [KO66], p. 467.

Laning and Zierler ‡
    One of the first systems to allow fairly normal mathematical expressions as input. It ran on the Whirlwind computer at M.I.T. at least as early as 1953.
    IV.2.1.3   [LA54], p. 301.

LDT (*L*ogic *D*esign *T*ranslator) †
    A language for writing logic equations for a computer from the information contained in the systems diagram and instruction repertoire of the machine.
    IX.2.3.3   [PC64], p. 695.

Lincoln Reckoner
: An on-line system on the TX-2 to do mathematical computations with high level matrix operations provided.
: IV.6.7   [SW66], p. 310.

LISP 1.5 (*LIS*t *P*rocessing)
: A sophisticated and theoretically oriented language for doing list processing. LISP 1 and LISP 1.5 differ significantly from LISP 2.
: VI.5   [BY66], p. 467; [WE67], p. 468.

LISP 2
: An ALGOL-like language which includes facilities and many concepts from LISP 1.5.
: VIII.6   [AH66a], p. 602.

LOLITA (*L*anguage for the *On-L*ine *I*nvestigation and *T*ransformation of *A*bstractions)
: An addition to one of the Culler-Fried systems to permit symbol manipulation.
: VI.9.3.4   [QK67], p. 470.

LOTIS (*LO*gic, *TI*ming, *S*equencing) †
: A language for describing a computer by describing the structure and behavior of its data flow.
: IX.2.3.2   [QY64], p. 695.

MAD (*M*ichigan *A*lgorithm *D*ecoder)
: A language for doing numerical computations which has a fast compiler.
: IV.5.2   [UM66], p. 308.

MADCAP
: A language for numerical mathematical problems and set theoretic operations, based on the use of a special typewriter which permits two-dimensional (i.e., natural) input of mathematical expressions.
: IV.7.3   [WS63], p. 312.

Magic Paper
: An on-line system on a specialized computer configuration for doing certain types of formal algebraic manipulations.
: VII.7.1   [CL66], p. 522.

MAP (*M*athematical *A*nalysis Without *P*rogramming)
: An on-line system (under CTSS) for doing numerical computation; it has certain higher level mathematical operations (e.g., integrate) and considerable dialogue with the user.
: IV.6.6   [KP66], p. 309; [KP66a], p. 310.

MATHLAB
: An on-line system for doing certain types of formal algebraic manipulation.
: VII.4   [EN65], p. 521.

762 LANGUAGE SUMMARY

MATH-MATIC (AT-3) ‡
 A language developed on UNIVAC, around the same time as FORTRAN, to do mathematical computations. Supplanted by FORTRAN.
 IV.2.2.1   [RR60], p. 301.

Matrix Compiler † ‡
 An early language to do matrix computations on the UNIVAC.
 IX.2.6.1   [MF57], p. 698.

META 5 †
 A language for syntax-directed compiling.
 IX.2.5.5   [OP66], p. 698.

MILITRAN †
 A discrete simulation language particularly oriented toward military applications.
 IX.3.1.6   [YC64], p. 700.

MIRFAC (*M*athematics *I*n *R*ecognizable *F*orm *A*utomatically *C*ompiled)
 A language for mathematical problems, based on a specialized typewriter to permit two-dimensional (i.e., natural) input of mathematical expressions.
 IV.7.4   [GK63], p. 312; [GK67], p. 312.

NELIAC (*N*avy *E*lectronics *L*aboratory *I*nternational *A*LGOL *C*ompiler)
 A language for doing numerical computation and some logical processes. The compilers are written largely in NELIAC.
 IV.5.1   [HS62], p. 307.

OCAL (*O*n-*L*ine *C*ryptanalytic *A*id *L*anguage) †
 A language for use in doing cryptanalysis.
 IX.2.6.2   [ED66], p. 698.

OMNITAB
 A very simple language containing some operations which are the same as those on a desk calculator and some which are at a high mathematical level (e.g., matrix inversion).
 IV.8.2   [HR66], p. 313.

OPS (*O*n *L*ine *P*rocess *S*ynthesizer) †
 A system under CTSS containing a discrete simulation language, among other facilities.
 IX.3.1.8   [YP65], p. 701; [YP66], p. 701.

PAT (*P*ersonalized *A*rray *T*ranslator)
 A small subset of APL.
 IV.6.8   [HH64], p. 310.

PENCIL (*P*ictorial *ENC*od*I*ng *L*anguage) †
 A language in an on-line system for use with simple data structures to display line drawings.
 IX.3.3.2   [VD67], p. 704.

PL/I
>   A language suitable for doing problems involving both numerical scientific computations and business data processing. It combines the most significant concepts from previous languages in the individual areas.
>   VIII.4   [IB66b], p. 601; [IB67d], p. 601; [IB67f], p. 601.

PRINT (*PR*e-edited *INT*erpreter)   ‡
>   An early language on the 705 for doing mathematical computations.
>   IV.2.1.6   [IB56a], p. 301.

Proposal Writing
>   An extension of FORTRAN to facilitate the preparation of proposals.
>   IV.3.6.1   [CT64], p. 302.

Protosynthex   †
>   A question-answering system whose data base is English text.
>   IX.3.2.7   [SE64], p. 703.

QUIKTRAN
>   An on-line version of FORTRAN with some restrictions, but with added facilities for debugging.
>   IV.6.3   [IB67e], p. 309.

SFD-ALGOL (*S*ystem *F*unction *D*escription—*ALGOL*)   †
>   An extension of ALGOL to permit descriptions of synchronous systems.
>   IX.2.3.7   [PN66], p. 695.

Short Code   ‡
>   Appears to be the first attempt at a higher level language for mathematical problems. Ran on UNIVAC I. Really allows a string of parameters for each operation.
>   IV.2.1.1   [RR52], p. 300.

SIMSCRIPT   †
>   A language for doing discrete simulation problems.
>   IX.3.1.4   [MA63], p. 700.

SIMULA (*SIMU*lation *LA*nguage)   †
>   An extension to ALGOL to do discrete simulation.
>   IX.3.1.7   [DH66], p. 701.

Simulating Digital Systems   †
>   A language with a flavor like that of FORTRAN, for describing the logical design of digital computers.
>   IX.2.3.4   [MZ65], p. 695.

SNOBOL
>   A string-handling and pattern-matching language.
>   VI.7   [FB66], p. 469; [GF68], p. 469.

SOL (*S*imulation *O*riented *L*anguage)   †
>   A language for doing discrete simulation problems.
>   IX.3.1.5   [KN64], p. 700; [KN64a], p. 700.

764 LANGUAGE SUMMARY

Speedcoding ‡
: One of the early attempts at a higher-level language for mathematical problems on the 701. Really allows a string of parameters following an operation code.
IV.2.1.2 [BS54a], p. 300.

SPRINT
: An approach to list processing which involves direct action on an operand stack.
VI.9.3.3 [KC67], p. 470.

STRESS (*STR*uctural *E*ngineering *S*ystems *S*olver) †
: A specialized language useful for solving structural analysis problems in civil engineering.
IX.2.2.2 [FE64], p. 694; [FE65], p. 694.

STROBES (*S*hared *T*ime *R*epair *O*f *B*ig *E*lectronic *S*ystems) †
: A language for communicating with the computer hardware for purposes of testing.
IX.2.6.5 [QU65], p. 699.

Symbolic Mathematical Laboratory
: An on-line system (under CTSS) to do formal algebraic manipulations, based on major use of a display screen and light pen.
VII.7.2 [ZB67], p. 522.

TMG †
: A syntax-directed compiling language.
IX.2.5.3 [MZ65a], p. 697.

TRAC (*T*ext *R*eckoning *A*nd *C*ompiling)
: A string manipulation language involving nested functions and macro facilities.
VI.8 [ME66], p. 469.

TRANDIR (*TRAN*slation *DIR*ector) †
: A syntax-directed compiling language.
IX.2.5.6 [DE64], p. 698.

TREET
: A list processing language which embodies many of the LISP concepts but in an easier notation.
VI.9.2 [HA67], p. 470.

UNCOL (*UN*iversal *C*omputer *O*riented *L*anguage)
: The concept of using a language intermediate between a programming language and machine language to minimize the number of compilers to be written.
X.2 [QR58], p. 719; [QR58a], p. 719.

UNICODE ‡
: A language (similar to MATH-MATIC) on the 1103 to do mathematical problems. Supplanted by FORTRAN.
IV.2.2.2 [RR59], p. 301.

NAME AND SYSTEM INDEX

This index contains names of people, organizations, systems, and languages. Page numbers shown in **bold face** represent a significant discussion, rather than just casual mention. Page numbers in *italics* for the major languages specify the location of the Sample Program for that language. Other *italic* page numbers indicate the existence of a bibliographic citation that is not referenced from the text or is not within an obvious subheading in the Reference Lists. Note that authors are *not* included in this index; they appear in the Author List in Appendix A.

A

A-0, 6, 12, 132
A-1, 6, 132
A-2 and A-3, 5, 6, 129, **132,** 134, 137, 316, 322, 753
A-3 (*see* A-2 and A-3)
Abrahams, P., 590
Abrams, P. S., 247
ACM, 173, 542, 638
   Collected Algorithms, 176
   Communications, 175, 176, 181, 335, 345, 472
   Programming Languages Committee, 176, 180
   SICPLAN Notices, 58, 341, 530, 548, *739*
   SICSAM, 65
ADAM (IBM), 718
ADAM (MITRE), **667–668,** 753

ADAPT, 606
AED, 244, 605, 641, **680–683,** 754
   AED-0, 651, 680–682
   AED-1, 682–683
   AEDJR, 682
Aerospace Industries Association, 606
AESOP, 461, **670–674,** 678, 754
AIMACO, 314, 323, **324,** 331, 754
Air Force, 525, 665
   Materiel Command, 324, 330, 605
ALCOR (*see* ALGOL, subsets)
ALGOL, 22, 39, 45, 48, 53, 56, 58, 75, 82, 84, 92, 103, 104, 134, 143, 144, 152, 153, 154, **172–196,** *178,* 205, 208, 229, 245, 294, 328, 329, 330, 335, 340, 344, 388, 400, 454, 501, 541, 542, 543, 544, 582, 583–589, 592–598, 621, 623, 625, 638, 651, 656–658, 680, 719, 723, 754

(ALGOL)58, 22, 143, **172–175**, 176, 179, 194, 196, 197, 199, 204, 205, 206, 215, 524, 525, 527, 635, 725
(ALGOL)60, **175–177**, *178*, 196, 406, 590, 598, 621, 658, 718
(ALGOL)60, Revised, **177–178**
(ALGOL)68, 178
(ALGOL)6X, **178**
(ALGOL)X, 194 (*see also* ALGOL 6X)
(ALGOL)Y, 194 (*see also* ALGOL 6X)
Bulletin, 174, 175, 177, 178, 180, 181
extensions, **194–196**
metalanguage, **53–56** (*see also* BNF; *see also* Metalanguage, ALGOL in Subject Index)
proposed ISO Standard, **178–192**
subsets:
    ALCOR, 175, 180
    ECMA, 180
    IFIP, 180
    SMALGOL, 180
ALGY, 472, **473–474**, 754
Allen-Babcock Corporation, 232
ALPAK, 502, 504, 505, 506
ALTAC, 42, 146, 153
ALTRAN, 472, **502–506**, *502* 636, 754
AMBIT, 387, **454–457**, *455,* 754
AMTRAN, 216, 217, 240, 245, 254, **258–264**, *259,* 754
AN/FSQ-7, 525, 526, 527
AN/FSQ-31, 525, 526, 527
AN/FSQ-32, 389, 406, 408, 491, 492, 525, 527, 591
Animated Movie language, **644–646**, 754
APEX, 245
APL, **620**, 621, 707, **712–715**, 725, 727, 754 (*see also* APL/360)
APL/360, 216, 217, **247–252**, *248,* 712, 754
APL (PL/I based), *465, 600*
APT, 7, 21, 33, **605–610**, 683, 754
Arden, B., 142, 205
ARITH-MATIC (*see* A-3)
Army Electronic Proving Ground, 198
ARPA, 590
ASA, 8, 46 (*see also* USASI)
Asch, A., 324
ASCII, 44
Ash, A., 135

ASP, *702*
AT-3 (*see* MATH-MATIC)
ATS, 684 (*see also* DATATEXT (IBM))
Auslander, M., 475
Autocoder, 6, 7, 11, 42, 717
AUTO-PROMPT, 606
AUTOSPOT, 606

B

B-∅, 5, 322 (*see also* FLOW-MATIC)
Babcock, J. D., 232
BACAIC, 5, 6, 7, 129, **133–134**, 755
Bachman, C., 376
Backus, J. W., 130, 143, 144, 174, 175, 176, 177 (*see also* BNF)
BALGOL, 174, 621
Barnett, J. A., 590
Barnett, M., 635
BASEBALL, **668**, 669, 755
BASIC, 216, **229–232**, *230,* 755
Basic FORTRAN, (*see* FORTRAN, (Standard) Basic)
Bauer, F. L., 174, 176, 177
Beeber, R. J., 143
Bell Telephone Laboratories, 244, 400, 436, 502, 542
BEMA, 47, 154, 345, 546
Bendix, 332 (*see also* CDC computers, G20)
Bennett, J., 417
Berg, H., 540
Best, S., 143, 417
BINAC, 129
BIOR, 6, 8
Blackwell, F., 464
BNF, **53–55, 175,** 408, 606, 638, 639
Bobrow, D. G., 590
Boeing Airplane Company, 133
Bond, E. R., 475
Book, E., 529, 590
Bosak, R., 709, 712
Bosche, C., 417
Bottenbruch, H., 174
Brackett, J., 240
Brayton, R., 406
Breed, L. M., 247
BRIDGE, 616
British Computer Society, 176
Broadwin, E., 135
Bromberg, H., 331, 340

NAME AND SYSTEM INDEX **767**

Brown, L., 540
Brown, W. S., 502
BUGSYS, **644–646,** 755
Burroughs, 330
Burroughs computers:
    Burroughs 220, 174, 198, 294, 389
    Burroughs 5000, 718
    Burroughs 5000/5500, 611, 656
    Burroughs 5500, 196, 261, *306,* 612
    Burroughs D825, 197, 527, 622

C

C-10 (MITRE), **664–665,** 755
C-10 (UNIVAC), 137
CADET, 683
CAL, 217
California, University of:
    Berkeley, 205
    Los Angeles, 253
    Santa Barbara, 253
Carnegie-Mellon University, 389, 583
Case Institute of Technology, 139
CDC, 332
CDC computers:
    CDC 160A, 198, 611
    CDC 1604, 198, 294, 389, 525, 526, 527, 611, 612, 622, 629, 678
    CDC 3100, 197, 436
    CDC 3200, 447
    CDC 3400, 612
    CDC 3600, 197, 393, 436, 525, 526, 527, 611, 612, 655, 718
    CDC 3800, 197, 655
    CDC 6400, 655
    CDC 6600, 655
    CDC 6800, 655
    CDC G20, 389, 583
CDL, **618–620**
CHARYBDIS, *522*
Chicago, University of, 417
Chipps, J., 139
CIB (*see* COBOL, Information Bulletin)
CITRAN, 217
Clapp, L., 511
Clem, P. L., Jr., 258
CLIP, 174, 196, 197, 215, 524, 529, 634, **635–636,** 755
Clippinger, R., 327, 712
CLP, 387, **461–462,** 755

COBOL, 20, 21, 23, 33, 35, 43, 46, 47, 53, 56, 58, 73, 78, 80, 82, 84, 85, 97, 106, 119, 121, 152, 153, 180, 193, 314, 324, 325, 326, 327, 328, 329, **330–376,** *336–337,* 536, 541, 542, 543, 544, 582, 634, 717, 723, 725, 756
(COBOL) 60, 324, 332, 333, 339
(COBOL) 61, 328, **332–333,** 344
(COBOL) 61 Extended, 328, **333**
(COBOL) 61 Required, 339
(COBOL) 65, **333–334, 345–375**
(COBOL) standard, **340–343** (*see also* X3.4.4)
Basic, 339
Compact, 339
Extended COBOL 61, 328, **333**
Information Bulletin, 333, 340, 341, 345
Maintenance Committee, **332–334,** 340
metalanguage, **53–56** (*see also* Metalanguage, COBOL, in Subject Index)
Short Range Committee, 323, 324, 325, 327, **330–333,** 339
CODAP, 678
CODASYL, 330, 340 (*see also* COBOL; Information Algebra)
    Executive Committee, 330, 331, 344
    Intermediate Range Committee, 330
    Language Structure Group (*see* LSG; *see also* Information Algebra)
    Long Range Committee, 330
    LSG, 709, 712, 719
    Short Range Committee (*see* COBOL, Short Range Committee)
COGENT, **638–639,** 756
COGO, 21, 247, **611–613,** 616, 618, 619, 756
COLASL, 79, **265–271,** *268,* 281, 756
COLINGO, **664–665,** 668, 756
COMIT, 56, 58, 68, 386, **416–436,** *418,* 438, 439, 447, 449, 453, 454, 455, 756
(COMIT) II, 417, *418,* 419–420, 423
compared with SNOBOL, **386–387**
COMMEN, *127*
Commercial Translator, 314, **324–326,** 327, 334, 375, 756
COMPASS (*see* Massachusetts Computer Associates)
COMPOOL, 526, 533, 536, 539

COMPROSL, 299–300
Computer Compiler, **623–624,** 756
Computer Design language, **623, 625,** 756
Computer Research Corp., 511
Computer Sciences Corp., 146, 327
COMTRAN, 324, 331 (*see also* Commercial Translator)
CONVERT, 388, 416, 435
Cooper, W., 417
CORAL, 387, **462–463, 674,** 680, 756
CORC, **294–296,** 461, 756
CORREGATE, 139
COSMOS, 283
Cousins, L., 316
Cox, J., 540
CPS, 58, 216, **232–240,** *233,* 689, 718, 756
CS, 132
CTSS, 216, 240, 244, 406, 408, 417, 419, 438, 475, 491, 611, 612, 660, **683,** 686, 688, **689–690,** *740*
Culler, G., 253
Culler-Fried system, 216, 217, 240, 245, **253–255,** 258, 464, 757

D

DAS, **631,** 757
Datamation, 42
DATA-TEXT (Harvard), **646–647,** 757
DATATEXT (IBM), 646, **684–686**
David Taylor Model Basin, 330
DEACON, **668–669,** 757
DEC computers:
    PDP-1, 406, 449, 510, 511
    PDP-5, 449
    PDP-6, 217, 223, 400, 498, 516
    PDP-8 and 8S, 449
Delaney, F., 132, 316
Della Valle, V., 135
DEMON, *379*
Department of Defense, 330
DesJardins, P. R., 232
DETAB 65, 315, 335
DETAB X, 315
Deutsch, P., 449
DIALOG, 216, 217, **255–258,** 674, 757
DIALOG (retrieval), *702*
DIAMAG, 195, **689–691,** 757
Dillon, G. M., 332

DIMATE, **647–649,** 757
Discount, N., 331
DITRAN, *303*
DM-1, *701*
Dobbs, C., 712
DOCUS, **678–679,** 757
DOL, 678, 679
DSL/90, 172, **632–633,** 757
Duncan, F., 175
Dunn, T., 226
DUO, 173
DuPont, 332
DYANA, **628,** 757
DYNAMO, **651–653,** 683, 758
DYSAC, **629–631,** 758
DYSTAL, 303, 387–388

E

Earley, J., 583
Eastwood, D., 450
ECMA, 156, 333, 340, 344
    ALGOL subset, 180
EDSAC, 4
Edwards, D., 406
Ellis, T. O., 217
Elmore, 4
Engelman, C., 491, 498
English, 52, 707, **715–717, 729–731,** 758
    (*see also* English-like; Natural English; Natural language; Query, all in Subject Index)
EPL, 542
ES-1, 684–685
ESI, 217
EULER, 194, 719
Evey, R. J., 475
Extended ALGOL, **196,** 758

F

473L Query, **665–667,** 669, 758
FACT, 314, 325, **327–328,** 334, 758
Falkoff, A. D., 247
FAP, 6, 244, 502
Farber, D. J., 436
FASE, *720*
FAST, 526
Feigenbaum, E. A., 391
Feldman, J., 641
Firth, D., 590

NAME AND SYSTEM INDEX

FLAP, 472, **506–510,** *507,* 758
FLOW-MATIC, 6, 314, **316–324,** 331, 332, 375, 758
FLPL, 388, 406
FMS, 148
Forgie, J., 245
FORMAC, 33, 105, 144, 170, 171, 195, 388, 472, **473–491,** *477,* 503, 758
 FORTRAN extension, **474–486**
 PL/I-FORMAC, 474–475, *477,* **486–490,** 554
Formula ALGOL, 194, 435, 473, 524, **583–589,** *583,* 641, 758
FORTRAN, 6, 14, 20, 21, 23, 33, 38, 45, 46, 47, 52, 58, 60, 73, 81, 100, 103, 114, 123, 132, 134, 135, 137, 138, 139, 141, **143–171,** *151,* 176, 180, 190, 194, 206, 208, 211, 216, 226, 229, 245, 269, 271, 294, 340, 344, 387, 388, 400, 472, 473, 474, 475, 476, 478–480, 498, 503, 525, 540, 542, 543, 544, 547, 582, 610, 621, 622, 628, 632, 634, 651, 657, 674, 675, 678, 687, 688, 717, 718, 725, 731, 732, 759
 (FORTRAN)I, **143–145,** 146, 164
 (FORTRAN)II, 42, 144, **146,** 147, 148, 150, 152, 153–154, 155, 156, 164, 170, 260, 503, 720
 (FORTRAN)III, 88, 147
 (FORTRAN)IV, 42, 146, 147, 148, 150, 152, 153–154, 155, 157, 164, 261, 475, 476, 478, 503, 540
 (FORTRAN)V, 541
 (FORTRAN)VI, 540
 (Standard) Basic FORTRAN, **150–165, 168–169,** 226, 228, 639 (*see also* X3.4.3)
 (Standard) FORTRAN, **150–157, 165–169** (*see also* X3.4.3)
 differences in standards, **168–169**
 extensions, **170–172**
FORTRANSIT, 7, **141–142,** 144, 150, 759
Fox, P., 406
Franciotti, R., 177
Franklin Institute, 5
Fried, B., 253 (*see also* Culler-Fried system)
FRINGE, 329
FSL, 583, **641–642,** 759

G

Galler, B., 205
GAMM, 173–174
GARGOYLE, *697*
GAT, 7, **142–143,** 206, 759
GATE, 139
GECOM, 314, **328–329,** 759
General Electric, 332, 376, 606, 668
General Electric computers:
 GE 225, 229, 284, 328, *379,* 419
 GE 235, 284
 GE 635, 231, 527
 GE 645, 244, 651, 661
 GE Datanet-30, 449
General Inquirer, 435
General Motors Laboratories, 628, 680
GIM-1, *702*
GIS, *702*
GOL, 678, 679
Goldberg, R., 143
Goldfinger, R., 324, 712
Gordon, G., 653
Gorn, S., 5
GPL, **195–196,** 759
GPM, 450
GPS, *466*
GPSS, 650, 651, **653–655,** 656, 759
GRAF, 172, **674–675,** 759
Graham, R., 142, 205
Graphic language, **677–678,** 759
Greber, M., 417
Green, B. F., Jr., 391
Green, J., 176, 177
Greenberger, M., 660
Greene, M., 135
Grisoff, S. F., 475
Griswold, R. E., 436
GUIDE, 541

H

Hansen, F., 417
Harper, M. H., 132, 316
Haines, E. C., 457
Halstead, M., 197, 198
Harvard University, 253, 646
Hawes, M., 316
Hawkinson, L., 590
Helwig, F., 417
Herrick, H. L., 130, 143

770 NAME AND SYSTEM INDEX

Hodes, L., 406
Holberton, F., 4
Honeywell, 327, 330, 334
Honeywell-800 business compiler (*see* FACT)
Honeywell-800 computer, 327, 379
Hopper, G. M., 4, 12, 132, 135, 144, 316
Hudson Laboratories, Columbia University, 258 (*see also* Klerer-May system)
Hughes, R. A., 143
Huskey, H., 197

I

IAL (*see* ALGOL 58)
IBM, 52, 134, 143, 152, 156, 247, 330, 334, 540–542, 546, 547, 606, 646, 684
  Hursley Laboratories, 52, 542, 548
  Vienna Laboratory, 52, 548
IBM computers:
  IBM 650, 7, 139, 141, 142, 146, 149, *301, 302*, 389
  IBM 701, 5, 6, *26*, 129, 130, 133, *300*
  IBM 704, 3, 6, *28*, 88, *126*, 137, 142, 143, 144, 145, 147, 149, 153, 170, 198, 205, 206, *302, 303, 307, 377*, 389, 406, 416, 606, 628, 651
  IBM 705, 5, 6, 129, 134, 137, 149, *301*, 324
  IBM 709, 42, 146, 147, 198, 204, *307*, 325, 406, 419, 524, 525, 636, 651
  IBM 709/90, 42, 147, 149, *303*, 378, 389, 417, 536
  IBM 1130, 261, 611, 612
  IBM 1401, 148, *307*, 664
  IBM 1410, 611, 665
  IBM 1620, *63*, 146, 149, 253, 258, *303*, 389, 406, 436, *468*, 611, 612, *696*, 718
  IBM 1800, 28
  IBM 7030, 146, 265, 267, 270, 458, *470*, 491, 667, 670
  IBM 7040, 207, 400, 462, 475, 677
  IBM 7044, 436
  IBM 7040/44, 226, 417, 419, 475, 502, 611, 654
  IBM 7070, 7, *28*, 146, 149, 198, 325, 611
  IBM 7080, 325

IBM computers (*cont.*):
  IBM 7090, 198, *307, 313*, 325, *377*, 475, 476, 506, 525, 526, 527, 606, 620, 631, 644, 651, 654
  IBM 7090/94, 63, 148, 205, 296, *303*, 326, 417, 419, 436, 445, 475, 502, 611, 612, 632, 646, 654, 655, 657
  IBM 7094, 197, 255, 400, 462, 516, 527, 542, 623, 645, 651, 683
  IBM 9020, 525, 526, 527
  IBM STRETCH (*see* IBM 7030)
  IBM System/360, 28, 29, *63, 126*, 197, 247, *303, 305*, 325, *377*, 400, 417, 419, 436, 447, 448, 449, 458, 476, 486, 526, 527, 542, *601*, 611, 615, 620, 651, 654, 655, 675, 683, 691, *693, 694, 700, 706*, 718, *719*
IBSYS, *63, 303*, 475, 476, 478
IBSYS/IBJOB, 148, 475
ICES, 605, 613, **615–620**, 759
ICETRAN, **617–618**
IDS, **376**, 760
IFIP, 175
  ALGOL subset, 180
  TC 2, 177
  WG 2.1, **177**, 178, 181, 194
IIT, 255, 606
Information Algebra, 707, **709–712**, 726, 727, *736*, 760
Information International, Inc., 590
Ingerman, P. Z., 177
IPL, 386
  IPL-I, 389
  IPL-II, 389
  IPL-III, 389
  IPL-IV, 389, 393
  IPL-V, 58, 386, **388–400**, *392*, 405, 406, 449, 462, 668, 718, 760
  IPL-VI, 389
  IPL-VC, 394, 718
IPSSB, **48**
ISIS, 217
ISO, 47, 48, 340, 341
IT, 5, 7, **139–143**, 760
Iturriaga, R., 583
Iverson, K., 217, 247, 712, 725, 727
Iverson's language (*see* APL; APL/360)

J

Jasper, R. B., 712

NAME AND SYSTEM INDEX 771

JCL, **691–693**
Jenny, A., 135
JOHNNIAC, 7, 217, 389
Johnson, R., 197
Jones, J. L., 324, 332
Jones, T., 316
JOSS, 216, **217–226,** *218,* 232, 258, 261, 689, 760
  (JOSS) I, **217–223**
  (JOSS) II, 217, **223–226,** *218*
JOVIAL, 33, 58, 174, 179, 194, 196, 197, 198, 205, 215, 523, **524–539,** *528, 541,* 542, 543, 634, 636, 639, 709, 760
  (JOVIAL) 1, 525, 536, 539
  (JOVIAL) 2, 525, 536
  (JOVIAL) 3, 525, 529, **530–539**
  Basic (JOVIAL) 529
  JS, 639
  JTS, 528, 639
  TINT, 528
JTS, 528, 639
JUG, *740*

K

Kahrimanian, H. G., 5
Kameny, S. L., 590
Kannel, M., 417
Kaplow, R., 240
Kapps, C. A., 462
Katz, C., 135, 174, 176, 177, 328, 329
Keating, W., 712
Keller, J., 226
Kemeny, J., 229
Kendrick, G. 712
Kenney, R., 475
Klerer, M., 284
Klerer-May system, 258, 265, **284–294,** *285,* 760
Knowlton, K. C., 400, 417, 644
Kogan, R., 177
Koschman, M., 139
Koss, M., 132
Krutar, R., 583
Kurtz, T., 229

L

L^6, 386, 389, **400–405,** *401,* 462, 645, 760
Landen, W., 197
Landin, P., 177

Laning and Zierler system, 5, 129, **131–132,** 760
LAP, 597
Larner, R., 540
Lathwell, R. H., 247
LDT, **621–622,** 760
Leagues, D. C., 502
LECOM, 419
Levin, M. I., 590
Lincoln Laboratory, 245, 462
Lincoln Reckoner, 216, 217, 240, **245–247,** 254, 258, 761
LIPL, 394
LISP, 60, 82, 96, 120, 386, **405–416,** *407,* 435, 449, 453, 457–461, 472, 491, 492, 494, 498, 506, 509, **589–598,** *591,* 725, 727, 761
  (LISP) 1, 405–406, 725
  (LISP) 1.5, 8, 82, 223, **405–416,** *407,* 506, 516, 589, 592, 598, 725
  (LISP) 1.75, 406
  (LISP) 2, 39, 195, 406, 410, 416, 458, 524, **589–598,** *591,* 725, 761
Logan, R., 129
Logic Theorist, 393
Logic Theory Machine, *466, 467*
LOLITA, 387, **464,** 761
Los Alamos Scientific Laboratory, 265, 271
LOTIS, **620–621,** 761
LUCID, *701*
Luckham, D., 406

M

MAC-360, 81, **264**
MAD, 142, 154, 174, 179, 194, 196, **205–215,** *207,* 216, 244, 644, 683, 688, 761
MADCAP, 265, **271–281,** *272*
MADTRAN, 154, 206
Magic Paper, 473, **510–514,** 761
Maling, K., 406
MANIAC II, 271
MAP, 216, 217, **240–245,** *244,* 254, 258, 733, 761
Martin, T., 540
Martin, W. A., 514
Massachusetts Computer Associates, 454, 542, 640
Masterson, K. S., Jr., 197

NAME AND SYSTEM INDEX

MATHLAB, 472, 473, **491–502**, *492*, 761
(MATHLAB) 68, 491, **498–501**
MATH-MATIC, 6, 88, 132, **135–137**, 139, 316, 322, 762
Matrix Compiler, 5, 6, **642**, 762
Matthews, G., 417
Mauchly, J., 129, 130
May, J., 284 (*see also* Klerer-May system)
McArthur, R., 197
McCarthy, J., 176, 405, 406, 590
McClure, R., 636
McGarvey, J. E., 132
McIlroy, M. D., 436, 450, 502, 541
Mealy, G. H., 391
Medlock, C. W., 540
META 2, 592
META 3, *698*
META 5, **638–640**, 762
Michigan, University of, 139, 205, 206
MIDAS, 627
MILITRAN, **657, 659–660**, 762
Miller, C., 611, 615
Minneapolis-Honeywell (*see* Honeywell)
Minsky, M., 590
MIRFAC, **281–284**, *282*, 762
M. I. T., 2, 129, 206, 240, 244, 405, 409, 417, 514, 542, 590, 605–606, 611, 612, 615–616, 651, 660, 661, 680, 683, 688 (*see also* CTSS; MULTICS)
Mitchell, L. B., 143
MITRE, 491, 664, 667, 670, 685
ML/I, *127*
MOBIDIC, 198, 400
Mooers, C., 448
Moore, R. D., 247
Morris, A. H., Jr., 506, 510
Morrissey, J., 226
MPL, 542
MPPL, 542
Mulder, M., 316
MULTICS, 244, 542, 634, 661
MULTILANG, 677
Myszewski, M., 475

N

9 PAC, 314
NAPSS, 299
NASA, 258
National Bureau of Standards, 296, 330
Naur, P., 175, 176, 177
Naval Research, Office of, 4, 5
Navy Electronics Laboratory, 197, 200
NCR, 332, 717
NCR 304 computer, 717
Nehama, I., 217
Nekora, M., 684
NELIAC, 58, 174, 179, 194, 196, **197–205**, *199*, 762
(NELIAC) BC, 197
Nelson, R. A., 143
Newell, A., 217, 388, 391
NICOL I, 542
NICOL 2, *600*
North Carolina, University of, 143
NPL, 542
Nutt, R., 143, 327

O

OCAL, **642–644**, 762
OCAS, 642
OLC, 253
OMNICODE, 5
OMNITAB, **296–299**, *297*, 762
ONR, 4, 5
OPS, **660–662**, 762
Orgel, S., 139
OS/360, 691 (*see also* JCL)

P

Packard Bell 250 computer, 198
PACT, 6, 8, 173
PACTOLUS, 627
Park, D., 406
PAT, 217, **252–253**, 762
Paul, M., 177
PDP, (*see* DEC computers)
PENCIL, **675–677**, 762
Pennsylvania, University of, 330
Perlis, A. J., 139, 174, 176, 583
Philco, 332
Philco computers:
 Philco 210–211, 207, 655
 Philco 2000, 146, 389, *466*, 473, 525
 Philco 2000/210, 526, 527
 Philco Basicpac, 198
 Philco CXPQ, 198

NAME AND SYSTEM INDEX **773**

Phillips, C. A., 330
PIL, 678, 679
PIL/I, 217
PL360, 9
PLANIT, *706*
PL/I, 20, 33, 37, 45, 52, 60, 73, 74, 81, 84, 100, 102, 103, 106, 119, 121, 125, 157, 179, 217, 232, 236, 472, 476, 486–487, 489–490, 523, 524, 530, 539, **540–582,** *544,* 610, 634, 639, 640, 644, 724, 725, 763 (*see also* CPS)
PL/I-FORMAC (*see* FORMAC)
Polonsky, I. P., 436
Porter, C. B., 197
Porter, S. W., 197
POSE, **299**
PRINT, 5, 6, 129, **134,** 763
PROJECT, 616
Proposal Writing, **170–172,** 763
Protosynthex, **669–670,** 763
PUFFT, *303*
Pulos, J., 130
Purdue University, 139

Q

Quarles, D., 130
QUIKTRAN, 12, 84, 170, 172, 216, **226–229,** *227,* 232, 611, **689,** 733, 763
QUIN, 691

R

Radin, G., 540
Ramo Wooldridge computers:
 RW 400, 254
 AN/UYK, 198
RAND Corporation, *61,* 217, 223, 389, 393, 655, *740*
Rapidwrite, 35, 338
RCA, 330, 332
RCA computers:
 RCA 501, 153, 332
 RCA 601/604, 436
RECOL, *701*
Reeves, V., **331**
Reinfelds, J., 258
Remington Rand, 12, 132, 135, 137, 144, 322, 323, 330, 332, 334 (*see also* UNIVAC (computers)

Rempel, R., 197
Report Program Generator, 11, 28
Revised ALGOL 60 (*see* ALGOL 60, Revised)
Ridgeway, R. K., 132
ROADS, 616
Roberts, L. G., 462
Rochester, N., 232, 486
Rockford Research Institute, 448, 453
Rogoway, H. P., 540
Rosenblatt, B., 540
Ross, D., 680
Rossheim, R., 316
RPG, 11, *28*
RPL, *706*
RUNCIBLE, 7, 139
RUSH, 217, 232, *309*
Russell, S., 406
Rutishauser, H., 5, 129, 174, 176

S

SACCS, 524
SAD SAM, 669
SAFARI, *470,* 685, *705*
SAINT, 410
Samelson, K., 174, 176, 177
Sammet, J. E., 330, 331, 474, 475, 712
SAP, 3, 6, 144
Saunders, R. A., 590
Sayre, D., 143
SC5, 48 (*see also* ISO; TC97)
Schmitt, W., 129
Schroeder, D., 232
Schwartz, J., 525, 529
SDC, 491, 524, 525, 528, 529, 530, 590, 634, 688 (*see also* AN/FSQ-7, 31, 32)
SDS 940 computer, 400, 406, 436, 467
Seegmüller, G., 177
Seitz, R. N., 258
Selden, W., 331
SFD-ALGOL, 195, **623, 625–627,** 763
SHADOW, 448, 605, *739*
SHARE, 173, 174, 206, 406, 417, 541, 547, 712, *740*
 Advanced Language Development Committee, 540–542, 548
 FORTRAN Committee, 147, 148, 540
Shaw, C. J., 58, 529
Shaw, J. C., 217, 389

Sheldon, J., 130
Sheppard, R. C., 541
Sheridan, P. B., 143
SHORT CODE, 5, 6, **129–130**, 763
Siegel, A., 417
Siegel, L., 130
SIFT, **42,** 150, **153–154,** 164
SIMSCRIPT, 33, 462, 650, 651, **655–656,** 661, 763
SIMULA, 195, 651, **657, 659–661,** 763
Simulating Digital Systems, **622–623,** 763
Sketchpad, 462, 678
Skillman, S., 130
SLANG, *697*
SLIP, *304,* 387–388, 644
SMALGOL (*see* ALGOL, subsets)
Smith, J. W., 139
SNOBOL, 68, 386, 421, 435, **436–448,** *437,* 454, 455, 644, 763
  SNOBOL3, **436–447,** *437*
  SNOBOL4, 436, **447–448**
  compared with COMIT, **386–387**
SOAP, 7, 141
SOCRATIC, *702*
SODAS, *741*
SOL, **656–658,** 661, 763
Somers, E., 316
Sparks, M. A., 258
Speedcoding, 5, 6, 129, **130–131,** 764
Sperry Rand (*see* Remington Rand; UNIVAC (computers))
SPLINTER, *600*
Springer-Verlag, 174
SPRINT, 387, **462–463,** 764
Standish, T., 583
Stanford University, 409, 542, 602
Stelloh, R. T., 197
Stern, H., 143
Stoller, G. S., 502
Stowe, A. N., 245
STRESS, **612–615,** 764
STROBES, **647–648,** 764
Strong, S., 240
STRUDL, 613, 616, 620 (*see also* STRESS)
Strum, E., 226
STUDENT, 664
Sullivan, D., 316
SURGE, 8, 314
Sutherland, I., 462, 678

Sylvania, 330
Sylvania computers (*see* MOBIDIC)
Symbolic Mathematical Laboratory, 473, **514–520,** *515,* 764

T

TABSOL, *28,* 329
TALL, *466*
TC97, 48 (*see also* ISO)
TELCOMP, 217, *308*
TELSIM, 627
TGS-II, 640
Tierney, G., 331
TINT, 528
TIPL, 393
TMG, 503, **636,** 764
Tobey, R. G., 474, 475
Tonge, F. M., 391
Tonik, A., 129
TRAC, 387, **448–454,** *449,* 764
TRANDIR, **640–641,** 764
TRANGEN, 640
TRANSIT, 616
TREET, 387, **457–461,** *458,* 764
TRW Systems, 253
Turanski, W., 176
TX-2, 462

U

Uncapher, K. W., 217
UNCOL, 539, 707, **708–709,** 764
UNICODE, 6, **137–138,** 764
UNIVAC (computers), 4, 5, 6, 129, 130, 132, 137, 139, *300, 301,* 322, 324, 334, 471, 642
  UNIVAC II, 6, 153, *301,* 324, 332, 334, 698
  UNIVAC 1103, *301*
  UNIVAC 1103A, 6, 137
  UNIVAC 1105, 137, 143, 255, *301,* 324, 389
  UNIVAC 1107, 207, 389, 527, 611, 612, 654, 655, 656, 657
  UNIVAC 1107/1108, 197, *307*
  UNIVAC 1108, 542, 655, 657, 683
  UNIVAC LARC, 7, 146
  UNIVAC M-460, 198
  UNIVAC M-490, 198, 655
  UNIVAC SS80, 146, 198

USASI, 8, **46–48,** 154, 339, 340 (*see also* X3; X3.4; X3.4.1; X3.4.2; etc.)
USE, 173
U. S. Steel, 332
Utman, R. E., 177

V

van der Poel, W., 177
van Wijngaarden, A., 176, 177
Vauquois, B., 176
VITAL, 641

W

Waite, 4
Washington State University, 419
WATFOR, *303*
Watt, J. B., 197
Wattenburg, W., 197
Wegstein, J., 174, 176, 177, 330
Weinstein, M., 417
Weissman, C., 590
Weitzenhoffer, B., 540
Wells, M. B., 271
Westinghouse, 339
Whirlwind, M.I.T., 2, 5, 7, 129, 131–132, *301,* 605
Wiesen, R. A., 245

Wood, L. H., 258
Woodger, M., 176, 177, 178

X

X3, **47,** 48, 341, 547
X3.4, **47–48,** 154, 180, 546–547
X3.4.1, **47**
X3.4.2, **47,** 180
X3.4.2C, 547
X3.4.3, **47,** 150, **154–156,** 340
X3.4.4, **47–48,** 333, **340–341,** 345
X3.4.5, **48**
X3.4.6, **48**
X3.4.7, **48,** 606
X3.4.8, **49,** 180
XPOP, 8

Y

Yang, G., 226
Yngve, V., 417, 420
Yntema, D. B., 245
Yu, L., 135

Z

Ziller, I., 143, 147
Zilles, S., 475
Zoeren, H. V., 139

SUBJECT INDEX

Page numbers shown in **bold face** represent significant discussion of the indicated subject, rather than just casual mention. Page numbers in *italics* indicate the presence of a bibliographic citation pertaining to this subject. This was done in only the few cases where the reference otherwise might be overlooked. The page references for all language names, system names, people, and organization names are given in the Name and System Index.

A

Addresses:
  regional, 2
  relative, 2
  symbolic, 3
Alphameric data (*see* Data types, alphanumeric)
Alphanumeric data (*see* Data types, alphanumeric)
Ambiguity, 38, 52
Application (*see also* individual applications listed by name):
  area, 21, 23, 24, 25, 31, **33–34,** 39, 45, 730, **733–734,** 736
  oriented languages, 19, **21, Chapter IX**
  package, **13–14**
Arithmetic:
  types of:
    complex, **99**
    double precision, **99**

Arithmetic (*cont.*):
    fixed point, **98**
    floating point, **98**
    integer, **98**
    logical, **99–100**
    mixed mode, **100** (*see also* Mode)
    mixed number, **98**
    multiple precision, **99**
    rational, **98–99,** *479*
    precision, 102 (*see also* types of, double; types of, multiple)
Arrays, 74, 95, 97, 100, 118, 124
Artificial intelligence, 383, 385, 389, 405, 409, *740*
ASCII, 70
Assembly language, 1, 2, 8, 14, 15, 16, 17, 18, 21, 23, 32, 42, 81, 86, 114, 120, 386, 389, 391, 604, 731
  symbolic, 2, 3, 5, 8, 11, 730 (*see also* Symbolic assembly program)

776

SUBJECT INDEX **777**

Assembly program (*see* Assembly language)
Assembly program using Information Algebra, 712
Atom (*see* descriptions of LISP 1.5 and LISP 2)
Automatic coding, 3, 5, **13**, 524, 717
Automatic programming, 4, **13**
  systems, 5–7

B

Backus–Naur form (*see BNF*)
Backus–Normal form (*see* BNF)
Backward pointers, 385
Blanks, 22, 70, 73, 77, **79** (*see also* individual language descriptions for specific rules)
Block, 84, 103 (*see also* Statements, compound; *see also* descriptions of ALGOL and PL/I)
BNF, **53–55, 175**, 408, 606, 638, 639
Bootstrapping, 40–41, 121–122
Bugs (*see* Debugging; *see also* description of L⁶)
Built-in names, 74 (*see also* description of PL/I)
Bundle (*see* description of Information Algebra)
Business data processing languages, 21, 33, **Chapter V**

C

Call by location (*see* Parameters, call by location)
Call by name (*see* Parameters, call by name)
Call by simple name (*see* Parameters, call by simple name)
Call by value (*see* Parameters, call by value)
Cambridge Polish notation, 416
CAR (*see* descriptions of LISP 1.5 and LISP 2)
CDR (*see* descriptions of LISP 1.5 and LISP 2)
Character(s), 22, 23, 72
  escape, **69**
  number of, 16
  punctuation, 70, 71, 72, 77, **78–79**, 82
  sets, 68, **69–70**, 264

Character(s) (*cont.*):
  ASCII, 70
  commercial, **69**
  sets for specific languages:
    ALGOL, **183**
    AMTRAN, **259**
    APL, **248**
    COBOL, **346**
    COLASL, **266**
    COMIT, **421**
    CORC, **294**
    CPS, **233**
    DIALOG, **256**
    Formula ALGOL, **584**
    FORTRAN, **69, 157**
    IT, **139**
    JOSS, **219**
    JOVIAL, **530**
    Klerer–May System, **285**
    LISP 1.5, **410**
    LISP 2, **592**
    MAD, **207**
    MADCAP, **273**
    Magic Paper, **511–512**
    MATHLAB, **493, 498**
    NELIAC, **200**
    PL/I, 70, **549**
    SNOBOL, **439**
    TREET, **458**
Civil engineering, 21, 603
  languages for, **610–620**
Closed shop, 24, 148
Code generation, 4, 322
Collating sequence, 37, 38, 338
Command and control, 27, 523, 525, 526, 530
Commands, **67,** 71, 77, 97, 99, 100, 105, 118, 121 (*see also* Statements)
Comments, 71, 72, 83, **86** (*see also* Programming languages, documentation, by using comments)
Compatibility, 25, 31, **36–42**, 43, 95, 335, 338 (*see also* Conversion)
Compiler, 11, **12,** 20, 24, 25, 38, 39, 43, 44, 50, 59, 87, 102, 111, 112, 113, 118, 119, 137, 708 (*see also* Implementation)
  criticism, 59
  debugging facilities, 114, **125**
  directives, 66, **68,** 71, **83, 119–120**

Compiler (*cont.*):
  efficiency, 21, 32, 61, 68, 119, 120, 724
    (*see also* Implementation efficiency)
  evaluation, 23, **59**
  independence, **38–39**
  language structure interaction with, **120–125**
  modification, 121
  syntax directed, 57, 605, 634, 635, 723
  writing application, **121–122**, 383, 530, 584, 603, 734
    languages for, **633–642**
Computational linguistics, 175
Computer, 102 (*see also* Hardware)
  configuration, 24, 25, 119
  errors, 110
  features, 112
  specific (*see* under the manufacturer's name in Name and System Index)
Computer aided design, 603
  languages for, **678–683**
Concatenation, 94, 104, 105, 112 (*see also* String)
Constants, 71, 75, 93 (*see also* Literal)
Constituents, 387 (*see also* description of COMIT)
Control cards, 35 (*see also* Control language)
Control language, **687–693**, 732–733
  (*see also* Time-sharing, control statements in; *see also* JCL in Name and System Index)
Conversion, 25, 31, **36–42**, 43 (*see also* Compatibility; Data types, conversion of)
  ease of, **16**, **41–43**
Core-Language, 31
Costs:
  hardware, 16, 36
  programming, 16, 36, 37, 38, 43
Cryptanalysis, language for (*see* OCAL, in Name and System Index)

D

Data:
  computation, types of, 93
  declarations (*see* Declarations)
  description, 66, 111, 114

Data (*cont.*):
  hierarchical, 74, **75–76**, **94**, 97, 100
    (*see also* Qualification)
  intermingling (*see* Data types, combining of)
  names (*see* Identifier)
  scope of, **103–104**
  structured (*see* hierarchical)
  types (*see* Data types)
Data base management (*see* Query)
Data names (*see* Identifier, types of, data names)
Data processing, 128, 134, 523, 539, 634, 717, 728 (*see also* Business data processing languages)
Data types, 66, **92–95**, 104, 121
  alphanumeric, 93
  algebraic (*see* formal)
  arithmetic, **93**, 101
  Boolean (*see* logical)
  character, **93**
  combining of, 77, 97, 100, 144
  complex, **93**
  conversion of, 100, **101**, **105–106**
  formal, **94**, 101, 111
  list, **94**
  logical, 77, **93**, 97, 101
  numeric (*see* arithmetic)
  pointer (*see* list)
  string, **94**, 97, 104, 105, 112
Debugging, 3, 15, 16, 18, 35, 59, 114, 125, 724 (*see also* Diagnostics; Compiler, debugging facilities)
Decision tables, 11, 315, 329 (*see also* DETAB 65 and DETAB X in Name and System Index)
Declarations, 66, **67–68**, 72, **83**, 84, 86, 87, 88, 89, 96, 97, 99, 114, **115–120** (*see also* Compiler, directives)
  default, 117 (*see also* description of PL/I)
  placement of, 88, 124
  types of:
    data description, 66, **116–117**
    dimension, 125 (*see also* Subscripts)
    environment and operating systems, 116, **119**
    file description, 116, **117–118**
    format, 116, **118**
    storage allocation, 116, **118**, 124
    subroutine, function, procedure, **119**

SUBJECT INDEX 779

Deconcatenable, uniquely, 55
Default (*see* Declarations, default)
Definition of programming languages, 8–11, 31, 41, 45, **48–59,** 724, 734
   administrative, 49
   by characteristics, **9–11,** 604
   by classification, **19–22**
   in glossaries, 8
   technical, 44, **51–57** (*see also* Metalanguage; Pragmatics; Semantics; Syntax)
   by users, (*see* User defined languages)
Delimiters, 66, **68,** 71, 77, 78, 80, **89,** 113
Diagnostics, 18, 59 (*see also* Compiler, debugging facilities; Debugging)
Differentiation, formal, 4, 61, 77, 104, 111, 472, 733
Dimensions, 118 (*see also* Arrays; Declarations, types of, dimension; Subscripts)
   two, for input, 81, 264, 732
Dispatcher (*see* description of COMIT)
Divisions in a language (*see* description of COBOL)

E

EBCDIC, 106
Editing generator, 4
Efficiency, 95
   of program, 9
Embedding, **92,** 111
Emulation, 41
English (*see* English-like; Natural English; Natural language; Query; *see also* English, in Name and System Index)
English-like, 82, 144, 314, 323, 325, 335, 343, 375, 715, 735 (*see also* Natural English; Natural language; Query; *see also* English in Name and System Index)
Environment, **36,** 68
   division (*see* description of COBOL)
Equipment checkout, languages for (*see* DIMATE and STROBES in Name and System Index)
Executable unit, 11 (*see also* Statements; Commands)
   smallest (SEU), **83–85**
Expressions, 102
   algebraic (formal), 95

Expressions (*cont.*):
   arithmetic, 74, 75, 95, 100
   creation and evaluation of, **100–103**
   evaluation of, **100–103**
   logical, 95
   mixed mode, 100, 144 (*see also* Data types, combining of)
Extensions (*see* Programming languages, extensions; Macro)

F

FIB, 691
File, 95, 107, 662, 664
   layout, 96
   handling systems, 376–377 (*see also* Query)
Fixed Words (*see* Key Words)
Flow of control, 107
Foreground initiated background, 691
Formal algebraic manipulation, 33, 383, 406, 471–472, 524, 582, 650, 733
   languages for, **Chapter VII, 583–589**
Format (*see also* Declarations, types of):
   file, 13
   fixed, 5, 11
   free, 10
Free list, 115, **384,** 385, 386, 476 (*see also* Garbage collection; List, processing)
Functions, **85–86,** 90, 92, 102, 111 (*see also* Subroutine)

G

Garbage collection, 115, 384, 385, 415–416, 598 (*see also* Free list)
General purpose languages, 523, 728 (*see also* Multipurpose languages)
Generator, 539, 709
Generators (See Code generation; Editing generator; Language generator; Report generator; Sort generators)
Global variable, **103** (*see also* Scope of data)
Glump (*see* description of Information Algebra)
Graphic(s), 70, 71, 582 (*see also* Character(s); Character(s), sets)
   application, 603
   display devices, 82, 244, 732

780 SUBJECT INDEX

Graphic(s) (*cont.*):
  languages, **674–678**

H

Hardware, 12, 15, 19, 36, 41, 50, 59, 95, 119, 730 (*see also* under manufacturer's name in the Name and System Index)
  compatibility, 16
  data units, **95**
  descriptions of (*see* Logical design; *see also* APL in Name and System Index)
  effective use of, 9, 34, 87, 732
  facilities, 9, 19, 37, 96, 113, 264, 732, 734
  implementation of programming languages, 707, **717–719**
  instructions, 9
  new, 16
  representation (*see* Hardware language)
  selection, 30, 31
Hardware language, 19, 22, **23**, 69, 70, 74
Hierarchy (*see* Data, hierarchical)
Higher level languages (*see* Programming languages)

I

Identifier, 24, 39, 71, 72, 725
  definition of, 69, **72–76**, 82
  types of, **72**
    data names, 72, 73
    program unit labels, 72, 73
    statement labels (*see* program unit labels)
    statement names (*see* program unit labels)
  Implementation, 24, 31, 32, 50, 51, 57, 59, 74, 95, 100, 113, 114, 115 (*see also* Compiler)
    efficiency, **25**, 74, **122–124**
Implicit multiplication (*see* Juxtaposition)
Information retrieval (*see* Query)
Interpreter, 5, 11, **12–13**

J

Juxtaposition, 73, 102, 131, 275, 512

K

Keyboard, 81, 82, 217, 512 (*see also* Typewriter)
Key punch, 69, 264, 732
Key words, 70, 71, **73**, 77, 80, 89, 108, 109, 124, 725

L

Labels (*see* Identifier, types of)
Language (*see* Assembly language; Machine language; Metalanguage; Programming languages; *see also* the individual languages in the Name and System Index)
Language generator, 725
Language L-like, **39**
Level numbers (*see* Data, hierarchical; *see also* descriptions of COBOL and PL/I)
Library routines, 4 (*see also* Subroutine, library)
Light pen, 82, 216, 217, 255–257, 510, 512, 516, 670, 671, 680, 682, 685, 691
Linguistic data processing, 383
List, 382, **383–385** (*see also* Free list; Garbage collection; Statements, executable, types of)
  processing, **383–385**, 386, 388, 650, 658
  languages for, **388–416**
  structure, 385
Literals, **71–72**, 80
Local variable, **103** (*see also* Scope of data)
Logical design, 33, 603
  languages for, **620–627**
Loops, 4, 84, **85**, 110 (*see also* Statements, executable, types of, loop control)
  parameters, 85, **109**
  range of, 84, **85, 109**
  termination of, 85

M

Machine code (*see* Machine language)
Machine independence, 10, 16, 37, 62, 96, 112, 331
Machine language, 1, 2, 8, 9, 10, 14, 18, 35, 88, 114, 120

SUBJECT INDEX 781

Machine tool control, 21, 33, 48, 603, 605–610
 languages for, **605–610**
Machines (*see* Computer; Hardware)
Macro (*see also* User defined languages):
 assemblers, 10, 121
 in PL/I, 579–581
 with programming languages, 86, 120, **121**, 196, 453, 524, 729
 with symbolic assembly program, 8, 14, 604, 729
Matrix computation, language for (*see* Matrix Compiler in Name and System Index)
Metalanguage, **53–57**, 635 (*see also* Compiler, syntax directed)
 ALGOL, **53–56** (*see also* BNF)
  used 191, 231, 409
 Backus Normal Form (*see* BNF)
 BNF (*see* BNF)
 COBOL, **53–56**
  used 234–235, 353–359, 364–365, 367, 368–374, 481–484, 557, 558–561, 619, 672–673
 metalinguistic formula, 53
 metametalanguage, 53
M-expression (*see* description of LISP 1.5)
Microprogramming, 232, 718–719
Mixed mode, **100**, 144 (*see also* Data types, combining of; Mode)
Mnemonics, 2, 3
Mode, **100, 102,** 106 (*see also* Data types)
Movie creation, languages for, **644–646**
Multipurpose languages, **Chapter VIII**

N

Names (*see* Identifiers)
Natural English, 82, 669–670 (*see also* English-like; Natural language; Query; *see also* English in Name and System Index
Natural language, 662–664, 715, **729–731**, 734–736 (*see also* English-like; Natural English; Natural language translation; Query; *see also* English, in Name and System Index)
Natural language translation, 386
Noise words, 74, **79–80**

Nonprocedural language, 19, **20,** 22, **726,** 734–736
Notation, 14, 39, 715
 better, 4
 formalism in, 24
 formalized, **52–57** (*see also* Metalanguage; Semantics; Syntax)
 natural, 4, 5, 16, 24, 31, 32, 35
 problem-oriented, **10–11,** 15, 87
Numerical analysis and computations, 128, 134, 216, 472, 490, 539, 540 (*see also* Scientific applications)
Numerical control (*see* Machine tool control)
Numerical scientific languages, 21, 33, **Chapter IV**

O

Object code, 11, 17, 125
 efficiency, 18, 21, 59, 120, 122, 123, 124
 errors, 125 (*see also* Debugging; Diagnostics)
 inefficiency, 18
Object program, **11,** 12, 32, 38, 59, 137
On line (*see* Time-sharing)
Open shop, 24, 148
Operating system, 3, 12, 35, 36, 37, 111, 113, 115, 603, 687, 732, 734 (*see also* Control cards; Control language; FIB; RJE; Statements, executable, types of, operating system interaction; Time-sharing; *see also* JCL, in Name and System Index)
Operation, high level primitive, 20
Operators, **66,** 71, **77,** 105 (*see also* Commands)
 computational, **66**
 graphic, 70
 infix, **67**
 logical, **66**
 post fix, **67**
 precedence rules for, **100**
 prefix, **67**
 relational, **66,** 71
 suffix (*see* post fix)
Overlays, 115, 118

P

Paper tape, 81, 82, 113, 131, 255, 510
Parallel processor, *306,* 732, *736, 737*

782 SUBJECT INDEX

Parameterized, 4, 13 (*see also* Parameters)
Parameters, 12, 78, 86, 89, 102, 119 (*see also* Functions; Procedure; Subroutine)
   call by location, **90–92**, 102, 160
   call by name, **90–92**
   call by simple name, 90
   call by value, **90–92**
   dummy arguments, **119**
   formal, **90**, 103, **119**
   input, 113
   for loops, (*see* Loops, parameters)
   passing, **90–92**, 113 (*see also* call by location; call by name; call by value)
Part programmer, 605, 606
Pattern matching, 68, 382, 386, 388, 582 (*see also* descriptions of COMIT and SNOBOL)
Payroll, 33
Phrase substitution (*see* Embedding)
Picture processing, 383
Plex, 195, 384, **680–681** (*see also* description of AED)
Pointer, 384 (*see also* Data types, list)
Polish notation, 67, 414, 416 (*see also* Cambridge Polish notation)
Polynomial manipulation, *302* (*see also* description of ALTRAN)
Pragmatics, 41, 44, **52**, 724
Precision, 4, 9, 37, 59, 102 (*see also* Arithmetic, precision)
Problem, one-shot, 17
Problem-defining language, 19, **21**
Problem-describing language, 19, **22, 726–727**
Problem-oriented language, 19, **21**, 603
Problem-solving language, 19, **22**
Procedure, **85–86**, 90, 107 (*see also* Subroutine)
Procedure-oriented language, **19–20**, 22
Production runs, 17, 18, 122, 229
Program:
   complete, **88**
   conversion, 16
   efficiency of, 9
   object (*see* Object program)
   self modification of, 120
   source (*see* Source program)
   structure, 66, **68**, 71, **82–88**, 103

Programming Languages, 1, 3, 8, 10, 11, 13, 16, 17, 23, 32, 35, 41, 42, 81, 87, 88, 96, 98, 101, 112, 113, 114, 120, 175, 604, 718, 723, 731, 732, 734, 736 (*see also* individual languages by name in the Name and System Index)
   advantages, **14–17**, 60, 113
   advantages and disadvantages, 1, **14–19**
   classifications of, 1, **19–23**, 34 (*see also* the following subentries where they appear as main entries)
   application oriented, 19, **21, Chapter IX**
   business data processing, 21, 33, **Chapter V**
   civil engineering, **610–620**
   compiler writing, **121–122**, 383, 530, 584, 603, **633–642**, 734
   computer aided design, **678–683**
   control, **687–693**, 732–733
   cryptanalysis (*see* OCAL in Name and System Index)
   equipment checkout (*see* DIMATE and STROBES in Name and System Index)
   formal algebraic manipulation, **Chapter VII, 583–589**
   general purpose, 523, 728 (*see also* Multipurpose languages)
   graphics, **674–678**
   hardware, 19, 22, **23**, 69, 70, 74
   information retrieval, (*see* query)
   list processing, **388–416**
   logical design, **620–627**
   machine tool control, **605–610**
   movie creation, **644–646**
   multipurpose, **Chapter VIII**
   nonprocedural, 19, **20**, 22, **726**, 734–736
   numerical scientific, 21, 33, **Chapter IV**
   problem-defining, 19, **21**
   problem-describing, 19, **22, 726–727**
   problem-oriented, 19, **21**, 603
   problem-solving, 19, **22**
   procedure-oriented, **19–20**, 22
   publication, 19, **22**, 69, 74
   query, 118, **662–674**
   reference, 19, **22**, 69, 70

Programming Languages (*cont.*):
  scientific applications, **Chapter IV, Chapter VII**
  simulation, **627–633, 650–662**
  social science research (*see* DATA-TEXT (Harvard) in Name and System Index)
  specialized (*see* application oriented)
  special purpose, 19, **21**
  string processing, **416–454**
  text editing, **684–686**
 consistency of, 32
 defined by users (*see* User defined languages)
 definition of (*see* Definition of programming languages)
 dialects, **39,** 42, 44, 45, 152, 179
 disadvantages, **17–19, 60–61**
 documentation, 25, 43, **57–58**
  ease of, **16,** 17
  by using comments, **86–87**
 efficiency of, 32
 English as a (*see* Natural English)
 evaluation, 19, 23, 31, 50, **58–61**
 extensions, **40–41,** 51
  self extensions, **121,** 723 (*see also* Macro)
 functional, 386
 generality, 31, 123
 hardware implementation of (*see* Hardware, implementation of programming languages)
 input form, **80–82**
 learning of, **14–15,** 17, 31, 32
 maintenance of, 45, **50–51**
 macro (*see* Macro, programming languages)
  ease of, **16,** 17
 modification, 8, **120**
 purpose, 31, **32–33, 49–50, 59–60**
 selection of, 31, 34, 36 (*see also* evaluation)
 standardization of, 31, **43–48** (*see also* USASI, X3, X3.4, and ISO in Name and System Index)
 subsets, (*see* Subsets of programming languages)
 suitability, 23–24 (*see also* evaluation; selection of)
 translating of, **42** (*see also* Compiler; Implementation)

Programming Languages (*cont.*):
 usage, 9, 13, 17, 18, 19, 20, 22, 23, 24, 25, 31, 32, 33, **34–36,** 58, **59–60,** 112, 730
 user defined (*see* User defined languages)
 users:
  types of, 24
  reactions of, 25
Pseudocode, 8, 129
Publication language, 19, **22,** 69, 74
Punched cards, 81, 82, 113
Punctuation:
 rules, 70
 symbols (*see* Character(s), punctuation)

Q

Qualification, **74–76,** 97
 prefixes, 76
 suffixes, 76
Query, 715
 application, 118, 603
 languages, 118, **662–674**
Question answering (*see* Query)

R

Rational arithmetic (*see Arithmetic, types of*)
Recompilation, 12
Recursion, **89–90,** 124
Recursive function theory, 386, 405 (*see also* description of LISP 1.5)
Reference counts, 385
Reference language, 19, **22,** 69, 70
Remote job entry, 689, 691
Report generator, 11, 20, 21, 118, 314 (*see also* Report Program Generator in Name and System Index)
Reserved words, **73–74,** 80, 89, 123–124
Ring (*see* description of CORAL)
RJE, 689, 691

S

Scientific applications, 122, 134, 148, 490, 523, 539, 540, 634, 728 (*see also* Formal algebraic manipulation; Numerical analysis and computations)

784 SUBJECT INDEX

Scientific applications (*cont.*):
  languages for, **Chapter IV, Chapter VII**
Scope of data, **103–104** (*see also*
    descriptions of ALGOL and PL/I)
Segmentation:
  automatic, 112, 137
  semiautomatic, 371 (*see also* Storage
    allocation)
Self modification of languages, **120** (*see
  also* Macro; Programming
  Languages, extensions)
Semantics, 41, 44, **51–52**, 71, 77, 125,
  635, 664, 716, 723 (*see also*
  Syntax; *see also* description of
  FSL)
Semiotics, *64*
Sentence (*see* Statements)
Sequence of control characters, 687
Set theoretic operations, 271, 279
S-expression (*see* descriptions of LISP 1.5
  and LISP 2)
Side effects problem, **102**
Simplification of expressions (*see* descriptions of FORMAC, Formula
  ALGOL, and MATHLAB)
Simulation, 33, 122, 582, 603
  of block diagrams, 627–633
  continuous, 650
  discrete, 627, 650–662
  languages for, **627–633, 650–662**
Social science research, 646
  language for (*see* DATA-TEXT
    (Harvard) in Name and System
    Index)
Sort generators, 4, 20, 21, 107, 322
SORT statement (*see* description of
  COBOL)
Source code, 12
Source program, 10, **11**, 12, 17, 43, 125
  inefficient, 18
Spaces (*see* Blanks)
Special purpose language, 19, **21**
Specialized languages (*see* Application,
  oriented languages)
Statement names (*see* Identifier, types of,
  program unit labels)
Statements, 10, 11, 38, 83–85, 88, 103,
  104–120 (*see also* Commands)
  compound, 84 (*see also* Block)
  declarative, 104 (*see also* Declarations)
  executable, types of, **104–115**

Statements (*cont.*):
  algebraic expression manipulation,
    111
  alphanumeric, **106–107** (*see also*
    conversion; editing; sorting)
  assignment, **104–106**
  computed GO TO, **107–108**
  conditional, 85, 107, **108–109**
  control transfer, 84, **107–108**, 119
  conversion, **106**, 118
  debugging, 112, **114**
  decision making, **107–111**
  editing, **106**, 118
  error condition, 107, **110–111**
  IF (*see* conditional)
  input/output, 112, **113**
  library referencing, 112, **113–114**
  list handling, **111–112**
  loop control, 107, **109–110**, 129
  operating system interaction **112–
    113**
  pattern handling, 112
  segmentation (*see* storage allocation)
  sequence control, **107–111**
  sorting, **106–107**
  storage allocation, 111, 112, **114–
    115**
  string handling, **112**
  switch control, **107**
  symbolic data handling, **111–112**
  imperative, **104** (*see also* Commands)
Storage allocation, **114–115**, 118, 383
  (*see also* Declarations, types of,
  storage allocation; Free list;
  Garbage collection; Statements,
  executable, types of, storage
  allocation)
Storage locations, 4
String, 385 (*see also* Statements, executable, types of, string handling)
  concatenation (*see* Concatenation)
  processing, 385, 386, 388, 650
  languages for, **416–454**
Structure (*see* Data, hierarchical)
Sublist (*see* List)
Subroutine, 4, 20, **85–86**, 90, 92, 103,
  113, 114, 115, 121, 124 (*see also*
  Parameters)
  body, 119

Subroutine (*cont.*):
  calling (*see* invoking)
  closed, 86
  invoking, 86, 88, 107, 119, 121, 388
  library, 4
  open, 86
  packages, 387–388, 604, 610, 729
  parameters (*see* Parameters)
  use of (*see* invoking)
Subscripts, 22, 24, **74–76**, 97, 125, 264
  (*see also* Dimensions)
Subsets of programming languages, 15, **40–41**, 45
  L-like extended, **41**
  nested, 40
  non-nested, **40**
Symbol manipulation, 383, 387, 406–407, 435, 436
Symbolic assembly program, 8, 10, 11, 14
Syntax, 41, 44, **51–57**, 77, 79, 89, 111, 125, 605, 664, 716, 723 (*see also* Compiler, syntax directed; Metalanguage)
Syntax directed compiler (*see* Compiler, syntax directed)

T

Table driven compiler (*see* Compiler, syntax directed)
Tables, 13 (*see also* Decision tables)
Tasking (*see* description of PL/I)
Teaching machine, 733
Terminal language, 687–691 (*see also* Control language; Time-sharing, control statements in)
Text editing, 603, 646
  languages for, **684–686**
Theorem proving, 382, 383, 389, 406

Threaded lists, 385
Thruput, 30
Time:
  for compilation, 17, 18, 123
  at compile, 118
  computer, 17
  elapsed, 17
  at object, 118
  for problem solution, 17
  turnaround, 3
Time-sharing, 3, 35, 36, 215–216, 729, 732
  control statements in, 87, 216, 603, 687–691 (*see also* FIB; Operating system; RJE; *see also* CTSS in Name and System Index)
Tokens, 68, **70–72**, 77
  system defined, **71**
  user defined, **71–72**
Tradeoffs, 122–124, 723, 725
Translator, 12, 539, 709, 724 (*see also* Compiler; Interpreter)
Typewriter, 69, 198, 216, 264, 732 (*see also* Keyboard)

U

Universal code, 5
Universal programming language, 723
User defined languages, 316, 605, **727–729**, 734–736 (*see also* descriptions of AED and ICES)

W

Words (*see* Key words; Noise words; Reserved words)
Word size, 37, 96
Workspace (*see* description of COMIT)

ST. MARY'S COLLEGE OF MARYLAND
　　ST. MARY'S CITY, MARYLAND

47826